Head and Neck

Enrico Marani · Ciska Heida

Head and Neck

Morphology, Models and Function

Springer

Enrico Marani
Department of Biomedical Signals and Systems, MIRA
University Twente
Enschede, The Netherlands

Ciska Heida
Department of Biomedical Signals and Systems, MIRA
University Twente
Enschede, The Netherlands

ISBN 978-3-319-92104-4 ISBN 978-3-319-92105-1 (eBook)
https://doi.org/10.1007/978-3-319-92105-1

Library of Congress Control Number: 2018941987

This Springer imprint is published by the registered company Springer Nature Switzerland AG
The registered company address is: Gewerbestrasse 11, 6330 Cham, Switzerland

Since interest by definition, connotes biased self-limitation, the scientific information gained remains incomplete, short of the comprehensiveness to which science in its professed universality aspires. So, how sure can we be that sheer analysis alone does not irretrievably destroy highly relevant data about nature? (Weiss 1967).

Preface

Peter Block in his “The answer to how is yes” tells the anecdote “What do you do when you find yourself in a hole?” The answer is: “The first thing you do is stop digging” (Block 2002). It seems foolish, but if you find yourself in a “hole” in anatomy, you start digging by performing extra preparations around and in the dissection area for identification of more structures. They help to solve your problem, because structures do have a basic topographical relation (see introduction on spatial coherence). Nowadays the question arises: Is anatomical “digging” still valuable for the sciences?

The most frequent question about the value of anatomy is: “Is anatomy ossified as a field of knowledge? Alternatively, is anatomy on the threshold of a renaissance? … Is anatomy education being threatened with extinction? Alternatively, is the teaching of anatomy continually evolving and reinventing itself?” (Eizenberg 2015). In other words, if science can be compared to a crossword puzzle does anatomy still have blank spaces? The answer is simple if one is involved in post-doctoral courses. You notice the lack of anatomical knowledge in medical and paramedical professions, mostly covered by arrogance or senseless nomenclature. Anatomy and neuroanatomy are an absolute need in both research and education (see, e.g. Chaps. 3 and 4 on development). Two research fields illustrate this: in the anatomical vascular research the concept of angiosomes (Taylor and Palmer 1987) and in neuroanatomy the cerebellar concept of longitudinal zones and compartments (Voogd et al. 1996; for overview see Ruigrok 2011). Both anatomical concepts totally renewed research in their fields. By the way, there is no Nobel Prize for Anatomy and only two neuroanatomists ever received a Nobel Prize, the one for physiology or medicine: Camillo Golgi and Santiago Ramon y Cajal in 1906. Seemingly valuable anatomical research does not earn a prize. There are also fields where anatomy loses track of developments. Chapter 7 gives an example concerning automated face recognition that no longer needs any anatomical information, quantitative or even qualitative.

In our current approach of anatomy or neuroanatomy, we prefer to look at the micro- and nano-levels as expressed in cell biology and molecular biology. In fact, we gain precision but lose perspective. Looking at the macrolevel of nature, to the properties of a group, like in histology or (neuro)anatomy, knowledge of unexpected properties of groups or populations can lead to study at the microlevel. For example, storage of fat molecules—think of cholesterol (see Sect. 14.2.1 on atherosclerosis)—occurs in the so-called smectic state (well-defined concentric layers of lipid molecules). If the substratum provides structure, all new fat molecules will follow these structural predestined constructions. If such an escorting substratum is missing, fat molecules will group in micels. They are still ordered structural patterns that are determined by the properties of the fat molecules. Although regularity and patterning of the micels are clearly recognizable and repeatable, each of the subparts are unique and no microsample is equal. The same holds for, e.g., glia cells: each glia cell is unique and the general pattern is the same everywhere in the same nucleus (see Weiss 1967).

Marani's introduction of a course on "kaakchirurgie" (oral and maxillo-facial surgery; 1988/1989/1990) together with Kostas Lekkas led to a series of clinical postdoctoral courses in the Leiden Academic Hospital (later named Leiden University Medical Centre, LUMC) on Head and Neck surgery, arm and hand surgery, leg surgery, pelvic surgery (by Marco de Ruiter, anatomy department) and craniotomy, organized by several clinical departments, together with Marani's neuroanatomical group. It started with Marani's regular courses in gross anatomy for medical and biomedical students in 1985/1986. At the end of each session, a clinician was invited to explain the importance of that days preparations to the students in the dissection room. It brought an interplay between the student interested in clinics and the clinician reviving and discovering anatomy. The clinician consequently wanted anatomical dissections to extend his knowledge often due to newly developed surgical techniques or to teach his younger colleagues in order to raise the quality of the department.

New techniques are developed by universities and firms. At the LUMC, a so-called HoloLens is used to project 3D pictures on the cadaver. It is called mixed reality and is also capable of instructing movements (LUMC 2017). Anatomage Table (info@anatomage.com, USA) or Sectra Table (www.3bscientific.com, Germany) are digital display equipments that react on the students' questions or make identification of structures possible. Navigator (Anatomage, USA) is a real-time volumetric inner anatomy tool with tracking stylus, visualizing cross sections or 3D anatomy. Digitizing and programming anatomical data goes rapidly and is effective in education.

Anatomy contains an enormous amount of data. Not all figures in this book are solely illustrations to the text. Some are a means of reducing text by the compilation of anatomical facts, e.g. Fig. 8.8 all blood vessels of the eyeball with the ophthalmoscopic view of the retina or subdivisions of the neck muscles Fig. 14.1. These illustrations allow the narrative to flow without too many interruptions.

Telling the story necessarily draws on ancient history, but also information from the recent past. It is interesting to note that various anatomical descriptions passed on from the Middle Ages and Renaissance are still valuable (see Historical panoramas in The Pelvis, Marani and Koch 2014). For example, ideas about symmetry in picturing of the human body have hardly changed, and the essence of topographic anatomy, picturing the exterior of the body for the projections of the different organs onto the surface, is still unaltered. "History determines our insight into the constituents and their interconnected functions in the human body. This perspective still defines our clinical approach in diagnostics and therapy" (Marani and Koch 2014). While it is a sad story with rather depressing results for the pelvis and its organs, in Head and Neck studies the results are far more optimistic, although in some areas there is a clear lack of knowledge and sometimes charlatans practice the (pseudo-)science.

"The sensory experience obtained by the anatomist during a dissection needs to be recorded by a medium outside his/her mind, in order to become accessible knowledge that can be theorized, discussed and disseminated" (Zwijnenberg 2004). It can be a drawing, painting, photograph and, of course, a model. Modelling arose also from the production of automatons, androids and robots. In arts, this anatomical, mechanistic view was expressed in engravings, but also vulgarized, e.g. for Head and Neck by Louis Poyet (1846–1913; Fig. 1). Modelling of the Head and Neck structures together with finite element methods has brought enormous progress in the understanding of functional morphology (think of applications in

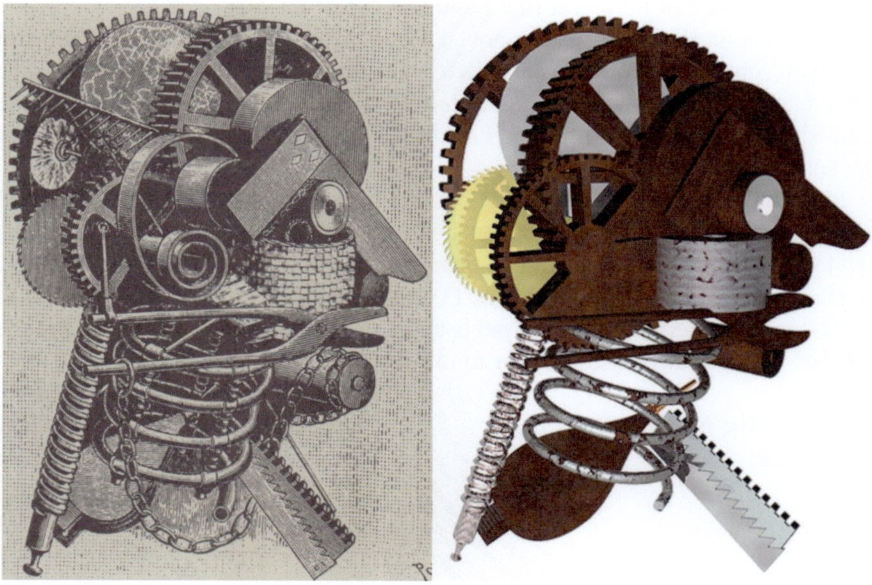

Fig. 1 Three-dimensional remake of "The Head and Neck of an Inventor", wood carving by Louis Poyet (1846–1913). *courtesy* E. Baas

robotics). In neuroanatomy, the cortico-basal ganglia thalamo-cortical map has been the basis for the study of Parkinson's disease. Direct and indirect pathways are discerned. The value of the subthalamic nucleus models for the verification of these neuroanatomical maps are indispensable and clearly contribute to the validation of deep brain stimulation (Heida et al. 2008) and thus to the quality of the motor behaviour of the body, including Head and Neck. Moreover, the direct electric stimulation of Head and Neck structures is evidently increasing: vagus stimulation in epilepsy, auditory stimulation for deafness, hypoglossus stimulation in obstructive sleep apnoea and Neuromodulation of the cervical dorsal root ganglion for arm pain, to give a few examples (see the journal *Neuromodulation* for more examples). Electric stimulation for monitoring is used for cranial nerves VIII and VII during operations. Tinnitus is the perception of sound in the absence of an external source. Vestibular–cochlear nerve stimulation should suppress this phantom hearing. For all these neurostimulation methods, new and additional anatomical data need to be gathered, as exemplified in the literature (see, e.g. Gharb et al. 2015 and for more examples see Marani and Lakke 2012) and modelling is in most cases a necessary prerequisite for human application. The scientists who collect these data can be grouped in lumpers and splitters. As an example, consider the types of the ansa cervicalis: 4 main groups by the older anatomists, 7 main groups by clinicians and 7 main groups with 21 subgroups by clinical anatomists. This last division clearly defeats the purpose (see Chap. 14), and it does not lead to basic concepts. It reminds us of Phil Patton's (1996) "Top this: Coffee-cup Lids". His description and study of the various types of plastic coffee-cup lids that piled up in his car (the data gathering) and finally the needed search for the original "Solo Traveler" coffee-cup lid patent (the basic concept of the lid).

This book on Head and Neck is not an anatomical handbook, you can find better and more complete ones. Instead it tries to show unexpected approaches and to give other basic views besides the classical ones. Head and Neck is written for the well-educated layman, scientist and clinician. It is for the reader to decide whether the authors succeeded.

Please note that citations in this book are in between quotation marks followed by the author(s) and the year of publication they are taken from. For short citations and text figures used or adopted (less than half a page) copyrights are not asked for, which is allowed by the copyright act. Figures from articles that have been used in the text are with permission of the publisher(s) or owners, unless they belong to the authors' own private possessions or to our department's property.

Enschede, The Netherlands — Enrico Marani
April 2018 — Ciska Heida

References

Block P (2002) The answer to how is yes. Acting on what matters. Berrett-Koehler Publishers, San Francisco

Eizenberg N (2015) Anatomy and its impact on medicine: will it continue? AMJ 8:373–377

Gharb BB, Tadisina KK, Rampazzo A, Hashem AM, Elbey H, Kwiecien GJ, Doumit G, Drake RL, Papay F (2015) Microsurgical anatomy of the terminal hypoglossal nerve relevant for neurostimulation in obstructive sleep apnea. Neuromodulation 18:721–728

Heida T, Marani E, Usunoff KG (2008) The subthalamic nucleus. Part II: Modelling and simulation of activity. Adv Anat Embryol Cell Biol 199, pp 1–88, Springer, Heidelberg

LUMC (2017), Leren van je bril, Department of Governance and Global Affairs, Leidraad, 2, p 47

Marani E, Koch WFRM (2014) The Pelvis, structure, gender and society, Chapter 4 and 5. Springer, Heidelberg, NY, Dordrecht, London

Marani E, Lakke EAJF (2012) Peripheral nervous system topics (Chapter 4: pp 82–140). In: Mai JK, Paxinos G (eds) The human nervous system 3rd edn. AP/Elsevier, Amsterdam

Patton P (1996) Top this: coffee-cup lids. I.D. Magazine May/June pp 40–47

Ruigrok TJ (2011) Ins and outs of cerebellar modules. Cerebellum 10:464–474

Taylor GI, Palmer JH (1987) The vascular territories (angiosomes) of the body: experimental study and clinical applications. Br J Plast Surg 40:113–141

Voogd J, Jaarsma D, Marani E (1996) The cerebellum, chemoarchitecture and anatomy. In: Björklund A, Hökfelt T (eds) Handbook of chemical neuroanatomy. Integrated systems of the CNS, Part III, Cerebellum, Basal ganglia and Olfactory system, Elsevier, Amsterdam

Weiss P (1967) $1+1 \neq 2$ When one plus one does not equal two. In: Quarton GC, Melnechuk T, Schmitt FO (eds) The Neurosciences a study programme pp 801–821, Rockefeller University Press, New York

Zwijnenberg R (2004) Project: philosophy, anatomy and representation. www.Narcis.Info/research

Acknowledgements

Financial support for charges on pictures and photographs (e.g. Figs. 1.3 and 9.6) and other expenses was given by The Department of Biomedical Signals and Systems, University Twente, Enschede, The Netherlands, and by Dental Passion, Amsterdam, The Netherlands, for the production of skull scans, dental overviews and Zirkonzahn prostheses of maxilla and mandible (see e.g. Fig. 6.3).

Albert van Helden is acknowledged for the correction of parts of the English text and the discussions on themes published in various chapters. Head and Neck scans have been made available by Carmen Vleggeert-Lankamp, Head of Spine Research, Department of Neurosurgery, Leiden University Medical Centre, The Netherlands, and she read critically some chapters, and by Maurits Joosse, Haaglanden Medical Centre, The Hague, The Netherlands. He was our vademecum for eye and orbita and corrected English. Gerard Kerkhof supported by discussions on sleep and on the sleep text. He helped with some sleep research carried out for, among others, this book too. Johan Frijns, ENT Department, Leiden University Medical Centre, made new publications available, checked text on cochlear implants and helped to find the representatives of cochlear implant firms. Hans ten Donkelaar made several figures from his book Clinical Neuroembryology (Springer) available for use in ours. Hans Landsmeer (1919–1999) made his unpublished text on Head and Neck already years ago available for use in Marani's lectures and publications, and permitted and supported the use of the figures. Janine Post from the Department of Developmental BioEngineering of the University of Twente is acknowledged for figures and reading the text on chondrocyte development. Alan Hobson is acknowledged for permission to use figures of "Ethology of the Sleep". Jan Voogd made several figures available and gave permission to use them. Jan Gevers Leuven allowed the use of his Fig. 1.2a and his thesis was a good introduction in the fatty acid metabolism. Figures from Marani's own lectures and from Marani's Neuroregulation group were allowed to be used by Leiden University Medical Center, by the Leiden Departments of Anatomy and Embryology and of Physiology, and by the Twente University (several departments). Ronald Bleys (University Medical Center Utrecht) is acknowledged for the check on the innervation of the soft palate muscles (Sect. 6.5.2). Martin Laman brought to our attention the hypoglossus paresis related to carotid dissections (Sect. 6.5.2). My daughter supported this book by the use of her facial photographs. Simone Ordelmans allowed us to use the introductory Fig. 1.3 of her thesis on vagus reflexes and part of her text on her p12. Marc Van Dijk is acknowledged for his help on Fig. 7.21.

Last but certainly not least, we thank Jantine Marani-Nijhof for her patience and support of the book by her critical remarks on several parts and her co-writing and permission of use of "Neurosurgical instruments pictured on icons and frescoes of the doctor-saints and healers: Panteleimon, Tryphon, and Cosmas and Damian" (*Progress in Neuroscience and Neuroregulation*, Vol II), in Sect. 2.1.5.2.1 Arabic and Byzantine trepanation and the treatment of head wounds.

Contents

Chapter 1
Introduction

The compartments of Head and Neck do contain spatial and structural relations. Paired and unpaired structures are present that are explained by processes involved in differentiation, morphogenesis and growth. The concept of spatial coherence will be used to elucidate these Head and Neck relations throughout this book. Attention is paid in this introductory chapter to personality, to Roman rhetoric and to head transplantation.

1.1 Compartments of Head and Neck

You cannot live without a head. Before the rifle, executions were performed by decapitation or by hanging. In both cases, main passage ways of the neck were destroyed or blocked.

"The only son of an artificer of six or seven years of age, went into the shop of a neighbour, who was his father acquaintance; in playing with the child, he put one of his hands under his chin and the other behind his head, and lifting him up into the air, told him, *he would shew him his Grandfather*, a mean and vulgar way of speaking. The child was no sooner lifted of the ground than he grew refractory, dislocated his head and died that instant. His father (who was immediately told of it) being transported with passion, ran after his neighbour, and not being able to overtake him threw after him a Saddler's Hammer, which he had in his hand; the sharp end whereof pierced what is called the pit of the neck, and cutting all the muscles, and penetrating the space between the first and second vertebra of the neck, cut the marrow of the spine, whereof he expired the same hour. Thus these two deaths happened almost after the same manner.

This way of playing with children is but too common with the vulgar, because they don't know the danger to which they expose them. Perhaps also, the child had not died if he had not been so refractory: for I do not at all doubt but his struggling was one of the principal causes of the dislocation of his head.

E. Marani and C. Heida, *Head and Neck*,
https://doi.org/10.1007/978-3-319-92105-1_1

T'is observed, in almost all hanged persons that the first vertebra of the neck is entirely separated from the second. It may be even the greatest and most proper cause of their deaths. This observation makes me believe the head and first vertebra of the neck are very difficult to be parted, and that generally when the head is thought to be dislocated t' is nothing else but a luxation of the first vertebra from the second" (Petit 1705, 1726).

Petit's observation is incorrect. Hanging breaks the isthmus of vertebra C2 and as a consequence the pinching off of the myelum occurs. It is called "Hangman's fracture". Moreover, separation of Head and Neck is caused by a head (C_0)–C_1 dislocation. C_1–C_2 connection is maintained, due to the strong ligament connections between C_1–C_2 and the weak C_0–C_1 connection is caused by the occipital condyles.

Besides the upper spinal cord, important pathways for air and food, a number of different structures and functions are incorporated in the Head and Neck of the body: Eye, ear, nose, mouth, the brain and the transition of the skull into the vertebral column that give humans knowledge of their environment by seeing, hearing, smelling, tasting and feeling by skin and mucous layers. The information thus obtained gives the brain and spinal cord the possibility to react adequately. Special structural and spatial organizations in this region are a prerequisite for storing all the organs effectively and let their different functions unharmed. To understand these relations, four main compartments are discerned (Fig. 1.1).

1. The neurocranium: the brain case with skull base, surrounding the brain, and including in its base the ear.
2. The viscerocranium or facial skeleton: the case containing the eye, nose and mouth.
3. The branchial system or primitive, developmental gill-arch system: mouth and nose can be considered extensions of the branchial system. The branchial system mainly makes up the collum of the neck and in it are organs like: pharynx, larynx, and thyroid and thymus glands.
4. The nucha constructed by vertebral column, within it the spinal cord and muscles around it.

The border between neuro- and viscerocranium is sharp: the skull base. At the opposite end, there is a clear overlap between the viscerocranium and the branchial system within the collum's upper part.

The border between collum (neck) and nucha (nape) is sharp and is determined by a fascia (a connective tissue blade): the fascia prevertebralis (layer of the cervical fascia), also called fascia colli profunda, placed just in front of the vertebral column.

These borders are clinically important: any damage below line 1 goes to the oto-rhino-laryngologist, ophthalmologist, or oral-maxillo-facial surgeon, above it is mainly the responsibility of the neurosurgeons. The same holds for line 2: behind it the neurosurgeons, in front of it the other specialists.

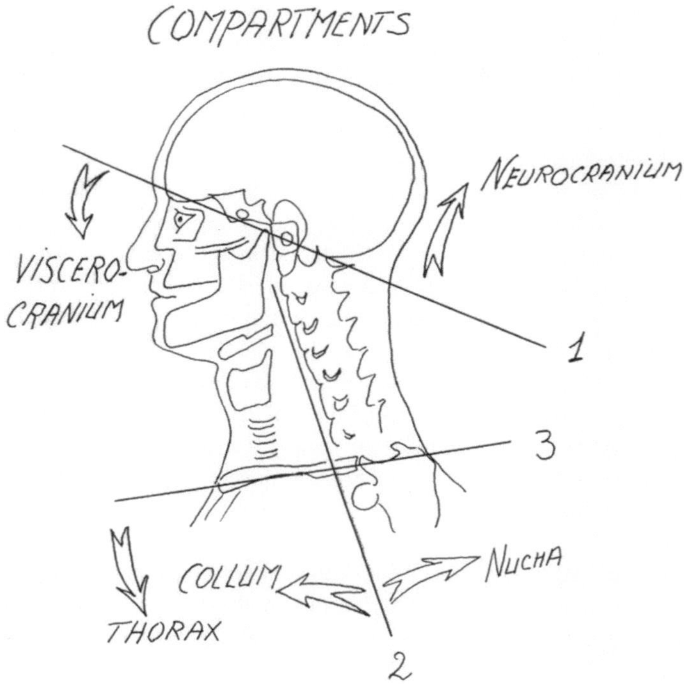

Fig. 1.1 Compartments in Head and Neck: 1 border between neurocranium and viscerocranium (the line between upper orbita border, the superciliary ridge, and ear entrance, meatus acusticus externus, parallels the skull base), 2 border between collum and nucha (the fascia prevertebralis or fascia colli profunda), 3 border between neck and thorax (thorax aperture) © E.Marani

1.2 Spatial and Structural Relations

The human embryo has acquired essential human characteristics by four weeks of gestation. The main developmental features are the four branchial vessel arches and the prosencephalon that has grown ventral of the pharynx. The progress of the facial structures are initiated together with the nasal structures. The interrelations between brainstem, pharynx, branchial arches and heart tube are increased. The relations between the developing eye and its surrounding mesenchyme of nose and face are established. Moreover, the relations of the otic vesicle in the area caudal to the bend of the rhombencephalon are determined. In the following weeks, this key stage is transformed into a creature with a Head and a Neck. Main characteristic of this part of the development is the lateral growth of Head and Neck, with the growing ear in a crucial position. This lateral extension leaves the primary relationships between brainstem, pharynx and skull base unaffected. In this lateral extension, the eyes and maxillary processes form the lateral border of the midregion that is dominated by the nasal structures. The mandible follows this lateral extension. A forward projection in the head region is advanced by the growth of the mandibular arch, the hyoid arch and the cervical sinus. This liberated these structures from the heart area. The widening of the cervical sinus could be seen as the formation of a short neck.

These spatial and structural relations of Head and Neck are indicated in Fig. 1.2:

a. The neurocranium has to encompass the brain, which increased in volume during development, but also underwent temporalization: the cortex curved and moved laterally, producing at maturity a spiralization like in an ammonite. The cerebellum has to stay below the cortex, due to the space-taking process of the brain hemispheres. The ear organ, included in the petrosal bone, part of the temporal bone, had to develop, as a consequence, in between cortex and cerebellum. The eye, as an extension of the central nervous system, had to take its place in front of it.
b. The skull base, also the roof of the orbit, needed to surround the cranial nerves and had to give access to the end of the brainstem in its continuation into the spinal cord. At the transition from the temporal lobe towards the facial bones, a special bony structure, the sphenoid bone, is developed and with its greater wing supports the temporal lobe of the brain. Cerebral veins found their way out of the skull.
c. The mandible had to connect to the skull to allow masticatory movements later in life and contains the muscular bottom of the mouth. Its prominence contributes to the maxillary part of the skull during development (see Section "Coherence Spatial").
d. The maxilla forms the base of the orbit and had to incorporate the nose. The start of the primitive digestive tract (tongue) and airways needed to find their internal and external connections in the viscerocranium. Pharynx and larynx developed their structural relations and needed connections to the skull base for hanging.
e. The muscles, mimic, skeletal and masticator ones, make the face and neck's outer appearance (here are indicated sternocleidomastoid and trapezius muscles that are important for turning and keeping the head upright).

1.3 Paired and Unpaired Structures

Structures present in the midline of Head and Neck are considered unpaired (normally they include an unpaired lumen), although they were generated embryologically as bilateral symmetric structures around it, e.g. pharynx, late development of the nose (see Section "Coherence Spatial"), mouth and larynx. Their bilateral innervation demonstrates overlap. It is therefore understandable that the twelfth cranial nerve (N. Hypoglossus; N XII) can be transplanted one sided, on other malfunctional neural connections, without serious dysfunction of the tongue (Malessy et al. 1999). This does not hold for the arytenoids, they are produced as paired structures. Paired and unpaired are also important in function. Although two temporomandibular joints are present, chewing is an unpaired process, like using the vocal cords by the arytenoids. Structures created in duplicate (eyes) are often partially independent in their function (you can close one eye and still function). To produce an integrated bilaterally functioning system, a complex control system in the nervous system is needed in such cases.

1.4 Coherence Spatial

In Head and Neck, the principle of the "*coherence spatial*" (spatial coherence: Landsmeer 1968) is especially valid. The spaces occupied by structures and organs are not only determined by their own form and function, but also by the form and function of surrounding structures. This principle is a good guide for understanding the development, but is also a good lead into the understanding of the mature Head and Neck. Within this principle, it is clear that some organs are more influential ("more equal") than others, e.g. in the case of the skull base these are: the eye, ear and brain (see Fig. 1.2a, b). If the coherence spatial is disturbed (e.g. by an abnormal development), the defect always has its origin in a shortness of space (encephalo-meningocèle) or abundance of space (hare lip).

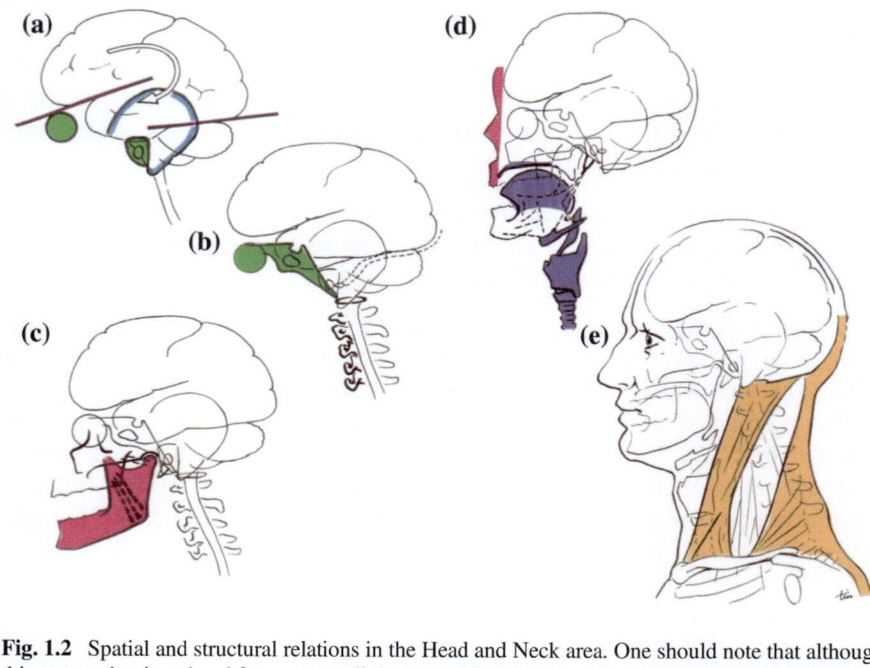

Fig. 1.2 Spatial and structural relations in the Head and Neck area. One should note that although this summation is ordered from a to e, all these spatial and structural relations more or less coincide during development. For explanation, see Section "Spatial and Structural Relations" (changed and adapted from J. M. F. Landsmeer[†])

Two examples will be treated here:

The encephalo-meningocèle (also called bifid cranium, cephalocèle or craniocèle) and chordomata are good examples of disturbance of the coherence spatial. Figure 1.3 shows an early embryological stage with three borders, where they can develop in the Head and Neck area:

1. Between future neurocranium and viscerocranium
2. Between collum and nucha
3. Between neck and thorax.

First note that these places coincide with the borders of the compartments as described above. Second, e.g. in border 1, between neurocranium and viscerocranium, the skull base develops. Everywhere within the skull base, the encephalo-meningocèle can originate. These encephalo-meningocèles, by their enlargements, deform other structures below and above the skull base, even to the extent that the skull vault cannot close and brain extends out of the skull (Fig. 1.3).

Thus, in normal situations structures or organs are *mutually dependent.* This dependence is not exclusively regulated by the human genome. The number of codons in the DNA is 10^8 for the entire body, while the brain alone has over 10^{16} connections. Therefore, the growth of the brain also has its own laws. This holds for most organs and structures in Head and Neck. These laws are grouped under the concept "Automation of Development" or "Cell Sociology" (e.g. see Chandebois and Faber 1983).

So, the construction of the spatial relations of the mature Head and Neck from early development till into maturity partially obeys its own laws and of course is also directed by genetics. For a better understanding of the "coherence spatial", one should notice that:

1. differentiation is producing cellular diversity
2. morphogenesis is organizing different cells into tissue and organs
3. growth is solely the increase in size.

The results of differentiation, morphogenesis and growth are organs that change relations by themselves in the bodily space to other changing organs: "The position of the elements in the process of ontogeny is also significant to their functions" …"The position of elements shifts greatly during ontogeny in order to carry out their activities" … "The specific position for the specific element is essential to carry out certain functions in a particular spatial arrangement" (Dutta 1982) to select just a few sentences from an overview of vertebral form and function discussion that swept through morphology from the 1960 to 1990 (see Dullemeijer 1974).

Processes that are involved in differentiation, morphogenesis and growth are:

- induction,
- fusion,
- cell death: apoptosis and necrosis,

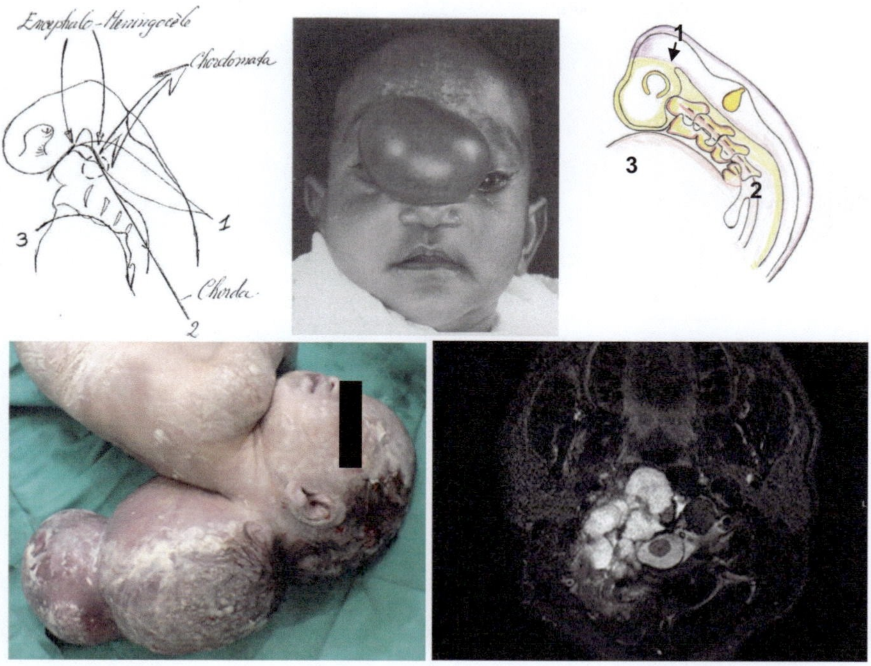

Fig. 1.3 Encephalo-meningocèle and chordomata: 1 border central nervous system, in its rostral part develops the skull; 2 chorda, later in development taken up in the vertebral column; 3 heart (primordium). Upper photograph: frontal skull base deviation; "Aboriginal infant with nasofrontal variety of frontoethmoidal meningoencephalocele. The facial defect lies at the junction of frontal and nasal bones, with the nasal bones attached to the inferior margin" (photograph and text from David, 1984, with permission); left colour plate shows an occipital encephalocèle at the backside of the head, figure taken from: (http://en.wikipedia.org/w/index.php?title=Encephalocele&oldid=662101223 Primary contributors: Revision history statistics)
Lower right side is a large chordoma having its origin in the clivus or the upper cervical spine. The tumour is hyperintense on this T2-weighted image, as is typical for a chordoma. (*Courtesy* C. Vleggeert-Lankamp with permission)

- patterning,
- segregation,
- migration,
- proliferation and
- pruning: fixing connections.

Several of these processes are functional in the production of the face (Fig. 1.2) of which the overview of the results, as published by Tapadia et al. (2005), is shown in Fig. 1.4 and supported by the following text:

"There are seven prominences that comprise the vertebrate face: the midline frontonasal prominence, and three paired structures, the lateral nasal, maxillary and mandibular prominences (Fig. 1.4d–f). These maxillary and mandibular prominences are derived from the first pharyngeal (branchial) arch, whereas the

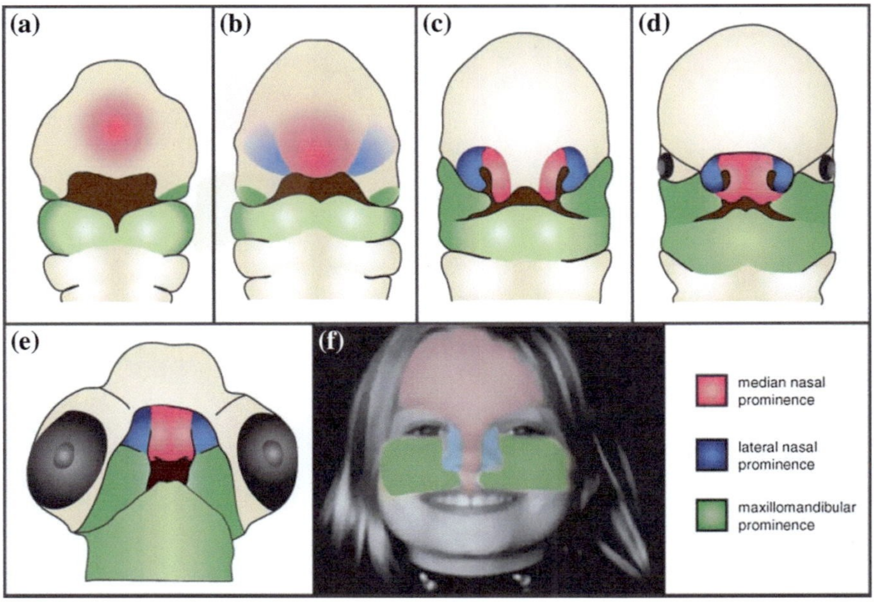

Fig. 1.4 Development of the craniofacial primordia. **a–d** Representations of frontal views of mouse embryos showing the prominences that give rise to the main structures of the face. The frontonasal (or median nasal) prominence (pink) gives rise to the forehead (**a**), the middle of the nose (**b**), the philtrum of the upper lip (**c**) and the primary palate (**d**), whereas the lateral nasal prominence (blue) forms the sides of the nose (**b**, **d**). The maxillo-mandibular prominences (green) give rise to the lower jaw (specifically from the mandibular prominences), to the sides of the middle and lower face, to the lateral borders of the lips and to the secondary palate (from the maxillary prominences). **e** Frontal view of a chick embryo, also showing which prominences give rise to different facial structures. **f** Frontal view of a human child, with different facial structures colour-coded to indicate the prominences from which each structure developed (Tapadia et al. 2005, with permission)

frontonasal prominence is derived from a midline primordium that forms on top of the forebrain. The frontonasal prominence contributes to the forehead, middle of the nose, philtrum of the upper lip and primary palate (Fig. 1.4). The lateral nasal prominence forms the sides (ala) of the nose; the maxillary prominences contribute to the sides of the face and lips, and the secondary palate; and the mandibular prominences produce the lower jaw (Fig. 1.4). Disruptions in the rate, the timing or the extent of outgrowth of any of these prominences will adversely affect the fusion process. Consequently, one can appreciate the wide variety of facial clefts that occur (Tessier 1976) and also why facial clefting is the most frequently occurring Head and Neck birth defect (Perrotin et al. 2001; Tapadia et al. 2005)".

Palatum clefting can be the consequence of a metabolic disturbance. The palatum originates from the fusion of two halves. Inhibition of glycosaminoglycans and glycoprotein synthesis inhibits the fusion of the two palatal shelfs. Palatum clefting is also determined by genes: a fully palatum cleft depends on other gene mutations than a partial anterior palatum cleft (Hillard et al. 2005).

1.5 What Is Said of Head and Neck and One's Personality?

Von Lanz and Wachsmuth (1955) published "Praktische Anatomie". The whole series is the anatomical standard for the clinics, and the series contains a separate part concerning the "Hals" (Neck). It starts with "Der Hals als Ausdruck der Persönlichkeit" (The neck as expression of personality). Few people would have expected that an anatomical handbook starts with personality expressions.

The anatomical text starts out by stating that few poets celebrate the neck, but painters and sculptors have to construct the neck with care. Here we restrict ourselves to the female. The ideal of female neck beauty has hardly changed through the ages. While Roman sculptors still used a small shorter female neck, the Middle Ages and Renaissance advocated the long and slender female neck also called the "swan neck" (Morris 2004): it became and to some extent remains the aesthetic archetype. The "collier de Venus" (Venus rings), rests of the fat traces of the child's neck contributed to this aesthetic ideal. Some expression of the thyroid gland supports the youth of the female.

Dante Gabriel Rosetti (painter, translator and poet, and a good exponent of the symbolistes) 1828–1882 lived in the nineteenth century and pictured a series of women by means of which the importance of the neck in relation to the stand of the head and eyes can be elucidated (Fig. 1.5). In the central picture, the neck continues into the "sea spell": 1877; right shoulder and the head are somewhat turned towards the eye direction. This combination forces the spectator to recognize reflection. The left upper figure is the model Alexa Wilding as Queen of Hearts. By her straight look and her raised head, she indicates attentiveness for the observer of the painting. Left lower picture is Ellen Smith who modelled for Rosetti several times. By the full rotation of the head that coincides with the direction of the eyes and the neck that is placed as a pillar for support of the head unhappy reflection is indicated. In the right upper picture by bending the head more downwards, as compared to Ellen Smith's position of the head, (active) participation is brought into the portrait.

The right lower picture has been described by the Tate Museum of Liverpool: "Aurelia (the name was presumably chosen for its classical connotations) exudes a powerful erotic appeal, emphasized in the picture by her red lips, flowing red hair and exposed shoulder and neck. Her dreamy expression and self-absorption render her entirely passive, the object of the artist's gaze". The visibility in the portrait of half the neck going over into the shoulder, the three quarters of the face and the staring eyes induces this self-absorption.

There are more relations between head, neck and eyes possible, e.g. neck straight and only inclining one's head indicates humility or sadness and an oblique head on a straight neck can indicate chastity. The angle between shoulder and head is important (see also Brücke 1891). In general, by Head and Neck movements you can notice someone's mood or state of mind, but hardly his/her character.

It cannot be a surprise that von Lanz and Wachsmuth's handbook after this personality introduction subdivides the neck's external form into three main

Fig. 1.5 Five pictures of Dante Gabriel Rosetti, English pre-Raphaelite painter, designer, translator and poet (1828–1882). Left upper figure, Regina cordium 1866; left lower figure Ellen Smith 1870–1872; central figure, a sea spell: 1877; right upper figure, Jane Burden; lower right figure, Fazio's mistress (Aurelia): 1863–1873. (All used figures are "libre de droits d'usage ou de distribution ou de modification". Figures are taken from Senex Magister and the Art Renewal Center, see http://hoocher.com/Dante_Gabriel_Rossetti/Dante_Gabriel_Rossetti.htm)

appearances for de male: athletic, leptosomic and eurosomic, and for the female neck: muscular, leptosomic and eurosomic. (Note that females could not be athletic in 1955.) These types of the human bodybuilding are scientifically no longer accepted, but still regularly used now with new terms: ektomorph (slender, leptosomic), mesomorph (muscular, athletic) and endomorph (inclination to obesitas, eurosomic).

1.6 Head and Neck and Quitilian's Rhetoric

At the end of this introduction, Head and Neck are regarded in relation to public speaking. Oratory functional style has been described already by the Romans. Cicero (106–43 BC) was famous for his oratory contemplations and the best-known hand-down books on this topic are from Quintilian (see Butler 1922). Marcus Fabius Quitilianus (35–96 AD) grew up in Spain, got his education in Rome and returned to Spain as a lawyer. He accompanied Sevius Sulpicius Galba (3–69 AD) back to Rome to be emperor. Galba was killed the same year. Quitilianus stayed and became professor in rhetoric and consul under Emperor Vespasianus (9–79 AD). At the end of his life, he wrote his Institutio Oratoria that was published in 95. It is the most famous collection of complete texts on rhetoric. After the detection of the complete work in 1416 in a convent in Sankt Gallen, it stayed renowned throughout all following ages. Here we cite some of the Quintilian instructions on Head and Neck posture and the expressions it has to produce to enhance the desired effect during public speaking:

"The head, being the chief member of the body, has a corresponding importance in delivery, serving not merely to produce graceful effect, but to illustrate our meaning as well. To secure grace it is essential that the head should be carried naturally and erect. For a droop suggests humility, while if it be thrown back it seems to express arrogance, if inclined to one side it gives an impression of languor, while if it is held too stiffly and rigidly it appears to indicate a rude and savage temper. Further, it should derive appropriate motion from the subject of our pleading, maintaining harmony with the gesture and following the movement of the hands and side" (Quintilian, Book XI, III, 68–70).

"The methods by which the head may express our meaning are manifold. For in addition to those movements which indicate consent, refusal, and affirmation, there are those expressive of modesty, hesitation, wonder or indignation, which are well known and common to all. But to confine the gesture to the movement of the head alone is regarded as a fault by those who teach acting as well as by professors of rhetoric. Even the frequent nodding of the head is not free from fault, while to toss or roll it till our hair flies free is suggestive of a fanatic" (Quintilian, Book XI, III 71–72).

To shorten the citations, one can easily reduce the text on the effects of the eyes: "But of the various elements that go to form the expression, the eyes are the most important, since they, more than anything else, reveal the temper of the mind, and without actual movement will twinkle with merriment or be clouded with grief" (Quintilian, Book XI, III, 75–76) on the effects of the eyebrows. "Complete immobility in the eyebrows is a fault, as also is excess of mobility or the tendency to raise one and lower the other, as in the comic mask…." (Quintilian, Book XI, III, 79–80) and on the lips and nostrils: "It is not often that the lips or nostrils can be becomingly employed to express our feelings, although they are often used to indicate derision, contempt or loathing" (Quintilian, Book XI, III, 80) the neck: "The neck must be straight, not stiff or bent backward. As regards the throat, contraction and stretching are equally unbecoming, though in different ways. If it be stretched it causes strain as well, and weakens and fatigues the voice, while if the chin be pressed down into the chest it makes the voice less distinct and coarsens it, owing to the pressure on the windpipe. It is, as a rule, unbecoming to raise or contract the shoulders. For it shortens the neck and produces a mean and servile gesture, which is even suggestive of dishonesty when men assume an attitude of flattery, admiration or fear (Quintilian, Book XI, III, 82–84).

One should note that for the Roman teachers in rhetoric Head and Neck are separate entities in their oratory posture employ. Within the head in succession eyes, eyebrows, and lips together with nostrils are the structures that can be applied to support the spoken words in persuading the audience.

1.7 Head Transplantation

This introduction started with a citation on head dislocation from 1705. It ends with a more recent unwarranted one on head transplantation, published 300 years later. This strongly contested surgery considers *total* head transplantation initiated by the Italian surgeon Canavero. This mention is taken over entirely from the *European Medical Journal*. One of the facts one should note is that regeneration of the spinal cord has not been functional at all, till now. A more extensive overview by Canavero himself can be found in *Surgical Neurology International* (2013).

Head Transplants: No Longer Science Fiction But a Step Closer to Reality? Reported by Alex Watt in the *European Medical Journal*, June 1, 2015

"HUMAN head transplants have, until recently, been nothing more than science fiction, but Dr. Sergio Canavero, Director of the Turin Advanced Neuromodulation Group, Turin, Italy, has announced plans to conduct the world's first human head transplant in a project named HEAVEN-GEMINI. Mr. Valery Spiridonov, a 30-year-old computer scientist from Vladimir, Russia, became the first person to volunteer for the procedure.

Mr. Spiridonov was diagnosed with Werdnig–Hoffman disease, a rare genetic muscle wasting disease, also known as Type 1 spinal muscular atrophy, at the age of 1. "I can hardly control my body now", said Mr. Spiridonov. "I need help every day, every minute. I am now 30 years old, although people rarely live to more than 20 with this disease".

HEAVEN-GEMINI is a controversial procedure and researchers have severely criticized its feasibility. The procedure, predicted to require 100 surgeons approximately 36 h to complete, will include spinal cord fusion (SCF), and the head from a donor body will be removed with an ultra-sharp blade, to minimize damage sustained by the spinal cord. "The key to SCF is a sharp severance of the cords themselves", explained Dr. Canavero, "with its attendant minimal damage to both the axons in the white matter and the neurons in the grey laminae. This is a key point".

The spinal cord of the donor body will then be merged with the recipient's head with chemicals called polyethylene glycol or chitosan available to stimulate SCF. The recipient will remain in a coma for 3–4 weeks, in which time the spinal cord will be subjected to electrical stimulation via implanted electrodes to bolster new nerve connections.

Dr. Canavero has suggested that the recipient may walk again within 1 year, supported by physical therapy. However, there are major challenges with the procedure, such as reconnecting the spinal cord and preventing the immune system from rejecting the head.

Mr. Spiridonov, however, remains hopeful in spite of the risks and suggested that the operation could take place as early as next year. "He is a very experienced neurosurgeon and has conducted many serious operations. Of course he has never done anything like this and we have to think carefully through all the possible risks", said Mr. Spiridonov, "but if you want something to be done, you need to participate in it".

1.8 Head and Neck Summary

Head and Neck are three-dimensional complicated structures containing the central and partially the peripheral nervous system, cranial and spinal nerves, glands, cartilaginous and bony structures, specialized muscles, blood vessels and specialized connective tissue. Tubes for food and air and the specialized sensory organs all have to find their place in such a way that the functionality of each organ, but also the total of functions stays efficiently operational. Missing links are manifold encountered in view of the complicated development, structure and function of the organs in Head and Neck. Surgery is complicated in these areas. Scientific fraud is also present in Head and Neck research and cannot be disregarded. This book intends to contribute to the indication and description of such problems, and where possible poses another view.

References

Brücke E (1891) Schönheit und Fehler der menschliche Gestalts. Wien (ISBN: 9783955073350)

Butler HE (1922) The institution oratoria of Quintilian, with English translation. Vol. 1–4, Harvard University Press, Cambridge Massachusetts

Canavero S (2013) HEAVEN: The head anastomosis venture project outline for the first human head transplantation with spinal linkage (GEMINI). Surg Neurol Int 4:s335–s342

Chandebois R, Faber J (1983) Automation in animal development. A new theory derived from the concept of cell sociology. In: Wolsky A (ed) Monographs in developmental biology. Karger, Basel

David DJ (1984) New perspectives in the management of severe craniofacial deformity. Ann Royal Col Surg Engl 66:270–279

Dullemeijer P (1974) Concepts and approaches in animal morphology. Van Gorcum Assen, The Netherlands

Dutta HM (1982) The form-function relationship of vertebrates: a selected review. Proc Indian Acad Sci 91:207–215

Hillard SA, Yu L, Gu S, Zhang Z, Chen YP (2005) Regional regulation of palatal growth and patterning along the anterior–posterior axis in mice. J Anat 207:655–667

Landsmeer JMF (1968) Les cohérences spatiales et l'équilibre spatial dans la région carpienne. Acta Anat 54:S1–84

Lanz von T, Wachsmuth W (1955) Praktische Anatomie Hals I, 2: Ein Lehr-und Hilfsbuch der anatomischen Grundlagen ärztlichen Handelns. Springer, Berlin, pp 549

Malessy MJA, Hoffmann CFE, Thomeer RTWM (1999) Initial report on the limited value of hypoglossal nerve transfer to treat brachial plexus root avulsions. J Neurosurg 91:601–604

Morris D (2004) The naked woman: a study of the female body. Jonathan Cape, London. ISBN 0312338538

Perrotin F, de Poncheville LM, Marret H, et al (2001) Chromosomal defects and associated malformations in fetal cleft lip with or without cleft palate. Eur J Obstet Gynecol Reprod Biol 99:19–24.

Petit, Jean Louis (1705) Traité des maladies des os (Petit, 1674–1750).English translation: John-Lewis Petit (1726) A treatise of the diseases of the bones containing an exact and complete account of the nature, signs, causes and cures. T. Woodward, London

Tapadia MD, Cordero RD, Helms JA (2005) It's all in your head: new insights into craniofacial development and deformation. J Anat 207:461–477

Tessier P (1976) Anatomical classification facial, cranio-facial and latero-facial clefts. J Maxillofac Surg 4:69–92

Chapter 2
The Skullcap, Cranial Vault or Calvarium

The landmarks of the skull are exemplified by studies concerning the Mona Lisa and the reconstruction of Dante's skull. Modern trepanation, which is still handicraft, is compared to ancient trepanation. Instruments, witch doctor trepanation in Africa and trepanation in the ancient Arabic world are treated. Skull studies performed in spaceflights, on fibroblast growth factor and on yellow bone marrow are discussed. At the end, the chapter returns to Vesalius's skull studies and Leonardo's "oldest" miniature-scale sculpture of the skull.

2.1 Introduction: Landmarks of the Skull

2.1.1 The Mona Lisa

Lilian Schwartz (1995) studied the Mona Lisa of Leonardo da Vinci (1452–1519) and used it in her computer experiments. She compared the Mona Lisa and its underlying Mona Lisa portrait of Isabella, Duchess of Aragon, with the self-portrait of Leonardo. By matching the two faces, feature by feature, she showed their hidden structural similarity. Lines drawn along forehead, for position of the eyes, nose and lips in the fitted portraits supported the hypothesis that in the absence of Isabella, and only possessing a picture of her, Leonardo used his own face to produce the Mona Lisa. For it, Leonardo needed a mirror and therefore changed the direction of the eyes of the Mona Lisa, used the sfumato technique (masking and fading borders) to shadow the eyes and reduced the contour of the round full face of Isabella to his cheek bones. The nostril of Leonardo is taken over in the Mona Lisa, and he enlarged her lower lip (Fig. 2.1).

There is criticism on the proposition that the Mona Lisa is in fact a self-portrait of Leonardo (see Schwartz 1995 and La Farge 1996). The underlying Isabella portrait does not fit, and the morphing of the three faces, although a good piece of

E. Marani and C. Heida, *Head and Neck*,
https://doi.org/10.1007/978-3-319-92105-1_2

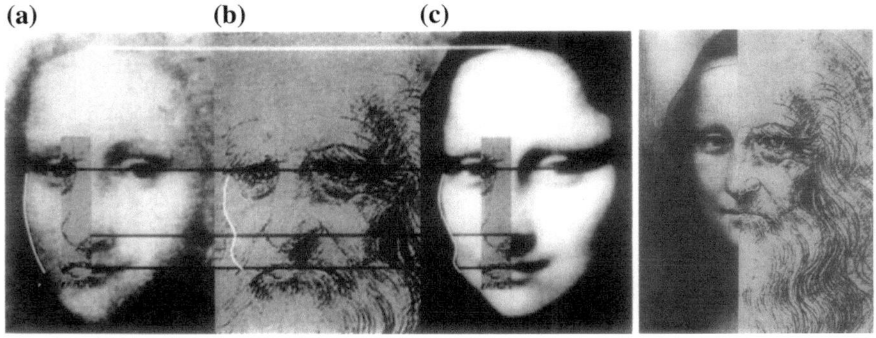

Fig. 2.1 Left: **a** Isabella, Duchess of Aragon, called the hidden Mona Lisa, **b** self-portrait of Leonardo da Vinci, **c** surface Mona Lisa. The cheek contour in **a** is changed in the cheek contour of **b**; see **c** Sfumato technique was used for eyes and mouth of the Mona Lisa. Right: picture composed of the morphed Mona Lisa half and the half of Leonardo's self-portrait. Note the good fitting of the face halves (Schwartz 1995, with permission)

fitting for that early computer period (1980s), can be criticized (see for newer techniques, e.g. the anatomical-based multilayer deformation (MLD) method, Zhang et al. 2006). The weakness of the approach is in the fitting: if anatomical skull landmarks could have been used on the faces, comparison would have been less criticized. It is a shame because Leonardo gave an excellent overview of the anatomy of the skull, produced the correct dental formula and detected more of the skull cavities than anybody else in that time (Gerrits and Veening 2013).

Clinical and anatomical landmarks on the skull are standardized nowadays (see, e.g. Deng et al. 2011) and they are needed in this book. Skull landmarks are extremely important for research as well as clinical purposes, as we will see later. A few landmarks are shown in the first anatomical figure, which will be used as a reference figure throughout the book (Fig. 2.2). Suture names are, in general, known and are indicated within the figures if necessary.

2.1.2 Dante Alighieri

Skull landmarks are also important in the technique of face reconstructions from skeletal remains. These applied methods are criticized because the thickness of the soft tissues (Stephan 2003), and, for example, colours of eyes are usually not known. Nevertheless, we still show the reconstruction of the face of Dante Alighieri (1265–1321) from his skull (of which the lower jaw is missing) to demonstrate the importance of skull measures and landmarks. Ancient portraits of Dante do exist but: "these ancient portraits gave rise to the universally known image of the poet, which has inspired all subsequent paintings of Dante and which has now become part of the collective imagination. However, it is important to remember that the

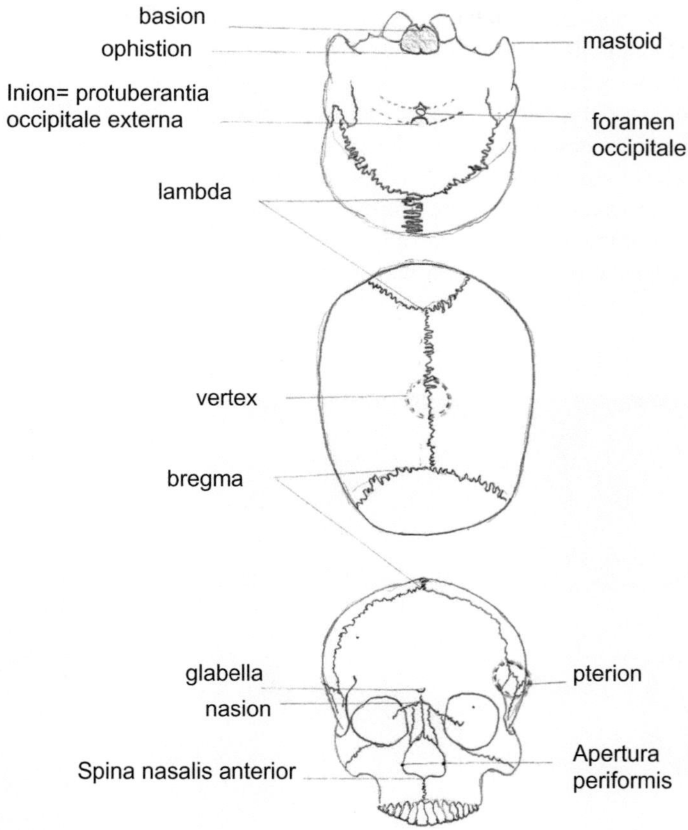

Fig. 2.2 Skull landmarks

representations of the facial features of the poet were mostly influenced by historical–cultural aspects rather than *anatomical considerations on Dante's skull* (italics ours). Moreover, the many death masks of Dante do not help recreate his true appearance since they are not considered authentic, i.e. taken directly from the cast of the dead poet's face, but rather artefacts" (Benazzi et al. 2009).

Dante's tomb was opened in 1921, on the occasion of the sixth centenary of Dante's death, and will be again unlocked in 2021 for the seventh centenary. Happily in 1921, an anthropological investigation of the skeletal remains was made and a plaster cast of the skull was made. Results were published in the literature and in a book: "This volume contained the description of the bones as well as 297 measurements and scale photographs of the cranium and postcranial skeleton" of course based on skeletal and skull landmarks, and "A mandible belonging to a cranium that was morphologically similar to that of Dante was selected" … "modified by virtual modeling to adapt it perfectly to Dante's cranium in terms of both size and level of correspondence in the temporo-mandibular joint and between the maxillary and mandibular alveolar process" (Benazzi et al. 2009 and see Fig. 2.3). Mandibular landmarks were used to change the 30-year-old museum mandible in one of a 50-year-old man.

The fitting of the facial remains like eyes, nose and lips was the end of the reconstruction process (Fig. 2.3). "The resulting facial model was provided with eyeballs whose size was dictated by the width of the orbital cavities, which are rather large in Dante's skull. The shape of the nose was influenced by the marked development of the nasal spine, the angulation of the floor of the nasal choanae and the convex course of the terminal part of the bony nasal profile, as well as by the

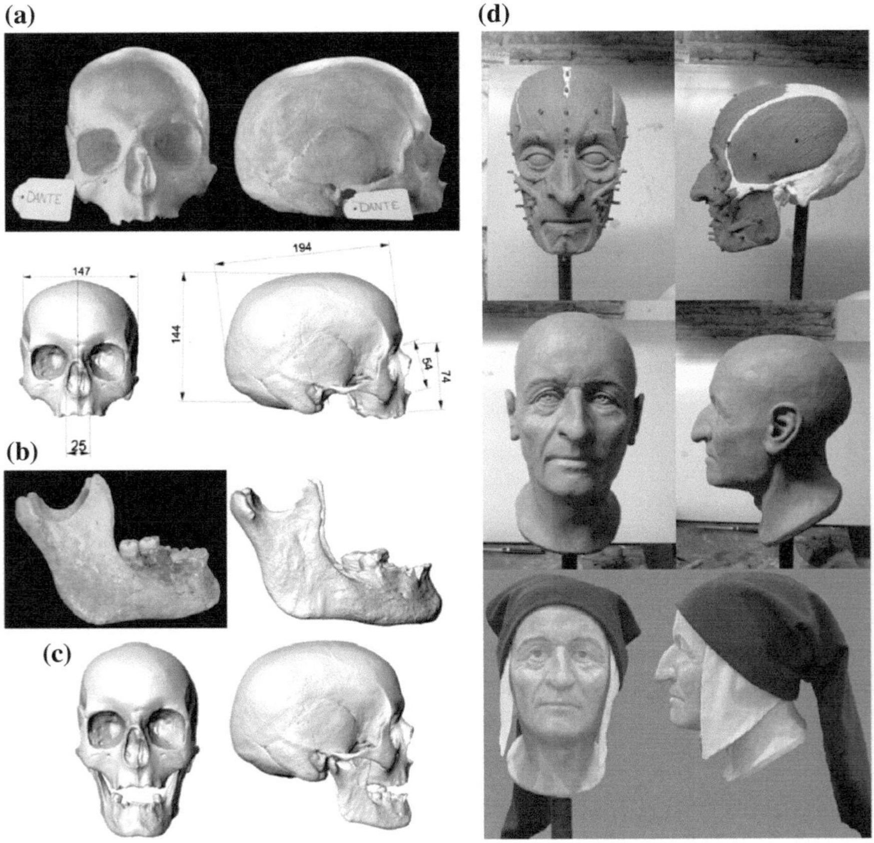

Fig. 2.3 **a** 1921 cast of Dante's skull with the 3D laser-scanned reconstruction and measures. For the measured distances, skull landmarks were used. **b** Photographs show reference mandible from the skeletal collection of the Museum of Anthropology, University of Bologna (in black), and its 3D laser-scanned reconstruction. **c** Joining skull 3D and 3D of the jaw. **d** Upper photographs: the soft tissue thickness is defined by the pegs over the modelled muscles (pegs are also placed in relation to skull landmarks), centre photographs: covering the face with the skin, bottom photographs Dante's face with cappa (Benazzi et al. 2009, with permission)

leftward deviation of the nasal septum and the protuberance on the right nasal bone" (Benazzi et al. 2009); by the way all skull landmarks.

The conclusion: "Obviously, we cannot claim that this is the definitive likeness of Dante, since too many variables were involved: apart from the uncertain nature of the mandible, some features are typically characteristic of the individual (e.g. the morphology and color of the eyes, the shape of the nose and mouth), while other factors concern more changeable aspects, such as the thickness of the adipose tissue. Hence, there is a certain amount of subjectivity in a practice that mixes science, technology and art" and "in this work we presented the recreation of the face that morphologically conforms most faithfully to the anatomical traits of Dante's skull, which remain the most objective evidence of the poet's physical appearance" (Benazzi et al. 2009; and see Fig. 2.3).

2.2 Trepanation

2.2.1 History and Research

Trepanation or trephination is the opening of a part of the human skullcap for medical or "religious" reasons in such a way that in the living, the dura mater (the outer brain membrane) stays unharmed. The instrument used is called a trepan or trephine (drill bit that cuts an annulus and leaves a centre core). The primitive surgical instrument can also be a saw, a drill, a chisel or a scraper.

The catalogue of surgical instruments of Arnold and Sons (West Smithfield, London, 1876), instrument makers by appointment to Her Majesty's Government; the Honourable Council of India; the Crown Agents for the Colonies; St. Batholomew's hospital; the Surgical Aid Society and the principal provincial and colonial hospitals, (truly an important firm) classes the trepan in the category: trephining, necrosis and resection instruments. You will find it next to a butcher's saw for excision and a hand chain saw. The trephines are offered small, medium and large. They can be bought in a case (containing a large and a small trephine) together with Hey's skull saw, elevator and skull forceps, strong scalpel (scale tang) and trephine brush: the set complete for £2.20. You can also move up to Arnold and Sons' complete case with butcher's saw with spare blades, amputating saw, extra forceps, chisel and two elevators and three amputating knives and some more instruments, for £14.00. It must be clear that the trepan around 1870 belongs in the West to the medical "butcher's shop" instruments.

Prehistoric trepanation was discovered by Broca (1824–1880) in 1867, by Prunières in tombs in Lozère and by de Baye in rock tombs in the Marne region (both France) in and around 1873. Broca's description (Broca 1867a, b, 1876) was of a skull from Cuzco (Peru, Fig. 2.4) sent in by E.G. Squier (1821–1888), US Chargé d'Affairs (you should call it now embassy's secretary) for Republics of Central America and later Member of the Mixed Commission for the relations between Peru and the USA. During his travels in Peru, he received the skull as a present. Since there was no agreement on the meaning of the four saw lines around the hole (Fig. 2.4) in the USA, Broca, as the most famous anthropologist of his time, was consulted. His conclusion of a neurosurgical intervention, containing signs of healing, started the study of trepanned skulls. The Cuzco burial fields later yielded several more trepanned skulls (Andrushko and Verano 2008) indicating that trepanation was a general neurosurgical practice in the Andean region. Over a 1000 ancient skulls that contain trepanation holes are now known, not only from Inca South America.

The neurosurgical success rate, as judged by the bone growth at the edges of the skull holes, is over 50%, some even indicate over 60%. In Inca times, a long-term survival rate of 78% was reached (Verano 2003).

Scientists searched for the reason to perform such a dangerous neurosurgical operation in ancient times. Causes like headache, migraine, epilepsy and demons or ghosts in one's head (indicative of insanity) are brought forward in literature (von Brunn 1928, Facchini et al. 2003). At least trepanation was seen as an intracranial decompression method during the early decades of the trepanation studies (Lucas-Championnière 1912).

Karl-Maria Heidecker rejected the demon–ghost theory and stressed a more natural scientific approach in the tracing of its causes. Comparing neurosurgical diagnoses from modern times with ancient written sources, this overview combined with mimicking of the production of the trepanation holes, strongly supported the conclusion that osteomyelitis (inflammation of bone marrow), damages of the skull due to fights or falls and other skull accidents are the causes for trepanned-like interventions (Heidecker 2009).

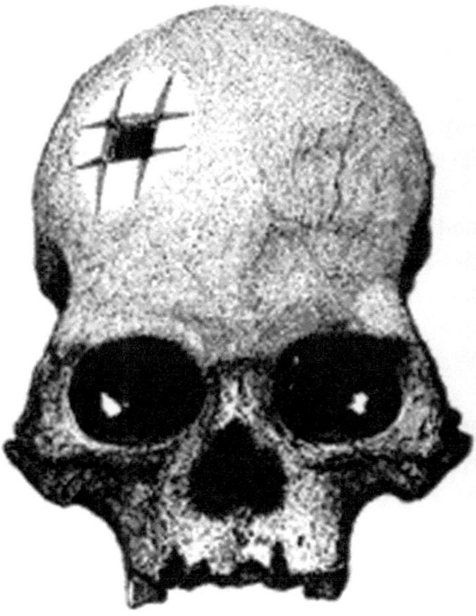

Fig. 2.4 Skull from Cuzco (Peru) sent in by E. G. Squier to P. Broca (1867a)

Sometimes, trepanned skull is found by accident, as was the case with the skull from Winsum in Friesland, in the North of the Netherlands (Inventory of the Friesch Museum, Nr 53,177). A right frontal and 1/3 left frontal bone with a trepanned hole of 13.5 mm is reported. Growth on its edges was found, so a survival after the operation was deduced. (Sypkens Smit 1943). The trepanned skull was found during the excavation of a *terp* (a man-made mound in tidal or flood-prone areas) in Winsum, in the layer that corresponds to Roman time. The assumption is that a Roman military doctor should have performed the trepanation, but since the age of the bones has not been determined, this is pure guesswork.

2.2.2 Human Migration Worldwide

Human mitochondrial DNA and other genetic markers of the human genome show that the spreading of humans started in Africa between 200 ka and 100 ka (1000 years = 1 ka) ago. (Anthropologists think this occurred 1000/1500 ka BC and disagree with this early "out of Africa" migration; see Gibbons 1992.) Spreading into the Arabic peninsula occurred at 80–60 ka BC and a further migration into Siberia around 60–30 ka BC. Europe and North America were occupied at 30–20 ka BC. South America was reached 20–15 ka years ago (Forster 2004).

The trepanned skulls are of a later date: Ukraine 10–4.8 ka BC, the oldest trepanation in Portugal 6 ka BC, Britain 4–2 ka BC and prehistoric South America 3 ka BC–1532 AD (Arnott et al. 2003). Thus, trepanation dates from the late palaeolithicum and early neolithicum, and at first sight, timeline agrees with the analogous trend of human spreading over the globe, but clearly later.

Such a comparison is undermined by recent studies that doubt several of the reported trepanations. They are considered the results of fights, wars and tumours (Bennike 2003). A re-evaluation was needed for material found in Denmark and in South Siberia. Osteoarchaeology has the means of determining the type of wounds caused by bladed weapons (Lewis 2008) and also the Scythian battle axe (*chekan*) wounds of the people of South Siberia. Several of the healed and unhealed wounds are now ascribed to fights and not to active trepanation. Murphy (2003) shows for the Scythian period (700 BC till 400 AD) twelve skulls with battle axe wounds of which four were healed. These skulls could easily be kept for trepanned skulls. Nevertheless in several primitive settlements, where archaeologists found cemeteries, signs of trepanation of one form or another were found.

2.2.3 Frequency of Trepanation

In general, trepanation occurs more in males than in females, and adolescents and youngsters were not excluded from this treatment in ancient times. Percentages are difficult to give since in 22.6% of the cases' gender of the skulls can not be determined; 64.5% were males and 12.9% females for the trepanations in British antiquity (4 ka BC–1.6 ka AD) (Roberts and McKinley 2003).

Two populations from ancient Egypt (graves of Thebes-West and Aswan) have been studied for the frequency of *skull trauma*. A total of 2048 human tomb remains were studied. About 0.6–0.7% of this ancient population showed perforating skull trauma. "This rate is rather high—in particular for the Theban population—since it can be assumed from archaeological findings that the individuals buried in those tombs belonged to a more advanced social status and had civil occupations" (workers, of course, have an higher chance of bodily accidents than administrators, remark ours; Nerlich et al. 2003a, b). If other facial traumas are added (e.g. nose, orbit and healed ones) 5.1% and in the other population 1.6%, *head traumata* could be calculated. In these ancient populations, head trauma reaches an estimated 2–3%, indicating the need of head surgery, including trepanation. For the Theban or the Aswan necropolis with over 100,000–1,000,000 (Thebe) inhabitants, this means that at least 2 or 3000 to 20,000 people must have had serious head trauma.

In 1862, Edwin Smith bought a papyrus scroll originating from Thebes, of which the content was deciphered later. It contains 48 clinical cases, among which head traumata, nose injuries, mandibular, ear or lip cases, throat and neck injuries, shoulder and thorax wounds (von Brunn 1928; Wilkins 1964). From the 27 head injuries, four are deep scalp wounds exposing the skull and 11 are skull fractures. So, skull traumata hold for 56% of the described medical cases, illustrating the prevalence of skull injuries. Since the papyrus scroll is not finished, it stops in the middle of a sentence; these figures should be taken with a grain of salt. One should note that trepanation is not mentioned in this papyrus scroll.

Severe skull trauma was already treated by surgical removal of fracture fragments to overcome intracranial haematoma. This type of haematoma leads easily to cerebral compression and often to rapid death. Already ancient Egyptian written sources described skull perforations and its distinct therapies (von Brunn 1928; Heidecker 2009; Nerlich et al. 2003a, b; Wilkins 1964).

Nowadays, 300–400 head traumas occur each year in a population of 100,000 of which 1–3% will get an intracranial haematoma. Today, the risk for males is twice that for females. By the way, these head traumas are considered mild in 80% of the cases (Vanpoecke 2004). Therefore, an estimate is that 60–80 patients are each year added with real skull trauma per 100,000 inhabitants, of which 9–12 will develop a haematoma.

2.2.4 Amulets and Postmortem Trepanation

The discovery of postmortem trepanation came late. Pseudotrepanations due to diseases or cranial weapon wounds were found early. Most cranial weapon wounds are on the left side, the aggressors being mostly right-handed. Luckily, entry wounds can be distinguished from exit wounds. Heads of defeated victims impaled on weapons are frequently reported from war situations, producing exit wounds that sometimes still are considered trepanations. Already Broca discerned postmortem trepanation from real trepanation in a Neolithic skull from France discovered by Prunières. Thus, postmortem trepanations were immediately noticed at the start of the study of trepanned skulls. Nevertheless, postmortem and pseudotrepanations were regularly taken to be real trepanations (Finger and Clower 2003; Donnabhàin 2003). Discoveries of skull parts with small holes and sometimes with inscriptions indicated that skulls and/or skull parts could be used as amulets, explaining some of the postmortem holes (Bennike 2003; Heidecker 2009).

2.2.5 *Modern Surgical Trepanation*

2.2.5.1 Modern Neurosurgery

Nowadays, neurosurgery and oral–maxillo-facial surgeons use modern techniques available to repair skull and facial wounds (see Fig. 2.5). Bone fragments that have been pushed into, for example, the sinus frontalis can be restored with plate osteosynthesis, in which frontal endoscopy can be a mean for minimally invasive reduction of outer table fractures (Mensink et al. 2009). The indications for and extensions of craniotomy became reduced as time went on, due to the neuronavigation techniques and skull base endoscopy in combination with MRI and (less) CT that produced an accurate location for tumours and a reduction of the needed operation territory. Sometimes, only a burr hole is needed because endoscopic techniques can be used (Seifert et al. 1999; Manwaring et al. 1994; Ciporen et al. 2010; Moshel and Schwartz 2010).

Nevertheless, trepanation is still used to make holes in the skullcap. They are handled as starting points to take out parts of the calvarium for access to the brain (Fig. 2.6). Various locations of the holes are used for the different approaches, like frontal or pterional. These so-called burr holes are connected by serrated edges

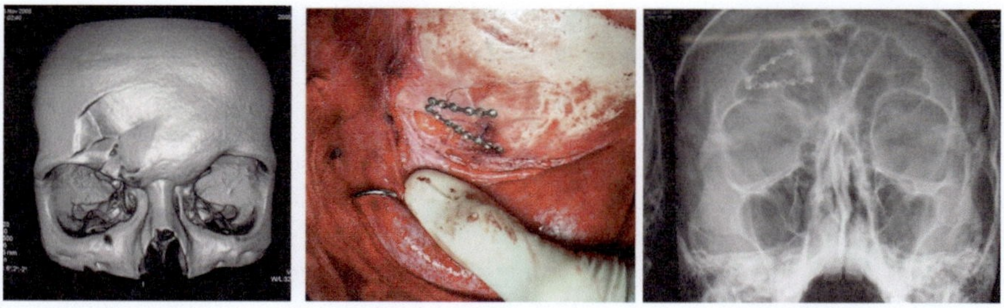

Fig. 2.5 Frontoorbital trauma (glabella and nasion not involved) together with its repair by plate osteosynthesis. Courtesy J. P. R. van Merkesteyn, LUMC, Leiden with permission)

using a hand Gigli saw. To lift the cranial vault part, it has to be freed from the upper dura mater layer, which adheres at the sutures of the skull (Fig. 2.6).

Recent studies on the healing and dry bone appearances in the present-day pathology of neurosurgical trepanation show a rather restricted regeneration of the needed bridged bone between walls of the serrated edges and of the holes. In eleven cases of removal and placing back of cranial vault parts in acute or survived (even up to 34 years) surgical trepanation, needed due to haematomas, meningioma and glioblastoma tumours or other causes, the conclusion is inevitable that trepanned skull wound closure is not self-evident (Nerlich et al. 2003a, b). Plate osteosynthesis, on the contrary (Fig. 2.5), that is used to restore the exact position of fractured skull fragments gives a far better restoration and healing (Nerlich et al. 2003a, b).

2.2.5.2 Witch Doctor Surgery in Africa

Articles on trepanation in Africa appear regularly in literature: "The traditional art of skull trepanations, traditionally passed down from generation to generation, is still practiced by Kisii tribesmen in the highlands of South Nyanza District of Kenya" (Grounds 1958). A feature titled "Skull Surgeon Who Never Went to Medical School" in one of the local daily newspapers in November 1982 (Mutahi 1982) focused on one well-known practitioner of trepanations, complete with his photograph and his traditional instruments. He reported that he had practiced the art since 1955, having learnt it from his grandfather, and had "operated on hundreds of patients who had initially sought treatment at hospitals without success" (Qureshi and Oluoch-Olunya 2010). And: "trephination in Africa has a long tradition originating in prehistory. The procedure, as practised in North Africa, originated from ancient Arabs and its practitioners included the extinct Guanches of the Canary Islands and the Tuareg Berbers, Chaouïas of Marocco, Libya, Chad and Algeria. The East African group included the Kisii and Tende in Kenya and Tanzania (see map Table 2.1). Tribes in both these geographical regions continue to perform the procedure today. The practice of trephination on the continent was not confined to these two regions, and examples can be found in Uganda, South Africa, Nigeria and Somalia (map in Table 2.1). Surprisingly, the early Egyptians did not routinely trephine patients" (Rawlings and Rossitch 1994).

Figure 2.7 shows the trepanation in Africa, as photographed by one of my students. During his three-month stay for medical training, he had the opportunity to see a witch doctor's trepanation.

It is now well established that the trepanation procedure is done for reducing headaches and not for any magical reason. There is no relation between brain function location and the trepanation place on the skull. For most, witch doctors claim that they underwent a local training in surgery and anatomy, mostly the son apprentice of the father or grandfather, within a setting of a restricted number of trainees in a tribe (Rawlings and Rossitch 1994).

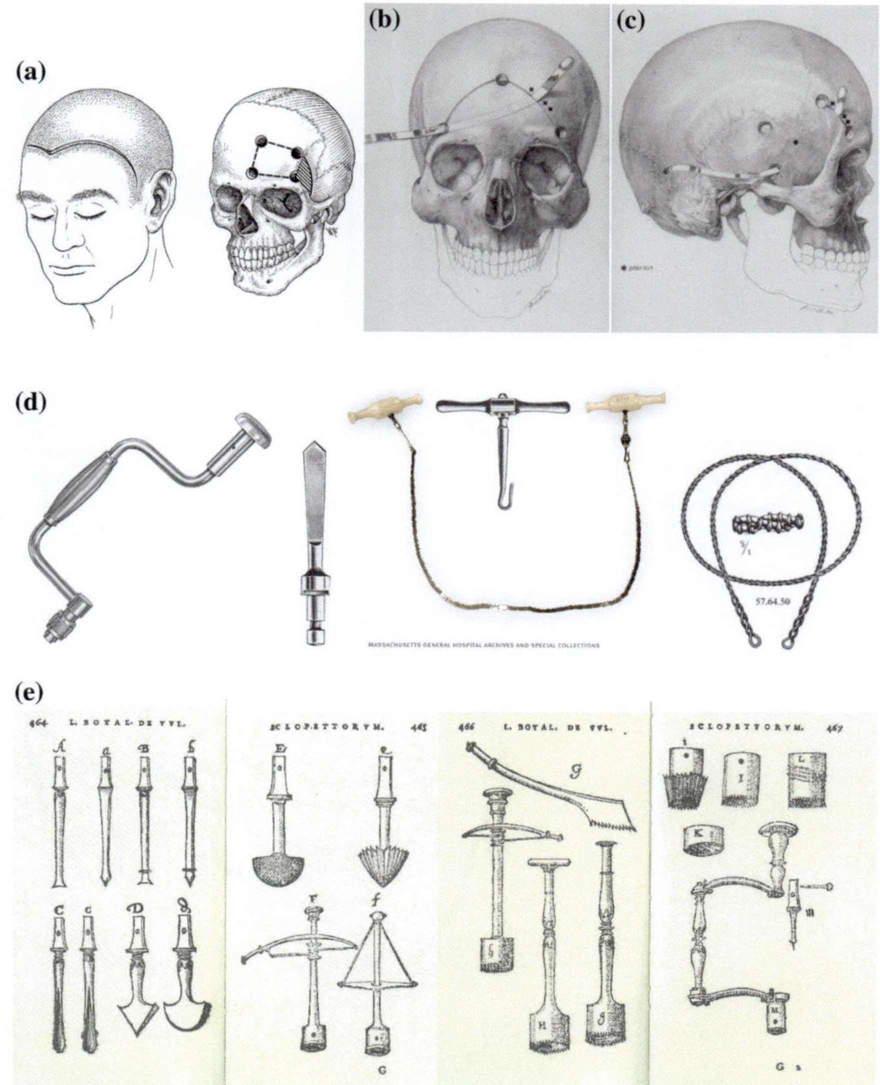

Fig. 2.6 Modern trepanation is still partly handicraft despite pneumatic and electric motor drills and saws: **a** skin cut and trepanation holes with lines for the intended removal of skullcap parts: basic craniotomy; **b** frontal approach; **c** pterional approach; **d** Hand drill to make burr holes; **e** Gigli saw with grip and rolled up saw (Courtesy M. Van Dijk, GUMC and RTWM Thomeer, LUMC with permission). For comparison, the pictures from the "Strumenti chirurgici: De curandis vulneribus sclopettorum" (1565) of Leonardo Botallo (1519/1530-1571/1588) are given below, pp. 464–467. Botallo's preference is the trepan bit indicated by L, because halfway down it contains a counter-screw thread preventing piercing of the skull and consequently preventing lesions of the meninges and brain ("per impedire l'affondamento oltre lo spessore del tavolato osseo e il conseguente pericolo di lesioni meningo-cerebrali", courtesy UTET 1981, with permission)

Table 2.1 Map of Africa with the tribes that still trephine patients (from Rawlings and Rossitch 1994 with permission)

Arabic and Byzantine Trepanation and the Treatment of Head Wounds

The roots of trepanation in Africa are found in the Arabic world, presumably before the first millennium, due to the Islamic dominance of the Nord of Africa that started in between seventh and ninth centuries. Arabic and consequently byzantine trepanations are studied in this part, byzantine being the transmitter of Greek medicine into the Arabic world and also being the intermediate between Arabic and

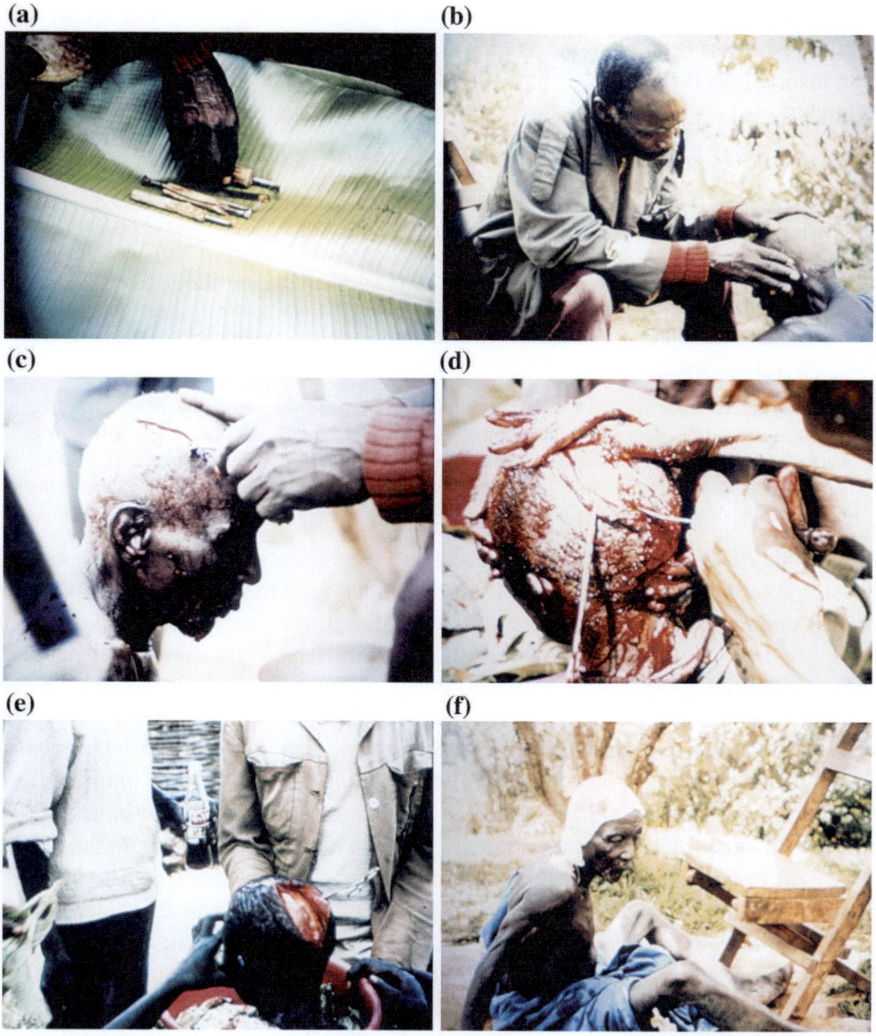

Fig. 2.7 Trepanation in Africa: **a** the scraping instruments; **b** patient inspection; **c** first cuts into the head skin; **d** progress of the trepanation; **e** patient drinks and one can look into the opened skin flaps; **e** patient after the trepanation. © E. Marani

Western medicine (see Duffy 1984; Lascaratos et al. 1998). This part is restricted to two famous Arab doctors: Rhazes and Albucasis.

Abū Bakrs Muhammad ibn Zakariyā al-Rāzī (called Rhazes in the West, 865–925 AD) lived at the start of the Arab Muslim Golden Age. As a physician and later on as the chief clinician of Baghdad's hospital, he was known with the works of Hippocrates (c460–c370 BC), Galen (129/130–200 AD) and Paul of Aegina (625–690 AD) due to earlier translations into Arabic, which he used in his practice and studies. He published over 150 books on several themes, but most on medicine. His Kitāb al-Mansūrī (known in the West as Liber medicinalis ad Almansorem or Liber Almansoris [see Aciduman et al. 2014]) contains a chapter on how to handle head wounds:

> *Chapter 26. On wound of the head and fracture of the head bone*
>
> When the head is wounded but the bone is not broken, aloe, myrrh, dragon's blood, starch of wheat, olibanum and coral should be sprinkled on the wound, since it may quickly heal. However, if the bone is dented or cranium is separated, confusion and spasm may occur and thereafter death promptly follows, [because of that] this condition should not be disregarded. In such case, broken bone(s) should be quickly taken out and the panniculus/ġishā (membrane) under it/them should be protected from tearing. A skilful and gentle physician/surgeon is necessary for this operation.
>
> We told our promise what we wanted to mention: the great danger of this illness must be learnt and the treatment is initiated immediately by removing the bone(s), before above mentioned evil symptoms, which are frightening and unavoidable, take place. Physician/surgeon should be very cautious for not to rupture the membrane which is located under the bone (Translation is from Aciduman et al. 2014).

The advice is along Hippocrates' and Galen's proposed treatment: stay away of the structures below the dura mater [(the superficial hard membrane enwrapping the brain) Aciduman et al. 2014]. Note that trepanation is not recommended in this part.

A good seventy-five years later, Abū 'l-Qāsim Khalaf ibn 'Abbās al-Zahrāwi called Albucasis (936–1013) published thirty volumes on medicine called: "Kitāb al-tasrif li-man 'ajiza'an al-ta'ālif" in the Arab Muslim world of those days. The surgical treatment is described in the three books "On surgery and instruments" (Albucasis 1000). In book III "On bone setting", both "Head injuries" and "Fractures and dislocation in general" are treated, indicating that head injuries were considered bone fractures. In its Chapter Two "On a fracture occurring in the head", three types of traumas are described: Ascial, a cut through the skull into the dura mater; Absolute avulsion, cut that is restricted to the bone; and Capillary, a fine crack within the bone concerned. Extra attention is given to falls or blows that will cause wounds with splinters especially in children.

The text continues with a clear warning. Do not treat the patient if he shows dangerous symptoms: "bilious vomit, convulsion, and mental derangement, loss of voice, fainting, high fever, protrusion and inflammation of the eyes. Leave the patient alone and do not treat him; for in the majority of cases these symptoms undoubtedly indicate the approach of death" (Albucasis 1000). The advice is understandable, since causing the death due to incompetent medical handling could be judicially punished with loosing one hand (Goodrich 2014).

"All the fractures are diagnosed by laying them bare and investigating them with probes and removing all the lacerated tissue over them. The "capillary" kind is diagnosed by laying bare the bone, wiping it clean and rubbing ink on it, when the fracture will appear black" (Albucasis 1000). The trick with Indian or Chinese ink for the capillary crack can also be found in the Latin translation (Ballester 1967) and must have been known in the Middle Ages. Moreover, inspection should be done with probes, thus with instruments that were specially developed by Albucasis.

"If the fracture has already reached the cerebral membrane with fragmentation of the bone and depression, then you must cut out the splintered and broken part as I shall now describe to you" (Albucasis 1000). So the dura mater is no longer the border for treatment or non-treatment. Splinters that stick into or through the dura can be removed.

The treatment is: shave the head of the patient. If you encounter an abscess or bleedings, treat these first with gauze bathed in wine and oil of roses. If these are under control, start the trepanation and take away splinters and bone pieces. Albucasis gives two methods:

Method 1: take chisels (strokes on the chisel should be gentle!) and start with the smallest and later on the broader ones for the bony parts removal. In case the bone is too hard, you make holes with drills called "non-sinking" drills (drills with locking edge are meant Fig. 2.8h). Since skullcap bone is different of thickness, you need different drills for various patients. Make several holes, cut with the chisel and remove the bone piece with your hands or forceps. Chisel or drill has to spare the dura mater. Roughness of the edges should be redressed.

Method 2: put beneath the severest part of the fracture but above the dura a lenticular chisel. Edges should be sharp, surface should be smooth. If dura sticks to the bone parts, "strip it off with the lenticular tip of the chisel and free it gently from it, for it will come away without harm or danger" (Albucasis 1000). Note that Albucasis knew that the dura adheres to the bone of the skull cap and the method of detaching is still the same in modern neurosurgery (see Fig. 2.6b, c).

If the cut is involving only bone, one should scrape the bony wound if there is roughness.

The text has been translated into Latin by Gerard of Cremona[1] (1114–1187) and stayed for ages the main medical source for medieval doctors. The instruments in surgery are also portrayed in Albucasis' "The practice". The two instruments used for trepanation are also taken from the Latin translation present in the library of the Montpellier University and from the Latin edition of 1594 of Gerard of Cremona (Fig. 2.8). It is clear that the Albucasis instruments are different from the Greco-Roman classical ones (see also the late byzantine instruments in Goodrich 2014). In Albucasis' description, the dura mater is no longer a border that can not to be crossed.

The medical profession got during byzantine times its own saints: Cosmas and Damian. Three couples of Cosmas and Damian exist that are recognized by the "synaxarion of the Greek church and with their own calendar day": Asian, Arab and

[1]Literature indicates that one century later, a second Gerard of Cremona, called originally Gerard de Sabloneta, translated Albucasis.

Roman Cosmas and Damians. "According to the Asiatic tradition, which is the oldest of the three, the saints died natural deaths and were buried in Pelusion, Egypt. The Roman and Arabian tales crown the life of the saints with the martyr's death, the latter version being the only one recognized by the Roman church" (Zimmerman 1936). Still these saints are the patron saints of surgery, and they are pictured with a box in their left hand and a surgical instrument in their right hand. The other doctor saint, Panteleimon: being a good physician, practicing in a philanthropic manner, like Cosmas and Damian did, healing diseases more by the "grace of Christ than by medical treatment". The amazing result of studying the surgical instruments of Panteleimon and of Cosmas and Damian is that they hold in some pictures the trepanation instruments of Albucasis and of the classical Greco-Roman ones (Fig. 2.9).

Undertaker

It is not generally known that the skullcap is important for the undertaker, as shown by the US Patent 4,275,490 dated June 30, 1981: during autopsy, the skullcap is sometimes removed to obtain access to the brain. The background of the invention reads in part:

"After the autopsy is performed, the corpse is taken by an undertaker who then has to prepare the corpse for the funeral. It then becomes the responsibility of the undertaker to resecure the cut and removed cranial cap skull section to the basal skull section, and this obviously is a very important responsibility where the corpse is going to be viewed by family, friends, acquaintances, and others". So, Mister Bivins, our undertaker, found a method and apparatus to overcome calvarium shift of the dead and to beat uneasiness of the mourners. In some cases, he has to cover up the total destruction of the presumably still present brain (parts).

The invention abstract is given here together with its first figure (Fig. 2.10): "the present invention relates to a method and apparatus for firmly securing a formerly cut and removed cranial cap skull section to the base portion of the cranium, after the cranial cap skull section has been cut and removed in the course of an autopsy. To accomplish this, a frictional plug or hook assembly is anchored in the cervical foramen cavity extending downwardly from the base of the skull, and another anchoring assembly is secured to the removed cranial cap skull section. A flexible coupler in the form of an elastomer is operatively attached between said frictional plug or hook assembly and said anchoring assembly anchored in said cranial cap skull section so as to hold with a tensional force the cranial cap skull section firmly in place about the cranium, thereby effectively completing the calvarium closure".

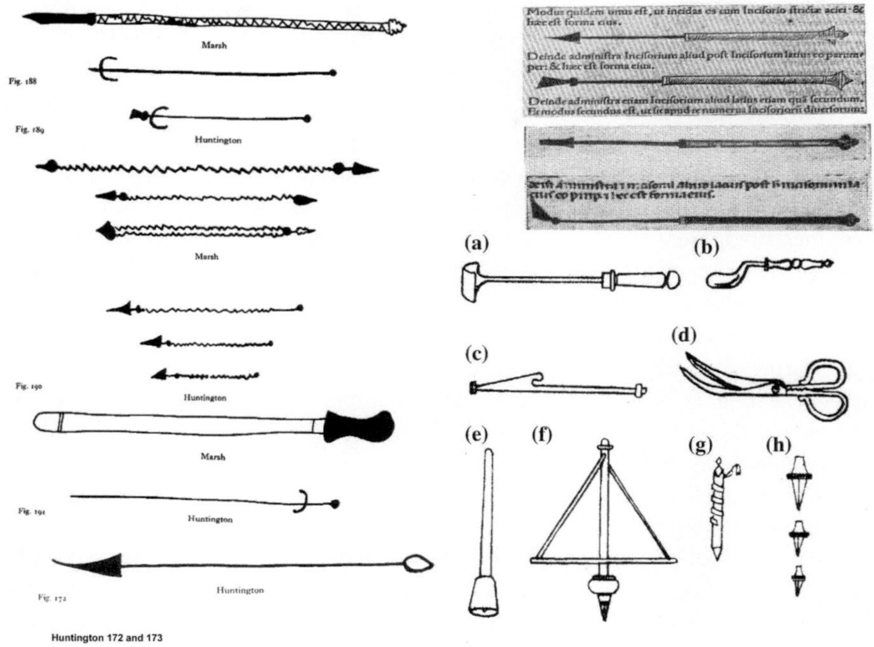

Fig. 2.8 Left side shows surgical trepanation instruments of Albucasis from the book specimens, called Huntington and Marsh, at the Bodleian Library. 188, fine blade chisels; 189, larger chisel; 190, large and small drills; 191, brain spatulas; 172/173, osteotome. Right side: demonstrates the reproductions of the surgical trepanation instruments as depicted in the Latin translations of Gerard of Cremona of Albucasis volumes (Ballester 1967, figures. are from "La trepanacion en España" (1967) ed. JM Lopez Piñero, Editorial Tecnica Española, Madrid). Right side demonstrates the surgical trepanation instruments as are known from the publications of Celsus (25BC–50AD) and Galen of Pergamon [(129/130–200), see Lopez Piñero 1967]: **a** xyster or scratcher; **b** meningophylax, controller of meninges; **c** phakatòs, lenticular knife; **d** koilískos, concave pincers tool; **e** modiolus, trephine; **f** and **g** terebra, trephine and drill; **h** trypana abaptista, trephine drill with locking edge

2.3 Calvarium Research

2.3.1 Spaceflights

We start with the most surprising research: spaceflights. During spaceflights, bones are unloaded, since gravity is strongly reduced, called microgravity. Bones that are normally loaded suffer severe reduction of bone mass when unloaded for a prolonged period.

> Physiological studies of the Skylab missions document increasing calcium and phosphate levels in both serum and urine of crewmembers, reflecting a gradual loss of bone mineral content in these astronauts. Some of these crewmembers continue to lose bone mass after returning to Earth, and those who experience the greatest bone mass decrements do not

fully recover even after five years. The correlation between bone loss and diminishing bone strength suggests a significant risk of fractures for astronauts (Zhang et al. 2013).

The calvarium, on the contrary, is not loaded; it does not bear body weight. The space shuttle flight STS-131 started on April 5, 2010, and returned on April 20, 2010, with on-board mice for calvarium research under microgravity. The results are unforeseen: "fifteen days of microgravity aboard the STS-131 mission induced adaptive changes in the non-weight-bearing murine calvaria. Of the three parameters measured, there was a statistically significant increase in bone volume, and a trend of increase in average cortical thickness. Tissue mineral density remained relatively unchanged" (Zhang et al. 2013). Thus, while your arms and legs become brittle by loosing bone minerals during microgravity, your skullcap increases its mass during spaceflights and keeps the bone mineral content stable. This result obtained within 15 days microgravity, although researched on mice, comes out of the blue for this author.

The other side of the picture is also present in the literature: osteoporosis, the loss of bone mass and density, thus a decrease of strength due to hormonal changes present during ageing also has influence on the cranial vault. The main results as obtained by Japanese researchers for sex differences on skull changes are summarized here: fracture load in cadavers was determined and the older the person the more brittle the skull cap was: "this indicates that the stiffness of cranial bones in Japanese individuals decreases with age, and the risk of skull fracture thus probably increases with age… Significant negative correlations between cadaver age and sample thickness, the ratio of the sample thickness to cadaver stature and the sample density were observed only among the female samples… These findings suggest that the skull is substantially affected by decreased bone metabolism resulting from osteoporosis" (Torimitsu et al. 2014). Criticism has been published earlier on osteoporosis effects on the calvarium. However, Obert et al. (2013) showed on statistical grounds that the skullcap osteoporosis effects are present in adult females.

2.3.2 *Sutures, Fibroblast Growth Factor Receptor and Craniosynostosis*

Craniosynostosis is the abnormal early growing together of some of the skull bones: abnormally early cap bone melting. "Craniosynostosis is important for two reasons. It provides a model system for studying the genetic and environmental factors in a pathway of developmental malformation, and it represents a significant medical problem, occurring in ~ 1 in 2500 individuals" (Wilkie 1997). "It is the premature loss and ossification of sutural growth centres and synostosis (fusion) of one or more sutures is accompanied by compensatory growth, both in other sutures and by remodelling (appositional growth) of other parts of the skull" (Morris-Kay and Wilkie 2005). Figure 2.11 shows the skull deformations if only one suture is involved in craniosynostosis.

Fig. 2.9 Left are Cosmas and Damian of a frescoe from the Nedelynsky monastery. Right upper picture is Cosmas and Damian from Asia Minor published by the Orthodox Synaxarion. Lower right is the doctor saint, Panteleimon in an icon from Chilandary monastery, Hagion Oros. Surgical instruments are enlarged: from left to right are noticed the surgical instrument of Albucasis twice and a phakatòs, the lenticular knife, in the hand of Panteleimon

Not only in skulls are deformations noticed, but in most syndromes broad thumbs and big toes, bony syndactyly (fusion of hand or foot bones) or connective tissue deviations are found. Some of the craniosynostoses were difficult to determine, but genetics showed that a series of these abnormalities like: Crouzon, Apert, Pfeifer, Muenke and achondroplasia syndromes can be reduced to mutations, among others, of fibroblast growth factor receptors (FGFR, Fig. 2.12).

How can one receptor be responsible for so many syndromes? It relates to the skull development, especially the bone edges at the sutures.

The edges of normal growing skull bones (Fig. 2.13) are characterized by bone producing cells of two types: differentiated bone cells (osteoblasts) and undifferentiated osteoblasts. Thus on the edges of developing skull bones, bone cells totally encapsulated by minerals are found and here the differentiation takes place. If the cells still get activation by FGFR2 receptors, they stay in the proliferating phase. Nearing the bone edge, their FGFR1 receptors are activated and the bone cells start to differentiate at the bony edges, taking up bone minerals and becoming encapsulated bone cells (Fig. 2.14).

It is the balance between proliferation and differentiation that produces a normal skull. In the Crouzon syndrome, there is strong activation of the receptor signalling

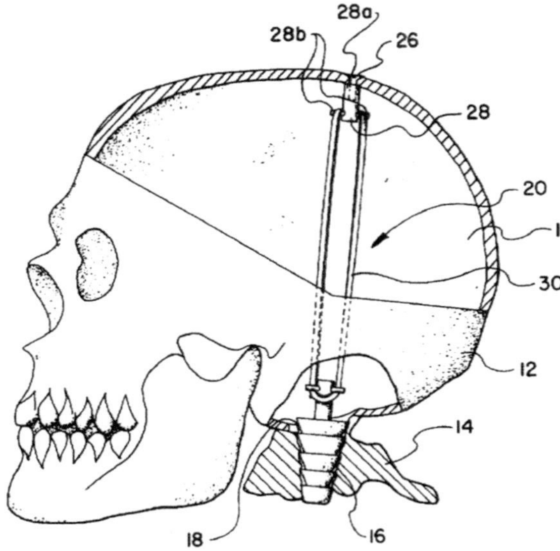

Fig. 2.10 Apparatus for securing calvarium skull section to basal skull section: 10, cranial cap section; 12, basal skull section; 14, plastic material; 16, a frictional plug (18) that is constructed of plastic or any other suitable material and which is adapted to be anchored within the cervical foramen (magnum, 16) cavity; 20, closure device; 26, a selected elastomer coupler 30 that when retained between the respective coupler attaching means 24 and 28, that the elastomer coupler 30 exerts a tension force that tends to pull the cranial cap skull Sect. 10 firmly against the basal skull, 28a hollow-threaded shaft, 28b a pair of pin studs or dogs (Bivins 1981)

because the receptor protein structure is changed and other abnormal types of activation work, which can cause strong growth and early closure of sutures, leading to a deformed skull. Early loss of the coronal suture (Figs. 2.11 and 2.13) due to strong development will be compensated for the domed skull, since the brain volume still increases. The not-yet-closed sutures with their bone plates have to absorb the volume increase of the brain by extra growth and late closure of its sutures. (Note that several other mutations also play a role in craniosynostosis, e.g. *TWIST1* and *EFNB1*; see Jezela-Stanek and Krajewska-Walasek 2013).

Surgical intervention in craniosynostosis has been effective in only 60–80% of the cases, but could be brought up to a 95% success rate (Schaller et al. 2012): "our study on 172 infants/children demonstrated that the functional and cosmetic outcomes in long-term follow-up are better after surgical treatment within the first 6 months of life. The cephalometric index of head circumference represents an excellent parameter to compare the postoperative outcome in this series and also will allow comparison with data from other centres of excellence. The operative risk is not higher in infants than in older children. SPECT studies underline the logic of treating infants with craniosynostosis with surgery within the first 6–8 months of life to give the best opportunity for the minimization of further cognitive sequelae" (Schaller et al. 2012).

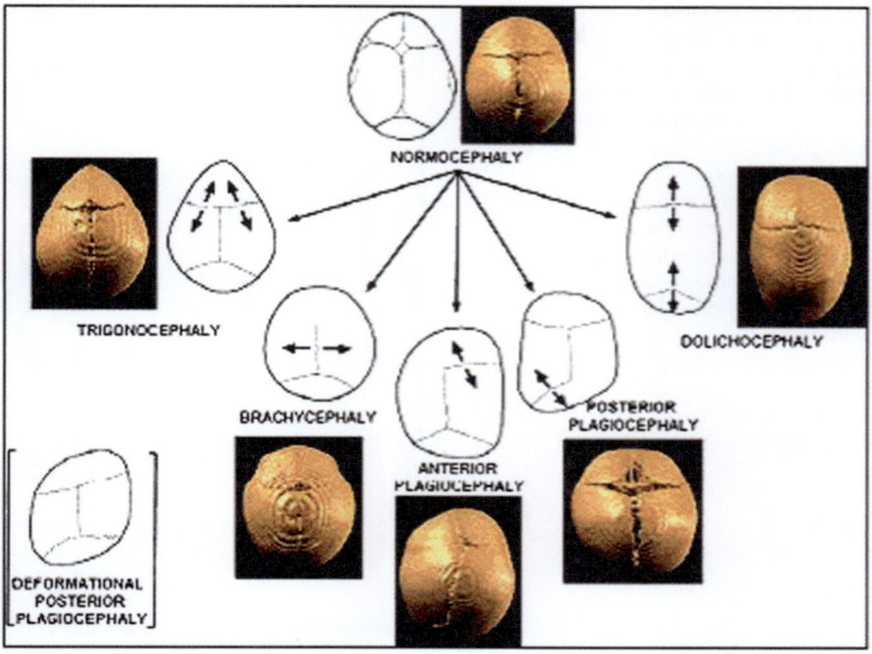

Fig. 2.11 Skull deformities that originate from craniosynostosis induced by one particular suture. Arrows indicate skull growth direction and 3D-CT reconstructions of the patient's calvarium deviations are presented (from Boyadjiev 2007, with permission). Trigonocephaly, triangular skull, premature fusion of metopic fissure; brachycephaly, flat, disproportional wide skull, premature closure of coronal (or lamboid) suture; plagiocephaly, asymmetrical distortion of the skull by partial premature closure of coronal or lambdoid fissure; dolichocephaly or scaphocephaly, long head, premature closure of sagittal suture

2.4 Skull Structure and Yellow Bone Marrow

The bony structure of the skull is developed by two processes: the calvarium arises from membranes and the skull base originates from cartilage. The cranial vault is made of two layers of bone, so-called tables, and in between there is bone marrow and cancellous bone. This diploic space is present from the age of 4 years, and at the age of 15, the bone marrow is totally replaced by fat. Bone marrow nearly gives no signal and fat a rather high signal in MRI (Fig. 2.15). The red to yellow, fat-containing, bone marrow transforms in a certain order: it starts in the facial bones (8.5 till 9.3 years of age), goes over to the frontal bone (9.7 years), reaches the parietal at 11.5 years and the change arrives at the occipital bone and skull base at 12 years of age (Simonson and Kao 1992). At the age of 15 years, all skull red bone marrow is changed into yellow bone marrow. Normally the yellow bone marrow can be noticed in the MRI as a small white line. Its form change is used to ascertain certain illnesses.

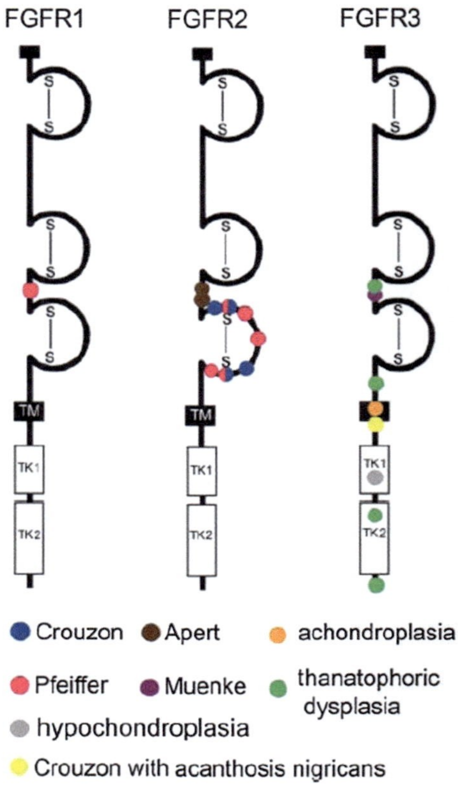

Fig. 2.12 "Structure of fibroblast growth factor receptor proteins (types 1–3), showing the position of some of the common mutations causing craniosynostosis, and (FGFR3 only) some of the mutations affecting long bone growth. Each receptor has three immunoglobulin-like domains (Ig) whose structure is maintained by disulphide bonds (s–s); TM, transmembrane domain; TK1, 2, tyrosine kinase domain" (Morris-Kay and Wilkie 2005 with permission)

Paget's disease is characterized by serious bone demolition and by a renewed increase of bone growth. The newly formed bone's construction is badly organized with the consequences of diminished structure. Skull involvement of Paget's disease is present in up to 65% of affected patients. The disease involves not only the calvarium but also the skull base (Fig. 2.15). The fat signal is increased due to the expansion of the calvarium and the clivius (Bag and Chapman 2013). The small white line in the bony skull changes into a cotton wool appearance of the yellow bone marrow.

2.5 Vesalius (1514–1564) and Skull Morphology

The birth of *scientific* anatomy in Europe is easily pinpointed. Andreas Vesalius (1514–1564) changed the direction of the anatomical science in favour of a real analysis of the human body, standing of course on the shoulders of others (Marani and Koch 2014). In his opus magnum: "*De humani corporis Fabrica libri septem*",

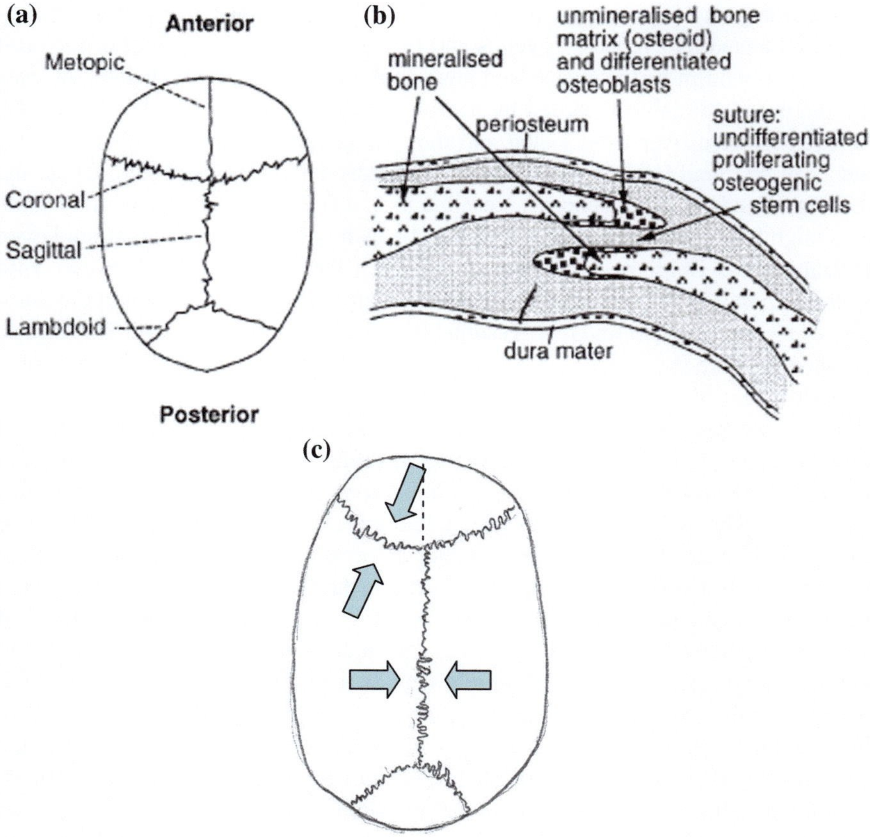

Fig. 2.13 "Normal cranial suture development. **a** View of child's skull from above, showing position of the major sutures. Coronal craniosynostosis leads to a short, broad skull; conversely, sagittal synostosis leads to a long, narrow skull. **b** Diagrammatic cross section through coronal suture. The skull bones overlap slightly. In craniosynostosis, the narrow space separating the bones is obliterated"; figure and text from Wilkie (1997 with permission). **c** The two main opposing juvenile growth directions indicated on a mature skull for the coronal and sagittal sutures. Metopic fissure normally disappears on the frontal bone. It is an expression of two bone growth centres that construct the frontal bone, which is a normal synostosis

Vesalius (1543) gave an extensive overview of the skull and spent several pictures to explain its bony structures.

As a child, Vesalius lived in Brussels in a house placed next to the "Helle straetken" (ruelle de l'Enfer or in English Hell Alley) behind which an open area with a hill, called "De Galgenberg" (gallows hill), was located. The body of the criminals that were hanged stayed on the gallows till the bones fell down. "With his friends he will have secretly climbed the hill and will have played with the bones and skulls, despite his parents ban" (Lindeboom 1964). As Vesalius indicated, this, together with the dissection of small animals, raised his anatomical interest.

According to Vesalius, the head was formed for the sake of the eyes, which nature protected by eyelashes, eyelids and frontal bone. The lower part of the orbit is saved from harm by the cheeks and jaws. In the middle, the nose shelters the eyes and the outer side is safeguarded by a strong bone. The idea of the head made for the eyes is taken over from Galen and the four unnatural cranial vault types (see later) can be found in Homer and Galen (Richardson and Carman 1998). Under the heading "Heads of other shapes", Vesalius lists: a Bolognese beggar with a square head slightly broader than long, a hydrocephalus boy in Genoa shown around even in Brabant and Belgium and deformed heads in the Styrian Alps. The two solid layers of bone of the calvarium with in between "fistulous or fungous bony substance" are already described by Vesalius (see Sect. 4.1), together with the sutures, the outside of the skull base, side view and interior of the calvarium and the interior of skull base.

The first extensive picture in the part "Why the head is so shaped, the number of different shapes" (his Chap. 5) concerned five skulls that were different in form and sutures: skull variability and malformations were already described by Vesalius (Fig. 2.16). The skull in Vesalius Fig. 1 "portrays the shape of a normal head or skull; it is like an elongated sphere slightly depressed on either side and swelling out at the front and back. Figure 2 demonstrates the first unnatural shape, in which the anterior eminence is missing. Figure 3 shows the second unnatural shape of skull, in which the anterior (anterior here is a print mistake of Vesalius and ought to be [posterior]) eminence is lost. Figure 4 depicts the third unnatural shape of skull, in which both swellings, anterior and posterior, are missing. In Fig. 5, we have illustrated the fourth unnatural shape of head, in which the two eminences are at the sides, not the front and back" (translation is from Richardson and Carman 1998).

In the Middle Ages, it was thought that the skull sutures are different in man and woman, based on Aristotle: "The human skull has the most sutures, and the male has more than the female". Vesalius denied it: "Men's skulls do not always differ from women's" (Richardson and Carman 1998).

Thus, Vesalius knew the general form of the sutures of the cranial vault: they form an H and he discussed the malformations: then they form a T.

Most remarkably, Vesalius noted that the outline of the sutures in the unnatural heads is clearly dissimilar, thus he recognized different malformations: "When the anterior eminence of the head is missing, the coronal suture is abolished but the lambdoid is preserved along with the sagittal that then extends between the eyebrows. When the posterior eminence is lacking, the lambdoid suture is lost but the coronal remains and the sagittal extends through the middle of the occiput all the way to the base of the skull". In the first case, the metopic suture remains present; we should say nowadays. Vesalius sees it as an extension of the sagittal suture, idem for the second case. The Sect. 2.2.2 explained what happened by fibroblast growth factor receptors' (FGFRs) malformations: early closure and thus losing, for example, the coronal suture. Vesalius described these phenomena in his figure (see Fig. 2.16) but could only use the presence, absence and extension of sutures to discern between variability and malformations.

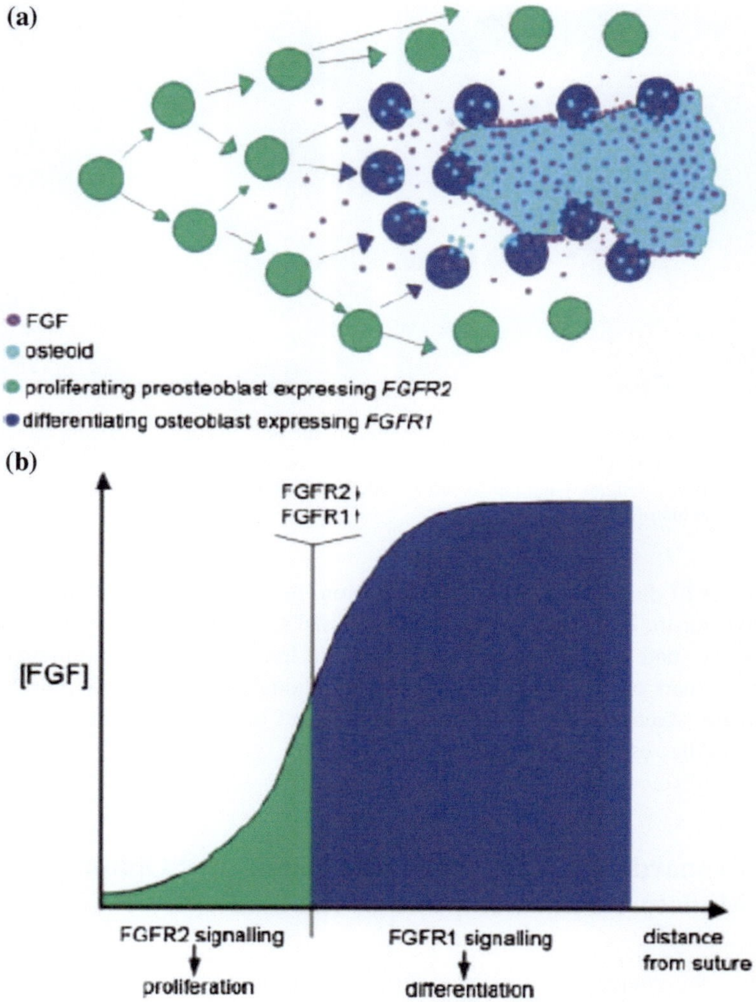

Fig. 2.14 "The relationship between FGFR expression and the progression from proliferating osteoprogenitor cells to differentiating osteoblasts at the edge of a bone in the mouse coronal suture. **a** Osteoblasts (blue) express FGFR1 and secrete bone matrix proteins (light blue) and FGF (pink); FGF diffuses into the region of proliferating osteoprogenitor cells (green). **b** Hypothetical scheme suggesting that a threshold of FGF concentration effects the change in gene expression from FGFR2 to FGFR1 and the change in cell behaviour from proliferation to differentiation". (Figure and text from Morris-Kay and Wilkie 2005 with permission)

Sutures are not a too constant phenomenon. Mosaic or wormian bones, small irregular, intrasutural bones, representing detached centres of embryological ossification in the skull, are not uncommon (see Fig. 2.17). In Vesalius' chapter on the skull, no indication of the recognition of mosaic or wormian skull bones was found.

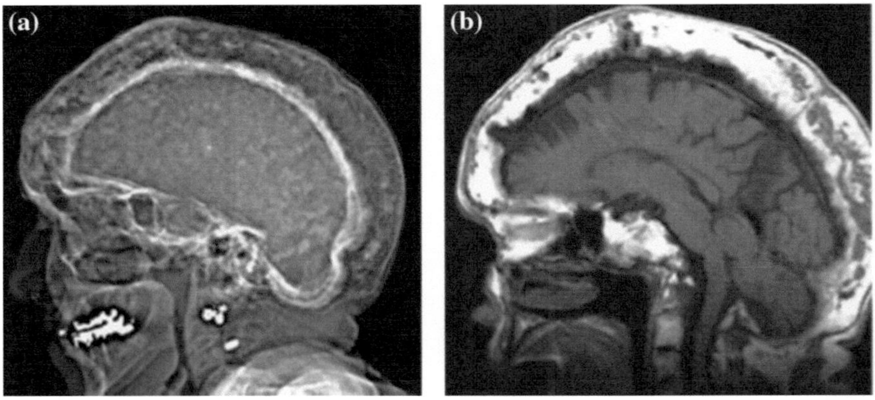

Fig. 2.15 "A 79-year-old female with Paget disease of the skull. **a** Lateral plain film demonstrates marked thickening of the skull base with cotton wool appearance. **b** Sagittal T1-weighted MRI reveals extensive fat signal within areas of expanded calvarium and clivus". Text and figures from Bag and Chapman (2013) with permission

A series of diseases show this phenomenon, for example, idiopathic osteolysis, hypothyroidism, Zellweger syndrome (autosomal recessive disease with absence of peroxisomes and thus a restricted break down of long fatty acids) and Prader–Willi syndrome (part of chromosome 15 is not functional leading to low muscle tone, incomplete sexual development, small stature and cognitive disabilities and insatiable appetite resulting in obesity).

2.6 Leonardo Da Vinci (1452–1519) and the "Oldest" Miniature-Scale Sculpture of the Skull

In writing this chapter, attention was attracted by an article in Le Monde of June 21, 2014: *Tempête pour un crâne: La sculpture miniature attribuée, début juin, à Léonard de Vinci est au centre d'un débat qui oppose les historiens d'art.* [Storm due to a skull: the miniature sculpture assigned to Leonardo da Vinci, since June, is the centre of a discussion between art historians]. (The title is for French people reminiscent of Victor Hugo's Les miserables by the chapter: Tempête sous un crâne.)

The original article was published in the Wiener Medizinische Wochenschrift by Stefaan Missinne (2014), a Belgian private researcher, who described the oldest globe that is also attributed to Leonardo da Vinci. The miniature skull was originally bought in 1987 in an antique shop by a couple and described by Saban and co-workers in 1996; it was also the subject of a thesis by Elisabeth Irene Ahner. While Saban and Salf (1996) and Ahner (2007) were rather careful in their wording in attributing the miniature of the skull, Stefaan Missinne made a big hit by attributing, beyond question, Leonardo as the maker of the miniature skull (see, e.g. Even, Le Monde 2014 or the Web). This is not the place to interfere in the discussion between art historians, but after reading the articles a more subtle and gentle consideration of the work of Elisabeth Irene Ahner by Stefaan Missinne should have occured. In short (according to Missinne), the miniature skull is of a more than 50-year-old male, contains filigree osteopathic details, has on its back side (os occipitale) most of the bone lines (linea nuchae superior and inferior and the external occipital protuberance) and has detailed eye sockets (see also Fig. 2.18 for miniature dimensions and below).

Our target, given this chapter, was to study the miniature skull as presented by the pictures of Missinne (2014), and we do it using the frontal pictures of the miniature skull and the drawing of the two skull halves by Leonardo at present in the Royal Collection at Windsor Castle, Great Britain. It is clear that in the miniature skull the lower jaw is missing and left cheekbone is presented, not the right one. Gerrits and Veening (2013), two Dutch anatomists also studied the Windsor Castle skull drawings (RL 19058v and RL 19057r) and some of their results are compared to those of Missinne (2014):

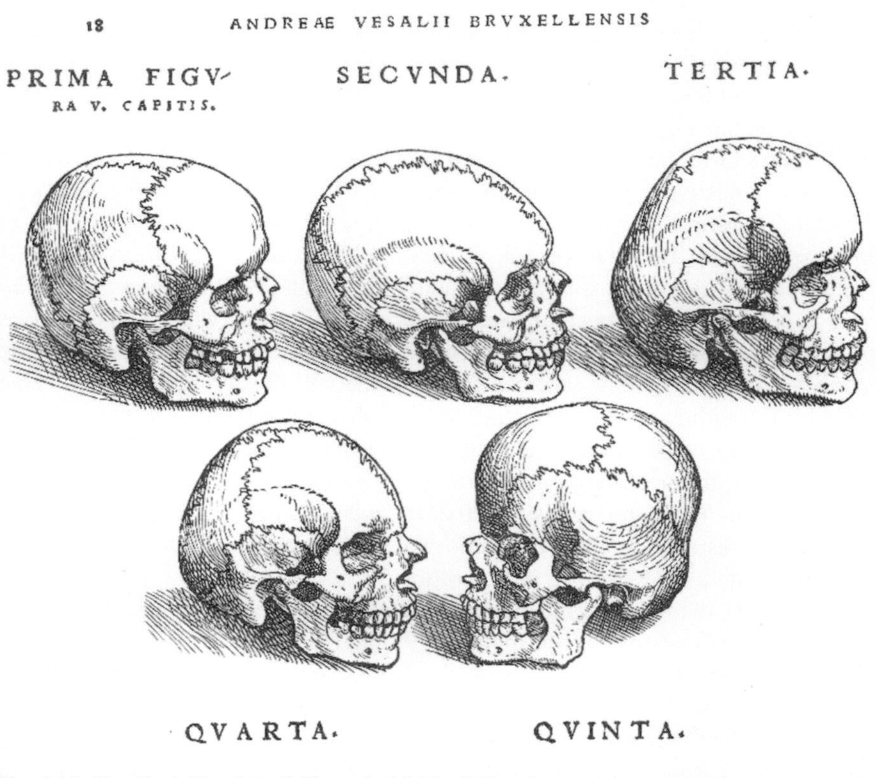

Fig. 2.16 Vesalius' Fig. 5.1 of Chap. 5 (1543): "Why the head is so shaped, the number of different shapes", compare to Fig. 2.11 (Author's collection)

(1). "The inferior orbital fissure (intraorbital suturae) is not shown either in the drawings by Leonardo or in this skull model" (Missinne 2014).
(2). "In both orbitae, the *fissurae orbitales* superior and *inferior* were delineated and the canalis opticus was positioned medially to these fissures" (Gerrits and Veening 2013).

Here, one encounters confusion: the fissura orbitalis inferior (=inferior orbital fissure) is not a suture, nor sutures. Within the orbita, several sutures are present between several bones. Indeed, Leonardo did not indicate the sutures within the orbita, since even for modern students these sutures are difficult to memorize. I (EM) at least cannot remember the sutures between: os nasale, os lacrimale, os ethmoidale, os sphenoidale, os frontale, os zygomaticus and os maxilla and regularly have to consult a handbook. However, it is surprising that the surface suture between cheek bone and maxilla (zygomaticomaxillaris) is not indicated in both. It is a clear suture and difficult to miss. The conclusion should be that fissures not related to the cranial vault are not reproduced (note that the sutura internasalis is connected to the frontal bone).

The canalis opticus with its foramen, the fissure orbitalis superior and inferior were absolutely correctly placed within the human orbita by Leonardo (Gerrits and Veening 2013) and this is confirmed by Missinne (2014) for the miniature skull. The soul or "senso comune" was placed by Leonardo just above the chiasma of the optic nerves after thorough skull, brain, cranial nerves and eye studies (see discussions on finding the soul by Leonardo; Del Maestro 1998).

The miniature skull has been attributed to Leonardo da Vinci. Let's use the good old Turner's brim index. "Turner's brim index stood the test of time, because it is expressed in relative terms so that it is applicable to pelves of every size from dwarf to giant" (Stewart 1984). Here we use it not for the pelvis, but for the skull to produce an index. Take the breadth, divide it by height and multiply it with 100. If we take these measurements from Lanz und Wachsmuth (1979), *always top cranial vault to bone edge of the maxillary teeth implants, as in the miniature skull*, it gives us 85, as Spalteholz (1932) does. Leonardo's pictures from the Windsor collection shows 87 or 88, depending whether one takes the half breadth twice or the whole breadth of both halves. The miniature-scale skull measure result, on the contrary, gives 105. Thus while Leonardo and the anatomical handbooks show the skull, over our measured distance, longer than broad, the miniature skull makes them equal thus too broad. The difference is 17–20 of the index. Is there anyone who believes that after all his skull studies, Leonardo would have sculptured a miniature skull with more than a 15% mistake in proportions?

The largest diameter of the orbita is obliquely placed (see, e.g. drawing of Leonardo and Lanz und Wachsmuth 1979). In the miniature skull, the largest diameter is nearly horizontal, a mistake Leonardo cannot have made. The index of the orbita, height divided by its breadth and multiplied with 100, gives 87 for the miniature skull. This value belongs to the hypsiconchy type (high shell orbita) known from Chinese and from Eskimos and is present in the Western population in a minority of cases (Lanz und Wachsmuth 1979). Since the miniature skull is said

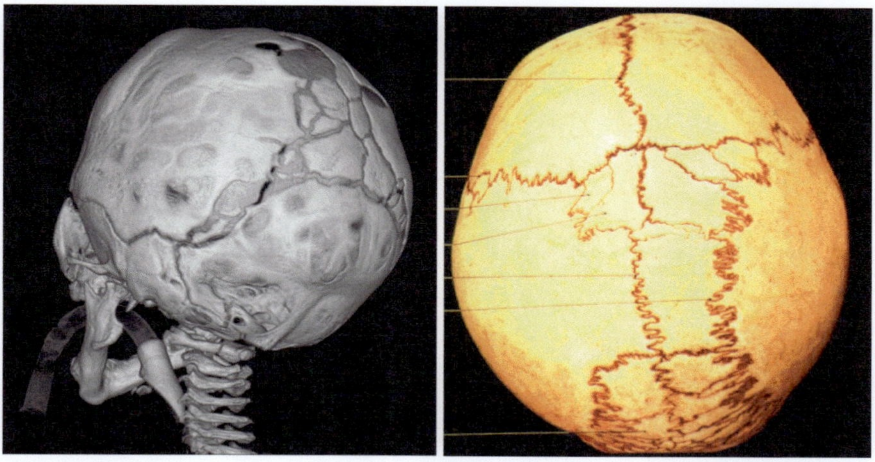

Fig. 2.17 Multiple mosaic bones: left 3D-CT study showing multiple intrasutural bones near the lambda and lambdoid suture in a patient with bicoronal synostosis. Note the widened sagittal suture (left figure from Carter and Anslow 2009; right figure from Lang 1981, with permission)

to represent an aged skull, one should also consider gerontology: ageing can increase the index but not too much, but ageing cannot change the oblique axis of the orbita into a horizontal one, even the reverse is noted since significant bone remodelling both supero-medially and infero-laterally occurs during ageing, reinforcing the oblique topography (Kahn and Shaw 2008) (Fig. 2.18).

The other argument is that the miniature skull shows all the topographical landmarks that only Leonardo had available before 1508. The check and comparison showed that most landmarks can be found in the miniature skull (see the comparison in Fig. 2.18 with my admiration for the maker of the miniature skull), unless the miniature skull is not from 1508 but from a much later date. Then the results of Vesalius and others could have been incorporated. Since the miniature skull has no date on it, stating its year of production is a gamble. We know that miniature skulls were produced later on, after Leonardo's death, in Italy, Germany and Venice (see Le Monde article). In short, Missinne (2014) has saddled art historians with a discussion, which is hardly anatomically supported due to a false skeleton construction: to broad skull and a square orbit (Fig. 2.18).

2.7 Epilepsy and Electrodes

If medication to overcome epileptic seizures does not work (it occurs in 20–30% of the patients), removal of the irritating part of the brain is considered. The problem is to find the irritating zone, especially if it is localized deep within the brain. Placing a

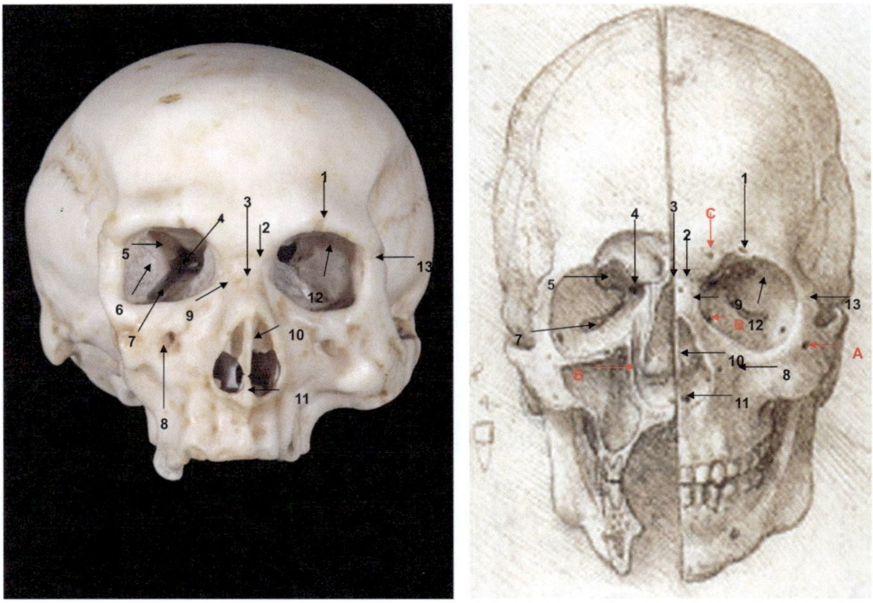

Fig. 2.18 Left: frontal view, from the miniature-scale model of an aged skull attributed to Leonardo da Vinci: measurements are: height 4.73 cm, width 4.98 cm length 5.83 cm (photograph is from Missinne 2014, with permission): 1, foramen supra orbitale; 2, sutura nasofrontalis; 3, sutura internasalis; 4 canalis opticus and lower part fissure orbitalis superior; 5 fissura orbitalis superior; 6, ala major; 7, fissure orbitalis inferior; 8, foramen infraorbitale; 9, sutura nasomaxillaris; 10, lamina perpendiculari ossis ethmoidalis; 11, vomer with spina nasalis anterior; 12, os frontale inside orbita; 13, sutura zygomaticafrontalis. The juga alveolaria, the bulging of the maxilla over teeth is clearly visible; the sutura intermaxillaris just below the vomer has been indicated as a bony edge not a suture. Right: drawing (drawing RL 19058v) made by Leonardo da Vinci. This drawing over traces of black chalk shows the anterior view of two halves of the human skull. Not found in the frontal picture of the miniature skull is the sutura zygomaticomaxillaris and the foramen frontale (C in red). In the miniature-scaled skull, it is indicated in the text that the foramen zygomaticofaciale (A in red) and the tear duct, ductus nasolacrimalis (B in red) are present

series of electrodes and stimulating such a series produce electrocorticograms that can be analysed (Lopez da Silva et al. 1977).

"The eleptogenic zone is elicited by deep intracranial electrodes' recordings during a spontaneous seizure targeting the resective surgery. Functional mapping using electrode stimulation is also performed to define eloquent cortex and to prevent postoperative functional deficits. That information is essential for taking final treatment decisions and depends greatly on the precise localization of deep intracranial electrodes".

"The *anatomical localization* of the deep intracranial electrodes after implantation in early stages of the EEG recording is a critical issue for the interpretation of neurophysiologic results and surgical planning that has great potential to affect

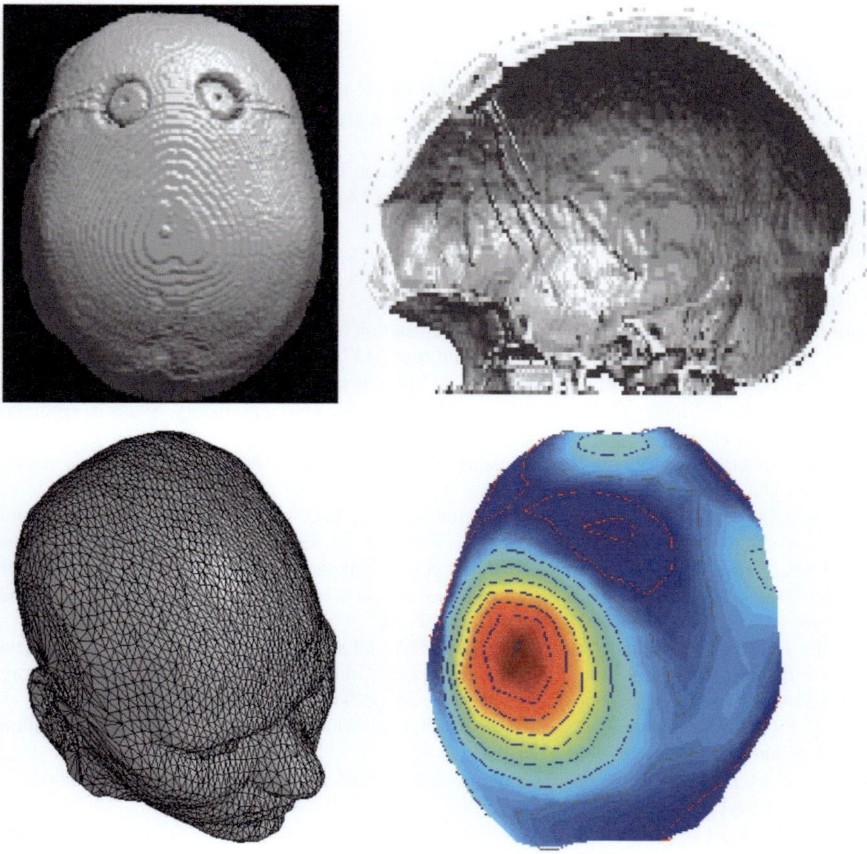

Fig. 2.19 Overview of the cranial vault with the two holes and both leads (left) and the cranial cap inside to show the subdural electrodes. The lower two figures show the mesh used in the finite element modelling and the detection by EEG/EMG (courtesy C. W. Van Veelen, Neurosurgery, University Utrecht and R. Peeters, the Dept. of Biomagnetics, University Twente, with permission)

outcome"(Princich et al. 2013; italics ours). The electrode localization at the end of the intracranial leads has to be accurate; otherwise, the calculated eleptogenic zone is wrongly placed (Fig. 2.19).

Developments in this area are progressing rather fast: multimodel three-dimensional neuroimaging can be connected to neuronavigation and is as such a help for the deep intracranial locations of electrodes to detect epileptogenic zones (Rodinov et al. 2013).

References

Aciduman A, Aşkit Ç, Belen D, (2014) Medieval times' influencing figure Rhazes' approach to head injuries in Liber Almansoris, World Neurosurgery. https://doi.org/10.1016/j.wneu.2014.02.031

Ahner EI (2007) Schädeldarstellungen der Renaissance in Italien - Ein Miniaturschädelmodell als historisch-anthropologische Quelle [Magisterarbeit]. Albert- Ludwigs-Universität Freiburg i. Br

Albucasis (1000) On surgery and instruments. A definitive edition of the Arabic text with English translation and commentary by Spink MS and Lewis GL (1973) Wellcome Institute of the History of Medicine, University Press, Oxford

Andrushko VA, Verano JW (2008) Prehistoric trepanation in the Cuzco region of Peru: a view into an ancient Andean practice. Am J Phys Anthropol 137:4–13

Arnold and Sons (1876) A catalogue of surgical instruments. London, Stationer's Hall

Arnott R, Finger S, Smith CUM (2003) Trepanation: history, discovery, theory. Swets & Zeitlinger Publ, Lisse, pp 1–408

Bag AK, Chapman PR (2013) Neuroimaging: intrinsic lesions of the central skull base region. Semin Ultrasound CT MRI 34:412–435

Ballester LG (1967) La obra de Albucasis, culminacion de la cirugia medieval. In: La trepanacion en España ed. JM Lopez Piñero. pp 51–61, texto 62–73, Editorial Tecnica Española, Madrid

Benazzi S, Fantini M, De Crescenzio F, Mallegni G, Mallegni F, Persiani F, Gruppioni G (2009) The face of the poet Dante Alighieri reconstructed by virtual modeling and forensic anthropology techniques. J Archeol Sci 36:278–283

Bennike P (2003) Ancient trepanations and differential diagnosis: a re-evaluation of skeletal remains from Denmark. In: Arnott R, Finger S, Smith CUM (eds) Trepanation: history, discovery, theory. Swets & Zeitlinger Publ Lisse, pp 95–116

Bivins CF (1981) Method and apparatus for securing calvarium skull section to basal skull section. US Patent 4,275,490, june 30, 1981

Boyadjiev SA (2007) Genetic analysis of non-syndromic craniosynostosis. Orthod Craniofacial Res 10:129–137

Broca P (1867a) Cas singulier de trepanation chez les Incas. Bull Soc Anthrop 2 (2nd series) 403-408

Broca P (1867b) Trépanation chez les Incas. Bull Acad Méd (Paris) 32:866–872

Broca P (1876) Sur les trépanations préhistoriques. Bull. Mém. Soc. Anthrop. Paris, 2e série 11: 236–256

von Brunn W (1928) Kurze Geschichte der Chirurgie. Julius Springer Verlag, Berlin

Carter RMS, Anslow P (2009) Imaging of the calvarium. Semin Ultrasound CT MRI 30:465–491

Ciporen JN, Moe KS, Ramanathan D, Lopez S et al (2010) Multiportal endoscopic approaches to the central skull base: A cadaveric study. World Neurosurg. 73:705–712

Del Maestro RF (1998) Leonardo da Vinci: The search for the soul. J Neurosurg 89:874–887

Deng Q, Zhou M, Shui W, Wu Z, Ji Y, Bai R (2011) A novel skull registration based on global and local deformations for craniofacial reconstruction. Forensic Sci Internat 208:95–102

Donnabhàin B (2003) Trepanations and pseudotrepanations: Evidence of cranial surgery from prehistoric and early historic Ireland. In: Arnott R, Finger S, Smith CUM (eds) Trepanation: history, discovery, theory. Swets & Zeitlinger Publ, Lisse, pp 79–94

Duffy J (1984) Byzantine medicine in the sixth and seventh century: Aspects of teaching and practice. Dumbarton Oak Papers 38:21–27

Even F (2014) Tempête pour un crâne. Culture & Idées p 2, Le Monde, 21 juin 2014

Facchini F, Rastelli E, Ferrero L, Fulcheri E (2003) Cranial trepanation in two skulls of early medieval Italy. Homo 53:247–254

Forster P (2004) Ice ages and the mitochondrial DNA chronology of human dispersals: a review. Phil Trans R SocLond B 359:255–264

Finger S, Clower W (2003) On the birth of trepanation: the thoughts of Paul Broca and Victor Horsley. In Arnott R, Finger S, Smith CUM (eds) Trepanation: history, discovery, theory. Swets & Zeitlinger Publ, Lisse, pp 19–42

Gerrits PO, Veening JG (2013) Leonardo da Vinci's "A skull sectioned": Skull and dental formula revisited. Clin Anat 26:430–435

Gibbons A (1992) Mitochondrial Eve: wounded but not dead yet. Science 257:873–875

Goodrich JT (2014) A medieval approach to head injuries from the Golden Era of Islamic Medicine from the writings of Abn Bakr Muhammad Ibn Zakariya al-Razi – Rhazes (865–925). World Neurosurgery (2014), https://doi.org/10.1016/j.wneu.2014.05.005

Grounds JG (1958) Trephining of the skull amongst the Kisii. East Afr Med J 35:369–373

Heidecker K-M (2009) Schädeltrepanationen in der Antike. In: Brockmann C, Brunschön W, Overwien O, De Gruyter W (eds) Antike Medizin im Schnittpunkt von Geistes- und Naturwissenschaften. Berlin

Jezelka-Stanek A, Krajewska-Walasek M (2013) Genetic causes of syndromic craniosynostoses. Eur J Pediat Neurol 17:221–224

Kahn DM, Shaw RB (2008) Aging of the bony orbit: a three-dimensional computed tomographic study. Aesthetic Surg J 28:258–264

La Farge A (1996) The bearded lady and the shaven man. Leonardo 29:379–383

Lascaratos J, Liapis C, Ionidis C (1998) Surgery on aneurysms in Byzantine times (324–1453 AD). Eur J Vasc Endovasc Surg 15:110–114

Lang J (1981) Klinische Anatomie des Kopfes- Neurokranium, Orbita, kraniocervikaler Uebergang. Springer, Berlin, Heidelberg, NY

Lanz von T, Wachsmuth W (1979) Praktische Anatomie. Kopf Teil B: J. Lang, Gehirn und Augenschädel. Springer, Berlin

Lewis JE (2008) Identifying sword Marks on bone: criteria for distinguishing between cut marks made by different classes of bladed weapons. J Archeol Sci 35:2001–2008

Lindeboom GA (1964) Andreas Vesalius (1514-1564) Een schets van zijn leven en werken. De Erven F. Bohn N. V. Haarlem

Lopez da Silva FH, Van Hulten K, Lommen JG, Storm van Leeuwen W, Van Veelen CWM, Vliegenthart W (1977) Automatic detection and localization of epileptic foci. Electroencephalo Clin Neurophysiol 43:1–13

Lopez Piñero JML (1967) La tradicion classica de la cirugia cranial. In: La trepanacion en España ed. JM Lopez Piñero. Editorial Tecnica Española, Madrid, pp 45–48

Lucas-Championnière JMM (1912) Les origins de la trépanation décompressive. Trepan. néolithique, Steinheil, Paris

Manwaring KH, Manwaring ML, Moss SD (1994) Magnetic field guided endoscopic dissection through a burr hole may avoid more invasive craniotomies: A preliminary report. Acta Neurochir 61:34–39

Marani E, Koch WFRM (2014) The pelvis, structure, gender and society. Springer, Heidelberg

Mensink G, Zweers A, van Merkesteyn JPR (2009) Endoscopically assisted reduction of anterior table frontal sinus fractures. J Cranio-Maxillafac Surgery 37:225–228

Missinne SJ (2014) The oldest anatomical handmade skull of the world c.1508: "The ugliness of growing old" attributed to Leonardo da Vinci. Wien Med Wochenschr. https://doi.org/10.1007/s10354-014-0282-0

Morris-Kay GM, Wilkie AOM (2005) Growth of the normal skull vault and its alteration in craniosynostosis: insights from human genetics and experimental studies. J Anat 207:637–653

Moshel YA, Schwartz TH (2010) Multiportal endoscopic approaches to the central skull base: A cadaveric study. World Neurosurg. 73:632–633

Murphy EM (2003) Trepanations and perforated crania from Iron age South Siberia: An exercise in differential diagnosis. In: Arnott R, Finger S, Smith CUM (eds) Trepanation: history, discovery, theory. Swets & Zeitlinger Publ, Lisse, pp 209–222

Mutahi W (1982) A skull surgeon who never went to medical school. Daily Nation, p 13, 26 Nov 1982

Nerlich AG, Zink A, Szeimies U, Hagedorn HG, Rösing F (2003a) Perforating skull trauma in ancient Egypt and evidence for early neurosurgical therapy. In: Arnott R, Finger S, Smith CUM (eds) Trepanation: history, discovery, theory. Swets & Zeitlinger Publ, Lisse, pp 191–201

Nerlich A, Peschel O, Zink A, Rösing FW (2003b) The pathology of trepanation: Differential diagnosis, healing and dry bone appearance in modern cases. In: Arnott R, Finger S, Smith CUM (2003) Trepanation: history, discovery, theory. Swets & Zeitlinger Publ, Lisse, pp 41–51

Obert M, Kubelt C, Schaaf T, Dassinger B, Grams A, Gizewski ER, Krombach GA, Verhoff MA (2013) Aging adult skull remains through radiological density estimates: a comparison of different computed tomography systems and the use of computer simulations to judge the accuracy of results. Forensic Sci Int 228:179.e1–179.e7

Princich JP, Wasserman D, Latini F, Oddo S, Blenkmann AO, Seifer G, Kochen S (2013) Rapid and efficiënt localization of depth electrodes and cortical labeling using free and open source medical software in epilepsy surgery candidates. Front Neurosci 7:260. https://doi.org/10.3389/fnins.2013.00260

Qureshi MM, Oluoch-Olunya D (2010) History of Neurosurgery in Kenya, East Africa. World Neurosurg 73:261–263

Rawlings C III, Rossitch E Jr (1994) The history of trephination in Africa with a discussion of its current status and continuing practice. Surg Neurol 41:507–513

Richardson WF, Carman JB (1998) On the fabrice of the human body. Book I: The bones and cartilages, Andreas Vesalius: a translation of De humani corporis fabrica libri septem. Norman Publ, San Francisco

Roberts C, McKinley J (2003) Review of trepanations in British antiquity focusing on funerary context to explain their occurrence. In: Arnott R, Finger S, Smith CUM (eds) Trepanation: history, discovery, theory. Swets & Zeitlinger Publ, Lisse, pp 55–78

Rodinov R, Vollmar C, Nowell M, Miserocchi M, Wehner T, Micallef C, Zombori G, Ourselin S, Diehl B, McEvoy AW, Duncan JS (2013) Feasibility of multimodal 3D neuroimaging to guide implantation of intracranial EEG electrodes. Epilepsy Res 107:91–100

Saban R, Salf E (1996) Une curiosité anatomique: un crâne miniature en marbre, témoin pathologique? Hist Sci Méd. 30:437–447

Schaller BJ FA, Merten HA, Buchfelder M (2012) Premature craniosynostosis -The role of skull base surgery in its correction. A surgical and radiological experience of 172 operated infants/children. J Cranio-Maxillo-Facial Surg 40:195–200

Schwartz LF (1995) Morphing the three faces of Mona: The decision-making steps Leonardo used to create his Mona Lisa. Comp Graphics 19:529–539

Seifert V, Zimmermann M, Trantakis C et al (1999) Open MRI-guided neurosurgery. Acta Neurochir (Wien) 141:455–463

Simonson TM, Kao SCS (1992) Normal childhood development patterns in skull bone marrow by MRI imaging. Pediatr Radiol 22:556–559

Spalteholz W (1932) Handatlas der Anatomie des Menschen, erster Band. Hirzel Verlag, Leipzig

Stephan CN (2003) Anthropological facial 'reconstruction' - recognizing the fallacies, 'unembracing' the errors, and realizing method limits. Sci Justice 43:193–200

Stewart DB (1984) The pelvis as a passageway. I. Evolution and adaptations; II The modern human pelvis. Br J Obst Gynaecol 91:611–617; 618–623

Sypkens Smit JH (1943) Een te Winsum (Fr) gevonden getrepaneerde schedel. De Vrije Fries 66–80

Torimitsu S, Nishida Y, Takano T, Koizumi Y, Makino Y, Yajima D, Hayakawa M et al (2014) Statistical analysis of biomechanical properties of the adult skull and age-related structural changes by sex in a Japanese forensic sample. Forens Sci Intern 234:185.e1–185.e9

UTET (1981) Leonardo Botallo: I doveri del medico e del malato. 41° convegno della società italiana di anatomia. Unione tipografico-editrice Torines, Torino

Vanpoecke M (2004) Hoofdletsels. Jaarverslag Rode Kruis België

Verano JW (2003) Trepanation in prehistoric South America: geographic and temporal trends over 2000 years. In: Arnott R, Finger S, Smith CUM (eds) Trepanation: history, discovery, theory. Swets & Zeitlinger Publ, Lisse, pp 223–249

Vesalius, A. (1543). De humani corporis Fabrica libri septem. (Facsimile, 1964) Reproduction by Culture et civilisation 115 Ave Grabriel Lebon. Bruxelles

Wilkie AOM (1997) Craniosynostosis: genes and mechanisms. Hum Mol Gen 6:1647–1656

Wilkins RH (1964) The Edwin Smith surgical papyrus. J Neurosurgery 21:240–244

Zhang B, Cory E, Bhattacharya R, Sah R, Hargens AR (2013) Fifteen days of microgravity causes growth in calvaria of mice. Bone 56:290–295

Zhang Y, Sim T, Tan CL, Sung E (2006) Anatomy-based face reconstruction for animation using multi-layer deformation. J Visual Lang Comput 17:126–160

Zimmerman LM (1936) Cosmas and Damian, patron saints of surgery. Am J Surg 160–168

Chapter 3
Malformation and Identity

One human live birth in fifty demonstrates an external malformation. Anomalies in the human central nervous system are 60%, while for bones muscles and skin it is 10%. The teratological aspects of the morphogenesis of the neuraxis, and thus also of head and neck, are inevitable. This chapter concentrates on conjoined twins, the relation of malformations to mythology, and treats acardiacus. The aspect of identity in relation to conjoined twins is part of this chapter. Neural crest malformations and the cephalic and trunk migration pathway are discussed together with the placode contributions for the production of ganglia. The chapter ends with craniofacial malformations, especially holoprosencephaly and its aberrant signalling pathway.

3.1 Introduction

In my collection of doctoral theses, I have a nearly forgotten one, by Jan Frederik Adolf Beins (1913–1970) written in the Department of Pathology of the University of Groningen, The Netherlands. Its title is "Misvorming en verbeelding" (malformation and imagination), and it is written in Dutch, so most of the foreign pathologists will have missed it.

The relation between mythology and teratology has been a theme of research from before the twentieth century in Western science. It has had its ups and downs. In 1929, Bolk (1866–1930) gave his view on mythology and teratology in the Netherlands, but it was hardly accepted within scientific circles. In 1948, Beins thesis appeared and he got a better reception. His thesis came also out as a hard cover book produced by the same printing house, indicating that the expected interest was enough to overcome the costs of a public edition. Indeed, the Web still shows a list of firms that sell the book and scientists that have commented on it. In 2010, Stahl and Tourame again gave an overview on malformations in relation to antique legends.

E. Marani and C. Heida, *Head and Neck*,
https://doi.org/10.1007/978-3-319-92105-1_3

Beins does not cite Bolk (both Dutchmen), and Stahl and Tourame are not aware of the Dutch contribution. Nevertheless, all three, of course, knew Etienne Geoffroy Saint-Hilaire (1772–1844), who founded Teratology, the science of malformations. The name Teratology was invented by Saint-Hilaire's son Isodore, who later became a professor of embryology. Saint-Hilaire and his disciple Serres (1786–1868) looked for systematics in malformations, and although the basic scheme that Saint-Hilaire and Serres produced for the divisions of malformations have been partially rejected, it is still the basis of modern ones (Leroi 2003).

"Scientific impetus was given to teratological research in this century following the observation of the Australian doctor Gregg in 1941 that German measles during early pregnancy could lead to severely deformed children. It was soon discovered that several other viral diseases including hepatitis, mumps, morbilli, poliomyelitis, varicella and AIDS can produce teratological effects. Teratological research received further stimulation as a result of atomic bombs that destroyed Hiroshima and Nagasaki in 1945, the thalomide catastrophe in Germany as well as in other countries from 1959 to 1962, and the Dioxin (2, 3, 7, 8-tetrachlorodibenzo-p-dioxin, TCDD) disaster of Seveso in Northern Italy in 1976" (Schumacher 2004). Knowledge of several causes for deformation of embryo and foetus has reduced the incidence of malformations, but is still calculated 2–3% of the total births, and "it *doubles* postnatal owing to the fact that many dysfunctions are not discernible at birth" (Schumacher 2004).

3.2 Terminology

"Variations are related to anomalies, which can be defined as marked deviations from an assumed normal standard. But, since the range included in such arbitrary standard cannot be rigorously defined, there are many deviations within the broad boundary zone between simple variations and 'true' anomalies. Anomalies are related to malformations. In these latter, however, defective or faulty development of structures is emphasized as the predominant characterization" (Kuhlenbeck 1973). In fact, there is no generally unifying terminology for malformations and the causes that can lead to teratologic deviations are summarized to be 19 different categories by Kuhlenbeck (1973). The monstrous pig in Albrecht Dürer's woodcut (Fig. 3.1) has his brother on his back, and the malformation is called cephalothoracoileopagus (*cephalo* = head; *thoraco* = chest; *ileo* = belly; *pagus* = attached to). In the first chapter on the calvarium, craniosynostosis has been treated, but the names for the malformation are rather different: Apert-, Crouzon-, Pfeiffer-, Muenke syndrome, and all are mutations in the fibroblast growth factor receptor genes. We will use the "pagus" nomenclature, since the involved bodily parts can directly be deduced from the name of the anomaly, with several morphological defects, the involved bodily parts can directly be deduced and syndromes (always signs and symptoms of a single condition) are mainly named after their discoverers.

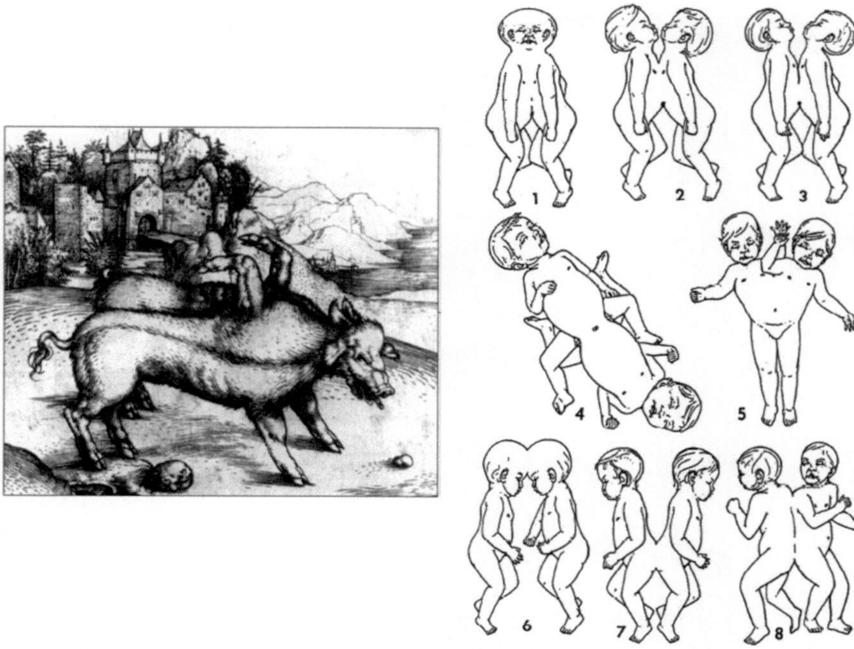

Fig. 3.1 Left: The cephalothoracoileopagus as pictured by Albrecht Dürer (1471–1528); Right: the proposed nomenclature for conjoined twins: "The eight types of conjoined twins: (1) cephalopagus, (2) thoracopagus, (3) omphalopagus, (4) ischiopagus, (5) parapagus, (6) craniopagus, (7) pygopagus and (8) rachipagus" (figure and text from Spencer 1966, with permission)

The discussion on terminology is especially important in conjoined twins (Fig. 3.1). It is used to indicate the right and left of the twins, thinking of separation of the twins (see Spencer 1966).

3.3 Head and Neck Malformations

The relation between teratology and mythology is not the subject of this chapter. The cephalothoracopagus (Fig. 3.2), or the so-called Janus-malformation, a two-headed malformation, was perhaps the origin of the Roman god "Janus" who also had two faces. Janus is the god of beginnings and transitions. The two faces, with the linked back of the heads, indicate looking to the future and to the past. The month of January was named by the Romans in his honour. An analogous story is the presence of more extremities, like or more arms and/or more legs: the four-armed Vischnu or eight-armed Japanese God that can be present in human malformations too (see Fig. 3.2). Nevertheless, whether the vision of the monstrous malformations is the cause of the manifestation of the external appearance of gods or not is not dealt with in this part.

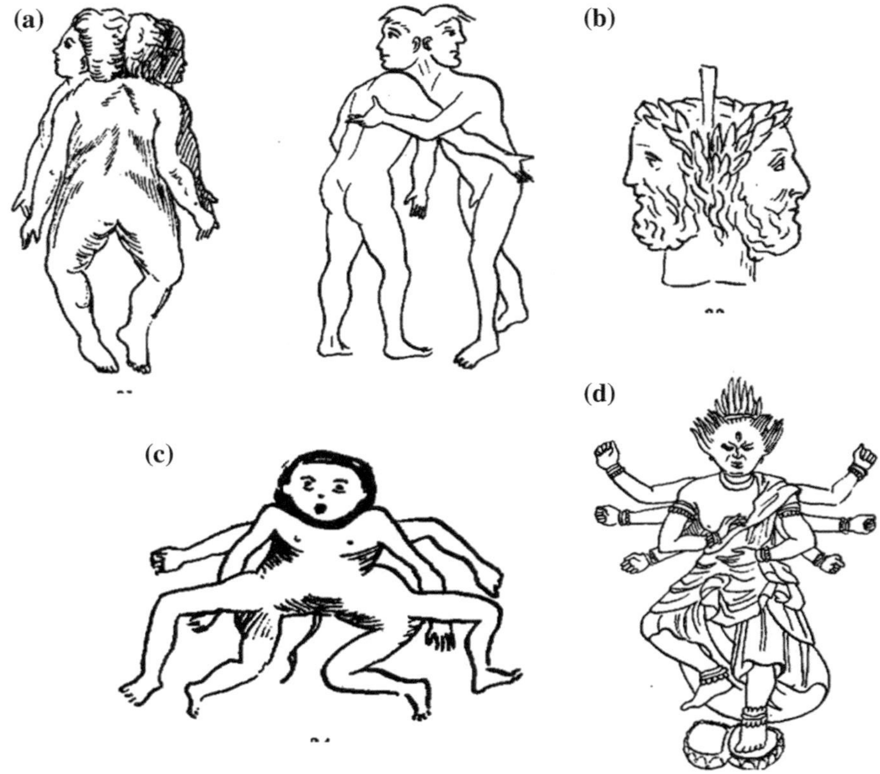

Fig. 3.2 **a** Cephalothoracopagus symmetros, double headed and with double thorax and four upper extremities, and next to it a double-headed monster "cephalopagus"; **b** Janus head; **c** cephalothoracopagus symmetros one headed with more extremities; **d** Eight-armed Japanese God (figures from Beins 1948)

3.3.1 Acardiacus

An acardiacus (literally without heart) is a deformed foetus having developed no heart (called the parasite or acardian twin), connected to another foetus (called the autosite) and staying alive by a connection to the blood circulation of the autosite or also called the pump twin. We know four types:

(a) Only a head is present, and rump and extremities are missing (acardiacus acornus), present in 5% of the arcadian twins (Fig. 3.3),
(b) A head on legs, rump is missing (acardiacus amorphus), present in 25% of the arcadian twins,
(c) A rudimentary head with an imperfect face, rest of the body is present (acardiacus anceps) in 8% of the cases,

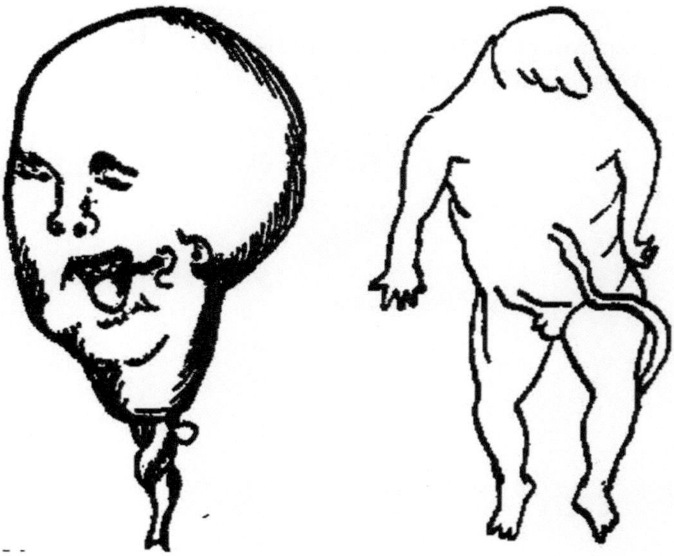

Fig. 3.3 Left, acardiacus acornus, only a head has developed and right, acardiacus acephalus, a body developed without head (figures from Beins 1948)

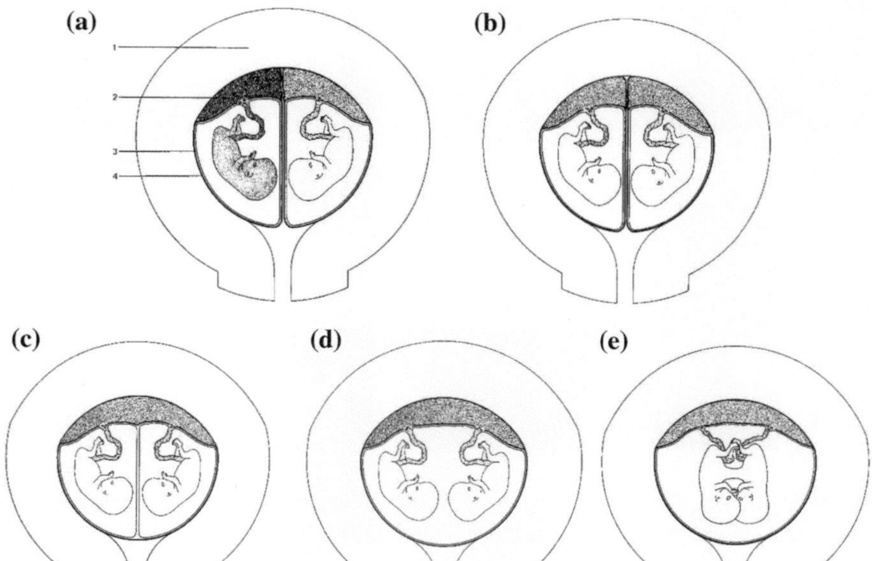

Fig. 3.4 **a** Dizygotic or fraternal twins with two amnions (biamniotic) and two placenta's (bichorionic); **b** monozygotic twins each with its own placenta (biamniotic and bichorionic); **c** monozygotic twins with one placenta (monochorionic) but with two amnions (biamniotic); **d** monozygotic twins with one placenta (monochorionic) and one amnion (monoamniotic), **e** monozygotic Siamese twins (monochorionic and monoamniotic): 1 uterus, 2 placenta, 3 amnion, 4 chorion (figures from Eskes 1977)

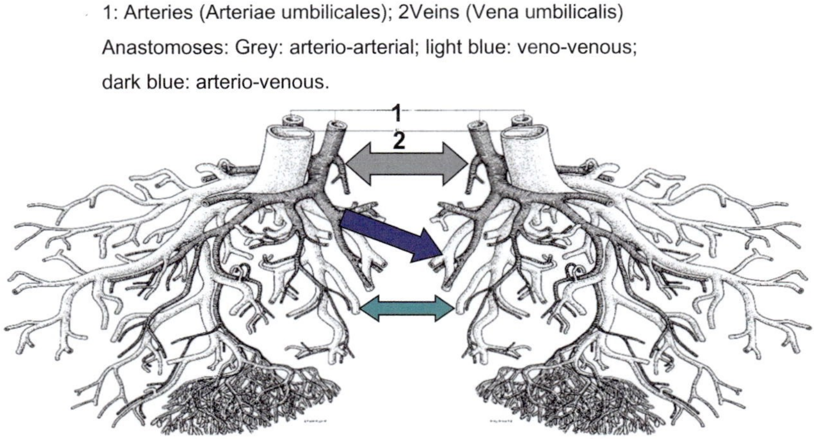

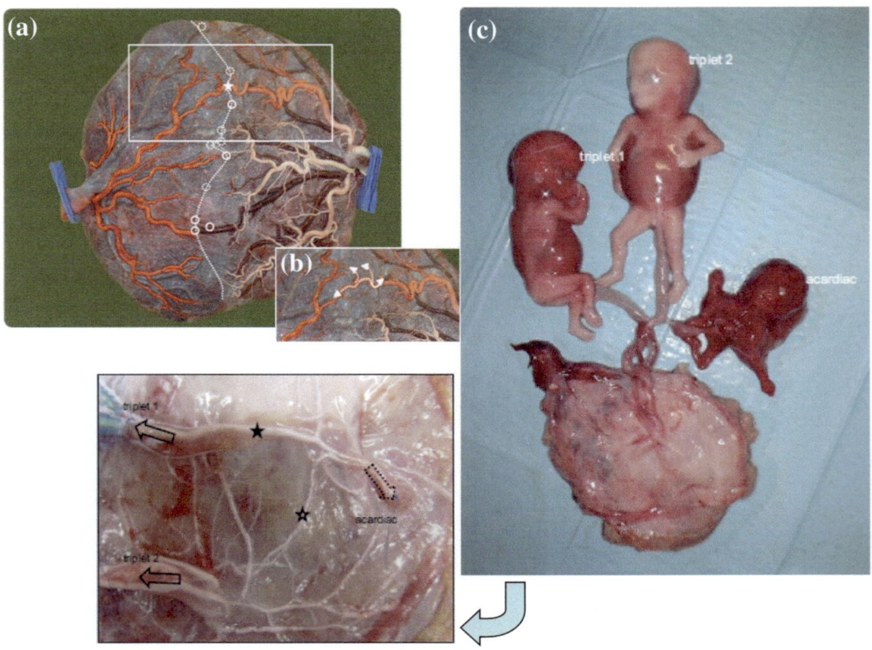

(d) A head is missing, and only rump with (rudimentary) or without upper limb extremities is present (acardiacus acephalus) in 62% (Fig. 3.3; see Chen 2005)

This brings us to twin studies. Twins developing in one and the same placenta are called monochorionic twins (for explanation, see Fig. 3.4). They form 20% of all twins and are subjected to extra death risks: 12% of them die before week 24 of pregnancy, as opposed to 2% for those twins that each have their own placenta (dichorionic twins, see Fig. 3.4).

◀**Fig. 3.5** Upper figure—three anastomoses between two placental territories for monochorionic diamniotic placenta; lower figure—"**a** typical monochorionic diamniotic placenta from uncomplicated pregnancy". Delivery was at 35 weeks of two healthy neonates of 2534 g and 2440 g. Placenta is equally shared. Each twin has its own individual placental territory (veins coloured blue for twin 1 and veins coloured brown for twin 2) defined by venous chorionic plate vessels of each twin (dotted line). There is one artery-to-artery anastomosis (star); five arteriovenous anastomoses (from twin 1 to 2) (open circles); and six oppositely directed venoarterial anastomoses (dotted circles). **b** enlargement of star area. "Each artery-to-artery anastomosis functions as flexible arteriovenous anastomosis. Depending on direction of flow, it can act as arteriovenous anastomosis from twin 1 to 2 (solid arrow), or as venoarterial anastomosis from twin 2 to 1 (dotted arrows)" (Lewi et al. 2013, with permission). **c** "Placenta of monochorionic monoamniotic triplet complicated by twin reversed arterial perfusion sequence and miscarriage at 15 weeks. Cords were entangled. Placenta shows 1 arterioarterial anastomosis (solid star) from cord of smaller triplet 1 (open arrow) and venovenous anastomosis (open star) to larger triplet 2 (open arrow). Triplet 1 was therefore pump triplet, whereas triplet 2 was draining triplet" (Lewi et al. 2013, with permission). Note the acardiac foetus

The monochorionic twins share their blood provision due to vascular anastomoses in the placenta (Fig. 3.5). "Another consequence of the shared circulation is that the well-being of one twin critically depends on that of the other. After the diagnosis of spontaneous demise of one of a monochorionic pair, the survivor has a 15% risk of death and a 25% risk of neurodevelopmental impairment" (Lewi et al. 2013).

The cause of the development of an acardiacus is not clear. "The two major theories of the embryologic origin of conjoined twins have been labelled 'fission' and 'fusion'. Proponents of the former suggest that incomplete fission of the blastocyst inner cell mass during the primitive streak stage, 13–15 days postfertilization, results in two centres of axial growth that retain a connection at some point. 'Fusion', in contrast, refers to two originally distinct inner cell masses that coalesce secondarily at a later stage. The most commonly advanced theory for the development of asymmetry between autosite and parasite after these events was first postulated by Donitz (1866). They proposed that vascular compromise causes tissue of the parasitic twin to become dependent on collaterals derived from the autosite. Selective ischaemic atrophy of the deprived portion of the parasite's body follows" (Sharma et al. 2010).

For a book on head and neck, another aspect is more important: what ever the acardiac cause will be. Seemingly, the head can be expanded independently, which is also stressed by the acardiacus acephalus. "Ironically, the presence of an element can best be comparatively studied, if in any species, or pathology, that element is absent" (Marani and Koch 2014).

Thus, the development of a head (and neck?) is an autonomous, self-governing process. If a differentiation programme is genetically started, it follows an automatic development (Chandebois and Faber 1983). The differentiating cell populations in head and neck are numerous. Still, their assemblage into a head entity indicates that the cell populations know their cohesion independently of their totally different developmental automation. (By the way, a conclusion for which no genetic explanation is yet available. It makes the coherence spatial even more mysterious).

3.4 Identity

Monochorionic and monoamniotic conjoined twins, also called Siamese twins, bring up the problem of identity. One could think that it is a problem created by philosophers (it is); however, it plays an important role in the surgical separation of conjoined twins. Conjoined twins join body parts. Being conjoined is not lethal as is known from the Rita-Christina case, born 1830 (Serres 1832) or from the Chang and Eng Burke conjoined twins, (Mitchell 2003) that subjoined "Siamese" to conjoined twinships (born 1811, arrival in the West in 1829, see also Dreger 1998 and Murray 2001 for more examples). Several conjoined twins were not separated because the parents objected the sacrifice of one of the two or the risk to loose both (sacrifice is by some authors considered an euphemism for killing, Dreger 1998). In several cases, separation has led to the death of both twins or to one survivor who could never leave hospital (Dreger 1998; Murray 2001).

Opposing results come from the Dalles hospitals that separated ten conjoined twins: 14 children survived and 6 died of which 3 before separation. Votteler and Lipsky (2005) studied the long-term follow-up of the separated ten conjoined twins. All survivors had school education and "the first four surviving twins have graduated from college. The 1981 twin completed high school. The remaining nine twins are in classes appropriate for their age" (Votteler and Lipsky 2005). The conclusion of the article shows: "In utero diagnosis of conjoined twins by ultrasonography compels physicians and parents early in the pregnancy to make decisions concerning possible surgical separation and the alternative of pregnancy termination. The case reports presented offer additional information for physicians and parents to make informed decisions at a crucial time by illustrating long-term physical and psychological outcomes of 14 surgical survivors" (Votteler and Lipsky 2005).

Psychological research brought up two things:

(1) "Partly merged bodies result in partly merged self" (Murray 2001). An answer to the identity problem?
(2) "Joined twins will surely not psychologically individuate like singletons, because even separate twins do not. Of course under this logic, even twins born separate are not psychological healthy or acceptable—a questionable conclusion in itself. Moreover, it is unclear how, if even separate twins suffer from these psychological 'problems', the separation of conjoined twins will solve this 'problem' (unless one dies)" (Dreger 1998).

Thus, the identity of conjoined twins is different from identity of singletons. Identity is made by the borders of the body as a separation from others and from its environment. Border information is skin information, since the skin is on the outside of the organism and does not permit outside substances and organisms (milieu exterieur) to penetrate into the inside (milieu interieur). Border information is brought from the skin by nerves to dorsal root ganglia (DRG), which in turn send this information via the spinal cord, into and along the brainstem, into the brain.

"There is one point that is hardly treated: the consequences of the DRG localization along the bodily segments. In fact, the somatotopic representation in the central nervous system originates in the segmental distribution of the DRG ganglia with the consequence that spinal segmental reflexes exist. In relation to this, the repetitive firing of the DRG neurons is responsible for generating background sensation of the awareness of the body scheme (Marani 2014). Awareness of the position of your toe being lower than your hip is organized by the positions of the DRG for the big toe (L5) being lower than the DRG of the hip (L1) along the vertebral column. These positions can be known by our brain due to the organized transfer of information from these DRG's towards brainstem and brain. Such consciousness is far less present for viscera" (Marani and Lakke 2012).

Thus, awareness and borders of your body are determined by layered DRG information towards the brain and the layers are also labelled higher or lower than the next layer. Now let us take a parapagus (Fig. 3.6), e.g. Rita-Christina. Although they had one front side of the pelvis, each had its own sacrum and own vertebral column, although both girls shared pelvic sexual organs (one of the set of pelvic sexual organs was degenerated) and had one vulva and one pair of legs. Abdominal internal organs were present in duplicate (Serres 1832; Leroi 2003; Marani and Koch 2014).

Before we go on, we look to the "original" Siamese twins Chang and Eng. They each married and had 10 and 12 children, respectively (Mitchell 2003). One week Eng (and Chang) was/were at the other house with the other's wife, and the next week they were at his house with his wife. "The fact that they engaged in sexual relations, even though they each had been married and monogamous, "shocked the moral sense of the community" (Dreger 1998), and it was considered by those outside the family "most immoral and shocking that the two should occupy the same material couch with the wife of one" even though one of the widows reported that "there never have been any improper relations between wives and the brothers" (Dreger 1998). Now, to put the identity question plainly: in bed during sex were there three persons or only two?

Back to Rita-Christina, they doubled up the same left and right legs. Thus, their body identity for the brain was either only with one leg or with both legs, which depends on the nerve innervation, reporting to their DRG's, entering this information into one vertebral column and spinal cord or into both. Other studies show that Siamese twins are each capable to control shared limbs. The argument "each twin has an appendage-like quality for their joined sibling, in which their body schema incorporates common areas of sensibility into an ambiguous body image" (Murray 2001) indicates that there must be an overlapping body identity in their brains.

Now let us go to the surgeon: if he separates a conjoined twin and conjoined twins do have ***overlap, thus two overlapping identities***, he is amputating part of the conjointment, thus changing the identity of both or of the surviving twin. Even worse if one of the twins agrees to be separated she or he condemns, mostly unknowingly, part of her or his identity to be changed and in case of death of the other always a renewed identity has to be built. Several authors see sacrificing one of the twins as murder. And in fact surgeons seek protection by asking permission from parents, and the courts, indicating that murder charge is indeed a real possibility. However, the identity problem has been introduced rather late, and little attention is paid to it. The "murder dilemma" is clearly deemed more important.

Is sacrificing or separating necessary? There are several examples where the conjoined twin identity is kept by not separating the twins and the twins survived: Yvonne and Yvette Mc Carther, craniopagus, they became 43 years old and had nursing degrees; Millie and Cristina, born to enslaved parents, survived well, could by their singing help the father to buy the plantation he worked on; Abigail and Brittany Hensel, born in 1990; "The sisters are dicephalic parapagus twins, sharing several organs but each with their own distinct personalities—and their own driver's licence. Brittany controls the right side of the body, while Abby steers the left (and it is her foot on the accelerator)" (from R. Murray, New York Daily News, 09 August 2012). Both nervous systems are connected with each other giving the possibility to control movements and to play piano. Moreover, each of the twins has its own identity, although they share one body. The need of splitting up Siamese twins can be doubted.

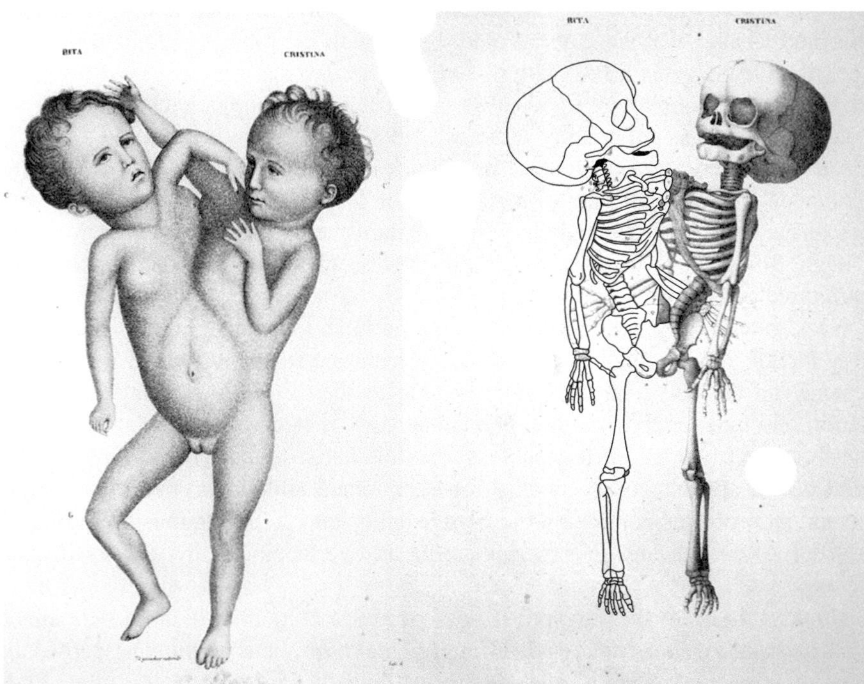

Fig. 3.6 Left is the impression of the conjoined twins Ritta-Christina. The right side is the skeleton of the conjoined twins after dissection. Plates are reproduced from Serres (1832), with permission from the Leiden University Library

3.5 Neural Crest Malformations

3.5.1 Neural Crest and Placodes

The earliest cells of the embryonic disc produce by its primitive streak (grooved line of cell mass producing mesoderm in an embryo of 15 days) intraembryonic mesoderm (precursors for bone and muscles) that come to lie in between the already formed upper layer (ectoderm) and the inner layer (endoderm) of embryonic disc and are also involved in the making of placodes (Streit 2004). Before the closure of the neural groove into a neural tube, which is caused by the rise of the neural walls, meso-ectoderm cells are already deposited into the embryo by both the epiblast and ectoderm (Fig. 3.7; Vermeij-Keers and Poelmann 1980; Smits van Prooije et al.

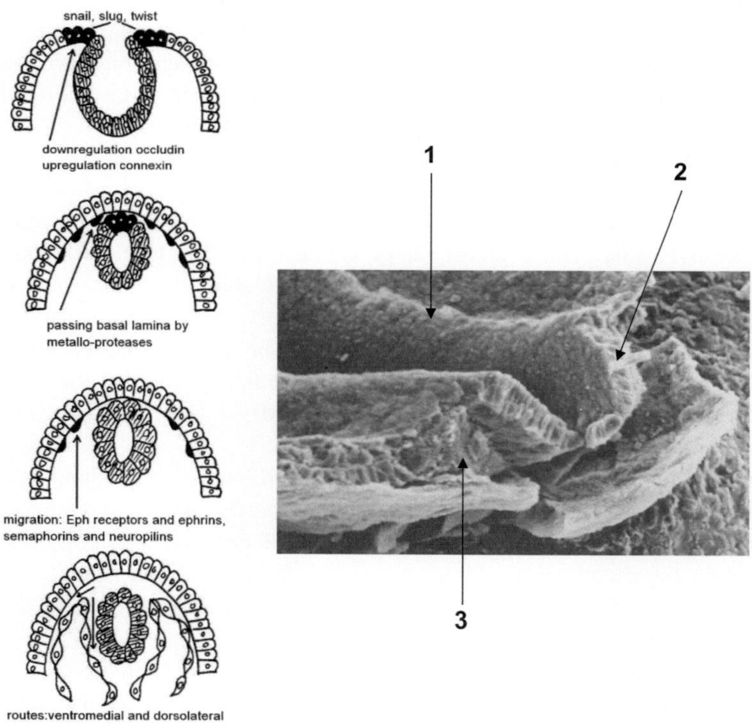

Fig. 3.7 Stages of neural crest development and migration; see also text. 1: neural crest, 2: neural wall, 3: early deposited mesoderm, scanning electron microscopy from E10 rat. Adapted and extended from Kuriyama and Mayor (2008) (see Marani and Lakke 2012)

1985a, b; for an overview of early neuroinduction and early neural patterning, see Stern 2001). The *neural crest* produces mesodermal (branchial arches: muscle and bone) and neuronal structures. *Placodes* produce neuronal structures and also contribute to the sense organs in the head, e.g. lens of the eye.

The neural crest is at the closure line of the neural tube and is produced by signals from the local epidermis, neural plate and underlying mesoderm. The boundaries of the neural plate are determined by bone morphogenic protein (BMP). This border determines together with *Wnts*, *fibroblast growth factor* and *retinoic acid* the change towards or neuronal plate cells or epidermal cells. The genes that are responsible for neural border cell determination are called *neural plate border specifier genes*. It is the set of *neural crest specifier genes*, under the influence of the first set that determines the neural crest cell type. This second set of genes is a prerequisite for survival of the neural crest cells. Migration and differentiation of the neural crest cells are regulated by a third group of genes, which are influenced by the first two sets, *called neural crest effector genes*. So a three-step cascade of genes organizes the neural crest cell type and their migration. Neural crest cells are capable of producing neurons, glia, mesoderm cells (cartilage, muscle) and endoderm cells (contributing to: head-neck, heart, peripheral nervous system and skin) (see Baker and Bronner-Fraser 2001; Kuriyama and Mayor 2008).

The migration of neural crest cells is only possible if they undergo a change called *epithelial–mesenchymal transition*. After this, the neural crest cells do have the possibility to leave the neural tube and use specific pathways to reach their destination. To migrate the premigratory neural crest cells lose their neural tube polarity, lose their tight junctions by down regulating *occluding* and produce gap-junctions by upregulating *connexin*. There is a loss of cadherin I and an increase in cadherin II enhancing cell movement. The downregulation of cadherin I is related to *snail, slug* and *twist* genes. Passing the basal lamina of the neural tube asks for the production of metalloproteases (MMP's) (Fig. 3.7; see Baker and Bronner-Fraser 2001; Kuriyama and Mayor 2008).

The *migration* of neural crest cells is determined by membrane-bound proteins. The receptor–ligand couples Eph receptors and ephrins, which play an important role in neural crest migration. The cranial neural crest uses in combination or separate for the rhombomeres Npn2, EphrinB2, EphA4/B1 and EphB1, while the trunk neural crest migration is determined by Slit/Robo, Eph/ephrins and semaphorins/neuropilins (Fig. 3.7).

The pathways used by neural crest cells for reaching their definite position have already been described ca. 1986 (Rickmann et al. 1985, chick; Smits van Prooije 1986; Smits-van Prooije et al. 1987, mouse and rat). A medial route through the somite, one between neural tube and somite, and a dorsolateral one are discerned. The neural crest cells passing through the medial route produce sensory ganglia and autonomic ones, Schwann cells and chromaffin cells. The dorsal route is restricted to melanocytes.

3.5.1.1 Cephalic Migration (Figs. 3.7 and 3.8)

The different Eph–ephrin combination regulates gap junction communication and motility and is responsible for the correct direction of groups of neural crest cells towards the second to fourth arch and inhibiting their intermingling (EphA4 and EphB1 third and third + fourth arch; ephrin B2 is expressed in the second arch).

Semaphorin proteins can bind to heterophilic receptors like neuropilin and plexin, changing the internal cell skeleton organization. They are important for the correct migration patterning in the head region. Axon guidance is determined by the interaction of semaforins and neuropilin 2 (Npn2). Moreover, the different kinds of semaforins can determine mesenchymal neural crest-free or invasion zones in the arches (Baker and Bronner-Fraser 2001; Kuriyama and Mayor 2008).

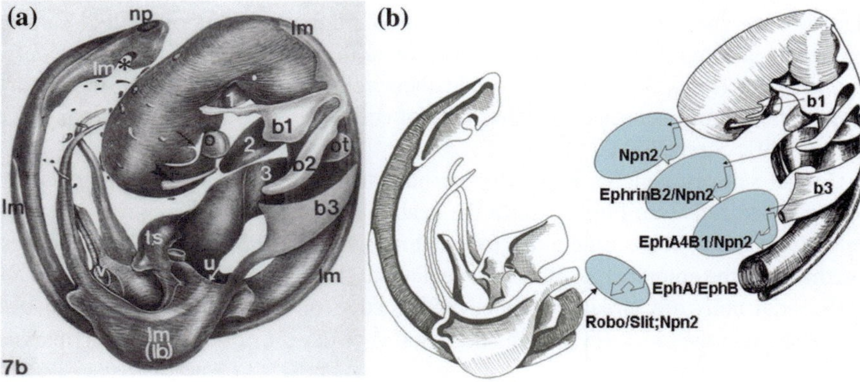

Fig. 3.8 Compare to Fig. 3.7. **a** Reconstruction of the neurectoderm and labelled cells in the mesodermal compartment of a 20-somite embryo that has been cultured for 30 h after injection of WGA-Au. Labelled cells are present all along its neural crest. The limb buds and caudal side of the transverse septum contain labelled cells. The arrow indicates labelled cells that presumably are deposited by the optic neural crest. The arrowhead indicates labelled cells that presumably are derived from the nasal placode. The labelled area of the third branchial arch is continuous with the transverse septum. Caudally of the transverse septum and alongside the umbilical veins lie the areas of the abdominal placode. The mesoderm underneath the vitelline artery is only labelled close to fts transition to the yolk sac. The asterisk indicates the depression the hindgut makes in the labelled primitive-streak-derived cells. The allantois is not reconstructed. Abbreviations: b, branchial arch; h, heart; lb, limb bud; lm, area of the mesodermal compartment containing labelled cells; rex, maxilla; np, posterior neuropore; o, optic vesicle; ot, otic vesicle; ts, labelled area of the transverse septum; u, umbilical vein; v, vitelline artery (from Smits van Prooije 1986). **b** Redrawn after Smits van Prooije. Grey ellipses envision sections through branchial arches 1–3 and through spinal cord. Arrows indicate migration routes only for neural crest cells at one side of each section together with their migration determining proteins (see Marani and Lakke 2012)

3.5.1.2 Trunk Migration (Figs. 3.7 and 3.8)

The trunk migration pathways can be described along the routes neural crest cells will take: ventromedial migration is controlled by Slit/Robo (Slit protein binds to Robo receptor) and also by semaphorins and neuropilins, the dorsolateral one by Eph–ephrins. The invasion of neurocrest cells is restricted to the anterior part of the somite. Entrance in one of the two main pathways is controlled by the same receptor–ligands, also called their bifunctionality. An explanation has yet to be found (see also Marani and Lakke 2012).

3.5.2 Placodes

Placodes are specialized transient parts of the columnar, pseudostratified surface ectoderm, later epithelia that are capable of adding their cells to the mesoderm compartment (Figs. 3.8, 3.9). They add their cells in the head region to sensory ganglia, producing neurons and other type of cells for the special sense organs (see Noden 1991; Streit 2004).

Cranial sensory ganglia, therefore, are built twofold, from neural crest cells and from placode cells (see Fig. 3.10). In the head area, the placodes that are present are named: olfactory, lens, trigeminal (ophthalmic and maxillo-mandibular) placodes, together with the epibranchial placodes: geniculate (VII), petrosal (IX) and nodose (X).

Placodal cells are already induced at the edge of the neural plate at the 0–1 somite stage in the chick (in humans this should be day 21 of development), marked

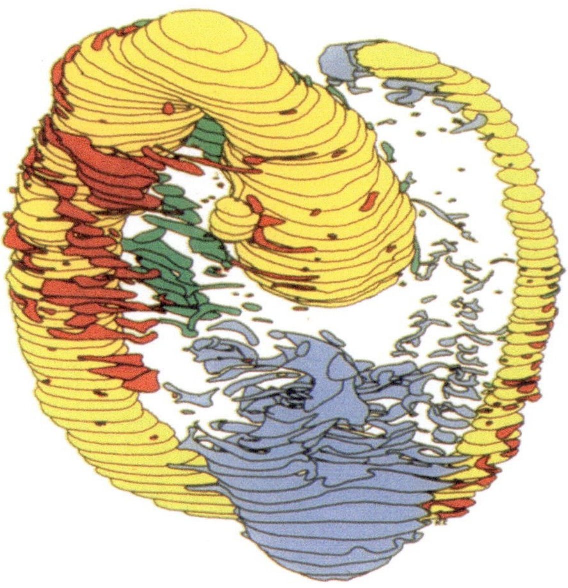

Fig. 3.9 Placodes (different colours) around the nervous system (yellow) (Reconstruction from the results of Smits van Prooijen by the Leiden Neuroregulation group)

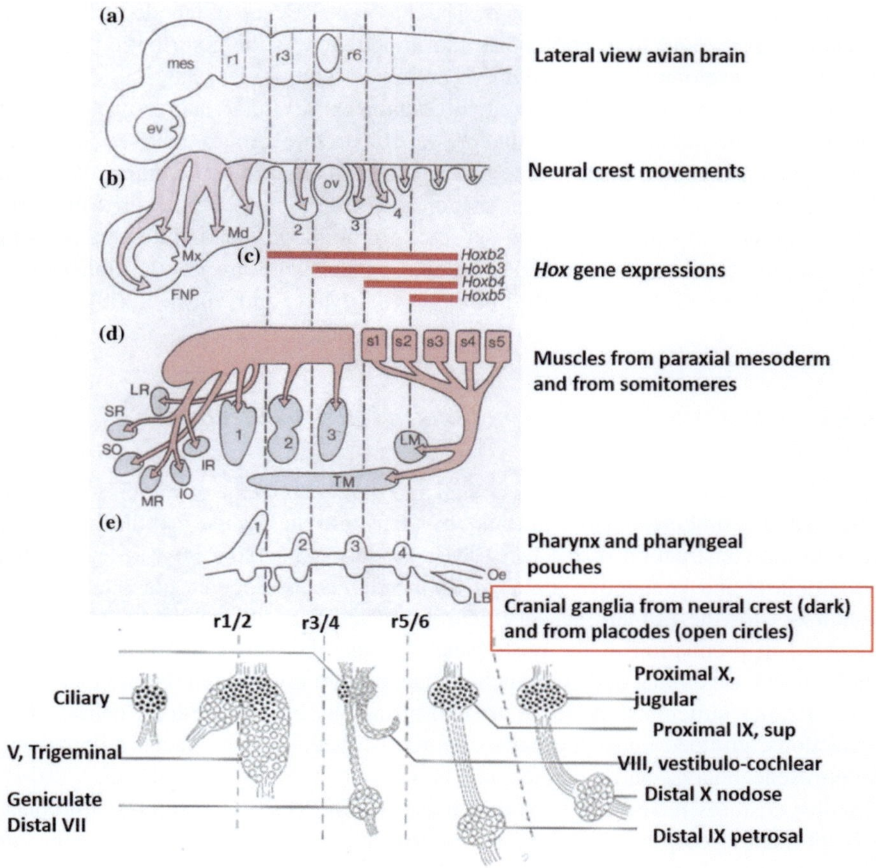

Fig. 3.10 Derivatives of the neural crest: cranial nerves, myogenic primordia and cranial ganglia. Figure is adapted from Ten Donkelaar et al. (2014) and from Noden (1991). ev eye vesicle, FNP frontonasal prominence, mes mesencephalon, IO inferior oblique, IR inferior rectus, LB long bud, LM laryngeal muscles, LR lateral rectus, Md mandibular prominence, MR medial rectus, Mx maxillary prominence, r1–r6 rhombomeres, SO superior oblique, SR superior rectus, TM tongue muscles, 1–4 pharyngeal arches and pouches)

by *Six, Eya, Id, Iro* and *Fox* genes, but the dispersed cells have to grade and group together to produce placodes. Inductors of placode cells are endoderm, future heart mesoderm and the neural plate. A role is played by BMP's as marker of the border of the neural plate, and *Iro-1* is regulated by BMP. *Sox, Dlx, Fox* and *Pax* genes are involved in determining placode cells. Only the transcripts *Six* and *Eya* are "to be maintained in all developing placodes, while being lost from interplacodal domains" (Streit 2004, for overviews see: Baker and Bronner-Fraser 2001; Streit 2004).

Using lectin-coated colloidal gold, Smits van Prooije succeeded already in 1986 to give a total overview of the localization of the neural crest and placode cells in the rat from the 5 till 41 somite stages. The massive presence of labelled cells in the branchial arches and limb bud shows the importance of the contribution of these cells to the development of the embryo and even foetus (Figs. 3.8, 3.9).

The migration routes are thought to contain extracellular matrix proteins like fibronectin, laminin and collagen that are capable to direct migration (see Sect. 4.1: chemotaxis and chemokinesis). The ECM pathways are unspecific highways, since these proteins can hardly be kept responsible for homing of the growth cone. Attracting and repellent factors are discerned for homing. Factors like glial-derived neurotrophic factor and netrin/deleted colon cancer gene are now established as neural crest attractants, especially in the gut (Kuriyama and Mayor 2008).

3.5.3 Craniofacial Malformations: Holoprosencephaly

Neural crest cells are present in the cranial and trunk part of the developing embryo. The cranial neural crest cells contribute by their, into mesectodermal differentiated, cells to the construction of the face. They produce precursors for cartilage, bone, odontoblasts and connective tissue of the head. The neurogenic placodes are the precursors for the primary sensory neurons in the head, the olfactory placode is involved in production of the nose (Sect. 7.2 and Fig. 7.16), and other placodes such as lens, trigeminal and epibranchial placodes are involved in the production of several other structures. A series of malformations are induced by neural crest deflections, grouped under the names neurocristopathies, craniofacial ciliopathies, holoprosencephaly and craniosynostoses (ten Donkelaar et al. 2014). Craniosynostoses have been treated in this book in Sect. 2.3.2, sutures, fibroblast growth factor receptor and craniosynostosis. It is impracticable to describe all neural crest craniofacial malformations. Clinical neuroembryology spent 50 pages to it (see ten Donkelaar et al. 2014). Therefore, only one is chosen: holoprosencephaly, because attention can be given to the proboscis.

Here, we generally follow the subdivision for holoprosencephaly as described by ten Donkelaar et al. (2014, see Fig. 3.11):

(1) Cyclopia: with single or double eye anlage and with or without a proboscis,
(2) Ethmocephaly with ocular hypotelorism and proboscis between the eyes,
(3) Cebocephaly with ocular hypotelorism and a single-nostril nose,
(4) Median cleft lip and palate (agenesis of the premaxilla) and ocular hypotelorism,
(5) Severe facial dysmorphism.

Holoprosencephaly can be caused by chromosomal abnormalities and teratogens and in general involves midline abnormalities caused by an interaction of genetic and environmental factors. For example, the diabetic mother has a 200-fold higher risk to induce holoprosencephaly, and in 1–2% of the newborns, this abnormality has been found. Although the counts are variable, its incidence is 0.44–1.2 per

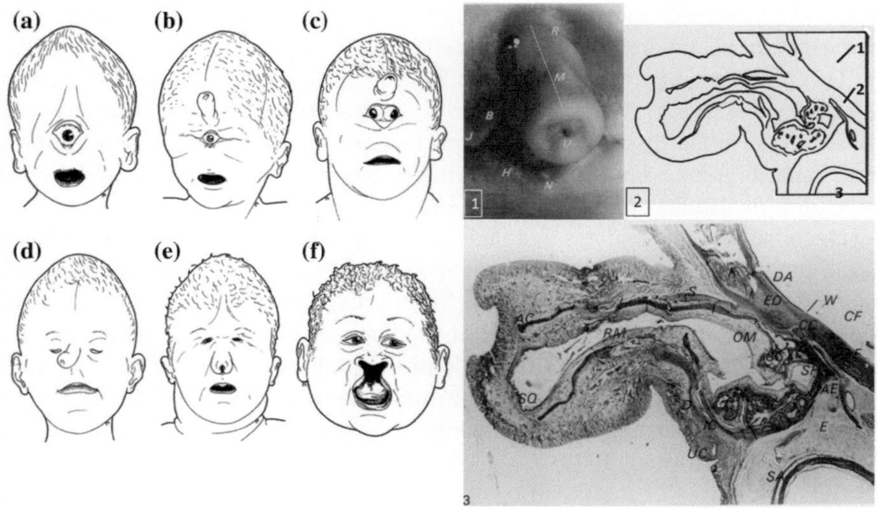

Fig. 3.11 Holoprosencephaly: general subdivision of its facial malformations. **a** cyclopia; **b** single median eye; **c** double median eye both with various degrees of doubling of ocular structures; **d** ethmocephaly; **e** cebocephaly; **f** median cleft lip with arhinencephaly (see Duhamel 1966; courtesy and permission of ten Donkelaar et al. 2014). 1: External view of proboscis: B, lacrimal part; H, lacrimal part of lower eye lid; J, lacrimal papilla of right upper lid; M, mobile distal part; N, median caruncle of orbital floor; R, fixed root; U, aperture. Part 2, scheme of part 3 with 1, anterior cranial fossa; 2, edge of orbital plate belonging to the left frontal bone; 3, the eye surrounded by sclera and orbital fat. Part 3, Midsagittal section of proboscis (see white line in 1). In its middle the cavity: at W the ethmoidal notch; CF, anterior cranial fossa; DA, dura-arachnoid; ED, extradural fibrous tissue; F, frontal bone; CC, cranial part of cartilage of base of proboscis; OF, edge of orbital plate of left frontal bone; AE, anterior ethmoidal nerve and artery; IC, inferior concha; OM, olfactory mucosa; I, persistent cartilage of nasal capsule; S, fibrous sheath; L, lateral nasal cartilage; RM, respiratory mucosa; AC, major alar cartilage; SQ, squamous epithelium; UC, superior canaliculus; D, nasolacrimal duct; E, orbital fat; SA, sclera (Figures 1, 3 and figure text are from McGrath 1992 and free for publication elsewhere)

10,000 newborns, but its rate increases to 40 per 10,000 in human abortions. Several genes have been identified that are involved in the abnormalities: SHH, SIX3, ZIC2, TGIF, PTCH, GLI2 and Nodal. SHH and SIX3 mutations produce major facial malformations, while ZIC2 mutation gives faint facial disturbances. Brain malformations appear accordingly that can go from alobar, complete absence of midline division containing one single ventricle and one undivided brain lobe, and to the lobar type containing a clear midline with still some continuity in its midline (see ten Donkelaar et al. 2014).

Hypotelorism (decreased distance between eyes) and proboscis (trunk) are characteristics of holoprosencephaly (Fig. 3.11). The area between the nasal placodes is either absent or too narrow (see Fig. 4 in book front matter) in holoprosencephaly. In normal situations, the medial nasal prominences produce the interplacodal area (red in Fig. 4 in book front matter). This area produces parts of the maxilla (premaxilla), the prolabium, the vomer and several other midline-related parts. The mesectoderm cells responsible for the production of these structures are derived from neural crest and placode cells.

The proboscis has been studied in human cyclopia (McGrath 1992). Since cyclopia arises due to disturbed midline structures of the nose and skull base and forming one orbita, its structural changes concern the nasal bone, vomer, frontal bone with its frontal sinus, the lesser wing of the sphenoid and the ethmoid bone structures: the lamina cribrosa and at its end the crista galli, containing the corridors for the olfactory bulb endings and its lamina perpendicularis. The ethmoidal notch is the submerged area of the lamina cribrosa. In cyclopia, this notch area enlarges and contains the base of the proboscis. Since it extends forward, it also organizes the roof of the orbit. At the backside of the roof are the fused lesser sphenoid wings, and they are also placed at the backside of the proboscis. The cavity of the proboscis is surrounded by tissue within it the nasal capsule. Histology showed that the lining of the cavity is by squamous epithelium, respiratory and olfactory mucosa (see Fig. 3.11; McGrath 1992).

Within the base of the proboscis olfactory fibres are found. They are clothed together. The anterior ethmoidal nerve and artery can be recognized. Branches of the ophthalmic artery are present in the ethmoidal notch in between the cranial and orbital parts of the proboscis base. A median neurovascular bundle can be discerned that is also determined as olfactory bundles, and an ethmoidal anterior nerve is present in this bundle and presumably continues as the internal nasal nerve. Cartilage curls are found at the base of the proboscis. The nasolacrimal duct is present.

The proboscis is the anterior and mainly superior part of the nose and its cavity. Due to the absence of its median structures, particularly the nasal septum, the distal part of the proboscis can extend and make a "freely mobile facial appendage" (McGrath 1992). The intricate relation between the ethmoid, sphenoid and nasal cartilage in holoprosencephaly has been left out, but is described in McGrath (1992). The proboscis is in fact a united nose and nasal cavity. In Chap. 7, the nose will be treated. The presence in all vertebrates of two cavities, placode-derived, separated by a septum demonstrates that this bicameral construction is a very conservative property. Thus, one wonders that a simple mutation in the identified genes can produce such major facial malformations. Research showed that mutations interact and extracellular effects are involved too.

For the understanding of Fig. 3.12, an explanation of the developmental involvement of the mutated genes in holoprosencephaly is given note that generally the gene produces a protein with the same name to get its effect: SIX3 is co-determining the development of the eyes and forebrain, SHH co-effects the patterning of the ventral part of the central nervous system, and ZIC2 is involved in the axis formation and dorsal brain development of the central nervous system. NODAL determines midline and laterality of the central nervous system. Remember that holoprosencephaly is characterized by midline aberrations and all summed up genes are midline or medio-lateral axis or dorso-ventral axis related.

Not only genes but chromosomal abnormalities can also be the base of holoprosencephaly. Moreover, research showed that changes in the cholesterol synthesis

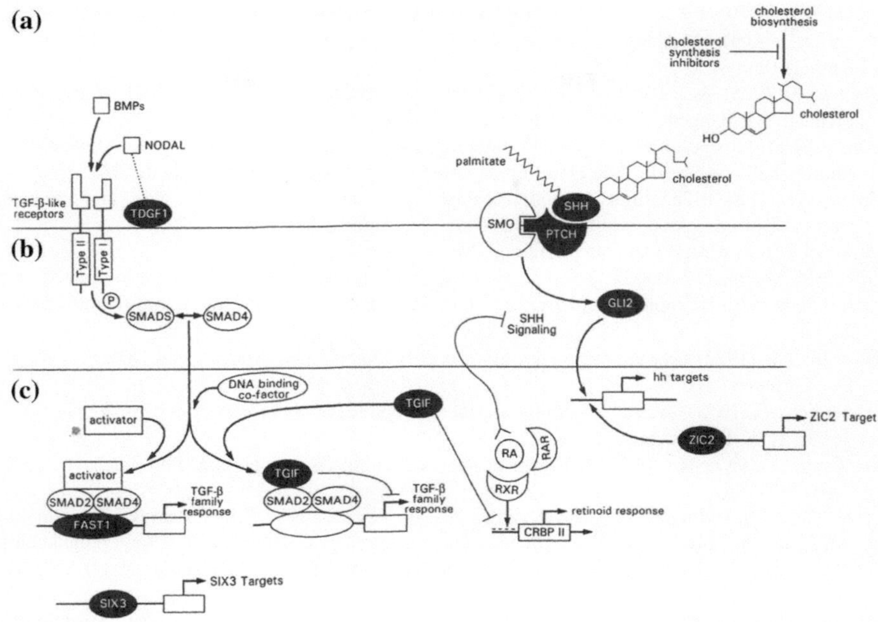

Fig. 3.12 Overview of signalling pathways involved in holoprosencephaly. Deviated functions of these cascades are subdivided in **a** extracellular events; **b** cytoplasmic cascade steps; and **c** nuclear effects (from Edison and Muenke 2003 with permission)

and uptake are an extracellular factor that can go wrong by SHH coupling for the uptake within the cell (Fig. 3.12). But the gene GLI2 can also contribute to the changed uptake of cholesterol, while ZIC2 (zinc finger transcription factor) is responsible for its nuclear anomaly. Since PTCH is a receptor for hedgehog (SHH) ligands, its mutation interferes also in cholesterol uptake. In the nuclear cascades TGIF, homeoprotein interacting with SMAD2 (mothers against decapentaplegic homolog 2) contributes after mutation to holoprosencephaly, because SMAD2 mediates the signal of the transforming growth factor (TGF)-beta, and co-regulates various cellular processes, such as cell proliferation, apoptosis and differentiation (see Fig. 4 in book front matter).

References

Baker CVH, Bronner-Fraser M (2001) Vertebrate cranial placodes. Part I. embryonic induction. Dev Biol 232:1–61

Beins JFA (1948) Misvorming en verbeelding. G.A. Van Oorschot, uitgever, Amsterdam

Bolk L (1929) Mythologie en teratologie. Rede Algemene Vergadering Akademie van Wetenschappen, Jaarboek 1929 der Akademie

Chandebois R, Faber J (1983) Automation in animal development. A new theory derived from the concept of cell sociology. In: Wolsky A (ed) Monographs in developmental biology, vol 16. Karger, Basel

Chen C-P (2005) Acardiac twinning (Twin reversed arterial perfusion sequence): a review of prenatal management. Taiwan J Obstet Gynecol 44:105–115

Donitz W (1866) Beschreibung und Erlauterung von Doppelmissgeburten, Dritte Abhandlung, Fünfter Fall. Archiv für Anatomie. Physiologie und Wissenschaftlige Medicin 534–544

Dreger AD (1998) The limits of individuality: ritual and sacrifice in the lives and medical treatment of conjoined twins. Stud Hist Phil Biol Biomed Sci 29:1–29

Duhamel (1966) Morphogenè pathologique. Masson, Paris

Edison R, Muenke M (2003) The interplay of genetic and environmental factors in craniofacial morphogenesis, holoprosencephaly and the role of cholesterol. Congenit Anom (Kyoto) 43:1–21

Eskes TKAB (1977) Gynaecologie en Obstetrie. In goede handen. Spruyt, Van Mantgem & De Does, Leiden

Kuhlenbeck H (1973) Overall morphologic pattern. The central nervous system of vertebrates, vol 3, Part II. S. Karger, Basel

Kuriyama S, Mayor R (2008) Molecular analysis of neural crest migration. Phil Trans R Soc B 363:1349–1362

Leroi AM (2003) Mutants: on genetic variety and the human body. Penguin Books, London

Lewi L, Deprest J, Hecher K (2013) The vascular anastomoses in monochorionic twin pregnancies and their clinical consequences. Am J Obstet Gynecol 208:19–30. https://doi.org/10.1016/j.ajog.2012.09.025

Marani E (2014) Dorsal root ganglion. Encyclopedia of the Neuroscience. Elsevier, Amsterdam

Marani E, Koch WFRM (2014) The pelvis, structure, gender and society. Springer, Heidelberg (40005-3 e-book). ISBN 978-3-642-40005-6

Marani E, Lakke EAJF (2012) Peripheral nervous system topics. In: Mai J, Paxinos G (eds) The human nervous system, 3th edn. Elsevier/Academic Press, Amsterdam

McGrath P (1992) The proboscis in human cyclopia: an anatomical study in two dimensions. J Anat 181:139–149

Mitchell S (2003) Exhibiting monstrosity: Chang and Eng, the "original" Siamese twins. Endaeavor 27:150–154

Murray CD (2001) The experience of body boundaries by Siamese twins. New Ideas Psychol 19:117–130

Noden DM (1991) Vertebrate craniofacial development: the relation between ontogenetic process and morphological outcome. Brain Behav Evol 38:190–225

Rickmann F, Fawcett WJ, Keynes RJ (1985) The migration of neural crest cells and the growth of motor axons through the rostral half of the chick somite. Embryol Exp Morphol 90:437–455

Schumacher GH (2004) Teratology in cultural documents and today. Ann Anat 186:539–546

Serres E (1832) Recherches d'anatomie transcendante et pathologique. Théorie des formations and deformations organiques, appliqué à l'anatomie de Ritta-Christina, et de la duplicité monstrueuse. J.B. Ballière, Paris

Sharma G, Nazarian Mobin SS, Lypka M, Urata M (2010) Heteropagus (parasitic) twins: a review. J Pediatr Surg 45:2454–2463

Smits-van Prooije AE, Poelmann RE, Gesink AF, Vermeij-Keers Chr (1985a) The cell surface coat during closure of the neural tube as revealed by concanavalin A. Verb Anat Ges 79:591–592

Smits-van Prooije AE, Vermeij-Keers C, Poelmann RE, Mentink MM, Dubbeldam JA (1985b) The neural crest in presomite to 40-somite murine embryos. Acta Morphol Neerl Scand 23:99–114

Smits-van Prooije AE, Vermeij-Keers Chr, Dubbeldam Mentink MMT, Poelmann RE (1987) The formation of mesoderm and mesectoderm in prosomite rat embryos cultured in vitro, using WGA-Au as a marker. Anat Embryol 176:71–77

Smits-van Prooije AE, Vermeij-Keers Chr, Poelmann RE, Mentink MMT, Dubbeldam JA (1986) The formation of mesoderm and mesectoderm in 5- to 41-somite rat embryos cultured in vitro, using WGA-Au as a marker. Anat Embryol 177:245–256

Spencer R (1966) Anatomic description of conjoined twins: a plea for standardized terminology. J Pediatr Surg 131:941–944

Stahl A, Tourame P (2010) De la tératologie aux monstres de la mythologie et des légendes antiques. (From teratology to mythology: ancient legends). Arch Pédiatr 17:1716–1724

Stern CD (2001) Initial pattering of the central nervous system: how many organizers. Nat Rev Neurosci 2:92–98

Streit A (2004) Early development of the cranial sensory nervous system: from a common field to individual placodes. Dev Biol 276:1–15

ten Donkelaar HJ, Lammens M, Hori A (2014) Clinical neuroembryology. Springer, Heidelberg

Vermeij-Keers Chr, Poelmann RE (1980) The neural crest: a study on cell degeneration and the improbability of cell migration in mouse embryos. Neth J Zool 30:74–81

Votteler TP, Lipsky K (2005) Long-term results of 10 conjoined twin separations. J Pediatr Surg 40:618

Chapter 4
The Development of the Bony Skull

This chapter treats the development of the bony neurocranium and the bony viscerocranium, and it runs from genetics to the three-dimensional enlargement of bony centres. The construction of head and neck, in fact of the whole body, is organized by the most conservative genes, called Hox genes. The fly Drosophila (the small red-eyed flies on your fruit) uses the same genes. In humans, these genes are employed to make the segmental borders in our body. Since the head exists of a fusing of several segments, the knowledge of the Hox genes clarified their borders. In the growth process, structures of head and neck are produced by mesoderm, neural crest and placodes, which ask for a thorough description.

4.1 Introduction

The presence of the conservative genetic code for the general blueprint of the vertebrate body and nervous system is exemplified by the Hox gene code. This gene code is responsible for a signalling stream of information in and outside the cell. The body uses not only genetic codes to compose the organization of the head, neck and nervous system for a specific species, but also by the so-called patterned migration. Once cells are born in a germ layer, they migrate towards their place of destination. Principles govern this migration: chemotaxis determines the direction of growth. A concentration gradient is needed, because such factors only work over a small distance. Chemokinesis, on the other hand, determines the growth velocity of, e.g. the axon. These factors are concentration-dependent, but a gradient is unnecessary.

Mechanical cues are also involved, mainly substrate alignment and galvanic fields. The pruning of contacts depends on the function of each contact. As a consequence, contacts or whole cells have to be removed, which in the case of cells happen by apoptosis, also called programmed cell death (see Lakke 1997).

E. Marani and C. Heida, *Head and Neck*,
https://doi.org/10.1007/978-3-319-92105-1_4

"In a homogeneous cell population, autonomous progression is an automatic process based on a continuous exchange of information which is effected through cell contact relations. The programme for this autonomous progression is not communicated to the individual cells of the population at the time of determination. Rather, the programme assigns to the cells a certain elementary social behaviour, which is then the basis for the structuring of the population" (Chandebois and Faber 1983). Induction changes this homotypic development into a heterotypic one: "in which one cell population, called the induced population, irreversibly engages in a new differentiation pathway under the influence of a neighbouring population, the inducer, which itself usually is not transformed" (Chandebois and Faber 1983). For example, the differentiation of certain brain segments (rhombomere borders due to Hox genes) is a necessary prerequisite for head placodes to produce sense organs. Seemingly, these induction interactions play an important role in the spatial coherence. Self-organization during the early human embryogenesis is considered the hallmark of the initial development, and this self-organization occurs in the absence of maternal tissue, supporting an automatic process (Shahbazi et al. 2016).

4.2 Breaks in Growth Velocity

According to literature, growth, from conception to senescence, should be from the beginning a steady exponential increase and followed by a steady decrease. In general, the increase (growth) is larger and faster than the decrease (ageing). The top of the growth increase is different for the various species, including man (Dobbing and Sands 1979). The turning points of increase to decrease can be differently placed on the time axis for various species, if the lines are expressed as first-order velocity curves.

During equal exponential growth, sudden changes in growth speed have been described by Goedbloed (1976, 1977, 1980, 1986) for the whole rat and mouse development and for various expanding organs and organ systems. Once the same growth rate changes were proposed for the brain and brain parts, in this case prenatal volume increase of the hypothalamus of the mouse (Fig. 4.1), journals stopped accepting their articles for publication. The argumentation, here taken from the referee reports of Roux's Archives of Developmental Biology for rejection, can be hostile:

(1) Overall morphometric measurements are of limited mechanistic significance in a complex organ containing different cell types whose behaviour may be quite distinct.
(2) The interpretation of this "breakpoint" in growth pattern of the hypothalamus at 13.4 days is purely speculative.

Statistics is statistics: if a change in growth speed (called breakpoint) can be statistically calculated then there is one, and of course measuring growth of brain

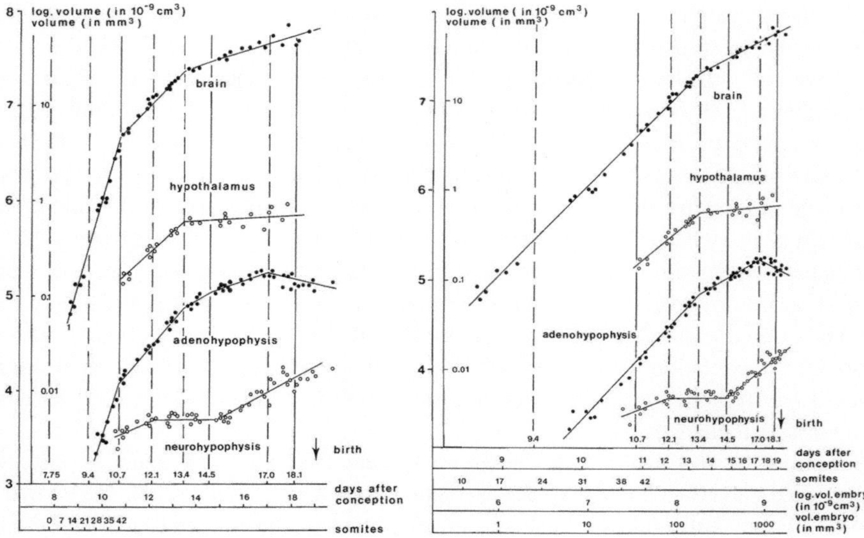

Fig. 4.1 Growth curves (left) and allometric (right) curves for the volume increase of the developing hypothalamus, adenohypophysis and neurohypophysis (unpublished, courtesy Goedbloed)

parts is important; look at all studies on human brain growth in the literature (see vertebrate overview Van Dongen 1998) and its importance for obstetrics and gynaecology: think of biometric ultrasound measurements of the unborn baby (consult, e.g. various articles in European Journal of Obstetrics & Gynecology and Reproductive Biology on foetal growth) or "Normal foetal growth is a critical component of a healthy pregnancy and influences the long-term health of the offspring. However, defining normal and abnormal foetal growth has been a long-standing challenge in clinical practice and research" (Zang et al. 2010), and its importance in paediatrics: "intrauterine growth restriction (IUGR) due to placental insufficiency affects 5–10% of all pregnancies, and it is associated with a wide range of short- and long-term neurodevelopmental disorders. Prediction of neurodevelopmental outcomes in IUGR is among the clinical challenges of modern foetal medicine and paediatrics" (Batalle et al. 2012). What is used to predict the brain function of the growing child: foetal brain measures and brain mapping with MRI.

The first indication that growth speed indeed differs in various parts of the developing body comes from the difference of areas in front of the primitive vertebral column (prechordal parts) and those behind it (postchordal parts) (see Herlin et al. 2011 and references therein): "the volumetric study of the skulls highlighted an asynchronous speed of growth between the pre- and postchordal parts of the skull base whose preferential growth is in the vertical and horizontal planes. We also found different speeds of growth in the prechordal part depending on the type of ossification (endochondral or membranous)" (Herlin et al. 2011).

In short, changes in growth velocity during embryological and/or foetal development are presumably present, but hardly accepted by the scientific world.

Postnatally, the situation is different. Brain weight correlates very well with skull circumference (Epstein and Epstein 1978). For the human skull circumference, growth spurts are found at 3–10 months, again at 2–4 years, 6–8, 10–12 and 14–17 years of age (Epstein 1974). The idea is every 4 years there is a growth spurt or a change in growth velocity. It could be confirmed by the increase in cerebral blood flow (Epstein 1999), total EEG data and cortical thickness changes (Epstein 1986). Postnatal changes in growth rates seem to be accepted, and smoking, alcohol and drug use are nowadays all negatively related to growth speed of the brain.

The consequences are: during the embryological and foetal development of the skull box, the relation with the growth of the brain is that both will be gradual. You should hardly notice any difference in growth speed. Postnatally, the skull knows growth speed changes.

Postnatally, the skull of the newborn is characterized by fontanelles, places where the calvarian bony distance is rather large. One should relate the growth spurts of the skull with the postnatal development of the fontanelles. Since their closure is after or at the end of the first growth spurt and the relation is not given in literature, growth spurts of the skull cap and skull base cannot be incorporated in the following part. Moreover, the earliest accepted postnatal growth spurt cannot be related to the earliest postnatal skull development due to the absence of information on their relation. Presumably, skull development is discontinuous but we cannot get a grip on it.

4.3 Principles of Skull Development

"The mammalian head is sharply marked off from the body and carried on a long and mobile neck. It is therefore a single "organ", whose structure has been moulded by its activities even more than has the head in fishes, amphibian and reptiles, which could be often said merely the anterior part of the body. Moreover, the shape of the head in mammals is usually less affected by considerations of frictional resistance during motion than is the case in aquatic or aerial mammals. Movement over the land surface is not often fast enough for the air resistance to be serious, and the head is therefore free to take on the different forms that we see in horse, elephant or man. In those mammals that have returned to aquatic life, there is usually a reintroduction of the "streamlined" form of the head and body. Our thesis is, then, that the shape of the head of any particular type of mammal is determined by the way its parts are used for the purposes: brain protection, carrying organs of special senses, carrying mouth and jaws for the entrance of the alimentary canal, jaws for offence or defence, passage for oxygen and communication expression, and we shall hope to be able to organize our knowledge about the various structures in the head in terms of these functions" (Young 1974, 1975).

Arguments for the head, with all its constituents, as an entity or organ that can develop independently have been given in Chap. 3: Malformation and identity. Genetics and among others patterned migration are important for the autonomous head development, which also holds for the bony skull developments.

Before the genetic base of bone formation can be treated (a section was already noticed in 2.3.2 on sutures, fibroblast growth factor receptor and craniosynostosis): the under-expression versus the overexpression of bone formation will be explained later, and now it concerns the pedestrian look on what the anatomist or embryologist notices. The principles that govern the development of the skull have to be well thought of:

(1) Concrescentia: fusing of bony centres to one bony element (occipital bone). A special case of concrescentia occurs in the sphenoid. The sphenoid is constructed from two different bony centres, next to the hypophysis. These two bony centres fuse behind the hypophysis and later in development in front. This organizes a craniopharyngeal channel, within it the hypophysal diverticulum. This development is not depending on the pouch of Ratke. This channel closes again in the third month of the human ontogenesis.
(2) Perforation: passage for structures like nerves and blood vessels. These passages can be present in the middle of bony centres (foramen opticum, foramen rotundum, canalis facialis in the otic capsule), but they can also be placed at the interfaces of two different bony elements (fissure orbitalis sup., the foramen lacerum for the carotis interna).
(3) Association: without any problem, the dermatocranium connects to the neurocranium and vice versa. At the edges, their connections are always late. The remaining openings are called fontanelles or fonticuli. Eventually, the closure lines form sutures. Note that the fonticulus major becomes bregma, the fonticulus minor is the lambda, and the fonticulus sphenoidalis ends up as the pterion (see Figs. 2.2 and 4.2). Association can also produce foramina (the foramen jugularis).
(4) Absence of ossification or late ossification present: ossification will not be present everywhere in the skull. The occipital condyles stay cartilaginous, which makes movements of the head with cervical vertebrae possible. The latest ossification (around 20 years of age) is the clivus.

4.4 Mesoderm, Neural Crest and Placode

Before the genetic, mesodermal, neural crest and placode items will be treated, the step of embryo to the neonate will show the direct relations between these items and the skull bony parts (Fig. 4.2). Seemingly, certain layers in the head and neck of the embryo are directly related to bony skull parts. The bony skull parts are still

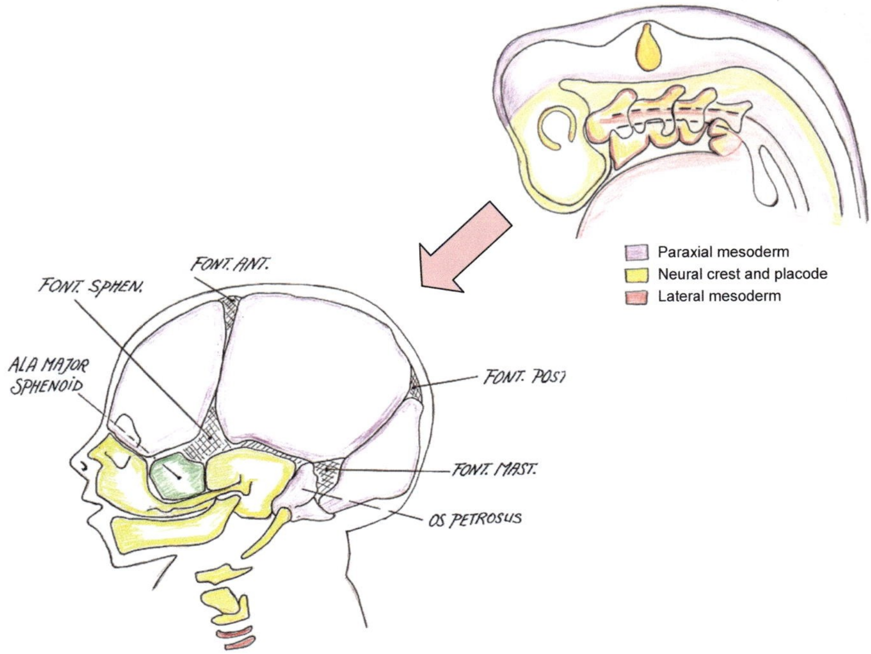

Fig. 4.2 Layering of the head and neck in an embryo of 10 mm by axial and lateral mesoderm and by neural crest and placode. Skull parts are given in relation to the paraxial mesoderm, neural crest and lateral mesoderm (compare to upper part of this figure in which the same colours are used) © E. Marani

developing (remember during labour the bony parts of the calvarium shift partly over each other upon entering the birth canal).

In the embryo, three longitudinal strips in the head and neck region can be discerned (Fig. 4.2):

(1) the paraxial mesoderm,
(2) the neural crest and the eye and otic placodes, and
(3) the lateral mesoderm.

The embryological situation can now be used to study the origin of the bony parts in a just born human. The paraxial mesoderm is responsible for the dermato-neurocranium (violet), and the neural crest and placodes make the maxilla, mandible and temporal bone, processus stylohyoideus, hyoid, thyroid and cricoid. The lateral mesoderm produces the first rings of the trachea. The origin of the great wing (ala major) of the sphenoid is still unknown. Firstly let's ask: the brain stem makes connections with the periphery by cranial nerves. They have to pass through the skull base to reach their end organs. How is this organized in development? Looking at the skull development, two methods of bone formation can be recognized: dermal (skin) ossification and cartilaginous ossification, of which the cartilaginous is the most involved in nerve passage. It brings us to the skull parts that originate from cartilaginous bone.

One can also discern three main constituent parts of the head and neck in another way:

1. Primordial neurocranium (skull base),
2. Splanchnocranium (bony elements of the viscerocranium around the primitive nose and mouth), and
3. Dermatocranium (bony elements of the rest of the neurocranium, roof of the mouth and the mandible).

In above-mentioned parts, 1 and 2 are constructed by cartilaginous ossification and 3 by dermal ossification (note Fig. 4.3). In fact, a sandwich construction of alternating desmal (intramembraneous) and cartilaginous ossification can be recognized in the head and neck areas. Some scientists wondered whether desmal and cartilaginous bone seamless connected to each other.

A few structures are lifted out of which the squamous temporal has the most complex relationship:

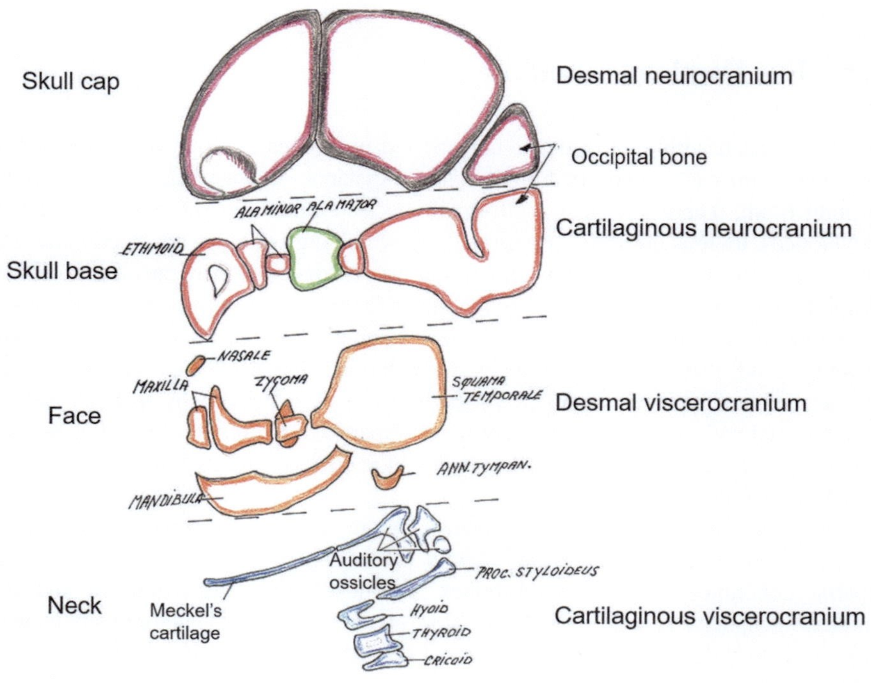

Fig. 4.3 "The skull can also be considered a big Mac". The layered construction of head and neck is shown in view of desmal and cartilaginous origin of its structures © E. Marani

(1) Both tooth-bearing bones (maxilla and mandible) do not constitute a joint early in development (Fig. 4.3). An articulation is formed late in development. Later on, the story on Meckel's cartilage will be told.
(2) The premaxillary and maxillary centres of ossification (in Fig. 4.3 indicated as two parts) fuse into one bone, the maxilla.
(3) The squamous temporal originally part of the viscerocranium is incorporated in the expanding neurocranium but without losing its visceral relationships: it contains a cerebral surface, an external surface, is involved in the otic capsule, participates in the formation of the roof of the middle ear cavity, forms the roof and posterior wall of the external auditory meatus and contributes to the formation of the mastoid process.

Thus, the skull can be considered a big Mac. The upper layer is dermal in origin. The second layer is cartilaginous, while the third layer is again dermal in origin. The fourth layer is again cartilaginous. The upper two layers produce the neurocranium (skull base included) and the lower two layers: the viscerocranium, of which the bony construction is rather difficult to comprehend, and branchial system that transforms heavily.

4.5 Eye Dominance and the Orbit

Two eyes each within an orbit are constructed in the head and placed at a distinct distance from each other. In humans, mean distance between the eye centres is around 6 cm. There are clear malformations: in anophtalmy (no eyes have been developed), there is no orbit, in cyclopedy in which only one eye has been matured, then only one orbit is present, and if three eyes develop there are three orbits. Buphthalmos or hydrophthalmos is the name for the development of an oversized eye; indeed, the orbit is also large and in microphthalmy with a small eye a small orbit is found. If two eyes develop next to each other, the orbits are also placed next to each other. Hypertelorismus is when the eyes are separated by distance larger than normal and indeed the orbits are farther apart too. Seemingly, the orbit has no tendency to differentiate by itself, but is in its localization and its form development totally dependent of the eyes.

An analogous reasoning can be given for brain and ear malformations, in which the skull bones are also deformed. The same holds for the nose and mouth: the spatial coherence between skull bones and head organs is extensive and its induction capacity is large as well (see Sect. 4.1 Introduction). The opposite is also true: if a skull deformity is present, one of the head organs is also malformed (see Van Limborgh 1967).

Remember that the same basic principle was already expressed by Galen and Vesalius in their medieval vocabulary: the head is formed for the sake of the eyes (see Sect. 2.5).

Cuvier (1769–1832), called the "dictator of biology" (Singer 1931), held strong opinions [think of his rude behaviour to the evolutionist Lamarck (1744–1829)]. Cuvier formulated the "principle of correlation of parts": "organs do not exist or function in nature as separate entities but as parts of organic living wholes. In these living wholes, certain relations are observed which are fundamental to their mode of life. Thus, feathers are always found in birds and never in other creatures. The presence of feathers is related to a certain formation of the forelimb with reference to its use as wing. This, in its turn, is related to certain formations of the collarbone and breast bone, with reference to the function of flight". This has consequences for the muscles and for breathing and so on. "So 'the principle of correlation' might be followed through the whole being of the bird, down to its minutest parts. Yes, and even to its psychology" (Singer 1931; see more recent Rudwick 1997; Buffetaut 2002).

This brings us to the rhetoric question: *Is the eye the "feather" of the skull?*

Recognition of humans by humans (not computer recognition!) occurs by their face, and the eyes are particularly important: "the concept of 'configural processing' (encoding of metric distances between features) is central to the study of face perception. It is widely held that viewers are sensitive to the relationship between facial components, and that they use this sensitivity to make perceptual judgements. This concept lies at the heart of many proposals concerning face identification" (Sandfort and Burton 2014). In the article, "Eye contrast polarity is critical for face recognition by infants", 7–8-month-old children were studied. "Just as faces share the same basic arrangement of features, with two eyes above a nose above a mouth, human eyes all share the same basic contrast polarity relations, with a sclera lighter than an iris and a pupil, and this is unique among primates" (Otsuka et al. 2013). The conclusion of the article is: "the results suggest that the contrast polarity relationships uniquely characteristic of human eyes, with the iris and pupil always darker than the white sclera, play a critical role in face recognition from early in development. Although the eyes may cover only a small *portion of the facial image (2.6%)*, it was only their local contrast polarity that affected face discrimination in these infants" (see Fig. 4.4; Otsuka et al. 2013).

There are racial differences in recognition: white people do have difficulties in recognizing black people, and vice verse (e.g. Hills and Pake 2013). Moreover, gender differences are well known (e.g. Palmer et al. 2013).

The enormous amount of literature on recognition is left aside in this chapter (see Chap. 7), because it is clear that the eye in recognition owns the "feather" capacity. The eye has an analogous position, due to its inducer function in the upper part of the face, also exemplified by the evolutionary differences between Homo sapiens and the other hominoids, based among others, on the form of the frontoorbital transition, belonging to the melting line between viscerocranium and neurocranium.

Remember that at the start of Sect. 4.4 it was noticed that several scientists wondered that desmal and cartilaginous bone seamless connected to each other: comparative anatomy found that in primitive species an *interorbital* septum can be present. This interorbital septum causes, if it is absent, what is called platybasic (broad-based) skulls (e.g. sharks and rays) and eyes in such skulls are placed far

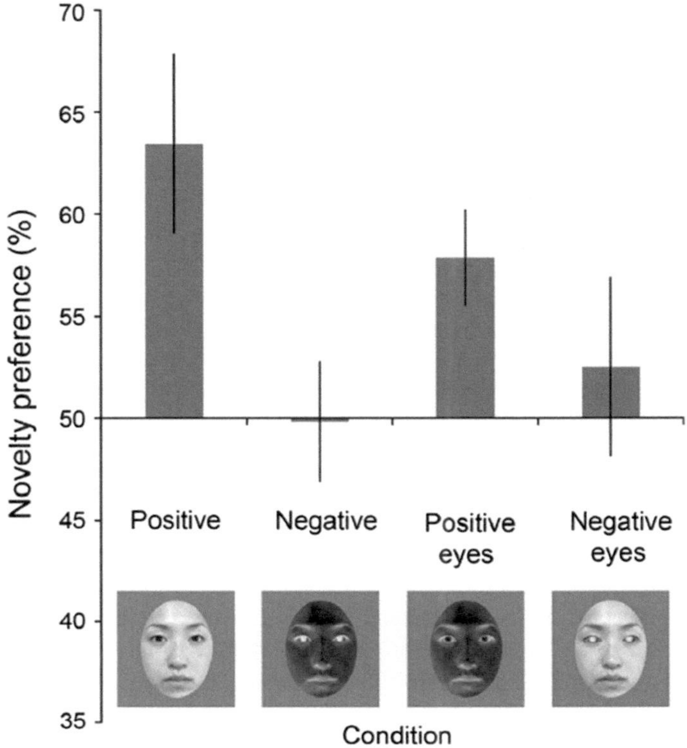

Fig. 4.4 Novelty preference scores (dividing each infant's looking time at the novel face by the total looking time over the two test trials and then multiplying this ratio by 100) obtained from each contrast polarity condition. Error bars represent 1 standard deviation (text and figure from Otsuka et al. 2013, with permission)

from each other. The space in between its orbits is often filled with brain. In cases of only the presence of this septum between eyes, they are placed nearer to each other in what is named tropidobasic (keel-based) skulls and orbits can border each other (most of the reptiles).

In mammals, including man, this interorbital septum is only present early in embryonic life and disappears later in its development. Therefore, mammalian skulls are called secondary platybasic. Now, it should be noted that only in platybasic skulls (secondary too) the neurocranium can join with the viscerocranium in an immovable skull fusion. In tropidobasic skulls, movement of skull parts is always possible and indeed kinesis (mobility) within the skull is observed in such cases (van der Klaauw 1941).

As clear as these theoretical considerations were before the Second World War, starting around the 1950s and later on, smell was brought into the interorbital septum description by the visual predation hypothesis: "orbital size and convergence necessarily reduced the size of the olfactory apparatus and resulted in an

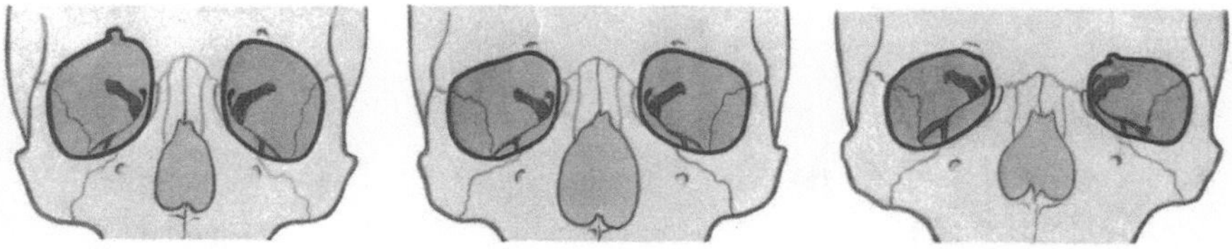

Fig. 4.5 Large (index 85, 0–>), middle (76, 0–84.9) and small (<–75, 9) human orbits (from Lanz und Wachsmut 1979)

interorbital septum". We know from holoprosencephaly (splitting of the brain in two halves is disturbed) that ocular convergence is related to the *Six 3* gene (Lozanoff et al. 2004) and it is this gene that presumably is responsible for eye placement. "Phylogenetic change should be driven by ocular convergence with frontal encephalization" (Lozanoff et al. 2004).

Indeed, frontal increase of the brain in man is extremely large, and our eyes are also converged for stereoscopic vision, but still with a mutual distance that leaves room for the nasal cavity to be in between both orbits.

In normal situations, the human orbit shows variability. A measure for the orbit is found by dividing the height by the width and multiplying the result with 100. One can discern a large (hypsiconchy), middle (mesoconchy) and a small (chamaeconchy) orbit (Fig. 4.5). This different typology of the orbits plays an important role in the study of, for example, metric traits and geographic distribution of the modern but also the Neolithic Chinese, Medieval Serbian and Modern American-Indian populations.

Now that orbit and eye have been placed in their shared influence sphere and the topography of the viscerocranium is given (see Fig. 2a, b in Introduction), the development of the skull base can be studied.

4.6 Foetal Movements and Skull Growth

The onset of foetal movements is variable and occurs somewhere in between week 6.5 and 8 (Fig. 4.6).

However, the skull base (and skull cap, lagging behind) is still in its developmental stages when movements of the foetus are present (Fig. 4.6): mandibular and maxillar processes are together with the otic placode noticed at the exterior of the embryo in between 23th and 30th day of development. Around the 37th till 40th day, the maxillary process fuses with the nasal processes (see Introduction and Fig. 5 in introduction). At day 46, the external ear develops, while at day 40 the cartilage next to the primitive vertebral column (parachordal cartilage) has already developed. From day 50 till day 90, most structures and passage ways for blood vessels and

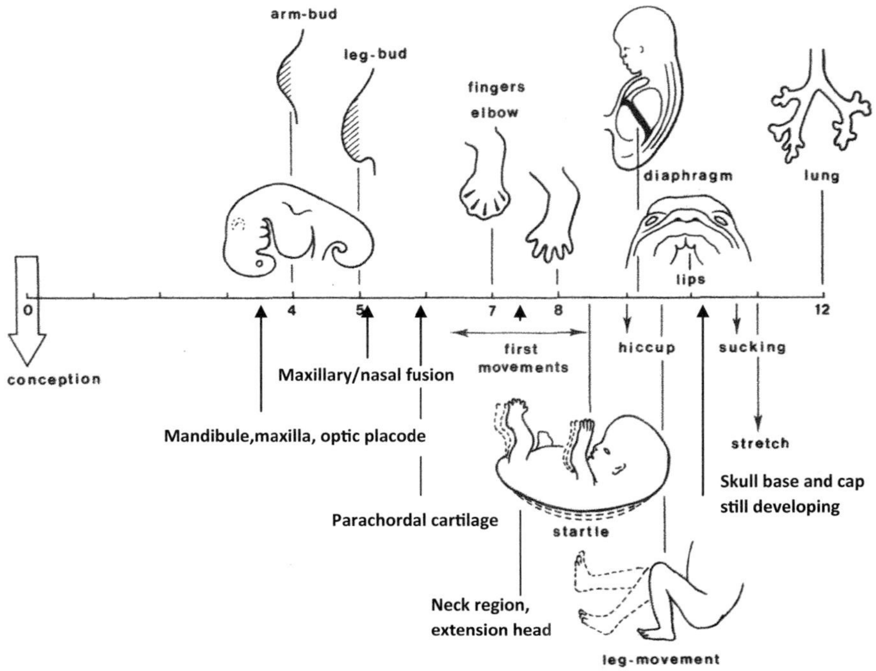

Fig. 4.6 Embryonal to foetal development indicating the weekly time frame for the first movements like startle, leg movements, hiccup, sucking and stretch in relation to skull cap and skull base development. © E. Marani

cranial nerves are already formed in the skull base. Extension of the head from the neck region occurs in between 50th and 60th day of gestation (Hamilton et al. 1959).

Minkowski (1921) used the caesarean operations done at the Maternité of the Hôpital of the Canton of Winterthur, to study the foetal movements by placing foetuses in lukewarm saline of 37 °C. (Note: operations were only carried out in cases that were [then] medical indicated). His youngest foetus was two months old and he observed, besides of course the movements of the extremities, head turning from one side to the other, and raising or bending of the head. These movements are slow, asymmetric, arrhythmic, disorderly, diffuse and amorphous, and of small amplitude. Movements can be of the head only or be in conjunction with extremities. Minkowski calls these movements "mouvements choreiathétostiques" [chorea-athetosis: is the occurrence of involuntary movements in a combination of chorea (also called St Vitus's Dance; irregular migrating contractions) and athetosis (twisting and writhing)], but can reach a "choreiform" tendency, especially in older foetuses. In the youngest foetuses "on n'observe pas de reflexes isolé, se limitant à l'extremité excite, mais *des reaction motrices qui irradiant plus ou moins sur d'autres extrémités, sur la tête et le tronc, et qui sont extrêmement variables"*; one does not notice isolated reflexes, restricted to the excited extremity, but motor reactions that more or less irradiate into other extremities, into the head and the trunk, and these are extremely variable (Minkowski 1921; italics are his). Cervical reflexes (foetus stays horizontal) indicate that movement of the head causes movement of the limb to which the head is turned. Turning the head to the right gives abduction of the right arm and an abduction of the left arm. Labyrinth reflexes can be activated by placing the foetus in a sitting position (bringing the foetus from horizontal to a vertical position). Symmetric bilateral movements of the extremities are provoked by these spatial head movements (Minkowski 1920, 1921).

Thus, while skull base and skull cap and brain are still in the middle of their development, the head can change its position, bring about reflexes to extremities and trunk, and in special cases the upper limb with hand can be brought towards the head.

Remember again that at the start of Sect. 4.4 it was noticed that several scientists wondered that desmal and cartilaginous bone seamless connected to each other. It is not so seamless. In birds and lizards, skull parts can move the so-called metakinetic skull (Bock 1960; Versluys 1912). In lizards, the movement is between frontal and parietal bones and between the ends of the zygomatic bone. Here, skull parts are not grown together. One of the consequences of mal-development, presumably evolutionary-driven, of the human skull connections can be schisis or cleft formation. During the embryonal and foetal development, grooves can enhance and close to form a tube. Grooves can also flatten ending with their disappearance. These processes occur frequently, but not only, at the transitions of desmal and cartilaginous bone. If during development one of these two processes goes wrong, dysplasia can occur. In fact based on embryology and on the known malformations, a classification became apparent (Vermey-Keers 1990; van der Meulen et al. 1990) that includes craniofacial dysplasias. Different types of dysplasia are distinguished. Some are only a schisis, others are characterized by a fibrous connection between bones caused by a non-neoplastic tumour-like congenital process. Dysplasia can be present over all bony body parts. It can concern one bone (monostotic) or involve several bones (polyostotic). Some are restricted to the viscerocranium and others to the frontal and lateral neurocranium (craniofacial dysplasia) but also to the mandible and maxilla alone (called cherubism) (Fig. 4.7).

A loop of these craniofacial dysplasia is described, forming a S-helix that relates to sutures and bony junctions (van Adrichem et al. 2008). Since our lead is the "spatial coherence" and, although we did not go into all developmental stages of the skull, the outcome of mutual dependence is here that seemingly developmental disturbances can be regrouped in a topographical craniofacial dysplasia map, because the causes and mal-growths produce a similar end result. Although the mandible is loosely bound to the rest of the skull, it still obeys the distribution map of dysplasia.

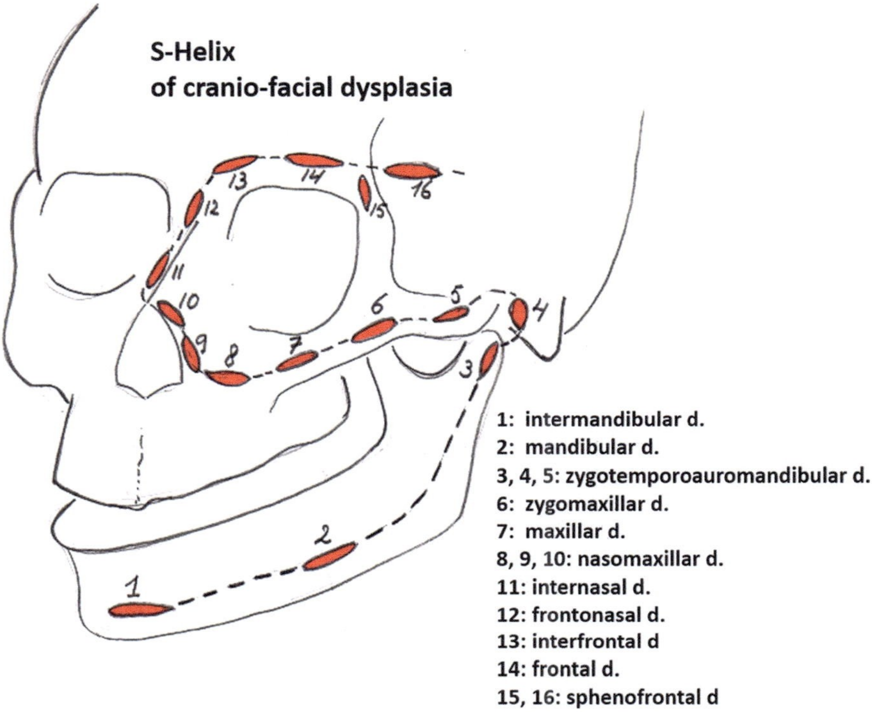

Fig. 4.7 Distribution of dysplasia (d) over the craniofacial region (see also van Adrichem et al. 2008)

4.6.1 The Mandible

The lower jaw of fishes has evolutionary less to do with our human (modern) mandible (see Figs. 4.8a and 6.6). In gnathostomes, the jaws are characterized by their loose structure. Jaws could be folded and during contact put forth pressure on each other. The gnathostome skull organized during evolution the demand for localization and support of jaws. The skull extended posteriorly behind the otic capsules, and the membranous skull became bony and extended ventrally, while the upper jaw elements fused (Fig. 4.8b). This extension and fusion of bony elements made the attaching of the hyobranchial (oropharyngeal and laryngeal) skeleton possible. Although a large diversity of cranial anchoring of the lower jaw is present in fossil and recent fishes, the place of the joint between skull and lower jaw became determined.

"Mammals are unique among vertebrates in having only the dentary bone in the mandible, the elements of the old jaw joint having been incorporated into the middle ear" (Gans 1988). In water, pressure waves can be noticed by innervated skin and thus predation is differently organized in aquatic animals compared to terrestrial ones. When living in an air surrounding, one needs specializations to

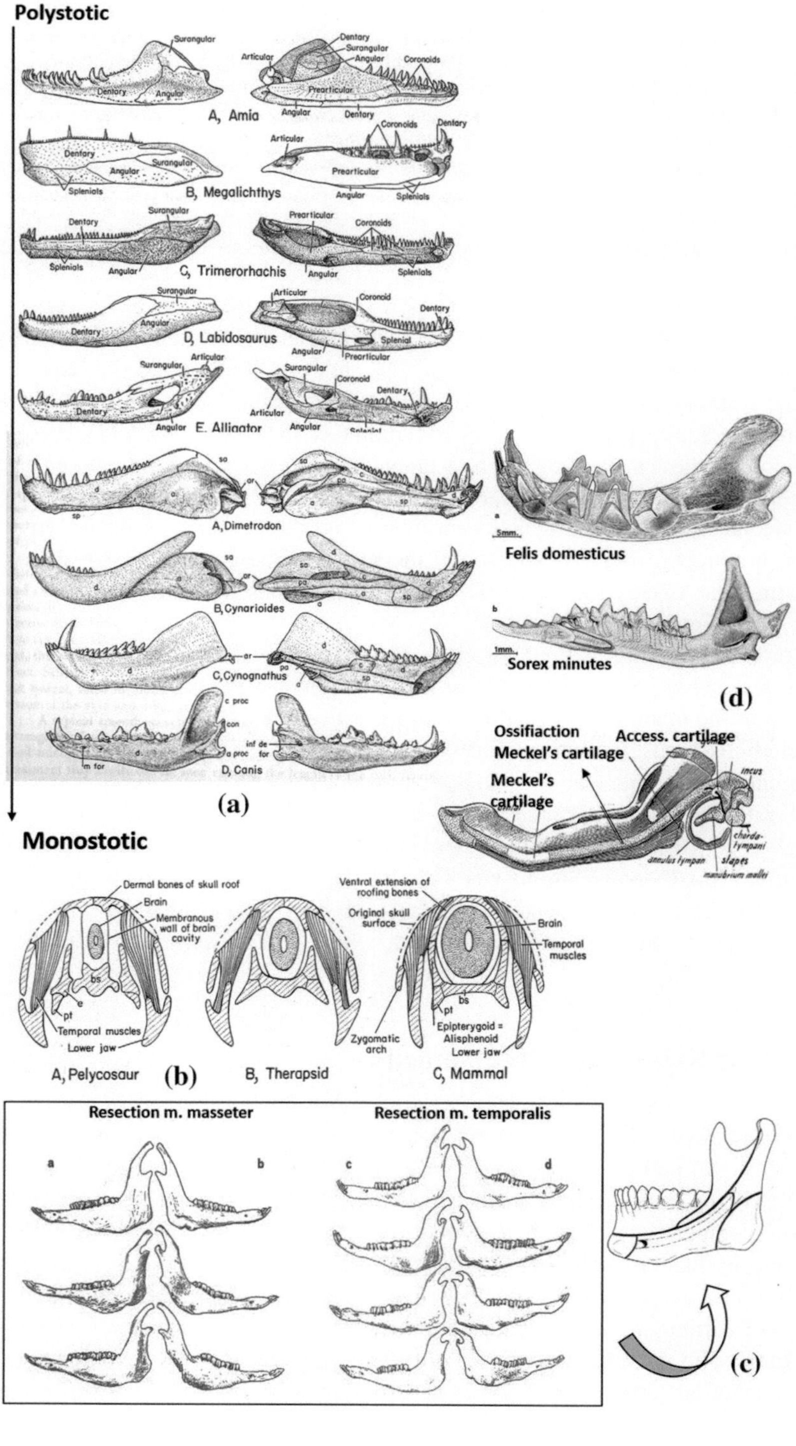
Polystotic
Surangular
Dentary
Angular
Articular
Coronoids
Prearticular
A, Amia
Splenials
B, Megalichthys
C, Trimerorhachis
Coronoid
Splenial
D, Labidosaurus
E. Alligator
A, Dimetrodon
B, Cynarioides
C, Cynognathus
c proc
con
a proc
inf de for
m for
D, Canis
(a)
Monostotic
Felis domesticus
5mm.
Sorex minutes
1mm.
(d)
Ossifiaction Meckel's cartilage
Access. cartilage
Meckel's cartilage
incus
chorda tympani
annulus tympan
stapes
manubrium mallei
Dermal bones of skull roof
Brain
Membranous wall of brain cavity
bs
pt
Temporal muscles
Lower jaw
Ventral extension of roofing bones
Original skull surface
Temporal muscles
Zygomatic arch
Epipterygoid = Alisphenoid
A, Pelycosaur
(b)
B, Therapsid
C, Mammal
Resection m. masseter
Resection m. temporalis
a
b
c
d
(c)

◂**Fig. 4.8** Developmental and evolutionary aspects of the lower jaw. **A** Left outer view, inner views at right of lower jaws from fish to dog. **B** Diagrammatic cross sections of the skull and jaws. **C**, **a** and **c** controls, **b** and **d** resection sides. **D** Meckel's cartilage and relation to the auditory ossicles. **E** Dark areas are X-ray transparent areas. Figures are adapted from Dullemeijer 1974, Romer 1962, Van der Klaauw 1941. Amia, ray-finned fish; Megalichtys, primitive crossopterygian; Trimerrorhachis, primitive labrynthodont; Labidosaurus, primitive reptile; Dimetrodon, primitive mammal-like reptile (pelycosaur); Cynarioides, primitive therapsid; Cynognathus, advanced therapsid; Canis, dog, example of a typical mammal. Abbreviations: a, angular; a proc, angular process; ar, articular/gonial; c, coronoid; con, condyle; c proc, coronoid process; d, dentary; inf de for, inferior dental foramen; m, mental foramen; pa, prearticular; sa, surangular; sp, splenial

detect sound. Although during a pop-concert you can feel the sounds if you are near the speakers, a few metres away the sounds cannot be bodily noticed anymore. The needed auditory specializations used the structures of the lower jaw of the fishes to produce the auditory ossicles during evolution (see Figs. 4.8d and 4.11). Placed one level lower, the cartilage of the first branchial arch was already involved in production of the jaw (Meckel's cartilage). Primitive jaws can contain several parts (e.g. dental, articular and Meckel's cartilage). Part of Meckel's cartilage is involved in the production of the malleus, and the rest integrates with the sole membrane structure that makes the bony mandible in mammals and humans, the dentary (Fig. 4.8d; Van der Klaauw 1941).

Comparative and functional morphology have studied the lower jaw frequently. Some researchers indicate that the articular or gonial part of the lower jaw is still present in the mammalian mandible (compare Fig. 4.8a, d). The effect of resection of mandibular muscles induced form and function relations for different parts of the mammalian mandible (see Fig. 4.8c; Schumacher and Dokladal 1968, sheep; Scott 1955, rat). The coronoid process is co-dependent of the attachment of the temporal muscle (Fig. 4.8c). Nature of bone has been studied in relation to the function of the mandible. Small jaws (Sorex) can function with mainly the compacta of bone; larger mandibles (Felis) need trabeculae (see Fig. 4.8e; Dullemeijer 1974). The mandible is of course a teeth-bearing part, which will be treated in Chap. 6.

4.7 The Skull Base Development

The skull base contains in early development several separated cartilaginous centres (Fig. 4.9). All cranial nerves are taken up in the skull base by perforation, except the N XII (by concrescentia), and NN IX, X, XI and the vena jugularis by association (compare Figs. 4.9a, b; see Sect. 4.3). It is the nerve or blood vessel that has the capacity to stop suturation or concrescentia. If a structure is absent, the closure goes on, and no passage remains. The consequence of spatial coherence?

These cartilaginous centres are organized by inductive collaboration between structures. The notochord, together with the hindbrain, induces several structures: the parachordal cartilage (Fig. 4.9), but also the basioccipital and the basisphenoid.

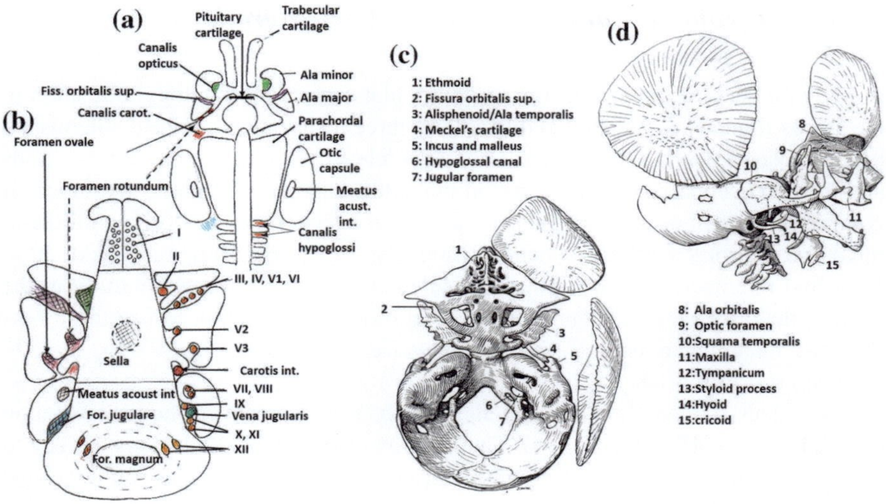

Fig. 4.9 Development of the skull base. Compare the left figures: **a** is earlier in ontogeny than **b** Both are schematic overviews for the development of foramina and fissures related to the cranial nerves and large blood vessels. **C** and **d** Show a more realistic overview with **c** a craniodorsal, and **d** a lateral view

Induction is also responsible for the origin of the squamosal, occipital, supraoccipital, parietal and frontal bones due to the actions of the posterior hindbrain and prosencephalon. The otic vesicle induces the otic capsule and also by induction the membrane bone in the maxillary and mandibular arches. Genetics controls skull skeletal development, e.g.: *Alx4* is co-responsible for nasal and maxilla ossification centres, the absence of *MSX2* can induce calvarial deflections, and *Bapx1* influences the development of basisphenoid and exoccipital (see Francis-West et al. 2003). Fusion of the developing centres finally produce the mature bones, which holds, e.g. for the occipital bone, the sphenoid and the mandible (see above). Induction and fusion are seemingly the two main processes that produce the skull base guided by genes, growth factors and transcription factors.

"Earlier in development, the cranial neural crest cell migration contributes to the formation of these centres of the facial skeleton: forebrain neural crest produces frontonasal mass, rostral midbrain gives periocular mesenchyme, caudal midbrain contributes to periotic mesenchyme, mandibular and maxillary primordia in their respective areas and the facioacoustic ganglion, while the rostral hindbrain still contributes to mandibular and maxillary primordia and the trigeminal ganglion. An analogous distribution is described for the mesoderm or the seven loose aggregations of these cells, called somitomeres" (Francis-West et al. 2003). The relations of the Hox genes and rhombomeres with the branchial arches are rather clear. Such a relation between subparts of the mes-, di- and telencephalon with the skull base is less clear because the early borders of these brain areas are under discussion.

4.7.1 Neuromere Borders and Crest Cell Migration

A study on the development of the rat ventricular surface morphology in relation to the sulcus limitans demonstrated that at least three neuromeres can be distinguished in the prosencephalon: neuromere I contains telencephalon, hypothalamic regions and the parencephalon anterius (ventral thalamus, its tegmental part), neuromere II comprises the parencephalon anterius (produces later on the sub- or ventral thalamus) and neuromere III the synencephalon (the dorsal thalamus; see Lakke et al. 1988 and references therein). Since the sulcus limitans terminates ventrally in the parencephalon posterius, there is enough segmental information to use these borders for mammalian neural crest and mesoderm migration from stage 9 (22–24 days) on (Fig. 4.10).

The sulcus limitans is in the limelight of the development of the diencephalon (see Lakke et al. 1988 for review). The location of the sulcus limitans is the key to the subdivision in alar (sensible) and basal (motor) areas. These areas can be related to a subdivision in functional components of somatic and visceral efferent areas. Several authors described the sulcus limitans ending at the mesencephalo–diencephalic border which has been agreed upon by Lakke et al. (1988) and Gribnau and Geijberts (1985). Other fissures subdivide the diencephalon into four longitudinal zones differing from the four functional zones mentioned above. These four diencephalic parts are called: epithalamus, thalamus dorsalis, thalamus ventralis and

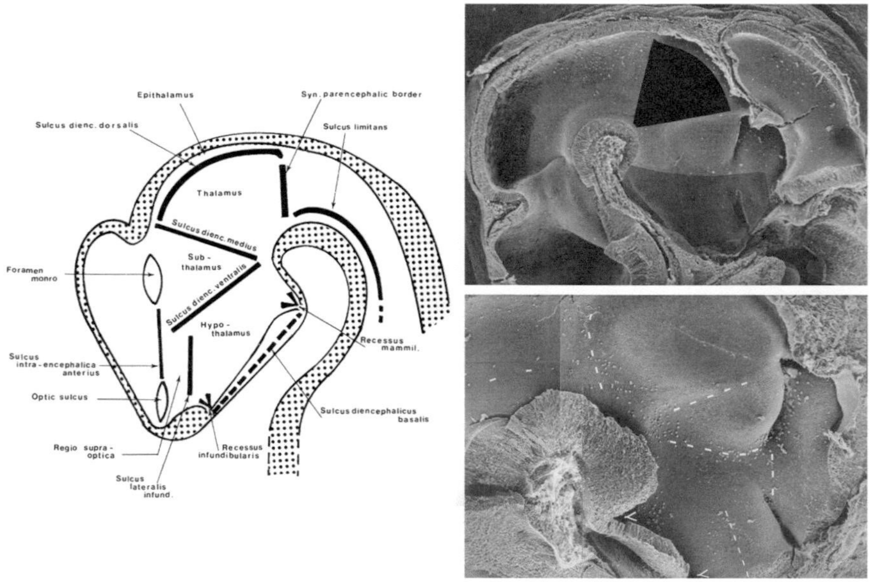

Fig. 4.10 Sulci and regions in the developing diencephalon of the rat at E13-14. Right side is mirrored to the left. Synencephalon/thalamus purple, parencephalon/subthalamus yellow and hypothalamic region blue

hypothalamus. They can be discerned by three sulci: the sulcus diencephalicus dorsalis, the sulcus diencephalicus medius and the sulcus diencephalicus ventralis (Fig. 4.10). During development, the diencephalic parts undergo differential growth and displacement also due to the extreme development of the cortex. The ventral thalamic area precedes the dorsal thalamus in its development. Two parts can be distinguished within the ventral thalamus: a dorsal or reticular zone and a tegmental zone. The reticular and tegmental zones are subjected to a lateral displacement and in later stages to stretching and flattening under the expanding dorsal thalamus (Papez 1940). These developmental extensions determine together with the flexures the basis and roof of the diencephalon and the craniodorsal and ventral form of the midskull and the midskull base. Optic nerve and infundibulum are structures that are skull base related and are topographically placed by the strong development of the thalamic areas.

4.7.2 The Development of the Tympanic Cavity and the Tympanic Ring

"Let us once and for all abolish one misconception that the tympanic cavity derives from the first pouch directly, simply by enlargement. Nothing is less true. The tympanic cavity is the result of a process of interaction of mesenchyme and epithelium in the region of the first and second arches, accompanied by obliteration of the first and second pouches. The region of the first and second arches is subject to a considerable development directed laterally and ventrally" (Landsmeer, unpublished). Three bones have to cooperate in space and structure to build the hearing apparatus: the inner ear, originating from the otic capsule in the petrosal bone, the middle ear with the ear bones with its connection to the pharynx (Eustachian tube) and the outer ear with the eardrum. The squamous temporal bone is the core of this construction. On one side, it has to contact the otic outgrowing capsule; on the other side, it has to take up the tympanic ring with its membrane (the eardrum). The petrous part makes a bowl, within it the otic structures. The squamous forms a roof over this bowl. A petro-squamous suture is the contact zone. At the other side, the squamo-tympanic suture originates.

To enclose the middle ear, the squamous temporal has to incorporate the petrous bone into the temporal bone, and its connection to the external auditory meatus with the eardrum (tympanic membrane) has to be constructed during development. Therefore, attention should be given to the tympanic region. The development of an epithelial plate from the first branchial groove occurs together with the extension of the pharynx (Fig. 4.11c). The epithelial plate and the epithelium of the future external auditory meatus fuse with a sheet of mesenchyme in between both, producing the tympanic membrane. This circular construct contains a bony structure at its border wall, the tympanic ring. The tympanic membrane adheres to this bony

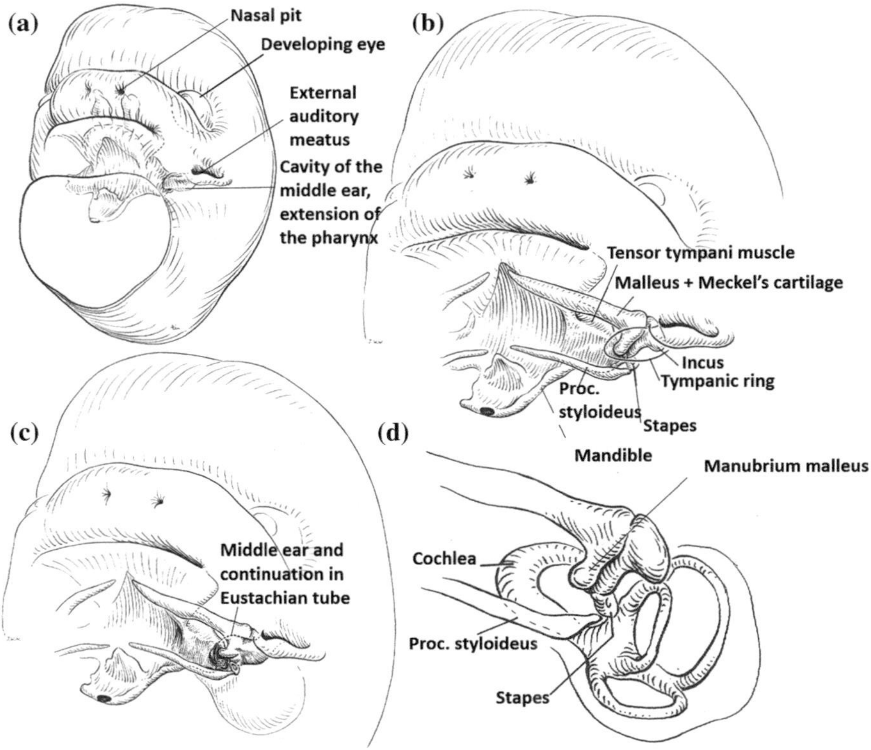

Fig. 4.11 **a**—An overview of an embryo reconstruction at the end of the second month. Translucent are the structures around the auditory apparatus drawn in (**b**) and (**c**). The labyrinth is pictured together with the tympanic structures of the middle ear in (**d**)

ring and can be clearly demonstrated in a dissection of a foetal specimen (Fig. 4.12).

The auditory ossicles together with the volume of the middle ear determine the sound one can hear. Thus, the mammalian ossicular chain, together with the ear-drum, overcomes the impedance mismatch between the surrounding air and the fluid within the inner ear better than in non-mammalian tetrapods. "Metrics expressing the morphological differences like mass of ossicles, moments of inertia, morphology of the connecting joint and differences in rotational motion correlates with variations in the frequency range of hearing across mammals" (Stoessel et al. 2016). Surprisingly for us humans, comparative anatomy of the auditory ossicles between great apes (*Pan troglodytes, Pan paniscus, Gorilla beringei, Gorilla gorilla* and *Pongo* sp.) and humans (*Homo sapiens*) shows that for the malleus, the incus and even the stapes of humans the difference to the great apes is obvious. The consequence must be that the hearing capacity of humans is species characteristic and even can be used to distinguish the hominin species during anthropological research (for techniques and mathematical methods, see Stoessel et al. 2016).

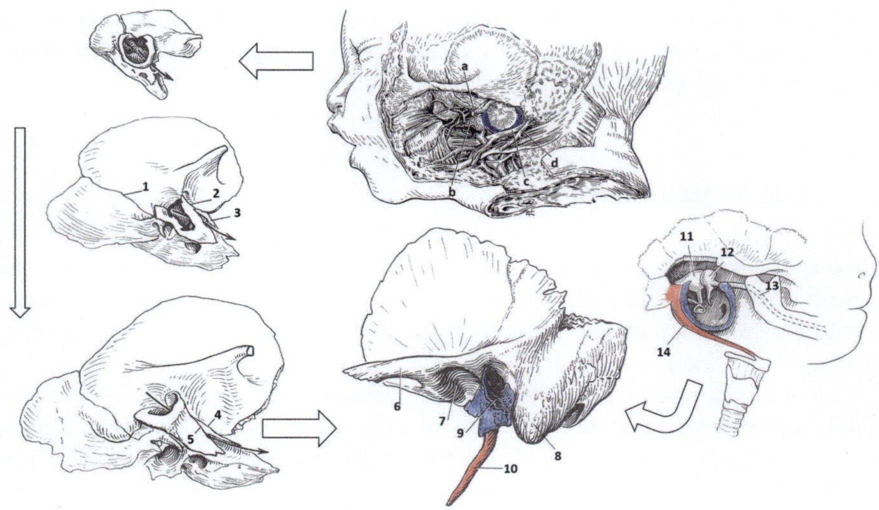

Fig. 4.12 Development of the tympanic ring. The ring is absent at the tympanic incisure. The relation to the carotid canal just below the tympanic ring is not indicated. Black arrows indicate the shifting direction of the external meatus from vertical towards nearly horizontal. 1, squamomastoid suture; 2, anterior tubercle of tympanic ring next to the postglenoid tubercle; 3, wall of the foramen of Huschke; 4, squamo-tympanic fissure and caudo-ventrally continuing into the petro-tympanic fissure; 5, fusion of tympanic walls; 6, zygomatic process; 7, mandibular fossa; 8, mastoid; 9, external meatus with tympanic plate in blue surrounding the styloid; 10, styloid; 11, incus; 12, malleus; 13, Meckel's cartilage; 14, Reichert's cartilage; a, spine of the sphenoid; b, chorda tympani; c, stylomuscles (stylohyoideus) and ligament, noting relation with m. digastricus; d, N VII

Earlier, the auditory ossicles have been developed medially behind the future tympanic membrane (Fig. 4.11). Meckel's cartilage plays an important role (see section mandible 4.6.1) in the production of the malleus, but also the incus, while the cartilage of the second branchial arch (Reichert's cartilage, red in Fig. 4.12) is hyoid and styloid related and is involved in the stapes ossicle production. The tympanic ring grows into the form of a funnel and tries to close the ring, which is occurring postnatally. This first succeeds halfway the funnel with the consequence that an extra hole exists below it, called the foramen of Huschke. Only in around 50% of the human population, this foramen closes later on. The funnel starts more vertically oriented, the tympanic membrane nearly horizontal and it displaces during the postnatal development towards horizontal, the tympanic membrane thus turning towards vertical (Fig. 4.12, note the direction of the arrows).

The delineation of the tympanic cavity that lies on the lateral aspect of the otic capsule and of the otic developing vesicle is important. It ends at the cartilaginous protrusion behind the facial nerve. The cavity has to extend behind the bulging cochlear part of the otic region for sound transmission that narrows into the Eustachian tube (Fig. 4.11b, c). The wall layer is continuous from the Eustachian tube into the auditory tube. In this area, two muscles develop: m. tensor veili palatini (palatum tightner) and m. levator veili palatini (palatum lifter). Therefore, the tympanic/otic region is during its development directly related to pharynx structures to keep the Eustachian tube open (see Fig. 6.6).

4.8 The Jugular Foramen

Another aspect of the coherence spatial is that one should be well aware of the typical relations present if they have to operate in or around the skull. Tumours originating from baroreceptors can easily reach the brain by the foramen lacerum (passage of the internal carotid).

Therefore, its place on the exterior side between the N VII, together with the stylohyoid process, and on the other side the hypoglossal canal is important. This is because the foramen jugulare with Nn IX, X, XI is just in front of the foramen lacerum.

However, the situation after passage of the skull is different. Here, N VII is localized next to N VIII, and Nn IX, X, XI concentrate towards the foramen jugulare (see Fig. 4.13).

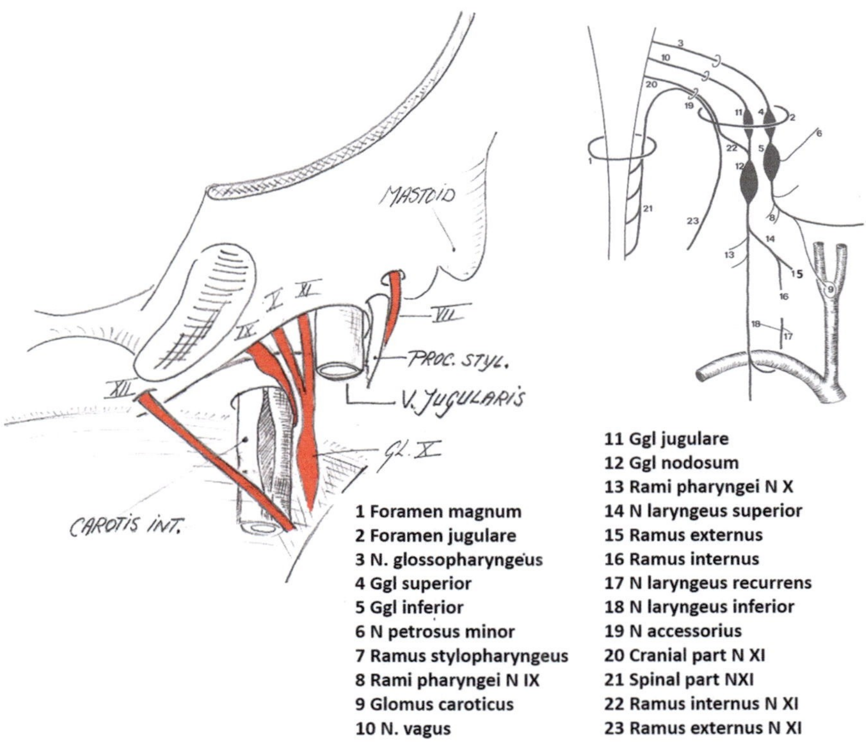

Fig. 4.13 Foramen jugulare and its cranial nerves in relation to the surrounding structures

References

van Adrichem NLA, Hoogeboom AJM, Wolvius EB (2008) Genetica van de ontwikkeling van schedel en aangezicht. Ned Tijdschr Tandheelkd 115:61–68

Baker CVH, Bronner-Fraser M (2001) Vertebrate cranial placodes. Part I. Embryonic induction. Dev Biol 232:1–61

Batalle D, Eixarch E, Figueras F, Munoz-Mureno E, Bargallo N, Illa M, Acosta-Rojas R, Amat-Roldan I, Gratacos E (2012) Altered small-world topology of structural brain networks in infants with intrauterine growth restriction and its association with later neurodevelopmental outcome. NeuroImage 60:1352–1366

Bock WJ (1960) Secondary articulation of the avian mandible. Auk 77:19–55

Bolk L (1910) De segmentale innervatie van romp en ledematen bij den mensch. De Erven Bohn, Haarlem

Buffetaut E (2002) Cuvier, Le découvreur de mondes disparus. Pour la Science, Paris

Chandebois R, Faber J (1983) Automation in animal development. A new theory derived from the concept of cell sociology. In: Wolsky A (ed) Monographs indevelopmental biology, vol 16. Karger, Basel

Dobbing J, Sands J (1979) Comparative aspects of the brain growth spurt. Early Hum Dev 311:79–83

Dongen Van PAM (1998) Brain size in vertebrates. In: Nieuwenhuys R, ten Donkelaar HJ, Nicholson C (eds) The central nervous system of vertebrates, vol 3. Springer, Berlin, pp 2099–2134

Dullemeijer P (1974) Concepts and approaches in animal morphology. Van Gorcum Comp, Assen, The Netherlands

Epstein HT (1974) Phrenoblysis: special brain and mind growth periods: I. Human brain and skull development. Dev Psychobiol 7:207–216. https://doi.org/10.1002/dev.420070304

Epstein HT (1986) Stages in human brain development. Dev Brain Res 30:114–119

Epstein HT (1999) Stages of increased cerebral blood-flow accompany stages of rapid brain growth. Brain Dev 21:535–539

Epstein HT, Epstein EB (1978) The relationship between brain weight and head circumference from birth to age 18 years. Am J Phys Anthropol 48:471–474

Francis-West PH, Robson L, Evans DJR (2003) Craniofacial development: the tissue and molecular interactions that control development of the head. Adv Anat Embryol Cell Biol 169:1–144

Gans C (1988) Craniofacial growth, evolutionary questions. Development 103:s3–s15

Goedbloed JF (1976) The embryonic and postnatal growth of rat and mouse. IV. Prenatal growth of organs and tissues: age determination and general growth pattern. Acta Anat 95:8–33

Goedbloed JF (1977) The embryonic and postnatal growth of rat and mouse. V. Prenatal growth of organs and tissues, general principles: allometric growth, absence of growth, and genetic regulation of the growth process. Acta Anat 98:162–182

Goedbloed JF (1980) The embryonic and postnatal growth of rat and mouse. VI. Prenatal growth of organs and tissues: age determination and general growth pattern. Acta Anat 106:108–128

Goedbloed JF, Smits-van Prooije AE (1986) Quantitative analysis of the temporal pattern of somite formation in the mouse and rat. A simple and accurate method for age determination. Acta Anat 125:76–82 cat

Gribnau AAM, Geijsberts LGM (1985) Morphogenesis of the brain in staged Rhesus monkey embryos. Adv Anat Embryol Cell Biol 91:1–63

Hamilton WJ, Boyd JD, Mossman HW (1959) Human embryology. Heffer, Cambridge

Herlin C, Largey A, deMattei C, Daures JP, Bigorre M, Captier G (2011) Modeling of the human fetal skull base growth: Interest in new volumetrics morphometric tools. Early Hum Dev 87:239–245

Hills PJ, Pake JM (2013) Eye-tracking the own-race bias in face recognition: revealing the perceptual and socio-cognitive mechanisms. Cognition 129:586–597

Hollyday M (1980) Organization of motorpools in the chick lumbar lateral column. J Comp Neurol 194:143–170
Keynes RJ, Stern CD (1984) Segmentation in the vertebrate nervous system. Nature 310:786–789
Klaauw van der CJ (1941) Skelet en spieren van den kop. In: Ihle JEW (ed) Leerboek der vergelijkende ontleedkunde van de vertebraten, sec. ed. I, ch V. Oosthoek's Uitgevers, Utrecht, pp 223–280
Kuriyama S, Mayor R (2008) Molecular analysis of neural crest migration. Phil Trans R Soc B 363:1349–1362
Lakke EAJF (1997) The projections to the spinal cord of the rat during development: a time-table of descent. Adv Anat Embryol Cell Biol 135:143
Lakke EAJF, van der Veeken JGPM, Mentink MMT, Marani E (1988) A SEM study on the development of the ventricular surface morphology in the diencephalon of the rat. Anat Embryol 179:73–80
Landmesser L (1978) The distribution of motoneurons supplying chick hindlimb muscles. J Physiol 284:371–389
Lanz and Wachsmuth (1938) Praktische Anatomie: Bein und Statik. Springer, Berlin, Heidelberg, NY
von Lanz T, Wachsmuth W (1979) Kopf-Gehirn- und Augenschädel: Kopf Teil Ib. Springer, Berlin
van Limborgh J (1967) Enige beschouwingen over de regeling van de embryonale schedelontwikkeling. Klin Lessen: Ned T Geneesk 111:1693–1698
Lozanoff S, Doll S, Hallgrimsson B, Neufeld E (2004) Prenatal growth in the interorbital septum in Macca mulatta. Acta Anat 186:435–442
van der Meulen JC, Mazzola R, Stricker M, Raphael B (1990) Classification of craniofacial malformations. In: Stricker M, van der Meulen JC, Raphael B, Mazzola R, Tolhurst DE (eds) craniofacial malformations. Churchill Livingstone, Edinburgh
Meyers PZ (1985) Spinal motorneurons of the larval zebrafish. J Comp Neurol 236:555–561
Minkowski M (1920) Ueber Bewegungen und Reflexe des menschlichen Foetus während der ersten Hälfte seiner Entwicklung. Schweizer Archiv fuer Neurologie und Psychiatrie Band VII, Heft 1. 1921
Minkowski M (1921) Sur les mouvements, les réflexes et les reactions musculaires du foetus humain de 2 à 5 mois et leurs relations avec le système nerveux foetal. Revue neurologique. Conférence Neurologigique Paris 7:1105–1118 et 1235–1250
Noden DM (1991) Vertebrate craniofacial development: the relation between ontogenetic process and morphological outcome. Brain Behav Evol 38:190–225
Otsuka Y, Motoyoshi I, Hill HC, Kobayashi M, Kanazawa S, Yamaguchi MK (2013) Eye contrast polarity is critical for face recognition by infants. J Exp Child Psychol 115:598–606
Palmer MA, Brewer N, Horry R (2013) Understanding gender bias in face recognition: effects of divided attention at encoding. Acta Psychol 142:362–369
Papez JW (1940) The embryological development of the hypothalamic nuclei in mammals. Res Publ Ass Nerv Ment Dis 20:31–51
Rickmann F, Fawcett WJ, Keynes RJ (1985) The migration of neural crest cells and the growth of motor axons through the rostral half of the chick somite. Embryol Exp Morphol 90:437–455
Romer AS (1962) The vertebrate body. W.B. Saunders Co, Philadelphia, London
Rudwick MJS (1997) Georges Cuvier, fossil bones and geological catastrophes. Univ. Chicago Press, Chicago/London
Ruigrok TJH (1984) Organization and morphology of motoneurons and primary afferents in the lumbar spinal cord of the turtle Pseudemys scripta elegans. Thesis University Utrecht, The Netherlands
Sandfort A, Burton AM (2014) Tolerance for distorted faces: Challenges to a configural processing account of familiar face recognition. Cognition 132:262–268
Schumacher GH, Dokladal M (1968) Ueber unterschiedliche Sekundärveränderungen am Schädel als Folge von Kaumuskelresektionen. Acta Anat 69:378–392
Scott JH (1955) Craniofacial regions. Dent Pract 5:206–214

Shahbazi MN, Jedrusik A, Vuoristo S, Recher G, Hupalowska A, Bolton V Fogarty NME, Campbell A, Devito LG, Ilic D, Khalaf Y, Niakan KK, Fishel S Zernicka-Goetz M (2016) Self-organization of the human embryo in the absence of maternal tissues. Nature Cell Biol 18:700–708

Singer C (1931) A short history of biology. Oxford at the Clarendon Press

Smits-van Prooije AE, Poelmann RE, Gesink AF, Chr Vermeij-Keers (1985a) The cell surface coat during closure of the neural tube as revealed by concavalin A. Verh Anat Ges 79:591–592

Smits-van Prooije AE, Vermeij-Keers C, Poelmann RE, Mentink MM, Dubbeldam JA (1985b) The neural crest in presomite to 40-somite murine embryos. Acta Morphol Neerl Scand. 23:99–114

Smits-van Prooije AE, Chr Vermeij-Keers, Dubbeldam Mentink MMT, Poelmann RE (1987) The formation of mesoderm and mesectoderm in prosomite rat embryos cultured in vitro, using WGA-Au as a marker. Anat Embryol 176:71–77

Smits-van Prooije AE, Chr Vermeij-Keers, Poelmann RE, Mentink MMT, Dubbeldam JA (1988) The formation of mesoderm and mesectoderm in 5- to 41-somite rat embryos cultured in vitro, using WGA-Au as a marker. Anat Embryol 177:245–256

Stern CD (2001) Initial pattering of the central nervous system: how many organizers. Nature Reviews Neurosci 2:92–98

Streit A (2004) Early development of the cranial sensory nervous system: from a common field to individual placodes. Dev Biol 276:1–15

Stoessel A, Gunz P, David R, Spoor F (2016) Comparative anatomy of the middle ear ossicles of extant hominids: Introducing a geometric morphometric protocol. J Human Evol 91:1–25

Teillet M, Kalcheim C, Le Douarin NM (1987) Formation of the dorsal root ganglia in the avian embryo: segmental origin and migratory behavior of neural crest progenitor cell. Dev Biol 120:329–347

Tosney K (1978) The early migration of neural crest cells in the trunk region of the avian embryo: An electron microscopic study. Dev Biol 62:317–333

Tosney K (1988) Proximal tissues and patterned neurite outgrowth at the lumbosacral level of chick embryo: partial and complete deletion of the somite. Dev Biol 127:266–286

Verbout AJ (1985) The development of the vertebral column Adv Anat Embryol Cell Biol 90 Springer, Berlin

Vermeij-Keers C (1990) Craniofacial embryology and morphogenesis: Normal and abnormal. In: Stricker M, van der Meulen JC, Raphael B, Mazzola R, Tolhurst DE (eds) Craniofacial malformations. Churchill Livingstone, Edinburgh

Vermey-Keers Chr, Poelmann RE (1980) The neural crest: a study on cell degeneration and the improbability of cell migration in mouse embryos. Neth J Zool 30:74–81

Versluys J (1912) Das Streptostylie-Problem und die Bewegungen im Schädel bei Sauropsiden. Zool Jahrb Anat Suppl 152:545–716

Weston JA (1963) An autoradiographic analysis of the migration and localization of trunk neural crest cells in the chick. Dev Biol 6:279–310

Young JZ (1974) An introduction to the study of man. Oxford University Press

Young JZ (1975) The life of mammals. Their anatomy and physiology. Clarendon Press, Oxford

Zang J, Merialdi M, Platt LD, Kramer MS (2010) Defining normal and abnormal fetal growth: promises and challenges. AJOG 202:522–528. https://doi.org/10.1016/j.ajog.2009.10.889

Chapter 5
Gustation, The Act or Sensation of Tasting

Taste seems to have two chief uses:
1. It invites us by pleasure to repair the losses which result from the use of life.
2. It assists us to select from among the substances offered by nature, those which are alimentary.
In this choice taste is powerfully aided by the sense of smell…; as a general principle, it may be laid down that nutritious substances are repulsive neither to the taste nor to the smell. It is difficult to say in exactly what the faculty of taste consists. It is more complicated than it appears. [Brillat-Savarin (1848), The physiology of taste: Meditation II, Taste, Definition of taste]

This chapter on taste starts with a trivial pursuit to approach the difficulty of the definition of taste with the consequence that the four tastes that are normally discerned should rather be considered a continuum of tastes. Attention is paid to the transmission of the taste signal, both its physiological and neuroanatomical aspects. Taste changes during pregnancy, diseases are treated, and clove, oregano and chilli are discussed. Taste is related to tongue reflexes and mechanics. Taste receptors are even found in the pharynx and start of the larynx. The chapter ends with the taste manipulations.

5.1 Introduction

Most people learned that humans discern four different tastes: sweet, salt, sour and bitter. Each of them is distinctly distributed over the tongue and located in taste buds (also called papillae, Fig. 5.1). Each taste bud contains around 50–150 taste receptor cells, and each of the receptor cells is connected by a cranial nerve branch of VII (facial nerve, its intermediate part), or IX (glossopharyngeal nerve), or X (vagus nerve) depending on its location on the tongue, palatum or pharynx and start of the larynx.

E. Marani and C. Heida, *Head and Neck*,
https://doi.org/10.1007/978-3-319-92105-1_5

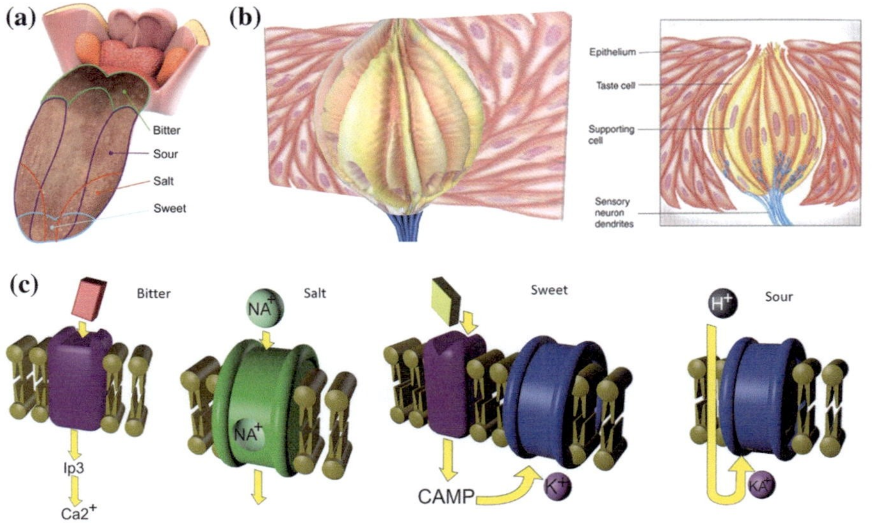

Fig. 5.1 **a** Distribution of bitter, sour, salt and sweet over the human tongue and **b** taste bud existing of taste receptor cells (containing type II and III) and supporting cells (type I and basal or progenitor cells). Note that the lifetime of taste bud cells is a few to 10 days. Nerve endings enter the inner side of the taste bud. **c** Binding of sweet, bitter, salt or sour to cell membrane receptors produces different cascades, inducing different neuronal signs to the brainstem and brain (figure is 3D adapted from R. Rhoades and R. Pflanzer's Human Physiology)

Under the heading "***Is taste alchemy or science?***" the start of this chapter is with a taste's trivial pursuit: questions to which you give the answers (see also Korsmeijer 1999):

(1) Lemonade

Let us take lemonade, containing only two tastes: sourness and sweetness. I put together the pure acids present in lemonade and its sugars in the same molar concentrations in water.

If I taste this solution, do I taste lemonade?

(2) Any taste can be analysed in the combination of the four categories: sweet, salt, sour and bitterness according to Hans Henning (1885–1946), who made a taste tetrahedron (1916). Lemonade sits about point a, Romaine lettuce at b and cabbage soup at c: to give you an idea of its application. Where should you place meat? (Fig. 5.2)
(3) Catching a cold

You caught a cold, and your nose is constantly sniffing. Do you taste lemon lemonade in the same way as you had not caught that cold?

Does the taste of lemon lemonade shift in Henning's taste diagram, having this cold?

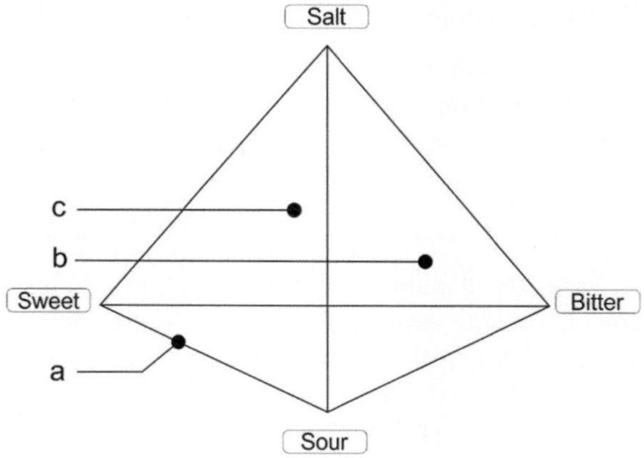

Fig. 5.2 Henning's taste tetrahedron (adapted from Korsmeijer 1999)

(4) Is your taste the same as mine?

The number of taste buds on the tongue and the number of taste receptors among the taste buds (Fig. 5.1a, b) vary tremendously from person to person.

Genetics determine whether you can taste certain substances: e.g. phenylthio-carbamide (PTC) is hardly noticed by some persons, others detect it, and some even dislike it. Substances like PTC are in broccoli: some like broccoli, and others hate it. You already noticed it by the abhorrence of your children.

Some cannot digest cow's milk, giving a different consumption pattern and thus different tastes. So is your taste the same as mine?

(5) The mix. Flavours mix such that taste sensations of the components are enhanced or reduced:

- Salt enhances sweetness (salt over strawberries).
- Sugar reduces sourness of acids (sugar in the acids of pickles).
- Caffeine, rather than blinding its bitterness with sour substances, increases their sourness.

"The alchemy in the mouth seems unpredictable".

(6) I look around me, sniff, taste and enjoy the restaurant.

At the first restaurant meeting, you as a male will not notice it, but at the second or third time you will for sure detect it: women look longer at the menu. Why?

- Do women count the calories?
- "If I eat that will I smell out of my mouth the whole evening?"
- "Can I eat this without dripping on my white shirt?"
- "There is nothing that stimulates my taste buds on the whole menu".

(Borrowed from the Dutch paper: de Volkskrant, Weekbijlage, 8 maart 2003)

(7) The phenomenology of taste, part I: What determines the taste perception?

Causal conditions of the body (differences caused by genetics).
The condition a person is in (hunger is the best sauce, thirst).
Cultural factors (a vegetarian can dislike meat, because an animal had to die for it).

(8) The phenomenology of taste, part II: What determines the taste perception?

Taste has attention outward to the object: you really dislike squid. In a restaurant you enjoyed the meal, until the cook tells you that it contained squid. You spit it out and expel the meal from your body.
The state one's body is in (e.g. diseases).
The pleasure or displeasure that is present in tasting (colours of cocktails signal to you that it is a pleasure to taste them).
So, is taste the detection of sweet, salt, sour and bitter?

5.2 Taste and Its Definition

Taste has been described for over two thousand years: whether it was based on the Yin–Yang dichotomy, or the five ways of Wu Xing with five tastes discerned, or Aristotle's description of seven tastes. The four basic gustatory tastes as described above: sweet, salt, sour and bitter are discussed. What is more problematic, a clear definition of each of these basic gustatory tastes is missing (Erickson 2008). To understand the problem, a small excursion into neurophysiology is necessary.

Different types of sensory receptors detect different types of sensory stimuli by differential sensitivities. Normally, each type of receptor is highly sensitive to one type of stimulus and receptors are non-responsive to normal intensities of the other types of sensory stimuli (Fig. 5.1c). To give an example, rods and cones of your eyes are sensitive to light but not to heat, cold, pressure on the eyeballs, or chemical changes in the blood.

We humans discern different types of sensations in our brains. These are called "*modalities*". However, although we experience different modalities, the nerve connection transmits only electrical signs (action potentials). Certain brain areas, to which nerves are connected, are responsible for the experience of that particular modality. Thus, if I could connect the acoustic nerve to the visual cortex, I could see what I heard!

Applied to the gustatory system this gives: the connection of a specific taste receptor to the brain, sending information towards a certain specific part of the brainstem and then relayed into specific brain parts, gives a gustatory sensation, gustatory being the modality. The *connection* of the sweet taste receptor cell with that part of the brain that gives us the sensation of sweetness is called the "*labelled line*" for sweetness. Thus, nothing else but sweetness information should go over this labelled line. Moreover, the train of action potentials over the "labelled line sweetness", of course, carries more information, e.g. how strong the sweetness of my ice cream is, mainly by the frequency of the action potentials.

You remember from the trivial pursuit questions that lemonade contains two basic tastes: sourness and sweetness. Each must have its own labelled line,each should be indifferent for the other basic taste, and *that is not supported by several studies* (e.g. for bitter, see Gilbertson et al. 2001; we come back to it in Sect. 5.10: reflexes, taste and tongue).

This was already demonstrated by Hanig in 1901. We owe to him the distribution of the four basic tastes over the tongue (Fig. 5.1a): "He demonstrated that the areas of sensitivity of the tongue to the basic tastes differ, with sweet largely in the front, bitter in the back, and salty and sour on the sides. Although their points of maximum sensitivity differ, their total distributions overlap greatly" (Erickson 2008). So, the discoverer of the tongue localization of the four basic tastes only indicated the maximum sensitivity and noticed serious overlap: no labelled line and truly receptor cells sensitive to more than one basic taste. Even worse, "another surprise is that taste receptors are not restricted to the gustatory organs but are expressed in many cell types and in some cases contribute to the ability to sense changes in temperature, light and olfactory cues" (Liman et al. 2014). It cannot come as a surprise that disappointment is present in the literature: "the idea of basic tastes has directed our research for over a century, we do not know what basic tastes are; we do not know the origins, rationale or definition of this idea; thus, we do not explicitly know how to test it" (Erickson 2008 and for more scientifically based arguments and articles).

So, here we are: Scientifically we don't know what taste is, but everybody tastes, presumably differently, and the number of cookbooks is enormous, all to enhance taste!

5.3 Henning Again

"Many scientists have been critical of the influence of our words on our understanding". Speaking about geological time periods, S. J. Gould (1994) discussed words used to invent categories.

"The human mind seems to work as a categorizing device. This deeply (perhaps innately) ingrained habit of thought causes us particular trouble when we need to analyse the many continua that form so conspicuous a part of our surrounding world. Continua are rarely so smooth and gradual in their flux that we cannot specify certain points or episodes as decidedly more interesting or more tumultuous in their rates of change, than the vast majority of moments along the sequence. We therefore falsely choose these crucial episodes as boundaries for fixed categories, and we veil nature's continuity in the wrappings of our mental habits. We must also remember another insidious aspect of our tendency to divide continua into fixed categories. These divisions are not neutral; they are established for definite purposes by partisans of particular viewpoints".

The close correspondence between Gould's comments and the probable continua in taste, including inflection points as the corners of Henning's tetrahedron (1916), is of note. Henning described the world of gustation as a continuum of tastes in three dimensions—"a tetrahedron, defined by "sweet, sour, salty and bitter" at its four corners" (Erickson 2008).

And somewhat further on, "beyond these, it might be useful to consider that even our most basic and secure word concepts, mass, time, distance and energy, are based on man's conception of the nature of things. They seem too good to be less than totally true in their general or particular usages, and quite real independent of naming them. But they are only our words, rubrics facilitating our understanding, and our communications (or miscommunications) with each other. Our understanding of even these "basic" physical concepts is changing; they are not as simple and real as these distinct words make them seem. In summary, it might be helpful to worry about allowing common usage words such as "sweet" and "sour" guide our research and understanding. These words could provide a problem rather than the solution" (Erickson 2008).

Thus, two main things come out of these reported texts:

(1) Taste is not a mixture of *four basic* tastes, but a *continuum* of tastes. There exist an infinite number of tastes or "many singular tastes" as Erickson (2008) wrote. At least we saved the cookbooks!
(2) Basic taste words do force us of thinking in the extremes of the tetrahedron: "sweet, sour, salty and bitter". There are no words for the many tastes. However, look into the descriptions of wines: fruity, with herb savour or "ce vin primeur aux parfums intenses de fruits rouge (framboise, fraise)" [this wine of early fermentation with intense perfumes of red fruit (raspberry and strawberry)], found on the backside of a Claret wine from Château Thieuley of the Bordeaux region, all indicate the continuum and not a mixture of the four basic tastes (see also L'énoncé de gout [phrasing taste] in Teil 2001).

Or read the description of the wine expert who with eight others tasted six Chardonnay wines in succesion that had been aged in oak barrels from different forests and/or species:

Darney (the tested wine, remark ours),

Robe: shiny greenish gold, subtle nose; the Chardonnay is easily identifiable; the wood is harmonious. Notes of smoked wood, vanilla, coconut, spices, honey, almonds. Distinct notes of toasted bread, butter and wood. Palate: the wood blends in remarkably well; excellent mouth feel; and no stiffening of mucous membranes. Leaves a clean taste of wine (from Sauvageot et al. 2006).

Imagine that a bottle label had indicated: 30% tannin bitterness, 20% fructose sweetness, 10% saltness and 40% acidic sourness! Thus in cooking and wines, humans do express the taste continuum in words, which also means that indeed we are aware of the continuum of tastes.

5.4 Changing Your Taste: A Puzzling Mystery

Your receptors can adapt to your previous taste experience: if you put your tongue for some time (15–30 s) in a sour solution, say lemon juice, you will notice that any other sour solution, for example vinegar, will taste less sour. Tasting, directly after the lemon juice, something salty, then you will notice that it tastes as usual: there is no cross-adaptation. The consequence of these sentences is also that you have a gustatory memory.

Glutamate presumably represents a separate taste, indicated with the Japanese word *umami* (also called savoury). The arcuate nucleus is a special hypothalamic nucleus since it misses the blood–brain barrier that is present in most other parts of the brain. The addition of the small molecule monosodium glutamate (also known as Vi-Chin as an additive in Chinese food) produces obese rats. Monosodium glutamate passes through the rat blood–brain barrier and destroys parts of the arcuate nucleus with taste neurons for glucose and fatty acids (Bunyan et al. 1976; Holzwarth-McBride et al. 1976; Marani and Rietveld 1982). This destruction occurs in rodents not in humans (high concentrations do produce headache and vomiting in humans). Chemicals that mix with salty tastes do not interfere with umami. Glutamate responses of taste neurons are different from the responses of other tastes, e.g. glucose and fatty acids neurons in the arcuate nucleus. Umami is declared (by some scientists) to be the fifth basic taste and is present in meat, (remember Sect. 5.1(2)), seaweed, mushrooms, peas and tomatoes.

The miracle berry of West Africa (Fig. 5.3) contains a protein, called miraculin that changes sweet receptors in such a way that they can be stimulated by acids. So, acids taste sweet (Barthosuk et al. 1974).

Fig. 5.3 Miracle berry (photograph with permission from H. Lyman, and is free in the public domain)

The other way around is also possible: extracts of Gymnema sylvestre make people insensitive to sweet tastes (Frank et al. 1992). Moreover, the substance theophylline reduces bitterness (Kodama et al. 1978) and people can report a sweet taste from *water* after they have eaten artichokes (DeSimone et al. 1980).

So, your modalities can be changed by substances that influence your taste receptors, which is different from enhancement or reduction of taste [see Sect. 5.1 (part 5)].

The last of these puzzling situations is a "taste illusion". (Of course, there is more; one can fill a book with these taste puzzles.) The tip of your tongue contains more sweet taste buds of a certain type (so-called fungiform ones) than both its adjacent sides for sweetness. Now, if you sweep with a cotton stick saturated with a sweet substance and you start, say at the left side of your tongue, your taste sensation is low but increases towards your tongue tip. One should expect that passing the tip and reaching the right side, containing fewer taste buds, a reduction of your taste for that substance should be noticed. However, you will have the same taste perception as on the tip of your tongue. Now, start at your right side of the tip of your tongue: indeed a low perception, going over the tip a high one and it stays after you passed the tip. This is called a taste illusion by touch: "taste sensations are not limited to the locations of taste buds, but rather to areas touched in the mouth. Thus during eating, taste sensations seem to originate from the entire inner surface of the mouth even though the taste buds are found only on certain loci" (Todrank and Bartoshuk 1991).

This part finishes with asking attention for the other function of the tongue: "besides taste, the surface of the tongue is also responsive to food texture, to heat and to pain. During mastication, it rolls the food round and round in the mouth, testing for lumps. When it judges that all sharp pieces have been crushed or rejected, it participates in the crucial act of swallowing" (see Sect. 6.3, Morris 2004).

5.5 Transmitting the Taste Signal

Taste receptor cells act differently if their membrane that can detect substances is activated by the basic taste substances (Fig. 5.1c). Electron microscopy showed that the taste bud contains three types of cells, of which the type II and type III cells are the taste receptor cells and the basal cells are progenitor cells. Type I is a supporting cell. Type II cells recognize bitter, sweet and umami stimuli. Type III cells are sour-sensing cells.

It is now generally accepted that one and the same *cellular* receptor (=a taste bud cell of type II or III) can detect several tastes (see also Sect. 5.10). For clarity reasons, we split its actions up into the four *membrane* actions present in each taste bud cell (Fig. 5.1c). Sweet substances have their effect via a receptor in the membrane of the taste cell, and the membrane receptor produces cyclic adenosine monophosphate (cAMP). cAMP closes the channels by which potassium ions can pass. The ion concentration on both sides of the membrane changes, which gives a change in the electrical potential over the membrane (here, it is a depolarization), and this change in membrane potential produces neurotransmitter release that is noticed by the nerve fibre that reacts with action potentials. Bitter uses a calcium ion cascade that finally releases neurotransmitters from the taste bud cell that activates the nerve fibre. Salt brings in sodium ions that change the membrane potential (depolarization), and sour substances also block the potassium channels (Fig. 5.1c). All these electrical membrane changes, causing transmitter releases, are noticed by the nerve fibre that reacts, in its turn, with production of action potentials: electrical signals that are transported over the nerve fibre towards the central nervous system. Sweet and bitter taste buds also can use adenosine triphosphate (ATP) as transmitter substance, while serotonin, glutamate, acetylcholine, norepinephrine and several neuropeptides are known as other taste bud neurotransmitters.

The genes for several membrane receptors are known and cloned. From G protein-coupled taste receptors (TRs), the so-called T2Rs, it is known that they have several genes in humans and that they detect various types of bitterness (Chandrashekar et al. 2000). "Importantly, each bitter-sensing cell co-expresses multiple T2R genes, suggesting that this cell type recognizes a diverse array of structurally distinct bitter compounds" (Sugita et al. 2013). The T1R receptors are responsible for umami and sweetness detection: "the single T1R2/T1R3 receptor can recognize diverse chemical structures of sweeteners such as simple sugars, a wide range of synthetic sweeteners, D-amino acids and even sweet proteins. Recent reports demonstrate the presence of multiple binding sites on the sweet taste receptor" (Sugita et al. 2013). Each of these receptors has their own cascade to produce release of neurotransmitters. In other words, Fig. 5.1c is too simple for modern research results, but still represents the receptor idea well. Therefore, the sour-detecting taste bud cell is extended using the overview of Liman et al. (2014). A proton channel is included at the top of the taste bud cell (Fig. 5.4).

"Entry of protons into sour cells produces cellular acidification, which may affect cell signalling. Notably, taste cells express several resting two-pore K+ channels, which may be blocked by intracellular acidification to produce further depolarization of the cell (Fig. 5.4). The idea that intracellular acidification activates sour cells is attractive, as it could explain why, at the same pH, acetic acid and other weak acids that penetrate the cell membrane taste more sour than strong acids, such as HCl, that do not penetrate the cell membrane" (Liman et al. 2014).

The nerve fibre part that connects to the taste bud cell is a branch. Thus, several branches at the end of the nerve fibre belong together, which means that over that single nerve fibre several action potentials induced by different taste bud cells are transported. For simplicity, the number of action potentials is cumulative over the nerve fibre. Thousands of fibres make the gustatory part of one cranial nerve like the VII (facial nerve), or IX (glossopharyngeal nerve) or X (vagus nerve).

The facial nerve (VII) brings the information of the frontal part of the tongue towards the brainstem. In general, sweet, salt and sour should pass over this cranial nerve. The back of the tongue is innervated by the glossopharyngeal nerve (IX), and thus, bitter should go by this nerve. The vagus nerve (X) innervates the palatum

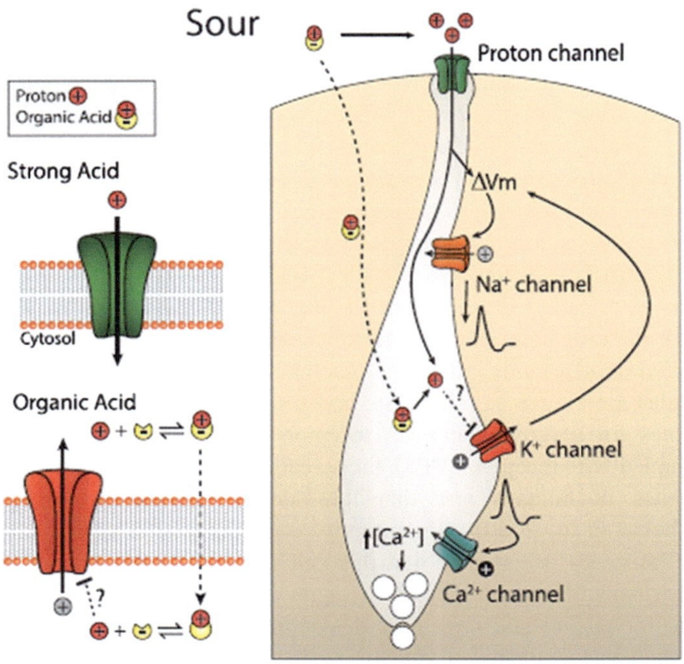

Fig. 5.4 "Sour taste in vertebrates is initiated when protons enter through an apically located proton-selective ion channel. Weak acids may also activate sour cells by penetrating the cell membrane and acidifying the cytosol, leading to closure of resting K+ channels and membrane depolarization", white balls are neurotransmitter vesicles, and release of the neurotransmitters is Ca^{2+}-dependent (figure is from Liman et al. 2014 with permission)

molle and pharynx (see Figs. 5.5, 5.6 and 5.9). The conclusion must be that if the tenth cranial nerve brings gustatory information to the brainstem that taste bud-(like) structures are present outside the tongue. In fact, they are found at the transition from palatum molle to durum and they are also present in the pharynx walls. One should note: "due to the lack of either human clinical or monkey experimental evidence, the innervation of the taste buds on the palate must be generalized from other species and cadaveric material" (Pritchard 2012).

Information of the taste buds travels over the nerve fibres of three cranial nerves, of which the cell bodies are grouped into ganglions (Figs. 5.6 and 5.9). They are called: nodose ganglion for vagus nerve, petrosal ganglion for the glossopharyngeal nerve and geniculate ganglion for the facial cranial nerve. The nerve fibres of these neurons pass the cell body and enter the brainstem into the gustatory part of the nucleus of the tractus solitarius. This happens in both the left and right part of the brainstem, and both nuclei of the tractus solitaries are connected at their caudal end (Figs. 5.7 and 5.9).

Comparative anatomy revealed that in primates, including man, this solitary nucleus extends much further than in rodents. This rostral extension projects

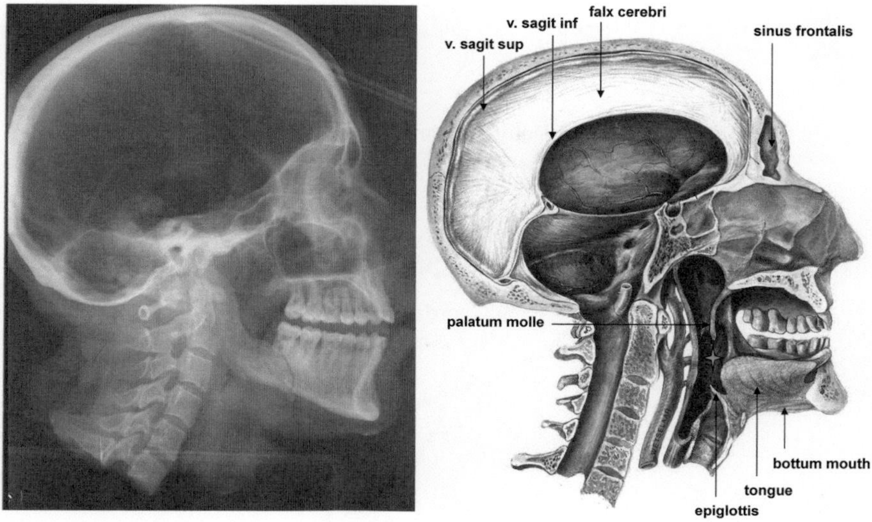

Fig. 5.5 Midsagittal section of the skull in an X-ray and a drawing. Note that the bottom of the mouth is made of tongue and mouth bottom muscles. There is a space between epiglottis and backside of the tongue. It belongs to the pharynx (space between uvula and epiglottis, blue star). The larynx is the space below the epiglottis to the entrance of the trachea. The border between nose cavity and mouth cavity is made by a bony structure (palatum durum) that passes into a non-bony structure (palatum molle with at its end the uvula). The trachea is in front of the oesophagus; during swallowing, food has to pass over the entrance of the airway (© E. Marani)

selectively to the thalamus and from there to the cortex of the insula, while in rodents another cortical (via the parabrachial nuclei) and other thalamic projection (also via the parabrachial nuclei) are present (see Voogd et al. 1998, Pritchard 2012; see also Fig. 5.7). Thus are primates more conscious of taste than rodents? At least their information stream is handled differently, and therefore, the results will be different. The consequences for this chapter are that we will not use most of the rodent results, since they presumably are too divergent from human results, of which we have, of course, less. Open questions remain.

Finally, the gustatory information will reach the cortex of the insula (the part of the lateral cortex overgrown by other opercular cortex; see Fig. 5.7). It is now well established that a topographic distribution is present for the basic different tastes within this gustatory cortex in rats with partial overlap of these cortical areas (Yamamoto et al. 1989; Chen et al. 2011).

Several clinical studies confirmed that in humans the anterior part of the insula is involved in gustatory awakening (see Fig. 5.7). Electrophysiological primate studies restricted the area: "only the granular insula located at the insular/opercular junction of the frontal lobe should be considered primary taste cortex" (Pritchard 2012). However, ".... There are a number of other reasons why it has been difficult to use the clinical literature to make a clear and compelling case for a question as fundamental as the location of gustatory cortex. First, patients with cortical injury

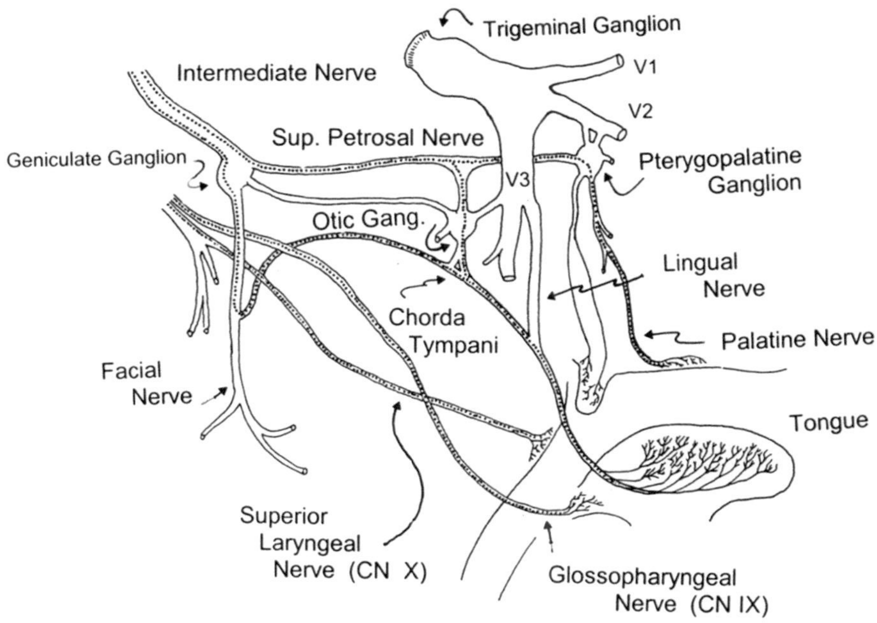

Fig. 5.6 Peripheral gustatory pathways: dotted lines are the taste connections towards the brainstem. Not only the chorda tympani of the nervus intermedius and the glossopharyngeal nerve, but also the connection to the palatum is indicated in this figure (from Pritchard 2012 with permission)

often do not notice and thus do not report gustatory impairment. Second, confusion between taste, smell, and even tactile sensations by most patients (and some physicians) has interfered with the proper diagnosis of taste deficits. Third, when gustatory deficits are reported, they may not be observed in the clinic due to inadequate or superficial testing. Fourth, when taste deficits are reported and detected in the clinic, they typically are given less attention than the motor and language deficits caused by coincident damage to the peri-Rolandic area. Fifth, coincident damage makes identification of the core gustatory region impossible unless data are available from several similar cases. Sixth, pure insular strokes are rare and, thus, seldom included in published studies. The clinical literature has provided important insights into the cortical organization of taste, but like all approaches, it has limitations" (Pritchard 2012).

There is one result that must be mentioned: Castriota-Scanderbeg et al. (2005) found that wine unilaterally activates the anterior insula and the orbitofrontal cortex in naïve subjects, but bilaterally in sommeliers.

Everyone knows the old batteries with one long and one short lip. To test its charge, as a child, you connected both lips with your tongue. The weak electric discharge still had a taste: "What electric stimulation of the tongue tastes like depends on various stimulus parameters. Anodal stimulation is usually

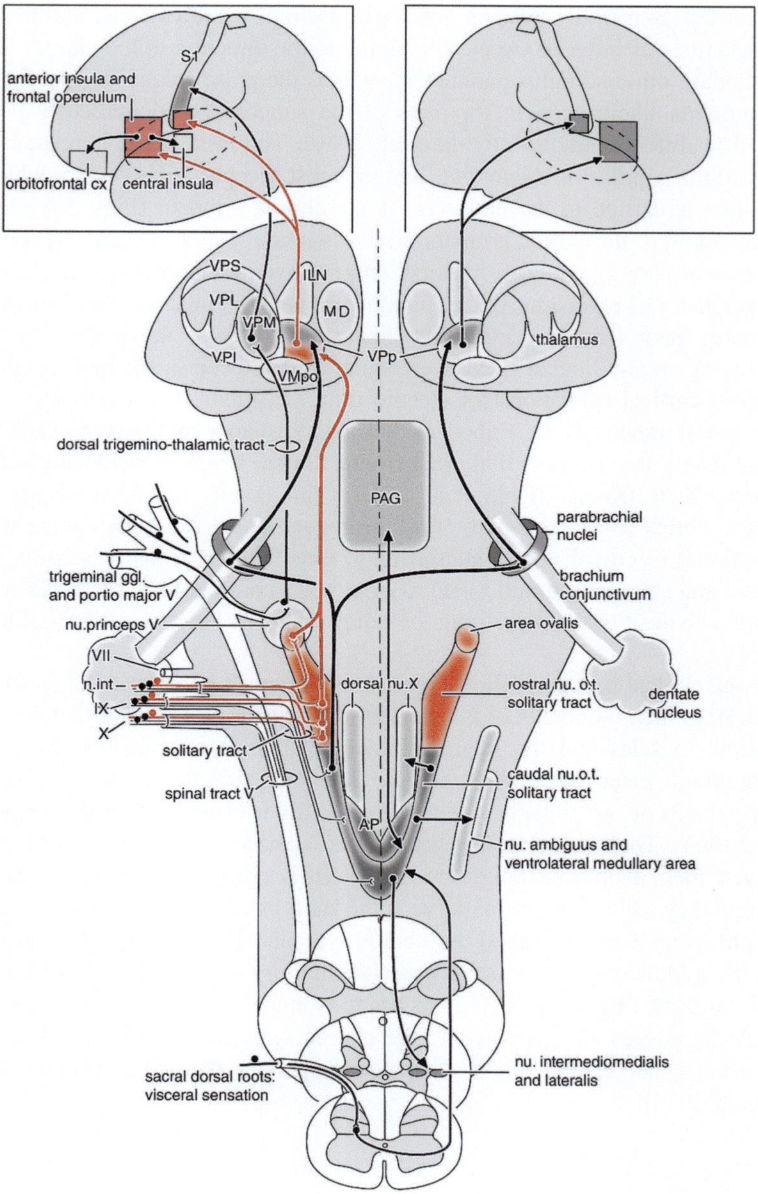

Fig. 5.7 Ascending central connections of the gustatory pathways. Note that inhibitory cortico-nucleus tractus solitaries connections are present, but not indicated in this figure (see Smith and Li 2000): VP, ventral posterior thalamic nucleus; PAG, periaqueductal grey; S1, gyrus centralis posterior; AP, area postrema (figure with permission and courtesy of J. Voogd)

characterized as metallic, sour or sour–salty tasting, while cathodal stimulation is often described as bitter or sweet. In addition to the direction of the current (anodal vs. cathodal), other stimulus parameters such as the pulse frequency and the locus of stimulation likely account for parts of the supposedly contradictory percepts reported in different studies" (Ohla et al. 2010). The taste noted is related to the density of the papillae on the tongue, and the most acceptable nerve over which this stimulation is carried to the brainstem is the chorda tympani (Figs. 5.6 and 5.9), since resection of the chorda tympani produces loss of electrical taste. Nevertheless in the case of very high pulses, information goes over the lingual nerve, a branch of the trigeminal (V) cranial nerve, presumably related to somatosensory information, presumably pain (see Fig. 5.6). Electrical stimulation of the tongue by which chorda tympani and lingual nerve were stimulated showed in the EEG results: "an overlap of cortical activations for electric taste and lingual somatosensory stimuli within areas previously described as key for gustatory processing, namely the bilateral insula and ventromedial orbitofrontal cortex, which were already activated within the first 100 ms" (Ohla et al. 2010). This connection from tongue to the gustatory cortex is unexpectedly fast. Moreover as to the differences between chemical and electrical taste, "interestingly, direct comparisons of sensitivities to electrical and chemical stimuli failed to evidence a correlation between electric and common chemical tastes, signifying the unique nature of electric taste" (Ohla et al. 2010).

For the clinical part that follows more attention must be paid to the facial nerve (Fig. 5.6). In fact, it consists of two nerves: the motoric facial nerve that innervates the mimic muscles and the intermedial nerve (nervus intermedius) over which (preganglionic parasympathetic) motoric innervation of the lacrimal glands and salivary glands occurs, but also the taste sensibility is brought from the tongue into the brainstem. Their neuronal cell bodies are localized in the geniculate ganglion. However, their path from tongue to brainstem is complicated. Viewed from the brainstem they follow, the facial nerve at the start, but after passage of the ganglion they split of as a branch called the chorda tympani. By its own bony channel, it enters the middle ear, passes in between the auditory ossicles, leaves it by its own fissure, finds the lingual nerve and reaches the tongue. *Thus, facial taste fibres pass through the middle ear.* Moreover, "the chorda tympani nerve is the closest to a pure gustatory nerve, but even its taste fibres respond to thermal and tactile stimuli" (Pritchard 2012).

5.6 Taste and Pregnancy

In fact, we come back to the various types of bitterness that can be registered with the taste bud receptors (Gilbertson et al. 2001) with no labelled line performance (see Sect. 5.2) of the bitter receptors. This plays an important role in pregnancy.

"Pregnancy alters taste responses and feeding patterns in women. Chief among these changes is increased sensitivity to bitter stimuli and feelings of nausea in response to bitter foods" (Breslin 2013). This field is subject to considerable speculation: thus, nausea and vomiting early in pregnancy are discussed in the light of the embryological production of the organs. This early developmental process is already influenced by low concentrations of toxins, of which most do undeniably contain a bitter taste. Several types of food do include toxins, and this taste-increase for toxins by the sensitivity increase for bitterness protects the embryo and early foetus. The negative relation between woman with nausea and miscarriages should support the hypothesis, and these women should have fitter and larger babies. "Pregnant women generally find a greater variety of flavours more aversive than do non-pregnant women" (Breslin 2013). But they have the same receptors for bitterness before and after pregnancy. The idea is that the hormonal change during pregnancy reaching the brain gustatory areas could be held responsible for the changed notion of bitterness in pregnant women.

5.7 Diseases and Taste

Inflammation of the middle ear cavity is called otitis media. Otitis media is often found in children. Parents should realize that children with severe otitis media undergo a change of taste or an increase in pain in the mouth, think of spicy food. "Damage to nerves is expected to produce loss of function; however, damage to taste nerves can produce unexpected oral sensory intensifications as well as phantoms (sensations in the absence of stimulation). Clinical as well as laboratory observations suggest that chorda tympani nerve (CN VII, Fig. 5.6) input can inhibit central areas receiving input from other nerves innervating the oral/nasal cavity; damage to the chorda tympani releases this inhibition leading to intensification of sensations from the previously inhibited areas as well as oral phantoms" (Bartoshuk et al. 2012), according to the introduction of the article on: "taste damage (otitis media, tonsillectomy and head and neck cancer), oral sensations and BMI" (body mass index). The other nerve of which less is known after destruction (that sometimes occurs after tonsillectomy) is the glossopharyngeal cranial nerve.

Anaesthesia applied to the middle ear, or partial, or total damage of the chorda tympani (thus facial innervation of the tongue: sweet, salt and less sour) both show that taste stays present but the part that normally is innervated by the glossopharyngeal nerve is intensified: in the old terminology bitter/partially sour is enhanced and, remarkably, at the contralateral side of the anaesthesia or damage. This phenomenon occurs mainly in supertasters: a supertaster is someone whose sense of taste awareness is far greater in intensity than the average, mainly because of a larger density of taste papillae. Some 35% of women and 15% of men are supertasters (the most sensitives are the tasters in coffee and tea industries; see Bartoshuk 1991). Thus, the chorda tympani, by its activity, inhibits or reduces the effect of the glossopharyngeal nerve. Seemingly, there is a balance between facial and glossopharyngeal activity and this is regulated in the central nervous system. To the best of my knowledge, it is not known whether this balance is in the brainstem nucleus or in the cortex (a good bet is both).

Although there are several notions of taste and diseases, here we report a few that are randomly chosen.

Phenylthiocarbamide, which we saw earlier in relation to broccoli in the trivial pursuit questions, has been suspected of being a trait marker of schizophrenia. Research by the Emory and the George Washington universities did not verify this relationship (Compton et al. 2013), but it remains persistent in discussions and the literature. Phenylthiocarbamide's evolutionary trend in non-tasters is rather interesting: "some humans and some chimpanzees cannot taste phenylthiocarbamide. Although these polymorphisms were thought to be ancient, predating the evolutionary split of these two species, Wooding et al. (2006) showed that the molecular basis for non-taster PTC alleles differs in the two species, and therefore the variants evolved independently" (Lieberman 2011).

Guillain–Barré syndrome is related to taste deficiency due to the affected myelin sheaths of the peripheral nerves, including the cranial VII, IX and X. However due to the clinical neglect of taste disturbances (see above Pritchard's citation), a restricted number of cases are reported in the literature (Soria et al. 1990).

Taste changes during ageing: "middle-aged Americans have higher obesity rates than any other age group, yet little is known about age-related changes in central taste function during this critical time" (Green et al. 2013). The "age by taste interaction" has been confirmed, and the insula and strange enough the basal ganglia are mainly involved in this change. Youngsters compared to middle-aged people show greater responses to hedonic results (sweet versus bitter) in insula and in parts of the basal ganglia. The last one is related to reward regions. The authors "speculate that these results might reflect early age-related differences in central taste processing that occur prior to deficits in gustatory function observed in old age, and this might have important implications for weight changes that occur during Middle Age" (Green et al. 2013).

5.8 Understanding Taste Information

One taste information system has already been described above: *the labelled line* (LL) model that originated from general receptor studies. The second approach is called *across fibre patterns* (AFPs), in which all information for a taste is registered and the overall coding of all gustatory cells activated determines the taste (Erickson 2008, his references herein). The relation between these two models has been described as: "the simplest version of LL theory is a special case of AFP theory, in which an arbitrary threshold causes one set of fibres to be described as 'on' and the others to be described as 'off'" (Katz et al. 2002). A third approach is called the dynamic and distributed taste model which is incorporated: adaptation over time periods, called time-varying gustatory responses, neuronal influences within a gustatory area and between gustatory areas, and mutual influence by other systems.

The conclusion for the dynamic and distributed model is: "we have attempted to justify the reasons for moving beyond static and unimodal models (LL and AFP are meant, note ours) of gustatory coding towards models in which processing occurs in time, is multimodal and involves interactions between neurons in the same and in spatially separate gustatory regions. We propose that the specifics of gustatory responses grow not only out of information ascending from taste receptor cells but rather from the cycling of information around a massively interconnected system" (Katz et al. 2002). Thus, one should study the interconnected systems.

Taste information (see Fig. 5.7) is relayed into the insular cortex and orbitofrontal cortex. The insular cortex considers the internal body conditions. It not only concerns taste. Heart rate, hunger and pain are also registered. The insular cortex and the orbitofrontal cortex are mutually connected and are crucial for reward processing and its evaluation. Orbitofrontal cortex and lateral prefrontal cortex are important for value-based decision-making and learning, especially in tasting, smelling and in food-related evaluations. "Ventral striatum, including the nucleus accumbens, has been shown to be critical for pleasure and reward, and a growing body of the literature demonstrates how important ventral striatum is in stress and food cravings in youth. The interrelations between insular interoceptive awareness, orbitofrontal cortex and ventral striatal pleasure systems are particularly important for better understanding brain mechanisms that contribute to obesity".

"Pleasant, delicious foods are sometimes referred to as "comfort foods" because of the emotionally appealing aspects of consumption. A growing body of the literature, predominantly in adult samples, has documented the role of pleasure centres in the brain, including ventral striatum, that respond to gustatory stimuli, including sucrose solution, chocolate milkshakes, colas and fruit juices. Eating in the absence of hunger is a related construct that is used to describe eating past the point of satiety. While there are likely many reasons that people eat in the absence of hunger, a strong body of research has emerged showing that brain-mediated emotional rewards may play a particularly important role in overconsumption" (Bruce et al. 2015). Preliminary results show that obese children are less effective in using their cognitive control centre (prefrontal cortex) if food logos (e.g. MacDonald's M) are shown (Bruce et al. 2013), indicating less antidote towards excessive food intake.

The cortical and basal nuclei are provided with gustatory information by the brainstem nucleus that belongs to the solitary tract. The gustatory information transmitted by the VII, IX and X cranial nerves ends in the rostral part of this nucleus (Figs. 5.7 and 5.9).

Undeniable trigeminal system is involved as was already described above. The trigeminal system sends pain information into its brainstem nuclei. It will be transmitted finally, passing the thalamus, into the cortex to make it a consciousness phenomenon (Usunoff et al. 1997). Oral pain during eating influences food intake, but also one's food taste. The interaction between trigeminus, and glossopharyngeus, vagus and facialis, and the nucleus of the solitary tract has been restrictedly studied in humans. The trigeminal nucleus projects to the solitary nucleus in experimental animals (Pritchard 2012). Figure 5.8 shows that fibres of the

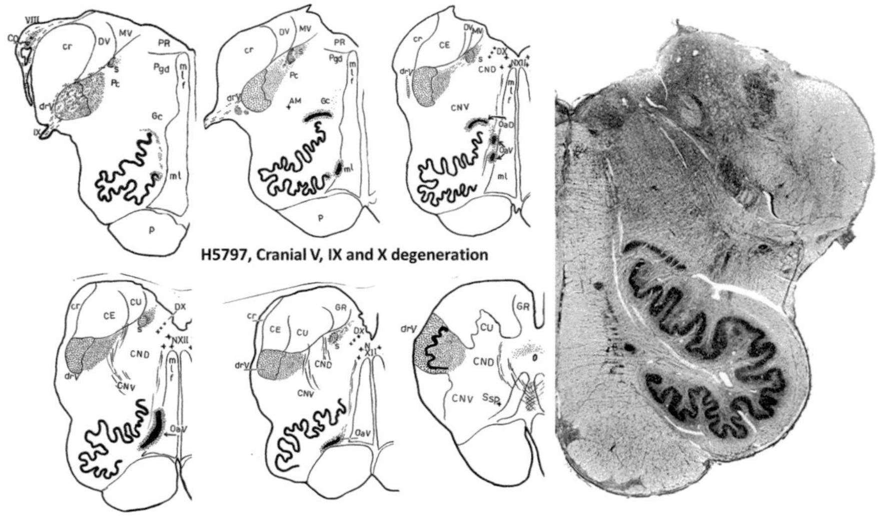

Fig. 5.8 Series H 5797: a 60-year old man developed 4 months before death disturbances of his cranial V, IX and X nerves by diffuse neoplastic infiltrations of the skull base. The degenerated fine glossopharyngeal and vagal fibre bundles (stippled) traverse the dorsal part of the spinal trigeminal complex and less pass through its ventral division. Proceeding dorsomedially the bundles converges to the solitary tract, and when reaching it, the fibres bend sharply caudally to join its nucleus. The solitary nucleus contained degenerated preterminal particles, diffusely spread between the dorsal small oval cells as well as between the ventral larger triangular neurons (from Schoen † unpublished)

glossopharyngeal nerve pass through the rostral spinal trigeminal nucleus in humans (Fig. 5.8). At the lower brainstem levels, where the glossopharyngeal projections are absent, still trigeminal projections can be found that enter the nucleus of the solitary tract (Fig. 5.8) supporting this trigeminal–solitary communication in humans.

The solitary tract has been described by Stilling in 1843 and by Clark in 1858. The tract had different names: slender longitudinal column, solitäres Bündel, bandelette solitaire or Respirationsbündel (due to the idea that the tract was a respiration regulator). The name “solitary tract has been taken on in the West”, although in different languages, because it is a pure topographic name. It lasted till 1900 before the role of IX and X cranial nerve was elucidated and the involvement of the facial cranial nerve by the chorda tympani accepted (±1909). The next step was the verification of the neurons around the tract in the so-called nucleus of the solitary tract. Not earlier than 1940–1950, the nucleus was firmly corroborated being the projection area of the solitary tract (Bradley 2007). The delay was caused by the false idea that the intercalate nucleus was the ending of the solitary tract. This is understandable since the intercalate nucleus parallels medially the solitary nucleus and solitary tract (see Paxinos et al. 2012).

The separation between rostral and caudal parts coincides exactly with the beginning of the ventricle free edge of the nucleus. The rostral nucleus is subdivided into four parts, each with its own functional importance: medial subdivision connects the caudal viscerosensory part of the solitary nucleus with the rostral, gustatory part. The rostral central part contains the largest input from the cranial nerves and represents an important part of the ascending gustatory information pathway. The rostral lateral part contains the oral tactile information. The ventral part projects to the other oromotor centres in the brainstem. In some animals, a small strip along the rostral central part is discerned. However, the cranial nerve endings avoid this part. There exists an intricate interaction between caudal and rostral parts of the solitary nucleus. Some rostral parts extend like a long column into the caudal part (medial but also the ventral part).

The neurotransmitter responsible for the information transfer of the primary afferents onto the gustatory solitary nucleus neurons is glutamate. Glutamate antagonists block the taste evoked response in solitary nucleus neurons. "The blocking action was the same for all tastants, indicating that all gustatory afferent fibre types employ glutamate at the first central synapse in the taste pathway" (Bradley and King 2007). The local interneurons of the rostral nucleus of the solitary tract use GABA as the inhibitory neurotransmitter. The GABA-A receptors are present on the postsynaptic membrane, while GABA-B receptors are placed on presynaptic membranes. Excitatory effects on solitary neurons are due to glutamate, while inhibition occurs mainly by GABA-A receptors.

Excitatory activity over the chorda tympani and over the glossopharyngeus nerve (see Fig. 5.9) summed up linear. Inhibitory activity summed up nonlinear. Moreover, long-term potentiation of gustatory inhibition is present in the solitary neurons that can change the transmission of the taste information. This is thought to be the basis for taste-related learning. In fact, gustatory inhibition in the solitary nucleus streamlines the incoming information before presenting it to the cortical areas: taste is not a mixture of four basic tastes, but a continuum of tastes [Sect. 5.3(2)].

5.9 The "Fusion" Kitchen: Clove, Oregano and Chilli

The Paris Match (No 1317/3) of August 1974 published an article "Des philters d'amour pour votre été" (love potions for your summer). Under the heading "La cuisine du diable" (the devil's kitchen), attention was paid to "The Perfumed Garden", written by Cheikh Nefzaoui (Umar Ibn Muhammed Al-Nefzawi). It is a book on all sexual performances and functions known in the Arab world in the fifteenth to sixteenth century. One of the suggestions is the use of egg yolk to increase energy for sexual potency. Another is to boil asparagus, fry them in oil and put them under egg yolk with ground herbs. In fact, dietary laws consider this a high protein approach that makes you tired, not energetic. Several other suggestions are: "if you want to acquire strength for the coitus, take fruit of the mastic tree, pound them and macerate them with oil and honey; then, drink of the liquid first in

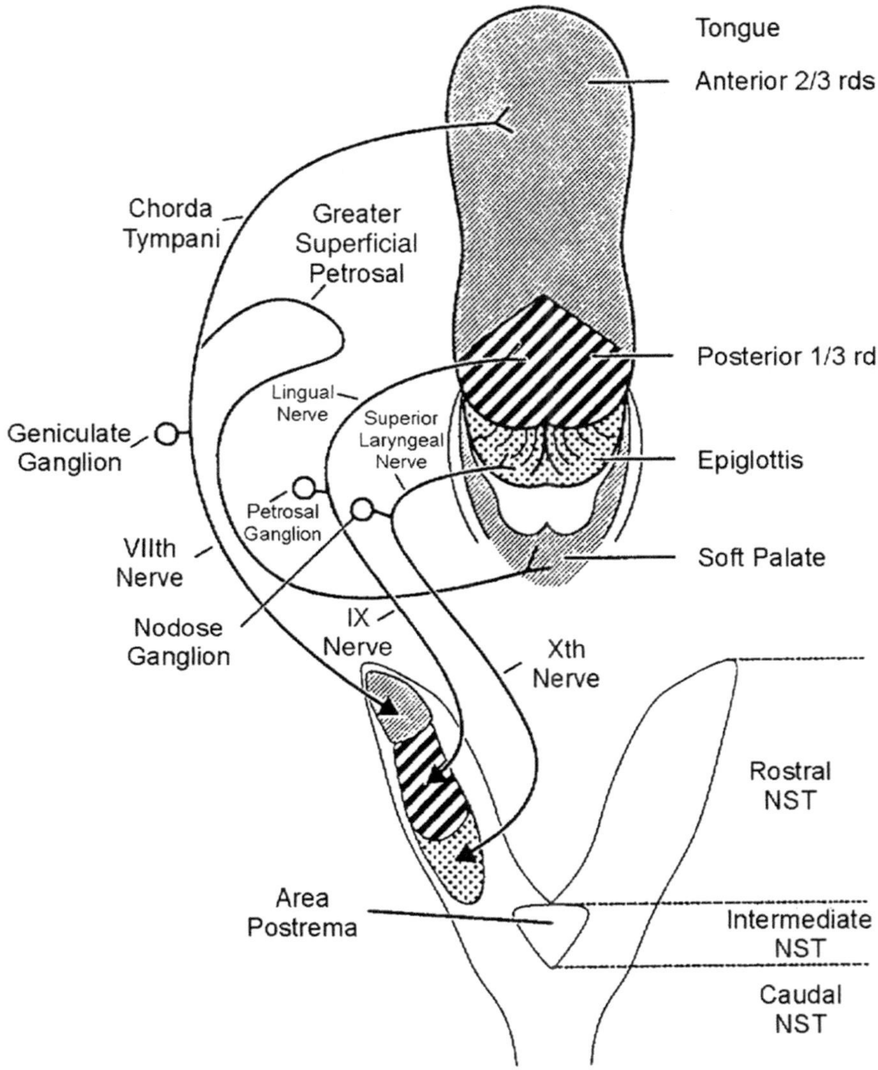

Fig. 5.9 Afferent sensory innervation of the oral cavity: tongue, soft palate and larynx and its projections onto the nucleus of the solitary tract. (From "Central Regulation of Autonomic Functions" by Loewy AD and Spyer KM eds. Oxford University Press, Oxford 1990 with permission)

the morning; you will become vigorous for the coitus; there will be abundance of sperm produced"; "he who feels that he is weak for the coition should drink before going to bed a glass full of very thick honey and eat twenty almonds and one hundred grains of the pine tree. He must follow this régime for three days. He may also pound onion seed, sift it, mix it afterwards with honey, stirring the mixture well, and take of this mixture while still fastening" (Al-Nefzawi (±1420).

Different, sometimes strange, aphrodisiac advice is known: soup of swallow nests, curry balls with sambal (spicy Indonesian chilli), dessert of herb bread made of clove, cinnamon and coriander followed by a tea made of saffron flowers and peppermint. Stimulating love potions or meals was known by their special taste, which indicated the "secret" actions and induced the placebo effects.

Special tastes from the Orient and beyond have been easily taken on in the West: "the globalization and our passion for visiting other cultures have led to development of new and exciting food styles, where exotic flavours from all over the world are combined with more traditional foods in new and exciting ways. The 'fusion' kitchen has provided us with greater food variety" (Reinbach et al. 2007).

Cloves were introduced by the Romans but got its spread in Europe in the seventeenth century by trade with the East Indies, reducing its price. *Oregano* (oros ganos, joy of the mountain) is known from Greek times. The Italian kitchen has been adopted by most Western countries. *Chilli* from East Indies or called "sambal" that was brought in from the Dutch Indies, now Indonesia, has been introduced in Europe as a result of the colonization. As a consequence, meal flavours have changed for a broad population in the twentieth century.

Although several more herbs and spices have had their influence on our appreciation of meals, and thus on our taste, only these three, clove, oregano and chilli, which are the best known in this respect, will be discussed here.

For the clove, the effective spice substance is eugenol [$C_{10}H_{12}O_2$; 2-methoxy-4-(2-propenyl) phenol, Fig. 5.10]. It is applied in foodstuffs, medicines and cosmetics. Clove oil was used in dentistry as an anaesthetic and was and is put into painful

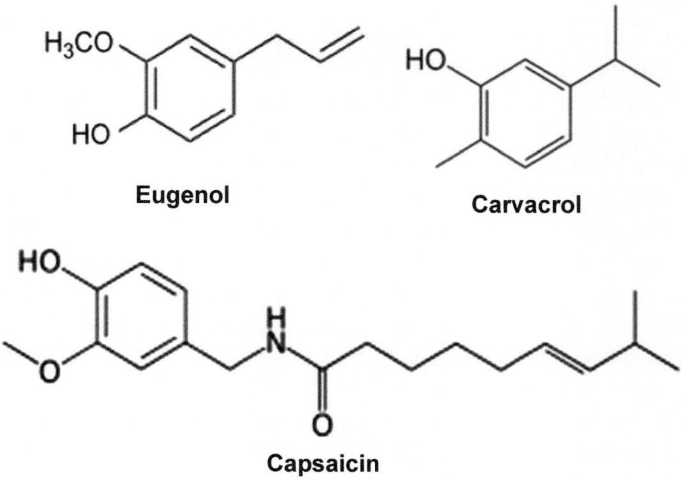

Fig. 5.10 Molecular formulas of effective substances of clove (eugenol), oregano (carvacrol) and chilli (capsaicin) are given

teeth by ordinary people to fight toothache. Clove oil is still used as a preanaesthetic for those who fear the prick of the needle.

Carvacrol the effective substance in *oregano* [$C_{10}H_{14}O$; 2-methyl-5-(1-methylethyl) phenol, Fig. 5.10] was recently found to inhibit growth and multiplication of bacteria, and low concentrations inhibit virulence and motility of pathologic bacteria. Contamination is seriously diminished for Bacillus cereus, supposedly also for Salmonella and Escherichia coli on meat, especially chicken fillet. It is the hydroxyl group that destabilizes the bacterial membrane (Ultee et al. 2002).

The actual substance in *chilli* is capsaicin [$C_{18}H_{27}NO_3$; (*E*)-*N*-(4-hydroxy-3-methoxybenzyl)-8-methylnon-6-eenamide; Fig. 5.10]. "It has in several human and animal studies been shown to stimulate digestion, to increase energy metabolism and to reduce energy intake and appetite" (Reinbach et al. 2007).

The human gene transient receptor potential cation channel, subfamily V, member 3, also known as Vanilloid Receptor-Like Protein 3 (TRPV3), that encodes a protein of the same name, is involved in both the effects of clove and oregano, and TRPV1 is also implicated (Klein et al. 2014). Since capsaicin is related to pain, its receptor distribution and its receptor effects were studied earlier than the effects of eugenol and carvacrol (for an early overview, see Szallasi 1994). Transient receptor potential vanilloid subtype 1 (TRPV1) receptor recognizes capsaicin. The receptor is present on dorsal root ganglion cells and by overstimulation with capsaicin the cytoplasm swells makes blebs, and finally overstimulation will induce necrosis. The cell swelling is the consequence of extra sodium ion influx and water transport after prolonged stimulation with capsaicin of the TRPV1 receptor (Pecze et al. 2013). The capsaicin receptor is also present in the taste buds of human circumvallate papillae. Co-localization of the capsaicin receptor with sweet (T1R2, T1R3, see Sect. 5.4) or bitter (T2R6) receptors in human circumvallate papillae was found (Jahng et al. 2010).

The venom of the tarantula or bird spider uses the capsaicin receptor. Hanatoxin from tarantula venom inhibits voltage-gated potassium channels. Various spider toxins are believed to partition into the membrane, bind to the voltage sensor part of the channel, which is formed by S1–S4 helices, and interfere with channel gating. Conformational changes of the voltage sensor that are required for opening the channel are presumably hindered by the toxin. "The toxins may bind to and stabilize the channel in the resting state, such that the channel becomes harder to open. In the presence of the toxin, the channel can still open but a stronger depolarization is required" (Fig. 5.11; Bohlen et al. 2010; Chen et al. 2012).

Hanatoxin favours the closed state of the potassium channel by its typical binding type to the capsaicin receptor. This activation of TRPV1 by the bird spider venom causes a robust inflammatory pain (Bohlen et al. 2010).

Artificial sweeteners (saccharin, aspartame and cyclamate) at low concentration give a pleasant sweet taste by T1R2/T1R3 sweet receptors, but at higher concentrations a bitter metallic taste arises. This bitter metallic taste originates from the involvement of the capsaicin receptor at high concentrations of these sweeteners (Riera et al. 2010).

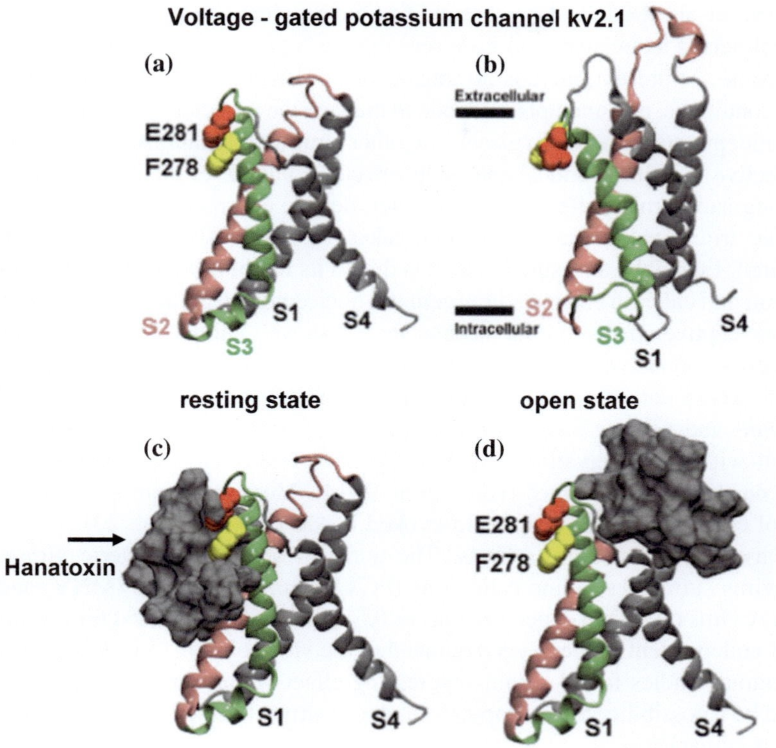

Fig. 5.11 Hanatoxin "interacts primarily with residues in the C-terminal half of the third transmembrane region (S3b), which together with S4 and the intervening extracellular protein receptor loop forms a flexible helix-turn-helix domain whose movements contribute to voltage-dependent gating. Hanatoxin appears to penetrate partially into the membrane, at the 278 and 281 positions, with its hydrophobic patch interacting with the binding groove near S3" (Chen et al. 2012, open access)

The capsaicin receptor is found in the human trigeminal ganglion and is related to pain. "Capsaicin-sensitive neurons are involved in nociception, are responsible for the neurogenic component of the inflammatory response and have efferent actions in peripheral tissues. The trigeminal capsaicin receptor is capable of activating the release of the neurotransmitters nitric oxide, CGRP and substance P from sensory nerves, and it contributes to central sensitization" (Hou et al. 2002). The consequence of organ sensitization is that pain in another organ occurs while the disease is not in the pain-inducing organ. Here, the activation of the trigeminal capsaicin receptor induces via the released neurotransmitters "pain in the brain", known as migraine (Hou et al. 2002), which is, of course, a pain caused by extracranial blood vessels or by the meninges. A specific type of oral pain, called "burning mouth syndrome, which is a chronic and intractable pain condition, which is most common in middle-aged and elderly women, affecting 1.5–5.5% of this population"

(Yilmaz et al. 2007) and "recent studies have pointed to dysfunction of several cranial nerves associated with taste sensation as a possible cause of burning mouth syndrome. At present, there is no satisfactory treatment for the idiopathic orofacial pain conditions, and attempts are made to manage the intractable pain using a variety of antidepressants, anticonvulsants or other drug combinations, all of which are ineffective and have undesirable side effects" (Yilmaz et al. 2007). Capsaicin's tingling and burning effect remain a secret, despite enormous research efforts.

The trigeminal capsaicin research takes us back to the clove and oregano research. Eugenol's anaesthetic effect is due to its inhibitory effect on voltage-gated sodium and calcium channels in trigeminal nociceptors. Carvacrol also accounts for antinociceptive effects, but its mechanism is less well known. "Many eugenol- and carvacrol-responsive units also responded to menthol, cinnamaldehyde and capsaicin. These data support a peripheral site for eugenol and carvacrol to enhance *warmth*- and noxious *heat-evoked* responses of trigeminal neurons and are consistent with the observation that these agonists briefly enhance warmth and heat pain on the human tongue" (Klein et al. 2014). Moreover, both eugenol and carvacrol cross-desensitized capsaicin-evoked irritation (Klein et al. 2013).

Carvacrol has antiviral effects. The antiviral activity on enteric viruses like norovirus surrogates, feline calicivirus (FCV), murine norovirus (MNV) and hepatitis A virus (HAV) in concentrations of 0.5–1% is nearly total and for certain virus types codependent of the oxygen demand of the viral solution (Sánchez et al. 2015).

Animal studies found a pain-suppressing effect of carvacrol in cancer, and this could be a possibility to develop new analgesic drugs for the management of cancer pain (Guimarães et al. 2014).

In conclusion, the phenomenology of the functions of these herbs in the gustatory process remains an enigma till now: "the alchemy in the mouth seems unpredictable" (see Sect. 5.1). Nevertheless, the research on clove, oregano and chilli seemingly opens an approach towards translational clinical solutions concerning important aspects of cancer, pain, viruses and sensitization.

5.10 Reflexes, Taste and the Tongue

5.10.1 Blowfly Snout Reflex

First, a "simple" system for reflex action will be considered: taste discrimination by the blowfly (Calliphora erythrocephala). The results are based on research done at the Biology Department of the Leiden and Groningen University in the Netherlands around the years 1970–1980 by van der Starre (1972) and van der Molen (1982).

The blowfly can taste with its legs. Its leg hairs on the tarsi contain taste neurons. Each taste hair contains five neurons that stick their dendrites into the hair pore that is covered by a sieve plate, through which water and diluted substances can reach the dendrites. Four of the neurons are chemosensitive, and one is mechanosensitive.

Their axons reach the suboesophageal and thoracic ganglia that react on the information by extending the snout (proboscis) of the blowfly towards, e.g., a sugar droplet. Since the structure of the leg is conservative, each hair of around 200 is genetically determined and physiology of one and the same hair in several blowflies can be studied, making averaging of the results possible and diminishing variability.

Each chemosensitive hair neuron can react on water, sugars and salts. Contact between a solution and the dendrites produces a spike train, which normally ends with adaptation of the chemosensitive neurons. The labelled line system is clearly not applied (see Sect. 5.2). Across fibre patterns is a good option (see Sect. 5.7). The long hairs show nearly no adaptation or reduction of the spike train, short hairs at the sides and upper part of the leg show total reduction, while hairs at the bottom of the leg react in between long and short hairs. Adaptation is concentration related. Salt and sugars can be discerned since the spike sequence for their spike trains is different. The neuron types do not differ in sensitivity for chemical substances but in their timely reaction on these substances: their temporal code is different. By the difference in temporal code, the neurons of the ganglia can discern the chemical substances. If to left and right leg different substances are offered (e.g. salt versus sugar), the snout will be stretched to one of the sides: the blowfly makes a choice.

5.10.2 Human Tongue Reflex

The Nobel Prize winner Ramon y Cajal described the tongue neuromuscular reflex related to feeding for the first time. Below we will discuss his reflex description, but we start with remarks of Bradley and King (2007): "Thus Cajal had already defined the basic reflex circuit from sensory endings in the dorsal tongue mucosa to the hypoglossal motor nucleus (right part Fig. 5.12).

Note that the sensory contribution of the facial nerve input is missing from Cajal's diagram (right side Fig. 5.12), because at the time he published his description, there was still some confusion regarding the contribution of the sensory part of the facial nerve to the solitary tract and nucleus (see Fig. 5.7). Cajal also failed to detail the contribution of gustatory and somatosensory input to the hypoglossal reflex. Functional details of the reflex circuits described by Cajal inspired a number of later investigators", Bradley and King (2007).

However, if one looks further (p. 733) into Cajal's Histologie du Système Nerveux de l'Homme et des Vertébrés, Volume I, one encounters Fig. 304 (left side of our Fig. 5.12). Thus, Cajal knew jolly well that the facial nerve contributed to the solitary tract (original figure text is kept in Fig. 5.12). He published his French Histologie in 1909, but first his Spanish original (1897) in between 1894 and 1897, in separate articles. His 1909 description was well known; e.g. the Dutch neurologist and neuroanatomist Winkler (1917) cites Cajal:

"Even within the N.glosso-palatinus (n.intermedius VII) centripetal (thin and partly unmyelinated), autonomic fibres are present that have influence on the respiration, as has been made probable by Ramon y Cajal. Along the tractus solitaries,

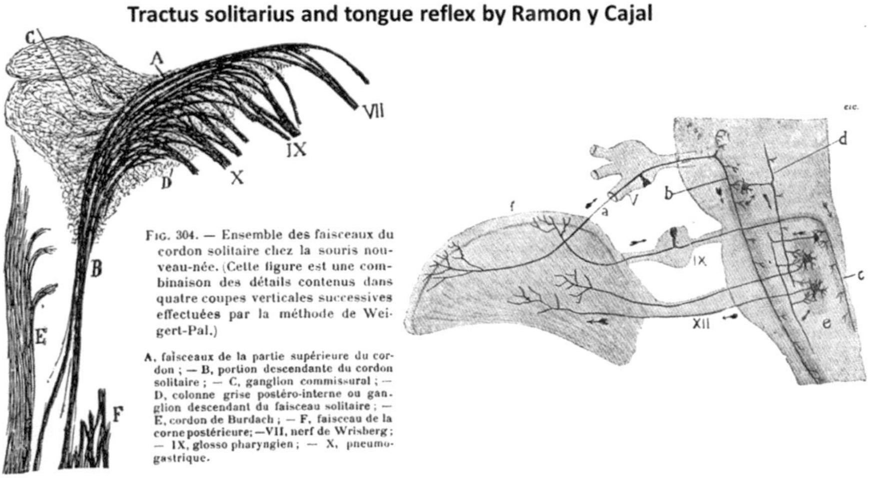

Fig. 5.12 Pictures of Ramon y Cajal (Figs. 294 and 304) on the composition of the solitary tract (304) and the tongue reflex. Text of Fig. 304 is: total of bundles of the solitary tract from the just born mouse. (This figure is a combination of the details found in four successive vertical sections processed with the Weigert-Pal method.) A, bundle of the upper part of the tract; B, descending part of the solitary tract; C, commissural ganglion (grey substance); D, grey posterior and internal substance or descendent ganglion of the solitary bundle; E, bundle of Burdach (fasc. cuneatus); F, bundle of the posterior angle (spinocerebellar fibres); VII nerf de Wrisberg (n. intermedius of VII); IX glossopharyngeus; X vagus nerve. Reflex abbreviations of Fig. 294 are: V, trigeminal (gasseri) ganglion; IX, ganglion d' Ándersch du glosso-pharyngien (petrosal ganglion); XII, hypoglossus; a, lingual nerve; b, end nucleus of the sensible trigeminus (ramus descendens N.V); c, tractus solitarius; d, central root of the trigeminus (nucleus princeps); e, hypoglossal nucleus; f tongue mucous surface; arrows indicate the direction of the stream

they can be followed into the nucleus commisuralis, although less in number. More important are the dorsal, centripetal autonomic fibres, which pass, being taste fibres, through the chorda tympani conducted into the N.glosso-palatinus" (note that n. intermedius was thought to be involved in respiration).

> Dat zich zelfs in den N.glosso-palatinus (n.intermedius VII) centripetale (dunne en ten deele merglooze) autonome vezels bevinden, die invloed op de ademhaling hebben is door Ramon y Cajal uiterst waarschijnlijk gemaakt. Zij zijn, langs den tractus solitaries heen, al is het dan minder aantal, tot in de nucleus commisuralis te volgen. Van meer gewicht echter zijn de dorsale, centripetale autonome vezels, die als smaakvezels door de chorda tympani naar den N. gloso-palatinus worden geleid.

Winkler's Chap. 6: "On the nervous system for the taste or the gustatory nervous system" does cite Cajal and others like Herrick, Luciani and Urbantschitsch, but also Krause and Cushing in relation to the clinical appearance of destruction of the chorda tympani: taste loss in the anterior part of the tongue, but pain and somatosensory information (trigeminus) stays present. Therefore, trigeminal contribution to the taste system was denied. Winkler already states that there must be, nevertheless, a correlation between trigeminus and gustatory system (see Sect. 5.7) that is present in Cajal's reflex diagram.

From Cajal's Histologie and from Winkler's description, one can conclude that Cajal knew the difference between rostral (gustatory) and caudal (visceral) solitary nucleus, and he described the commissural part of this nucleus for the first time (see Cajal remarks in Histologie (1909), pp. 732–737) and was well aware that the cranial nerves VII, IX and X contributed to the solitary tract (his Fig. 304, left in Fig. 5.12). Cajal found the tract to descend till the pyramidal decussation and perhaps into C1. Collaterals and their arborizations are described (see Cajal's Fig. 305). From page 847 on Cajal's description on the "Nerf intermédiaire De Wrisberg" starts. It is undeniable that Cajal here repeats its relation to the solitary tract and to the geniculate ganglion. He gives its trajectory and indicates that collaterals are seldom found. Its relation with the gustatory part or nucleus as described by Nageotte (1906) is reproduced with the solitary nucleus receiving the three cranial nerves.

Cajal did not indicate the VII or the X cranial nerve in his diagram of the tongue reflex (see Fig. 5.12). One should note that Cajal was well aware of the contribution of cranial nerve X (confirmed by Bradley and King 2007). Cajal used only the IX cranial nerve to show the reflex system. This was because the gustatory contribution of the IX cranial nerve was the most important in the eyes of the scientists of those days, and it was educational rather clever because this approach avoids discussion on his diagram.

Now, we got rid of a historical error, the diagram can be studied. Tongue mucosa activation sends by its ganglion its information into the brainstem, into the solitary tract with its nucleus. The afferent leg of the reflex has been confirmed electrophysiologically by Sherrington in 1916 and 1917 (see Bradley and King 2007). The solitary cells project to the hypoglossal nucleus, of which the motoneurons can contract the tongue muscles (Fig. 5.12). The interaction between the trigeminal sensory system and the hypoglossal nucleus is also shown. The diagram shows the correlation between gustatory and pain/somatosensory information via the hypoglossal nucleus. Note that the trigeminal system is not involved in the gustatory pathway. In his text, Cajal indicates that both circuits (the trigeminal and the glossopharyngeal one) "mediate chewing, swallowing, sucking and so on".

One learned above that trigeminal information projects to the solitary nucleus in humans (area ovalis). The difference between human and animal gustatory organization as explained above in Sects. 5.5 and 5.7 also involves the relation gustatory system and trigeminal input. In rat and monkey, a different input in the solitary nucleus has been found. Moreover, Nageotte's result (1906): "trigeminal axons terminate throughout the prefacial extension of the nucleus solitaries, which places the human data add odds with the experimental evidence from other animals, including primates…. *Although this issue may seem esoteric, the organization of the remainder of the central gustatory system hinges upon it*" (Pritchard 2012), which has already been explained in Sect. 5.5.

Here, we repeat: "besides taste, the surface of the tongue is also responsive to food texture, to heat and to pain. During mastication, it rolls the food round and round in the mouth, testing for lumps. When the tongue judges that all sharp pieces have been crushed or rejected, it participates in the crucial act of swallowing" (Morris 2004). Thus, while the stretching of the proboscis in the blowfly is a simple reflex action on the added taste information, the human tongue reflex is modified by mechanosensitive, taste, pain and heat, larynx and pharynx information, that is V, VII, IX and X cranial nerve information (add both figure parts of Ramon y Cajal in Fig. 5.12) influencing together with cortical descending information the tongue reflex. It is therefore understandable that the hypoglossal area in the brainstem is characterized by steering of premotor neurons that correlate with the oral phase of swallowing (mastication), and is involved into the second phase of swallowing and in the esophageal phase (ambiguous nucleus). Higher centres (caudal pons, cortex, cerebellum and striatum) also relay to the hypoglossal and ambiguous area and can act on jaw-opening motoneurons of the trigeminal nucleus. The steering of the tongue reflex is complicated.

5.11 Manipulation of Taste

One of my grandchildren together with his friend went on family holidays. Carrots were on the menu that evening. To my surprise, the friend could not believe that carrots tasted that way. After questioning, it became clear that he already ate his whole life carrots from tins. The flavour enhancers caused a totally different taste.

Some use cyclamate as a replacement for sugar (above we noticed already that its sweetness effect is concentration dependent), and industries add glutamate to get a better umami taste. The negative reaction is by people that go for life-reform-food: "Lebensreform ('life reform') was a social movement in late nineteenth-century and early twentieth-century Germany, Austria and Switzerland that propagated a back-to-nature lifestyle, emphasizing among others health food/raw food/organic food, nudism, sexual liberation, alternative medicine and religious reform and at the same time abstention from alcohol, tobacco, drugs and vaccines" (Citation from Peter Crawford 2013), for macrobiotic food ("macrobiotics is a system that can be used to create extraordinary health and make adjustments to food and lifestyle choices that support health improvement", Kushi institute, the Netherlands) and for biodynamic food ("biodynamics is a holistic, ecological and ethical approach to farming, gardening, food and nutrition", Biodynamic Association, USA). Here, we try to stay away from the discussion on cancer induction by additives, but you will notice that it is impossible. In fact, getting an E-number should mean that the additive is safe in the concentrations and in the food types prescribed.

Let us repeat the monosodium glutamate story (see start Sect. 5.4). "Rats provide the most spectacular fat storage deficiency disease known. Ve-Tsin, which is in fact sodium glutamate (E-621, see Table 5.1), in rats selectively destroys a hypothalamic nucleus, known as the arcuate nucleus. Severe obesity arises in these animals. In our Western society, obesity is also a threat, as it induces diabetes. Ve-Tsin is of course present in Chinese food. Fortunately, our arcuate nucleus is not

Table 5.1 E-numbers and their significance

	Sweeteners		Flavour enhancers
E-number	Chemical substance	E-number	Chemical substance
E-950	Acesulfame	E-620	Glutamine acid
E-951	Aspartame	E-621	Sodium glutamate
E-952	Cyclamate	E-622	Potassium glutamate
E-954	Saccharine	E-623	Calcium glutamate
E-957	Thaumatine	E-626	Guanyl acid
E-959	Neochesperidine	E-627	Sodium guanylate
E-967	Xylitol	E-630	Inosine acid
E-420	Sorbitol	E-631	Sodium inosinate
E-421	Mannitol	E-633	Calcium inosinate

Shows two types of additives: sweeteners and flavour enhancers. By their E-number the substance is allowed in the European Community and its use is restricted to certain maximum levels and to the indicated types of food

destroyed, but the typical headache, though not the diarrhoea, is caused by sodium glutamate. The arcuate nucleus is a special hypothalamic nucleus since it misses the blood–brain barrier that is present in most other parts of the brain. The small molecule monosodium glutamate passes through the rat blood–brain barrier and destructs parts of the arcuate nucleus (Bunyan et al. 1976; Holzwarth-McBride et al. 1976; Marani and Rietveld 1982). Large molecules can also pass into the surroundings of the arcuate neurons due to the absence of the blood–brain barrier. The arcuate nucleus, therefore, plays an important role in fatty acid sensing, which is in its turn important for the control of energy homeostasis. Neurons sensitive to fatty acids are present in the lateral hypothalamus, and fatty acids are capable to change the firing rate of neurons in the arcuate nucleus (Migrenne et al. 2006, 2007). Fatty acids regulate the membrane, neuronal conductance of a "wide variety of ion channels" (Migrenne et al. 2006, 2007). Studies showed a differential effect on arcuate neurons if oleic acid (an essential fatty acid, people have to take it from plants) was administered. Three distinct subpopulations of arcuate neurons could be discerned that reacted differently on oleic acid if low or high concentrations of glucose were present. Therefore, more than one population of arcuate neurons is involved in the neural control of insulin secretion by fatty acids (Migrenne et al. 2006, 2007). Seemingly, a balanced effect of these subpopulations of arcuate neurons is present depending on the glucose level noticed in the arcuate nucleus and this effect is via the hypothalamic network and hormones passed towards the pancreas, regulating insulin delivery, producing fine tuning of energy levels. Insulin co-determines the amount of energy-rich substances present in the body that can be converted to fat (Sainsbury and Zhang 2010; Könner et al. 2009). Thus, arcuate fatty acid sensing is present: "a local increase of fatty acids in the brain triggered changes in insulin secretion and action (involving in particular the hepatic glucose output), with or without food intake modifications"" (see Marani and Koch 2014).

Thus, monosodium glutamate as additive does have effects on the fat/glucose metabolism. Both phenomena, the headache and the overweight production, are regularly encountered in alternative food groups.

The aspartame case shows another dimension of the food additive discussion. In 1974, aspartame has been approved by the Food and Drug Administration (FDA). From then on, criticism on the decision persisted. It started with claims that the producer falsified and/or withheld negative information. In the "60 min programme" (an American newsmagazine television programme broadcast on the CBS television network) of 1996, the claim was discussed that aspartame causes brain tumours in humans and the criticism on aspartame FDA approval was repeated. The producer offered two of the FDA aspartame administrators a job that they accepted and indeed the initial research had flaws. It all contributed to the so-called conspiracy theory. Conflict of interest was used as argument to block the 1981 approval. Moreover, false allegations that consuming aspartame causes multiple sclerosis, skin lesions (lupus), blindness, spasms, shooting pains, headaches, depression, memory loss, birth defects and death have been e-mailed around (Edell 1998; see also http://www.fda.gov/ohrms/dock, last visit 28-10-2015). The controversy forced the US government into action. The USA Government Accountability Office studied the FDA approval and concluded that the official protocol was followed. The "fuss making" forced several governmental institutions over the world to study aspartame again. The general political outcome is that restrictions for the use are given, e.g. label on tins, or some firms removed aspartame from their products, due to the damaging antiaspartame PR. Research by the European Food Safety Authority (EFSA) and by the British Food Standards Agency supported the approval or indicated that the public health was not endangered by aspartame in 2013. Italy has its own story. The European Ramazzini Foundation of Oncology and Environmental Sciences published two studies (Soffritti et al. 2006, 2007, see also La Repubblica 07-03-2015, p. 6, Il Ramazzini convince la Pepsi "Via l'aspartame") that indicated that aspartame should be restudied. One of the claims was that it could induce cancer. The judgement of the FDA and the EFSA was destructive for the Ramazzini studies; e.g. high incidence of possibly carcinogenic infections could also explain the Ramazzini study results. In short, the flaws were enormous and the check of the Ramazzini pathologic slides did not support the researchers' claims (see: the truth about aspartame & Nancy Markle, http://www.xpress.com/bheal, last visit 28-10-2015). Aspartame is mud-plastered by healthcare frighteners, although thorough research does not support the fear for aspartame.

These considerations of food additives allow to decipher the problem of taste manipulation in a way that advances the debate and stimulates refinement of current healthcare ideas.

References

Al-Nefzawi UIM (±1420) The Perfumed Garden. A manual of Arabian erotology (16th century). English translation (1886), Imprimerie Veuve Félix Guy et Cie

Bartoshuk LM (1991) Sweetness: history, preference, and genetic variability. Food Technol 45:108–113

Bartoshuk LM, Gentile RL, Moskowitz HR, Meiselman HL (1974) Sweet taste induced by miracle fruit (Synsephalum dulcificum) Physiol Behavior 12:449–456

Bartoshuk LM, Catalanotto F, Hoffman H, Logan H, Snyder DJ (2012) Taste damage (otitis media, tonsillectomy and head and neck cancer), oral sensations and BMI. Physiol Behavior 107:516–526

Bohlen CJ, Priel A, Zhou S, King D, Siemens J, Julius D (2010) A bivalent tarantula toxin activates the capcaisin receptor, TRPV1, by targeting the outer pore domain. Cell 141:834–845

Bradley RM (2007) Historical perspectives. In: Bradley RM (ed) The role of the nucleus of the solitary tract in gustatory processing. Frontiers in Neuroscience. CRC, Taylor and Francis, Abindon, UK

Bradley RM, King MS (2007) Neurotransmitters and receptors expressed by rNST neurons. In: Bradley RM (ed) The role of the nucleus of the solitary tract in gustatory processing. Frontiers in Neuroscience. CRC, Taylor and Francis Abindon, UK

Breslin PAS (2013) An evolutionary perspective on food and human taste. Curr Biol 23:R409–R418

Brillat-Savarin (1848) Physiologie du gout. Gabriel de Gonet, Paris (The physiology of taste; translation by F. Robinson, version 2014) French edition BNF

Bruce AS, Lepping RJ, Bruce JM, Cherry JBC, Martin LE, Davis AM, Brooks WM, Savage CR (2013) Brain responses to food logos in obese and healthy weight children. J Pediatr 162:759–764

Bruce AS, Lim S-L, Smith TR, Cherry JBC, Black WR, Davis AM, Bruce JM (2015) Apples or candy? Internal and external influences on children's food choices. Appetite. https://doi.org/10.1016/j.appet.2015.04.061

Bunyan J, Murrel EA, Sah PP (1976) The induction of obesity in rodents by means of monosodium glutamate. Br J Nutr 35:25–39

Castriota-Scanderbeg A, Hagberg GE, Cerasa A, Committeri G, Galati G, Patria F, Pitzalis S, Caltagirone C, Frackowiaket R (2005) The appreciation of wine by sommeliers: a functional magnetic resonance study of sensory integration. NeuroImage 25:570–578

Chandrashekar J, Mueller KL, Hoon MA, Adler E, Feng L, Guo W, Zuker CS, Ryba NJ (2000) T2Rs function as bitter taste receptors. Cell 100:703–711

Chen R, Robinson A, Chung S-H (2012) Binding of Hanatoxin to the voltage sensor of Kv2.1. Toxin 4:1552–1564

Chen X, Gabitto M, Peng Y, Ryba NJP, Zuker CS (2011) A gustotopic map of taste qualities in the mammalian brain. Science 333:1262–1266

Compton MT, Ionescu DF, Broussard B, Cristofaro SL, Johnson S, Haggard PJ, Potts AA, Wan CR, Walker EF (2013) An examination of associations between the inability to taste phenylthiocarbamide (PTC) and clinical characteristics and trait markers in first-episode, non-affective psychotic disorders. Psychiat Res 209:27–31

Crawford P (2013) Lebensreform. www.gusto-graeser.info/S

DeSimone JA, Heck GL, Bartoshuk LM (1980) Surface active taste modifiers: a comparison of the physical and psychophysical properties of gymnemic acid and sodium lauryl sulfate. Chem Senses 5:317–330

Edell D (1998) Beware the E-mail hoax: the evils of Nutrasweet (Aspartame). Health Central, 18 Dec 1998

Erickson RP (2008) A study of the science of taste: on the origins and influence of the core iseas. Behavior Brain Sci 31:59–105

Frank RA, Mize SJS, Kennedy LM, de los Santos HC, Green SJ (1992) The effect of Gymnema sylvestre extracts on the sweetness of eight sweetners. Chem Senses 17:461–479

Gilbertson TA, Boughter JD Jr, Zhang H, Smith DV (2001) Distribution of gustatory sensitivities in rat taste cells: whole-cell responses to apical chemical stimulation. J Neurosci 21:4931–4941
Gould SJ (1994) The persistently flat earth. Natural History 3:12–19
Green E, Jacobson A, Haase L, Murphy C (2013) Can age related CNS taste differences be detected as early as middle age? Evidence of fMRI. Neuroscience 232:194–203
Guimarães AG, Scotti L, Scotti MT, Mendonça Jr FJB, Melo NSR, Alves RS, De Lucca Jr W, Bezerra DP, Gelain DP, Quintans Jr LJ (2014) Evidence for the involvement of descending pain-inhibitory mechanisms in the attenuation of cancer pain by carvacrol aided through a docking study. Life Sci 116:8–15
Hanig DP (1901) Zur psychophysik des geschmacksinnes. Philosophische Studien 17:576–623
Henning H (1916) Die qualitatenreihe des geschmacks. Zeitschrift Psychologie 74:203–219
Holzwarth–Mc Bride, Sladek JR, Knigge KM (1976) Monosodium glutamate induced lesions of the arcuate nucleus Anat Rec 186:197–206
Hou M, Uddman R, Tajti J, Kanje M, Edvinsson L (2002) Capsaicin receptor immunoreactivity in the human trigeminal ganglion. Neurosci Let 330:223–226
Jahng JW, Moon YW, Lee J-H (2010) Immunohistochemical detection of capsaicin receptors in taste cells of human circumvallate papillae. Asian J Oral Maxillofacial Surg 22:193–197
Katz DB, Nicolelis MA, Simon SA (2002) Gustatory processing is dynamic and distributed. Curr Opin Neurobiol 12:448–454
Klein AH, Carstens MI, Carstens E (2013) Eugenol and carvacrol induce temporally desensitizing patterns of oral irritation and enhance innocuous warmth and noxious heat sensation on the tongue. Pain 154:2078–2087
Klein AH, Joe CL, Davoodi A, Takechi K, Carstens MI, Carstens E (2014) Eugenol and Carvacrol exite first and second order trigeminal neurons and enhance their heat-evoked responses. Neuroscience 271:45–55
Kodama J, Fukushima M, Sakata T (1978) Impaired taste discrimination against quinine following chronic administration of theophylline in rats. Physiol Behavior 20:151–155
Könner AC, Klöckener T, Brünner JC (2009) Control of energy homeostasis by insulin and leptin: targeting the arcuate nucleus and beyond. Physiol Behav 97:632–638
Korsmeijer C (1999) Making sense of taste. Cornell Univ. Press, London
Lieberman DE (2011) The evolution of the human head. Belknap Press, Harvard University Press, Cambridge, England
Liman ER, Zhang YV, Montell C (2014) Peripheral coding of taste. Neuron 81:984–1000. https://doi.org/10.1016/j.neuron.2014.02.022
Marani E, Rietveld WJ (1982) Monosodium glutamate effects on arcuate cells at puberty. In: Bradford HF (ed) Proceedings of the NATO conference Transmitter interaction and comparmentation. Plenum Press, NY, pp 573–576
Marani E, Koch WFRM (2014) The Pelvis. Structure, gender and society. Springer, Heidelberg. NY
Migrenne S, Cruciani-Guglielmacci C, Kang L et al (2006) Fatty acid signaling in the hypothalamus and the neural control of insulin secretion. Diabetes 55(S2):S139–S148. https://doi.org/10.2337/db06-S017
Migrenne S, Magnan C, Cruciani-Guglielmacci C (2007) Fatty acid sensing and nervous control of energy homeostasis. Diabet Metabol 33:177–182
Molen van der JN (1982) A study on coding and variability in taste responses of the blowfly. Thesis University Groningen
Morris D (2004) The naked woman: a study of the female body. Jonathan Cape, London
Nageotte J (1906) The pars intermedia or nervus intermedius of Wrisberg and the bulbo-pontine gustatory nucleus in man. Rev Neurol Psychiat 4:472–488
Ohla K, Toepel U, le Coutre J, Hudry J (2010) Electrical neuroimaging reveals intensity-dependent activation of human cortical gustatory and somatosensory areas by electric taste. Biol Psychol 85:446–455
Paxinos G, Xu-Feng H, Sengul G, Watson C (2012) Organization of brainstem nuclei. Ch. 8. In: The human nervous system eds JK Mai, G. Paxinos, Elsevier, Academic Press, Amsterdam, Boston
Pecze L, Blum W, Schwaller B (2013) Mechanism of capsaicin receptor TRPV1-mediated toxicity in pain-sensing neurons focusing on the effects of Na^+/Ca^{2+} fluxes and the Ca^{2+}-binding protein calretinin. Biochim Biophys Acta 1833:1680–1691

Pritchard TC (2012) The gustatory system. In: Mai J, Paxinos G (eds) The human nervous system. Springer, Heidelberg, pp 1176–1200

Ramon y Cajal S (1909, 1972) Histologie du systeme nerveux de l'homme et des vertébrés. Instituto Ranon Y Cajal Madrid

Reinbach HC, Meinert L, Ballabio D, Aaslyng MD, Bredie WLP, Olsen K, Møller P (2007) Interactions between oral burn, meat flavor and texture in chili spiced pork patties evaluated by time-intensity. Food Qual Prefer 18:909–919

Riera CE, Vogel H, Simon SA, Damak S, le Coutre J (2010) The capsaicin receptor participates in artificial sweetener aversion. Biochem Biophys Res Com 376:653–657

Rhoades R, Pflanzer R (2003) Human Physiology, 4th edn. Thomson Books/Cole, Australia

Sainsbury A, Zhang L (2010) Role of the arcuate nucleus of the hypothalamus in regulation of body weight during energy deficit. Mol Cell Endocrinol 316:109–119

Sánchez C, Aznar R, Sánchez G (2015) The effect of carvacrol on enteric viruses. Intern J Food Microbiol 192:72–76

Sauvegeot F, Urdapilleta I, Peyron D (2006) Within and between variations of texts elicited from nine wine experts. Food Qual Prefer 17:429–444

Smith DV, Li CS (2000) GABA-mediated corticofugal inhibition of taste responsive neurons in the nucleus of the solitary tract. Brain Res 858:408–415

Soffritti M, Belpoggi F, Esposti DD, Lambertini L, Tibaldi E, Rigano A (2006) First experimental demonstration of the multipotential carcinogenic effects of aspartame administered in the feed to Sprague-Dawley rats. Environ Health Perspect 114:379–385. https://doi.org/10.1289/ehp.8711

Soffritti M, Belpoggi F, Tibaldi E, Esposti DD, Lauriola M (2007) Life-span exposure to low doses of aspartame beginning during prenatal life increases cancer effects in rats. Environ Health Perspect 115:1293–1297. https://doi.org/10.1289/ehp.10271

Soria ED, Candaras MM, Truax BT (1990) Impairment of the taste in Guillan Barré syndrome. Clin Neurol Neurosurg 92:75–79

van der Starre H (1972) Tarsal taste discrimination in the blowfly. Thesis University Leiden, Calliphora vicina Robineau-Desvoidy

Sugita M, Yamamoto K, Hirono C, Shiba Y (2013) Functional dissection of sweet and bitter taste pathways. J Oral Biosci 55:66–72

Szallasi A (1994) The vanilloid (capsaicin) receptor: receptor types and species differences.Gen Pharmac 25: 223–243

Teil G (2001) La production du jugement esthétique sur les vins par la critique vinicole. Sociol Travail 43:67–89

Todrank J, Bartoshuk LM (1991) A taste illusion: Taste sensation localized by touch. Physiol Behavior 50:1027–1031

Ultee A, Bennink MHJ, Moezelaar R (2002) The phenolic hydroxyl group of carvacrol is essential for action against the food-borne pathogen *Bacillus cereus*. Appl Environ Microbiol 68:1561–1568

Usunoff KG, Marani E, Schoen JHR (1997) The trigeminal system in Man. Adv Anat Embryol Cell Biol 136:1–126

Voogd J, Nieuwenhuys R, van Dongen PAM, ten Donkelaar HJ (1998) Mammals. In: Nieuwenhuys R, ten Donkelaar HJ, Nicholson C (eds) The central nervous system of vertebrates, vol 3. Springer, Berlin, pp 1637–2219

Winkler C (1917) De bouw van het zenuwstelsel. Deel 1: het zenuwstelsel voor den reuk, voor het zien, voor het algemeen gevoel en voor de smaak. De Erven F. Bohn, Haarlem, pp 1–447

Wooding S, Bufe B, Grassi C, Howard MT, Stone AC, Vazquez M, Dunn DM, Meyerhof W, Weiss RB, Bamshad MJ (2006) Independent evolution of bitter-taste sensitivity in humans and chimpanzees. Nature 440:930–934

Yamamoto T, Matsuo R, Kiyomitsu Y, Kitamura R. (1989) Taste responses of cortical neurons in freely ingesting rats. J Neurophysiol 61:1244–1258

Yilmaz Z, Renton T, Yiangou Y, Zakrzewska J, Chessell IP, Bountra C, Anand P (2007) Burning mouth syndrome as a trigeminal small fibre neuropathy: increased heat and capsaicin receptor TRPV1 in nerve fibres correlates with pain score. J Clinl Neurosci 14:864–871

Chapter 6
Oral Cavity: Tongue, Palate and Teeth

This chapter starts with the topography of the spaces discerned in head and neck, which is needed to understand their relations with the oral cavity. The oral cavity morphology is described (see what one can observe by the cheeks alone Fig. 6.1), and a short history of carious teeth is given. Part of this chapter is devoted to tooth eruption. An extensive description of the tongue and the supra- and infrahyoidal muscles together with their innervation is provided. Relations of the hypoglossal nerve are researched. Substantia nigra and the oral cavity are described in relation to oro-facial dyskinesia. Attention is paid to the soft palate muscles and their function in speech and swallowing. Palate myoclonus and its relation to olivary hypertrophy are treated.

6.1 Intermezzo: Topography of Spaces in Head and Neck

Before reading this part, notice the following. A space for most people is empty, not for the anatomist. Thus, the masseter muscle fills a space in the head called fossa masseterica. Now be aware of the following definitions: *Fossa is normally a groove,* but it is also the virtual space taken by a muscle or a group of muscles with the same function. The word is homologous with lodge for extremity muscles. *Spatium* is a real space present in head or neck, by which structures pass to reach organs.

The five areas that are discussed here are from inside to the outside: spatium parapharyngeum, fossa pterygopalatina, fossa infratemporale, fossa temporalis and the fossa masseterica (see Fig. 6.2).

In order to explain the spatial relations, it is useful to refer to the muscles of mastication. The temporal muscle and masseter muscle are well noticed from the lateral side. In fact, the fossa temporalis continues to the fossa masseterica, below the zygomatic arch. However, both spaces are discontinuous if filled with the muscles. The temporal muscle occupies the temporal fossa beneath the strong temporal fascia, a layer that is attached to the bony zygomatic arch. The temporal

E. Marani and C. Heida, *Head and Neck*,
https://doi.org/10.1007/978-3-319-92105-1_6

Fig. 6.1 Photographs of the foot of Maria statue, originally one of two opposite blocks in the crypt of the église de Saint Amant de Boixe, Charente (±1350), moved to the "chapelle ossuaire" (chapel for bone rudiments), now chapel for the victims of the 1914–1918 war. On both sides of the central piece, the glutton and the soaker are sculptured as they are vulgarly called. The soaker should suck at a bottle with the bottle neck in its mouth. However, his cheeks swell out indicating blowing. For indicating sucking, his cheeks should have been pictured hollow

muscle inserts into the coronoid process of the mandible (Fig. 6.5), exercising a fast force during biting. The masseter arises from the zygomatic arch and is inserted into the angle of the mandible, occupying the fossa masseterica. It exercises a lasting force to hold food. Both temporal and masseter muscle can retract and protrude or pull laterally and protrude the mandible, respectively.

The arrangement of the pterygoid muscles is more complicated. They originate from the pterygoid process, which is a bilaminar pillar, going downward from the sphenoid bone. The lateral pterygoid muscle arises from the lateral part and protrudes the condyle. The medial pterygoid originates in between both pillars and ends on the inside of the angle of the mandible. So the posterior part of the mandible is hold within a sling made up by the masseter muscle at its outside and medial pterygoid muscle at the inside. The digastric muscle is a depressor of the mandible, but can also elevate the hyoid due to its intermediate tendon that adheres to the hyoid (see Sect. 6.3 for hyoid stabilization).

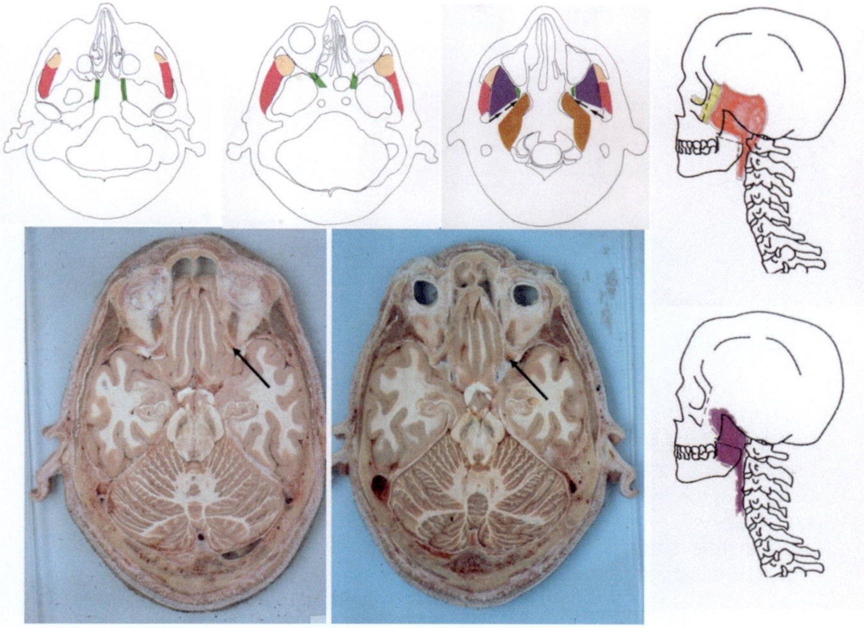

Fig. 6.2 Top of this figure shows three horizontal sections through the head. Orange: insertion M. temporalis, red: M. masseter, green: fossa pterygopalatina, brown: spatium parapharyngeum, blue/purple: fossa infratemporale. To demonstrate these structures, they are shown in horizontal sections through the head, from orbita to the zygomatic bone. Right side: These spaces are projected on the skull and the cervical vertebral column, purple: spatium parapharyngeum (deepest), red: fossa infratemporale, yellow: fossa pterygopalatina (© E. Marani). In the middle are two sections pictured that pass through the upper part of the human head (courtesey G. Maat, Dept. Anatomy, LUMC), compared to other figures. Arrows indicate the fossa pterygopalatina. The fossa temporale can be discerned by the presence of the temporal muscle, directly below the skin in front of the ear. By the way, note that these sections through the human head both show the substantia nigra (see Sect. 6.4)

The fossa pterygopalatina gives way to high dorsal structures towards the eye (Fig. 6.2), especially via the fissura orbitalis inferior and medially into the nasal cavity and palatum (Fig. 6.3). Its exponent is the A. maxillaris that enters this fossa and by its branches passes through bony channels and openings to inner parts of the skull. The plane on which this part of the skull in the drawing rests in Fig. 6.3 is now turned 180°, so one sees the back of this plane. The relation of the fossa temporale (with the temporal muscle) and infratemporale (with the lateral pterygoid muscle) and the fossa pterygopalatina becomes clear. The fossa pterygopalatina is the corridor to pass from the infratemporal fossa to the orbit (see Chap. 8).

Wharshafsky et al. (2012) discussed the imaging of the deep neck spaces. In summarizing the key points of their article, it says: "there are many opinions about what defines neck spaces and how to best organize them into useful clinical tools. A solid understanding of the anatomy and relationship of the various neck spaces

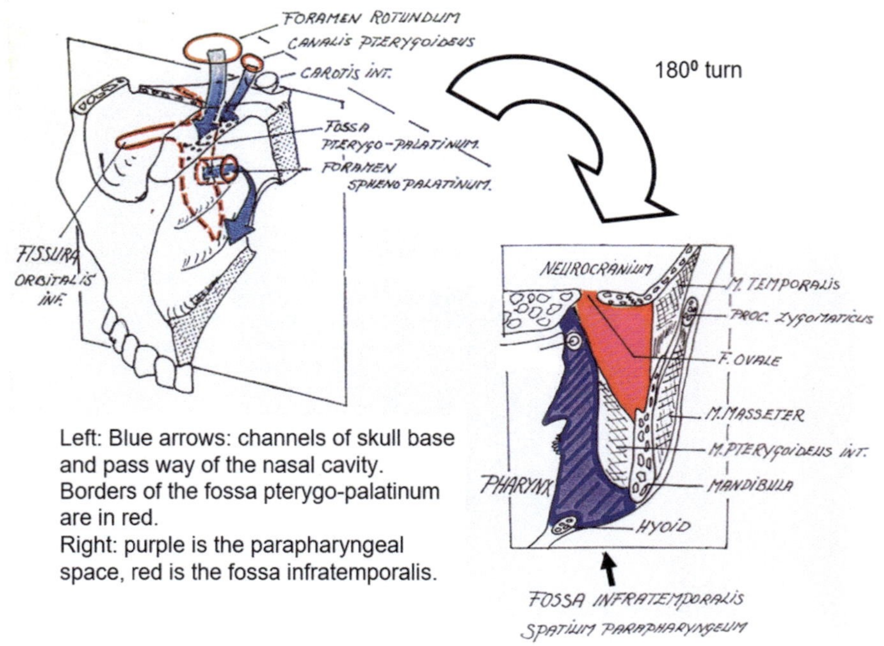

Fig. 6.3 Section through maxillary quarter of the skull, cut off at the orbit. Left, the spatium parapharyngeum is shown in red, with its entrances from the skull base. Note the pass way of the carotid artery (see Sect. 12.7 for cranial nerves that enter this area). By turning the section surface a 180 degrees, one sees the lateral relations of the fossa infratemporalis (red) and the parapharyngeal space blue/purple. Note that this parapharyngeal space reaches to the hyoid

is valuable in diagnosing and treating diseases of the neck". The same subdivision is supported and was used already far earlier (see Chap. 14).

6.2 Teeth

The oral cavity is variable, covered totally by mucous membranes and bordered by the lips at the outside. The roof of the mouth is the palatum durum (palate) and distinguishes the nasal cavity from the oral cavity. Its lateral part is by the bony maxilla and its teeth (Fig. 6.6) covered by gum. The bottom of the oral cavity is named the diaphragma oris, mainly built by muscles, bony mandible and teeth. The space between cheeks, lips and teeth is termed vestibulum oris, and the real inner part within the half circles of teeth is called the real cavity of the mouth, cavitas oris proprium. The passage space or exit, from mouth to the pharynx, is called the isthmus of the fauces (throat) or the oropharyngeal isthmus. It is surrounded by part

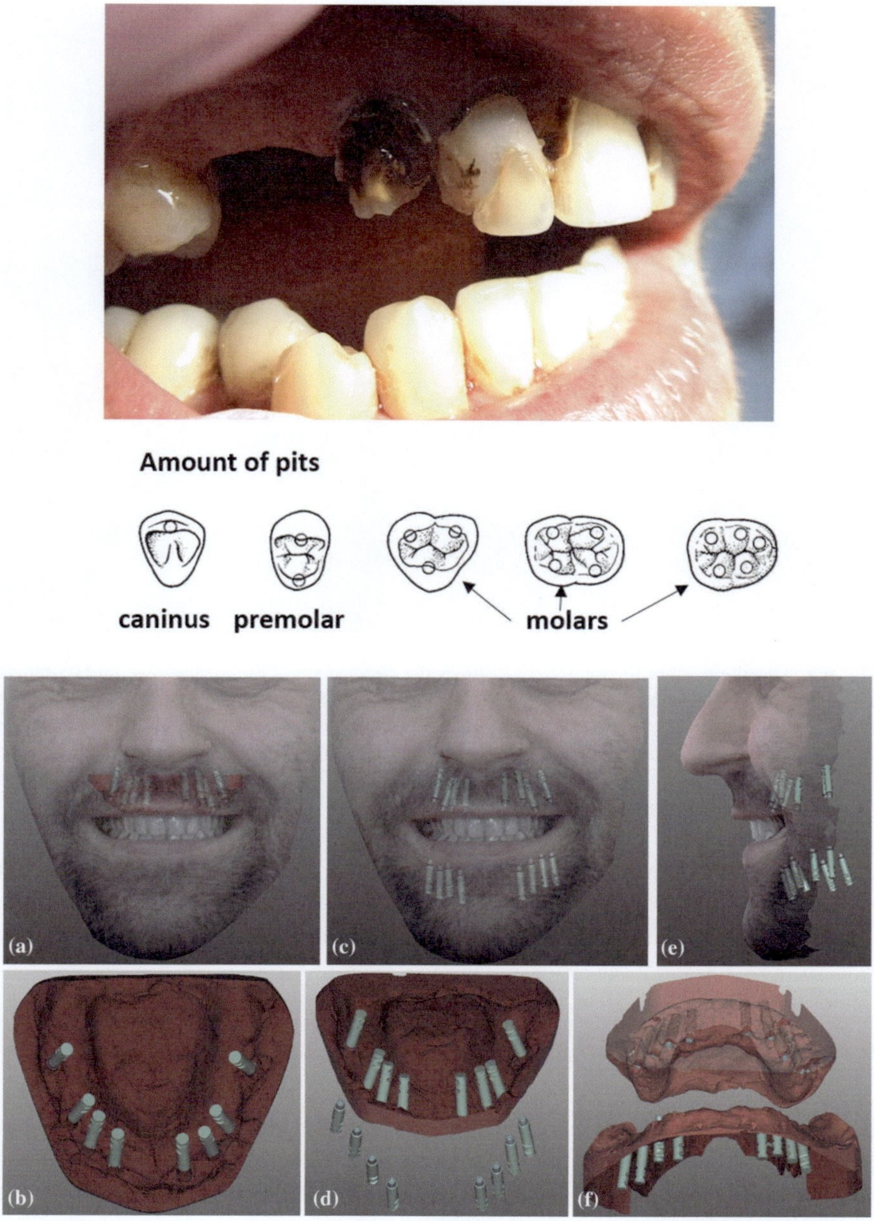

Fig. 6.4 Photograph of a 29-year-old patient demonstrating caries rot on the caninus and more loss. (Courtesy V. Thierry, with permission). The amount of pits is indicated for the various types of teeth. The lower part of the figure shows maxillary implants in (**a**) and view on these implants with reconstructed gingiva (**b**). Upper and lower jaw implants are shown in (**c**) and (**d**). The lateral view on the implants is demonstrated in (**e**) and **16** upper and lower jaw implants with reconstructed gingiva in (**f**) (part figures **a**–**f** courtesy G. Agosti with permission)

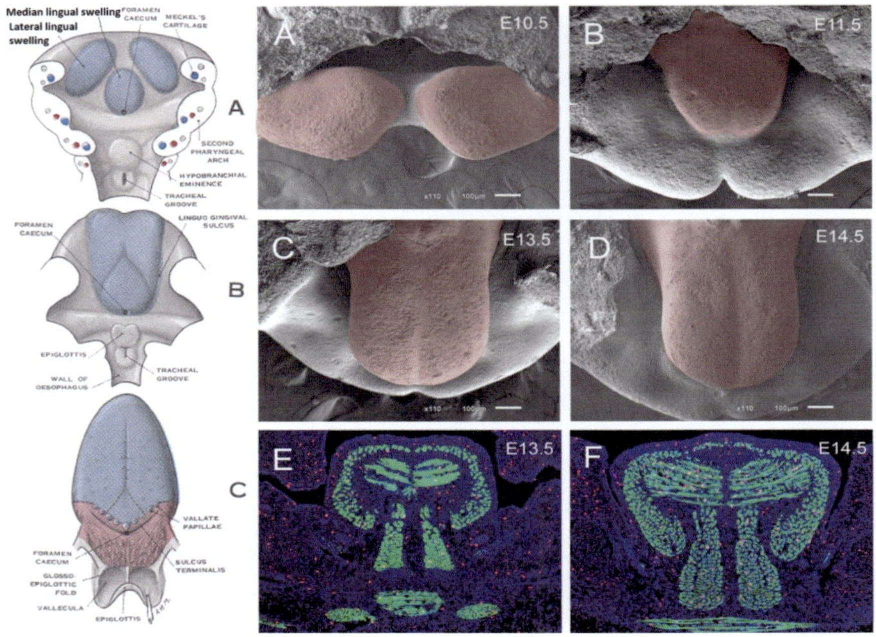

Fig. 6.5 Development of the tongue. Left shows the overgrowth of the median swelling (compare **a** and **b**) and the contribution of the hypobranchial eminence (**b** and **c**; adapted from Hamilton et al. 1946). Right panel demonstrates scanning electron microscopic pictures of the bilateral swellings and their fusion (**a**–**d**). Myosin heavy chain immunocytology (green) shows the developing musculature of the tongue (compare to Fig. 6.7). Red stips are proliferating cells (figure is from Parada and Chai 2016 with permission)

of the oropharynx directly behind the mouth cavity, bounded superiorly by the soft palate (palatum molle), laterally by the palatoglossal arches, and inferiorly by the tongue. Food uses this route to reach the pharynx and oesophagus.

6.2.1 *Awareness of the Oral Cavity: Carious Toothache*

For most people, the oral cavity is not earlier consciously present until one has toothache. To concur with this pain awareness, caries effects on teeth will be studied with some respect to history, passing through ages and going from the north, Spitsbergen, to the south of Africa.

Let us start with Spitsbergen. In the seventeenth and eighteenth centuries, the Dutch settled at a whaling station on the island, called Zeeusche Uytkyck (Zealand's Look-out). The Groninger expedition of 1980 studied fifty man buried on the island with a total of 1397 collected teeth. We do not go into the determination of age and results of most other teeth, but only give the results on the molars (Maat and Van der Velde 1987). Caries develops especially within the fissures and

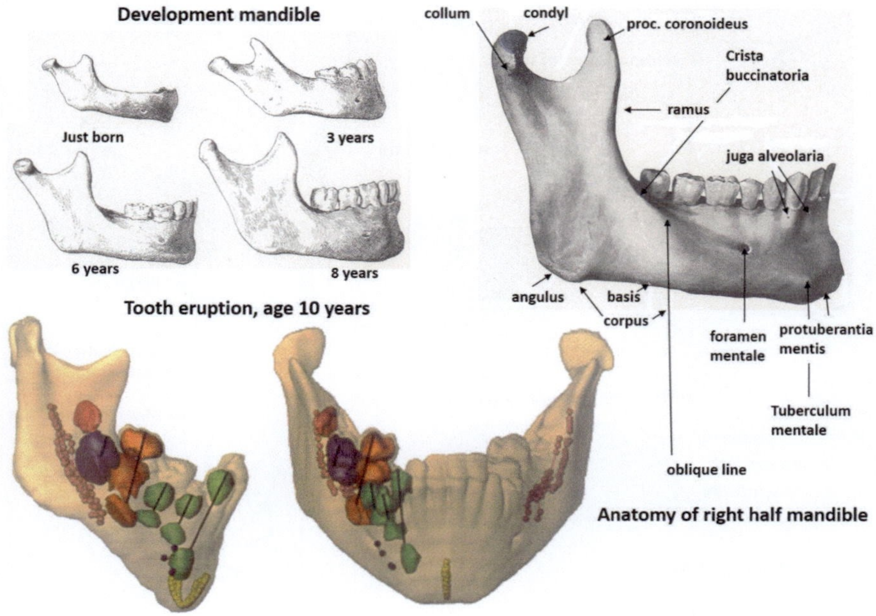

Fig. 6.6 Development of the mandible is demonstrated from just born to 8 years together with the mature main anatomical structures of the mandible outside (Spalteholz 1932). Lower coloured figures show mandible of 10-year-old child with superimposed the teeth of ages 0, 1, 7, 10 years. Black lines are the eruption paths: note the mandibular canals. All structures are aligned to the symphysis menti (Krarup et al. 2005 with permission). Purple, permanent third molar; blue, permanent second molar; red, permanent first molar; green, permanent premolars and canine; dark blue, the mental foramen; yellow, symphysis mentis; pink: mandibular canals

pits of the teeth of which molars are renowned (Fig. 6.4, middle part). Caries is overcome if the rate of attrition is faster than rate of caries progression. Antemortem teeth loss was 6.75% for the Dutch whalers, but for the molars the highest loss was maxillary M1 17% and mandibular M1 14%. The overall incidence of caries was 13.4%, and molars had the highest incidence: M1 33–37%; M2 38–39% and M3 25–35% (figures not corrected for the postmortem loss of teeth). Occlusal caries was exclusively seen on molars (64.2%), and neck caries was found in 35.8% of the caries cases. Note that 44% of the men showed intact dentation.

In the Middle Ages, the caries rate was low nearly 3–6%, while during the scientific revolution (1500–1700), it slowly increased to 10–13%. Its cause is unknown, but it surely is not produced by an increase of sugar addition to the food. Presumably, the finer flours, due to sieving after grinding, reduced the substances responsible for attrition and thus reduced its rate, increasing the caries effect. Indeed, the decrease during this time in dental wear with 20 years supports this idea (see Maat and Van der Velde 1987).

In South Africa, it concerns migrant labourers that worked in Kimberley during the late nineteenth century. Diamond mining started in 1871 on Colesberg Kopje

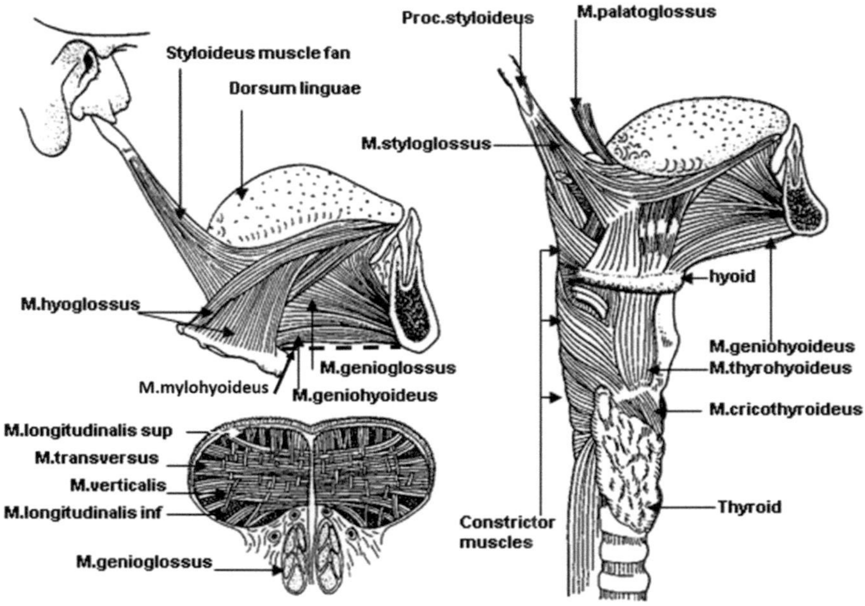

Fig. 6.7 External and internal muscles of the tongue (left). Right side shows the tongue muscles in relation with pharynx and larynx muscles and structures (figures are adapted from Samandari and Mai 1995 with permission). Note that the mylohyoid muscle is a very thin muscle and is indicated with a stippled line. The transverse section of the tongue is depicted with the largest muscle beneath: the M. genioglossus

(finally it became De Beers Consolidated mines), and by 1899, the population around the mines had increased from a few hundred till 16,300 Europeans and 28,200 black individuals. "Several complete skeletons were exhumed from damaged, unmarked graves alongside the fenced Gladstone cemetery in Kimberley, South Africa in 2003. Several skeletal lesions suggestive of scurvy, tuberculosis, treponemal diseases, violent trauma and congenital abnormalities were observed during the investigation of the remains" (Van der Merwe et al. 2010). Caries was demonstrated in 56% of the males and in 46.2% of the females and with an average of 2.7 carious teeth per head in males and 3.8 in females. Antemortem dental loss was found in 29% of the cases. Molars were most damaged. In this population, the highest carious molar was M2. "However, the highest prevalence of carious lesions was found on the second molar and not on the first. This can most likely be explained by the pattern of antemortem loss in the Gladstone sample, where the first molars were mostly affected, skewing the distribution of dental caries" (Van der Merwe et al. 2010). 31.2% of the molars were carious, and no statistical difference was noted between males and females. Periodontal granulomata ("abscesses") were observed in 17.7% of males and 15.4% of females, and periodontal disease affected 40% of those investigated.

The results demonstrate that the authentic meals without sugar or without machine ground maize induce lower caries frequency than those meals in which carbohydrates are easily available for the saliva of the mouth. The low carious teeth per mouth in the Gladstone population are presumably caused by the short stay of these rather young mineworkers at the Kimberley mines. The appearance of serious bone damage and fistula of the jaws as a consequence of caries is frightening realizing the pain aspect, since up to 40% of the skeletal cases were with periodontal disease.

Thus, the comparison between the 13.4% molars at Spitsbergen and 31.2% at Gladstone, Kimberley, shows more than a doubling of the molar caries. Moreover, dental loss in Spitsbergen was antemortem 6.75% and at Gladstone 29% a fourfold increase. Thus, caries increased within 200 years enormously and at the same time the toothache paralleled it of course. And in the young Gladstone population with a restricted time to carbohydrate-rich mineworkers food, the jaw bone damage was 40%. One should realize that the jaw bones started to rot to use a non-scientific word.

The reaction on caries and on these periodontal diseases has been the development of dentistry that has assumed enormous proportions up to the use of implants, bone replacements and total prostheses. Two factors have stimulated the development of dental implants: jaws are bony structures that accept inert implants, and gingiva always closes around jaw structures even if these structures are artificial. Prosthodontics (replacement of teeth by artificial structures) induced the development of new substances that are accepted by the jaws (e.g. zirconia or metals, like titanium, for dental implants and crowns), application of 3D techniques, of laser techniques and new ones for the production of artificial structures including implants (Fig. 6.4a–f). "Zirconia has interesting microstructural properties and is the material of choice for the "new generation" of implants, thanks to its biocompatibility, osseoconductivity, tendency to reduce plaque accumulation, and interaction with soft tissues, which leads to periointegration. Moreover, osseointegration of zirconia dental implants may be comparable with that of titanium implants. Zirconia particles may have the potential to improve initial bone healing and resistance to removal of torque" (Hisbergues et al. 2009; Özkurt and Kazazoğlu 2011). However, long-term zirconia results are still absent. Of course, aesthetics is an important aspect of dentistry and orthodontics (e.g. Gürel 2003), and often application of aesthetics is an expensive hobby.

The idea that caries has been conquered by our oto-laryngeal knowledge and dental health care developed in the Western world is a fairy tale. Only one example is given here: Fig. 6.4 shows a photograph of a 29-year-old patient. The caninus has been totally destructed by caries. Note that this tooth is a hard nut to crack by caries possessing only one pit and restricted superficial fissures. The dental care costs are in most countries hardly insured or the price is rather high, excluding poor(er) people from dental treatment. The overview of the WHO oral health studies shows: worldwide, 60–90% of school children and nearly 100% of adults have dental cavities. Severe periodontal (gum) disease, which may result in tooth loss, is found in 15–20% of middle-aged (35–44 years) adults. Globally, about 30% of people aged 65–74 have no natural teeth. Oral disease in children and adults is higher among poor and disadvantaged population groups (see WHO's media center, fact sheath 318, April 2012).

6.2.2 Tooth Eruption

The oral cavity widens in relation to the pharynx during early development, which is essentially linked to tooth eruption and thus to the formation of the alveolar ridges that contain the dentition. The vestibule or space outside of the alveolar ridges permits the mobility of the cheeks and lips and is important for articulation, deglutition and mastication. Understanding of the eruption of teeth is important for dental disciplines. The development of the vestibule of the mouth and the dental development are intimately related. Both developments are based on the inclusion of epithelial membranes. Its primary and secondary developments finally give dental crests. Cell nests in the dental crest are the start of the milk teeth. Deeper placed cell nests are the anlagen of the permanent teeth. Till 2005, three-dimensional studies of these growths were absent (Krarup et al. 2005), presumably due to separate studies of the developments of the teeth and the mandible. The mandible development shows large morphologic changes (see Fig. 6.6), which makes it difficult to compare the different age results. The results due to metallic implants showed: assurance that mandible palpable and structural points could be used without implants and research of tooth eruption in longitudinal studies of facial growth is possible. The mandibular structures that can be used for age comparisons are: "(1) the anterior contour of the chin; (2) the inner contour of the cortical plate at the lower border of the symphysis menti (suture line of both mandible halves enclosing caudally the protuberantia mentis); (3) any distinct trabecular structure in the symphysis menti (at the protuberantia mentis Fig. 6.6); (4) the contour of the mandibular canal and (5) the lower contour of mineralized tooth germs before root development begins" (Krarup et al. 2005). For X-ray comparison, one needs the head placed according to the Frankfurt horizontal which is in babies and children a real problem (for the solution, see Krarup et al. 2005). Roentgencephalometry is a two-dimensional technique that was used earlier, but replacement by CT solved this problem. CT gives the possibility to make 3D compositions. Growth could be registered by the mathematical method of least squares estimation of the best fit between 3D landmarks representing homologous points in pairs of landmark datasets. Although children with Apert syndrome (see Chap. 2) were studied, these mandible results can be generalized for all children. The symphysis mentis showed bone apposition at its posterior side, which gives antero-posterior growth of the mandible. The mandibular canals stayed stable in their position viewed from lateral. Still they increased in length. However, the mandibular canals changed position in frontal view towards lateral (Fig. 6.6). Bone apposition was found at: the condyle, coronoid process, alveolar process, posterior aspect of the ramus and buccal surface of the mandible. To change the mandible form, bone resorption is also needed.

It was present at: anterior aspect of the ramus, underneath the posterior part of the angle and lingual aspect of the mandible body, especially the coronoid process, the ramus and condyle. The surprising result is the straight line of the position of the molars during development and eruption despite the large morphological changes that occur during the first 8–10 years of age (Fig. 6.6).

6.3 The Tongue

The tongue looks an unpaired system but is built from paired muscles covered by an aponeurosis that creates this unpaired impression. The migration of the cranial neural crest cells (see Sects. 3.5.1 and 4.4) that contact the ecto- and endoderm constitutes the base of the tongue development and induces swellings. The patterning information by Hox-negative distal-less homeobox code and the intra-arch polarity contributes to the axes necessary for the tongue development. This polarity organizes mandibular and maxillary differentiation into their respective prominences (see Fig. 1.5 in Introduction). The mandibular prominence produces 2/3 of the tongue, among other structures, of course. The tongue originates of different contributions of all branchial arches. It concerns a median lingual swelling, which will be overgrown by the lateral lingual swellings during the fusion of these lateral swellings (Fig. 6.5 left). Together, these swellings are responsible for 2/3 of the tongue. The rest, at its proximal end, is produced by the hypobranchial eminence (see Fig. 6.5).

Four main stages are discerned within the development of the tongue:

(1) Invasion of myogenic progenitor cells into the tongue primordia that is regulated by the molecules tyrosine-protein kinase Met (c-met), growth factor receptor-bound protein 2-associated protein 1 (Gab-1), ladybird homeobox protein homolog 1 (Lbx1).
(2) Proliferation and differentiation of these myogenic progenitor cells within the tongue primordia by Pax3 and Pax7 (a family of nuclear transcription factors).
(3) Determination and further differentiation occur by myogenic regulatory factors: myogenic factor 5 (Myf5), muscle-specific regulatory factor (MRF4) and myoblast determination protein (MyoD) together with myogenin.
(4) Finally, myoblast fusion occurs (Fig. 6.5 right; Parada and Chai 2016).

The tongue is not related to the pharynx as a cavity wall structure during development; hence, it would have been a (half) ring-like structure as is the case, for example, with the larynx constrictors placed at pharynx dorsal wall (see Fig. 6.7). Moreover, the two halves of the tongue are mutual structurally and functionally related due to their fusion. It is the organ that belongs to the most mobile structures of the body, and its four functions can be summarized as: chewing, swallowing, speaking and tasting.

The tongue rests free on the oral bottom for movement and is only connected at its back that starts as an elevation in the medial part of the oral floor. The floor and the tongue anterior are connected with a small median string called frenulum linguae. At its base, the outlets of the submaxillary glands are found. If this frenulum is too short (ankyloglossia), it has to be cut (frenulectomy) to return to a normal function of the tongue. Not only tongue mobility after frenulectomy but also the masseter, anterior temporalis, upper and lower orbicularis oris, and submental muscles increased their muscle activity and returned to normal (Tecco et al. 2015). Seemingly, the movement restrictions of the tongue induce action decrease in several other muscles in and around the oral cavity. Could this indicate the presence of not only a spatial coherence but also a functional coherence?

One should distinguish the tongue muscles that are explicitly for mobility of the tongue from the suprahyoid muscles, all involved in movement of the hyoid. By the way, the tongue resides within the borders of the hyoid bone. The floor of the mouth is constructed by the myelohyoid muscle, a rather flat and thin muscle. Its continuity with the pharyngeal wall is not assured due to its free border. The tongue muscles are subdivided into those that start their connections from outside to the tongue inside, called extrinsic muscles and the muscles that make all their connections within the tongue, the intrinsic ones. The tongue's proper structure is mainly muscular and is build by several intrinsic muscles: longitudinalis superior and inferior, the transversus linguae and the verticalis linguae (Fig. 6.7). "The intrinsic muscles run into three directions, vertically, horizontally and longitudinally. This complex is also responsible for the astonishing way in which the tongue can change shape, becoming wide and flat, narrow and thick, or rolled up laterally to form a gutter shape. The latter shape is impossible to achieve for a small number of people, and this inability is genetically determined. It is of no practical importance but try it yourself" (Moffat 1993). Moffat's statement on practical importance is not totally true. Tongue rolling, as making the gutter shape as it is called, has been related to left-handedness and several other physiological conditions. The proportion of tongue rollers in women is higher among right-handed than left-handed females. Azimi-Garakani and Beardmore (1979) found a greater incidence of tongue rolling ability among undergraduate students majoring in the biomedical sciences than among those studying arts (see for more references Kothtari et al. 2011). Tongue rolling has been studied in relation to rehabilitation of oral functions in patients with difficulty in swallowing due to brain damage. "Relearning of motor skills is an important part of rehabilitation for patients with acquired brain damage. The human tongue is involved in various essential and complex functions. A rapid and effective control of the tongue musculature is required in many daily oral functions like chewing, swallowing and speech. Motor task training causes a reversible functional reorganization in the human motor cortex. This phenomenon is called cortical plasticity" (Kothari et al. 2011). Tongue training schemes for rehabilitation are effective, but the applied methods are too complicated for the patients. Question was whether tongue rolling could be used as a more simple approach for tongue training and thus for easier rehabilitation or not. The outcome was that it did not influence tongue-training performance (Kothari et al. 2011).

Macroglossia, a too big tongue, contains the risk of blocking the airway. It disturbs oral hygiene, produces difficulties with speech and can be related to malocclusion. Operative reduction of the tongue is then carried out by anterior resection of tongue parts. On both sides of the midline tongue, tissue is removed sparing the neurovascular pedicle (Georgiou et al. 2013). One should expect sensory loss, but a case report states: “no sensory loss of any category tested (static tactile threshold, vibration sense, static 2-point discrimination, pain threshold and taste) was observed after tongue reduction” (Matsumoto et al. 2014).

Only recently a family with verbal dyspraxia (impaired ability to carry out the coordinated movements that are needed for speech) has been genetically studied to define their chromosomal mutation. The mutation was related to the long arm of chromosome 7 (7q31). The mutated gene was *FOXP2*. *FOXP2* plays an important role in the development of structures related to speech. This gene mutation utters deficiencies in speech performance (Varga-Khadem et al. 2005). Speech needs a highly coordinated cooperation between different anatomical areas innervated bilaterally by six different cranial nerves (V, VII, IX, X, XI, XII). Moreover, respiratory activity has to be synchronized. This asks for the involvement of paired spinal nerves (Neef et al. 2011). Auditory system, motor and somatosensory cortex, cerebellum, basal ganglia and the thalamus all are involved in speech. Moreover, speech perception is linked to speech production (for overviews, see, e.g. Guenther and Hickok 2016; Hickok and Poeppel 2016).This coordination of the involved neuronal networks occurs mainly in central nervous system and is beyond the scope of this book.

Surprisingly, persistent stuttering is related to abnormal cortical activity of the tongue motor area (Neef et al. 2011). Cortical tongue areas remain to astonish: “recently, it has been suggested that the tongue deviates towards the language-dominant cerebral hemisphere on full protraction. This suggestion is based on the important role the tongue plays in articulation and on the close anatomical relationship between the cortical tongue motor area and the motor speech area. It was proposed that this phenomenon could serve as a reliable and simple method for language brain lateralization. However, this hypothesis is still open for verification”. And the answer of this research is: “we conclude that, at least in Jordanians, tongue deviation cannot be considered as a reliable indicator for language lateralization” (Ghaida et al. 2014), as one could have expected! A more acceptable suggestion is that the tongue is motoric related to mental concentration. Children intensely concentrated on their (class) work protrude their tongue, while the football player shooting with his right leg protrudes his tongue even into the direction of the shooting leg. This is called synkinesis or associated automatic movement. It can be caused in other cases by a wrongly wiring of the nerves after an accident (see Sect. 12.4. Crocodile tears).

The tongue is bilateral innervated by the hypoglossal cranial nerves. One-sided damage of the hypoglossal nerve has nearly no influence on speech, but results in a flaccid paralysis of the affected side with atrophy. First, a difficulty should be mentioned: the neuroanatomical studies indicate one-sided contralateral innervation of the hypoglossal motor neurons by the pyramidal tract (Voogd 2008). Older neurophysiology suggests bilateral pyramidal innervation. Focal transcranial magnetic cortex stimulation of the tongue motor cortex shows that both the ipsilateral and contralateral tong parts are activated: the contralateral part more than the ipsilateral tongue part. After stimulation of the hypoglossal nerve main unilateral effects are noticed, while cross-talking was due to "muscle fibres traversing in a lateral direction through the tongue. Following focal stimulation over the lateral motor cortex of one hemisphere, the sizes of the ipsilateral responses ($n = 21$ subjects) were on average 77% of the contralateral responses. This gives evidence for true ipsilateral corticofugal projections when the significantly smaller amount of cross-talk (*P,* 0.01, Wilcoxon test) after unilateral electrical nerve stimulation is considered" (Meyer et al. 1997). Problem has been solved one should think, only ipsilateral pyramidal projections on the hypoglossal motor neurons are present. However, neonatal rat studies indicated that the end branches of the hypoglossal nerve cross the midline towards the other tongue half, thus peripheral contralateral innervation (Fukushima et al. 2015). This fact was already known for years by anatomists for the human situation (see Braus 1929).

6.4 Supra- and Infrahyoidal Muscles

The best known of the tongue external muscles is the *genioglossus* muscles with an "empty" areolar space between them. The genioglossus muscles originate from the mandible inside by bony elevations known as the genial tubercles and move into the tongue. (Note that the *geniohyoideus* belongs to the suprahyoidal muscles.) The M. hyoglossus makes a direct connection between the hyoid and the tip of the tongue. The M. styloglossus starts at the processus styloideus, goes towards the lateral side of the tongue and from the lateral side reaches the tip of the tongue (Fig. 6.7). The M. palatoglossus starts at the border soft to hard palate and bends downward to the tongue base. It adheres to the M. transversus linguae fibres. These bilateral palatoglossus muscles form a closed ring around the tongue base. These muscles are responsible for the palatoglossal ridge at the fauces, where it forms the anterior pillar and co-constructs with the palatopharyngeal ridge, the tonsillar fossa.

All these intrinsic and extrinsic tongue muscles are innervated by the hypoglossal nerve except the palatoglossus muscle.

The hyoid can be stabilized due to the presence of the main suprahyoid muscles (digastric and stylohyoid) and infrahyoid muscles (omohyoid and sternohyoid). Their concerted action can open the mouth provided stabilization of the hyoid and of the head on the vertebral column is established. Stabilization of the thyroid can be managed by action of the thyrohyoid and the sternothyroid. The sternohyoid, omohyoid, thyrohyoid and sternothyroid are all innervated by the ansa cervicalis (C1–C3) loop (see Sects. 14.1/6.2). The stylohyoid and the posterior belly of the digastric muscle are innervated by the cranial nerve VII, while the anterior belly of the digastric muscle is innervated by the trigeminal cranial nerve, by its mylohyoid nerve branch.

6.5 Substantia Nigra and the Oral Cavity

For many years it has been known that the substantia nigra is involved in oral movements and in oro-facial dyskinesias (impairments of voluntary movements; Mora et al. 1977; De Deurwaerdere and Chesselet 2000). The trigeminal nuclei receive afferent inputs from an unexpectedly large number of brain nuclei (for reviews, see Usunoff et al. 1997; Lazarov 2000; Waite and Ashwell 2004). The substantia nigra in rats sends projection fibres to the reticular formation around the trigeminal motor nucleus, but the axons of this non-dopaminergic pathway do not enter the territory of this nucleus. Rather, the substantia nigra influences this nucleus via a multisynaptic pathway (Yasui et al. 1992, 1997). Evidence that afferents on the sensory mesencephalic trigeminal nucleus neurons originate from cells in dopaminergic areas, i.e. in the substantia nigra, ventral tegmental area and the medial hypothalamus, was provided by Copray et al. (1990). A direct bilateral connection between the substantia nigra the sensory trigeminal nucleus, a unique collection of pseudounipolar proprioceptive primary trigeminal afferent neurons was found located within the brainstem (reviewed in Lazarov 2000). Until now, it has been believed that the substantia nigra influences the trigeminal motoneurons via a multisynaptic pathway (Yasui et al. 1992, 1997). The results of our study (Marani et al. 2010) provide strong evidence that the substantia nigra also directly innervates the proprioceptive trigeminal neurons and, thus, both the motor and sensory neurons controlling jaw muscles involved in mastication. Since pseudounipolar mesencephalic trigeminal nucleus neurons send axons to the pontine and spinal trigeminal nuclei, it appears that the entire trigeminal nuclear complex is profoundly influenced by the substantia nigra. Therefore, it can be inferred that inputs from this area possibly modify, modulate or interact with outputs from all these nuclei to control the masticatory behaviour.

6.6 The Soft Palate

For most people, including surgeons working outside the head and neck area, the projections of the neck structures onto the pharynx and oral space remain problematic. The first arcus of the pharynx (arcus palatoglossus) is located exactly below the lower edge of the atlas, which means that the palate is located at the atlas, thus at C1 height. The mylohyoideus muscle in the new born is projected at C3 and in mature situations somewhat lower at border C4/C5 (Figs. 6.8 and 6.9). There is a slight position difference between adult male and adult female. It concerns a half

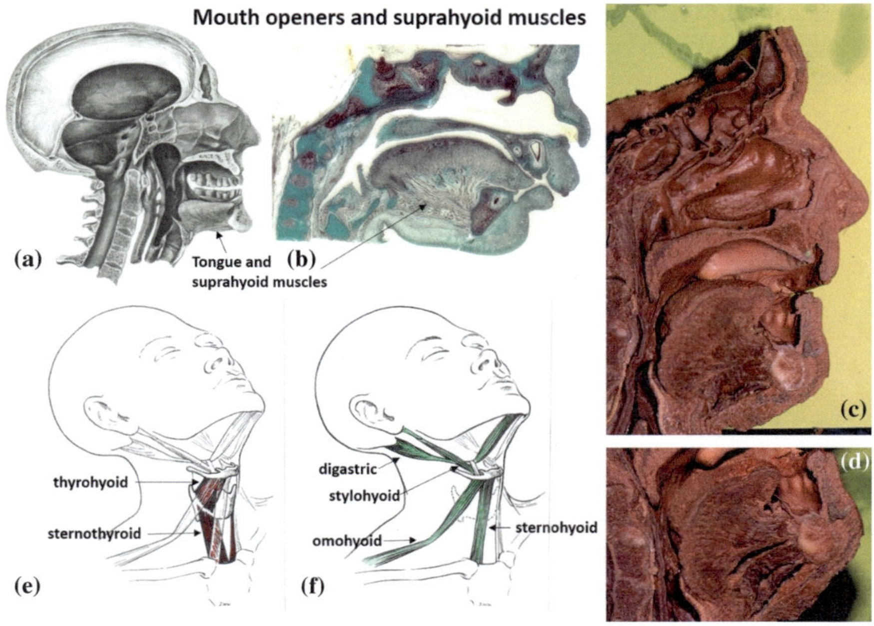

Fig. 6.8 Suprahyoidal muscles (**a**–**d**) and the hyoid stabilizers and mouth openers (**e**–**f**). A shows the position of the tongue and the suprahyoidal muscles in a semiskeletal picture of head and neck. **b** Is a histological section through the oral area of a 4-month-old human specimen. Arrow indicates the geniohyoideus muscle. **c** and **d** are midsagittal dissections of the mature human nasal and oral areas (made by W. Reychard, courtesy Dept. Anatomy and Embryology, LUMC with permission). Note that in **d** are present the tongue, directly related to the genioglossus muscle and below and separated the geniohyoideus. The mylohyoideus is present underneath. Note the different direction of its fibres. Within the subcutaneous fat, the fibres of the platysma can be discerned. **e** and **f** Demonstrate the main stabilizers and mouth openers and their relation to the hyoid and thyroid

vertebra height that projects its borders higher in the adult female (Lanz and Wachsmuth 1955). While the palate is placed horizontally in the human, it starts during development in a vertical position. The antero-posterior and the oro-nasal axes are organized by a series of genes: *Msx1, Efnb1, Shox2, Fgf10* and *Bmp4* are limited to the anterior region, whereas *Barx1, Fgfr2, Meox2, Mn1, Pax9* and *Tbx22* belong to the posterior region, and in the oro-nasal axis *Smad1/5/8, Dlx5, Fgf7, Fgfr2* and *Osr2* are related to the nasal side, and *Fgf10, Gli1, Osr1* and *Osr2* are expressed on the oral side. The bony palate is determined by bone morphogenetic protein (*BMP*) at the nasal side and the soft palate by *SHH* at the oral side (see Parada and Chai 2016). Although the genes in palate development involved are now known, the general development has been described adequately earlier in relation to harelip and facial clefts (see also Fig. 4.7 and Sect. 4.6).

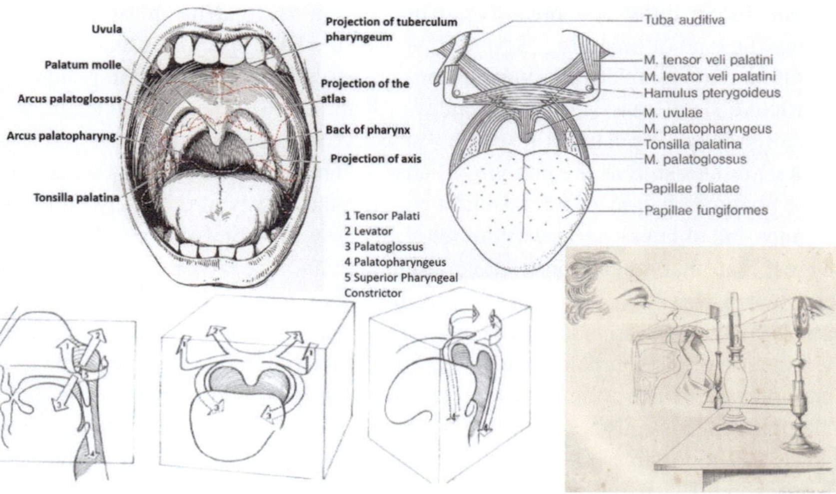

Fig. 6.9 View into the mouth cavity. The oropharyngeal isthmus is located directly behind the mouth cavity, which is bounded superiorly by the soft palate, laterally by the palatoglossal arches and inferiorly by the tongue. Note that the arcus palatopharyngeus is officially placed outside the mouth cavity. In between both arches, the tonsil is localized. The muscles involved in the soft palate (Samandari and Mai 1995, with permission) are indicated together with the forces exerted by these muscles (see cubes, Fritzell 1969 with permission). The "Laryngoskopie" figure is from Czermak's "Monographie" (1860)

6.6.1 Anatomy and Function of the Soft Palate Muscles

The view into the mouth (Fig. 6.9) shows part of the back of the upper pharynx. However, its lower pharyngeal part with the vocal cords can only be visualized by laryngoscopy. Czermak (1828–1873) popularized the "Kehlkopfspiegel" (the pharynx mirror, Fig. 6.9) by a series of publications and a monography (1860) that drew attention to laryngoscopy. Since the function of the soft palate is codetermined by several pharyngeal structures (e.g. pharyngeal constrictor muscle, Fig. 6.9), especially in speech, the development of the laryngoscope technique also extended the possibility to study the velopharyngeal functions (see Lanz and Wachsmuth 1955, pp. 313–316). An adequate velopharyngeal closing and opening mechanism is essential for clear speech. It involves elevation and dorsal movement of the soft palate, medial movement of the lateral pharyngeal walls and ventral bulging of the posterior pharyngeal wall (see cubes in Fig. 6.9).

One has to realize that the superior pharyngeal constrictor muscle contains several parts (pars pterygopharyngeal, buccopharyngeal, mylopharyngeal and glossopharyngeal) that differently contribute to its function of narrowing or stiffening the pharynx and forward bulging of the posterior pharyngeal wall (Fig. 6.9). Beside speech, the muscle contributes to swallowing and to airway closure during deglutition. Respiratory-related superior pharyngeal constrictor activity was found present during quiet oral breathing, while others say that the muscle is inactive during this type of breathing. Several investigators have speculated that this muscle activation may stimulate pharyngeal airway narrowing during sleep in patients with obstructive sleep apnoea (see references in Kuna 2000). These problematic results are partly solved by Kuna's research: "selective activation of the pharyngeal constrictor muscles stiffens the pharyngeal airway. The results also show that activation of the pharyngeal constrictor muscles constricts the airway at relatively high airway volumes but dilates the airway at relatively low airway volumes. These results suggest that pharyngeal constrictor muscle activation at the end of an apneic episode, when airway volume is relatively low, may help restore airway patency in patients with obstructive sleep apnoea". The pharyngeal muscles that contribute to the function of the soft palate are researched, but the outcomes are from time to time contradictory. Since "the fibres of the superior pharyngeal constrictor and the palatopharyngeus muscles are often so hopelessly entangled in the human subject that it is impossible to say where one muscle begins and the other ends" (Townshend 1940) indistinctness increases. This contributes little to the understanding of the function of the soft palate in mastication, swallowing, deglutition and speech.

The muscles of the soft palate, included the palatopharyngeus, are described below. The palatopharyngeus muscle, located behind the tonsil (Fig. 6.9) and its longitudinal fibres that run into the Eustachian tube cartilage, called the salpingopharyngeus muscle, displaces the soft palate dorsocaudally at contraction, the lateral pharyngeal walls are shifted medially and the posterior pharyngeal wall directs forward and upward. The palatoglossus muscle lowers the palate and is the antagonist to the levator veli palatini.

The levator veli palatini muscle can only be discussed in relation to the uvular muscle. If the levator muscle contracts, the medial portion of the soft palate is moved up and backwards (Fig. 6.9). The function of the uvular muscle was mysterious till the 1980s. Due to the cleft palate surgery attention arose for the uvular function in speech and swallowing in both normal and cleft palate humans. The convexity of the dorsal surface of the velum is essential for normal velopharyngeal function. This convexity is mainly kept by the levator veli palatini and also, but less, by the musculus uvulae. This muscle in fact consists of two muscle strands and is paired due to a septum (Azzam and Kuehn 1977; Landes et al. 2011; Older references doubted the bilateralism of the muscle). A concave or flat nasal velar surface is related to symptoms of velopharyngeal dysfunction including hyper nasality and nasal air emission (see also part 7.4.2; Inouye et al. 2016 and references herein). The uvular muscle adds stiffness to the velum and contributes to rotate the velum towards the posterior pharyngeal wall. 3D model studies of the uvular muscle showed: it acts as a space-occupying structure and a velar extensor. "As a space-occupying structure, the uvular muscle helps to nearly triple the midline velopharyngeal contact length. As a velar extensor, the musculus uvulae acting alone without the levator veli palatine muscle decreases the velopharyngeal distance 62%. Furthermore, activation of the musculus uvulae decreases the levator veli palatine muscle activation required for closure almost threefold, from 20% (without uvular muscle) to 8% (with uvular muscle)" (Inouye et al. 2016).

The tensor veli palatini muscle has a peculiar position since his tendon coils around the hamulus of the pterygoid. The Hamulus functions as a hypomochlion (the fulcrum of the lever) for the muscle. Dimensions of the Hamulus known are length is 7 mm, and distance between the tips of the hamuli is nearly 40 mm (Krmpotić-Nemanić et al. 2006). Function of the tensor veli palatini is large during deglutition and swallowing, but its function during speech is debated still. By contraction of the tensor veli palatini, the palate aponeurosis is pulled tight and the Eustachian tube opens. This function has been denied by Fritzell (1969) and Brescia (1971), while Van Gelder (1965) supports the opposite of narrowing the entrance of the tube. Recent research involves more muscles in opening of the Eustachian tube. They are the tensor veli palatine, levator veli palatine, the superior pharyngeal constrictor muscles and even the buccinator muscle (Okada et al. 2017). The palatoglossus muscle is the continuation of the transverse fibres within the tongue (Fig. 6.7) that ascends to the lateral side of the palatal aponeurosis. It organizes a mucomuscular fold seen at the surface as the arcus palatoglossus (Figs. 6.9 and 6.10c). Palatoglossus action lowers the palate.

6.6.2 *Innervation of the Muscles of the Upper Pharynx and Soft Palate*

The innervation of the soft palate knows few studies that investigate these muscles in humans. The review article of Logjes et al. (2016) weakens the information of animal studies for understanding human soft palate innervation: "studies on the motor nerves to the soft palate muscles in animals had very conflicting results and seem to be less useful as a model for the human situation" (Logjes et al. 2016). Four to five cranial nerves have to be discussed in relation to the innervation of the soft palate: branches of the trigeminus nerve, of the facial nerve, of the glossopharyngeus nerve, of the accessory nerve and of the sympathetic superior cervical ganglion. Since sympathetic ganglion connections exist with the hypoglossal nerve (Pick 1970), this nerve will be included since the nerve is used as a step to sympathetically innervated palatine structures. The general pattern for the innervation of the palate can be found in most study books and on the Web: sensory innervation goes over greater and, one or more, lesser palatine nerves (N. palatinus major and NN. palatini minores), while the glossopharyngeal nerve also contributes to the innervation of the soft palate. The motor innervation of the palate muscles is by the accessory cranial nerve except the tensor veli palatine, which is taken care by the mandibular nerve.

However, some doubt on this scheme exists. It is generally accepted that the tensor veli palatine muscle is innervated by motor fibres in a mainly sensory maxillary branch of the trigeminus nerve. The N. pterygopalatinus or also named

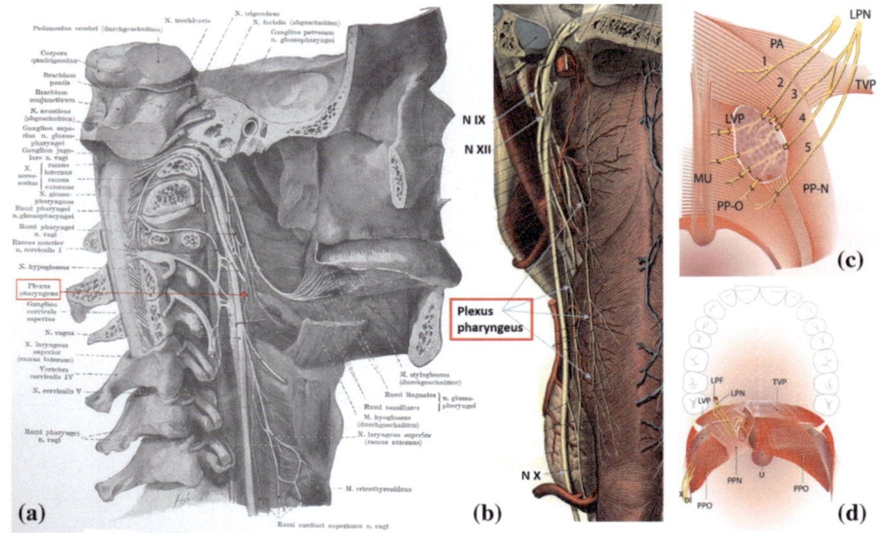

Fig. 6.10 Innervation of the soft palate. **a** Is the dissected laryngopharynx to show the nerves. The pharyngeal plexus (red) is indicated in relation to the glossopharyngeus and vagus nerves and the near located sympatethic trunk. **b** Preparation of posterior part of the pharynx to demonstrate the plexus pharyngeus: revealed are the hypoglossus nerve, the vagus nerve and the anastomoses with the glossopharyngeus nerve. **c** "Superior view on soft palate with the innervation by the lesser palatine nerve. A part of the LVP is removed for showing the five branches of the lesser palatine nerve. These nerves run underneath the palatine aponeurosis and the nasal part of the PP and penetrate the inferior-velar part of the LVP on its lateral surface close to the insertion of the LVP in the midline of the velum. LPN lesser palatine nerve, LVP levator veli palatini muscle, PA palatine aponeurosis, PP-N palatopharyngeus muscle nasal strand, PP-O palatopharyngeus muscle oral strand, TVP tensor veli palatini muscle, MU uvulae muscle". **d** "The dual innervation of the soft palate by the LPN and the pharyngeal plexus. View of the plastic surgeon on the soft palate; the pharyngeal plexus penetrates the superior-extravelar part of the LVP on the lower lateral border. The lesser palatine nerve runs through the lesser palatine foramen and runs over the palatine aponeurosis of the TVP and the nasal part of the PP to enter the inferior-velar part of the LVP on its lateral surface. Here, the LPN innervates the small inferior-velar part of the LVP and the anterior part of the oral part of the PP, together referred to as the anteromedial region of the soft palate muscles. LPF lesser palatine foramen, LPN lesser palatine nerve, IX glossopharyngeal nerve, X vagus nerve, TVP tensor veli palatine muscle, LVP levator veli palatini muscle, PP-N palatopharyngeus muscle nasal strand, PP-O palatopharyngeus muscle oral strand, U uvula". (**a** is from Spalteholz (1932), **b** excision from figure of Sobotta (1907) and the part figures **c** and **d** are, together with text parts, taken from Logjes et al. (2016) with permission)

N. sphenopalatinus passes the ganglion pterygopalatinum and takes up parasympathetic fibres from it. Its continuation partially continues in the hard palate as the N. palatinus major and penetrates in its trajectory through the hard palate by a canal that finishes in the great palatine foramen. In the canal, the lesser palatine nerves branch off and arrive by their own small canals at the lesser palatine foraminae (Fig. 6.10c, d). The amount of lesser palatine foraminae is variable from 1 to 5 or 6. In most cases, however, 1 or 2 are found (Lanz and Wachsmuth 1985). The lesser palatine nerves are divided into medial and posterior ones. The posterior goes along the Hamulus pterygoideus and turns back and also medial to innervate the oral part of the soft palate and the arcus palatopharyngeus. The sensory trigeminal and motor fibres of the ***facialis*** nerve present in the lesser palatine nerve also reach the palatopharyngeus, levator veli palatine and uvulae muscles. The lateral palatine nerve is mainly sensory and responsible for the proprioception of lateral parts of the soft palate and the tonsil area (Braus 1940; Podvinec 1952; Sedlácková 1967; Sedlácková et al. 1973).

(Early branches of the chorda tympani contain the motoric fibres of the M. levator veli palatini and uvulae. They leave the facialis nerve, pass through the ganglion oticum and by the nervulus sphenoidalis reach the nervus canalis pterygoidei (N. vidianus). Finally, by the N. petrosus superficialis major they find the ganglion sphenopalatinum and through this ganglion reach the NN. palatini minors [translated from Braus 1940]).

This involvement of the facial nerve leads to the dual innervation hypothesis that during speech the facial nerve and during swallowing the vagus nerve is active for the levator veli palatini muscle (Podvinec 1952; Sedlácková 1967; Sedlácková et al. 1973).

This brings up the involvement of the vagus nerve in the innervation of the soft palate muscles. In Fig. 6.10a, b, the intimate relation between the three nerves (NN. IX, X, XI) leaving the foramen jugulare (see Fig. 4.13) is also shown. The connection between the accessory and vagus cranial nerve by the ramus internus of the accessory cranial nerve brings accessory motor information into the vagus cranial nerve (Fig. 4.13). Since the plexus pharyngeus contains intimate connections between the glossopharyngeus and vagus nerve and this plexus contributes to the innervation of the soft palate, accessory–vagus information reaches the soft palate also by the glossopharyngeus (Fig. 6.10a–d). This was renewed and confirmed in the publication of Logjes et al. (2016). "Both the levator veli palatini and palatopharyngeus muscles receive motor fibres from the accessory nerve (through the vagus nerve and the glossopharyngeal nerve) and also the lesser palatine nerve" (Logjes et al. 2016). The constrictor pharyngeus superior muscle is innervated by the upper branches of the pharyngeal plexus (see Fig. 6.10a, b). The pharyngeal plexus takes up branches of the glossopharyngeus and vagus cranial nerve as indicated above. The topographic localization of the plexus branches makes reaching the palatopharyngeal and the levator veli palatine muscles possible at their lateral side (Fig. 6.10d). Therefore, the dual innervation of the soft palate by the lesser palatine nerves and the pharyngeal plexus is strongly supported by Logjes et al. (2016).

Attention to the hypoglossal nerve in relation to the superior sympathetic ganglion is given by Pick (1970): "the ganglion is connected with the glossopharyngeal, vagal and hypoglossal and with the upper 3 or 4 cervical spinal nerves, and it sends off muscular and osseous rami; it also provides visceral branches for the pharynx, the larynx (usually via the superior laryngeal nerve), to the thyroid gland

and to the carotid arteries, especially the ramifications of the external carotid". And somewhat further on: "the superior cervical ganglion connects with the vagus nerve and in front of the ganglion with the hypoglossal. These two nerves are also connected with each other by several strands".

Our interpretation is that the plexus pharyngeus contains sympathetic fibres, which is well known (Tubbs et al. 2015), and seemingly also encompasses hypoglossal contributions. In this respect, the hypoglossus paresis by carotis dissection (present in nearly 5% of all dissections and producing complaints) is interesting. The supposition is or expansion of the internal carotid that should oppress the hypoglossus nerve in the neck, or a disturbance of the microcirculation of the hypoglossus (Mokri et al. 1995; Hafkamp et al. 2004; Hennings et al. 2014).

6.6.3 Palate Myoclonus

Palate myoclonus, also called palatum myorithms, is a non-epileptic myoclonus and is characterized by rhythmic and complex synchronic striated muscle contractions of the palate with a frequency of 120–140/min (Hagen 1969). It was described for the first time by Müller in 1837 and Politzer in 1862 according to Frenken (1977) and it was Spencer (1886) according to Hagen (1969). Palate myoclonus was published in the literature with large- or small-time intervals and with a variety of names (Hagen 1969; Frenken 1977: rhythmic myoclonus, nystagmus du voile, Gaumensegelnystagmus and Hirnnervenmyorhythmie).

Palate myoclonus has been regularly related to olivary hypertrophy (for overviews, see Hagen 1969; Frenken 1977). In short, lesions or haemorrhages of the brainstem or cerebellum are always related to olivary hypertrophy, which is generally considered a degeneration of the inferior olive. First, detection of the relation between typical lesions and olive hypertrophy was already described in 1882 by Meyer. A series of publications restricted the lesions to damage of the tegmentum pontis with homolateral inferior olive hypertrophy or to cerebellar lesions, always involving the dentate nucleus, producing contralateral inferior olive hypertrophy. Bilateral inferior olivary hypertrophy was described after a shot wound involving lesions of the right-sided dentate nucleus and right-sided tegmentum pontis (Uemura 1917) clearly supporting the earlier findings. The correlation between palate myoclonus and these types of lesions and inferior olive hypertrophy came late. From around 1930 on several tens of articles have been published relating palate myoclonus to inferior olivary hypertrophy. The uncertainty between the connection of inferior olive hypertrophy and myoclonus around the 1930s has been well described by Guillain (1934). Sinnige (1938) studied the connections of the cerebellum in the dog and undeniably treated in his part the connections of the inferior olives (pp. 136–140) after cerebellar lesions. Towards the end, Sinnige still had to explain his "pseudohypertrophic degeneration of the inferior olives". He considered it a regenerative phenomenon: "how should we explain that this regeneration is only present by total extirpation and not by a one-sided one. According to me, the possible explanation is: during one-sided cerebellum extirpation, functional compensation is present by the sound homolateral olive cells of the degenerated ones of the other side, as we regularly encountered in other one-sided degenerations. The need of regeneration of the degenerated cells disappears. However, at total extirpation, at which the possibility of compensation has disappeared, an excitation of regeneration originates" (translated from Dutch). Sinnige's explanation hardly contributes to the understanding of olivary hypertrophy since one-sided cerebellum extirpation (including the cerebellar nuclei) should produce olivary hypertrophy.

The experimental approach of inferior olivary hypertrophy started with the research of Verhaart and Voogd (1962). Verhaart published in 1949 "The central tegmental tract", as a continuation of his 1936 paper. This tract was exclusively related to the red nucleus. The rubrospinal tract in humans was found inconstant and of "slight dimension", and "through the development of the frontal part of the red nucleus and the regression of the caudal part, this nucleus has changed its character from a pre-eminently motor nucleus to a nucleus connected to the olivocerebellar system" (Verhaart 1938). Within Verhaart's group, the rubro-olivary system was established and it was the basis for Verhaart and Voogd's (1962) experimental results. Olivary hypertrophy is judged a combined retrograde and antegrade transneuronal reaction, because the direct cerebello-olivary projections, the dentate-rubral projections and the Darkschewitsch-olivary projections coincide in the hypertrophic olivary areas (Fig. 6.11).

Olivary hypertrophy has been considered a degenerative phenomenon. Slowly, the notice arose that olivary hypertrophy could be caused by hyperactive cells and that it is primarily not a degenerative cellular type. The idea of hyperactivity has been suggested by studies on human material (Ben Hamida 1968; Rondot and Ben Hamida 1968). The support of the hyperactivity theory was based on the increase in diameter of the nucleolus and of the cell volume and on the absence of neuronal degeneration in the hypertrophic areas up to 524 days in the cat. An accumulation of dark, osmiophilic balls within the rough endoplasmic reticulum also supported the hyperactivity concept (Voogd and Boesten 1976). Since biochemistry is difficult to perform due to the small area involved and the borders of the hypertrophic areas are variable in different experimentally induced olivary hypertrophy, histo- and cytochemistry provides the solution (see Fig. 6.11). Histo- and cytochemistry showed that the osmiophilic accumulated substance was PAS negative and was negative in the Millon reaction for proteins and also negative for Sudan black staining. The acid anhydride method (Pearse 1980; Stoward and Burns 1968) coloured the osmiophilic balls red indicating high concentration of protein-bound carboxyls (Marani 1986). Low molecular proteins or proteins rich in terminal amino acid groups are identified by this method (Barrnett and Seligman 1958).

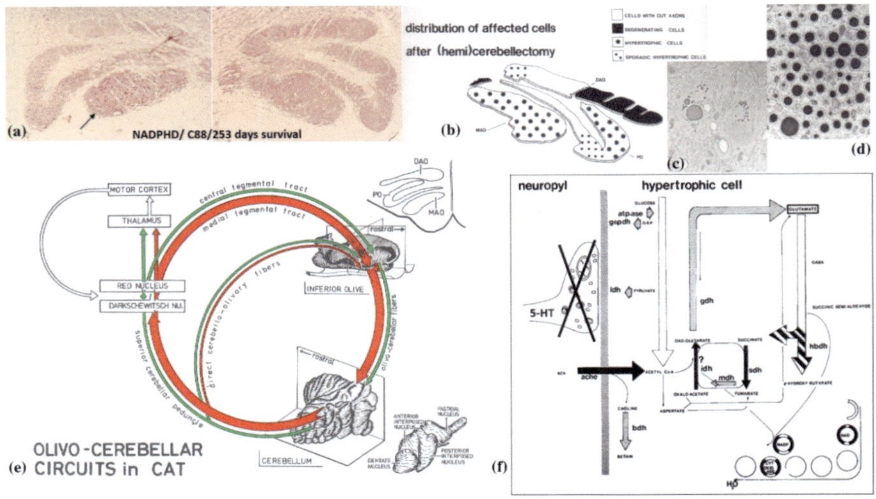

Fig. 6.11 Experimentally induced olivary hypertrophy in the **cat**. **a** Increased enzymatic activity of NADPHD after a survival time of 253 days is shown after a hemicerebellectomy present in the cat rostro-lateral medial accessory olivary (MAO) hypotrophic neurons, left side. **b** Next to it is the overview of the areas containing hypertrophic olivary neurons together with sporadic hypertrophic areas in the inferior olive. **c** Lightmicroscopical section treated for the anhydride technique. Positive dense globular inclusions are noted within the hypertrophic neurons indicating that accumulation of protein-bound carboxyls could occur within the hypertrophic cells in the cat (Marani 1986). **d** The electron micrograph shows the osmiophilic globular inclusions within the cat extended granular endoplasmic reticulum (courtesy AJP Boesten, see Boesten and Voogd 1985). **f** The neuronal connections involved in the inferior olivary complex: the inferior olive (top right) project to the cerebellar cortex as climbing fibres (called the olivocerebellar tract). Intrinsic cerebellar circuitry relays the signals through the cortex into the cerebellar nuclei (bottom right). The output system of the cerebellar nuclei is the superior cerebellar peduncle, which terminates in the contralateral red nucleus and Darkschewitsch nucleus. The cerebellar nuclei also have a direct projection back to the inferior olive. A closed loop exists because both red nucleus and Darkschewitsch nucleus project back to the inferior olive by the central and medial tegmental tract, respectively. The output of this closed loop is through axons of the cerebellar nuclei that project into the thalamus (courtesy J. Voogd). **f** The summary diagram of enzymatic and histochemical results (Marani 1986) shows the main pathways for the hypertrophic inferior olivary cells. Enzymes tested and scheduled in this diagram are stippled when enzyme reaction products did not differ. Black arrows indicate increased enzyme reaction products. Enzymes of both the ketone cycle and of the respiratory chain react in hypotrophic olivary cells with an increase of reaction products. The neuropil contained an increase in acetylcholinesterase activity and a decrease in serotoninergic innervation

Enzyme histochemistry demonstrated normal distribution of monoamine oxidase and an increase in acetylcholinesterase distribution in hypertrophic cells and neuropil. Esterases can be involved in the conversion of choline, and the last step of this conversion of choline to betaine showed no increase in hypertrophic cells or in the neuropil. β-Hydroxy-butyrate shows an increase in reaction product formation in exclusively the hypertrophic neurons. The enzyme diverts acetyl CoA derived from fatty acid or pyruvate oxidation. The lytic enzyme acid phosphatase or changes in the glucose catabolic pathway (lactate dehydrogenase) or pentose phosphate pathway (glucose-6-phosphate dehydrogenase) were not increased in hypertrophic cells or its surrounding neuropil. The same was found for tricarboxylic acid cycle (iso-citric dehydrogenase and malate dehydrogenase). Except succinic, dehydrogenase was increased in the perikaryal cytoplasm of the hypertrophic neurons. The glutamate pathway (glutamate dehydrogenase and succinic semi-aldehyde dehydrogenase) showed no increase of the reaction products. In the respiration chain, three enzymes were found to be raised compared to controls: NADH tetrazolium reductase, NADPH tetrazolium reductase and cytochrome oxidase (see Fig. 6.11; see Marani 1986, with permission). These results indicate that ketone utilization occurs in the hypertrophic olivary cells (β-hydroxy-butyrate and the Krebs cycle enzymes NADH, NADPH tetrazolium reductases and cytochrome oxidase were increased). "Recent studies in man and animals have shown that the brain in vivo turns in special circumstances to β-hydroxy-butyrate and aceto-acetate as it main substrates. The ketone bodies serve not only as source of energy derived from their oxidation to CO_2 and H_2O, but also contribute carbon residues to constituents of brain normally dependent on glucose metabolism, for example, phospholipids, cholesterol, and the amino acids glutamate, glutamine, aspartate and γ-aminobutyric acid" (Sokoloff 1973). This hyperactivity hypothesis has been sparsely supported by others (Welsh et al. 1998). The detection of the **reduction** of serotonin in the hypertrophic inferior olive supports its hyperactivity. Reduction of serotonin causes olivary neurones into rhythmic activity. This activity presumably contributes to certain forms of rhythmic myoclonus (Plancantonakis et al. 2000). The detection of this serotoninergic effect opened a new medical approach for myoclonus. "In fact, it has been stated that the discovery of the antimyoclonic action of the serotonin precursor 5-hydroxytryptophan was nearly as momentous as the discovery of the anti-Parkinson action of levodopa" (Welsh et al. 1998).

References

Azimi-Garakani C, Beardmore JA (1979) An association between tongue-rolling phenotypes and subjects of study of undergraduates. J Biosoc Sci 11:193–199

Azzam NA, Kuehn DP (1977) The morphology of musculus uvulae. Cleft palate J 14:78–87

Barrnett RJ, Seligman AM (1958) Histochemical demonstration of protein-bound alpha-acylamido carboxyl groups. J Biophys Biochem Cytol 4:169–179

Ben Hamida MEM (1968) Contribution à l'ètude anatomique du couple olivo-dentèle. Thesis, Paris

Boesten AJP, Voogd J (1985) Hypertrophy of neurons in the inferior olive after cerebellar ablations of the cat. Neurosci Lett 61:49–54

Braus H (1929) Anatomie des Menschen, Band I, Bewegungsapparat. J.Springer Verlag, Berlin

Braus H (1940) Anatomie des Menschen. Periphere Leitungsbahnen II. Band IV. Verlag J Springer Berlin

Brescia MJ (1971) Cleft lip and palate. In: Grabb WC, Churchill J (eds) Surgical, dental and speech aspects. A Publ, London

Copray JCVM, Liem RSB, Ter-Horst GJ, van Willigen JD (1990) Dopaminergic afferents to the mesencephalic trigeminal nucleus of the rat: a light and electron microscope immunocytochemistry study. Brain Res 514:343–348
Czermak JN (1860) Der Kehlkopfspiegel und seine Verwerthung fuer Physiologie und Medizin. Verlag W, Engelmann, Leipzig
De Deurwaerdere P, Chesselet MF (2000) Nigrostriatal lesions alter oral dyskinesia and c-Fos expression induced by the serotonin agonist 1-(m-chlorophenyl)piperazine in adult rats. J Neurosci 20:5170–5178
Frenken CWGM (1977) De niet-epileptische myocloniëen. Thesis Nijmegen University, The Netherlands
Fritzell B (1969) The velopharyngeal muscles in speech. An electromyographic and cinéradiographic study. Acta Otolaryngol Suppl 250:1+
Fukushima N, Yokouchi K, Kawagishi K, Karasawa M, Kuroiwa M, Moriizumi T (2015) Two specific tongue regions receive bilateral hypoglossal innervation: a study in neonatal rat pups. Arch Oral Biol 60:432–438
Gelder Van (1965) Het zachte gehemelte bij de spraak. Thesis, University Amsterdam, Publ. Erven F. Bohn NV, Haarlem, The Netherlands
Ghaida JA, Hani SB, Mustafa A, Eldwairi Q (2014) Deviation of the fully protracted tongue: is it a reliable indicator for language cerebral dominance? Med Hypot 83:270–272
Georgiou R, Dulguerov N, Ayliffe P (2013) Reduction of the tongue: how we do it. Technical note. Br J Oral Maxillofacial Surg 51:e13–e14
Guenther FH, Hickok G (2016) Neural models of motor speech control. Neurobiology of language. Elsevier. http://dx.doi.org/10.1016/B978-0-12-407794-2.00058-4
Guillain G (1934) L'olive bulbaire. C. Doin & Cie, Paris
Gürel G (2003) The science and art of porcelain laminate veneers. Quintessence Publ, Co Ltd, London
Hafkamp HC, van der Goten A, Manni JJ (2004) Unilateral spontaneous dissection of the internal carotid artery presenting as hypoglossal nerve palsy. Eur Arch Otorhinolaryngol 261:405–408
Hagen CJ (1969) Over hypertrofe olijfdegeneratie. Thesis University of Amsterdam
Hamilton WJ, Boyd JD, Mossman HW (1946) Human embryology. Heffer & Sons, Cambridge
Hennings JM, Hoehn D, Schoemann-Spaeth E, Weber F (2014) Painless hypoglossal palsy as an isolated symptom of spontaneous carotid dissection. J Stroke Cerebrovas Dis 23:1988–1990
Hickok G, Poeppel D (2016) Neural basis of speech perception. Neurobiology of language. http://dx.doi.org/10.1016/B978-0-12-407794-2.00025-0
Hisbergues M, Vendeville S, Vendeville P (2009) Zirconia: Established facts and perspectives for a biomaterial in dental implantology. J Biomed Mater Res B Appl Biomater 88:519–529
Inouye JM, Lin KY, Perry JL, Blemker SS (2016) Contributions of the musculus uvulae to velopharyngeal closure quantified with a 3D multi-muscle computational model. Ann Plast Surg 77(Suppl 1):S70–S75
Kothari M, Svensson P, Basic A, Christiansen B, Vigsø M, Truc L, Baad-Hansen L (2011) Influence of the ability to roll the tongue and tongue-training parameters on oral motor performance and learning. Arch Oral Biol 56:1419–1423
Krarup S, Darvann TA, Larsen P, Marsh JL, Kreiborg S (2005) Three-dimensional analysis of mandibular growth and tooth eruption. J Anat 207:669–682
Krmpotić-Nemanić J, Vinter I, Marusić A (2006) Relations of the pterygoid hamulus and hard palate in children and adults: anatomical implications for the function of the soft palate. Ann Anat 188:69–74
Kuna ST (2000) Respiratory-related activation and mechanical effects of the pharyngeal constrictor muscles. Resp Physiol 119:155–161
Landes CA, Weichert F, Steinbauer T, Walckzak L, Hasenfus A, Veith C, Schröder A, Fritsch H, Theegarten D, Wagner M (2011) Histology and function: Analyzing the uvular muscle. Cleft Palate-Craniofacial J 48:639–645
Lanz T, Wachsmuth W (1985) Kopf: Uebergeordente systeme. Lang J, Springer, Heidelberg

Lazarov NE (2000) The mesencephalic trigeminal nucleus in the cat. Adv Anat Embryol Cell Biol 153:1–103

Logjes RJH, Bleys RLAW, Breugem CC (2016) The innervation of the soft palate muscles involved in cleft palate: a review of the literature. Clin Oral Invest 20:895–901

Maat G, Van der Velde EA (1987) The caries-attrition competition. Int J Anthropol 2:281–292

Marani E (1986) Topographic histochemistry of the cerebellum. Gustav Fischer Verlag, Stuttgart, NY

Marani E, Lazarov NE, Heida T, Usunoff KG (2010) Nigro-subthalamic and nigro-trigeminal projections in the rat. In: XII MediterConf Med Biol Engineer Comp, MEDICON 2010, Chalkidiki, Greece, pp 184–198

Matsumoto K, Morita K-I, Jinno S, Omura K (2014) Sensory changes after tongue reduction for macroglossia. Oral Maxillofacial Surg 117:e1–e2

Meyer B-U, Liebsch R, Röricht S (1997) Tongue motor responses following transcranial magnetic stimulation of the motor cortex and proximal hypoglossal nerve in man. Electroencephalogr Clin Neurophysiol 105:15–23

Meyer P (1882) Ueber ein Fall von Ponshaemorraghie mit secundären Degenerationen der Schleife. Arch f Psych u Nervenkrankh 13:63

Moffat DB (1993) Lecture notes on anatomy. Blackwell Scientific Publications

Mokri B, Silbert PL, Schievink WI, Piepgras DG (1995) Cranial nerve palsy in spontaneous dissection of the extracranial internal carotid artery. Neurology 46:356–359

Mora F, Mogenson GJ, Rolls ET (1977) Activity of neurons in the region of the substantia nigra during feeding in the monkey. Brain Res 133:267–276

Neef NE, Paulus W, Neef A, Wolff von Gudenberg A, Sommer M (2011) Reduced intracortical inhibition and facilitation in the primary motor tongue representation of adults who stutter. Clin Neurophysiol 122:1802–1811

Okada R, Muro S, Eguchi K, Yagi K, Nasu H, Yamaguchi H, Miwa K, Akita K (2017) The extended bundle of the tensor veli palatine: anatomic considerations of the dilating mechanism of the eustachian tube. Auris Nasus Larynx. https://doi.org/10.1016/j.anl.2017.05.014

Özkurt Z, Kazazoğlu E (2011) Zirconia dental implants: a literature review. J Oral Implantol 37:367–376

Parada C, Chai Y (2016) Recent advances in palate and tongue development. Kaufman's atlas of mouse development supplement: with coronal sections. Doi: © 2015 Elsevier Inc. http://dx.doi.org/10.1016/B978-0-12-800043-4.00017-8

Pearse AGE (1980) Histochemistry theoretical and applied: analytical technology, 4th ed. Churchill Livingstone, London

Pick J (1970) The autonomic nervous system. J.B. Lippincott Co, Philadelphia

Plancantonakis DG, Schwarz C, Welsh JP (2000) Serotonin suppresses subthreshold and suprathreshold oscillatory activity of rat inferior olivary neurones in vitro. J Physiol 524:833–851

Podvinec S (1952) Physiology and pathology of the soft palate. J Laryngol Otol 66:452–461

Rondot P, Ben Hamida MEM (1968) Myoclonies du voile et myoclonies squélettiques. Etude clinique et anatomique. Revue neurol 119:159

Samandari F, Mai JK (1995) Funktionelle Anatomie für Zahnmediziner Band I und II. Quintessenz Verlag, Berlin

Sedlácková E (1967) The syndrome of the congenitally shortened velum. The dual innervation of the soft palate. Fol Phoniat 19:441–450

Sedlácková E, LašŤovka M, Šram F (1973) Contribution to knowledge of soft palate innervation. Folia Phoniat 25:434–441

Sinnige JLM (1938) Anatomisch onderzoek over de verbindingen van de kleine hersenen bij de hond. Noord Holl Uitg N.V. Mij Amsterdam

Sobotta J (1907) In: McMurrich P (ed) Atlas and text-book of human anatomy, vol 3. W.B. Saunders Company, Philadelphia and London

Sokoloff L (1973) Metabolism of ketone bodies by the brain. Ann Rev Med 24:271–280

Spalteholz W (1932) Handatlas der Anatomie des Menschen, vol 1. Verlag S. Hirzel, Leipzig

Spencer HR (1886) Pharyngeal and laryngeal "nystagmus". Lancet 2:702, 758
Stoward PJ, Burns J (1968) Some evidence confirming the specificity of Barnett and Seligman's technique for demonstrating side-chain carboxyl groups in proteins. Histochemie 13:7–10
Tecco S, Baldini A, Mummolo S, Marchetti E, Giuca MR, Marzo G, Gherlone EF (2015) Frenulectomy of the tongue and the influence of rehabilitation exercises on the sEMG activity of masticatory muscles. J Electromyogr Kinesiol 25:619–628
Townshend RH (1940) The formation of Passavants bar. J Laryngol 55:1554–1561
Tubbs RS, Shoja MM, Loukas M (2015) Anatomy of the glossopharyngeus nerve. Nerves Nerve Injuries 1:373–381. (Elsevier, Amsterdam, NY). http://dx.doi.org/10.1016/B978-0-12-410390-0.00027-5
Uemura H (1917) Pathologisch-anatomische Untersuchungen ueber die Verbindingsbahnen zwischen dem Kleinhirn und den Hirnstam. Schweiz Arch Neurol Psychiatr 1:151
Usunoff KG, Marani E, Schoen JHR (1997) The trigeminal system in man. Adv Anat Embryol Cell Biol 136:1–126
Van der Merwe AE, Steyn M, Maat GJR (2010) Dental health of 19th century migrant mineworkers from Kimberley, South Africa. International Journal of Osteoarchaeology. https://doi.org/10.1002/oa.1143
Varga-Khadem F, Gadian DG, Copp A, Mishkin M (2005) FOXP2 and the neuroanatomy of speech and language. Nat Rev Neurosci 6:131–138
Verhaart WJC (1936) Die Zentralen Haubenbahn bei Affen und menschen. Schw Arch Neur u Psych 38:270–283
Verhaart WJC (1938) The rubrospinal system with monkey's and man. Psych Neurol Bl 42:335–343
Verhaart WJC (1949) The central tegmental tract. J Comp Neurol 90:173–192
Verhaart WJC, Voogd J (1962) Hypertrophy of the inferior olives in the cat. J Neuropath exp Neurol 21:92–103
von Lanz T, Wachsmuth W (1955) Praktische Anatomie, Hals. Teil I, 2, Springer Verlag, Berlin
Voogd J (2008) Motor systems. In Nieuwenhuys R, Voogd J, van Huijzen C (eds) The human central nervous system. Springer, Heidelberg
Waite PME, Ashwell KWS (2004) Trigeminal sensory system. In: Paxinos G (ed) The rat nervous system, 3rd edn. Academic Press, San Diego, pp 1093–1124
Welsh IP, Chang B, Menaker ME, Aicher SA (1998) Removal of the inferior olive abolishes myoclonic seizures associated with a loss of olivary serotonin. Neurosci 82:879–897
Wharshafsky D, Goldenberg D, Kanekar SG (2012) Imaging anatomy of deep neck spaces. Otolaryngol Clin N Am 45:1203–1221
Yasui Y, Nakano K, Nakagawa Y, Kayahara T, Shiroyama T, Mizuno N (1992) Non-dopaminergic neurons in the substantia nigra project to the reticular formation around the trigeminal motor nucleus in the rat. Brain Res 585:361–366
Yasui Y, Tsumori T, Ono K, Kishi T (1997) Nigral axon terminals are in contact with parvicellular reticular neurons which project to the motor trigeminal nucleus in the rat. Brain Res 775:219–224

Chapter 7
Cut off Your Nose to Spite Your Face: Face, Nose, The Paranasal Sinuses and Smell Related to Taste

This chapter considers facial attractiveness and face physiognomy. It studies human face detection that is compared to automated face recognition. The relation between facial skin colouration and blushing or flushing is defined. The nose is portrayed in art, in anatomy, in evolution and in development. Blood vessels of the face, nose bleeds, nose fractures and medieval rhinoplasty are described. Attention is given to: smell related to taste, stereo smelling, sinuses, the nasal cavity with its airflow and fast olfactory uptake of therapeutics by bulk flow. The chapter summarizes the immunology of mucosal epithelium.

7.1 Facial Attractiveness, Physiognomy and Its Consequences

The question "What makes a face attractive" (Rhodes 2006; see also Thornhill and Gangestad 1999) has been studied frequently. Three main characteristics are emerged: averageness, symmetry and sexual dimorphism. There are of course other characteristics [pleasant expression, youthfulness, liking a person (Rhodes 2006)], but most studies looked for attractiveness, assuming that there is general aesthetic blueprint for it. Averageness is the intersection of facial characteristics of a population. It was found that the more average a face is, the more it is perceived attractive. In other words, the more it is "common" face for that population is more attractive. There are indications that this also holds for non-Western populations and no difference has been found between men and women in this respect. Symmetry contributes to attractiveness, and significant gender or race differences are absent. Sexual dimorphism is the fact that female faces are different from male ones, induced genetically and hormonally. Feminine characteristics (small chin, full lips and high cheekbones) are attractive and are race-independent. Masculine facial characteristics do contribute to attractiveness but clearly less than facial femininity. "Particularly relevant are the results showing that the perception of human beauty appears *largely*

E. Marani and C. Heida, *Head and Neck*,
https://doi.org/10.1007/978-3-319-92105-1_7

shared by people different in culture, ethnicity and age, and thus supposedly data-driven. These findings are a rationale in computer techniques that attempt to emulate the human perception on the base of objective facial features" (Laurentini and Bottino 2014). These objective features are also a rationale for the plastic surgeon to produce a general accepted attractive face in which the nose plays an important role. Since the nose (co)-determines the beauty of the human face (Glaser 2002), the consequent extra attention for the nose by plastic surgeons is expressed in their "Rhinoplasty Society": an organization of those plastic surgeons that have specialized in human nose surgery. There is no square mm of the human nose that has not been studied by these plastic surgeons.

The Nose: A profile of sex, beauty and survival by Gabrielle Glaser (2002) is described as: "a fascinating tour of its subject through history and biology, art and culture, sex and sensibility, sickness and health". However, it is a voyage through *Anglo-Saxon* olfactory research and statistics with lively examples. Her chapter "The Sinus Business" on sinusitis says that in the 1993–1998 periods there were at least 37 million cases of chronic sinusitis in the USA and sinusitis was responsible for 12 million doctor's visits and 1.2 million hospital admissions. Nasal surgical and hospital costs are 3 billion dollar and 15 billion on over-the-counter-care. Sinusitis causes 73 million lost workdays. Surgical interventions by ear, nose and throat surgeons are "often" useless, but still carried out. "The revision rate on sinus surgery is estimated to be anywhere from 20 to 50%". At the other side of the spectrum is fragrance. Fragrance retail sales have increased from $250,000 in 1950 to $6 billion in 2000 in the USA, and "the notion of scented products has drifted into household cleaning products". These data demonstrate, at least, the clinical and scent-smelling importance of the nasal cavity. A few general data concerning the nose and olfaction are: the total surface area of the human nasal cavity is 150 cm^2, and the olfactory region is 10 cm^2 (in dogs, it is 150 cm^2). The sensory neurons of the olfactory epithelium have a lifetime of 30–90 days, averaged to 40 days. A constant neuro-genesis is responsible for the renewal of the sensory neurons. This property is only present in the olfactory epithelium. Presumably, we do have over 1000 different odorant genes coding at least for 1000 different protein receptors. Together, these receptors can discern 10,000 different odours using 5 million olfactory receptor neurons (a dog has over 200 million). In 2004, Richard Axel and Linda Buck were awarded the Nobel Prize in Physiology or Medicine for their description of the odorant receptor genes. One is hardly engaged in daily life with these scientific aspects. Our perception of the nose as a bodily structure is mainly determined by cultural aspects and by its beautiful form or ugly appearance, but hardly by its anatomical and physiological features, unless we caught a cold or get a sinusitis.

7.1.1 Face Philosophy and Emotions

The human face, as it is perceived by us, is mainly determined by its bony and cartilage structures. Although soft tissue parts can differ per individual, they have a minimal effect on the relations between the separate parts of the face in healthy people. Although all mammals do contain a face, one does not deny that ants or bees possess a face. Fish, lacking a nose, is perceived as having a face. Seemingly, the entity "face" is species independent. Even a 45-year-old woman with a shotgun blast into her face, destroying the nose, midfacial skeleton, lower eyelids and partly the upper lip is still perceived having a face. Seemingly, a face stays a "face" even if parts are missing. **So, what is a face?**

Cartooning needs understanding of the relations of the face parts, eyes, nose and mouth. Instructions for placing simple points within an oval or circle induce face images in our brains.

It is clear from the first row of the middle section (Fig. 7.1) that the types of signs used for eye, nose and mouth are irrelevant, as is the surrounding circle, square or triangle in the second row. The third row shows that missing parts do not influence the brain imaging of a face. Even a sign of a nose alone in a circle is perceived as a face, and the lowest row shows that bordering the eye, nose and mouth is unnecessary. However, turning the structures upside down suddenly denies the face perception (fourth row, right side).

If an animal notices a part of a prey, his brain will code for the whole prey. If we hear the first notes of a piece of music, we can sing the whole song. If we smell the perfume of our love, we can notice his/her whole body. The question of how the visual, auditory and other areas of your brain influence one another to produce a combined brain image of a single object part is called the **binding problem**. Binding of a perception depends on precisely simultaneous activity in various brain parts. The relative weight of the activation of a neuron involved in binding depends on its place within the neuronal network. A network with four neurons and five activation levels (the activation level can represent, for example, firing rate or firing

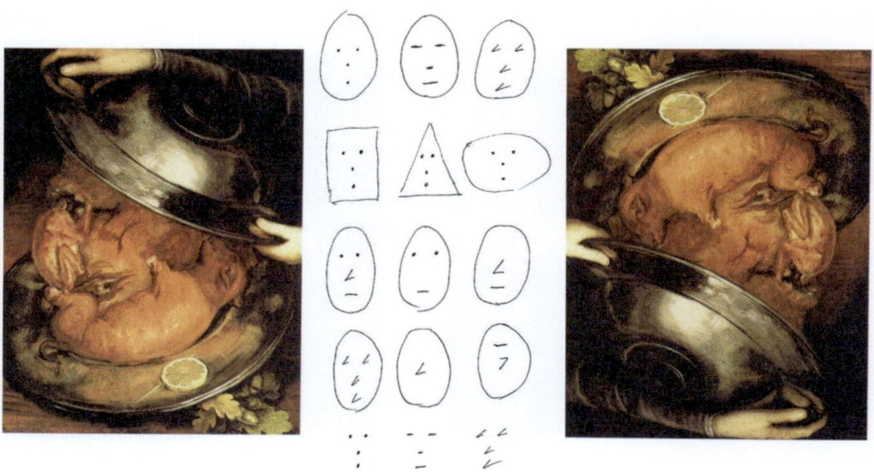

Fig. 7.1 Cartooning the face. Middle section, cartooning instructions given by J. Hamm (1991) that have been exercised in the relations of facial structures. Left and right panel from Arcimboldo: 1527–1593, so-called Mooney pictures (Mooney 1956)

pattern) has $5^4 = 625$ combinations. Cortex of 1 mm^3 contains 10^5 neurons and 10^9 synapses. So, the binding capacity is unlimited.

The famous Arcimboldo's (1526–1593) painting of the head and face of "autumn" is constructed by fruits (not shown). The arrangement of the fruits indicates a head and highlights certain parts of a head, e.g. the eye/nose. The process of binding produces a representation of a head into your brain. Structures put upside down inhibit this binding by the brain. Mooney faces are named after the investigator who used the comparison with inverted pictures for the first time (Mooney 1956). Flipping over the left panel (Fig. 7.1) shows you a face that you did not recognize the first time (left versus right panel, Fig. 7.1). Your cortex barely responds at the first picture, but at the second one the cortex produced activity at a frequency of 30–80 action potentials per second: known as gamma waves. You used binding to see a face in the grilled pig, quail and turkey (see also Rossion et al. 2011).

Now, one can ask: do we indeed notice a face? Or do we notice a configuration, say relations between several structures: the triangle formed by the eyes and nose, the triangle of eyes and mouth and for a nose alone you need a bordering of the space (here a circle). As long as these configurations are somehow remained intact, binding organizes the image or idea of a face in your brain. There is a strong analogy with automated face recognition. Later on (part Sect. 7.1.6), we will see that automated face recognition uses a mathematical type of "binding".

Special brain structures are involved in face recognition. This is not the place to describe extensively these brain structures, but see Fig. 7.2 for a neuroanatomical overview of face and of object recognition. The fusiform area is localized in the lower edge of the temporal lobe of the brain and is selectively involved in face recognition (Kanwisher et al. 1997), although others propose a more general function (gaining of expertise) to this structure (for the anatomy of the temporal lobe, see Kiernan 2012). The inferior occipital gyrus and the posterior part of the superior temporal sulcus all are involved in face recognition in adults. One should

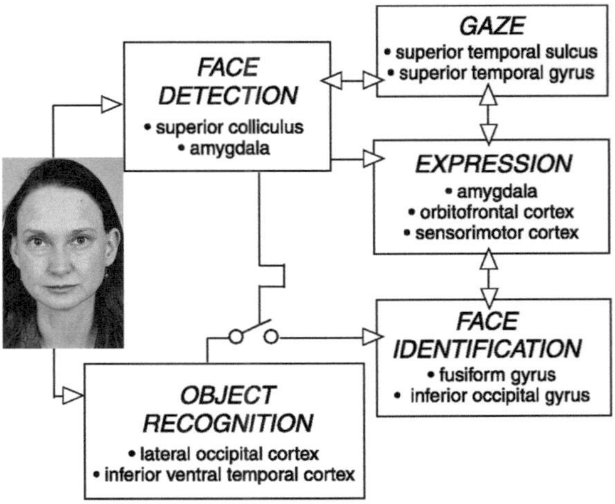

Fig. 7.2 "A schematic presentation of brain circuits for face detection and recognition, emphasizing the subcortical route via superior colliculus and amygdala". (Redrawn from de Gelder et al. (2003) and adapted from Hari and Kujala 2009)

note that the cortical representations of faces are not solely in the fusiform area, but are distributed both spatially and temporally over the whole brain. This coincides with other object representations. Facial attractiveness modulates the cortical activity. “The involvement of the orbitofrontal cortex co-varied linearly and the activity of amygdala nonlinearly with judgments of facial attractiveness” (Hari and Kujala 2009, Fig. 7.2).

Face memory is spectacular. Even when one has not seen a person for over 30 years, instantly identification will occur. Lesions of the fusiform area only impair the ability to recognize faces (face blindness). This condition is known as prosopagnosia in neurology: prevalence rate of non-syndromic hereditary prosopagnosia is 2.47%. By the way, those who suffer from prosopagnosia instantly recognize people by their voices.

One large area of research on facial expression, the one on emotion, is difficult to summarize. Questions like “Do facial expressions convey universal emotions?” and “How do we interpret facial expressions?” (Freitas-Magalhães 2012) are not treated in this book. Nevertheless, basic emotions are expressed by the facial muscles in certain patterns and they are universal: corners of the mouth raised are recognized as joy, cheeks raised, nose wrinkled and the upper lip curled represent disgust.

Emotions like anger, contempt, fear, surprise and sadness are recognized due to our experiences and messages given by others during our earliest education. Each of these emotions can be influenced by culture, age and gender. However, the face cannot express two emotions at the same time, presumably because different neuronal networks are involved. Women show basic emotions quicker and more intensively than men do. An expression can last half a second to 4 s (called macro-expression), but there are also microexpressions of less than half a second that normally go unnoticed. Our brain needs only 350 ms for the global perception of an emotional interpretation (see overview of Freitas-Magalhães 2012). Smiling and weeping are special research areas in which women and men confirm that a smiling face induces positive emotions and that smiling can have therapeutic effects. “Criminals who smile are more favourably perceived and seen as less responsible for the crimes they have committed than criminals with a neutral expression, especially among women” (Freitas-Magalhães 2012). Weeping is a reactive and compensatory mechanism for both women and men. Women cry more frequently than men. These facial expressions are made by subtle contractions of the facial skin muscles that are treated in Sect. 7.1.5.

Face disfigurement induces avoidance by others. The range goes from acne via cleft palates to facial birthmarks, and it is found to hamper “interpersonal relationships and inequities in employment, education and healthcare settings” (Ryan et al. 2012). One seemingly treats people with facial disfigurement as though they have an infectious disease (Ryan et al. 2012). The facial expression due to pain is used to assess and document relief of pain in critically ill non-communicative patients (Arif-Rahu and Grap 2010). Moreover, deviations of normal foetal facial expressions using 4D ultrasound for its evaluation are expected to indicate specific neurological disorders (Kanenishi et al. 2013). These results bring up the question: Where did this facial research originate?

7.1.2 Physiognomy

The face plays an important role in human communication. We easily pass a judgement after seeing somebody's face for the first time. Remarks like "an honest face" and "the look of his eyes is sinister" represent revealing a person's character or mindset. Over the whole globe, in China (Mar 1974), in India and in the West (Braus 1929), people studied the face in order to extract the other person's nature. These studies are grouped under the name physiognomy, gleaning the countenance features.

Marcus Antonius Polemo (AD 88–144) wrote "*Physiognomy*". The various manuscripts that survived via the Arabic route—the Greek original is lost—have been translated in "*Seeing the Face, Seeing the Soul*" (Swain 2007). Physiognomy (firāsa in Arabic) "tells what a person is like from the way he or she looks". Several other Greek philosophers paid attention deducing a person's nature or character from his face and/or body. Aristotle is the most important among them, but Galen too pays attention to the bodily relations with the soul in "*The Powers of the Soul Depend on the Mixtures of the Body*" a text in Galen's "*Scripta Minora*".

To get an idea of what Greek–Roman physiognomy means, two short texts of the reconstructed Polemon's Physiognomy are given here:

"Twenty-fourth chapter: on the physiognomy of the chin and its signs. If you see that the man has a long chin, you will find that he has little evil, suppresses fury and begins speaking at the wrong time. In his nature, there is laxity and weakness. Shortness of the chin is a sign of bad, you will find their owners entering into evil, and it also indicates boldness and killing. Shortness and roundness of the chin indicate weakness and effeminacy. If you see that the chin has four edges, judge for their owner boldness and strength. If you see that the beginning of the chin is long and separated, as if it had two heads, this should please you, for it indicates cunning and complicity. If its beginning is not separated very much, there is some joining, that is a sign of much desire for sexual intercourse" (The Leiden Polemon; Swain 2007).

"Twenty-ninth chapter: on the physiognomy of the ears and their signs. Largeness of the ears indicates indifference towards things. Smallness of the ears indicates treachery, and wicked and hidden desires. If they are extremely small, as if pared, they indicate folly and stupidity. The evenly proportioned ear, having the form of a square, indicates alertness in things and power. If you see that the ears have flaps, this indicates knowledge and understanding. If they are without this attribute, judge for them envy, much evil, and the practice of it" (The Leiden Polemon; Swain 2007)

7.1.3 Natural Portraying and History of Physiognomy

Physiognomy came into its own when portraits became realistic representations of body or face. Earlier statues and paintings pictured important persons with altered faces to express their nobility and superiority. Painters like Zograf Vasilie (curriculum vitae unknown; Saint Nicola temple, Boyana, Bulgaria 1259), or Cimabue

(1240–1302; e.g. Florence crucifixion 1287–1288; Basilica di Santa Croce, Italy) started the Pre-Renaissance natural portraying. A short excursion to the less known Vasilie fresco portraits made in 1259 (Fig. 7.3) in the Boyana Church what is nowadays Sofia residential area illustrates the renewed facial expressions that are at the base of applying pictorial physiognomy. The Boyana Church is small and contains two floors: the ground floor is dedicated to Saint Nicolas and the upper floor to Saint Panteleimon (see Sect. "Arabic and Byzantine Trepanation and the Treatment of Head Wounds". Sofia had a strategic position between the West and Asia and also between both parts of the Bulgarian Kingdom (632–1396). So, the tsars appointed directly related family, mostly brothers, as rulers of the area. These so-called sebastokrators financed the embellishment of their churches (see Dimitrov 2009). The result was Pre-Renaissance pictures of Christ, Maria and Saints in this small Boyana Church. These frescoes are a culmination of art, beauty and professional skill and of realistic human faces (see Fig. 7.3).

Cimabue's apprentice Giotto di Bondone (1266/67–1337), abbreviated to Giotto, is also known for the dramatic expressions of his painted figures. Giotto was well

Fig. 7.3 Parts of the frescoes of the Boyana Church to demonstrate the Pre-Renaissance picturing of 1259. Left upper fresco is the Holy Mandylion (cloth upon which a miraculous image of Jesus's face is present); left lower figure contains the faces of the Holy Virgin and Child; upper right figure shows the face of St. Nicholas of Myra, and the lower right is the face of St. Ekaterina. Note their realistic portraying and their facial expressions (with oral permission)

acquainted with physiognomy. Around 1300 renewed interest arose for the Greek and Roman literature on physiognomy, parallel to the realistic portraying of faces. A series of over 1000 manuscripts has survived, indicating the importance of physiognomy. Giotto used the current fourteenth-century physiognomy for the expression of the characters he painted. Giotto knew Pietro d'Abano (1257–1315), writer of "*Liber compilationis physonomiae*" (book on physiognomic compilations), which is a collection of texts from other physiognomic books and contains physiognomic translations from the Arabic "*Almansor*" of Rhazes (854–925). It appears that the shapes of faces and the nature of the characters they indicated were well known, even by unlettered people (for an extensive review, see Steinke 1996).

Giambattista della Porta (1535–1615) wrote "De *humana physiognomonia libri III*" (on human physiognomy in three books) in 1586. Although the Greek and Roman works already indicated the resemblance between human faces and animal ones, Della Porta's book stressed the similarity. In Fig. 7.4a, b, two examples of resemblances for the nose are given. Since characteristic properties are given to animals (e.g. the lion symbolizing force), the resemblance also indicated that these animal properties are present in the nature of man. The importance of Della Porta's work on physiognomy should not be underestimated: both Lavater and Lombroso (see below) were strongly influenced by Della Porta's ideas.

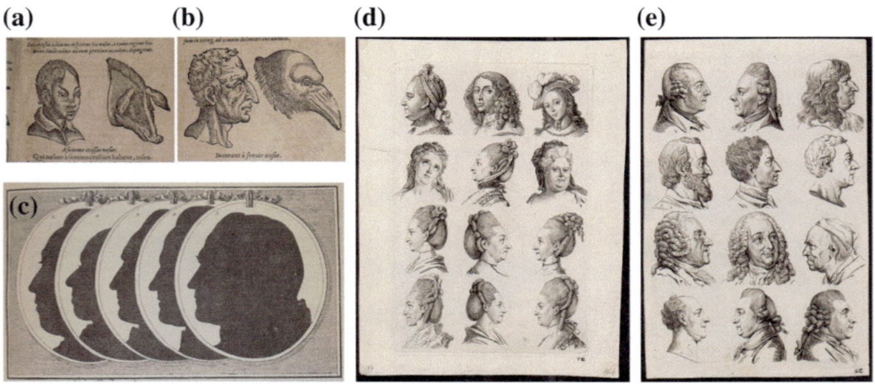

Fig. 7.4 **a** and **b** are figures from Della Porta's work on physiognomy, indicating comparable face structures between man and animal. **c** Contains the silhouettes of faces from Johann Casper Lavater: from left to right, 1 is the silhouette of Johan Martin Miller (poet, 1750–1814) described as: forehead and eye of a soft-feeling poet, weak lower lip and weak chin; 2, Asmus, pseudonym of Matthias Claudius (poet, 1740–1815): not weak, nor clever, honourable, credulous and direct, truth, man of feeling, modesty and tranquillity; 3, Friedrich Heinrich Jacobi (cited by Edward Allwill's Briefsammlung (1776): sharp angle of the forehead and the characteristic upper lip to chin indicates searching, wisdom and by its bent face weakness or poetry; 4 and 5 are compared: 4, critical mind, nose and forehead indicate poetic forces; 5, more active brain, nose indicates wisdom and is more beautiful, and mouth and chin relate to wisdom and tranquillity, modesty. **d** Zwölf Köpfe unbekannter berühmter Frauen verschiedener Nationen und Zeitalter, zu Lavater (twelfth heads of unknown famous wives (men in **e**) of different nationalities and ages for Lavater); **d** and **e** are made by Daniel Chodowiecki for Lavater's books showing the eagerness of the bourgeoisie to be incorporated in his books, but were not published. (All figures courtesy Leiden University Library with permission)

Before attention can be paid to Lavater and his "Essays on physiognomy", first notice should be given to fear and anxiety in the eighteenth century. The eighteenth century was marked by a radical change: "the world picture became mechanized and secularized" (Luyendijk-Elshout 1990).

The eighteenth-century plays and books were repeatedly dealt with people who were socially frightened. Clearly, an excess of fear and anxiety was enough to declare you insane. People, impaired by their fears, regularly consulted physicians. The doctors concluded that it produced hypochondria and hysteria. The Incubus, a ghostlike monster, was held responsible for pressure on the thorax, because it sits on the thorax and abdomen of the patient. "In this world, eighteenth-century man tried to be "absolute", that is, free from the chains of mystery and fantasy, obedient to reason and understanding. In this world, there was no place for human feelings, which had to be corrected and kept hidden, or, later in the century, exaggerated by the heroes and heroines of Romanticism" (Luyendijk-Elshout 1990). Being "absolute", in its psychological meaning, induced suppressed emotions that appeared as fear and anxiety. It is also an unconscious drive to look for certainty and Lavater with his physiognomy pretended to give a real true insight into the one's fellow man, giving security and certainty in social relations.

Johann Casper Lavater (1741–1801) claimed physiognomy to be a science. His four volumes called "*Physiognomische Fragmente zur Beförderung der Menschenkenntnis und Menschenliebe*" (Essays on physiognomy designed to promote the knowledge and the love of mankind, 1775–1778) produced serious controversies in literary and scientific circles. The pros and cons were examined, to name a few, by Johan Wolfgang Goethe (1749–1832) and by Friedrich Schiller (1759–1805), both having doubts about physiognomy. The physicist Georg Christoph Lichtenberg (1742–1799), who thought its popularity, showed "a lack of self-examination and knowledge of human nature" among the Germans. The success of Lavater's physiognomy books was based on his claim that physiognomy scientifically studied the inner and outer selves. Two arguments have been used: one borrowed from semiotics, the study of the meaning of a sign and its process relations: "the set of roles that facial traits occupy within a given sign relation". Thus, the relations of facial signs contain a meaning. The other is that instinctively we all think at the first appearance of a person that we can deduce his nature, due to our life experience: unconscious judgements or snap judgements (see Highfield et al. 2009). Nowadays, physiognomy claims to be a type of psychology. Some call it morphopsychology and others face reading (see Louis Corman [1901–1995]), and as such physiognomy had a great appeal to the public and to scientists and authors in the eighteenth century. Lavater was good in external communication: he often appeared in public giving speeches, exploiting the new fashion of silhouettes (Fig. 7.4c) and his appearance, both socially and personally thrilled people, just as a pop idol does in our time. For his next volume, he invited successful authors and scientists for their silhouettes or portraits that increased fame to both sides (Fig. 7.4c–e). "Beyond its scientific claims and social modishness, physiognomy also spread like literary wildfire" (Shookman 1993). Many German, French, Swiss and English authors used physiognomy to describe the nature and characteristics of their personages. Lavater studied portraits of several authors, among them the one of Goethe. "Goethe's whole 'self' could be found in studying his nose: The Nose, Lavater wrote, was full of expression and productivity—taste and love—that is of poetry. When Lavater here remarks that Goethe's nose is full of poetry, he exposes himself to utter ridicule" (Shookman 1993; see also Siegrist 1993). Nevertheless, his influence was great, partially due to the portraying of important persons (Fig. 7.4d, e). It should be remarked that Lavater and all of his predecessors considered physiognomy a non-mathematical science.

Early French Romanticism profiles went in another direction as can be seen in the paintings of Théodore Géricault. Best known is "Le radeau de la Méduse" (the raft of the Medusa) for which studies of the facial and bodily expressions of dead and sick people were performed. Géricault studied dead corpses, did interviews with the survivors and his friend Theodore Lebrun, who was seriously ill, and had to stand to be painted for the Meduse: "Mais j'étais beau pour le peintre (Géricault) qui cherchait partout de la couleur de mourant" (I was perfect for the painter, who looked everywhere for the colour of the dying) wrote Lebrun (Cailler 1947). The same holds for the series of portraits of the insane, in total ten different portraits, of which only five survived. La Monomane de l'envie (insane jealous woman), woman with a gambling mania, the kidnapper, man with delusions and the kleptomaniac were all painted from patients that were studied by his friend Dr. Étienne-Jean Georget, a pioneer in psychiatric medicine. Thus, facial characteristics were studied almost scientifically, and their change induced by altered life situations was noted and described. Physiognomy in France was more scientifically based, but still it stayed a qualitative approach (for an extensive overview, see *Géricault* 1996).

Franz Josef Gall (1758–1828) invented phrenology. You could define it as physiognomy by the skull cap alone. Gall supposed that parts of the brain were specialized in steering special human functions. Cortex areas exist for hearing, for speech but also for music, language and so on. Moreover, moral functions were localized in brain areas. He proposed that in specialized cases (e.g. musicians or scientists) these cortical areas could grow larger than other areas and as a consequence the skull at these places protruded. Words like "Muziek knobbel" (music bump or knot) and "wiskunde knobbel" (maths-knot) are still used in Dutch to characterize people who excel in music or mathematics as relics of phrenology. By qualitatively studying the skull, you could predict the nature of the person. Family games were produced with small dolls. You could feel all over the doll's skull and compare it to skulls of your friends or family during the party (see Fig. 7.5). However, the human skull does not deform because of the extra subarachnoidal space, brain functional areas are organized on the microscale, and the quality of functioning is not solely depending on the extent of the cortical areas, but mainly by the effective use of the existing networks. Although Gall's descriptions of a number of the functional localizations in the brain were wrong, it initiated studies to find the actual localization of the functional areas of the cortex. It gave neuroanatomy and neurophysiology a boost.

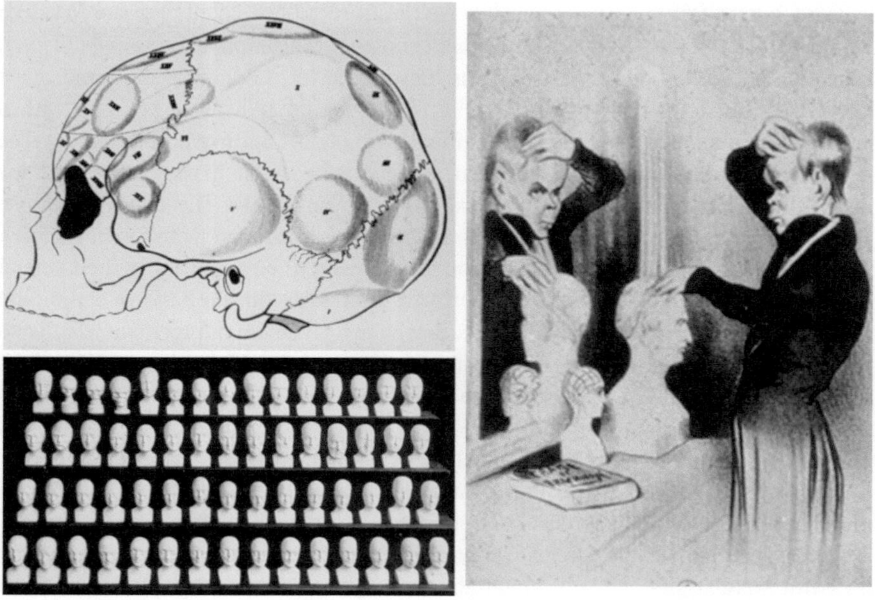

Fig. 7.5 Phrenology. Skull with extended bone areas related to specific brain areas and functions. Round game with all types of skulls for comparison and caricature of the phrenology application (courtesy Leiden Anatomical Museum 1982 with permission)

Cesare Lombroso (1835–1909) argued that criminality was mainly inborn. One could distinguish criminals by anthropological properties that showed anomalies in comparison with normal persons. The criminal was at a lower moral level and returned to an ape-like situation, or returned to an earlier evolutionary stage of man. Anatomical characteristics of the criminal are: sloping forehead, strange ears of unusual size, prognathism (chin and cheekbones protrude, and eyes are widely separated), asymmetry of the face and/or of the skull. These characteristics were called stigmata. By these anatomical deviations, even different types of criminals could be discerned. Several moral properties mark the criminal: cruelty, abnormal vanity, a lack of moral sense and excessive tattooing (!).

Lombroso's approach of criminality failed, and one should have pity with those who were sent to prison or condemned to death on Lombroso's judgement. Lombroso's criminal hypotheses were the direct consequence of physiognomy and phrenology.

The police composite sketch is the reversed approach: victims tell their story and what they recall of the face of the criminal. The police artist distils the portrait of the criminal from the victim's information. In several cases, criminals are arrested due to these composite sketches. The basic sketch is made with the knowledge of the facial proportions diagram and the head shape reference points diagram. In fact, a quantitative and a qualitative model are used in the police composite sketch (Mancusi 2010).

The Chinese face reading (Mar 1974) is far more sophisticated than our Western counterpart. Restricted to the nose in this chapter, Chinese physiognomists discern basically the three main divisions as described by Leonardo da Vinci (see Fig. 7.14). The long nose (diritto/A, Fig. 7.14) should indicate a person "with a conservative personality, often an individual who is disdainfully proud and capable of intellectual, artistic and/or technical achievement. Such a person is unfit for business enterprises, for the commercial world" (Mar 1974). The concavo/C nose (Fig. 7.14), indicated as a short nose, is "usually associated with open-minded, optimistic, outgoing individuals. Such persons are apt to be loose about their sexual morals. They also dislike detail and perform their work on the basis of emotional impulses. This type of individual works best when given encouragement, that is "a pat on the back"" (Mar 1974). The convesso/D nose (Fig. 7.14) is typed as: "the skinny and bony nose, which is a sign the individual lacks the power of concentration. If the nose is too high in proportion to the face as a whole, the person is likely to be proud, haughty and difficult to get along. Such a person, in authority, invariably has problems with subordinates. This individual is lonely and is only on tolerable terms with his spouse" (Mar 1974).

During pregnancy, the face appearance changes. Pregnancy glow originates by the extra small blood vessels under the skin (your surroundings notice it often earlier, signalling you are pregnant), while melisma or chloasma is the extra pigmentation of the facial skin during pregnancy. Facial expressions are also different from normal in cases of disease. The best known is the illness of Basedow (Fig. 7.6). Physiognomy therefore has had an important application in medicine and was controversially used by the German psychiatry during the Second World War. De Crinis (1889–1945), a German physician, psychiatrist and national socialist before and during the Second World War, was co-responsible for the euthanasia of children and mentally disordered people. He published a booklet on the human facial expression in 1942. Facial clinical implications are the main theme. In Fig. 7.6, eight of his facial recognition of diseases are given. In total, from facial paralysis to alcoholism, 35 diseases are considered. The booklet of De Crinis (1942) encountered opposition outside national socialist circles. It was considered controversial due to its possible involvement in euthanasia: it could be used for the prediction of insanity with its national socialist "consequences". Since the pictures are static, it remains difficult to diagnose the diseases in most cases. Morphopsychology or face reading alleges to be dynamic (video, film) in its study of the face and therefore is more reliable.

7.1.4 Facial Skin Colouration

The human perception of a face is influenced by skin colour or skin condition, called skin texture. Skin colour is used by the viewer to determine: health, attractiveness and age. Skin quality, colour and pigmentation are observed and are also used in judging. Increased redness, increased melanin and carotenoids in the skin

Fig. 7.6 Facial expressions of patients. Diseases should be evident from the facial characteristics. Upper row left to right: Basedow (thyroid disease), mental deficiency with muscle dystrophy, schizophrenia, epilepsy. Lower row from left to right: scrofula (lymphadenitis or king's evil), early multiple sclerosis, late multiple sclerosis, disease of Winniwarter-Bürger (blood vessel closure by thrombus) (De Crinis 1942)

(within limits, of course) enhance the perceived health of faces. Skin colour is co-determined by the facial blood vessels.

The face gets its blood from two branches of the external carotid artery (A, facialis and A. temporalis superficialis) and one that comes from inside the orbit (A. ophthalmica), a branch from the internal carotid artery (Fig. 7.7). There are two small foramina above and below the orbit (Fig. 7.15) by which two small arteries pass: the A. supraorbitalis (branch of the internal carotid artery) and the A. infraorbitalis (branch of the external carotid artery). The A. temporalis superficialis supplies blood to the sides of the forehead, and the eye areas are supplied by the A. ophthalmica and mainly the supraorbital artery. Thus, the whole face is supplied by the external carotid artery except the eye region and the forehead above the orbit, which is done by the internal carotid and its branch the supraorbital artery (Fig. 7.7).

In places where blood vessel bifurcates, e.g. the division of the abdominal aorta into both its common iliac arteries and the division of carotid communis into its internal and external branches, blood vessels undergo extra stress promoting the formation of atherosclerotic plaques. For the carotid arteries, this results in stenosis

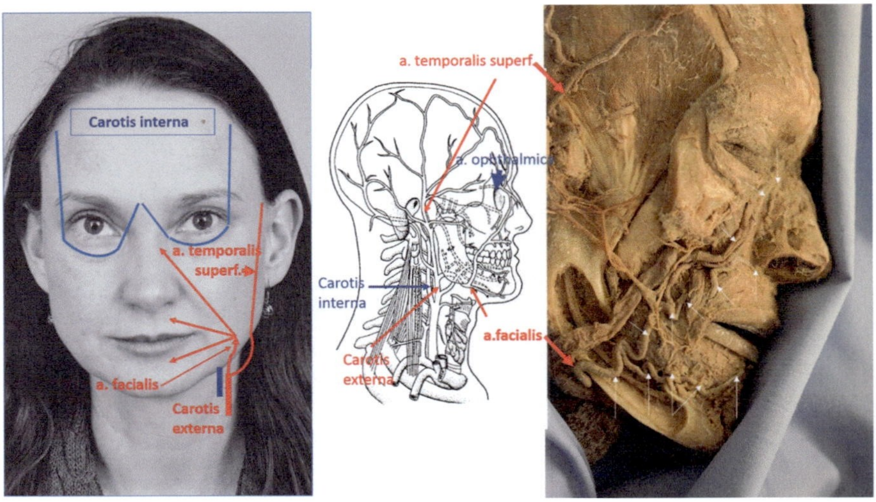

Fig. 7.7 Arterial blood vessels of the head. In red are the vessels that originate from the external carotid artery, and blue indicates the blood supply by the internal carotid artery (middle figure adapted from Samandari and Mai 1995; right figure courtesy © R.G.E Notenboom, Dept. Anatomy & Embryology, LUMC, Leiden with permission)

of the external and/or internal carotids. Occlusion of the internal carotid artery causes hemispheric brain problems and amaurosis fugax (momentary blindness). By removal of the plaque area and external endarterectomy, replacing the vessel on either the external carotid or the common carotid artery (Walker et al. 1994; Sterpetti et al. 1988) hemispheric and eye problems are in general solved.

The facial artery can be palpated along the edge of your mandible: a small bump on the mandible at 1/3 from the chin. You can feel it beat. The facial artery after branching from the external carotid artery grooves into the back and upper part of the submandibular gland, is pressed to the inside of the mandible and at its edge turns around the mandible. It brings blood to a wide triangle from the chin towards the nose root and is responsible for blood supply of the chin, lips and the nose. At the medial eye corner, it ends and normally has contacted the superficial branches of the ophthalmic artery with its end branch (A. angularis, see Fig. 7.7). The nose also receives blood vessels of the internal carotid artery by a branch of the ophthalmic artery, called the dorsal nasal artery. The deeper layers of the nose receive blood from branches of the ethmoid artery (belonging to the ophthalmic artery), and passing through the lamina cribrosa its nasal cavity is reached. These branches divide for reaching the septum, the olfactory region and a ramus nasalis externus that reaches the nose dorsum. Clearly, the exterior of the nose is provided by both branches of the internal and external carotid arteries. All these branches are accompanied by veins. Venous drainage of the nose thus can reach the neurocranium.

Epistaxis or nose bleeding normally is harmless. Mostly, it is the plexus of Kiesselbach that causes the bleeding (see Fig. 7.21) and is called an anterior epistaxis, often a consequence of hypertension. Posterior epistaxis (10% of cases) can also be caused by hypertension, but is considered far more dangerous due to the involvement of larger blood vessels, especially because the internal carotid can be involved. In cases of skull trauma, an aneurysm of the internal carotid vessel can be the cause of this nose bleeding with a lethal outcome (Nooij et al. 2015).

Discussing blushing and flushing gives an uneasy feeling: it is still characterized as "The puzzle of blushing". It has remained an elusive phenomenon, lacking sustained investigation" (Crozier 2010). A clear distinction between flushing and blushing is not found in dictionaries or in scientific articles. A good neurophysiological explanation for "face" reddening is absent, which is caused by:

(1) The different physiological types of reddening (e.g. with or without blood pressure changes).
(2) Reddening can be caused by many substances (e.g. alcohol, nicotine).
(3) Reddening can be induced by various situations (e.g. emotion, carcinoma, menopause).
(4) Several humoral mediators exist for reddening (e.g. prostaglandins, substance P, histamine).
(5) Oral cavity heating can induce reddening (e.g. ingesting hot fluids) (see Cutlip and Leary 1993 and references herein).

In order to arrange the following explanation, two definitions are proposed here that can be met with criticism. Let us start with Fig. 7.8. Using our definitions (see below), the right series of face photographs shows by the upper two flushing (forehead not red) and by the lower two blushing (reddening of the whole face). Reddening of the face is related to postganglionic vasomotor nerves (upper scheme Fig. 7.8) that ascend mainly in the cervical sympathetic trunk. They reach their targets using various pathways preferentially along blood vessels. The sympathetic trunk is also responsible for the innervation of the carotids (Fig. 7.8), in which are involved the superior and middle cervical ganglion and ganglia of the vertebral artery plexus. The branches to and the plexus of the external carotid artery contain glandular, pseudomotor and afferent fibres (Baljet and Groen 1985; Groen 1986; Groen et al. 1987). It should be noted that the variability in these connections is large.

Flushing is defined by us as face reddening, restricted to the external carotid supported area of the face (forehead stays free from reddening), caused by descending rami of the lower part of the sympathetic superior cervical ganglion, and rami of the cervical media and inferior ganglia and/or stellate ganglion, and activating the so-called sympathetic external carotid plexus. Due to activation of the concomitant fibres of the sweat glands, it is also called wet flushing.

Blushing is reddening of the whole face and neck, eventually arm and upper breast, by vasodilator chemicals in the bloodstream reaching the wall receptors of both internal and external carotids and/or by activation of both the external and internal carotid plexuses, involving among other ganglia the whole sympathetic

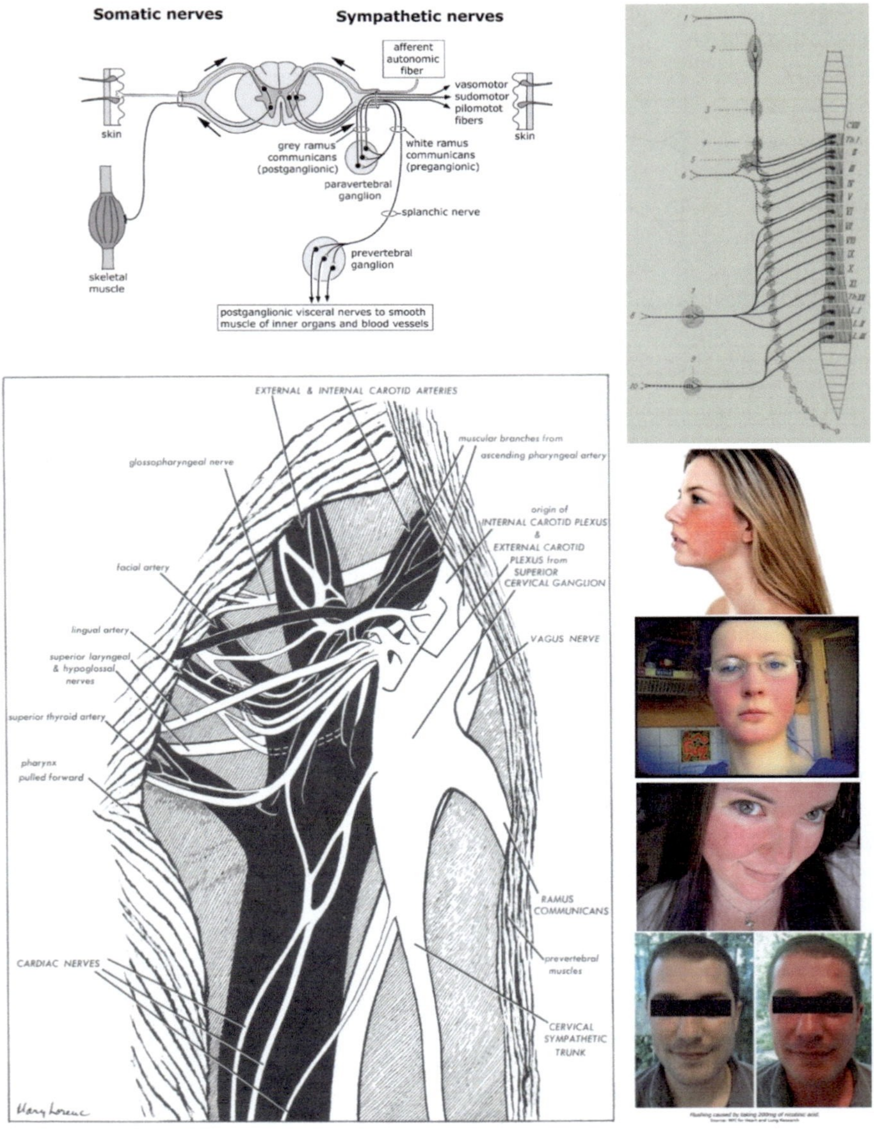

Fig. 7.8 The left upper figure shows the somatic versus the sympathetic pathways at spinal cord levels (Marani and Koch 2014). The right upper figure shows the sympathetic cord and its cervical ganglia (Villiger 1946): 1, eye, glands; 2, superior cervical ganglion; 3, medial cervical ganglion; 4, inferior/caudal cervical ganglion; 5, stellate ganglion; 6, heart, lungs; 7, ganglion/plexus coeliacus; 8, internal viscera; 9, plexus hypogastricus; 10, bladder, colon sigmoideus and genitals. The photographs at the lower right are taken from the Web showing flushing upper two, blushing third and the effect of 200 mg nicotinic acid (bottom; to the best of our knowledge, they were free to use on 23-06-2016). The left lower figure shows the external and internal carotid plexus in relation to the superior cervical ganglion, head and neck cranial nerves, the related branches and blood vessels (taken from Pick 1970; owner of copyrights is unknown to the last seller, i.e. Wolter Kluwers Health)

superior cervical ganglion. Vasodilator chemicals need not to activate the sweat glands. One can speak of dry flushing.

The spinal motor neurons of the sympathetic areas in the head are present in the spinal cord from T2 (some say from T2–T4) upwards (Fig. 7.8; Pick 1970; Marhold et al. 2005). Their axons extend into the cervical sympathetic cord as white rami. Presumably, only cervical level C8 still contributes to the sympathetic cord. Sympathetic rami from the upper thoracic and C8 sympathetic cord intrude in the spinal nerves, sometimes by jumping to the nearby spinal nerve to enter into the next lower or upper spinal nerve. Kuntz ramus (or the nerve of Kuntz) originates at T2 and enters T1 (Kuntz 1927; Cho et al. 2005; Singh et al. 2005; Fig. 7.9).

If all other possibilities (psychology and/or medication) are exhausted, sympathectomy can be considered (Licht et al. 2006, 2012). The idea is to cut the sympathetic cord at T2 to inhibit all sympathetic information arriving in this cord and thus vasodilatation, and therefore, reddening of the face either by flushing or blushing is overcome. The same occurs with heavy sweating of the hands (called sudoresis or hyperhidrosis), and here, the nerve of Kuntz comes up. Sweating of the

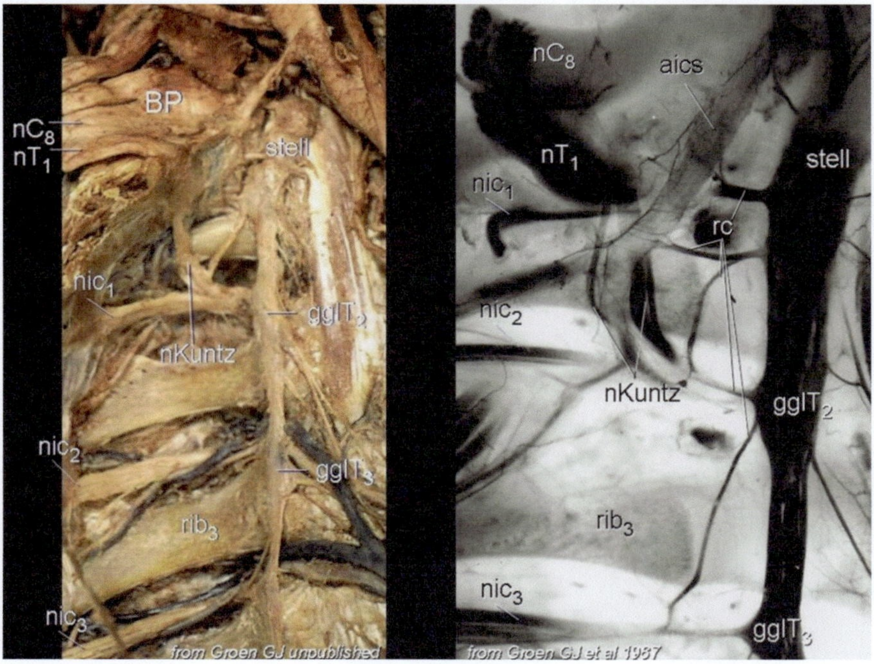

Fig. 7.9 Localization of the Kuntz nerve in a mature macroscopic human sample and a foetal acetylcholinesterase coloured overview sample. Aics, superior intercostal artery; BP, brachial plexus; ggIT, sympathetic thoracic cord ganglia 2 and 3; nC8, eighth cervical nerve; nic, intercostal nerves 1, 2 and 3; nKuntz, nerve of Kuntz; nT1, first thoracic spinal nerve; rc, rami communicantes; rib, third rib; stell, stellate ganglion, combination of the first thoracal sympathetic ganglion and the inferior cervical ganglion (courtesy G. J. Groen, with permission)

hands goes over the brachial plexus (C5–T1), and sympathectomy at T2 overcomes it. However, if rami communicantes from T2 can jump to the first thoracic nerve, escaping the surgeon's knife, the operation is useless. We spare you the literature on the nerve of Kuntz but present Fig. 7.9. While anatomists can find the nerve of Kuntz (Fig. 7.9; Cho et al. 2005; Ramsaroop et al. 2004; Singh et al. 2005; Pick 1970; Kimmel 1959; Kuntz 1927), surgeons do have difficulties with it. The literature series of the "best thoracic evidence topics" contains the article "Does the nerve of Kuntz exists?". The answer is: there is a strong prevalence of its presence in up to 60% of the literature (McCormack et al. 2011).The surgeon should perform "a lateral diathermy along the second rib lateral to the sympathetic chain" (Fig. 7.9). Anatomical studies suggest that a nerve of Kuntz or other rami communicantes towards T1 would be cut between 30 and 70% of the time (McCormack et al. 2011).

7.1.5 *Muscles of Facial Expression*

The mimic muscles originate in two layers: one is superficial and named after the platysma muscle, and the other is a deep layer and named after the sphincter colli muscle (Fig. 7.10). During the development, the two layers are telescoped into each other largely effacing their distinction. These muscles can take on two different

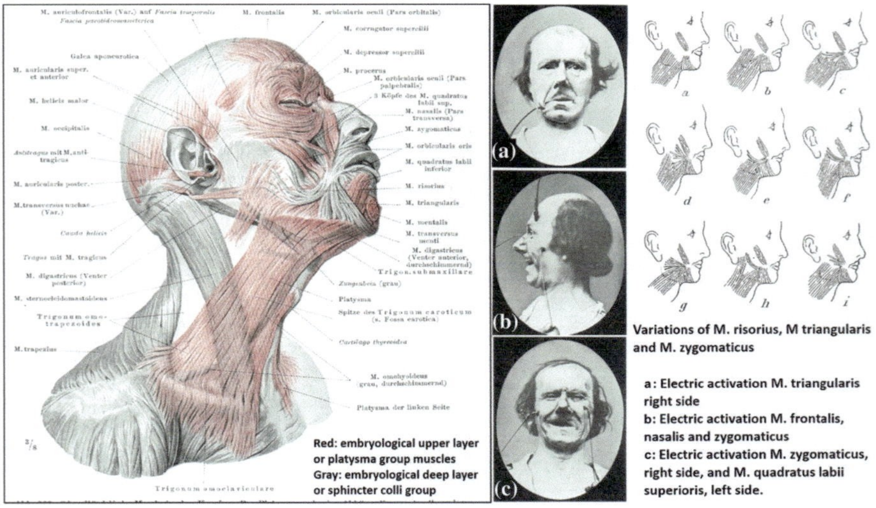

Fig. 7.10 On the left, an overview of facial muscles in head and neck with some neck muscles (sternocleidomastoideus, trapezius and digastricus) is presented, and on the right activation patterns of the facial muscles controlling the corner of the mouth are presented and are taken from Braus (1929). The photographs **a**, **b**, **c** are from Duchenne's (1806–1875) "Le Mécanisme de la Physionomie humaine" (1862) illustrated by photographs. (**a** is Duchenne's plate 43, **b** plate 39, **c** plate 48; see Cuthbertson 1990)

forms, circular or straight. The circular ones are placed around the eyes (m. orbicularis oculi, superficial layer origin) and around the mouth (m. orbicularis oris, deep layer origin, Fig. 7.10), and the straight ones (e.g. platysma, m. zygomaticus, m. nasalis, m. frontalis) are spread over the face.

Around the corner of the mouth, the risorius straight muscle, closely related to the triangular muscle, shows a rather large variation (Fig. 7.10). The initiation of laughing features and the facial expression of laughing are the responsibility of the risorius, triangular and zygomaticus muscles (7.10c); they produce the laugh furrow next to the mouth (Fig. 7.13 black arrows). Botulinum injections into the masseter muscle can reach this muscle causing asymmetric smiles. In Melanesian Papuans, this risorius muscle is absent and the muscle contributes to the rictus grin or risus sardonicus, the spastic grin due to tetanus. The number of facial muscles around the mouth (ten are discerned) plays a role in eating, but also small children use them for breastfeeding and later on for taste-testing in their oral phase. Expressions of the mouth play an important role in signalling happiness and sorrow. Kissing with the lips is a sign of greeting for humans, but also for apes. Presumably, it is required for communication to be effective. Likely, it stems from the suckling procedures of infancy. The orbicularis oris is present in tarsiers, but absent in lemurs. In monkeys, all possessing the orbicularis oris, the triangular muscle is powerful and other mouth muscles start to subdivide into smaller distinct muscles. The similarity between human and monkey/ape facial expressions is great (Waller et al. 2006, 2008).

The buccinator muscle has a special position. It arises from the maxilla and mandibule behind the corner of the mouth, passes beneath the masseter muscle and has its insertion at the raphe pterygomandibularis. The muscle fibres cross in such a way that those arising from the mandibule reach the upper lip and those from the maxilla reach the lower lip. These fibres also constitute the deep layer of the m. orbicularis oris, which is known as the trumpet muscle due to its capacity to blow air through the lips. The corners of the mouth will be pulled laterally by its action; it closes the mouth and pushes the lips against the teeth. The parotid duct penetrates the muscle at the level of the second upper molar.

Expressions are made by single facial muscles or groups of these facial muscles. The muscles that express attention, the medial frontalis and lateral frontalis muscles (the skin of the front of the head is pushed superiorly, and eyebrows are pulled upward), are contrary to the effect of the muscle of reflection (depressor supercili; the medial corner of the eyebrow is lowered, in combination with the depression of the root of the nose). In general, each facial muscle has its counter-force muscle and they are grouped according to their emotional expressions: muscles of aggression (m. procerus), of pain (m. corrugator supercilii), of joy and benevolence (m. zygomatic major), of lasciviousness (m. nasalis), of sadness (m. triangularis), of weeping and whimpering (m. zygomaticus minor, m. levator labii sup.), and of fright and terror (platysma; see Waller et al. 2006). The first grouping of facial muscles along emotional expressions has been described by Duchenne (1862). Duchenne's subdivisions are still right and nearly identical to Waller et al.'s (2006) summary (see also Cuthbertson 1990).

The Facial Action Coding System (FACS) describes human facial expressions. It is a standard to systematize and categorize emotions expressed by the human face. FACS can support computed automated face recognition (see Sect. 7.1.6). Each emotion can be split into action units that are contractions or relaxations of single facial muscles or groups of muscles.

7.1.6 Automated Face Recognition

From 1800 onwards, Anatomy and Physical Anthropology collected the measures of the skull and its parts, face, teeth, nose, ears and eyes. These quantitative results (only one example is given, Fig. 7.11) are summarized in Lanz und Wachsmuth (1985). The measures are studied during development into adulthood and ageing (Fig. 7.11).

It is a labour lost for, e.g., the two fissures (superior and inferior) of the orbit, because left and right orbits are always different. A total of 48 measures are needed for their developmental description, not accounting all other structures of the orbit. These quantitative results, produced for the whole body, have been important for the medical profession and are extended in plastic surgery (see Fig. 7.14 lower pictures). It was and is the basis for replacement of bony parts and joints

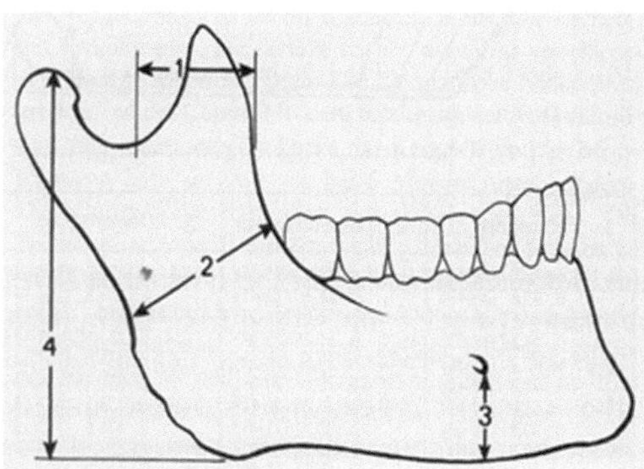

1 Proc. coronoideus, Breite	2 Ramus mandib., Breite
0- 2 J. re. 14,8 (13-16)	0- 2 J. re. 23,1 (21-25)
li. 14,3 (13-15)	li. 22,5 (20-24)
2- 4 J. re. 15,25 (12-17)	2- 4 J. re. 24,37 (20-26)
li. 14,83 (13-18)	li. 23,63 (19-25)
5- 6 J. re. 15,9 (13-19)	5- 6 J. re. 24,0 (22-26)
li. 15,2 (13-18)	li. 24,4 (23-26)
7-10 J. re. 16,0 (13-19)	7-10 J. re. 26,96 (23-34)
li. 16,45 (14-21)	li. 26,77 (23-34)
Erw. re. 20,41 (14-25)	Erw. re. 31,30 (20-39)
li. 20,30 (14-26)	li. 31,07 (20-39)
Gr. re. 19,25 (16-25)	Gr. re. 27,73 (12-33)
li. 19,22 (14-24)	li. 27,56 (12-34)

3 For. ment., Abstand zu Basis	4 Ramus mandib., vertik. Höhe
0- 2 J. re. 7,3 (6-8)	0- 2 J. re 29,6 (25-36)
li. 7,9 (5-9)	li 29,6 (26-34)
2- 4 J. re. 8,91 (7-10)	2- 4 J. re 30,6 (25-38)
li. 8,83 (7-10)	li 31,9 (25-39)
5- 6 J. re. 8,6 (6-10)	5- 6 J. re 28,8 (26-32)
li. 8,7 (7-10)	li 29,4 (26-33)
7-10 J. re. 10,17 (8-12)	7-10 J. re 35,8 (32-53)
li. 10,07 (8-13)	li 38,6 (33-58)
Erw. re. 14,11 (11-17)	Erw. re 56,7 (38-75)
li. 14,32 (11-18)	li 56,2 (42-74)
Gr. re. 13,06 (8-15)	Gr. re 52,8 (37-68)
li. 13,06 (9-15)	li 53,2 (37-66)

Fig. 7.11 Mandible and its post-natal changes: 1, Processus coronoideus, post-natal breadth development; 2, ramus mandibulae, post-natal breadth development; 3, foramen mentale, post-natal upward displacement; 4, caput mandibulae, post-natal upward displacement relative to the basis mandibulae; J, years; Erw, adult; Gr, aged (taken from Lanz und Wachsmuth 1985)

(best known are hip, knee and ankle replacements), it is needed in craniofacial trauma, and 3D reconstructions of the face for skull casts (see Fig. 2.3), mandible, maxilla and even whole face reconstructions favoured from it. These measures were initially also used in face recognition.

These days automated face recognition is explored and applied in safety services and is used in passport fraud, identifying missing children and minimizing identity fraud. In the 1960s, the first semi-automated method for face recognition was developed. It needed features like eyes, ears, nose and mouth on the photographs. The calculated distances and ratios were brought to a common reference point, which were then compared to reference data. Subjective markers such as hair colour and lip thickness were added and used in the 1970s to automate the recognition. "In 1988, Kirby and Sirovich applied principal component analysis, a standard linear algebra technique, to the face recognition problem. This was considered somewhat of a milestone as it showed that less than one hundred values were required to accurately code a suitably aligned and normalized face image" (NSTC). Principal component analysis (PCA) transforms the data (signals or images) to a new coordinate system with basis vectors derived from the variance in the data, the greatest variance is projected on the first coordinate, i.e. the first principal component (or eigenvector with the largest eigenvalue), the second greatest variance is projected on the second coordinate and so on. The use of PCA led to the "eigenfaces approach": any face can be recalculated in terms of weight factors describing the contribution each eigenface makes to that image, with an eigenface representing a principal component created by using PCA on a large set of images from different human faces. This approach brought a new dimension and a serious reduction of data needed for automated recognition.

Nowadays in automated face recognition, two methods are in use: geometric (feature-based) and photometric (view-based), and three mathematical algorithms are applied: principal component analysis (PCA), linear discriminant analysis (LDA) and elastic bunch graph matching (EBGM). Like PCA, LDA is a linear transformation technique. LDA tries to increase or to maximize the separation between groups of faces. As a consequence, the face discriminatory information stands out more clearly.

"Elastic bunch graph matching relies on the concept that real face images have many nonlinear characteristics that are not addressed by the linear analysis methods discussed above, such as variations in illumination (outdoor lighting vs. indoor fluorescents), pose (standing straight vs. leaning over) and expression (smile vs. frown). A Gabor wavelet transform creates a dynamic link architecture that projects the face onto an elastic grid. The Gabor jet is a node on the elastic grid (see Fig. 7.12), which describes the image behaviour around a given pixel. It is the result of a convolution of the image with a Gabor filter, which is used to detect shapes and to extract features using image processing. Recognition is based on the similarity of the Gabor filter response at each Gabor node" (NSTC).

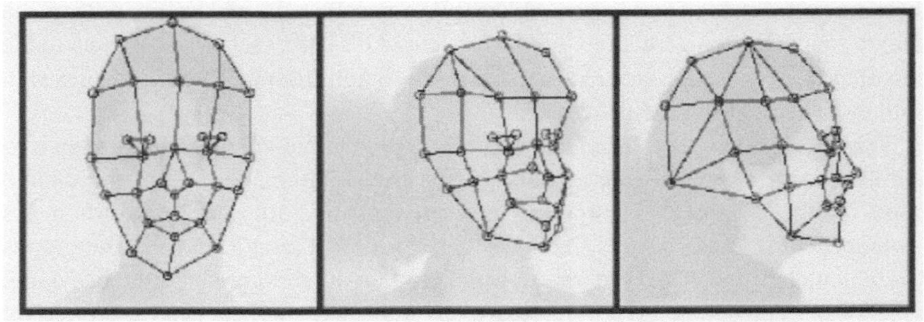

Fig. 7.12 Elastic bunch graph matching; Gabor wavelets are marked by circles on the face images (free figure is taken from the National Science and Technology Council (NSTC) subcommittee of Biometrics)

Let us try to understand the above-quoted sentences using some basic information:

(1) A wavelet is a short vibration that can be written down by a pencil on a paper like in a seismograph. It has an amplitude, it starts at zero and makes the wave, and after extinction it returns to zero. This signal can also be digitally registered and can be recognized in its digital form. The most used example is: a wavelet with the same frequency as the music note C with a short duration can be used to find all the C's in a piece of music. In the same way, it can be used to compare obtained information from pictures.
(2) Dennis Gabor invented a certain type of wavelets. This type of wavelets is in its function extremely good in edge detection. Gabor wavelets help to recognize identity despite variations of pose, illumination or expression of the face. The method is robust despite variations in hairstyle or the presence of glasses.

The conclusion is that face recognition does not need any anatomical qualitative or quantitative information using these mathematical methods. You need, e.g., a series of Gabor nodes, reference points, and you can place them anywhere on the face. You can use the point of the nose, but instead you can also use the wing of the nose. As long as you can put a network of Gabor nodes over the face, you can start automated face recognition. So, imagine anatomists and physical anthropologists studied and measured over 200 years the face and its skeleton and they are bypassed by the mathematicians in fifty years (note: the mathematicians started in the 1960s).

We come back now to the pictures of Fig. 7.6 as made by De Crinis (1942). Schizophrenia indeed shows different facial emotional expressions as studied by face recognition (Young Lee et al. 2015). The same holds for Parkinson due to the mask face, a consequence of the imperfect voluntary control of the facial musculature (Marneweck and Hammond 2014). Studies on Huntington disease do indicate that facial emotional expressions are impaired (Trinkler et al. 2013). Seemingly, the facial expression in a series of neurodegenerative diseases is altered, and presumably, they can be pictured two dimensionally. Whether the impaired facial expressions between the neurodegenerative diseases differ too remains unknown.

In addition to face recognition, various techniques are tested to automatically identify emotional visage characteristics of humans. In two-dimensional (2D) pictures, recognition of emotions on the human face is hardly possible and is highly dependent on pose and illumination (Fig. 7.13 and see Sect. 7.1.6). It is necessary to use a consistent facial pose in 2D images. With the use of three-dimensional (3D) pictures, the results improve: while lip pucker and jaw clencher are difficult to discern in 2D, the 3D approach does recognize such subtle changes (Sandbach et al. 2012). Spontaneous expressions are still difficult to recognize but are feasible if 4D (3D over time) images are used. Using 2D pictures, a 3D image can be made using the "3D morphable model methodology". Although the method is not very robust, under good lightning conditions it works well in the recognition of facial expressions like smiles. Using encoded coloured light patterns that are projected on the face, a 3D image can be made. Only limited changes of the face can be noted, and most emotional expressions are characterized by large changes of the facial surface. The multiview stereo technique needs several cameras and, therefore, will not be the first choice since it is expensive and the technique is slow (Sandbach et al. 2012).

Another approach uses 3D muscle structures that are embedded in a finite element continuum of the face and neck. The mechanical contribution of muscles is evaluated and mapped onto the facial computational domain. Muscle forces are brought into this facial overview. "Muscle meshes can be constructed independently from the facial mesh giving two main advantages: (i) the muscle geometries can be refined independent of the facial computational domain, and (ii) it is not required for the computational domain to conform to complex topology of muscle structures" (Wu et al. 2013). In fact, the facial muscles are truthfully portrayed on a finite element mesh of head and neck and their activity implemented (Wu et al. 2013). Realistic contraction of facial muscles is studied with the VICON movement capture system. Using the VICON system, normal patterns for the movements of wrinkling the forehead and frowning, smiling and opening and closing the eyes are described (Jorge et al. 2012). Best feature detection (Li et al. 2012) shows that facial expression features are better recognized using points of local expression regions than points in the facial contour (see Fig. 7.13).

The use of retrieval methodology using automatic detection of specific facial landmarks gives a descriptor that contains topological and geometric information from 3D face scans using dynamic time wrapping technique. Topological and geometric information are both needed since happiness and surprise produce changes in mouth topology but anger, fear and sadness cause only geometric changes (Danelakis et al. 2016). Several methods are employed nowadays: convolutional neural networks (Texeira Lopez et al. 2016) and pose-specific nonlinear mapping for facial expressions (Jampour et al. 2016). The series of techniques applied are still not ended, and among them are of course the eigenface technique (De et al 2015) and the Gabor technique (Zhang et al. 2014; see Sect. 7.1.6).

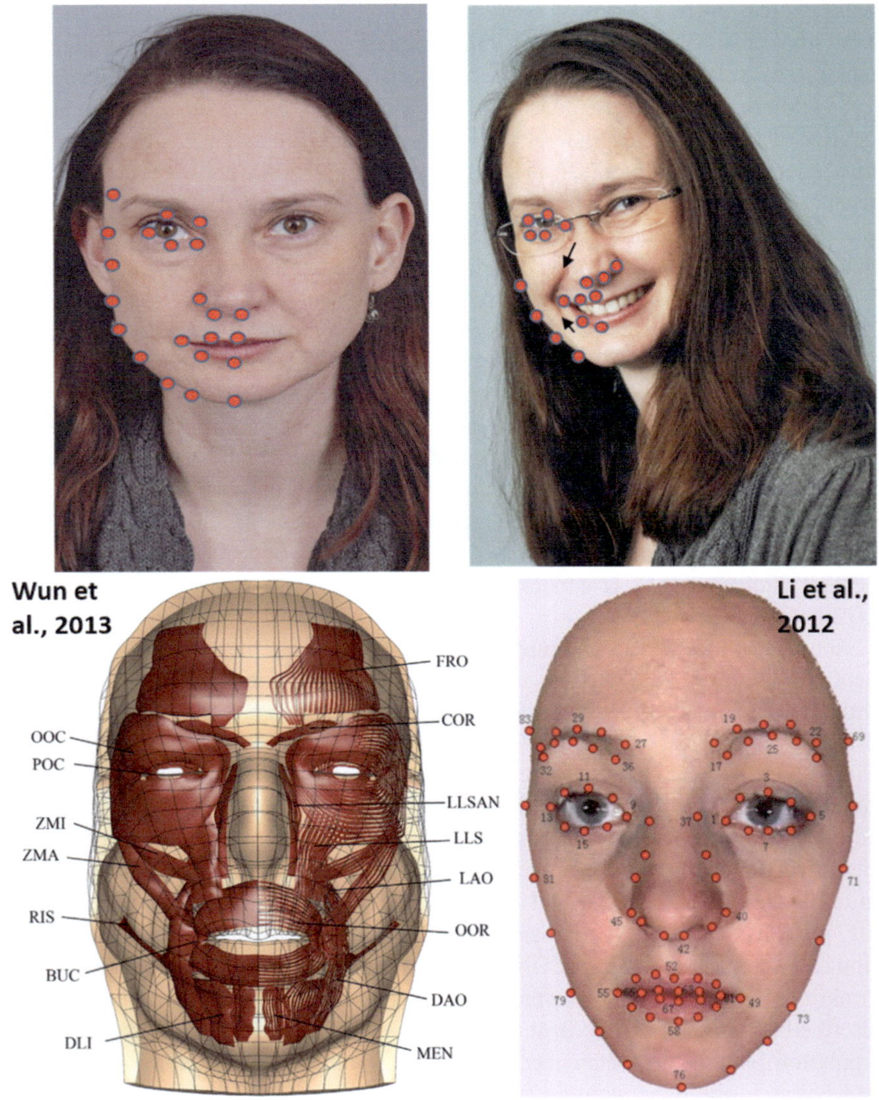

Fig. 7.13 During smiling subtle changes are also noted around the eyes, mouth, face edges and nose: compare two upper photographs. Wearing spectacles disturbs even more the recognition of facial emotions. Therefore, a series of subtle changes has to be detected that needs registration of facial changes around eyes, mouth nose and circumference of the face. Techniques in producing a finite element mesh with the realistic face muscles projected in it (Wu et al. 2013 with permission) are studied. The basic research on circumferential information versus local changes in the face is shown by Li et al. (2012 with permission); thus where should we look for the emotional expression? M. risorius effect has been indicated by black arrows. Smiling is due to the risorius effect and thus easy to detect

7.2 The Nose: In Arts, In Anatomy, During Development and in the Animal World

In arts, the study of the nose was necessary, obliged and taught regularly. Here, we give a part of Leonardo da Vinci's approach and skip the Greek and Roman classic canons, and also Albrecht Dürer's vision. Moreover, the Fibonacci sequence[1] seemingly does not hold all too well for the face (Laurentini and Bottino 2014), as is also true for the "golden ratio",[2] since the literature shows "little correlation between the asserted ideal proportions (golden ration applied on human faces) and the beauty scores given by human raters" (Laurentini and Bottino 2014). Nevertheless, some plastic surgeons favour both rules (Ricketts 1982).

For the study of the nose, Leonardo discerned three main forms in his Trattato Della Pittura, part 285;[3] "on the making of a human portrait and profile after having seen it only once, we will say first something on noses there are three kinds, straight, concave and convex ones. There are only four types of the straight one: long, short, high ones with a point and low ones. The concave ones are of three kinds, and some have the concavity in the upper part, some in the middle and some in the lower part. Finally, the convex ones differ in three ways, that is to say some have the hump on the upper part, some in the middle and others at the bottom. The protrusion that causes the hump in the middle of the nose varies in three ways, or they are straight, or concave or convex" (Fig. 7.14). The modern view is that the nose is basically symmetric in form, but contains small declines from the vertical straight line, a wry nose in essence (present in man in 82%). In normal cases, this inclined construction does not cause problems. A wry nose will always be accompanied by a slight or serious nose septum deviation depending on the severity of the asymmetry. In general, four deflecting types are nowadays discerned (Fig. 7.14): the straight long nose, hawk nose and both the bony and cartilaginous saddle nose. Lang in Lanz and Wachsmuth (1985, 2004) studied 119 students from Würzburg (Germany) and found the main subdivisions of Leonardo: straight (36% in males, 34% in females), concave (male 24%, female 38%) and convex (males 40%, females 28%). Their subtypes, which only partly follow Leonardo's subdivision, have been neglected in these percentages.

The anatomical structure of the nose starts below the glabella and is indicated in Fig. 7.15, left side. Measurements of the nose are made comparable by the nose index: the distance between both nostril outsides (Alaria) multiplied with 100 and divided by the nose height (from nasion to the subnasal angle). The nose index varies slightly during development, with a dip around the fifth/sixth year of age (at 3 years, the nose index is around 75; at 5/6 years 73 (males) and 69 (females); and later going down again for adolescents: 63 (females) and 66 (males)). The nose becomes longer and less broad towards maturity. Comparisons between people in colder regions and in warmer areas show that the elongation and narrowing of the face together with reduction of the nose are climate related and correlate to heat loss (see Newman 1953).

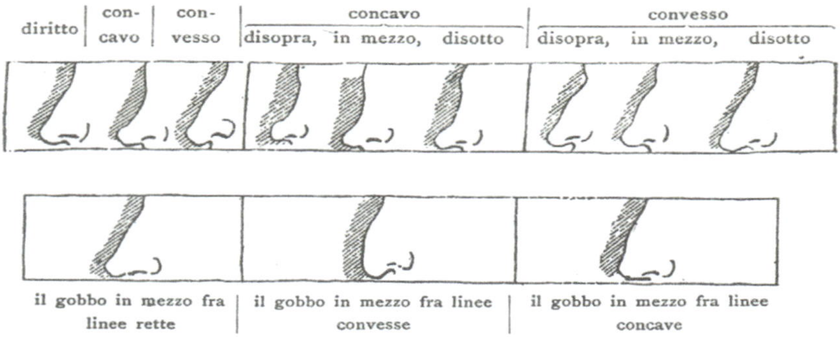
diritto
con-cavo
con-vesso
concavo
disopra, in mezzo, disotto
convesso
disopra, in mezzo, disotto
il gobbo in mezzo fra linee rette
il gobbo in mezzo fra linee convesse
il gobbo in mezzo fra linee concave

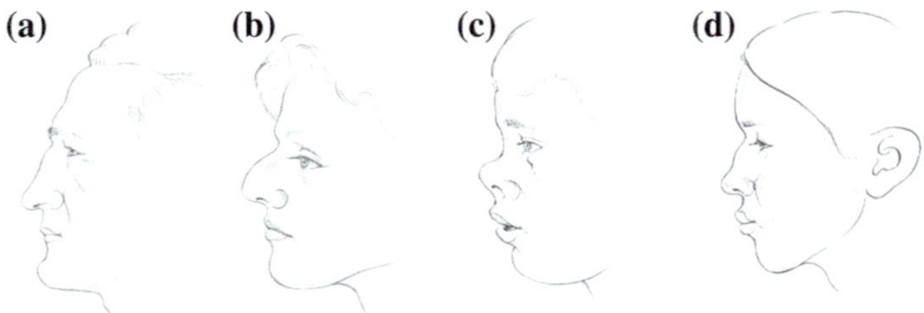
(a)
(b)
(c)
(d)

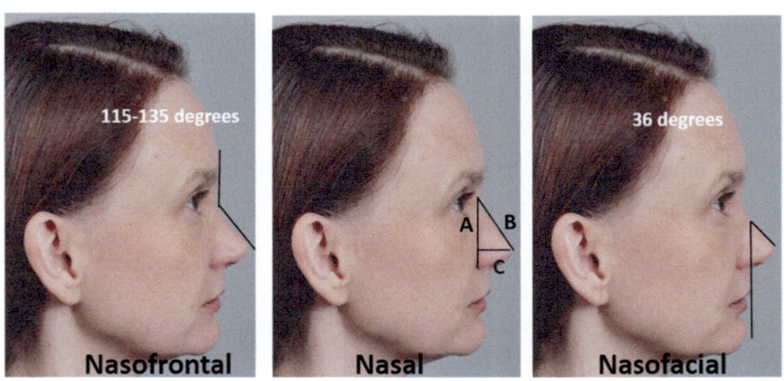
115-135 degrees
Nasofrontal
A
B
C
Nasal
36 degrees
Nasofacial

◄**Fig. 7.14** Upper part: nose forms as described by Leonardo da Vinci (see text). Middle part: **a** long nose; **b** hawk nose, **c** bony saddle nose; **d** cartilaginous saddle nose. The four types **a–d** can be found on the medallion showing four Italian composers: Nicola Zingarelli [1752–1837] (**d**), Giuseppe Sarti [1729–1802] (**c**), Giacomo Tritto [1733–1824] (**b**) and Giovanni Paisiello [1740–1816] (**a**) (Milan, Civica Raccolta Bertarelli). Lower part: measures of the nose showing nasal, nasofrontal and nasofacial angles (taken from Steiger's Facial Plastic Surgery PR, no answer on our permission demands over 2 years)

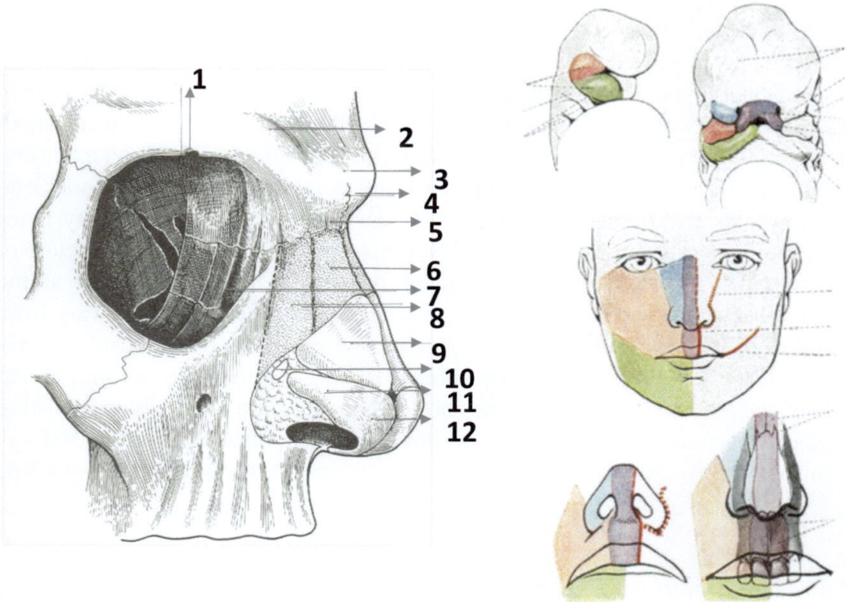

Fig. 7.15 Left: 1, supraorbital foramen; 2, upper rim of the orbit; 3, glabella; 4, sutura metopica ending in nasion; 5, nose root; 6, os nasale; 7, lacrimal rim; 8, mounting apophysis of the maxilla; 9, triangular cartilage; 10, sesame bone; 11, wing cartilage; 12, forming the nostril (adapted from Legent et al. 1976). Right (see also Fig. 5 in Introduction), purple, central frontonasal process with nasal pits at its border; green mandibular process; blue, lateral nasal process of maxilla (apophysis); red towards pink, maxillary process. By red lines are given the locations of facial clefts. Right lower figure shows relation of os nasale and os incisivum over the septal cartilage, and the left lower one shows this relation along the subnasal angle (from Braus 1934)

"Nasal facial angle is formed by the intersection of two lines. One line is drawn from nasion to pronasalae and another line drawn from nasion to pogonion. This angle ideally approximates 36°. The nasofacial angle defines nasal projection from the face (Fig. 7.14). A more acute angle bestows the appearance of increased nasal length, while more obtuse angles contribute to the appearance of a shorter, more projected nose. The nasofrontal angle is the obtuse angle between a line tangent to the glabella and a line tangent to the pronasalae, with both lines originating at the nasion. This angle is ideally between 115 and 135° (Fig. 7.14). The nasolabial angle is the angle formed between the plane of the columella and upper lip as seen in profile. Ideally, it measures between 105 and 115° in females and 90 and 105° in males" [see also Ahmed et al. 2014; note that Ahmed et al. found for women that "a rotation of 106° (columellar facial angle) was found to be the most aesthetic". Their results were published in most main Western journals, discussing the female nose contributing to facial beauty].

In reptiles, amphibians, birds and mammals, the nose is present with normally two holes and a septum for the separation of two cavities for smelling and breathing. The spout-hole in whales starts during embryonic development at the front of the head, nose-like, but is displaced during foetal development towards the top of the head. Whales do have two nose holes, but in the toothed whale one of the holes and cavity have been changed into an echolocation system. The spout-hole in whales is directly connected to its trachea, which allows whales to open their mouth underwater that prevents drowning. The trunk in elephant and tapir is an extension of the nose and upper lip. The elephant's trunk is constructed from muscles and membranes. Its nasal cartilages "serve only as valves at entrance to bony nares (nose openings) or turbinals (conchae). The trunk is a tapering muscular structure of nearly circular in cross section, extending proximally from attachment at the anterior nasal orifice and ending distally in tip or finger. The length may vary from 1.5 to 2 m or longer depending on the species and age. Four basic muscle masses (the radial, the longitudinal and two oblique layers) and the size and attachment points of the tendon masses allow the shortening, extension, bending and twisting movements accounting for the ability to hold and manipulate loads up to 300 kg" (Rasmussen 2006). And indeed, two holes are present in the elephant's trunk tip.

7.2.1 *Nose Evolution and Nose Development*

How did the capacious nose arrive evolutionary in the human face? In Sect. 7.1.1, it was stated that the human face is mainly determined by its bony and cartilage structures. Figure 7.15 shows the bony and cartilaginous parts that constitute the human nose. Soft tissues hardly influence the relations between the various parts of the face. This helps evolutionary Anthropology to base its results on archaic skulls. As long as one stays within the archaic hominids, a comparison with Homo sapiens is scientifically sound. The main differences noticed between Homo sapiens (early modern human, indicated by S,) and Homo neanderthalensis (as an example of archaic humans, 40,000–70,000 BC, indicated by N) are:

(1) Long low skull (N) versus rounded skull (S),
(2) Big brow ridges (N) versus vertical forehead with small brow ridges (S),
(3) Big projecting face (N) versus small retracted face (S),
(4) No chin (N) versus chin (S, see Lieberman 2013).

The Neanderthal face is prognathic. The jaws are more retracted in modern man. You could say that the Neanderthal still has some weak characteristics of a snout. A consequence is a more flattened nose compared to Homo sapiens' protruding nose. However, the most exciting result of the comparison is the association of canine form with the face structure. Teeth co-determine the face form, which is obvious. Remember your parents or other family members when the dentist removed all their teeth (often in the 1950–60s) and a denture was placed. You had to adapt to the change of their faces. Evolutionarily, food determines your teeth and your teeth co-determine your face form. Before the line Homo habilis (2 million), Homo erectus (1.5 million) and Homo sapiens (0.5 million years ago) are discussed, the food of the Australopithecus must be considered. While the Neanderthal is still in the line of the hominids, the Australopithecus branched off the hominid line 3.5 million years ago and was extinct 1 million years ago. The food of the Australopithecus was hard fibrous, and therefore, they needed big thick molars. Moreover to grind that type of food, one needs big, strong chewing muscles. Thus, the m. temporalis was large with a big belly inducing bony crests to anchor it. The zygomatic arches were large, bowed and thick to give space to the thick temporalis belly. They were also thick to adhere the large masseter muscle (Lieberman 2013). The consequence is a broad face (the heavy-duty jaws of the Australopithecus got them the nickname "Nutcracker Man", Lieberman 2013).

The canine length limits the gape (the median margin to margin length of the open mouth). Smaller canines seemingly are associated with smaller gapes and more powerful chewing (Lieberman 2013). Since the hominid species changed their food (they ate more fruit), a series of changes are noticed:

(1) Homo habilis showed an expansion of his brain and lost its snout.
(2) Homo erectus reduced its cheek teeth, and chewing muscles were reduced, finally reducing the lower face in the genus Homo sapiens.
(3) This all resulted in a large vertical projecting face, a low forehead, big orbits and a capacious nose to inhale air.
(4) The big enigma remains the chin ("an upside-down T-shaped projection of bone at the base of the lower jaw" Lieberman 2013). Only humans possess a chin.

The nose with its cavities is seemingly a very conservative property that is basic to inhale, but can evolve in various ways in the vertebrate kingdom. During development, the nose starts with the so-called nasal pits or nasal placodes. Placodes are transient parts of the columnar, pseudostratified surface ectoderm, later epithelia, that are capable of adding their cells to the mesoderm compartment. They add their cells in the head region to sensory ganglia, producing neurons and other type of cells for the special sense organs, like the olfactory areas. In the head area, the placodes that are present are named: olfactory, lens, trigeminal (ophthalmic and maxillo-mandibular) placodes, together with the epibranchial placodes producing geniculate (VII), petrosal (IX) and nodose (X) ganglia. Placodal cells are induced already at the edge of the neural plate at the 0–1 somite stage in the chick, marked by *Six, Eya, Id, Iro* and *Fox* genes, but the dispersed cells have to sort out and group together to produce placodes (see Sect. 3.5.2; Marani and Lakke 2012). "By the form of the epithelia, namely conspicuously higher than the surrounding ectoderm, placodes are easily detected in the foetal situation. Moreover, these cells show properties that are unusual for neuroectodermal cells. These nasal pits therefore can easily be identified on a cross section of the fore brain region, just at its anterior extremity (Fig. 7.16). The mesoderm developing from the neural crest, near the brain flexure, becomes arranged in a characteristic fashion (Fig. 7.15, right part, purple). It is deposited between and in front of the eyes, a region which can be reached by passing over the optic stalks. If it were to develop in this way, i.e. between and in front of the eyes, there would be a kind of mass protruding in ventral or ventro-caudal direction. The nasal pits would then float on this mass of mesenchyme. But it appears that the nasal pits are able to resist this displacement and stay in place, so that the mesenchyme bulges around them, burying the nasal placodes in deep, blind pockets, the nasal pits (Fig. 7.16 a–c). This development is soon followed by another formation from the same source, deposited in a column of mesenchyme beneath the eye, the maxillary process (Fig. 7.15, blue and pink). These mesenchymal formations at the base of the fore brain obviously concern the formation of the face and nasal cavities. The process in its simple form does not seem difficult to follow as it is one of the inverse mouldings: instead of there being a mass of clay in which the nasal passages are moulded, one must imagine a growing mass enclosing blind alleys such as the nasal passages. The mesenchymal formation, a dense mass of tissue, "moulds" the external contours, and those processes can be seen on the inner aspect of the nasal passages. There is in fact a

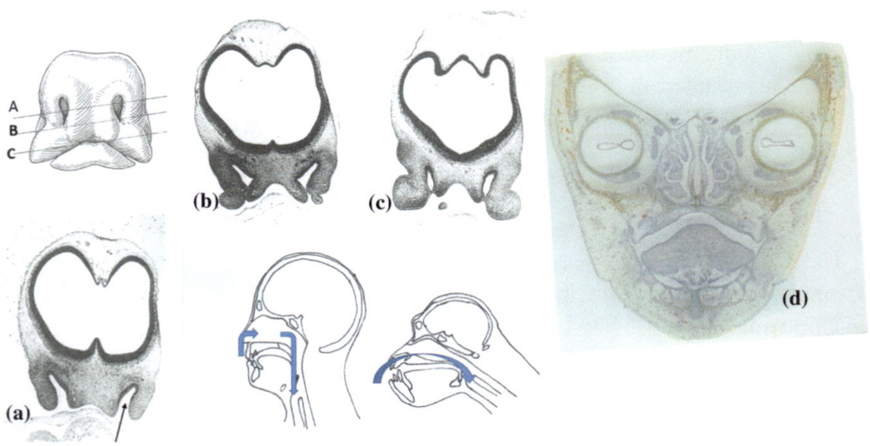

Fig. 7.16 Left upper drawing shows the ventral aspect of the face of a 10–11 mm (±40–47 days) human embryo with medial and lateral nasal fold around entrances of the nasal pits (arrow) and below it is the mouth cleft. **a–c** show consecutive sections through the head dominated by the large telencephalon. In **d**, a frontal section through a 5-months-old foetus shows the labyrinth of the nasal cavities with nasal septum and palate. The lower drawing indicates the airway through nasal nostrils, nasal cavity and the pharynx–larynx complex in man and chimpanzee

continuous mass, but due to external folding and contours, and in anticipation of future development, the mesenchymal mass is distinguished as facial processes or swellings" (Landsmeer†, unpublished).

Since the embryonic brain induces the grouping of the placodal cells into the head placodes that form sense organs and the frontal part of the brain, the telencephalon induces the olfactory placode, and attention is paid here to "induction". The role of induction is often misinterpreted. The nature of the inducer is less important. Often different tissues can mimic the same inductive phenomenon. The inductive effect is far more dependent on the individualities of the responding cells. There exists an induction time window. Before or after this window, that specific type of induction is impossible. Within that window, only a restricted amount of cells has to be induced. The activated cells will force the others in the population in the same direction of development (Chandebois and Faber 1983).

The genes and growth factors that play an important role in facial outgrowth are known. They are endothelin-1, SHH (sonic hedgehog), BMPs (bone morphogenetic proteins), Wnt (wingless-related integration site) and FGFs (fibroblast growth factors, see Sect. 2.3.2 sutures). The frontonasal mass outgrowth is also mediated by retinoid signalling. The epithelial–mesenchymal interaction, as described above, is dependent on Msx1 transcription factor. Cell migration, cell survival and cell differentiation (see Introduction) in the facial area need the transcription factors Pax6, Cart1 and Dlx5, respectively (Francis West et al. 2003). Four months later, this hollow tube within its end the olfactory epithelium has changed into a labyrinth due to the conchae extending into the tube's space. The nasal septum separating both nasal cavities has joined the palate (Fig. 7.16d). If the fusion between the nasal septum and both left and right palatal processes that can make the palate go wrong, anomalies of lip and palate (harelip and cleft palate) can occur. Note that the paranasal sinuses (see Sect. 7.4, below) appear late in the foetal development and early post-natal life. They start as small spaces that grow during childhood to their mature expansion in adulthood.

The construction of the nose and its airways is mainly explained by their capability to warm and moisturize the incoming air by the mucous layer. Inspiratory flow (inspiration lasts around 2 s) is generally considered as laminar flow, while expiratory flow has more characteristics of turbulent flow and requires more energy expenditure (expiration lasts approximately 3 s). Air that passes through a straight tube has a laminar flow. Near the edges of the *straight* tube, this laminar flow nears zero, while in the middle the highest air velocity is found. Since the mucous membrane of the nasal cavities covers the wall of the tube, nearly no exchange between wall and air is produced, for its flow is nearly zero. If turbulent flow is organized, its related circulation is less ordered, velocity is lowered relative to pressure, and exchange time is increased. The incoming air is moisturized and warmed. In the chimpanzee, the airway is a slightly bowed tube due to the presence of a snout and the form of the midface. In humans, a snout is absent, the nose is placed perpendicular to the midface, and due to the upright placed head the trachea is also at about a 90° angle to the nasal cavity (Hahn et al. 1993; Churchill et al. 2004). In fact, the airway contains two 90° angles increasing turbulent flow

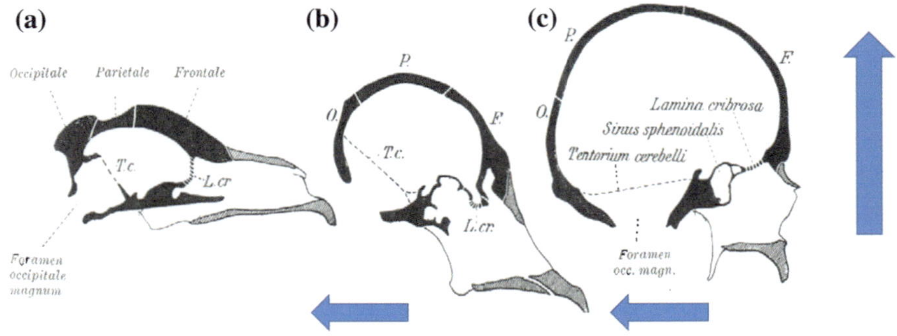

Fig. 7.17 Midsagittal sections of the skull of dog (**a**), ape (**b**) and man (**c**). The level of the tentorium cerebelli is stippled and the lamina cribrosa striped (Braus 1934). Abbreviation explanations are indicated in the figure. Arrows indicate changes in length

enormously. Warming and humidifying the air, also due to the conchae channels, increase proportionally (Fig. 7.16; see also Lieberman 2011). Since on the one hand the inspiration is considered to be a laminar flow and on the other turbulent flow is advocated for moisturizing, the air unanimity is absent.

The nose position is determined by the frontal skull bone. Increase of the brain causes a dome-shaped skull roof. The anterior part of the frontal bone pushes over the eye, producing a reduced eyebrow wall. Compared to other hominid species, this upper rim of the orbit is tilted towards the front in humans. The orbit roof even bulges further out anteriorly to give space to the frontal lobe of the brain in humans. The consequence is also that the relative distance between occipital bone and teeth implant in the maxilla decreases, while the height of the distance between parietal bone and the maxilla implant line increases. An index of this skull height multiplied with 100 and divided by the skull length shows: dog $\pm$ 35, ape $\pm$ 75 and man $\pm$ 90, indicating that the increase in skull height in relation to skull length in man is large, reaching nearly equal measures. Moreover, the relative length of the skull decreases, the most in ape and human compared to the dog. Between ape and human, there is less difference (see Fig. 7.17).

7.3 Taste and Smell Interaction

Now, the nose has been culturally and anatomically placed within the face, its blood vessels have been described, the nose development has been extended (see Sects. "Coherence Spatial" in the Introduction and Sect. 7.2.1), and smell has to be studied. Smell is part of a multimodal perception called flavour. Flavour is in the brain, and it is caused by combined, brain relayed, actions of taste, smell and somatosensory systems (mainly trigeminal-relayed information) (see Small 2012). Thus, smell alone is difficult to describe, due to its integration with taste and sensory inputs. We leave out here the interaction between flavour perception and hearing and minimalize flavour and visual cues (see the very critical book review by Spence 2012, on neurogastronomy and references herein)

In her book "Making Sense of Taste", Carolyn Korsmeyer (1999) treated food and philosophy. Her description of the pleasure of food (P) in the chapter Science of Taste has been put into a formula by these authors:

$$P = 3B + \frac{H \cdot C + 0}{T}$$

in which "B is causal factors of the body that are unchangeable (liking sweet or ability to taste PTC); 3 is a constant different for each person; H is bodily conditions at the time of ingestion (I am hungry, thirsty, stuffed, nauseated) and its variability can be manipulated; C contains cultural factors that enter into the selection of food and drink as palatable, good, edible and tasty and vary by society, and they are manipulated by education, experience and practice; T is the intentional object of taste that directs attention to the state of one's body; O is the intentional object considered as the item tasted"(see Korsmeyer 1999).

What is missing is smell. Korsmeyer treated smell in her book on several places to the interaction between taste and smell. One can distinguish two main types of reasoning:

(1) Taste in fact is determined by smell, without smell "no gourmets, only consumers of nutriments".
(2) Taste relies on touch of the tongue and mouth by the food, because taste receptors are everywhere in the mouth and are mainly activated by chewing.

Some avoid this discussion: "other researchers speak of a 'mouth sense' that combines taste, olfaction, chemical sensitivity, temperature and touch" (Korsmeyer 1999). In her chapter Philosophies of Taste, the following is noted: "philosophers from Aristotle onwards commonly observe that we do not call the objects of any senses except hearing and sight 'beautiful', since the bodily senses provide sensuous pleasure rather than the contemplative delight occasioned by beauty" or citing Bailie from Korsmeyer: The *Eyes* and *Ears* are the only *Inlets* to the sublime. *Taste, Smell* nor *Touch* conveys anything that is great and exalted, and from Burke, "smells and tastes have some share too, in ideas of greatness; but it is a small one, weak in its nature, and confined in its operations. I shall only observe that no smells or tastes can produce a grand sensation, except excessive bitters and intolerable stenches".

Thus, the prejudiced conditions are that taste and smell, called the chemical senses, can only give sensuous pleasure, smell is a necessary prerequisite for taste, and without activated touch receptors taste does not work. Any study of the interaction between smell and taste therefore seems biased. Nevertheless, Korsmeyer (1999) ends here philosophical approach with "In the next chapter I shall present reasons to suspect that some of the traditional assumptions about this sense exaggerate both the degree to which it is a natural disposition to respond with pleasure or displeasure and its stubbornly subjective character. Unravelling the philosophical prejudices in regard to taste appropriately takes a turn here in the direction of scientific studies of this sense". Here, one is redirected to this Chap. 5: Taste. Nevertheless, studies on the interaction between smell and taste need the utmost care, since bias is around the corner.

Why this attention for the philosophical discussion on taste and smell. Normally, good research elucidates the problems raised (and mostly not solved) by philosophers. Seemingly, psychological circumstances do affect your taste and smell perceptions. Korsmeyer's only example is given: "in three experiments, we found that interpersonally unfair treatment at work, defined as treatment that violates an individual's sense of dignity and respect and triggered disgust emotions over and above anger which subsequently related to stronger taste and smell reactions to gustatory and olfactory stimuli. This effect was observed for pleasant and unpleasant tasting products, for agreeable and malodorous scents, and among both mistreated victims and third parties" (Skarlicki et al. 2013). Therefore, research on taste and on smell and on its interaction is fiendishly difficult, because the psychological situation one is in can hardly be excluded in human research. We all know that dinner in a pleasant environment increases the enjoyment of the meal. In the laboratory setting, dinner will be less enjoyable. During development from baby to maturity and during ageing, your reaction to taste and smell changes. In 1999, a workgroup studied the effects of ageing on smell and taste: increase in age is related to decrease of smell, taste loss is minimal during ageing, unless drug intake increases, and decreased energy expenditure relates to decreased food intake caused by decreased appetite, but is often related to olfactory dysfunction and chewing problems in the elderly interference with taste sensations (Finkelstein and Schiffman 1999). Thus, while grandma loved the Indian restaurant, her ageing and her medication cause taste and smell changes and finally she avoids the Indian meals to surprise her children. And everybody knows that children have to learn to appreciate food other than the home-made cuisine.

Back to the interaction between taste and smell. Texture of food as noticed by touch and vision of food does interplay with taste and smell interaction. All information shows that a direct interaction between taste, smell, touch and vision does not occur at relay stations present before their cortical areas (Fig. 7.18). The direct cortical end stations of taste, smell, touch and vision even do not interact directly. In front of the taste's insula and frontal operculum, a cortical area is present called the orbitofrontal cortex. It is kept responsible for the interaction of taste, smell, vision and oral texture (touch) information. Some of its neurons are simply reactive to smell or taste information, others to both types of information and others to more input of different origin (temperature is also a factor). Thus, not earlier than information of smell, taste, touch and vision has arrived at this cortical station their interaction is possible (Rolls 2015). Everyone knows by passing the fish and chips stand that the smell makes the mouth watery and you can taste the food already by memory. It is the amygdala nucleus, which is at the start of the hippocampus, that organizes your smell and taste memory. It does it not only for pleasant feelings but also for aversive tastes and smells (Fig. 7.18; see Rolls 2009a, b, 2015; Poothullil 1995).

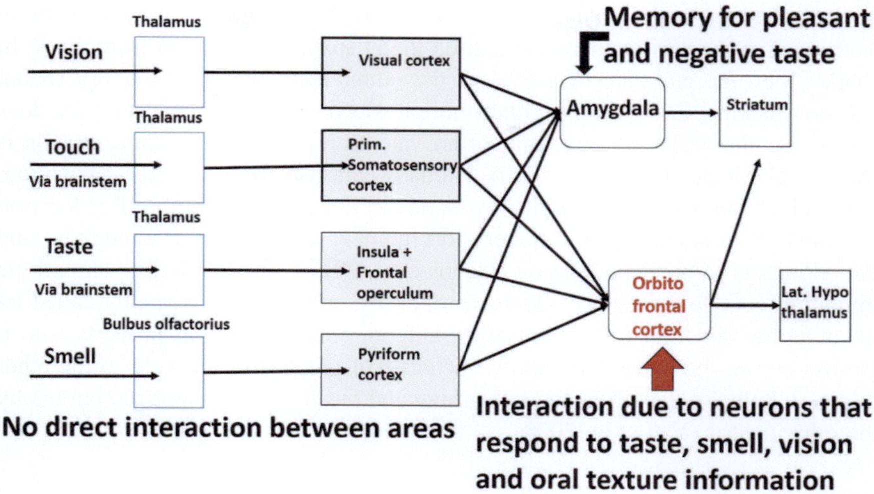

Fig. 7.18 Interaction between vision, touch, taste and smell (adapted from Rolls 2009a, b)

7.3.1 Can We Smell in Stereo?

The bilateral construction of the eyes and ears is used for stereo vision and for stereo acoustics, respectively. It goes without saying that the bilateral symmetry of the nose brings up the question of stereo olfaction. Results on insect larvae support the enhancing of signal-to-noise ratio by dual odour inputs. For several vertebrates, stereo smelling has been proven in the laboratory: sharks, moles and rodents. The odour information of each nostril remains at the start of the olfactory circuitry on the same side of the brain from the ipsilateral olfactory epithelium into the ipsilateral olfactory bulb towards the ipsilateral cortex. Halfway contralateral connections originate (see Fig. 7.21). Odorant receptors, group for the same odour, are organized in distinct antero-posterior directed zones. The same receptors project to a specific pair of glomeruli in the olfactory bulb (Johnson and Leon 2007). Thus, a stereo-typed, mirror-symmetric glomerular map is present for both olfactory systems. The connection between left and right olfactory system goes via a part of the anterior olfactory nucleus. This nucleus is a portion of the olfactory cortex and sends its odour information along the anterior commissure towards the specific iso-functional odour glomeruli on the other side (Fig. 7.21). Interbulbar connections are inhibitory due to their GABAergic neurotransmitter use. “Interbulbar reciprocal connections may enhance the difference in inputs between the two nostrils, allowing the ‘stereo’ sensation of odours” (Imai and Sakano 2008). This mechanism has been found in rats, in which the anterior olfactory nucleus neurons can discern between right and left position of an odour source. These nucleus neurons reference signals from the two nostrils (Kikuta et al 2010).

Stereo odour sensation has also been related to navigation and to olfactory learning. Cross-nostril comparison is used to get spatial information from smell. In sharks, the time and side of arrival of the odour determine the olfactory spatial decision-making. Even if a low concentration was first, a higher concentration does not change the decision. Thus, time of arrival and not the odorant concentration is pivotal (Gardiner and Atema 2010). Humans can also use odour for navigation. Once a location has been identified by its odour, humans can return to that location by odour information alone. A rather short moment (one minute) is enough to map the odour and to navigate later on (Jacobs et al. 2015). The interbulbar circuitry is important for odour memory. Odour learning by one nostril is normally called by the other nostril. After transection of the anterior commissure, this property is lost. Destruction of the anterior olfactory nucleus that projects exclusively to the other olfactory bulb did disturb olfactory learning and recall of the memorized odours by the other nostril (Yan et al. 2008).

7.4 Sinuses

One should note in advance that scientists and clinicians do not know the function of sinuses and its morphology shows a strong evolutionary, developmental and mature variability that makes any problem impossible. Development of the ethmoidal sinus is retarded in 10% of the studied patients, and the same holds for the maxillary sinus in 4%. Of patients with inflammatory sinuses, 20% showed anomalies of the sinusoidal structures and in 55% the middle concha (important for air flow) showed deflections. To it, one can add: deviations of supraorbital recess in 6% of the cases, concha bullosa (**middle turbinate pneumatization**) in 30% and pneumatization of the anterior clinoid processes in 12%. Asymmetry is nearly normal in sinuses, and sometimes an entire sinus is absent. In evolution, the Neanderthal skull is heavily pneumatized and scientists have attributed it to cold adaptation. However, Eskimos contain smaller frontal sinuses and Japanese apes living in the colder north do have smaller frontal sinuses compared to those living in the south of Japan. Here, one notices again the spatial coherence idea: "this is consistent with a model in which the final size is determined by the available space, and the sinuses are excavated in response to lack of bone loading" (see O'Higgins et al. 2006).

7.4.1 Midface Construct

Scientific American of December 1979 published "*The Neanderthals*", written by Trinkaus and Howells. Both are experts in Anthropology, and their approach placed the Neanderthals back into the row of ancestors to *Homo sapiens* and out of the ape-like line. Here, two citations from their text follow: "the Neanderthal face is unique. A prominence down the midline brings the nose and the teeth farther forward (with respect to the vault of the skull) than they are in any other human fossil, either older or younger. The cheek arches slope backward instead of being angled, as they are in modern 'high cheekbones'. The forehead slopes instead of rising abruptly as it does over the tucked-in face of modern man". Somewhat further on, after stating that the Neanderthal face is large, but smaller than in man, "the nasal cavity and the rounded eye sockets are capacious. The sinus cavities are also large. For example, the frontal sinuses fill the brow ridges from above the nose out to the middle of the eye sockets with multichambered 'cauliflower' cavities. They do not, however, reach up into the frontal bone above the brows, as is the case in earlier members of the genus *Homo*. In modern skulls, the frontal sinuses are flattened, often extend above the brows and are quite irregular in size and shape. In order to explain the Neanderthals' large projecting face, scholars have invoked a variety of causes, most often adaptation to cold. Carleton S. Coon and others have suggested that the Neanderthal midfacial prominence was such an adaptation. The nasal cavities were placed well away from the temperature-sensitive brain, and at the same time the enlarged cavities may have provided additional space for warming inhaled air". The authors end up by indicating that the Neanderthal facial morphology is presumably a multifactorial result, among which the chewing apparatus and cold are important. In this chapter, one regularly encounters the temperature and humidity-related function of mucous covered spaces in the head. It is a narrow entrance to these nasal cavities (Fig. 7.19), and airflow studies do not show turbulences within the sinuses (see below). To show that several other functions are discussed, the citation from the article on cooling the brain is given: "the paranasal (frontal, ethmoid, maxillary and sphenoid) sinuses are essentially mucosa-lined airspaces within the bones of the face and skull, the physiology of which has been the subject of much research. Multiple theories of function exist. These include the functions of warming/humidification of air, assisting in regulation of intranasal pressure and serum gas pressures, contributing to immune defence, increasing mucosal surface area, lightening the skull, giving resonance to the voice, absorbing shock and contributing to facial growth. We suggest that all of these functions are part of a bigger, more involved picture than is now apparent and they are likely an important component of a selective brain cooling system. Intracranial temperature measurements have demonstrated that the rostral base of the brain can be cooled by conduction to the nearby upper respiratory tract. A very thin layer of bone separates the posterior ethmoid air sinus from the subarachnoid space, and only thin plates of bone separate the sphenoidal sinuses from internal carotid artery and cavernous sinuses, so that adjacent brain lobes and vessels could be cooled by conduction. The cooling may also occur through convection process caused by evaporation of mucus and the transfer of the cool venous blood from paranasal mucous membranes to the brain. These may demonstrate the very substantial potential for cooling of the brain by paranasal sinuses" (Irmak et al. 2004). The midface is characterized by spaces: paranasal sinuses, nasal cavity and orbit (see Fig. 7.19). Pneumatization is present by the frontal, ethmoid, sphenoid and maxillary sinuses, and within the mastoid analogous spaces or cellules are present.

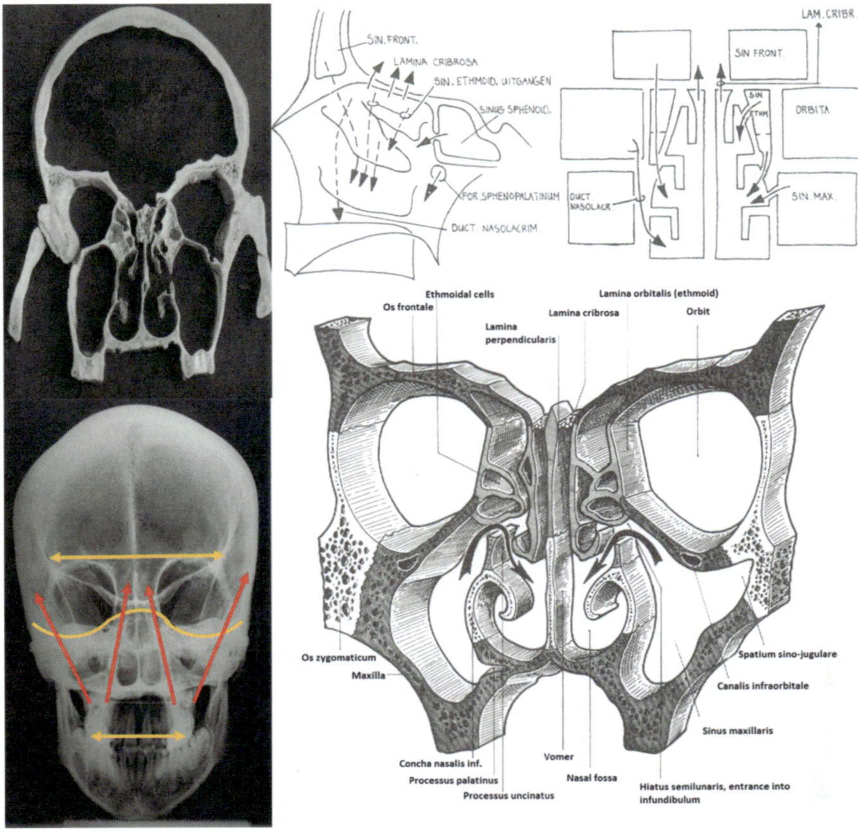

Fig. 7.19 Nasal sinuses: left transverse cut through human skull with its explanations to the right. Lower X-ray shows the horizontal (yellow) and vertical (red) buttresses by arrows. (The left figures are from the Neuroregulation Group, LUMC, Leiden. Right part is adapted from Legent et al. 1976)

The consequence is that the midface misses a solid construction. Two bone pillars are vertically placed on each side: the anterior medial nasomaxillary buttress and the lateral zygomatic-maxillary buttress. Three horizontal reinforcements are discerned: supraorbital frontal bar, the infraorbital rims and the maxillary alveolar process (Fig. 7.19; see Hardt and Kuttenberger 2010).

Facial trauma causes fractures that are commonly known, and standard classifications are used, e.g. Le Fort fractures or Wassmund fractures. These fractures taught us that a fracture over the nose root, separating it from the frontal base, and through both zygomas (Le Fort II fracture), mobilizes the midface (maxilla, nose and ethmoid) (see Hardt and Kuttenberger 2010, their Fig. 3.6). In other words, the midface is clicked in between skull base and mandible. In those mammals that have a snout (e.g. dog, rabbit), the nasal bone, the mes-ethmoid (extra extension of ethmoid to lateral), vomer, palatine and premaxilla all are enlarged and contribute to the snout. The dog and rabbit have a trumpet bell (visceral cranium) with a real tube (the snout) with consolidations of the trumpet wall. Ours is a disc, flattened without a real tube to the bell (see Fig. 7.17). A lateral X-ray of the human head demonstrates that the area between anterior skull base and dental part of the maxilla is poor in bone construct. It looks as if a gap is present where the midface is located. So, the question is how is the midface organized as a lightweight block and contains for its subdivisions thin bony walls to separate the midface spaces? Let us first put a series of arguments that are derived from evolutionary studies:

(1) The content of the neurocranium increases from the Australopithecus with 400–500 ml to the modern man containing 1400 ml. It means that the weight of the brain increased nearly 300%.
(2) Rough calculations show that our head weight is 5–6 kg, while the contribution of the human brain is 1.4 kg. The brain is responsible for 25% of the head's weight. Due to the extension of the brain, extra skull cap bone weight should be added to this calculation.
(3) Australopithecus africanus (brain content 440 ml) and A. robustus (brain content 530 ml) were characterized by strong masticatory muscles that even forced a crest on the skull cap for anchoring. Moreover, their bony skull contained bow ridges, reinforced zygomas and strong frontal directed maxillas. The Australopithecus head was typified by bone consolidations. The hominid head is characterized by reduction of zygomas, reduction of bow ridges and reduction of the chewing muscles by change of the hominid diet.
(4) A reduction in the shoulder and neck muscles is also noticed (mm. splenius, semispinalis, trapezius) if compared to chimpanzee (our closest living relative, see Lieberman 2011). Although still a nuchal ligament can be discerned in humans, it is greatly reduced.

By summing up these facts, one should conclude that reduction of masticatory muscles and bony adherences for these muscles in Homo sapiens, which is in fact reduction of weight, coincided with increase in brain weight. Moreover, due to human head orientation with the vertebral column perpendicular to the "middle" of the skull base, head balance is forcing the centre of gravity directly above the vertebral column, and reduction of the neck and shoulder muscles is then possible.

Now, look into the hominid line. The differences between hominid skulls and Neanderthal skulls have been studied by morphometric analysis of growth trajectories (Ponce de Léon and Zollikofer 2001; Zollikofer and Ponce de Léon 2004; see also Lieberman 2011). Within the skull depository and resorptive, growth fields can be discerned. These growth fields show a strong resorptive area within the human midface (Dutterloo and Enlow 1970; Enlow 1990) that is supported by the morphogenetic studies of Zollikofer. The morphogenetic studies concluded that "early ontogenetic modification of a small set of growth parameters is a major source of evolutionary novelty during hominid evolution" (Zollikofer and Ponce de Léon 2004). Moreover, in the earlier publication of Ponce de Léon and Zollikofer (2001) they state that "Homo neanderthalensis and Homo sapiens represent morphologically discrete, separate species, which belonged to distinct evolutionary lineages". The bony reduction of the midface therefore should be considered a unique development restricted to Homo sapiens. A second fact is that there is hardly any relation between the skull morphology and masticatory muscles in Homo sapiens, while such a relation is present in apes, Australopithecus and Neanderthals. Studying this, only the human temporalis muscle could fulfil such a relation. The temporalis muscle and perhaps the masseter muscle cross-sectional area seem to be weakly positively correlated with facial width (Weijs and Hillen 1986; Toro-Ibacache et al. 2015).

The largest strains during biting are present in the midface and of course at the biting point. This holds for molar bite and incisor bite; during incisor bite, strains are more to the frontal maxilla and palate and at molar bite more to lateral of the maxilla up to the orbit (Toro-Ibacache et al. 2015). Thus, although the midface has a lightweight bony construction, it still can withstand bite strains, which should be reassuring for orthodontists replacing teeth by implants.

7.4.2 Blood Vessels, Airflow and Nasal Drug Uptake in the Interior of the Nose

An aspect of the nose that has not yet been treated is the airflow in the nasal cavity. The nasal space is subdivided into two halves by the septum, and each halve is further subdivided into the anterior aperture at the facial side. Next, the nasal vestibule is discerned. The following space is divided into an upper area, related to the upper turbinate with the olfactory epithelium, and lower area, related to the two other turbinates and called the respiratory region. The outlet is the posterior nasal aperture ending into the rhinopharynx. There is a direct relation between airflow and uptake of substances by the mucous layer and the blood vessels of the interior of the nose: think of cocaine snorting. Cocaine is placed on a mirror, chopped, placed in lines and next snorted. The inhaling of the cocaine asks for a strong airflow within the nasal cavity. The effect after long cocaine use is wearing through the mucous membrane and perforation of the nasal septum. One becomes "high" due to the uptake of the cocaine by the blood vessels that do have a close topographical relation to the brain (AA and VV ethmoidalis and sphenopalatine, see Fig. 7.21). Or the cocaine has first to be transported to the heart and from there pumped into the brain vessels to have its effect in the brain (the systemic pathway to the brain), or it can be taken up by the olfactory epithelium and enter the brain (nasal olfactory pathway, see also below).

Before discussing nasal airflow, a few remarks are needed. Due to the anatomical structure of the interior of the nose (Figs. 7.19 and 7.20), any measuring by instruments within the nasal cavity will disturb the airflow. Measuring is replaced by modelling, but modelling of the nasal interior needs (over)simplification inducing inaccuracies. Therefore, translation of the model results to the real airflow has shortcomings (for an overview of these problems, see Doorly et al. 2008). The

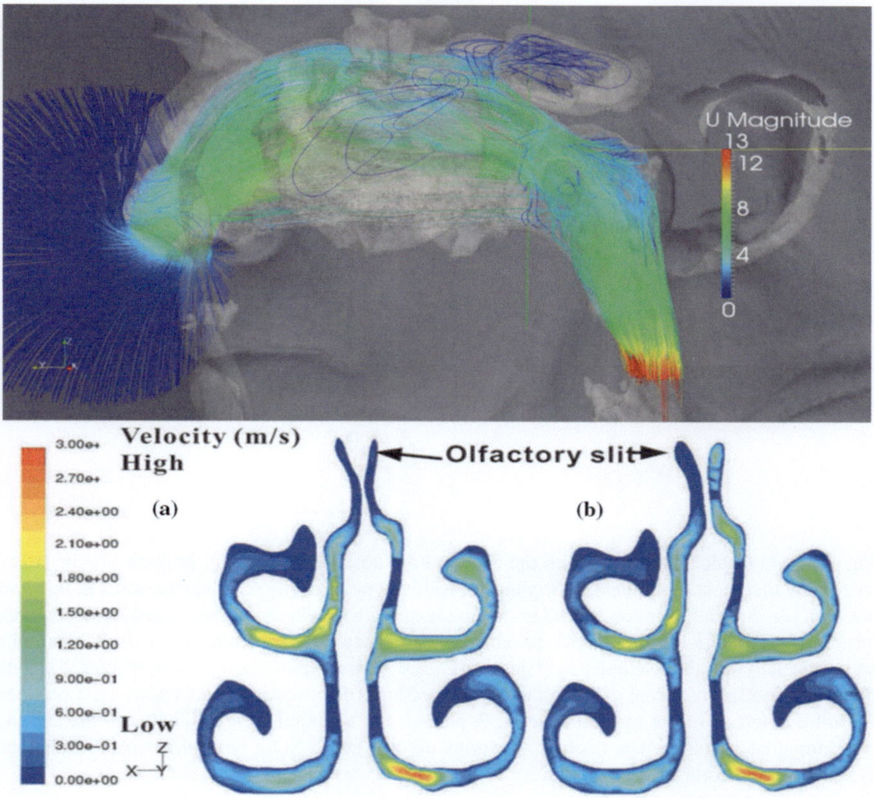

Fig. 7.20 Pictures of airflow in the nasal cavity. Upper figure shows three-dimensional view of the nasal cavity flow, equivalent to a steady inspiration. Flow is illustrated through the trajectory of massless Lagrangian tracers, released in front of the head and colour-coded with the local velocity, measured in m/s (Quadrio et al. 2016, with permission). Lower figure is coronal sections through the nasal cavity. These contour plots of velocity magnitude 6.2 cm from the tip of the anterior nostrils. Airflow is simulated by applying atmospheric pressure at the external nares plane and a steady pressure of 15 Pa at the nasopharynx. **a** Narrowed olfactory slit. **b** Original airway model generated from CT scans (Zhao et al. 2004 with permission)

pictures of Fig. 7.20 are derived from model studies and should be considered with care. Upper part of Fig. 7.20 (steady inspiration) shows the change of velocity around the turbinates as expressed by the colour coding that is proportional to the magnitude of the velocity vector. Upward stream reaches the smell area below the cribriform plate. Moreover, the highest velocity is noticed in the nasopharynx. Lower part of Fig. 7.20 demonstrates in coronal sections that the highest velocities within the nasal cavity are found below the inferior turbinate and below the middle turbinate (the respiratory region), a result that has been supported by several articles (e.g. Doorly et al. 2008; Chung and Kim 2008). Narrowing the nasal cavity just at the olfactory slit inhibits airflow at the smell area (lower part Fig. 7.20). Moreover, the upper part of Fig. 7.20 "shows the presence of a recirculating region in

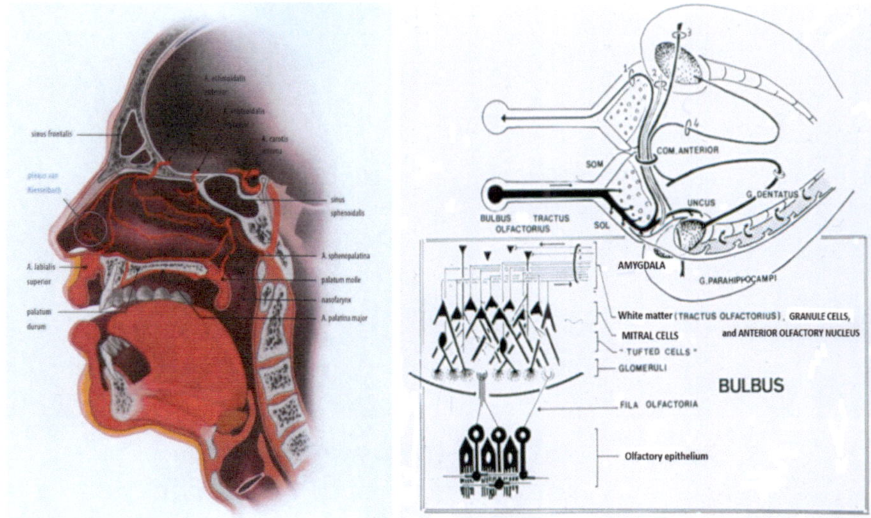

Fig. 7.21 Left: blood vessels within the inside of the nose (fossa nasalis), projected on its lateral side. Note that most blood vessels originate outside the nasal cavity. The anastomoses in this area are multiple. Special attention should go to the plexus of Kesselbach for nose bleeding: a posterior bleeding is mainly a bleeding of the proximal parts of these arteries (courtesy to Dr. M. van Dijk and especially to the Nederlands Tijdschrift voor Geneeskunde, see: 2015;159:A9352 with permission). Right: scheme of the olfactory connections: commissural connections exist between 1, Bulbi; 2, left and right pyriform cortex; 3, parts of the temporal cortex; 4, between the left and right amygdala nucleus. The tractus olfactorius divides into a SOL: stria olfactoria lateralis and SOM: stria olfactoria medialis (courtesy J. Voogd)

correspondence of the maxillary sinus and—more importantly— on the uppermost region of the nasal cavity, at the upper boundary of the so-called olfactory cleft where most of the olfactory receptors are located" (Quadrio et al. 2016; see also Doorly et al. 2008).

The question whether intranasal volume influences olfactory function leads to controversies. Hong et al. (1998) state that there is no correlation between smell ability and size of the nasal and sinus structures. A series of publications say that shape and volume of the nasal cavity do influence olfactory function (see Damm et al. 2002 and references herein). Hong et al. (1998) stated from their results that left smell receptors are used for preference, while Damm et al. (2002) found a right-sided predilection for odour thresholds and volumes (see also Sect. 7.3.1. Can We Smell in Stereo?). Moreover, "total, not partial, opacity of the left olfactory cleft, frontal recess or ethmoidal infundibulum was correlated with decreased sense of smell" according to Hong et al. (1998), while others do find olfactory changes by reduction of the olfactory slit (see Fig. 7.20b; Zhao et al. 2004).

A more fundamental question is brought up by Hong et al (1998): "there may be olfactory receptors inside the sinuses, and there may be a relation between the trigeminal receptors in the sinuses and the olfactory system". Indeed olfactory sensory neurons are found in the frontal sinus of the dog (Skinner et al. 2005), and the presence of olfactory neuroblastomas in sphenoid, maxilla and ethmoid sinuses (Holmes et al. 2016) could indicate the presence of olfactory neurons within the mucosa of the sinuses.

The epithelial/mucous layers in the nasal cavity are known for their underlying fine network of small blood vessels. This holds for the septum, the turbinates and the sinuses. This network is needed for the production of mucus. The mucous layer produced by the epithelial cells is shifted with a displacement of 1–10 mm, which means that within half an hour this layer is totally removed from the sinuses. Within the respiratory epithelium, four cell types are recognized: non-ciliated columnar epithelial cell with microvilli, goblet cells that produce the mucous layer by their mucus granules, the basal cell and the ciliated columnar cell. They are placed on a basal connective tissue membrane. Mucus can have different compositions but in general in the nose and sinuses a low viscosity sol layer is present at the level of the cilia and a viscosity gel layer is present on top of it. The olfactory epithelium is differently organized. In between sustaining cells, olfactory sensory receptor cells are present containing tight junctions, while at the basement membrane basal cells are located. The olfactory receptor cells form a knob with cilia at the nasal lumen side and send a non-myelinated axon that passes the basal membrane and forms together with other axons small bundles that enter the olfactory bulb (see Fig. 7.21). The olfactory receptor neurons have a restricted lifetime (30–90 days). A constant neurogenesis takes place, and ageing of this renewal system is unknown (Illum 2000). The bundles of olfactory axons are surrounded by olfactory glial cells. These cells, important in neurogenesis, are known for their ability to enhance axonal regeneration and are applied in spinal cord injury (Skinner et al. 2005). Mucus that covers the olfactory receptor layer is produced by Bowman's glands that are placed in between the receptor and sustaining cells. The production of mucus is different in the respiratory region (individual goblet cells) and the olfactory region (clustered mucous cells in a gland). The olfactory epithelium also contains in between their cells sensory trigeminal axons. The meninges are innervated along their blood vessels by sensory trigeminal axons with main neurotransmitters: substance P, Neurokinin A, CGRP and, the often overlooked autonomic innervations, the parasympathetic innervation from the ganglion sphenopalatine with main neurotransmitters: acetylcholine, VIP, NO and sympathetic innervation from the cervical superior ganglion with main neurotransmitters acetylcholine, neuropeptide Y and ATP. One should understand that there exists an intricate series of actions between these three systems and with mast cells (see Aubineau et al. 1980; Delépin and Aubineau 1997; Chaldakov et al. 2014).

The uptake of substances by the olfactory epithelium depends on the properties of the substance involved. The uptake is determined by molecular weight (small molecules are taken up more easier than large molecules), lipophilicity (strong lipophilicity forces uptake in the systemic route, and moderate lipophilicity supports olfactory uptake) and the degree of dissociation. The pathways for olfactory uptake are across the epithelium by moderate lipophilic substances and through tight junctions, present between the epithelial cells, by more hydrophilic drugs. The olfactory uptake pathway passage is fast (cocaine euphoria occurs within 3–5 min), and drugs pass rapidly into the cerebrospinal fluid and olfactory bulb. The olfactory pathway seems to be used for substances that cannot pass the blood–brain barrier via the systematic route, but viruses are also known to take this olfactory passage (Illum 2000). The nasal drug delivery route is common in our lives. Nasal sprays for sinusitis and cold are the best known, but they only attack within the nasal cavity. The pharmaceutical industry detected this route, and nasal sprays for allergic rhinitis, migraine, treatment of severe pain (fentanyl spray), for smoking cessation and for menopausal symptoms are on the market. These sprays have their effect within the neurocranium. Enhancers of the uptake, called nasal promotors, benefit the uptake of larger molecules like peptides and proteins. Nasal peptide products that are on the market now are, e.g., calcitonin and oxytocin (low systemic levels of these substances already have effect), and the expectation is that with a nasal promotor their uptake effectivity will increase. However, the lack of these nasal promotors forces the industries to find molecules that will fulfil such a role to administer peptide drugs via the olfactory pathway. Nasal delivery by nano-encapsulated drugs is being investigated now (Illum 2012; McDonough et al. 2010).

The optimism about the use of the nasal uptake route is qualified by the study of 73 publications on 82 compounds: "outcomes of this analysis indicate that efficiency of brain delivery by the nasal route differs widely between the studies and does not correlate with the drug's physicochemical properties. Particle and gel-based delivery systems offer limited advantage for brain drug delivery in comparison with the intranasal administration of drug solution. Nevertheless, incorporation of specialized reagents (e.g. absorption enhancers, mucoadhesive compounds, targeting residues) can increase the efficiency of drug delivery to the brain via the nasal route" (Kozlovskaya et al. 2014). Note that the efficiency of peptide absorption by the nasal route is below 1% (see Illum 2012).

7.4.3 Bulk Flow Transport of Therapeutics

The effect of several compounds taken up via the nasal route is noticed within minutes. Comparison between blood vessel injection and nasal application of compounds shows that for most substances the appearance within the cerebrospinal fluid is earlier for the nasal administration (e.g. cocaine, progesterone, zidovudine [AIDS inhibitor], see Illum 2000). Most substances that are brought into the nasal cavity not only will be taken up by the olfactory epithelium, but also by the fine network of blood vessels in other parts of the nasal cavity. It means that the systemic route could still operate. A second problem concerns the uptake itself. Cocaine needs to reach special brain areas to have its effect. Animal studies confirm that cocaine is rapidly taken up, but the special brain areas are reached differently and high concentrations are present in the olfactory bulb, olfactory tract and less in the specific brain areas. Since the "nasal" systematic route also works for cocaine, it remained doubtful at the start of the research whether it was the nasal route that provided the specific brain areas with low concentrations of cocaine or that this was the consequence of the systemic route administration of cocaine. Diazepam administered in animal experiments showed uptake and after 10 min reached the olfactory tract, indicating that the olfactory brain system is involved in uptake of compounds via the nasal route. Neuroanatomical tracers (radioactive, enzymatic and fluorescent substances are used) were added into the olfactory bulb. From these earlier studies with tracer injection in brain parts, one knows the projection areas of the olfactory system (Fig. 7.21). So, the expectation is that radioactive or fluorescent dyes (often lipophilic) or other tracer substances administered into the olfactory epithelium, thus using the nasal route, should reach the olfactory target areas rapidly. This does not occur. [^{125}I]-labelled interferon-β1b is found after 30 min. in high concentrations in the olfactory bulb, the trigeminal branches V1 (ophthalmic) and V2 (maxillary), and after 60 min. in the basal ganglia of monkeys (Lochhead and Thorne 2012). The basal ganglia are not a target of the olfactory system (see Fig. 7.21; Shipley et al. 1996). An analogous result has been obtained by intranasal administration of chloramphenicol acetyltransferase (Appu et al. 2016). These results are supported by insulin-like growth factor-1 (brain reached within 30–60 min), insulin, nerve growth factor, benzoylecgonine (polar metabolite of cocaine), morphine, testosterone (Dhuria et al. 2010), exendin (9–39 min) (an antagonist of glucagon-like peptide-1) and galanin-like peptide (Lochhead and Thorne 2012).

This rapid delivery of compounds into the nervous system within 1–2 h cannot be explained by diffusion (over 10 h are needed) or axonal transport (4–5 h or more are needed) (Lochhead and Thorne 2012). The explanation given in the literature is bulk flow. Bulk flow is driven by fluid pressure gradients. The brain cells are surrounded by interstitial fluid that at brain's edges connects with the surrounding cerebrospinal fluid, and these fluids do have a (slow) flow. This is comparable to flow through a saturated porous substance (in this case the brain cells with their intercellular spaces) that can be described by Darcy's law. By the way, Darcy (1803–1858) detected this effect during his construction of the aqueduct of 12, 7 km with sand filters in and around Dijon to get proper drinking water into the city. The description of the effect is ultimately useful for geohydrology e.g. for these authors the maintenance of the Dutch dikes. Darcy's law defines that flow through a saturated porous substance (dike or brain) is proportional to the difference in head of cerebrospinal fluid and inversely related to the length or distance of the flow.

$$V = Kh/l$$

in which V = Velocity of cerebrospinal fluid, K = Permeability of brain tissue, h = Head of cerebrospinal fluid (head: difference in elevation between two points in a body of fluid) and l = Length of flow.

The presence of added substances within the olfactory bulb due to intranasal administration and within the cerebrospinal fluid causes these fluid pressure gradients with the consequence of bulk flow that is faster than diffusion and only needs 1–1, 5 h to reach the back of the experimental monkey brain. These bulk flows follow neuronal connections outside the brain, but within the neurocranium, like the trigeminal nerve branches to reach the brainstem and use tracts inside the brain (Abbott 2004). Perivascular spaces around the blood vessels are seemingly also involved as channels for bulk flow. But the brain vessel walls contain blood–brain barrier, which seriously restricts the amount of substances, especially larger proteins, which can pass, and therefore, the transport must be related to interstitial perivascular flow (see Abbott 2004). Since the area perforata, between olfactory tracts, contains a myriad of small blood vessels that penetrate from below into the basal ganglia, it is understandable that bulk flow mainly transports into the basal ganglia. Bulk flow seemingly shows perivascular space distributions and not axonal connections. Bulk flow also called convection-enhanced delivery has been used in clinical applications. Inserting an infusion cannula near the desired brain part causes targeting of the added therapeutics by the bulk flow. The neurological disorders that are researched concern lipofuscinosis, Gaucher and Parkinson's disease, multiform glioblastoma and intrinsic pontine glioma (Lonser et al. 2015).

7.4.4 Spatial Aspects of the Walls of the Nasal Cavity

During the earliest development of the skull, the frontal basal part of the neurocranium contains two main cartilaginous structures: on both sides a lateral plate, called the nasal capsules, and a midsagittal construct named trabeculae. In front and above these structures is the frontal bone centre. More posterior are placed two cartilaginous structures known as the orbitosphenoids (see Chap. 4 for development of the primitive neurocranium). Below these structures are present the premaxillary and maxillary centres that fuse and produce the maxilla. Later on zygoma and pterygoid relate to the maxilla. These structures have to construct the walls of the nasal cavity and its cavity dividing septum. The oral cavity is in connection with the nasal cavity since the palatum still has to grow out from both the maxillary sides and must fuse with the midsagittal septum.

The nasal capsule ossification leads to the formation of the ethmoid. The ethmoid after being developed into its final position contributes to both the nasal septum and to the nasal lateral walls (see Fig. 7.19). The vomer, one of the bony components of the nasal septum, is membranous in origin, while the perpendicular part of the ethmoid is cartilaginous in origin. These two structures fuse, forming a T-like structure producing roof and septum of the nasal cavity together with the vomer that originates from membranous ossification. The vomer is considered a spacer between the central part of the sphenoid and the hard palate, and the palate can be considered a spacer between the two maxillae. The ethmoid ossification and outgrowth produce also the superior and medial concha. The inferior one is an independent developing bony structure (Fig. 7.22).

The pterygoid is related to the nasal cavity due to its position behind and aside of the maxilla and to the oral cavity (Fig. 7.22) by its contribution to adhere two masticatory muscles (medial and lateral pterygoid muscles,). The pterygoid finally becomes an integral part of the sphenoid. The palatine part of the nasal cavity floor,

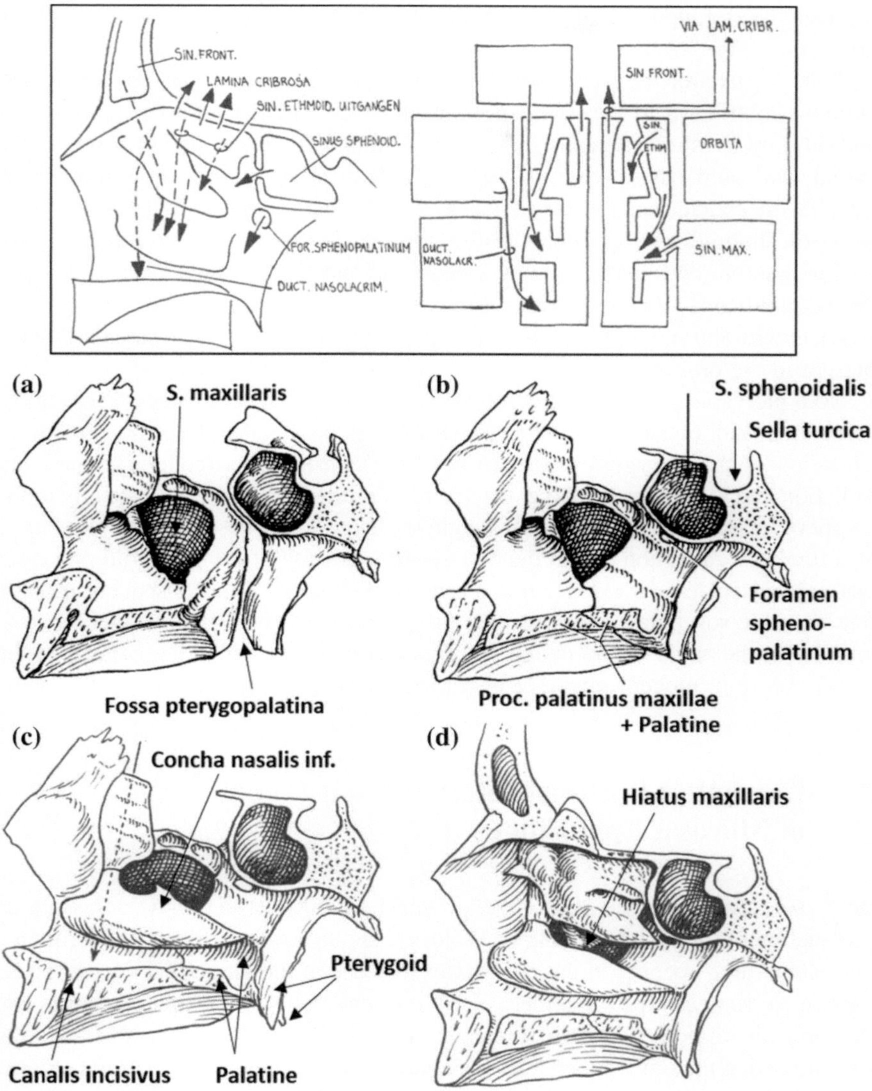

Fig. 7.22 Four consecutive views of the lateral wall inside of the nasal cavity (**a**-**d**). Upper figure shows the sinuses in a sagittal and a frontal scheme. Stippled arrow indicates the position of the ductus nasolacrimalis (tear duct), upper figure courtesy N. Gerrits, and lower four are from Landsmeer, Leiden Dept. of Anatomy

localized behind the palatum, extends upward along the pterygoid and reaches the sphenoid. The palatine has an L form with the vertical part contributing to the lateral wall. One extension of the palatine is the so-called sphenoid process. The second is the orbital process. These two extensions enclose the sphenopalatine foramen, also called incisura sphenopalatine, the deep end of the fossa pterygopalatine (Fig. 7.22).

Thus, a junction is established between maxilla, ethmoid and sphenoid. The ethmoid contributes to the lateral nasal cavity wall and separates the nasal cavity and orbit by its ethmoidal cells (Fig. 7.19). The maxilla forms the lower part of the lateral wall of the nasal cavity (Fig. 7.22). As a consequence, the ethmoid and maxilla do contact each other at the medial border of the maxilla and contact between ethmoid and sphenoid is also formed, but more postero-dorsally, producing a suture between both bony pieces. The palatine's orbital process contacts the ethmoid too, and the palatine extends dorsally to contact the sphenoid. Its contacts with the maxilla, ethmoid and sphenoid are expressed as sutures that are present in the orbital wall.

Both pterygoid processes are strong bony pillars and therefore capable of energy absorption and of support of the posterior end of the maxilla. The downward extension of the pterygoid processes has been considered a modern human constellation and affects the caudal extension of the maxilla. This relates directly to the height of the nasal cavity and to the height of the posterior nasal orifices or nares. It also implies a retraction due to the downward extension of the pterygoid processes that organizes a less far extension in a backward direction. The consequences are: a free movable soft palate and it favours the proper position of the medial pterygoid muscles at the angle of the mandible. This angle supports the enlargement of the mouth cavity in modern man (see also Landsmeer, unpublished).

7.5 Combat of Bacterial Infections and Immunology of Mucosal Epithelium

Here, we start with the most unexpected fight against bacteria, although an explanation of the lymphatic ring of Waldeyer should go first (Fig. 7.23). From the Eustachian tube opening in the pharynx to the base of the tongue lymphatic tissue, forming a ring can be discerned. Two concentrations of lymphatic tissue are anatomically discerned within this ring: near the Eustachian tube opening, the adenoid and at the oral wall end, the palatine tonsil. At the back side of the tongue, another concentration of lymphatic tissue is present. Micro-organisms that enter are fought by tonsillitis and adenoiditis.

"We previously observed that the human bitter taste receptor T2R38 is an important component of upper respiratory innate defence because it detects acyl homoserine lactone (AHL) quorum sensing molecules secreted by gram-negative bacteria. T2R38 activation in human sinonasal epithelial cells stimulates calcium and nitric oxide signals that increase mucociliary clearance, the major physical respiratory defence against inhaled pathogens" (Lee et al. 2014). The bitter taste receptor not only functions in taste, but is also involved in detection of bacteria. These taste receptors are expressed in the sinonasal cavity and detect quorum sensing molecules of bacteria (Lee and Cohen 2014). The production of these taste

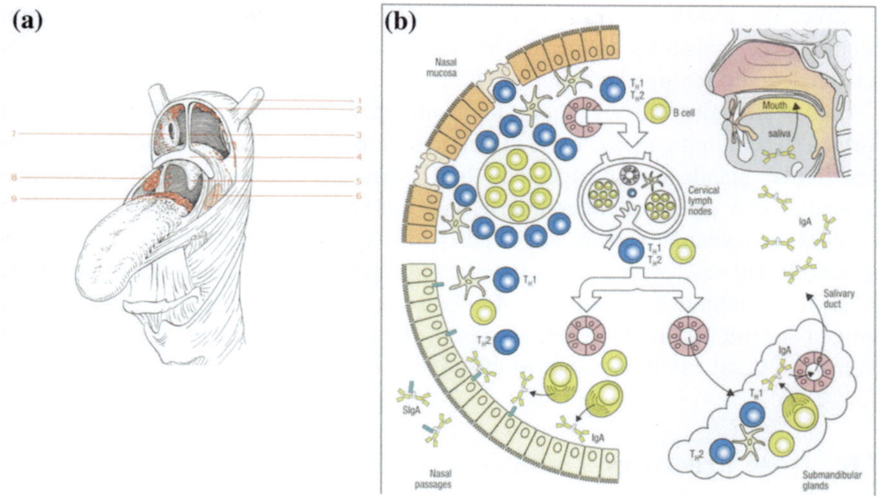

Fig. 7.23 A shows the lymphatic ring of Waldeyer adenoid at the naso-oral transition towards the pharynx: 1, Eustachian tube; 2, adenoid or pharyngeal tonsil; 3, nasal septum; 4, palatum; 5, uvula; 6, left palatine tonsil; 7, opening Eustachian tube; 8, right palatine tonsil; 9, lingual tonsils (courtesy J. Tinklenberg). **b** Gives an overview of the nasopharyngeal–oral mucosal immune system (with permission from Smith et al. 2013). M cells placed between the epithelial cells are capable to ingest viruses and bacteria to expose them to dendritic cells and T cells. Antigen activation results in stimulation of T and B cells producing IgA. Via the epithelial venula cervical lymph nodes are co-activated and submandibular glands reached too. These glands increase the production of IgA by the activated T and B cells that accessed these glands by the venula

receptors is seemingly within the brush cells and in lining cells of the duct entrances of mucosa producing glands (Krasteva-Christ et al. 2015; Perniss et al. 2016). Moreover, the release of acetylcholine from ciliated and secretory mucosal cells contributes to the activation of the innate defence mechanisms. Next to the taste receptor activation, the local auto-/paracrine function of acetylcholine release contributes to the mucociliary clearance, regulation of macrophage function and modulation of sensory nerve function (Kummer et al. 2008).

The nasopharyngeal and oral immune system (Smith et al. 2013) are subdivided into mucosa-associated lymphoid tissue (MALT) and nasopharynx-associated lymphoid tissue (NALT). MALT contains B cells, T cells, dendritic cells and antigen-presenting cells but lacks afferent lymphatics, and no capsule is present, while NALT exists of lymphatic tissue and scattered isolated lymphoid follicles (the ring of Waldeyer exists of NALT). Diffuse pockets of M cells, responsible for the uptake and delivery of foreign macromolecules and micro-organisms, are present and bring these substances from the luminal surface into mucosal lymphoid tissue. Inherently, this is a danger, because pathogens can enter the body too. Thus, antigen sampling is done by these cells. NALT, being lymphoid tissue, contains an epithelium with M cells, a follicular zone with B cells and a parafollicular area with T cells and dendritic cells. Although the IgA is the main effector antibody molecule in the mucosa, most tonsillar B cells are surface IgG positive and surface IgM positive. The tonsillar B cells are characterized by immediate antibody production against M cell taken up antigens. Tonsils and adenoids are equipped with M cells for inhaled antigen uptake (Fig. 7.23).

Saliva and gingival crevicular fluid add IgA, IgG and IgM to the mucosal surface. The dominant iso-type of immunoglobulin-producing cells in the submandibular glands is IgA. IgG and IgM types are also present. T cells in the submandibular glands produce interferon-γ and/or interleukin 5 or 6.

Viral or bacterial vectors are needed to activate the oral and nasal immune system. It will react with antigen-specific mucosal and systemic immune responses. It is the sublingual region of the oral cavity where the initiation of the antigen-specific reaction occurs. However, induction of antigen-specific immunity within the nasal cavity is the most effective (Smith et al. 2013).

7.6 Nose Fractures, Damage and Infections

There are of course disadvantages to have a human nose: the broken nose, which can occur in and around the cartilaginous or bony midsagittal septum, or an upward blow that induces an deleterious effect from the nose bones onto the crista galli (bony spine at the frontal brain base, on which the falx cerebri, double leaf of the dura mater with main venous vessels, adheres) and onto the lamina cribrosa (frontal perforated plate of the skull base) with its olfactory bulbs. Frontomaxillary and frontonasal fractures induce frequent disturbances of smell. Higher and more extensive fractures are present itself with an increase in unconsciousness. “Consequently, impairment of smell can be attributed to the fracture itself, but also to a cerebral lesion located more proximally” (van Damme and Freihofer 1992). Although the Van Damme and Freihofer article is old, it has not lost its value. More recent studies from the Taste and Smell Clinic in Washington DC showed that of their already diagnosed 5183 studied patients, taste loss was found in 62% and smell loss in 87% of the cases.

Nose infections can easily spread into the brain base, due to the venous drainage of the nose into the skull base. The best-known historical disease of the nose is syphilis. Degeneration of the nose was well known due to this infection disease (e.g. see portrait of the painter De Laresse by Rembrandt with a clear saddle nose due to congenital syphilis; note that 20% of the population of Venice was infected with syphilis in the eighteenth century).

7.6.1 *Medieval Rhinoplasty*

Medieval surgery went so far that skin flaps were put on the degenerated nose parts to restore the patients face. First, the text of Hugo and Theodore of Lucca (1205–1298) from Book 2 Chap. 11 is given, where Theodore deals with treating injuries of the nose by weapons.

"*On the injuries of the nose:* If the nose has been cut by a sword or a similar weapon, in such a way that it hangs wobbly and in the case it is trusted to you, before it dies off, you have to be careful and cautious. Place the nose back, as good as possible at the correct position, after you have first, if necessary, the parts brushed up a little, to reunite them better. Then, with sutures you stitch them together. After ending the suturing and the nose is at the correct place, one takes gauze or rolls of soft flax, which are rolled slowly in gauzes and plunge them in very hot wine. Then after you have them squeezed well, put them into the nostrils. They have to be of proportions and length that they cover the whole wound, in such a way that the spongy parts of the nose and its external parts are kept well in place and the patient can endure them without too much pain" (Tabanelli 1965).

In 1442, Branca da Branca developed the first rhinoplasty. It was applied but kept secret in the family for two generations. The Branca family used the cheek or forehead skin, called the Indian technique, where it was discovered, but also developed the pedicle flap using the arm skin. The same technique was described by Heinrich von Pfolspeundt in 1460. Leonardo Fioravanti (1517–1588) travelled through Calabria, after the last crusade. Here, he learned the Branca technique from the Vianeos brothers, being local barber surgeons. Finally, Gaspare Tagliacozzi (1545–1599) brought it to fame (by binding the patient's arm to the nose. He used the skin flap for recovery) and published it (1597), but the technique was forbidden on religious grounds (Santoni-Rugiu and Mazzola 1997; Marani and Koch 2014).

7.6.2 *Again Leonardo Da Vinci*

Leonardo's paintings are so subtle in the representation of faces and bodies that clinicians can clearly make the diagnosis if diseases are present. Here, an example of skin diseases of the nose and face is shown. Bacteria or fungi infections of the skin are ascertained producing deleterious effects. Some terminology is explained below in parentheses.

Leonardo's "the Adoration of the Magi" ('L' Adorazione dei Magi') has been studied by Marchetti and Panconesi (1997) for the pathological conditions, including common cutaneous lesions and dermatologic diseases recognizable on this painting. The whole painting has been scrutinized for pathological properties of the persons painted, but here the restriction is to the figure in the centre of the adoration (Fig. 7.24), presumably an adolescent boy, which is painted with various lesions.

"Nose. The nostrils are ulcerated (the right one is practically non-existent) and covered with blackish scabs. A blackberry-like vegetant papillomatous lesion (comprised of four or five hemispheric elements) on the left nostril, presuming a relationship (contagion?) with the female figure on his left described above, suggests a diagnosis of ulcero-vegetant pyoderma (chronic cellulitis of the skin producing granulation, ulceration and necrosis) (phagedenic on the right nostril).

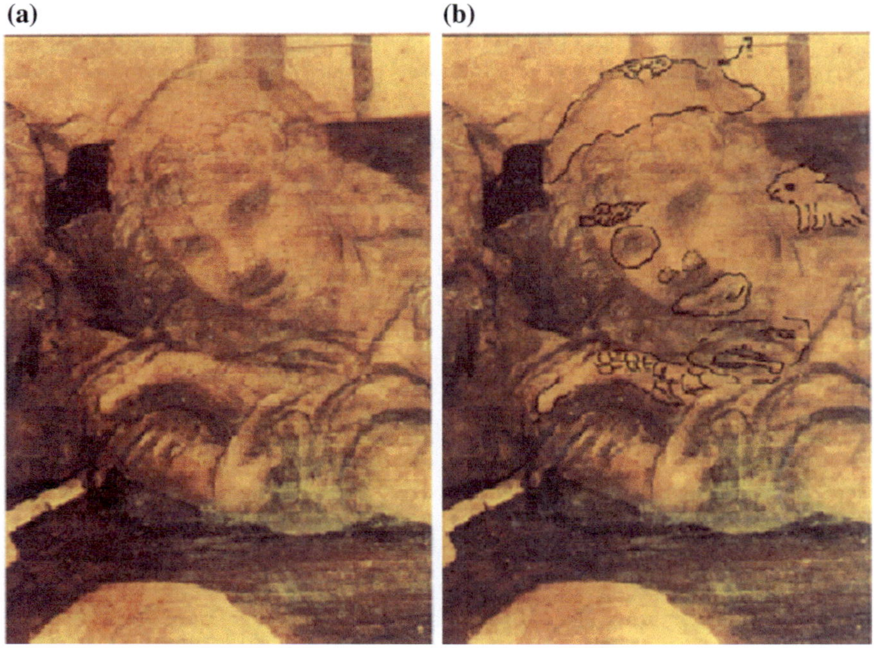

Fig. 7.24 (Figure 4 of Marchettia and Panconesi 1997 with permission): an adolescent boy with various lesions. "Observe closely in (**a**) the areas marked here with dotted lines or shading in **b** the hands, lower lip, upper sides of nostrils, right eye, right part of forehead, hair on left side near the ear, occipital part of head (in the latter area note also the curious feature of the nail in the head)"

Eyes. The boy is probably blind. The right eye is missing altogether, and the left one appears fixed in a void, like a glass eye: LD (in relationship with the infectious pathology prospected above) destructive staphylococcic ophthalmia (inflammation of the eye by bacteria).

Chin. The chin is small, and the skin appears as a smooth cheloid-like (inflammated) mass: LD presumably the outcome of a destructive process, with a cheloid-like scar (overgrowth of scar tissue producing contraction deformity), presumably resulting from a destructive infectious (again pyogenic? [pus formation]) process.

Head. The head has a crown of curly hair surrounding an area of alopecia (baldness) with a central triangular-shaped squamo-crusted mass: LD impetiginoid favus (skin disease by staphylococcus), or in consideration of LD impetiginoid favus, or staphylococcic kerion (a boggy suppurative mass of the scalp associated with ringworm (fungus infection) of hair). Observing the hair area, we also note on the right side of the forehead a depression with clear-cut margins: LD scarring, presumably due to the tinea (ringworm infection) or the pyogenic infection that destroyed the eye on the same side" (Marchetti and Panconesi 1997).

The diagnosis made of this boy painted by Leonardo (Fig. 7.24) shows an aspect of the facial skin that is nowadays absent in the Western world: bacterial infections that destroy facial organs or organ parts like the eye, the nostrils of the nose, the chin and the hair coverage of the head. Due to antibiotics and the concentration on the beauty of the face, the early detection of blemishes on the facial skin surface prevents distortion of facial skin.

Notes

1. The Fibonacci sequence is every number in the row (after the second) is the sum of the previous two numbers (1, 1, 2, 3, 5, 8, 13,).2. Golden ratio is the ratio of each number in the Fibonacci sequence to the one before it: 1/1 = 1, 2/1 = 2, 3/ 2 = 1, 5 and 5/3 = 1.666, 8/5 = 1.6, 13/8 = 1.625 ending with φ, $\varphi = 1.6180339887$.
2. Golden ratio is the ratio of each number in the Fibonacci sequence to the one before it: 1/1 = 1, 2/1 = 2, 3/2 = 1.5 and 5/3 = 1.666, 8/5 = 1.6, 13/8 = 1.625 ending with $\varphi = 1.6180339887$.
3. Diremo prima de'nasi, I quali sono di tre sorta, cioè dritti, concavi e convessi. De'dritti non ve n'è altro che quattro varietà, cioè lunghi, corti, alti con la punta, e bassi. I nasi concavi sono di tre sorta, de'quali alcuni hanno la concavità nella parte superiore, alcuni nel mezzo ed alcuni nella parte inferiore. I nasi convessi ancora si variano in tre modi, cioè alcuni hanno il gobbo nella parte di sopra, alcuni nel mezzo ed altri di sotto; gli sporti che mettano in mezzo il gobbo del naso si variano in tre modi, cioè o sono dritti, o sono concavi, o sono convessi (285. Del fare un'effigie umana in profilo dopo averlo guardato una sola volta: Trattato Della Pittura, Leonardo da Vinci, ristampa CodexUrbinas Latinus Vaticanus 1270, 1996).

References

Abano Pietro d' (1295) Liber compilationis physonomiae. Ms. Mailand, B Ambros Z; Edition Pierre Maufer, Padua 1474

Abbott NJ (2004) Evidence for bulk flow of brain interstitial fluid: significance for physiology and pathology. Neurochem Int 45:545–552

Ahmed O, Dhinsa A, Popenko N, Osann K, Crumley RL, Wong BJ (2014) Population-based assessment of currently proposed ideals of nasal tip projection and rotation in young women. Jama Facial Plast Surg 16:310–318

Appu AP, Arun P, Krishnan JKS, Moffett JR, Namboodiri AMA (2016) Rapid intranasal delivery of chloramphenicol acetyltransferase in the active form to different regions as a model for enzyme therapy in the CNS. J Neurosci Methods 259:129–134

Arif-Rahu M, Grap MJ (2010) Facial expression and pain in the critically ill non-communicative patient: state of science review. Inten Crit Care Nursing 26:343–352

Aubineau P, Sercombe R, Seylaz J (1980) Parasympatheticomimetic influence of carbachol on local cerebral blood flow in the rabbit by a direct vasodilator action and an inhibition of the sympathetic-mediated vasoconstriction. Br J Pharmacol 68:449–459

Braus H (1934) Anatomie des Menschen, zweiter band: Eingeweide. J Springer, Berlin

Baljet B, Groen GJ (1985) Anatomical considerations of the autonomic nervous system. In: Gelderman PW (ed) Symposion on pathological blushing and sweating. Acta neurochir 74:148–157
Braus H (1929) Anatomie des Menschen, vol I. Bewegungsapparat. Julius Springer, Berlin
Cailler P (1947) Géricault, raconté par lui-même et par ses amis. Vésenaz, Genève
Chaldakov GN, Fiore M, Ghenev PI, Beltowski J, Rančić G, Tunçel N, Aloe L (2014) Triactome: neuro-immuno-adipose interactions. Frontiers immunol, Implications in vascular biology. https://doi.org/10.3389/fimmu2014.00130
Chandebois R, Faber J (1983) Automation in animal development. In: Wolsky A (ed)Monographs in developmental biology 16. Karger, Basel
Cho HM, Lee DY, Sung SW (2005) Anatomical variations of rami communicantes in the upper thoracic sympathetic trunk. Eur J Cardiothorac Surg 27:320–324
Chung SK, Kim SK (2008) Digital particle image velocimetry studies of nasal airflow. Respir Physiol Neurobiol 163:111–120
Churchill SE, Shackelford LL, Georgi JN, Black MT (2004) Morphological variation and airflow dynamics in the human nose. Am J Hum Biol 16:625–638
Crozier R (2010) The puzzle of blushing. The psychologist 23:390–393
Cuthbertson RA (1990) The mechanism of human facial expression. Cambridge University Press, Ed Maison Sciences de l'Homme, Paris
Cutlip WD II, Leary MR (1993) Anatomic and physiological bases of social blushing: speculations from neurology and psychology. Behav Neurol 6:181–185
Damm M, Vent J, Schmidt M, Theissen P, Eckel HE, Lötsch J, Hummel T (2002) Intranasal volume and olfactory function. Chem Senses 27:831–839
Danelakis A, Theoharis T, Pratikakis I, Perakis P (2016) An effective methodology for dynamic 3D facial retrieval expression. Pattern Recog 52:174–185
De A, Saha A, Pal MC (2015) A human facial expression recognition model based on Eigen face approach. Procedia Comput Sci 45:282–289
De Crinis M (1942) Der menschliche Gesichtsausdruck und seine diagnostische Bedeutung. Georg Thieme Verlag, Leipzig
De Gelder B, Frissen I, Barton J, Hadjikhani N (2003) A modulatory role for facial expressions in prosopagnosia. Proc Natl Acad Sci USA 100:13105–13110
Delépin L, Aubineau P (1997) Plasma protein extravasation induced in the rat dura mater by stimulation of the parasympathetic sphenopalatine ganglion. Exp Neurol 147:389–400
Dhuria SV, Hanson LR, FreyII WH (2010) Intranasal delivery to the central nervous system: mechanisms and experimental considerations. J Pharmacol Sci 99:1654–1673
Dimitrov B (2009) St. Nicholas and St. Panteleimon Boyana Church. Unicart, Sofia
Doorly DJ, Taylor DJ, Schroter RC (2008) Mechanics of airflow in the human nasal airways. Resp Physiol Neurobiol 163:100–110
Duchenne G-B (1862) Mécanisme de la physionomie humaine or analyse électro-physiologique de l'expression des passions. J. Renouard, Paris
Dutterloo HS, Enlow DH (1970) A comparative study of cranial growth in Homo and Macaca. Am J Anat 127:357–368
Enlow DH (1990) Facial Growth, 3rd edn. Saunders Co, Philadelphia
Finkelstein JA, Schiffman SS (1999) Workshop on taste and smell in the elderly: an overview. Physiol Behavior 66:173–176
Francis West PH, Robson L, Evans DJR (2003) Craniofacial development: The tissue and molecular interactions that control development of the head. Adv Anat Embryol Cell Biol 169:1–143
Freitas-Magalhães A (2012) Facial expression of emotion. In: Ramachandran VS (ed) Encyclopedia of human behavior. Elsevier/Academic Press, pp 173–183. ISBN 978-0-08-088575-9
Gardiner JM, Atema J (2010) The function of bilateral odor arrival time differences in olfactory orientation of sharks. Curr Biol 20:1187–1191

Géricault (1996), Louvre conferences et colloquies, Tome I and II, La documentation Francaise, ed. Régis M. ISBN Fr 2-11-003327-4
Glaser G (2002) The nose: a profile of sex, beauty and survival. Atria Books, NY
Groen GJ (1986) Contributions to the anatomy of the peripheral autonomic nervous system. University of Amsterdam, Thesis
Groen GJ, Baljet B, Boekelaar AB, Drukker J (1987) Branches of the thoracic sympathetic trunk in the human fetus. Anat Embryol 176:401–411
Hamm J, (1991) Cartooning the head and figure. Grosset & Dunlap Inc, NY. Dutch Translation
Hahn I, Scherer PW, Mozell MM (1993) Velocity profiles measured for airflow through a large-scale model of the human nasal cavity. J Appl Physiol 75:2273–2287
Hari R, Kujala MV (2009) Brain basis of human social interaction: From concepts to brain imaging. Physiol Rev 89:453–479
Hardt N, Kuttenberger J (2010) Craniofacial trauma. Springer, Berlin, Heidelberg
Highfield R, Wiseman R, Jenkins R (2009) How your looks betray your personality. New Sci 2695:11 feb
Holmes M, Su SY, Bell D (2016) Ectopic primary olfactory neuroblastoma of the maxillary sinus. Ann Diagn Pathol 22:45–48
Hong SC, Leopold DA, Oliverio PJ, Benson ML, Mellits D, Quaskey SA, Zinreich SJ (1998) Relation between CT scan findings and human sense of smell. Otolaryngol Head Neck Surg 118:183–186
Illum L (2000) Transport of drugs from the nasal cavity to the central nervous system. Eur J Phamaceut Sci 11:1–18
Illum L (2012) Nasal drug delivery - Recent developments and future prospects. J Control Release 161:254–263
Imai T, Sakano H (2008) Interhemispheric olfactory circuit and the memory beyond. Neuron 58:465–467
Irmak MK, Korkmaz A, Erogul O (2004) Selective brain cooling seems to be a mechanism leading to human craniofacial diversity observed in different geographical regions. Med Hypotheses 63:974–979
Jacobs LF, Arter J, Cook A, Sulloway FJ (2015) Olfactory orientation and navigation in humans. PLoS ONE. https://doi.org/10.1371/journal.pone.0129387
Jampour M, Lepetit V, Mauthner T, Bischof H (2016) Pose-specific non-linear mappings in feature space towards multiview facial expression recognition. Image Vision Comp http://dx.doi.org/10.1016/j.imavis.2016.05.002
Johnson BA, Leon M (2007) Chemotopic odorant coding in a mammalian olfactory system. J Comp Neurol 503:1–34
Jorge JJ, Pialarissi PR, Campos Borges, Squella SAF, de Fatiima de Gouveia M, Saragiotto JC, Ribeiro Gonçalves V (2012) Objective computorized evaluation of normal patterns of facial muscles contraction. Braz J Otorhinolaryngol 78:41–51
Kanenishi K, Hanaoka U, Noguchi J, Marumo G, Hata T (2013) 4D ultrasound evaluation of fetal facial expressions during the latter stages of the second trimester. Int J Gynecol Obstetr 121:257–260
Kanwisher N, McDermott J, Chun MM (1997) The Fusiform Face Area: A module in human extrastriate cortex specialized for face perception. J Neurosci 17:4302–4311
Kiernan JA (2012) Anatomy of the temporal lobe. Epilepsy Res Treatment. https://doi.org/10.1155/2012/176157
Kikuta S, Sato K, Kashiwadani H, Tsunoda K, Yamasoba T, Mori K (2010) Neurons in the anterior olfactory nucleus pars externa detect right or left localization of odor sources. PNAS 107:12363–12368
Kimmel DL (1959) The cervical sympathetic rami and the vertebral plexus in the human fetus. J Comp Neurol 112:141–161
Korsmeyer C (1999) Making sense of taste. Cornell University Press, Ithaca, London
Kozlovskaya L, Abou-Kaoud M, Stepensky D (2014) Quantitative analysis of drug delivery to the brain *via* nasal route. J Control Release 189:133–140

Krasteva-Christ G, Soultanova A, Schuetz B, Papadakis T, Weiss C, Deckman K et al (2015) Identification of cholinergic chemosensory cells in mouse tracheal laryngeal glandular ducts. Internat Immunopharmacol 29:158–165

Kummer W, Lips KS, Pfeil U (2008) The epithelial cholinergic systems of the airways. Histochem Cell Biol 130:219–234

Kuntz A (1927) Distribution of the sympathetic rami to the brachial plexus: Its relation to sympathectomy affecting the upper extremity. Arch Surg 15:871–877

Lanz von T, Wachsmuth W (1985, 2004) In: Lang J, Kopf Teil A (eds), Uebergeordenete systeme. Springer, Heidelberg

Laurentini A, Bottino A (2014) Computer analysis of face beauty: a survey. Comp Vis Image Unders 125:184–199

Lavater JC (1775–1778) Physiognomische Fragmente zur Beförderung der Menschenkenntnis und Menschenliebe. Weidmann und Reich, Leipzig

Lee RJ, Chen B, Redding KM, Margolskee RF, Cohen NA (2014) Mouse nasal epithelial innate immune responses to Pseudomonas aeruginosa quorum-sensing molecules require taste signaling components. Innate Immun 20:606–617. https://doi.org/10.1177/1753425913503386

Lee RJ, Cohen NA (2014) Sinonasal solitary chemosensory cells "taste" the upper respiratory environment to regulate innate immunity. Am J Rhinol Allergy 28:366–373(8)

Legent F, Perlemuter L, Quéré M (1976) In: Waligora J, Perlemuter L, Anatomie (eds) Nerfs crânies et organs correspondants, vol 3. Masson Paris

Leonardo da Vinci, Trattato della pittura (Ristampa, Edizione integrale, 1996) I Mammut. Grande Tascabili Economici, Newton & Compton, Roma

Li X, Ruan Q, Ming Y (2012) A remarkable standard for estimating the performance of 3D facial expression features. Neurocomp 82:99–108

Licht PB, Ladegaard L, Pilegaard HK (2006) Thoracoscopic sympathectomy for isolated facial blushing. Ann Thorac Surg 81:1863–1866

Licht PB, Pilegaard HK, Ladegaard L (2012) Sympathectomy for isolated facial blushing: A randomized clinical trial. Ann Thorac Surg 94:401–405

Lieberman DE (2011) The evolution of the human head. Belknap Press, Harvard University Press, Cambridge, MA, London

Lieberman D (2013) The story of the human body. Evolution, health and disease. Penguin Books. London

Lochhead JL, Thorne RG (2012) Intranasal delivery of biologics to the central nervous system. Adv Drug Deliv Rev 64:614–628

Lonser RR, Sarntinoranont M, Morrison PF, Oldfield EH (2015) Convection-enhanced delivery to the central nervous system. J Neurosurg 122:697–706

Luyendijk-Elshout A (1990) Of masks and mills: the enlightened doctor and his frightened patient. In: Rousseau GS (ed) The languages of psyche. Mind and body in enlightenment thought. University California Press, Berkeley, pp 186–230

Mancusi S (2010) The police composite sketch. Human Press/Springer Science + Business Media, NY

Mar TT (1974) Face reading. The Chinese art of physiognomy. Dodd, Mead & Co, NY

Marani E, Lakke EAJF (2012) Peripheral nervous system topics. In Mai JK, Paxinos G (eds) The human nervous system. Academic Press, Elsevier, Amsterdam, Boston

Marani E, Koch WFRM (2014) The Pelvis, structure, gender and society. Springer, Berlin, Heidelberg

Marchetti C, Panconesi E (1997) Leonardo da Vinci: beauty or human suffering in the world. Notes on pathological cutaneous alterations in the 'Adoration of the Magi'. J Eur Acad Dermatol Venerol 8:101–111

Marhold F, Neumayer C, Tschabitscher M (2005) The sympathetic trunk and its neural pathway to the upper limb: review of literature. Eur Surg 37:114–120

Marneweck M, Hammond G (2014) Voluntaryof facial musculature in Parkison's disease. J Neurol Sci 347:332–336

McCormack AC, Jarral OA, Shipolini AR, McCormack DJ (2011) Does the nerve of Kuntz exist? Inter Cardiovasc Thor Surg 13:175–178

McDonough J, Dixon H, Ladika M (2010) Nasal delivery of micro- and nano-encapsulated drugs. In: Kulkarni VS (ed) Handbook of non-invasive drug delivery systems. Elsevier, Amsterdam, pp 193–208

Mooney CM (1956) Closure with negative after-images under flickering light. Can J Psychol 10:191–199

National Science and Technology Council, Subcommittee of Biometrics (2006) Face recognition. www.biometrics.gov

Newman MT (1953) The application of ecological rules to racial anthropology of the aboriginal New World. Am Anthrop 55:311–327. In: Bresler JB (ed) Human ecology, 1966. Addison-Wesley, Massachusetts

Nooij RP, van Dijk M, Groen RJM, Uyttenboogaart M, Eshghi OS, Korsten-Meijer AGW (2015) Epistaxis na een schedeltrauma. Ned Tijdschr Geneeskd 159:A9352

O'Higgins P, Bastir M, Kupczik K (2006) Shaping the human face. Int Congr Ser 1296:55–73

Perniss A, Kummer W, Bufe B, Krasteva Christ G (2016) A bacterial signal peptide increases mucociliary clearance in explanted mouse trachea. In: Poster 128, 111th annual meeting anatomische gesellschaft, Goettingen

Pick J (1970) In: Lippincott JB (ed) The autonomic nervous system. Toronto, Philadelphia

Ponce de León MS, Zollikofer CPE (2001) Neanderthal cranial ontogeny and its implications for late hominid diversity. Nature 412:534–538

Poothullil JM (1995) Oral satiation and regulation of intake. Physiol Behav 57:349–352

Porta IB (1586) De humana physiognomonia. Cornelii Sutorii, Rofae, France

Quadrio M, Pipolo C, Corti S, Messina F, Pesci C, Saibene AM, Zampini S, Felisati G (2016) Effects of CT resolution and radiodensity threshold on the CFD evaluation of nasal airflow. Med Biol Eng Comput 54:411–419

Ramsaroop L, Singh B, Moodley J, Partab P, Satyaoal KS (2004) Anatomical basis for successful upper limb sympathectomy in the thoracoscopic era. Clin Anat 17:294–299

Rasmussen LEL (2006) Chemical, tactile, and taste sensory systems. In: Fowler ME, Mikota SK (eds) Elephant biology, medicine and surgery. Blackwell Publ Co, Oxford

Rhazes (Abū Bakr Muḥammad ibn Zakarīyā ar-Rāzī) (before 903): Kitāb al-Manṣūrī fī aṭ-Ṭibb, (translation 1175 by Gerhard von Cremona, text called *Almansor)*

Rhodes G (2006) The evolutionary psychology of facial beauty. Annu Rev Psychol 57:199–226

Ricketts RM (1982) The biologic significance of the divine proportion and Fibonacci series. Am J Orthodont 81:351–370

Rolls ET (2009a) Cortical processing of the reward value of food. Encyclopedia Neurosci. Elsevier, Amsterdam, NY, pp 189–195

Rolls ET (2009b) Flavor physiology. Encyclopedia neuroscience. Elsevier, Amsterdam, NY, pp 227–232

Rolls ET (2015) Taste, olfactory and food reward value processing in the brain. Prog Neurobiol 127–128:64–90

Rossion B, Dricot L, Goebel R, Busigny T (2011) Holistic face categorization in higher order visual areas of the normal and prosopagnosic brain: toward a non-hierarchical view of face perception. Front Hum Neurosci. https://doi.org/10.3389/fnhum.2010.00225

Ryan S, Oaten M, Stevenson RJ, Case TI (2012) Facial disfigurement is treated like an infectious disease. Evol Hum Behav 33:639–646

Samandari F, Mai JK (1995) Funktionelle Anatomie für Zahnmediziner Band I+II. Quintessenz Verlag, Berlin

Sandbach G, Zafeiriou S, Pantic M, Yin L (2012) Static and dynamic 3D facial expression recognition: a comprehensive survey. Image Vis Comp 30:683–697

Santoni-Rugiu P, Mazzola R (1997) Leonardo Fiorvant (1517–1588): A barber-surgeon who influenced the development of reconstructive surgery. Plast Reconst Surg 99:570–597

Shipley MT, MCLean JH, Zimmer LA, Ennis M (1996) The olfactory system. In: Swanson LW, Björklund, Hökfelt T (eds) Handbook of chemical neuroanatomy, vol 12. Integrated systems of the CNS, Part III. Elsevier Science, Amsterdam

Shookman E (1993) Pseudo-science, social fad, literary wonder: Johann Caspar Lavater and the art of physiognomy. In: Shookman E (ed) The faces of physiognomy: interdisciplinary approaches to Johann Caspar Lavator. Camden House, Drawer California, pp 1–24

Siegrist C (1993) "letters of the devine alphabet"-Lavater's concept of physiognomy. In: Shookman E (ed) The faces of physiognomy: interdisciplinary approaches to Johann Caspar Lavator. Camden House, Drawer California, pp 1–24, 25–39

Singh B, Ramsaroop L, Partab P, Moodley J (2005) Anatomical variations of the second thoracic ganglion. Surg Radiol Anat 27:119–122

Skarlicki DP, Hoegg J-A, Aquino K, Nadisic T (2013) Does injustice affect your sense of taste and smell? The mediating role of moral disgust. J Exp Soc Psychol 49:852–859

Skinner APC, Pachnicke S, Lakatos A, Franklin RJM, Jeffrey ND (2005) Nasal and frontal sinus mucosa of the adult dog contain numerous olfactory sensory neurons and ensheating glia. Res Vet Sci 78:9–15

Small DM (2012) Flavor is in the brain. Physiol Behav 107:540–552

Smith PD, MacDonald TT, Blumberg RS (2013) Principles of mucosal immunology. Garland Science/Taylor & Francis Group London, NY

Spence C (2012) Book review: 'Neurogastronomy: how the brain creates flavour and why it matters by Gordon M Sheperd. Flavour 1:21–24

Steinke H (1996) Giotto und die Physiognomik. Zeitschr Kunstgesch 59:523–547

Sterpetti AV, Schultz RD, Feldhaus RJ (1988) External carotid endarterectomy: Indications, technique, and late results. Vasc Surg 7:31–9)

Swain S (2007) Seeing the face, seeing the soul. Polemon's physiognomy from classical antiquity to medieval Islam. Oxford University Press

Tabanelli M (1965) La chirurgia italiana nell'alto medioevo. Bibliotheca della "rivista di storia delle scienze mediche e natural, Vol XV, parts I and II, L.S. Ol`schki, Florence

Texeira Lopez A, de Aguiar E, De Souza AF, Oliveiro-Santos T (2016) Facial expression recognition with convolutional neural networks: coping with few data and the training sample order. Pattern Recogn. https://doi.org/10.1016/j.patcog.2016.07.026i

Thornhill R, Gangestad SW (1999) Facial attractiveness. Trends Cogn Sci 3(12):452–460

Toro-Ibacache V, Munoz VZ, O'Higgins P (2015) The relationship between skull morphology, masticatory muscle force and cranial skeletal deformation during biting. Ann Anat. https://doi.org/10.1016/j.aanat2015.03.002

Trinkaus E, Howells WW (1979) The Neanderthals. Scientific American, Dec 1979, 94–105

Trinkler I, Cleret de Langavant L, Bachaud Lévi A-C (2013) Joint recognition-expression imparmaint of facial emotions in Huntington's disease despite intact understanding of feelings. Cortex 49:549–558

Van Damme PA, Freihofer HPM (1992) Disturbances of smell and taste after high central midface fractures. J Cranio-Maxil-Facial Surg 20:248–250

Villiger E (1946) Gehirn und Rückenmark. Benno Schwabe & Co Verlag, Basel

Walker PJ, May J, Apris JP, White GH, Halliman J (1994) External carotid endarterectomy for amaurosis fugax in the presence of internal carotid artery occlusion. Austr NZ Surg 64:48–52

Waller BM, Parr LA, Gothard KM, Burrows AM, Fuglevent AJ (2008) Mapping the contribution of single muscles to facial movements in the rhesus monkey. Physiol Behav 95:93–100

Waller BM, Vick SJ, Parr LA, Bard KA, Smith Pasqualini MC, Gothard KM, Fuglevent AJ (2006) Intramuscular electrical stimulation of facial muscles in humans and chimpanzees: Duchenne revisted and extended. Emotion 6:367–382

Weijs WA, Hillen B (1986) Correlations between the cross-sectional area of the jaw muscles and craniofacial size and shape. Am J Phys Anthropol 70:423–431

Wu T, Hung AP-L, Hunter P, Mithraratne K (2013) Modelling facial expressions: a framework for simulating nonlinear soft tissue deformations using embedded 3D muscles. Finite Elem Anal Des 76:63–70

Yan Z, Tan J, Qin C, Lu Y, Ding C, Luo M (2008) Precise circuitry links bilaterally symmetric olfactory maps. Neuron 58:613–624

Young Lee S, Bang M, Ran Kim K, Kyung Lee M, Young Park J, Young Sun Y, Kang JI, Lee E, An SK (2015) Impaired facial emotion in individuals at ultra-high risk for psychosis and with first episode schizophrenia, and their associations with neurocognitive deficits and self-reported schizotypy. Schizophrenia Res 165:60–65

Zhang L, Tjondronegorob D, Chandranb V (2014) Random Gabor based templates for facial expression recognition in images with facial occlusion. Neurocomp 145:451–465

Zhao K, Scherer PW, Hajiloo SA, Dalton P (2004) Effect of anatomy on human nasal air flow and odorant transport patterns: implications for olfaction. Chem Senses 29:365–379

Zollikofer CPE, Ponce de León MS (2004) Kinematics of cranial ontogeny: Heterotopy, heterochrony, and geometric morphometric analysis of growth models. J Exp Zool (Mol Dev Evol) 302B:322–340

Chapter 8
Eye and Orbit: The Eye Sees What You Do not Spot

"More is seen than can be remembered" implies two things: a memory limit that is a limit on the (memory) report and it infers that more information is available. (G. Spencer [1921–2014], 1960)

Since the eyes are placed on the surface of the face, infections, poisoning vapours and etching substances from outside can easily reach the eyes. Especially punches, blows and stabs are producing sever eye injuries, because the orbit and eye are in relation with the weak midface structures. Eye and orbit contain their own vessels and muscles, constructed by a peculiar embryology. Innervation of orbital structures, function and morphology of the optic motor systems is special. Perception and oculomotor steering are greatly interwoven. Diseases or genetic deviations, intrinsically or extrinsically related to the eyes, are greatly different in their phenotype. The cellular photo-transduction, on–off processing and receptive field studies are not treated.

8.1 Introduction: The Orbit's Bony Construct

In Chap. 7, it was discussed that the midface is characterized by cavities (see Sect. 7.4.1, Fig. 7.19). The consequence is that the midface misses a solid construction, as demonstrated for the orbit medial wall after trauma in Fig. 8.1 left. Thus, the interrelation between the eye and the bony structures asks for a special descriptive approach in which the "weakness" of the orbita walls has to be included.

Earlier holoprosencephaly has been treated with cyclopia, single median eye and a double median eye in Chap. 3, part 3.5.3. In general, the skull and orbita adapt to the eye situation (see Sect. 4.5). Often, the sphenoid is reduced and the ethmoid can be absent in cyclopia and other holoprosencephalic deflections. The cyclope has been used by Belgian artists. Along the motorway E17 Gent-Kortrijk (Belgium) at the height of Deerlijk, a statue of a cyclope is present (April 2014–April 2018, last visit). The artist's collective had to adapt the front of the skull to the single eye

E. Marani and C. Heida, *Head and Neck*,
https://doi.org/10.1007/978-3-319-92105-1_8

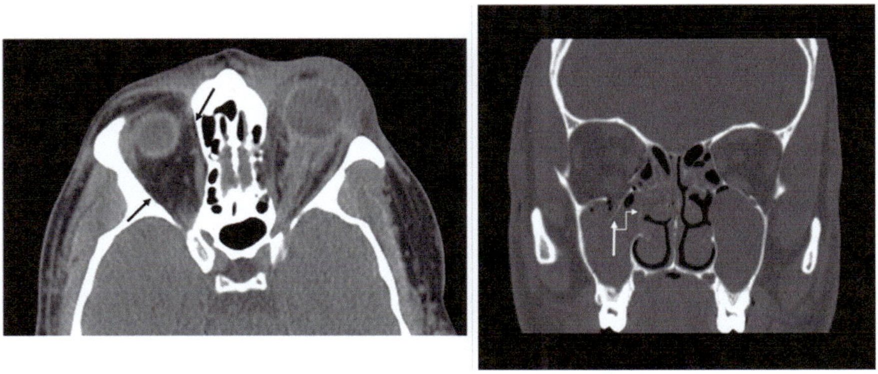

Fig. 8.1 Left picture: Medial communitif orbit wall fracture, left, and opticopathy, left, (both black arrows and note prolaps of m. rectus medialis) caused by maltreatment trauma due to multiple fist blows. Visus was 20%, after treatment with high doses of methylprednisolon (3 days, 1000 mg iv) visus was restored to 90%. Right picture shows a blowout fracture, left, of a 42-years-old male after fist blows on his left eye. Note the descent of the bony bottom (straight arrow), the hypersensibility of the n. infraorbitalis regio and the tissue changes of the conchae and blood within the nasal cave (hooked arrow) (with permission and courtesey of Dr. M. Joosse, Haaglanden Medical Centre, The Hague, The Netherlands)

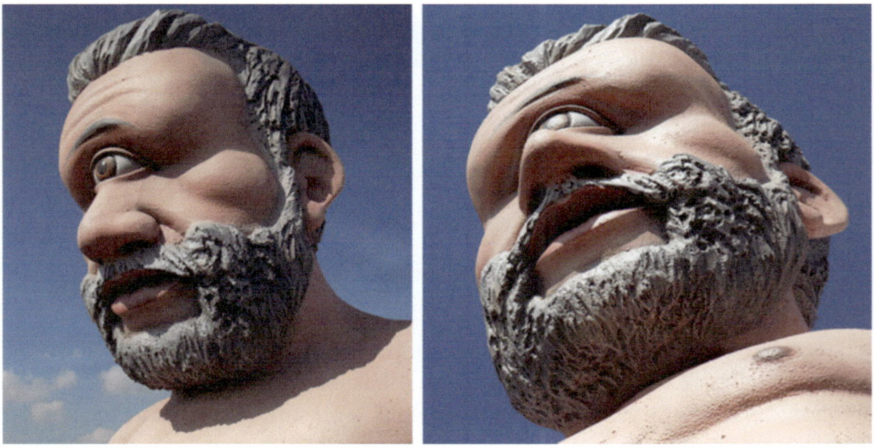

Fig. 8.2 Face of the Belgian cyclope: An artists collective called 100 Titans from Flanders (see www.100titanen.be) placed the cyclope on the 1 April 2014 along the road (with permission and courtesey of Hervé Missiaen)

(Fig. 8.2). The frontal skull part is a shed over the eye, nearly a Neanderthal-like big brow ridge (see Sect. 7.4.1). One should note how deep the lateral walls of the orbit are placed posteriorly (backwards?) in the Belgian cyclope. Normally, only the eye's anterior chamber extends out of the bony borders of the orbita. Thus, the normal aditus orbitae or outer open surface of the orbit extends far to the front of the eye, the orbita encapsulating most of the eyeball. A strongly reduced sphenoid together with the maxilla and zygomatic bones in the Belgian cyclope form a large recess in which the connection with the frontal bone has to be made. Although it is a good artist's solution, such a strong exophthalmus seems impossible in the *natural* situation.

The orbit complex is formed at the transition of cranial bones and facial bones. The orbit is constructed by seven different bony parts (Fig. 8.3). Their contact areas are sutures of which at least 6 are present within the orbit. The orbit's form is a quadrangular pyramid with its top posterior and the oblique rectangular base anterior. In Sect. 7.4.4, a junction has been described between maxilla, ethmoid and sphenoid. Contact between ethmoid and sphenoid is formed, producing a suture between both bony pieces. The palatine's orbital process contacts the ethmoid too and the palatine extends dorsally to contact the sphenoid, both with sutures (Fig. 8.3 left). Orbit's contacts with the maxilla, ethmoid and sphenoid are expressed as sutures that are present within the orbit wall. Several other bony structures with their sutures are found within the orbit: lacrimal bone, frontal bone

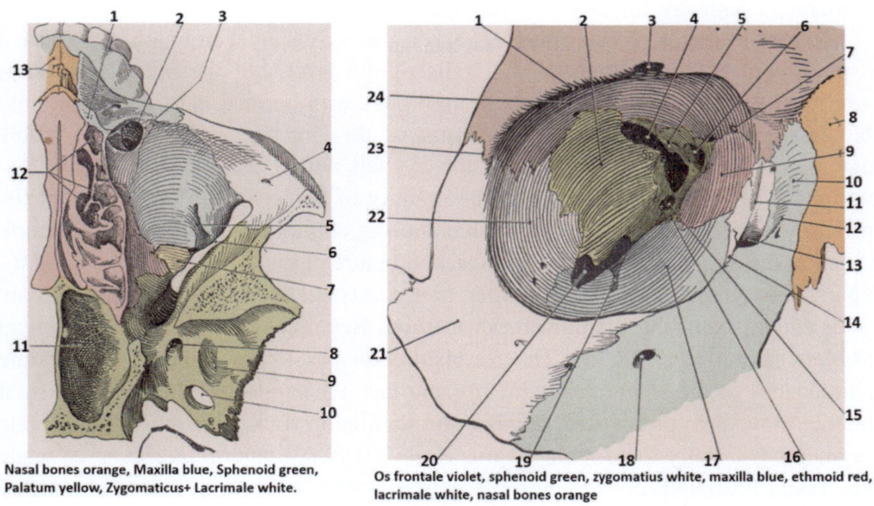

Fig. 8.3 Left, inferior part of the orbit: 1, frontal part maxilla; 2, nasolacrimal duct; 3, hamulus lacrimalis; 4, zygomaticofacial foramen; 5, infraorbital sulcus; 6, inferior orbital fissure; 7, orbital process of palatum; 8, foramen rotundum; 9, cerebral surface of sphenoid; 10, foramen ovale; 11, sphenoid sinus; 12, ethmoid cells; 13, nasal bones. Right is a frontal view of the orbita: 1, supraorbital line; 2, sphenoid surface; 3, incissura orbitalis superior; 4, superior orbital fissure; 5, ala parva of sphenoid; 6, optic foramen; 7, anterior foramen of ethmoid; 8, nasal bone; 9, ethmoid (lamina papyracea); 10, frontal part maxilla; 11, posterior lacrimal crista of lacrimal bone; 12, anterior lacrimal crista of maxilla; 13, fossa sacci lacrimalis; 14, hamulus lacrimalis; 15, foramen rotundum; 16, orbital processus of paslatum; 17, orbital surface of maxilla; 18, foramen infraorbitale; 19, canalis infraorbitalis; 20, inferior orbital fissure; 21, cheek surface of zygomaticum (facies malaris); 22, orbital surface of zygomaticum; 23, zygomaticofrontal suture; 24, orbital surface of frontal bone, (figures adapted from Corning 1919)

and zygomatic bone (Fig. 8.3). Sometimes the knowledge of these sutures and bony structures are advantageous in a totally different context (see Sect. 2.6. Leonardo da Vinci (1452–1519) and the "oldest" miniature-scale sculpture of the skull). From Fig. 8.3, one should conclude that the medial wall trauma in Fig. 8.1 concerns the lamina papyracea of the ethmoid (nr. 9 in Fig. 8.3 right).

The sphenoid contains the optic canal, the superior orbital fissure and the foramen rotundum. At the border of sphenoid, zygomaticus and maxilla the inferior orbital fissure and the canalis infraorbitalis (between zygomaticus and maxilla) are constructed within the orbit (see Fig. 8.3). The sphenoid is in the centre of the bony orbit and participates in all passage structures needed to enter the orbit from the back. Its ala minor contains the optic canal, while its ala major borders both orbital fissures. The bony structures are covered with a periost layer, called conjunctive membrane, which is directly connected to the superficial layer of the dura mater around the superior and inferior orbital fissures.

The upper part of the orbita (portion of the frontal bone) knows two indentations: lateral the one for the lacrimal gland and medial for the trochlea along which the tendon of the superior oblique eye muscle functions. Since the roof of the orbita is the floor of the anterior cranial fossa and the frontal bone is very thin, this is an endocranial pathway for tumours that originate in the orbit.

Blowout fractures of the orbit (Fig. 8.1, right, mostly by a frontal blow) cause a displacement of one or two of the walls of the orbit, but the orbital rim stays unharmed. Orbital floor displacement often goes with herniation of soft tissue into the maxillary sinus. Common symptoms of this fracture are diplopia (double vision), enophthalmos (retraction of the eyeball within its orbit), dystopia (displacement of the eye), paresthesia (abnormality of perception or sensation) of the infraorbital nerve, and soft tissue incarceration or entrapment, leading to restriction of ocular movements. Due to the intricated structure of the orbit floor (Fig. 8.3 left), reparation occurs not always with bone transplants but several other techniques are used, among them the titanium mesh method. Repair of orbit walls is a surgical problem (Lima et al. 2015). The reconstruction has to accomplish the accurate maintenance of the anatomical structure and thus the support of the orbit and by it the free movement of the eye. "Joining and stabilizing small, thin and delicate bone fragments is usually impossible" (Lima et al. 2015). A titanium mesh can be made into a bent form and screwed onto the orbit rim after repositioning of the herniated soft tissue (Lima et al. 2015). After damage of both the zygomatic bone and the orbita wall (lateral blow), the eye will retract posteriorly and enophthalmos occurs. Bony volume expansion and atrophy of soft tissue are thought to be the main causes of enophthalmos (Lee 2010). Treatment includes osteotomy and an increase in the volume of orbital tissue. Using diced cartilage as graft for the increase of orbit volume, a rather high success rate in a restricted series has been accomplished to reduce enophthalmos (Lee 2010).

8.2 Look and Estimate

From anatomy to perception seems a large jump, but both will come together later in this chapter in Sect. 8.7 oculomotor steering. You learned that a scientist has to make correct observations with or without instruments. My students thought, that after their bachelor's part of the study, they could observe well. In order to test this, a classic method was used in which the students had to verify their observations by estimating dimensions or distances (see Fig. 8.4). The dimensions of the Notre-Dame de Chartres were used (see A. Prache 2003 for cathedral information).

The first student question asked was "Estimate the height of the right plain pyramid tower at the street side". Followed by the second question that concerned the round rose window. It has a diameter of 12,885 meters. How often goes the length of the A arrowed window in the diameter of the rosette? A: 2, 5 x, B: 3,5 x, C: 4,5 x, D: 5,5 x. And how often this B arrowed window? A: 14 x, B: 17 x, C: 20 x, D: 23 x. And what is the relation of the length of the arrowed windows? The longest (A) is x times the smallest (B): a: 4 x, b: 5 x, c: 6 x, d: 7 x. A labyrinth is present in the floor of the cathedral, made to get in a fixed area, here a circle, as much running meters by a fixed pathway. The labyrinth has the same diameter as the round window. Chartres's pathway is 34 cm in width with an interpath distance of 8 cm. "If you follow the path with your finger, you will see that we first traverse the inside five rings on the left half, and then those on the right. The path continues right across the top of the sixth to join the remaining outer five circles, first passed on the left and then on the right: it is a clear and consistent pattern. There is a symmetry from one

Fig. 8.4 Cathedral Notre Dame of Chartres has been built mainly in the twelfth–thirteenth century. Its interior contains the labyrinth, and in this picture, the beautiful thirteenth century north transept rose window is depicted (figures are taken from brochures of the cathedral with oral permission)

side to the other, and from the inside to the outside" (James 1977). You start at the right side, at the opening in the notched line (red arrow). Now, you have to reach the centre. The labyrinth's diameter is 12,885 meters. How many meters path do you need to reach the centre: A: 101.5 m, B: 165.3 m, C: 261.5 m, D: 305.6 m. Of course you will say this is not observation, this is calculation (by the way, it is not since you can solve it by estimation. You need not to calculate).

During these lectures, given each year, barely 30% of the answers were correct. This nears the gambling percentage, indicating that our distant surrounding is hardly noticed in correct measures and that humans are not well trained in estimating these distances.

Correct answers on the questions:

105 meters high
A-window is B
B-window is D
B in A-window is D
Path length is C

The circumference of the eye is 8 cm. If the projection of the 105 m tower uses something more than half the retina, hence we have a reduction of 2600 times, and often its projection on the retina is less, resulting in even a higher reduction. However, the apple on the fruit dish has a lower reduction and the eye of the needle nearly none. The eye of the needle equals the tower image on the retina. Single or monocular retina projections per se do not help in accurately estimating distances or depth, unless we know from experience the size of the object. The retina projects via the optic nerve and optic tract to brainstem areas, to the lateral geniculate thalamic area and finally into the visual cortex via the optic radiation. Intrinsic and extrinsic eye muscle information arrives in brainstem areas and its eye nuclei. All these structures and more are involved in depth perception. "The possibility to see the world in three dimensions remains quite fascinating considering the following peculiar fact: images arrive on the retina in two dimensions. Spontaneously, no one would doubt that there is a physical world that includes a third dimension (but see Russell 1980, remark ours). Nevertheless, the brain has to build this third dimension" (Grondin 2016). Absolute distances are difficult to estimate (see above), but relative distances, to estimate whether an object is closer to us than an other object or not, are extremely well assessed. Binocular convergence differences are measured by the kinesthetic movement information of the receptors of the extraocular muscles. The binocular convergence gives data on the absolute distance. This information is more accurate when objects are close to the observer within 5–7 m, while inaccuracy is beyond this distance. Moving parallax shows relative distances. Having two normal eyes, with their mutual distance, causes that two partially different images are made. This is called retinal disparity and is the basis of stereoscopy. Stereoscopy organizes in the brain the perception of depth. There are monocular cues that also contribute. Accommodation of the lens to get a sharp image on the retina by the lens muscles is noticed by the nervous system by its muscle receptors for assessing distance. Linear perspective contributes too: in the Netherlands, the ditches are mainly parallel over long distances; two parallel ditches seem to end in one point, which gives the idea of depth, and the same holds for, e.g., train rails. Interposition of objects, one object partially obscuring another and relative height (below the horizon line, the higher in that visual field the farther away and above the horizon line vice versa), relative brightness (shadow can create depth impression) and texture, the linear perspective combined to the relative size of objects (density of objects and their size vary with distance), all contribute to three-dimensional perception. Thus, large absolute distances are difficult to estimate, depth can be noticed by the brain due to different retina images and by receptor information of intrinsic and extrinsic muscles, while the image itself does introduce cues that are also used by the brain to produce a 3D image of our external world (see Grondin 2016).

Object recognition starts in the primary visual cortex, proceeds into the association visual cortical areas (indicated by V2, V3 and V4) and is also partial present in the lateral occipital cortex. From here on, information reaches the inferior ventral temporal cortex including the fusiform gyrus for face recognition (see Fig. 7.2). Colour recognition also occurs in this inferior ventral temporal area. This series of connected areas for object recognition is named the ventral stream to discern it from the dorsal stream of visual cortical areas related to motion. This ventral stream eventually forms the visual foundation for our cognitive life. Within the ventral temporal lobe, different areas are involved in the various types of recognitions: face recognition was in the fusiform area and visual presented words are recognized in another area. This also holds for images of places and for body parts. Object recognition also has its own temporal area. Incoming visual information in the primary optical area has to be sent to their distinct regions within the temporal cortex via the association visual cortex: a topographical map of object form is seemingly present within the ventral temporal cortex. The question is what characteristics are used to determine any object's localization within the temporal lobe? Low-level properties are held responsible for this choice, which is also saying that one category of objects does have the same low-level properties. The semantic or categorical significance is added later to the object after being classified by its brain topographical localization. Using scrambled pictures, that still convey low-level characteristics but not the semantic significance, showed indeed that the visual choice is done by low-level characteristics. The rose window and labyrinth do contain the same low-level characteristics that bring them to their topographic object area (say all round structures) in the temporal lobe, where the semantic is added and the structures are named rose window and labyrinth (see Coggan et al. 2016a, b). Thus, assigning a perceived object to its temporal category is done on other grounds than properties like height, length or thickness that belong to the object. Our speculation is that therefore, distance or dimension estimations of objects are lousy at the outset. Training will increase distance estimations, think of primitive hunters.

8.3 Intelligence, Tears and the Eye

Somehow, one cannot escape Johann Caspar Lavater. In his "Physiognomische Fragmente" (1778), an overview of eight eyes is presented in the part Genius. The eyes are portrayed "in profile" (Fig. 8.5a). Following his German text in a quick, only partial, translation it says: "Of none, as they appear here, could I say with certainty: A great genius. Nevertheless, 3 and 5 show geniality of different kind: 3 is a fast powerful strong genius and 5 is a genius by experience. 1 is clearly no genius, 2 has more brains—but too little energy to reach genius".

Lavater goes on in describing the eyes: 4 is nearly designated being a blockhead. 6 is susceptible to genius. 7 does have something of a genius even more if the back corner was less pointed. 8 is also not really a genius by the neglect of the back portion of the eye. The remark at the end of this screed is that everyone can endorse this kind of meaning of the eyes by him or herself. However, pupil, iris, limbus and sclera are not applied by Lavater to decide on genius.

Nevertheless, Lavater brings up the structure of the outside of the eye and eyelids. The environs of the eye are characterized by the two corners of the eye (anguli oculi) containing two connective bands (raphe and ligamentum palpebrae), and within the medial one, the lacrimal caruncula (Fig. 8.5b) is present. It is a rosy, globular nodule at the inner corner of the eye and is made of skin covering sebaceous and sweat glands. Note that in dog and cats, the caruncula has a tear function, but not in man. The eye's skin, fat and the m. orbicularis oculi are placed over a

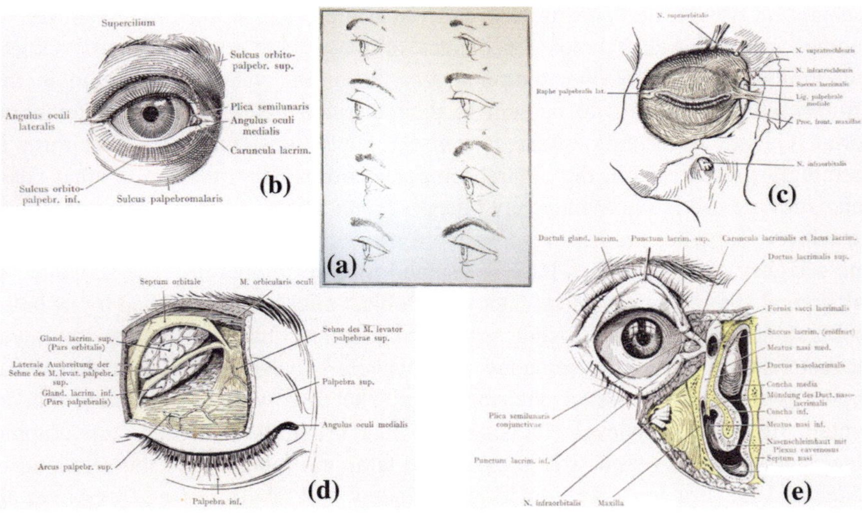

Fig. 8.5 **a** Overview of Johann Caspar Lavater's figure for genius and the form of the eye. **b** View on the exterior of the right eye. **c** Septum of the orbit, contributing to the tarsi placed beneath the skin of the eye lids. **d** tear gland in situ. **e** conjunctiva sac and the channels for drainage of tear liquid (Part figure **a** from Physiognomische Fragmente 1778, courtesy Leiden University Library with permission, part figures **b–e** from Corning 1919)

connective layer, the septum of the orbit (Fig. 8.5c) that contains still a superficial layer of fat, beneath the muscle. This septum makes the connection of the eyelids to the rim of the orbit and continues in both eyelids to produce its deepest layer. They are called tarsus superior and inferior, and at their rim, hair-associated glands are present. At the tarsi's ocular sides, the Meibomian gland and the lacrimal glands of Krause and Wolfring are present. The Meibomian gland is a holocrine gland that, by suicide of the cell, delivers lipid to the surface of the eye. It changes the tear viscosity, evaporation of tear fluid is diminished, and increases sliding of the eyelids. The tear gland (see Fig. 8.5d) produces the tear fluid that contains IgA, and also antimicrobial peptides like lactoferrin, defensins, cytokines (mostly interleukins), transforming growth factor-β, surfactant proteins and mucins. Production, spreading over the eye surface, collecting and draining away pathways (Fig. 8.5e), all these areas contain associated lymphoid tissue: lacrimal gland-associated lymphoid tissue, conjunctiva-associated lymphoid tissue (CALT), tear duct-associated lymphoid tissue (TALT/LDALT), which by their plasma and T cells add these substances. B cells are mainly found in the CALT and in the TALT, not in the lacrimal gland. By an unknown way, fluid is taken up again for lymphocyte recirculation, presumably for antigen sampling (Smith et al. 2013; Schechter et al. 2010).

Dry eye disease or keratoconjunctivitis sicca is a frequently occurring ocular surface problem. In the USA, over 5 million persons over the age of 65 do have problems that are diagnosed dry eye or are dry eye related. The majority of dry eye patients are women (65% versus 35% males). The disease can be due to deficiency of tear fluid or to strong evaporation. Tear deficient is Sjögren's syndrome, alacrima, which can be congenital or acquired, lacrimal obstruction and neurogenic causes (neuroparalytic keratitis, contact lens related or VIIth nerve palsy). Sjögren's syndrome is an autoimmune disease with inflammation of the lacrimal and salivary glands. Antigens related to Sjögren's syndrome are α-fodrin (also named SPANT-1; actin binding protein at the plasma membrane that is important for alignment of exocytose vesicles), Ro and La extractable nuclear antigens and other antinuclear antibodies associated with ribonucleoproteins. Evaporation concerns deficiency of lipids by absence of the glands, inflammation of the eyelid rim (Blepharitis) or dysfunction of the Meibomian gland. Deflections at the eyelid can produce dry eye including aperture abnormalities to eyelid incongruity. Normally, the embryonic folds of the eyelids fuse and keep the epithelial layer. Persistence of this layer helps the formation of the epithelial structures within the eyelids. Opening of the eyelids by splitting this layer produces a precise fit of the two free borders overcoming eyelid incongruity. The surface changes like xerophthalmia (desiccation of the connective tissue of the eye) at the start induces dry eye, but can end in destruction of the cornea. Approaches that are current, but have their disadvantages, are: "(1) artificial ointments and eye drops, which provide only short-term relief; (2) punctal plugs, which reduce the tear drainage into the nasal cavity and may improve symptoms, but which also prevent clearance of the inflammatory proteins responsible for corneal desensitization; (3) tarsorrhaphy, a surgical procedure to restrict eyelid opening, which also reduces the visual field of view; (4) cyclosporine, an anti-inflammatory drug that provides mild improvement in symptoms, but is plagued with irritating side effects; and (5) heat therapy, to clear blocked Meibomian gland secretions" (Brinton et al. 2016). One cannot be surprised that dry eye solutions are also found in electric stimulation of the lacrimal gland in one way or another. The experimental animals for tearing studies are cat and rodents (mouse, rat and rabbit). The rabbit lacrimal apparatus is different from the human one. While in humans, one lacrimal gland is present lateral above the eye (Fig. 8.5d), and the rabbit has two (Fig. 8.6 left). One gland has the same position as in man, but the accessory lacrimal gland is located at the inferior rim of the eyeball and is nearly three times as large as the "real" lacrimal gland (Fig. 8.6; Van Dyck 2015). Moreover, not only the anatomy is different, but several characteristics are at variance: difference in serous and mucous cells in the acini, variations in the content of proteins produced, absence of mast cells in human acini, the enzyme lactoperoxidase is hardly produced in the rabbit, but is a prominent product in humans. "In humans, reflex stimulation accelerates the secretion of the secretory component domain of polymeric immunoglobulin receptor and IgA, but not as much as it accelerates the production of fluid, so the concentration of secretory component and IgA in the fluid film actually decrease" (Schechter et al. 2010). This is not the case in the rabbit, its concentrations stay at level. There is a difference in innervation pattern concerning the sympathetic and parasympathetic system between man and rabbit, and gender differences in the rabbit are large. As expressed by Schechter et al. (2010) in the title of their overview paper: "A lacrimal gland is a lacrimal gland, but rodents and rabbits are not human". Given these remarks, care should be considered reading the following electric stimulation results. The electric stimulation of tear secretion increased the tear production after stimulation of the facial nerve (cat; Yasui et al. 1997), of the superior cervical ganglion (rabbit; Botelho et al. 1976), and of the afferent branches of the trigeminal nerve (cat; Yasui et al. 1997; see also Brinton et al. 2016). Brinton and co-workers implanted bipolar platinum foil electrodes near the accessory lacrimal gland of rabbits and a monopolar electrode near the afferent ethmoid nerve (see Fig. 8.6a, b). Scopolamine is a blocker of the neuronal pathway part that originates in the brainstem and projects to the sphenopalatine ganglion, which innervates the rabbit's acini (see Fig. 8.6). Using scopolamine, the bipolar stimulation of the lacrimal gland had no effect: only the efferent parasympathetic pathway that originates from the sphenopalatine ganglion is blocked. Prazosin, an efferent $\alpha 1$ sympathetic inhibitor, had no effect. Hexamethonium (see Fig. 8.6) did not affect bipolar stimulation tearing at the gland, but did affect tearing by monopolar stimulation of cornea and ethmoid, thus blocking the effect of afferent nerve stimulation. Stimulation of the afferent ethmoid nerve produced the same amount of tear fluid as stimulation of the accessory lacrimal gland, however, at a lower current and shorter pulse duration. Stimulation of the lacrimal nerve in the upper orbit, and thus stimulation of the upper lacrimal gland, gave tearing by far lower stimulation parameters (Kossler et al. 2015). The conclusion should be that afferent nerve stimulation is more effective compared to (accessory) lacrimal gland stimulation, which is understandable by the immense branching of the parasympathetic nerves

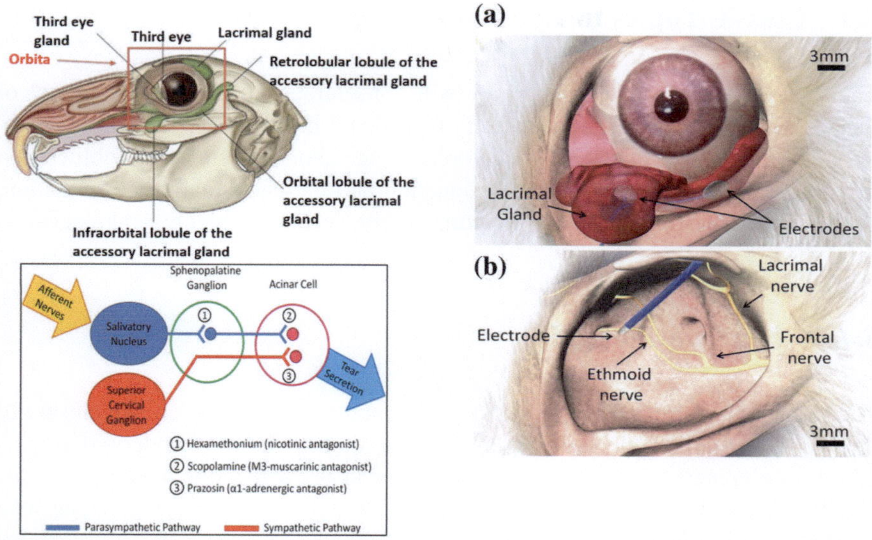

Fig. 8.6 Left, anatomy of the orbit and lacrimal glands of the rabbit (taken from Van Dyck 2010 with permission). Right, electrode placement in acute stimulation experiments: **a** bipolar electrode implantation within the orbit near the accessory lacrimal gland, **b** monopolar electrode placed on the ethmoid nerve (from Brinton et al. 2016 with permission)

over the whole gland, while electrode stimulation activates a restricted area of the lacrimal gland (see also Sect. 12.4.2 and Figs. 12.17 and 12.18).

To the best of our knowledge, electric stimulation of the human lacrimal gland has not been published till now (Sept. 2016), which is understandable by the noticed differences between rodent and cat lacrimal gland characteristics compared to the human situation.

At the end of lacrimal gland studies, one should note that various experimental approaches have been undertaken or are still in progress. Here, we sum up a restricted series: (1) lacrimal gland stem cell approach. It is possible to force stem cells into an epithelial gland cell, because the hormones and factors to change stem cells are known (Hirayama et al. 2016). (2) Gland germ transplantation. Very young developing acini can be transplanted and brought to function (Hirayama et al. 2013). (3) Extra attention for the cornea and conjunctival epithelia and glands: Possibility that the missing of the lacrimal gland can be surmounted by the resting epithelia and lacrimal glands of Krause and Wolfring within the tarsi; e.g. removal of the main lacrimal glands in the rabbit did not reduce tearing (Stevenson et al. 2016).

8.4 Classic Topics in Eye Embryology

The treated topics in eye embryology will concentrate on a restricted amount of themes. Still, not all can be extensively considered in this chapter part: (1) evagination of the optic vesicle, (2) eyecup polarity, the choroidal artery and obliteration of vessels around the eye, (3) lens production and mesenchymal space formation, (4) iris and ciliary formation and (5) retinal artery, canal of Schlemm and glaucoma.

8.4.1 Eye Cup Formation

The formation of the optic stalk with vesicle starts early, even before the closure of the neural tube. The optic vesicle is in constant contact with the ectoderm during its initial development. The ectoderm becomes thickened and forms the lens placode. The term lens placode can be used inappropriately. "There is nothing wrong with this term if it is applied to any ectodermal thickening. However, if it is implied that the cells in the placode are endowed with qualities derived from the neuroectoderm, then the lens placode is not a placode in the proper sense of the term" (Landsmeer, unpublished). The optic vesicle changes into a cup by the outgrowth of the borders of the vesicle and proliferation of the surrounding mesenchyme is a prerequisite too. The growing lens anlage therefore can be directed into the interior of the optic cup (Fig. 8.7a, c). Recent research shows that the ectoderm that is in contact with the outgrowing eye before being transformed into the lens placode (prelens ectoderm) is responsible for cup formation. Removal of the prelens ectoderm inhibits cup formation. Therefore, the optic cup is formed even after removal of the lens placode, as long as prelens ectoderm has been present, indicating that the lens placode is not the inducer of the cup formation. Remarkably, retina development occurs also without cup formation (Hyer et al. 2003).

The transcription factor *Pax6* is needed to induce cup formation (Valeria Canto-Solor and Adler 2006), but cooperates with eye field transcription factors (EFTFs). The EFTF genes include *Six3, Lhx2, Rax (Rx), Tbx3, Optx2 (Six6), Nr2e1 (Tlx)* and *Pax6*. These factors influence each other, and therefore, cup formation is not solely induced by *Pax6*. Co-expression of the other factors is needed for optical vesicle and optic cup formation (Shaham et al. 2012). Sonic hedgehog (*Shh*) is important in midline developmental aspects. It reduces *Pax6* effects. If *Shh* effects are reduced, *Pax 6* effects increases and the eyes fail to separate causing cyclopia (see Sect. 3.5.3 Holoprosencephaly).

The outgrowth of the optic vesicle, even before double walled cup formation, is arrested at one point of the stalk by a blood vessel. It arrives from the surrounding mesenchyme and enters the optic cup. A choroidal fissure originates from it within the cup stalk (Fig. 8.7b). Seemingly the blood vessel is responsible for the local arrest of growth. The developing retina within the cup needs the fissure to contact the visual nervous system parts by axons of the retinal ganglion cells. "Hence, if

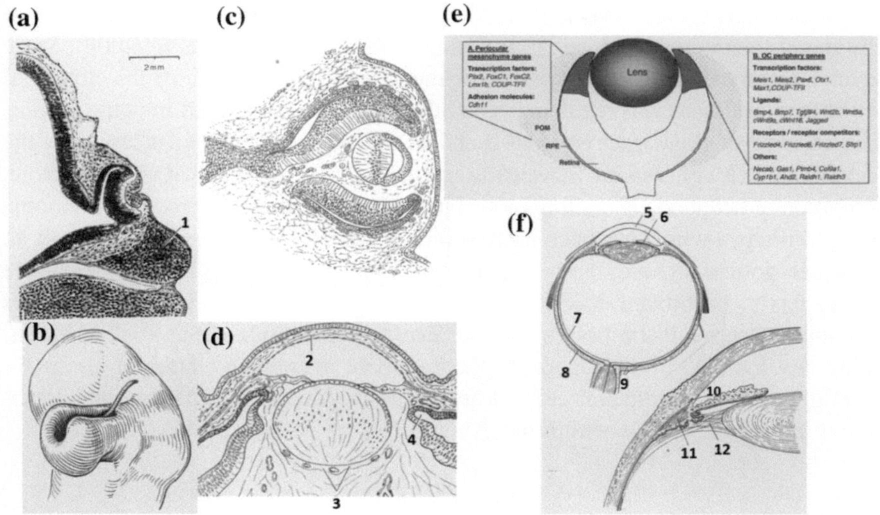

Fig. 8.7 Various stages of eye development. **a** Section through the early outgrowth of the eye cup, always in relation to the ectoderm; **b** 3D view on fissure formation within the cup by the choroidal vessel; **c** Mesenchym penetration within the cup, separation of the lens vesicle from overlying ectoderm, but surrounded by vascular mesenchym of the choroidal artery; **d** Disappearance of the vessels induces "necrosis" or is it apoptosis of tissue inducing wide spaces, cup rim connects to the lens; **e** Genes, transcription factors, ligands, receptors and adhesion molecules involved in iris development: OC, optic cup, POM, periocular mesenchyme, RPE, retinal pigment epithelium (figure is from Davis-Silberman and Ashery-Padan 2008 with permission); **f** Chambers and free iris border together with the system of crypts and ridges in the ciliary body. Canal of Schlemm has been indicated (10) (part figures **a**, **b**, **c**, **d**, **f** Courtesy Dept. Anatomy and Embryology, Leiden University Medical centre). 1, maxillary process (see Fig. 4 in book front matter); 2, multilayered cornea (note that conjunctival sac and eyelids are not drawn); 3, lens mesodermal capsule with vessels of the hyaloid artery; 4, lumen of the optic cup; 5, choroid and entering bloodvessels; 6, cornea with epithelium; 7, iris; 8, retina; 9, sklera; 10, canal of Schlemm; 11, ciliary muscle; 12, zonal fibres

there were no choroidal fissure, the nerve fibres would be obliged to cross the free border of the optic cup, a ridiculous suggestion, especially when we consider that the free border of the cup eventually forms the movable border of the iris. Thus, the choroidal fissure is a most ingenious solution to a serious problem" (Landsmeer, unpublished). This fissure closes towards distal, if the growth of the border of the cup stops. A network of intercellular channels presents within the embryonic retina continues into the optic stalk. Before the differentiation of the retinal ganglion cells, these channels are noted and are present in the route the axons of the retinal ganglion cells will take to leave the eye cup along the edge of the fissure (Silver and Robb 1979). If the choroidal fissure does not close the deviation that occurs is called coloboma. Coloboma disturbs the development of various regions in the eye. It comprises the iris, the best-known deflection, and also cornea, ciliary body, retina, pigment layer and the optic nerve. It sometimes is related to microphthalmia, anophthalmia and anterior chamber effects. Coloboma is responsible for 3–11% of childhood blindness worldwide (see Chang et al. 2006; Hornby et al. 2000). Since several genes are involved in eye development, it stayed unclear how the closure of the choroidal fissure was regulated. The SoxC genes are important in oculogenesis. The absence of SoxC factors *Sox4 a* and *b* produces coloboma. The cause is the elevated levels of *Indian Hedgehog* signalling by absence of *Sox4* factors. Inhibition of hedgehog signalling by its inhibitor cyclopamine overcomes coloboma in the zebrafish, which has been used as model system for these studies (Wen et al. 2014). In normal closure, the basement membrane of the cells at both sides of the fissure has to be broken down. Periocular mesenchyme derived endothelial cells that migrate through the fissure into the eye cup and produce the hyaloid vasculature are involved in destruction of the basement membrane (James et al. 2016): seemingly, these endothelial cells close the fissure door after entering and establishing the blood vessels within the eye cup.

8.4.2 Eye Cup Polarity

The eye cup has a dorsal to ventral polarity. This is due to the start of the eye cup formation at the more ventral side of the outgrowing stalk. It progresses mediodorsally. This dorsoventral polarity is induced by signals from midline structures. *Pax2* is a marker for the ventral and *Tbx5* for the dorsal polarity of the eye cup (Uemonsa et al. 2002). An anterior posterior polarity in the developing eye cup is noticed by the blood vessels, in which the hyaloid vessels are located in the anterior of the eye, mainly for blood circulation of the developing lens, while the choroidal vessels are posteriorly located. In the meantime, the chorioid artery has supplied the mesenchyme around and in the cup. After the lens development, the hyaloid artery degrades and the circulation within the cup is abolished. Presumably, retinoic acid also inhibits the mesenchyme development around the eye cup. A transparent medium originates within the cup after ending of the blood circulation (Fig. 8.7c towards d). It also makes mobility of the iris and the formation of the chambers possible. The proximal part of the choroidal artery persists and supplies the retinal layer of the eye cup. This becomes the central artery of the retina. It will branch through the retina and these vessels can be seen by ophthalmoscopy (Fig. 8.8). The retinal artery is a cerebral artery (internal carotid artery—ophthalmic artery—retinal artery) and by its sole entrance within the eyeball and sole supply of the optic nerve fibres and ganglion layer, its dysfunction means blindness. The blood vessel network within the retina is made by specialized endothelial cells, so-called tip cells. "Tip cells are motile, invasive and extend numerous filopodial protrusions sensing growth factors, extracellular matrix and other attractive or repulsive cues in their tissue environment. Axonal growth cones and endothelial tip cells also respond to signals belonging to the same molecular families, such as Slits and Roundabouts, Netrins and UNC5 receptors, Semaphorins, Plexins and Neuropilins, and Eph receptors and ephrin ligands"

Table 7.1 Blood supply of the eye ball.

	arteries	veins
eye ball (total)	a. ophthalmica	vv. ophthalmicae (superior and inferior)
retina	a. centralis retinae	v. centralis retinae
	lamina choroidocapillaris	vv. vorticosae
Uvea	aa. ciliares posteriores brevis	vv. vorticosae
	aa. ciliares posteriores longae	vv. ciliares anteriores
	aa. ciliares anteriores	
conjunctiva	aa. ciliares anteriores	vv. ciliares anteriores
	aa. palpebrales	vv. palpebrales

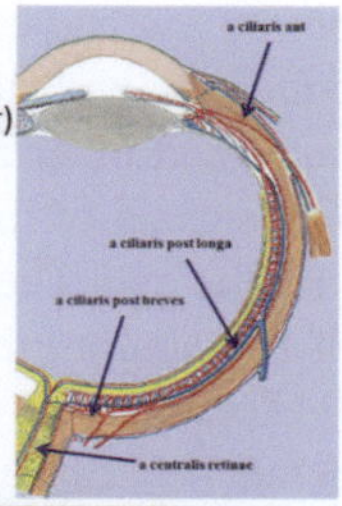

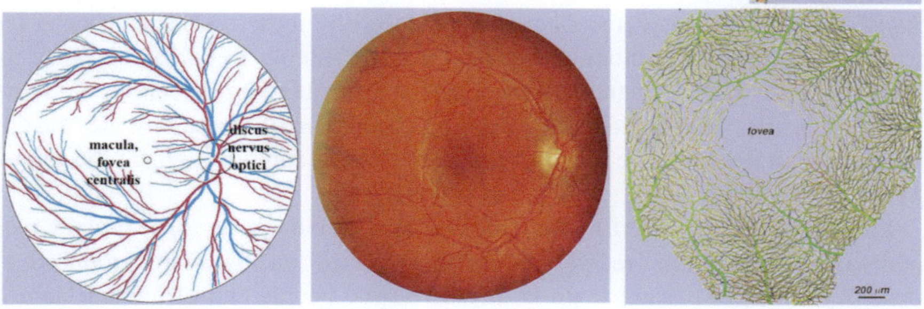

Fig. 8.8 Blood vessels of the eye: Table 8.1 gives the main arteries and veins related to the eye areas, together with the localization of the ciliary arteries in the midsagittal section. The distribution of the retinal artery branches in relation to fovea and discus is pictured. The distribution of the capillary networks in avoiding the fovea is presented in the right lower picture. Table 8.1 is adapted from Imhof and Bleys (2008) with permission and figs are adapted from first digital version of Sobotta (15th ed.), except blood vessels around the monkey fovea which is from Snodderly et al. 1992 with permission

(Adams and Eichmann 2010, see also Eichmann et al. 2005). Vessel branching is done by the transmembrane ligand Delta 4 by inhibiting tip cell formation (Suchting et al. 2007). Not only the predetermined genetic control produces the blood vessel network (see Table 8.1 together with Fig. 8.8), but also the physiological function of the vessels in the organ (Le Noble et al. 2005). Whether the physiological function determines the fine tuning or not within, the retina has still to be underpinned in the one way or the other.

8.4.3 Iris Development

The development of the iris starts with the increase in length and the thinning of the rim of the optic cup (Fig. 8.7c, d). The lens is a necessary prerequisite to form iris, ciliary body and cornea. Removal of the lens or changes in expression of lens genes induces eye anterior segment dysgenesis (disturbed development). Before the iris development, the retina differentiation has started and progresses towards the cup rim. It stops near the margins of the cup. The compartmentalization of the ocular cup in a neuronal area and a non-neuronal peripheral cup rim area is present long before iris and ciliary body development. The growth in length of the cup margins coincides with the production of smooth muscle cells of ectodermal origin. The outer and inner layers of the iris become pigmented, while the ciliary body is not. The stroma of the iris originates from the periocular mesenchyme, which migrates into the eye. The periocular mesenchyme is suspected to contain also contributions of the neural crest. The factors and genes involved are summarized in Fig. 8.7e. *PAX6* is certainly essential for the growth and development of the iris, since *PAX6* deficiency causes iris defects. Its downstream gene targets are *Foxc1*, *Tgfβ2* and *Bmp4*. Presumably, iris defects are also produced by these downstream gene targets: *Bmp4* expression inhibition changes the inner and outer epithelial layer of the iris, and *Tgfβ2* deficiency induces corneal defects, patterning defects in the retina and serious iris stroma reduction and in patients *Foxc1* mutations are directly related to complete absence of iris tissue and to ocular anterior segment dysgenesis (Wang et al. 2017).

Iris colour is determined by variable amounts of eumelanin (brown/black melanin) and pheomelanin (red/yellow melanin). These melanins are produced by melanocytes. More of eumelanin is in brown-eyed people and of pheomelanin in blue- and green-eyed people. Melanocortin-1 Receptor Gene regulates eumelanin production and is located on chromosome (MCIR) 16q24.3. Several point mutations in the MCIR gene alleles are commonly found in light skinned and blue/green-eyed people (Metzelaar-Blok et al. 2001; Valverde et al. 1995).

8.4.4 Glaucoma

The eyesight of Galileo (1564–1642) progressed to blindness, and finally, he became blind. He reported bad vision of his left eye. Portraits showed that he suffered from a mucocoele of the right frontal sinus around his age of 60. "A mucocele of a frontal sinus is an accumulation of mucoid secretion and desquamated epithelium within the sinus with distension of its walls and is regarded as a cyst-like expansive and destructive lesion. If the cyst invades the adjacent orbit and continues to expand within the orbital cavity, the mass may mimic the behaviour of many benign growths primary in the orbit" (Agarwal et al. 2012). Although a serious affection, it will not lead to blindness. Based on a series of arguments angle-closure (irido-corneal angle-closure) glaucoma should be considered the cause of Galileo's blindness at the end of his life (Marmor 2009).

All the above summed up *Pax6*-related downstream genes (*Foxc1*, *Tgfβ2* and *Bmp4*) are also inducers of glaucoma. These genes change the iris, but also the irido-corneal angle. This is exactly the location of the canal of Schlemm with its interwoven trabecular meshwork to take up the eye aqueous humour. Eye liquid is produced by the ciliary processes, streams from the posterior chamber via the iris margin and pupil into the anterior chamber and is removed by the trabecular meshwork into the canal of Schlemm to end in the normal blood stream via the scleral plexuses. Whether an acute or gradually change of the balance between production and removal, an increased amount of eye humour will induce glaucoma.

"Chinese and East Asians have a higher risk of developing primary angle-closure glaucoma [also known as acute glaucoma, remark ours], a major cause of blindness. There is increasing evidence that the iris plays a crucial pathophysiologic role in the development of angle closure and its glaucoma. Irises that are more voluminous (e.g. thicker, larger, and more convex) are associated with angle closure. In addition to the anatomic characteristics, the physiologic dynamic responses of the iris during pupil dilation have been implicated in primary angle-closure glaucoma. By using anterior segment optical coherence tomography imaging, eyes with previous acute angle-closure attacks do not loose as much iris volume during pupil dilation as that of healthy eyes" (Chua et al. 2016). Primary angle closure is generally caused by relative pupillary block, which is the increased resistance to flow of aqueous humour through the pupil from the posterior chamber to the anterior chamber, leading to anterior bowing of the peripheral iris over the trabecular meshwork and to angle-closure glaucoma. And one of the other causes is seemingly the volume of the iris during dilatation. The iris contains so-called crypts: Holes present in the middle circle (collarette) of the front of the iris. The collarette is just above the ciliary region. Insufficient volume decrease of the iris, thus a relative thick iris during dilatation, closes also the irido-corneal angle. The iris can be pushed to the trabecular meshwork and hinders the removal of eye humour. "Irises with more crypts were significantly thinner and lost more volume on pupil dilation. In view that the latter feature is known to be protective for acute angle-closure attack, it is likely that the macroscopic and microscopic composition of the iris is a contributing feature to angle-closure disease" (Chua et al. 2016).

8.4.5 Blood-Retinal Barrier

The exterior of the eye is protected by lymphoid tissue (see Sect. 8.3). The interior is still equipped with means to control inflammatory immunity (for the history of ocular immunology see Margo and Harman 2016). However, the requirement is, of course, that the visual function cannot be disturbed. In the centre of the optic nerve, the artery and vein for the retina is localized (a.v. centralis retinae). The rest of the wall of the eyeball gets its blood from the a. ciliares (see Table 8.1 in Fig. 8.8), branches of the a. chorioidea. Both main vessels enter the eye from the back. The outer retina gets its perfusion from the chorioidea (80% of the perfusion) and the internal retina from the retinal artery (20%). Rods and cones therefore are supplied by the ciliary arteries and the ganglion and plexiform layers mainly by the retinal artery. The network of blood vessels does not enter the area of the fovea (Fig. 8.8). Several capillary layers are distinguished: radial peripapillary capillaries (nerve fibre layer), an inner layer (ganglion cell layer) and an outer layer (inner and outer plexiform layer) that contains transitional areas in between (Snodderly et al. 1992). The ocular immune privilege (a mechanism of immunological ignorance by a

Table 8.1 Congenital malformations of the human eye

Inherited eye diseases in man				
Gene	Function or required for	Disease	Inheritance	Phenotypes
Panocular defects				
PAX6	Induction of lens, retina, iris, cornea	Aniridia	AD	with foveal or optic nerve hypoplasia, cataract, glaucoma, corneal dystrophy
CHX10	Retina, optic nerve development	Microphthalmia	AR	microphthalmia, cataract, coloboma
Anterior segment defects				
MITF	Melanin synthesis retina	Waardenburg 2	AD	Iris pigment defects, hearing loss white forelock
FOXC1, -F2	Development cornea iris	Axenfeld-Rieger anomaly	AD	Minor: glaucoma, iris hypoplasia
FOXQ1				Major: iridogoniodysgenesis post. Embryotoxon
PITX2	Development ocular mesenchyme	Rieger syndrome		
		Iridogonialdys gen	AD	Rieger anomaly, dental hypoplasia, facial dysmorphism
PITX3	Development lens placode	Anterior mesenchym. dysgen. congenital cataract		Defects in all tissues of anterior chamber
Posterior segment defects				
PAX2	Development optic cup, retinal fissure	Pappilorenal syndrome	AD	Optic disk dysplasia
HESX1	Forebrain, optic vesicle, nasal placode	Septo-optic dysplasia	AR	Optic disc hypoplasia midline defects
Photoreceptor abnormalities				
CRX	Photoreceptor outer segment, transactivation rhodopsin	Cone-rod dystrophy or Leber congenital amaurosis retinitis pigmentosa	AD/AR	Degeneration cone, than rod or absence of funct. Photoreceptors
NRL	Cotransactivator rhodopsin		AD	Rod receptor degeneration

AD, autosomal dominant; AR, autosomal recessive; Amaurosis, total blindness; Axenfeld-Rieger anomaly, abnormalities anterior eye chamber like, thin iris, multiple pupils, abnormal cornea, late development of glaucoma; iridogoniodysgenesis, abnormalities of the anterior eye chamber due to abnormal migration or terminal induction of neural crest cells; papillorenal syndrome, underdevelopment (hypoplasia) of the kidney and colobomas of the optic nerve; Rieger syndrome, midface abnormality with development of glaucoma; retinitis pigmentosa, retina degenerative condition of both eyes with pigment formation on the retina which progresses to blindness; Waardenburg 2, among other deviations enlargement of the eyeball (table is adapted from ten Donkelaar et al. 2014) different immune response) is co-organized by a blood-retinal barrier due to tight junctions between the endothelial cells. The ocular fluids contain TGFβ, also somatostatin, VIP, and αMSH (inhibitory neurotransmitters) and retinoic acid, while the tissues express FasL, PDL-1, thrombospondin, galectins (all inhibitory cell surface molecules). The consequence is that antigen presenting cells are made less active and produce an inhibitory environment for immune cells from the blood circulation. Activation of antigen presenting cells, due to anterior eye chamber infection, gives strange enough spleen activation, inducing $CD4^{+}$and 8^{+}-T_{reg} cells. Nevertheless, direct activation of T cells within the eye is also possible (Taylor 2009; Smith et al. 2013). Removal of the superior cervical ganglion, removal of sympathetic innervation of the eye, abolishes immune privilege, presumably by a serious reduction of TGF-β in the aqueous humour (Vegal et al. 2009). While intrinsic factors diminish the activity of antigen presenting cells, extrinsic factors like the neurotransmitters substance P (SP), vasoactive intestinal peptide (VIP), calcitonin gene-related peptide (CGRP) and neuropeptide Y (NPY) do increase trophic and immune functions at the ocular surface, partially by their relation with mast cells and recruiting immune cells. These monocytes and neutrophils are also stimulated by the neurotransmitters in "their innate immune functions by increasing the production of arachidonic metabolites (i.e. prostaglandin E2 and thromboxane B2) and their secretion of granule constituents", which holds especially for SP (Sabatino et al. 2016). CGRP induces IL-10 secretion and diminishes inflammation reactions as VIP does. NPY increases TGF-β and prevents autoantigen processing and presentation and contributes to immune privilege. How these different and contradictory functions of the neurotransmitters contribute to a healthy ocular surface is still a matter of research.

The goldfish has been frequently used for studying the neurotransmitters in the optic system in the seventies and eighties of the previous centuries. Damage of the goldfish optic nerve induces a fast increase in acid phosphatase within 2 days in the optic nerve and all the projection areas (see Marani and Ruigrok 1981). A lysosomal reaction follows the injury of the optic nerve and seemingly not an immune reaction. However, allografts are readily rejected by the goldfish immune system. Hence the presence of ocular immune privilege was studied in goldfish eyes by injecting allografts and autografts within the anterior eye chamber. The results showed that autografts survived, but allografts were rejected within 8 days. "Thus, immune privilege does not exist in the goldfish eye, implying that immune privilege in the eye (or central nervous system) may be an evolutionary adaptation acquired by higher vertebrates" (Jiang et al. 1994).

8.4.6 *Congenital Malformations of the Eye*

Congenital malformation is due to the genetic control of the eye development and is related to their transcription factor genes. These human congenital malformations are subdivided in panocular defects, anterior segment defects with congenital aphakia, posterior segment defects and photoreceptor anomalies. Table 8.2 gives an overview of these malformations.

Since we hardly know the sequence of gen activations for the development of the eye and the interrelation between several transcription factors, e.g. eye field transcription factors (Sect. 8.4.1), syndromes can only be coupled to a transcription factor but they cannot be coupled to the order of appearances and mutual effects of these transcription factors. It remains a medley of syndromes and of transcription factors.

8.5 Spectacles

The eye is of course part of the visual system. Already Felix Platter (1536–1614) showed that the retina is responsible for vision and not the lens (Milos and Akert 1982). Pupil function was earlier detected by Leonardo da Vinci (1452–1519; Capra 2007). By the way, Leonardo used spectacles. Lens function, turning of the image on the retina, remained still inexplicable, since the adaptation of the brain to the turned image was unknown. Thus, one discerns an optical component and, already Leonardo, a neuronal component of the visual system. Present day, the eye as part of the optical component has been compared to a camera. Advantages of the eye over a camera are superb lenses, self cleaning, auto-focus and auto-tracking, steady cam or good stabilization, and the eye is functional at widely different light levels. The last four advantages include a neuronal component.

Dysfunctions of the eye lenses are called far-sighted or hyperopia (=hypermetropia) and near-sighted or myopia, while the normal function of the lenses is indicated by emmetropia. Presbyopia, the fact that during aging the lens becomes more rigid and can hardly adapt to project images exactly on the retina, was considered a medical problem, not an optical one in the Middle Ages. An mediaeval answer could be found if an optical solution was looked for, which indeed occurred. It is a written sermon of the monk Giordano da Rivalto (dated 23 February 1305) that says: “It is not yet twenty years since the discovery of the art of making eyeglasses for good vision, one of the best and most necessary arts that the world has” (Willach 2008). Therefore, and of course on more grounds, the first appearance of eyeglasses is estimated to be somewhere between 1285 and 1290 (Willach 2008; Ilardi 2007). A rock crystal plate with an oval shape, the Nimrud lens, stems from an Assyrian town excavation and is determined from the period 900–700 BC. The Lothar’s rock crystal is from the Carolingian time 855–869 AD and has lens properties. Both are rock crystals proving that lens grinding was a common technique throughout these ages (Willach 2008). This is also supported by the round precious stones in religious art treasures and if glass-like they could be used as magnifying glasses of 30–40 diopters. However, they cannot be used as an ancillary lens, because they do miss the positive refracting power to compensate the rigid aged lens (Willach 2008). Venetian glassmakers solved the problem of the hygroscopy of glass (water destructed easily ancient, potassium based, glass) by the use of soda (Na_2CO_3). Non-hygroscopic glass could be produced, opening the possibility to make permanent lenses. The first spectacles were rivet spectacles of which the halfs could be slipped over each other and thus were easy to transport. The first rivet-spectacle picture stems already from 1352 (fresco of French cardinal Hugh of St Cher, monastery of San Nicoló in Treviso). The primitive bridge, to put the spectacles on the nose, arrived later. The materials varied from, bone, wood, horn and leather and this variety was applied a hundred years after the spectacle invention. The steel spring bridge combined with the leather one originated in the eighteenth century. "This improvement lightened the frame and provided a spring-like action for better grip on the nose" (Ilardi 2007). One should note that metal or wire rivet spectacles were already produced from 1387 on. Till so far the presbyopia, but the myopic "patients" were hardly helped, because the spectacles in daily life would easily fall down. They used concave lenses with a handle. Leonardo da Vinci sketched spectacles with strings attached, presumably to wind them around the ears. It is then a small step to add legs to the bridge spectacles. By trial and error, like now at the optician, the correct lenses for the spectacles were chosen. Optic laws were detected at the end of the sixteenth to the start of the seventeenth century (Thomas Harriot [1560–1621] in 1601–1602; Willebrord Snell van Royen called Snellius, [1580–1626] in 1621, both (re-)discovering the law of refraction of light). Long-distance viewing devices were developed "for military and naval purposes, distant views of landscapes for painting, topographical maps and closer observation of nature" in the sixteenth century (Ilardi 2007). Camera obscura studies improved images by understanding lens function and mirrors: "equipped with convex lenses and/or concave mirrors placed over its aperture to project distant images on a screen, glass or piece of paper suitably placed opposite the aperture" (Ilardi 2007). Thus, lenses were three ages in production before optical laws were applied and the telescope (combination of convex objective lens and a concave eyepiece) arrived not earlier than in 1608.

It is rather astonishing that spectacles that are placed for decennia on the nose skin with its underlying nasal bone of the bearer, hardly shows bone deformation, nor bedsore. Bacterial infections do occur and best-known skin infections are impetigo (cold sores) and herpes zoster (shingles) ophthalmicus (involved are the trigeminal branches V1 and V2).

8.6 Mature Orbit and Its Muscle Content

Eye movements belong to the fastest movements one can make. The motor unit contains the smallest amount of muscle fibres related to one motor neuron, in fact 5–10 extrafusal muscle fibres. The amount of muscle spindles is the highest in the eye muscles. Moreover, for fixation slow muscle fibres are also present. Eye muscles are a special type of striated muscles and can be considered the ADHD one among the striated muscles (by the way, Ritalin© need not, but can induce glaucoma). These eye muscles cover partially the topographical relationships between

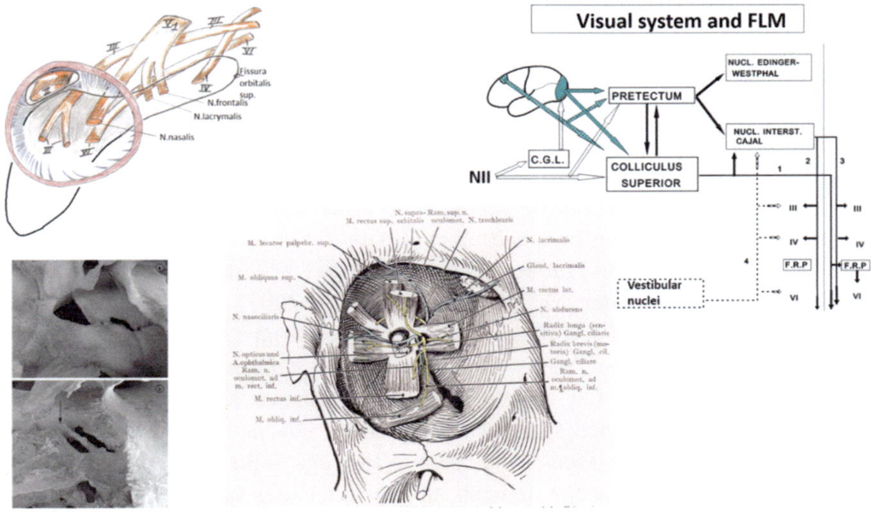

Fig. 8.9 Left upper(© E. Marani) and middle figures give the relation of the annulus of Zinn or common tendinous ring for the eye muscles and the fissura orbitalis superior with the cranial nerves and their branches. Note that the caudal part of fissura orbitalis superior lies exterior of the tendinous ring and no structures pass through it to enter the orbita. Left down shows the anterior lacerate foramen (white arrow) of the two Guiana dolphin species: a riverine species, b marine species. Black arrow shows subdivision of this fissure in the marine species (figure from Fettuccia et al. 2009 with permission). Right shows the steering of the eye nuclei, formatio reticularis pontis and the relation to the fasciculus longitudinalis medialis (FLM). 1 = FLM Tectospinal (contralateral), 2 = FLM Interstitiospinal (ipsilateral), 3 = FLM Interstitiospinal (contralateral), 4 = FLM Vestibulo-mescencephal (ipsi and contralateral), CGL Corpus geniculatum laterale, FRP Formatio reticularis pontis: horizontal gaze and saccades. (Figs are from lectures the Neurosurgical Depart. Leiden University Medical Centre and from Corning 1919)

the optic foramen, the superior orbital fissure and the common tendinous ring (annulus of Zinn). The four rectus muscles originate from this common ring and various structures enter the orbit through this ring, thus, are located within the cone of these muscles (Fig. 8.9 left and middle).

They are the optic nerve, the ophthalmic artery (both passing through the optic foramen), the oculomotor nerve, the nasalis (nasociliary) nerve, branch of the ophthalmic trigeminal nerve, and the abducens nerve to the lateral rectus muscle (Fig. 8.10). Structures passing outside the ring through the lateral part of the superior orbital fissure are the frontal nerve and the lacrimal nerve, branches of the trigeminal ophthalmic nerve, and the trochlear nerve to the superior oblique muscle (Fig. 8.9). The innervation of the rectus muscles (except rectus lateralis) and the inferior oblique and the superior palpebral muscle by the oculomotor cranial nerve (N III) is plain (see Fig. 8.10 for cranial nerve innervation of eye muscles).

Presynaptic parasympathetic fibres, originating in the Edinger–Westphal nucleus, travel along the N III towards the ciliary ganglion for the parasympathetic innervation of the eye. Orbit entrance is by the back of the eye through the annulus

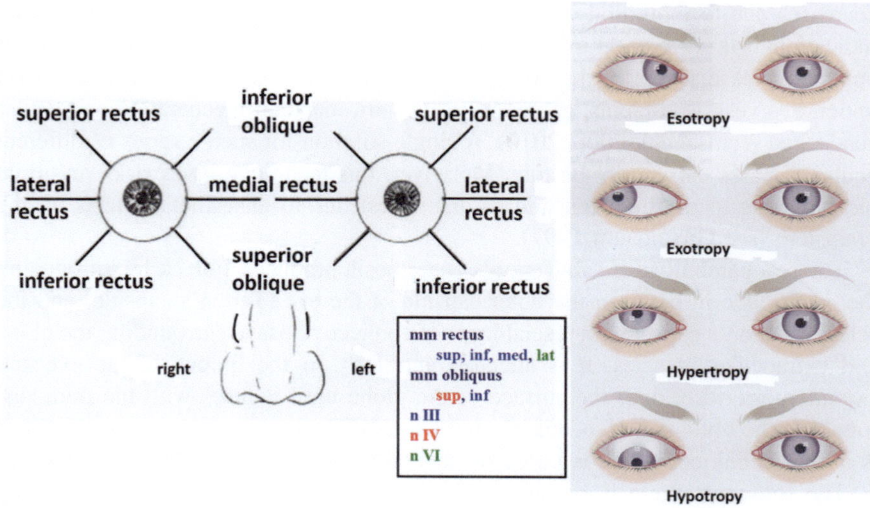

Fig. 8.10 Left: Directions of the eye movement caused by eye muscle function, right: types of strabismus, and middle: cranial nerve innervations of the eye muscles. (right picture from M. Joosse, Haaglanden Medical Centre, The Hague with permission)

of Zinn. Its ganglion ciliare is located within the rectus muscle cone. Small nerves from this ganglion (nn ciliares breves) reach the eyeball. These postganglionic parasympathetic fibres innervate the m. ciliaris and m. sphincter pupillae. Among these small fibres, sympathetic and sensible axons are hiding. The sympathetic innervations can pass, together with the cranial abducens nerve or by the ophthalmic artery (originating from the internal carotic plexus), also through the annulus, thus are located within the rectus muscle cone (Fig. 8.9).

The nervi ciliares longi contain sensible fibres of the n. nasociliaris. The nasociliary nerve brings with it the sensible parasympathic nerve fibres, called ramus communicans cum ganglio ciliari or radix longa. It splits from the nasociliary nerve directly after entering the orbit and connects to the back of the ciliary ganglion.

"Neural therapy is a form of alternative medicine in which a local anaesthetic is injected into certain locations of the body in an attempt to treat chronic pain and illness. There is no good evidence that the therapy works and it carries some risk; it has not been found to be of overall benefit" (Wikipedia UK, 3 December 2016). However, peribulbar injections reaching the ciliary ganglion with a local anaesthetic are still used in neural therapy to suppress inflammation and chronic eye diseases. By the way, it is labelled as a parabulbar one in neural therapy but it is a peribulbar injection (Weinschenk 2008, see below sub-Tenon's anaesthesia). Procain (1%; 1–2 ml) will be injected through the lower eyelid (left at 5 h, right 7 h) at a depth 1, 5–2 cm. Mydriasis (dilatation) of the pupil and ptosis (paralysis of m. levator palpebrae, hanging of the upper eyelid) is the indications that the anaesthetic reached the ganglion ciliare, which is placed behind th eyeball. Blocking by procaine of the N III gives double vision by exotropy (Fig. 8.10). Eye pressure decreases due to inhibition of both parasympathetic and sympathetic innervation. Indications for it are wide in neural therapy, to name a few: conjunctivitis, zoster ophthalmicus, diseases of the cornea, uveitis, iridocyclitis, cataract, neuritis nervi optici, eye vessel problems, glaucoma at its start, macular degeneration and ocular headache (Weinschenk 2008, 2010). A single solution for such a series of different ocular diseases can hardly be true. Moreover, this technique gives risks on orbita bleeding due to the injection and cardiac arrest due to anaesthetic effects on the brainstem (see Greenbaum 1997).

The real parabulbar or sub-Tenon's anaesthesia produces a block by introducing the anaesthetic in between the outer capsule of the eye (Tenon's capsule) and the sclera. "Tenon's capsule is a fascial layer of connective tissue surrounding the globe and extraocular muscles. It is attached anteriorly to the limbus of the eye and extends posteriorly over the surface of the globe until it fuses with the dura surrounding the optic nerve" (Canavan and Dark 2003; see also Greenbaum 1997). It is a safe technique, which is based on several studies (see Canavan et al. 2003)

The sensory system is organized by the ophthalmic nerve and the maxillary nerve. The nervus nasociliaris (nervus nasalis, V_1) enters the orbita, located cranial of and placed on the ophthalmic artery (the artery is in 98% responsible for the perfusion of all orbita structures, the other 2% is by the meningeal medial artery), and passes over the optic nerve with the ophthalmic artery, to reach the medial side of the orbita. It gives the ethmoidal branches that enter the nasal cavity and ends as the nervus infratrochlearis in the skin of the upper part of the nose. The lacrimal nerve uses the lateral wall of the orbita to reach the tear gland, the upper eyelid and the conjunctiva. The frontal nerve, placed cranial of the levator palpebrae muscle, passes through the roof of the orbita to split into the supraorbitalis nerves and supratrochlearis nerve. Partially parallel to the nerves, the veins of the nose and forehead skin (v. supraorbitalis, v. nasofrontalis) are branches of the superior ophthalmic vein, located in the orbit, which is also responsible for the drainage of the eyeball (together with the inferior ophthalmic vein). This vein leaves the orbit by the most upper lateral part of the superior orbital fissure. The inferior ophthalmic vein connects to the superior one, before leaving the orbit. Infections of the skin can reach the inside of the orbit and the eye by this venous drainage and even can reach the sinus cavernosus. By the connection of the superior ophthalmic vein (v. angularis) with the branches of the facial vein, infections can reach the fossa pterygopalatine (by v. faciei profunda, plexus and pterygoid vein) and of course the fossa nasalis (vv nasalis externae). After the description of most of the internal structures of the orbita attention is payed to the entrances at the back of the orbita. The *cranial* orbital part knows three entrances:

(1) The optic canal (optic nerve and ophthalmic artery)
(2) The annulus of Zinn entrance, located over the middle of the superior orbital fissure (n. nasociliaris, n. abducens, n. oculomotorius)
(3) The lateral opening of the superior orbital fissure (n. trochlearis, n. frontalis, n. lacrimalis and v. ophthalmica superior)

It does not mean that once a structure is in the orbit it is targeted for the ocular/visual system. Orbital pathways are also used to reach the upper viscerocranial skin and parts of the nasal cavity, even the fossa pterygoidea by venous connections (see also Sect. 6.1).

Within the human *caudal* orbital part, the inferior orbital fissure contains a sulcus (infraorbitale) that directs towards the foramen infraorbitale. Its infraorbital nerve, main branch of the human maxillary nerve and placed on the bottom of the orbit, uses this pathway to reach the teeth, the lower eyelid, cheek, lower nose skin, anterior temple and upper lip partially. Thus, the orbit walls are not only a separation for the ocular system cave, but also a hindrance for structures that need to reach areas that are not ocular related.

One should remember Sect. 4.7 on the development of the skull base. Yes, the cranial nerves impose the foramina during development. The superior orbital fissure is produced at the closure edges of two different bony elements, while perforation makes the passage for foramen opticum and rotundum (see Sect. 4.3). Nevertheless, the intriguing question is not posed in topographical handbooks. Of the superior orbital fissure, the caudal one third is not used for passage of structures (Fig. 8.9 left upper part), which holds for nearly the whole inferior orbital fissure, since the nervus infraorbitale has its own sulcus, except for some small infraorbital vessels, the zygomatic nerve and a few small twigs of the pterygopalatine ganglion. "Moreover, a thin layer of non-striated muscle, the orbitalis (or muscle of Müller, remark ours), bridges the inferior orbital fissure; its precise function is uncertain" (Gray's anatomy 1973). What, *in a topographical sense*, forces the human orbita to keep such a large cleft? Spatial coherence says that structures and organs are not only determined by their own form and function but also by the form and function of surrounding structures (see Sect. "Coherence Spatial" in Introduction). Is it the orbita wall's own structural property or is it a demand from surrounding structures? Before going on: "A direct comparison of the massive skull of the extinct hominid Pithecanthropus with that of a small gibbon led some authorities to the conclusion that the former was really a giant gibbon. This can hardly be regarded as valid (Le Gros Clark 1960). Statistical comparisons, by ignoring allometric phenomena, have grossly overemphasized the taxonomic significance of differences in shape" (Le Gros Clark 1960, in Kuhlenbeck 1967). After this caution, part of the solution for the posed question can seemingly be found in the dolphin skull structure. The dolphin's skull is characterized by telescopy, lengthening of the maxilla and mandible in such away that a snout evolves. The neurocranium becomes shorter and broader. Risso's dolphin (*Grampus griseus*) has a robust and well-ossified cranium. In this dolphin, the foramen **infraorbitale** was composed of several holes, not a single hole as in humans. The infraorbital *nerves* run through these foramina (Yamagiwa et al. 1999). Thus, lengthening of the front of the face into a snout causes the infraorbital branches of the maxillary nerve to use several channels to innervate the longer maxilla surface. The peripheral nervous system adapts to the changed maxilla and mandible. The changed neurocranium contains the foramen orbitorotundum also called anterior lacerate foramen or cranial hiatus (dolphin nomenclature is confusing, because anatomical names are inconsistent (Yamagiwa et al. 1999). A large hiatus is present for the abducens (VI), ophthalmic (V_1), maxillary (V_2), trochlear (IV) and oculomotor nerves (III). Adjacent to the anterior margin of the cranial hiatus, the canalis opticus for the optic nerve (II) is present. "Some openings are very large and irregular and are referred to as hiatuses. The cranial hiatus of the dolphin and manatee may include the following nerve openings: optic foramen, orbital fissure (anterior lacerate foramen), and oval foramen—plus the openings for vessels between the last two" (Rommel et al. 2009). In the bottlenose dolphin (*Tursiops*), the cranial nerves II, III, IV, V_1 and VI use the lacerate foramen, and sometimes, this foramen also contains V_2 (Rommel 1990). In the Guiana dolphin (*Sotalia* spec.), with a marine (*S. guianensis*) and a riverine (*S. fluviatilis*) species, the riverine one has an open lacerate anterior foramen, while the marine dolphin separates this foramen by a bone chip (called a spike-shaped projection, Fig. 8.9; Fettuccia et al. 2009). The subdivision of the lacerate foramen is believed (on weak grounds, by the way) a neotenic property (Neoteny is the retention of juvenile characters in the adult, also known as pedomorphosis, see Sect. 10.1.2).

In fact, the foramen orbitorotundum or anterior lacerate foramen or cranial hiatus is comparable to the superior + inferior orbital fissure of Man. Note that the maxillary nerve V_2 (infraorbitalis is its main part in humans) is localized in it. Seemingly both human orbital fissures, constituting one hole, are the orbit's back wall own property, which can be freely varied as seen in the dolphins. One should note that the bone parts forming the orbit's back wall are differently organized in the dolphin compared to humans (see Rommel et al. 2009).

There is one more remarkable fact concerning the back of the orbit. The skull has two main points that produce forces on the skull base: the tempero-mandibular joint and the occipital condyles. These stresses are distributed over a large area of the skull side walls towards the cranial vault. "In this way, less stress is exerted upon mechanically less resistant regions of the skull base, namely the middle part of the skull cavity permeated by the numerous foramina, and the anterior skull cavity with the fragile roof of the nasal and orbital cavities" (Demes 1985).

8.7 Oculomotor Steering

Seemingly, the smallest nuclei in the central nervous system are involved in important basic functions: Onuf's nucleus for bladder and colon steering, Barrinton's nucleus for the miction control and the eye nuclei for oculomotor functions. For example, the trochlear nucleus (origin of cranial IV and innervating m.obliquus sup) contains 2000–3500 neurons, while the retina has 120–140 million rods, 5–6 million cones and 1 million retinal ganglion cells.

Francis Crick (from Crick and Watson's DNA molecular structure) studied the "mystery of consciousness" by putting most weight and considerations on the visual system to decipher the problem, with very limited success. Emphasis is laid by Crick on short-term memory: "Consciousness in general, and visual awareness in particular, obviously incorporate into their processes much that we have already stored in long-term episodic and categorical memories. What concerns us more is very short-term memory, since it is plausible to argue that if we lost *all* forms of memory for new events we would not be conscious. However, this essential form of memory needs only last a fraction of a second or perhaps a few seconds at most" (Crick 1994). Attentional operations within the central nervous system have been studied in the previous millennium mostly from a philosophical point of view and Crick's hypothesis played an important role in it (see Churchland 1986). An experiment can explain the short memory: A very short flash of light of around 50 milliseconds will be noticed by a test person. If this short light flash contains a series of letters a few will be memorized, but not all. However, if the letters are organized in rows and the test person is asked after the 50 milliseconds flash to repeat the letters of the, e.g. middle row, he is capable to do so partially. Thus, this so-called George Sperling experiment shows that the letters are noticed but seemingly this visual trace is rapidly decaying. This is called iconic memory, is related to the retina and it is concluded that "*a certain minimum time is needed for adequate visual processing* "(Crick 1994). The sufficient time is somewhere between 100–200 ms. This means for oculomotor steering that the eye needs a minimum time, let us say 200 ms, to keep a constant position for visual processing. The motor neurons of the eye muscles therefore need a minimum steering constancy of its muscles of about 200 ms. There are several more conditions the motoneurons have to fulfill of which none of the underlying arguments are reproduced here. It is enough to know that several premotor networks force different types of eye movements, and gaze holding:

1: saccades. Flicks of the eyes: fast, jump-like, voluntary, conjugated changes in direction of vision, resulting in a new fixation point. Flicks can be made max. 5 times/s, which is 200 milliseconds. This is the time needed for adequate visual processing (see above).

2: vestibulo-ocular reflex. Stabilisation of the image on the retina needed during movements of the head. Excitatory connections link each vestibular canal to a set of two eye muscles that move the eyes in the same plane as the canal. Inhibitory connections link the vestibular canal with their muscular antagonists (see Fig. 8.12).

3: optokinetic response. It contains a reflex and a nystagmus. In a moving surrounding, the eyes will directly find a new fixation point by this reflex. Large visual fields (stimulus fields) that move over the retina induce reactions of the retinorecipient accessory optic nuclei and optic tract nucleus that activate vestibulo-oculomotor circuits producing eye movements. Optokinetic nystagmus: slowly following the movement, till the eye cannot follow anymore, a saccadic movement brings the eye back to a new position and slow following can start again.

4: smooth pursuit. In fact, these are scanning movements: moving objects stay projected on the foveae of both retinae. It is the consequence of the dorsal stream of visual cortical areas that notices motion and sends information to the vestibulocerebellum.

5: vergence. Vergence movements cause the eyes to rotate in different directions. These are the only disjunctive eye movements in the visual system. The reticular formation below and dorsal of the ocular nuclei sends these commands.

6: gaze-holding. Involves the velocity of the position integrator. It has its own neuronal circuits and uses the cerebellum to affect the ocular motor nuclei.

How do eye muscles work: (1) any contraction of an eye muscle needs activation of its antagonist for, e.g., horizontal movements (Fig. 8.11); (2) activation of the antagonist is relayed via the proper contralateral eye nucleus by the ipsilateral nucleus involved; (3) the semicircular ducts note by angular movements the turning velocity of the head; (4) from this, the slow compensatory phase of the vestibulo-ocular reflex must be generated (or nystagmus, or saccades); (5) eye motor neurons need not only turning velocity but also a position signal; (6) this neuronal integration of position and velocity happens in the paramedian reticular formation.

An eye motor neuron is bombarded with synaptical information directly and indirectly by the six premotor networks, indicated above, and it has to produce the correct answer to steer precisely the eye muscle concerned. The eye motor nuclei and vestibular nuclei are distantly localized in the brainstem. To connect these nuclei, the fasciculus longitudinalis medialis (FLM) is present (see Fig. 8.9), which is divided in subconnections. Several areas in the brainstem (colliculus superior and

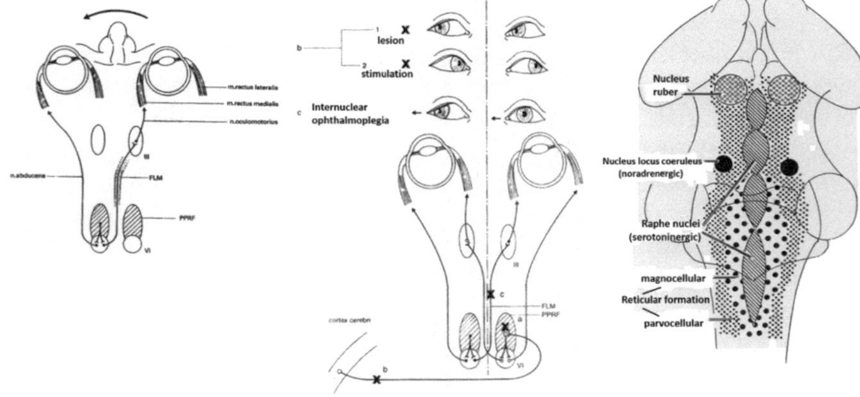

Fig. 8.11 Overview of eye movements. Left shows that simultaneous horizontal movement of the eyes needs contralateral antagonist activation. Middle shows the effects of interventions of the paramedian pontine reticular formation (**a**), of the involved cortex (**b**) and of the FLM (**c**). Crosses in the map indicate the location of the interventions, crosses in the upper part the brain stem side. Right figure shows parvo and magnocellular parts of the reticular formation (courtesey J. Voogd). FLM, fasciculus longitudinalis medialis; PPRF paramedian pontine reticular formation

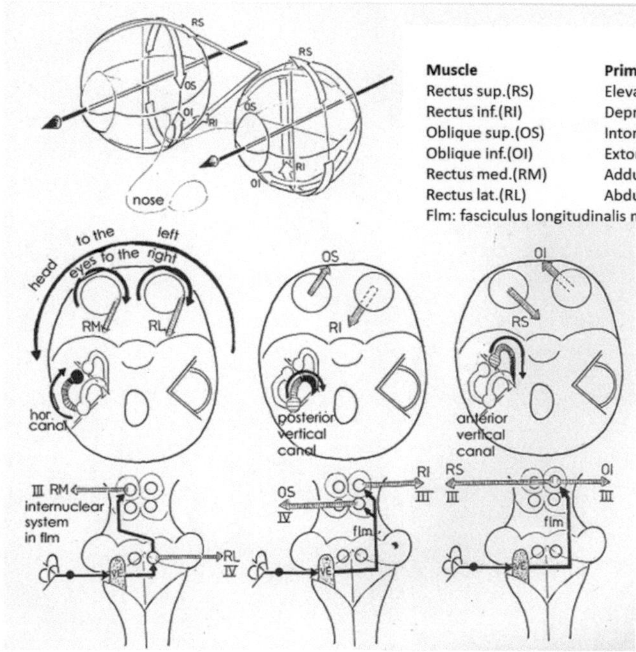

Muscle	Primary action	Secondary action
Rectus sup.(RS)	Elevation	Intorsion, adduction
Rectus inf.(RI)	Depression	Extorsion, adduction
Oblique sup.(OS)	Intorsion	Depression, abduction
Oblique inf.(OI)	Extorsion	Elevation, abduction
Rectus med.(RM)	Adduction	
Rectus lat.(RL)	Abduction	

Flm: fasciculus longitudinalis medialis

Fig. 8.12 Vestibulo-ocular reflex regulating the interaction of head movements and eye movements. Upper part shows eyes with the direction of eye muscle actions. Middle figures demonstrate the movements of the head, represented by the actions of the vestibular canals, and muscle actions for stabilisation of the image on the retina during these movements. Lower series gives the brainstem with action of the vestibular nuclei (see also Fig. 8.13) on the eye motor nuclei for the different head movements (courtesy J. Voogd with permission)

pretectum) are essential for the generation of the correct answers to the orientation of the eyes, to the coupled head and neck position, and, of course, to the whole body, towards the point of interest. Moreover, one should notice that the visual occipital cortex can influence both colliculus superior and pretectum. The paramedian reticular formation is part of the reticular formation, which is an interneuron pool. Three interneuron types can be discerned: Large (magnocellular) interneurons, small (parvocellular) interneurons, midline interneurons. The reticular formation is subdivided according to the developmental parts of the brainstem: rhombencephalon, mesencephalon, diencephalon.

Each area is in its turn subdivided in a lateral small (parvo) and paramedian large (magno) interneuronal population (Fig. 8.11). The mes and diencephalic reticular formation contains mainly small interneurons: the paramedian reticular formation around eye motor nuclei relays on mainly small interneurons (parvocellular area). The midline area is serotoninergic and is excluded from these reticular eye subdivisions.

I. What happens during stimulation of the caudal part of the paramedian pontine reticular formation? In the caudal area of the pons lies nucleus VI, which is

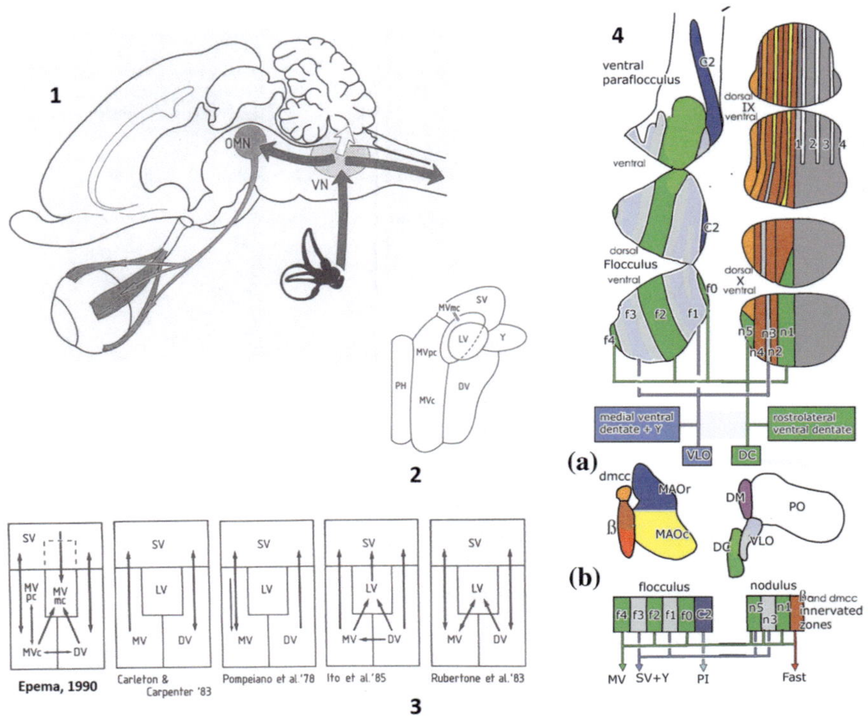

involved in horizontal gaze. The caudal paramedian reticular formation projects unilateral on nucleus VI, but the nucleus VI projects onto the contralateral eye nucleus III. Thus, conjugated horizontal deviation of the eyes to the same side as the stimulation is carried out.

II. What happens after stimulation of the rostral paramedian reticular formation? The rostral paramedian reticular formation is related to the eye motor nuclei III and IV. Thus, vertical eye movements will arise after stimulation. Note that vertical eye movements involve always both eyes.

III. What happens after stimulation of the colliculus superior? Stimulation of the colliculus superior gives contralateral stimulation only of the caudal paramedian reticular formation. Thus, contralateral VI will be activated, and thus conjugated horizontal deviation of the eye to the other side.

IV. How is the relation between forebrain and eye movements? Forebrain areas are involved in voluntary eye movements including visual guided saccades. The best-known area is the posterior part of the middle frontal gyrus, just in front of the premotor cortex, also called frontal eye field (see Fig. 8.9). Inputs originate especially from the posterior parietal eye field. This frontal eye field is especially related to saccades. It is the saccade triggering area. Connections are directly to the brainstem but mainly via the colliculus superior. Thus, always contralaterally towards VI (see Fig. 8.9).

◂**Fig. 8.13** 1: Vestibular nuclei (VN) and their cerebellar and ocular effect (OMN: Ocular Motor Nuclei; see Fig. 8.12). 2: The vestibular nuclei are subdivided in several subnuclei, which not all are represented here (DV: descending vestibular nucleus; LV: Lateral vestibular nucleus; MV: Medial vestibular nucleus; mc: magnocellular; pc: parvocellular; c: caudal; SV: Superior vestibular nucleus; Y: group y). 3: Research gave the increasing detected interactions between the various vestibular subnuclei (see Epema 1990, with permission). 4: Vestibular cerebellar effects produce longitudinal zones in the "vestibulo-cerebellum" containing nodulus, uvula and flocculus. PH, the nucleus prepositus hypoglossi is not considered a vestibular nucleus, but does contribute with a floccular projection. **a** Olivocerebellar climbing fibre projections to Purkinje cell zones of the flocculus and the nodulus-uvula of the rat. Illustrated are the zones on the ventral and dorsal surface of the flocculus and the adjacent ventral surface of the ventral paraflocculus. In the right column the ventral and dorsal surface of the nodule (X) and the uvula (IX) are shown. In the right halves of these lobules the zebrin-positive bands 1–4 are indicated in lobule IX. The cortex of lobule X is uniformly zebrin-positive. The medial (MAO) and principal olive are shown as flattened maps. The dorsal cap (DC) projects to the f4, f2 and f0 zones of the flocculus and the n1 and n5 zones of the nodulus. The ventrolateral outgrowth (VLO) projects to the f1 and f3 zones of the flocculus and the n3 zone of the nodulus. The other zones of the nodulus that extend into the zebrin-positive zones of the uvula are innervated by the group beta and the dorsomedial cell column (dmcc). The caudal medial accessory olive (MAOc) and the dorsomedial group (DM) project to the zebrin negative zones of this lobule The most medial C2 zone of the flocculus and the paraflocculus receives climbing fibres from the rostral medial accessory olive (MAOr). B. Efferent projections of the Purkinje cell zones of the flocculus and the nodulus. The f0, f2, f4 and the n1 and n5 zones project to oculomotor relay cells for rotations around a vertical axis in the medial vestibular nucleus (MV). The f1, f3 and n3 zones project to oculomotor relay cells in the superior vestibular nucleus (SV) and the group Y for rotations of the eye around a horizontal axis co-linear with the anterior horizontal canal. The c2 zone projects to the posterior interposed nucleus (PI). The group beta and dmcc innervated zones project to the fastigial nucleus. (Katayama, paraflocculus 1988; Voogd et al. 1998; Ruigrok 2003; Ruigrok et al. 1992; with permission and courtesy J. Voogd)

This description of horizontal eye movements can easily be repeated for activation of the other vestibular canals (see Fig. 8.12). The eye motor nuclei are guided by the brainstem nuclei and its reticular formation, the cerebellum (Voogd et al. 1998; Ruigrok 2003; Ruigrok et al. 1992; Thunnissen 1990) and the cortex (Horn and Adamczyk 2012). Thus although the eye motor nuclei are small, the nervous system areas that contribute to their function force the eye motor nuclei to function as an integrator of all the different incoming information.

The vestibular canals gather information for angular acceleration along the three main axes, while otolith sense linear acceleration during roll tilt of the head about the longitudinal axis (utricule) and during pitch about the intra-aural axis (saccule).

The vestibular cranial nerve projects to various vestibular nuclei that do interact with each other and by a separate branch into the cerebellum (Barmack 2016). These vestibular nuclei project also to the cerebellum. Vestibular incoming information within the "vestibulo-cerebellum" (nodulus, uvula, flocculus and paraflocculus) is projected into longitudinal zones over the cerebellar lobules (indicated by f for floccular zones, or n for nodular ones). These zones are determined by the olivocerebellar (climbing fibre) input. The output modules of these zones are the Purkinje cells that project onto cerebellar nuclei (in this case dentate nucleus and vestibular nuclei).

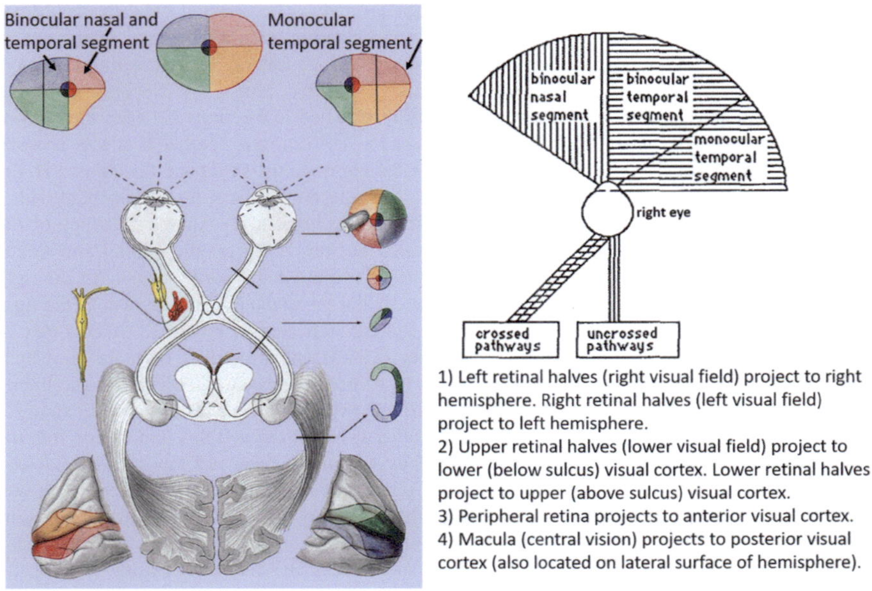

Fig. 8.14 Visual fields in humans. Binocular nasal and temporal segments with the monocular temporal segment are indicated in an overview of crossed and uncrossed visual pathways (left figure is adapted from first digital version of Sobotta (15th ed.), right figure is from Joosse (1999, with permissions)

These efferent projections of the Purkinje cell zones of the flocculus and the nodulus connect back to vestibular nuclei. The f0, f2, f4 and then 1 and n5 zones project to oculomotor relay cells for rotations around a vertical axis in the medial vestibular nucleus (MV, Fig. 8.13). The f1, f3 and n3 zones project to oculomotor relay cells in the superior vestibular nucleus (SV) and the group Y for rotations of the eye around a horizontal axis co-linear with the anterior horizontal canal (Fig. 8.13). Thus, steering of the vestibular nuclei by the cerebellum produces indirect cerebellar control over the ocular motor nuclei (Fig. 8.13). This cerebellar steering is used to guide and refine motor behaviour and to ascertain spatial orientation as described above (Roth et al. 2016).

8.8 Monkeys and Squint

One learns from students, especially if they have done part of their study abroad. Maurits Joosse went to the Anatomy and Cell Biology Department of the Emory University in Atlanta, USA, which has an integrated cooperation with the Yerkes Regional Primate Research Centre. Did you know that monkeys could be squint? Squint was thought a deflection of the human race, which is incorrect since already Hubel and Wiesel (1965) used cats for strabismus studies.

"Hubel and Wiesel showed that the connections degenerate as a result of visual deprivation, such as monocular deprivation, binocular deprivation, and squint, also known as strabismus. Monocular deprivation reduces the percentage of cells that can be driven by the deprived eye, and binocular deprivation reduces the specificity of the receptive fields of single cells in the cortex for orientation and direction of movement, and squint reduces the percentage of cells that receive binocular input" (Daw 2009). Joosse's study (1999) concerned the southern pig-tailed macaque (Macaca nemestrina). These monkeys do have naturally occurring convergent strabismus. Thus, squint monkeys do exist with view directions of the eyes that cross in front of the macaque. Their monocular fields (see Fig. 8.14) extend from 45° nasally to 90° temporally. The eye visual field is subdivided into binocular segments (nasal and temporal) and a monocular segment (see Fig. 8.14). In cats, it was found that the nasal field changed due to strabismus. In the squint monkeys, like in strabismic amblyopia in humans, the change of the nasal field was absent. In strabismus, a preferred eye and a non-preferred eye is distinguished. Everybody knows a child with a sealed off preferred eye to activate the non-preferred one. In the non-preferred eye of few of the squint monkeys, changes in the temporal monocular segment were found, but hardly in strabismic amblyopia in humans. "The results of our study, when combined 'with previous studies of visually deprived animals and humans' with amblyopia, show that monocular field deficits can be of graded severity" (Joosse 1999, see also Wilson et al. 1989; Boothe et al. 1990; Joosse et al. 1990).

The final remark concerns the iris. Ocular manifestations of syphilis can come about at every stage of the disease. "Whether an oral localisation of the primary chancre increases, the risk of ocular involvement is unknown". The first description of a translucent vascularised iris tumour in a patient with ocular syphilis has now been published (De Jong et al. 2017).

References

Adams RA, Eichmann A (2010) Axon guidance molecules in vascular patterning. Cold Spring Harb Perspect Biol. https://doi.org/10.1101/cshperspect.a001875

Aggarwal SK, Bhavana K, Keshri A, Kumar R, Srivastava A (2012) Frontal sinus mucocele with orbital complications: management by varied surgical approaches. Asian J Neurosurg 7:135–140

Barmack NH (2016) Vestibular nuclei and their cerebellar connection. In: Gruol DL, Koibuchi N, Manto M et al (eds) Essentials of cerebellum and cerebellar disorders, pp 69–78

Brinton M, Chung JL, Kossler A, Kook KH, Loudin J, Franke M, Palanker D (2016) Electronic enhancement of tear secretion. J. Neural Eng 13.https://doi.org/10.1088/1741-2560/13/1/016006

Botelho SY, Martinez EV, Pholpramool C, Prooyen HC, Janssen JT, De Palau A (1976) Modification of stimulated lacrimal gland flow by sympathetic nerve impulses in rabbit. Am J Physiol 230:80–84

Boothe RG, Quick MW, Joosse MV, Abbas MA, Anderson DC (1990) Accessory lateral rectus orbital geometry in normal and naturally strabismic monkeys. Invest Ophthalmol Vis Sci 31:1168–1174

Canavan KS, Dark A (2003) Garrioch MA (2003) Sub-Tenon's administration of local anaesthetic: a review of the technique. Br J Anaesth 90:787–793

Capra F (2007) The science of Leonardo. Doubleday, NY

Chang L, Blain D, Bertuzzi S, Brooks BP (2006) Uveal coloboma: clinical and basic science update. Curr Opin Ophthalmol 17:447–470

Chua J, Thakku SG Tun TA, Nongpiur ME, Lee Tan MC, Girard MJA, Wong TY, Min Quah JH, Aung T, Cheng CY (2016) Iris crypts influence dynamic changes of iris volume. Ophthalmol 123 2077–2085

Churchland, Smith P (1986) Neurophilosophy. Toward a unified science of the mind-brain. MIT Press Cambridge

Coggan DD, Liu W, Baker DH, Andrews TJ (2016a) Category-selective patterns of neural response in the ventral visual pathway in the absence of categorical information. NeuroImage 135:107–114

Coggan DD, Baker DH, Andrews TJ (2016b) The role of visual and semantic properties in the emergence of category-specific patterns of neural response in the human brain. http://dx.doi.org/10.1523/ENEURO.0158-16.2016

Corning HK (1919) Lehrbuch der topographischen Anatomy für studierende und Ärzte. Verlag J. F, Bergmann, Wiesbaden

Crick F (1994) The astonishing hypothesis. The scientific search for the soul. Simon & Schuster, London

Davis-Silberman N, Ashery-Padan R (2008) Iris development in vertebrates; genetic and molecular considerations. Brain Res 1192:17–28

Daw NW (2009) The foundations of development and deprivation in the visual system. J Physiol 587:2769–2773

De Jong Y, Haverkort ME, Van Sorge AJ, Jansen C, Luyten GPM, Josse M (2017) A translucent vascularised iris granuloma in a patient with secondary syphilis. The Lancet Inf 17:255–256

Demes B (1985) Biomechanics of the primate skull base. Adv Anat Embryol Cell Biol 94:1–59

ten Donkelaar HJ, Lammens M, Hori A (2014) Clinical neuroembryology. Springer, Heidelberg

Eichmann A, Le Noble F, Autiero M, Carmeliet P (2005) Guidance of vascular and neural network formation. Curr Opin Neurobiol 15:108–115

Epema AH (1990) Connections of the vestibular nuclei in the rabbit. Thesis Erasmus University, Rotterdam

Fettuccia DC, da Silva VMF, Simöes-Lopes PC (2009) Non-metric characters in two species of Sotalia (Gray, 1866) (Cetacea, Dephinidae) Braz. J Biol 69:907–917

Gray's Anatomy (1973) In: Warwick R, Williams PL (eds) 35th ed. Edinburgh, UK

Greenbaum S (1997) Ocular anesthesia. Saunders WB Co, Philadelphia

Grondin S (2016) Psychology of perception. Springer Nature Intern Publ, Switzerland

Hirayama M, Kawakita T, Tsubota K, Shimmura S (2016) Challenges and strategies for regenerating the lacrimal gland. Ocular Surf 14:135–143

Hirayama M, Ogawa M, Oshima M, Sekine Y, Ishida K, Yamashita K, Ikeda K, Shimmura S, Kawakita T, Tsubota K, Tsuji T (2013) Functional lacrimal gland regeneration by transplantation of a bioengineered organ germ. Nat Comm. https://doi.org/10.1038/ncomms3497/www.naturecom

Horn AKE, Adamczyk C (2012) Reticular formation; Eye movements, gaze and blinks. In: Mai JK, Paxinos G (eds) The human nervous system, Academic Press/ Elsevier Amsterdam, Sydney, pp 328–366

Hornby SJ, Gilbert CE, Rahi J, Sil AK, Xiao Y, Dandona L, Forster A (2000) Regional variation in blindness in children due to icrophthalmos, anophthalmos and coloboma. Ophthal Epidem 7:127–138

Hubel DH, Wiesel TN (1965) Binocular interaction in striate cortex of kittens reared with artificial squint. J Neurophysiol 28:1041–1059

Hyer J, Kuhlman J, Afif E, Mikaw T (2003) Optic cup morphogenesis requires pre-lens ectoderm but not lens differentiation. Develop Biol 259:351–363

Ilardi V (2007) Renaissance vision from spectacles to telescopes. Amer Phil Soc Memoirs, vol 259. Philadelphia

Imhof SM, Bleys RLAN (2008) Oogheelkunde. Eds. Stilma en Voorn, Praktische Huisartsgeneeskunde, 2e herziene druk, BSL, Houten

James A, Lee C, Williams AM, Angiliri K, Lathrop KL, Gross JM (2016) The hyaloid vasculature fascilitates basement membrane breakdown during choroid fissure closure in the zebrafish eye. https://doi.org/10.1016/j.ydbio.2016.09.008i

James J (1977) The mystery of the great labyrinth, Chartres Cathedral. Studies in Comparative Religion, vol 11, no 2. World Wisdom, Inc. www.studiesincomparativereligion.com

Jiang LQ, Streilein JW, McKinney C (1994) Immune privilege in the eye: an evolutionary adaptation. Develop Comp Immunol 18:421–431

Joosse M (1999) Visual fields in strabismus suppression and amblyopia. Thesis Erasmus University, Rotterdam

Joosse MV, Wilson JR, Boothe RG (1990) Monocular visual fields of macaque monkeys with naturally occurring strabismus. Clin Vis Sci 5:101–111

Katayama S, Nisimaru N (1988) Parasagittal zonal pattern of olivo-nodular projections in rabbit cerebellum. Neuroscience Res. 5:424–438

Kossler AL, Wang J, Feuer W, Tse DT (2015) Neurostimulation of the lacrimal nerve for enhanced tear production Ophthal. Plast Reconstr Surg 31:145–151

Kuhlenbeck H (1967) The central nervous system of vertebrates, vol 1. Propaedeutics to comparative neurology. Karger S, Basel, NY

Lee J-W (2010) Treatment of enophthalmos using corrective osteotomy with concomitant cartilage graft implantation. J Plastic Reconstr Aest Surg 63:42–53

Le Gros Clark WE (1960) The antecedents of man. An introduction to the evolution of the primates. Quadrangle Books, Chicago

Le Noble F, Fleury V, Pries A, Corvol P, Eichmann A, Reneman RS (2005) Control of arterial branching morphogenesis in embryogenesis: go with the flow. Cardiovas Res 65:619–628

Lima LB, Miranda de Miranda RB, Furtado LM, Rocha FS, Parreira da Silva MC, Silva CJ (2015) Reconstruction of orbital floor for treatment of a pure blow-out fracture. Rev Port Estamotol Med Dent Cir Maxilofac 56:122–126

Marani E, Ruigrok TJH (1981) Enzyme histochemical changes in some optic projection areas of thegoldfish after optic nerve lesions. Neurosci Lett 23:233–238

Margo CE, Harman LE (2016) Autoimmune disease: Conceptual history and contributions of ocular immunology. Surv Ophthalmol 61:680–688

Marmor MF (2009) The enigma of Galileo's eyesight: some novel observations on Galileo Galilei's vision and his progression to blindness. Surv Ophthalmol 54:630–640

Metzelaar-Blok JAW, ter Huurne JA, Hurks HM, Keunen JE, Jager MJ, Gruis NA (2001) Characterization of melanocortin-1 receptor gene variants in uveal melanoma patients. Invest Ophthalmol Vis Sci 42:1951–1954

Milos G, Akert K (1982) Die neurobiologischen Grundlagen der binokulären Raumwahrnehmung. Historische Übersicht über die Entwicklung der Vorstellungen und Erkenntnisse. Schweiz Arch Neurol Psychiatr 130:39–67

Prache A (2003) Notre-Dame de Chartres. Image de la Jérusalem céleste, CNRS editions Paris. ISBN 2-271-05012-X

Rommel S (1990) Osteology of the bottlenose dolphin. In: Leatherwood S, Rees RR (eds) The bottlenose dolphin. Academic Press, San Diego, NY, London, pp 29–49

Rommel SA, Pabst DA, McLellan WA (2009) Skull anatomy. In: Perrin WF, Wursig B, Theivissen H (eds) Encyclopedia of marine mammals, 2nd ed. Acad Press, San Diego, CA, pp 1033–1047

Roth MJ, Lindner A, Their P (2016) Visual circuits. In: Gruol DL, Koibuchi N, Manto M et al. (eds) Essentials of cerebellum and cerebellar disorders, pp 89–100

Ruigrok TJ (2003) Colateralization of climbing and mosy fibers projecting to the nodulus and flocculus of rat cerebellum. J Comp Neurol 466:278–298

Ruigrok TJ, Osse RJ, Voogd J (1992) Organization of inferior olivary projections to the flocculus and ventral paraflocculus p mofg the rat cerebellum. J Comp Neurol 316:129–150

Russell B (1980) Appearance and reality. The problems of philosophy. Oxford University Press, Oxford, pp 1–6

Sabatino F, Di Zazzo A, De Simone L, Bonini S (2016) The intriguing role of neuropeptides at the ocular surface. Ocular Surface. https://doi.org/10.1016/j.jtos.2016.10.003

Schechter JE, Warren DW, Mircheff AK (2010) A lacrimal gland is a lacrimal gland, but rodent's and rabbit's are not human. Ocul Surf 8:111–134

Shaham O, Menuchin Y, Farhy C, Ashery-Padan R (2012) Pax6: A multi-level regulator of ocular development. Prog Ret Eye Res 31:351–376

Silver J, Robb RM (1979) Studies on the development of the eye cup and optic nerve in normal mice and in mutants with congenital optic nerve aplasia. Develop Biol 68:175–190

Smith PD, MacDonald TT, Blumberg RS (2013) Principles of mucosal immunology. Garland Sci, London, NY

Snodderly DM, Weinhaus RS, Choi JC (1992) Neural-vascular relationships in central area of macaque monkeys (Macaca fascicularis). J Neurosci 12:1169–1193

Sobotta, Atlas of human Anatomy. vol 1, Head, neck, upper limbs and skin. Ed Staubesand J, Digitalized 15th ed. Urban&Schwarzenberg, Elsevier

Spencer G (1960) the information available in brief visual representations. Psychol Monogr Gen Appl 74(11):whole 498, 1–29

Stevenson W, Pugazhendhi S, Wang M (2016) Is the main lacrimal gland indespensible? Contributions of the corneal and conjuctival epithelia. Survey Ophthalm 61:616–627

Suchting S, Freitas C, le Noble F, Benedito R, Bréant C, Duarte A, Eichmann A (2007) The Notch ligand Delta-like 4 negatively regulates endothelial tip cell formation and vessel branching. PNAS 104:3225–3230

Taylor AW (2009) Ocular immune privilege. Eye 23:1885–1889

Thunnissen I E (1990) Vestibulocerebellar and vestibulo-oculomotor relations in the rabbit. Thesis, Erasmus University, Rotterdam

Uemonsa T, Sakagami K, Yasuda K, Araki M (2002) Development of Dorsal-Ventral polarity in the optic vesicle and its presumptive role in eye morphogenesis as shown by embryonic transplantation and in Ovo explant culturing. Dev Biol 248:319–330

Valeria Canto-Soler M, Adler R (2006) Optic cup and lens development requires Pax6 expression in the early optic vesicle during a narrow time window. Dev Biol 294:119–132

Valverde P, Healy E, Jackson I, Rees JL, Thody AJ (1995) Variants of the melanocyte − stimulating hormone receptor gene are associated with red hair and fair skin in humans. Nat Genet 11:328–330

Van Dyck R (2015) Een retrobulbair abces bij een gezelschaps konijn (oryctolagus cuniculus). Masterproef and thesis, Veterinary Faculty, University Gent

Voogd J, Gerrits NM, Ruigrok TJ (1998) Organization of the vestibulocerebellum. Ann NY Acad Sci 781:553–579

Vegal JL, Keino H, Masli S (2009) Surgical denervation of ocular sympathetic afferents decreases local transforming growth factor β and abolishes immune privilege. Am J Pathol 175:1218–1225

Wang X, Shan X, Gregory-Evans CY (2017) A mouse model of aniridia reveals the in vivo downstream targets of Pax6 driving iris and ciliary body development in the eye. Biochim Biophys Acta 1863:60–67

Weinschenk S (2008) Parabulbäre Injektion an das Ganglion ciliare. Komplementäre un Integrative Medizin 49(8):39–43

Weinschenk S (2010) Handbuch Neuraltherapie. Diagnostik und Therapie mit Lokalanästhetica mit Zugang zum Elsevier-Portal. Urban & Fisher/Elsevier, München

Wen W, Pillai-Kastoori L, Wilson SG, Morris AC (2014) Sox4 regulates choroid fissure closure by limiting Hedgehog signaling during ocular morphogenesis. Developm Biol 399:139–153

Willach R (2008) The long route to the invention of the telescope. Amer Philosoph Soc, Transaction series, 98, vol 5. Philadelphia, pp 1–116

Wilson JR, Lavalle KA, Joosse MV, Hendrickson AE, Boothe RG, Harwerth RS (1989) Visual fields of monocularly deprived macaque monkeys. Behav Brain Res 33:13–22

Yamagiwa D, Endo H, Nakanishi I, Kusauagi A, Kurohmaru M, Hayashi Y (1999) Anatomy of the cranial nerve foramina in the Risso's dolphin (Grampus griseus). Ann Anat 81:293–297

Yasui T, Karita K, Izumi H, Tamai M (1997) Correlation between vasodilatation and secretion in the lacrimal gland elicited by stimulation of the cornea and facial nerve root of the cat. Invest Ophthalmol Visual Sci 38:2476–2482

Chapter 9
Vestibule, Cochlea and Implants

The direct observation of a natural phenomenon or its analysis, even with the most sophisticated techniques, never leads to absolute certainty. R. Chandebois (1983)

9.1 Introduction

The start of modern ear research is difficult to establish (see Helidonis 1993 for overview). Deafness was earlier fought by so-called ear or speaking trumpets and hearing tubes (Fig. 9.1c–e). The Paris horns, which even could be carried as a diadem, helped both ears (Fig. 9.1b). Joseph Toynbee's "Disease of the ear" (1860) has been taken as the starting point of ear research in this book. The concomitant invention of the auriscope by John Brunton (1826–1899) in 1860 (Fig. 9.1a) opened the possibility to enlighten the auditory tube and to inspect the eardrum with artificial light. It became the start of operations for deafness of the ear.

Toynbee described ankylosis (stiffening and immobility of joints, here of the auditory ossicles) of the stapes and otosclerosis (an excessive growth of the middle ear ossicles) of the stapes's footplate producing deafness, in 1883. Nowadays, persons with otosclerosis often benefit from a hearing aid. The other approach, surgical replacement of the stapes, is a common procedure to improve conductive hearing problems. During stapedectomy, the stapes is removed and replaced with a stapes prosthesis. The operation is carried out under local anaesthesia. Surgery is done on only one ear at a time. Waiting time is nearly a year between procedures. Hearing improvement reaches its top about four months after the surgery with a success rate of over 80%. While stapedectomy has been carried out from the 1880s on after Toynbee's description and some success has been noted, it stopped around 1908–1910 due to primitiveness of light sources, of instruments and of magnification instruments. Moreover, the lack of standardized and reproducible audiometry made comparisons before and after operations difficult. Benefit of the operation was more decided by the patient's mood than by measured reduction of deafness. Tuning forks, although present, were hardly used to check ear functions. "Perhaps the greatest handicap facing early otologists in their quest of a surgical cure for

E. Marani and C. Heida, *Head and Neck*,
https://doi.org/10.1007/978-3-319-92105-1_9

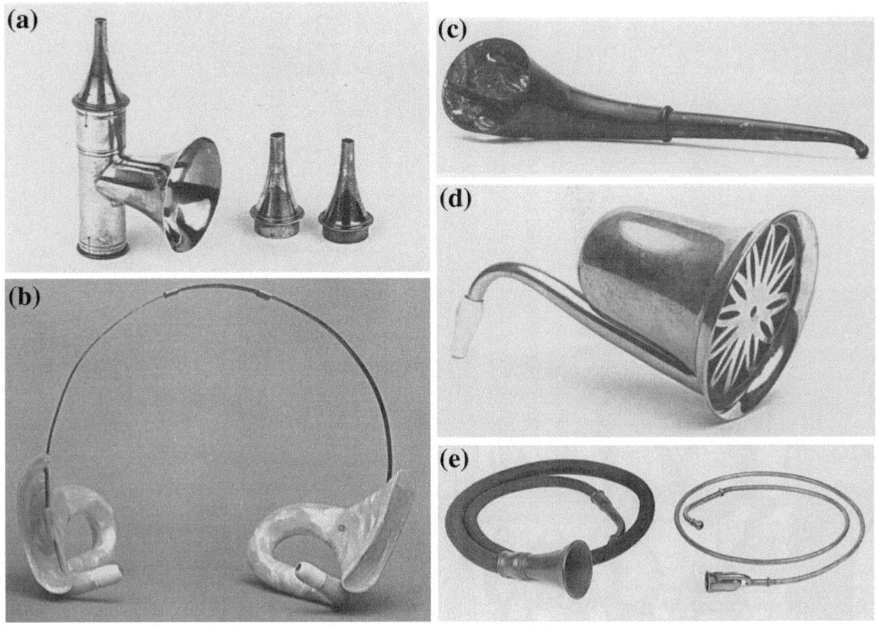

Fig. 9.1 Auriscope and speaking trumpets. **a** Auriscope of John Brunton (1826–1899) this specimen was produced in 1890; **b** So-called Paris horns, which could be carried as a diadem on the head; **c** Trumpet horn of tortoise; **d** Cupper trumpet, so-called casing and rod puller clamp (trekpot in Dutch); **e** Hearing tube of Tröltzsch (left) and cupper conversation tube (Griffioen 1988; Courtesy and permission Reinier de Graaf Gasthuis, Delft, The Netherlands)

deafness was their apparent lack of understanding of the middle ear transformer. Their ignorance of ossicular chain function was demonstrated by their not infrequent performance of Kessel's operation, removing the tympanic membrane and ossicles to treat deafness. Even in the 1900s, the significance of the ossicular chain was not clear" (Hillel 1983). It lasted till 1950 before the operation was restored, and problems were overcome co-based on the knowledge of the function of the ossicular chain.

Audiology developed due to the detection of tonotopy by von Békésy (1899–1972; Nobel Prize in 1961), by understanding bone conduction and the middle ear function, model studies on cochlear waves, the propagation of transients in these models and cochlear input impedance by Tonndorf (1914–1989; see Zwislocki 1981)

In his article "The audiology revolution" (1973), Kershaw gives an overview how the science of hearing developed during and after the Second World War with special attention to children's deafness. Two ladies should be named, but first attention is paid to the impact of antibiotics. In general, deafness has been seriously reduced due to the fencing in of otitis media with its consequence of long-lasting infections before the application of antibiotics. "For the child health doctor, acquired deafness was "out" and congenital deafness was "in"" (Kershaw 1973).

The consequence was the so-called glue ear, in which unnoticed chronic otitis in the preschool years still could cause deafness. Edith Whetnall (1910–1965) started her crusade insisting that "'no child is totally deaf". She specialized in ear, nose and throat clinics and (co-)organized and became director of the Royal National Throat, Nose and Ear Hospital in London (UK) in 1947. She found out that early recognition of children's deafness and with training of the mother, children could learn to speak despite their deafness. In fact, auditory training in young deaf children due to an early diagnosis overcomes muteness. Mary Sheridan (1899–1978) was a paediatrician and public health officer. She noticed that late diagnosis of hearing and visual problems produced serious retardation of children's education. Due to her publications on speech and language delay, attention to the general child development arose. "The ingenuity of Mary Sheridan and others has made it possible to devise a series of simple screening tests which can be carried out by health visitors on children of between 6 and 8 months old and, it is fair to say, will detect those children who have a sufficient hearing defect to justify concern and more detailed investigation" (Kershaw 1973). The consequence of these developments was also the erection of special schools for deaf children to support their education and training.

Due to the discovery of electricity and the microphone, a series of electrical hearing aids have been developed. The carbon microphone played an important role in this so-called telephone age (1900–1920) "Dr. Ferdinand Alt (1867–1923), working at the Politzer Clinic in Vienna, demonstrated an electric hearing aid at the Austrian Otological Society in 1906. The hearing device used a magnetic earphone connected to a carbon granule microphone that was placed on top of a small battery box. This instrument was of little use if the speaker was more than two feet away".

The vacuum tube age was from 1920 to 1940. At the moment that the triode vacuum tube was invented by De Forest (1873–1961) in 1906, it was also used in hearing aids. The other problem was solved by Helmut Sell, of the German firm Siemens, since he invented the first carbon amplifier that could be placed separately from the microphone in the hearing aid in 1925. "This amplifier is a magnetic receiver coupled to a carbon microphone with a common diaphragm powered by the same battery supplying the current for amplification". Carbon amplifiers, or "boosters", were used widely in the 1930s. Application of bone conduction was affected by Greibach of Sonotone by his "reaction" type bone receiver in 1933. "The design, a major improvement over earlier bone conduction hearing aids, is still in use today".

"Transistor hearing aids were introduced and began replacing vacuum tube hearing aids around the 1950s. The smaller transistor size and lower battery voltage requirements permitted a dramatic miniaturization of hearing devices. The transistor also allowed other styles of hearing aids to be developed, such as the behind-the-ear model and the hearing aid built within eyeglass temples" (the citations above are from "Timeline of hearing devices and early deaf education" 2005–2009). The transistor hearing age took place from 1950 to 1980. The digital hearing age is from 1980 to 2000 and brings us to the second millennium with digital renewals and cheap hearing aids.

Restoration of hearing by technical aids was an early target of functional electrical stimulation. In 1950, Lundberg (see Gisselsson 1950) stimulated the auditory nerve for the first time. The patient could only notice noise. Stimulation of the acoustic nerve in a deaf patient was successfully carried out by Djourno and Eyries (1957). The implanted induction coil was connected to an indifferent electrode placed in the temporal muscle on one side and the other side was connected to an inner ear electrode. A second coil was placed over the first to transmit the signal from a microphone by induction. The patient could hear simple words. The hereupon following unpleasant criticisms of neuroscientists and clinicians that continued over 20 years are skipped here. It lasted till 1984 before the Food and Drug Administration allowed the use of cochlear implants in adults.

Hearing aids paralleled electrical discoveries from 1900 onwards. Anatomical knowledge together with paediatric expertise fought deafness in young and old during and after the Second World War, exemplified by the restricted illustrations above. Here we left out the deafness due to military actions, like bombing and explosions during both world wars. To be honest, the application of the physics of sound contributed as much as anatomy, perhaps even more. Here we start with the developmental and anatomical topics of the external, middle and inner ear. To rub up the anatomical relation of the petrous bone to other skull parts, the following description is given.

The pars tympanica of the os temporale is the bony part of the external auditory canal and consists of a tube-like bended plate containing a cleft at its upper side. It is surrounded by the ear auricle and is open at its lateral side, while closed medially by the eardrum or membrana tympani (for its development see Fig. 4.12). This pars tympanica is enclosed by the pars squamosa, which constitutes part of the skull, the mandibular fossa and the zygomatic process. Other parts of the temporal bone are the mastoid process, styloid and its petrous portion. The temporal bone contains foramina: carotid canal, foramen lacerum, external auditory canal, stylomastoid foramen and the internal acoustic canal. The petrous portion contains medially the cochlea and laterally and caudally the semicircular canals. The ossicle chain is placed laterally of cochlea and semicircular canals. The division into external ear, middle ear and inner ear is mainly related to the petrous bone, except the lateral half of the external auditory canal (see Chap. 4, Sect. 4.7.2. The development of the tympanic cavity and the tympanic ring).

9.2 The Otic Vesicle Development Topics

The otic vesicle is the result of placode activity at the rhombomere 5 by four weeks of gestation. The middle ear derives from paraxial mesoderm and neural crest cells (see Sect. 4.4 and Fig. 4.2). The lining of the middle ear is, of course, endodermal in origin (Fekete 1999). The otic vesicle expands into three subparts: the endolymphatic sac, the vestibular and cochlear parts (Bruska et al. 2009). Clefts in the vestibular part produce three flattened discs that contain a lumen. By fusion of the disc walls in their centres, semicircular canals originate. The outcome consists of three circular canals embedded in the periotic mesenchyme or periotic connective tissue (Streeter 1906). Figure 9.2 shows clearly that one of the discs (the horizontal/lateral one) is independent, while the other two (the superior and posterior one) share a common crus. The origin of the common crus is shown in the lower part of Fig. 9.2, in which the arrows indicate the two places of imminent dehiscence. "The two canals clearly share a common portion, so that only three openings in the utricle give access to the two canals. As a result, five openings in the utricle lead to the three semicircular canals" (Landsmeer unpublished; week 8–14th of gestation, Dechesne and Sans 1985). "It is not known when the organs reach adult size, but it has been proposed this occurs when the bony labyrinth reaches maturity between 17 and 19 week of gestation" (Lim and Brichta 2016).

The otic mesenchyme will produce the otic capsule. The capsule has to fulfil two demands. It has to ensure the accessibility of the various related organs and at the same time guarantee their firm fixation to the capsule and through the capsule to the head. "To obtain reliable information about the position of the head in a gravitational field and about any acceleration to which the head is subjected, it is essential that the position of the sensory apparatus is itself under rigorous control. The same applies, mutatis mutandis, to the organ of hearing.

One can argue that, if the sensory conditions are restricted in the sense of a proper tuning of the sensitive elements, there would be no objection to making the

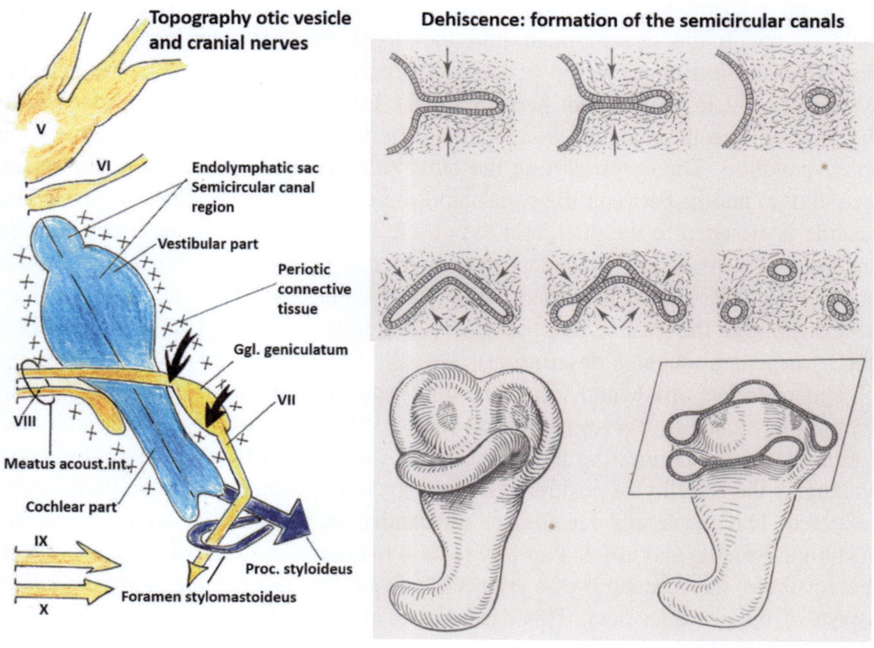

Fig. 9.2 Otic vesicle and its relations to the cranial nerves. The development of the labyrinth is explained by the fusion of epithelial walls of the developing discs (© E. Marani, left, and Dept of Anatomy LUMC, right)

apparatus movable in the body, provided the motion was under proper control. In fact, this is affected by the body itself, in the sense that the head carrying these organs is movable on the body. It is, however, essential that the organs of the two sides cannot move relative to one another. Discrimination, based on the bilateral positioning of the sensory systems, would possibly be more ambiguous, or at least unduly complicated, if the two systems could move in relation to each other" (Landsmeer unpublished). This makes it comprehensible that both the human vestibule–cochlear systems are centred within the petrous bones, which are firmly connected to the temporal bone and by this bone immovable to the skull. Moreover, it demonstrates the importance of the otic capsule and its mesenchyme for the development of both vestibule and cochlea (for their bony developments see Jeffery and Spoor 2004).

Since the otic vesicle originates next to the rhombencephalon and caudal of the pontine flexure and cranial nerves enter the otic vesicle, the spatial relation between vesicle and developing brainstem is obvious. Three directions of development of the otic vesicle can be discerned: the extension into the anteromedial direction paralleling the pharyngeal axis, an expansion in the dorsolateral direction and a lengthening dorsally into the areolar space around the rhombencephalon (Fig. 9.2 left). The epithelial cochlear extension has a triangular form and starts to coil at the 8–9th week of gestation. This spiralization goes on till two and a half turns have been made (see Fig. 9.3b; Kim et al. 2011). This epithelial development makes the cochlear duct. At its start, the tube side near the rhombencephalon will develop into the organ of Corti. The epithelium within the cochlear duct exists of high columnar cells of which the development into the organ of Corti has been pictured in Fig. 9.3c, d. Note that the top of the coil finally is perpendicular to the nervous system (Fig. 9.3a, inset is not perpendicular). Cells that separate themselves from the duct will produce ganglion cells along the whole spiralized tube, forming the spiral ganglion. The outgrowth on the other side produces the semicircular canals (see above) and in between the canals and the cochlea two vesicles originate. The saccule is attached to the duct, and the utricle is the vesicle that broadens into the canals. In between these vesicles a small duct is present the so-called endolymphatic duct that widens into the saccus endolymphaticus. The duct has two connections one to the saccule and the other to the utricle (see Raphael and Altschuler 2003 for general cochlea description).

The ampullae, utricle and saccule contain sensory epithelium called maculae that should be reached by nerves and blood vessels. "The accessibility is ensured by the continuity of the connective tissue spaces, but the satisfaction of this requirement endangers the stability of position. Hence, a specific solution to the problem can be expected. The utricle and saccule are surrounded by a connective tissue space, the vestibule. Saccule and utricle partially remain in contact with the otic capsule. From the vestibule, the formation of a connective tissue space proceeds along the apical aspect of the cochlear duct. This extension which crawls along the cochlear duct like a hollow snake is called the scala vestibuli. At the same time, another cavity begins to develop de novo on the basal aspect of the coil, and this also creeps towards the tip of the coil. This space is termed the scala tympani (the tympanic staircase). The two scalae, vestibular and tympanic, fuse around the tip and this is the helicotrema (inset Fig. 9.3b). From the vestibule, the space also extends around the semicircular canals. They are completely surrounded by the space except at the sites of the cristae ampullares" (Landsmeer unpublished). This space is of course the perilymphatic space that enwraps the vestibule, the scala vestibuli, the scala tympani and the spaces around the semicircular canals. The footplate of the stapes closes of the vestibule.

The accessibility is a prerequisite for the perception of sound and equilibrium and as such enables that nerves and vessels can reach the different parts of the otic capsule. The nerves enter the otic capsule from the medial side, i.e. brain. The entrance within the capsule is called the internal acoustic meatus (Fig. 9.2). This meatus starts at the medullary side containing the facialis, with its intermedius part, and statoacoustic cranial nerves and a branch of the middle cerebellar artery that gives the internal auditory artery. This total is enwrapped by the dura mater. Within the meatus, the facial nerve finds its way towards the middle ear and places its ganglion geniculatum at the end of the canal, still branching of the superior and inferior petrous nerves. Halfway the meatus canal, the vestibular part of the statoacoustic cranial nerve splits in an inferior and a superior vestibular nerve. Just before this division, the vestibular ganglion (also named ganglion of Scarpa) is present in the middle of the meatus. The facial nerve continues dorsal over the otic capsule in between the cochlear and canalicular parts. This is over the roof of the vestibule. At the lateral side of the vestibule, the nerve makes a 90° turn (black arrow Fig. 9.2 left). It is in this turn that the geniculate ganglion is present. The facial nerve continues below the horizontal canal and turns again since it enters a furrow organized by a lateral protrusion of the otic capsule. This lateral protrusion is prolonged into the styloid process (see Fig. 9.2 left).

Since the otic capsule on both sides is enclosed in the so-called chondrocranium, cartilaginous structures will encompass the developing cochlea and vestibular organ parts contributing to the stability of the otic system and to a fixed position of the sensory apparatus. This otic system can in this way be kept under rigorous control (see above). Outgrowth of the first and second arches interacts with the otic development and with the development of the tympanic cavity. The tympanic cavity ensues from the inclusion of epithelial folds of pharyngeal arch developments. The tympanic cavity is in fact a dorsolateral directed wing of the pharynx. Inherently, the tensor tympanic muscle, handle of the malleus, vertical process of the incus and the stapes are developmental consequences of the secondary folds of this pharynx wing. The handle of the malleus, being in direct contact with the cavum epithelium, will co-organize the expansion of such a fold into the tympanic membrane. The cartilaginous development starts at the same time in the otic capsule, future chondrocranium and branchial arches. The final position of the aperture into the pharynx is related to the development of the secondary palate. Lengthening and relative narrowing of the medial part of the primitive tympanic cavity cause the Eustachian tube. This formation of the auditory Eustachian tube goes along with the appearing of the levator palatini and tensor palatini muscles, both inserted into the soft palate and located near the auricle side of the auditory tube (see Sect. 6.6 the

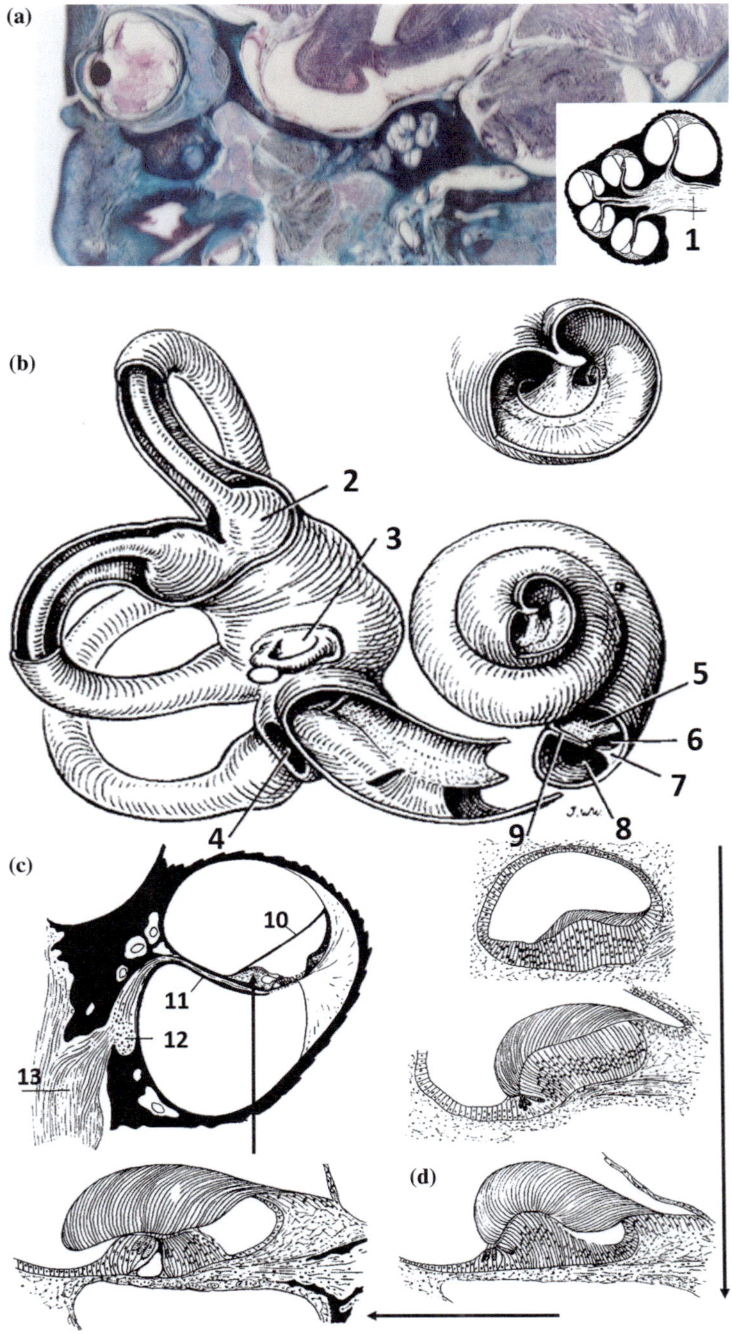
(a)
1
(b)
2
3
4
5
6
7
8
9
(c)
10
11
12
13
(d)

◀**Fig. 9.3** Cochlea and organ of Corti. **a** Section through a 7-month-old foetus for the position of the cochlea in relation to skull base, brain and intern carotid artery, inset shows diagrammatic view through the cochlea with the branches of the cochlear nerve (1). **b** Cochlea with semicircular canals, inset helicotrema, 2: crista with ampulla, 3: stapes covering oval window, 4: round window, 5: vestibular duct (scala vestibule), 6: cochlear duct (ductus cochlearis), 7: spiral ligament, 8: tympanic duct (scala tympani), 9: modiolus bone segment. **c** Transverse section through the cochlea, 10: vestibular membrane of Reissner, 11; bony spiralized lamina (lamina spiralis ossea) with fibres of the cochlear nerve, 12: ganglion spirale, 13: start of the development of the n. cochlearis. **d** Development of the organ of Corti (follow arrows). Finally, the internal spiralized sulcus and the internal tunnel with the space of Nuel (not separated in the figure) are developed in the middle of the hair cells, covered by the tectorial membrane

soft palate). (Several developmental topics have not been treated; see e.g. Whitfield 2015; Raphael and Altschuler 2003).

9.3 Hair Cell Development

Hair cells of cochlea and vestibulum are mechanoreceptors that are capable of transmitting auditory and vestibular stimuli into the related brainstem parts. Their stereociliary bundle of microvilli on the upper, luminal surface of these cells is able to register small movements or pressure waves. Mechanosensory transduction channels will react on such pressures and open with the consequence of influx of positive ions. The depolarization of the hair cell changes the neurotransmitter release at the base of the hair cell. This activates the synaptic end of the ganglion cell, which change its transmission rate of action potentials towards the brainstem.

The hair cell developmental characteristics are mainly obtained in experimental animals with the mouse being the most known for cochlear research. This research has been carried out in various experimental animals, and differences are present. The otic vesicle is a placodal structure. Placodal cells are induced already at the edge of the neural plate at the 0–1 somite stage in the *chick*, marked by Six, Eya, Id, Iro and Fox genes, but the dispersed cells have to sort out and group together to produce placodes. A role is played by BMP's as marker of the border of the neural plate, and Iro-1 is regulated by BMP. Sox, Dlx, Fox and Pax genes are involved in determining placode cells. Only the transcripts *Six* and *Eya* are "to be maintained in all developing placodes, while being lost from interplacodal domains" (Streit 2004, for overviews see: Baker and Bronner-Fraser 2001; Streit 2004). Wnt is explicitly involved in otic placode induction, increasing Notch-1 and refining ventral boundaries.

Within the otocyst the ventral epithelia change into the so-called prosensory area by the factors FGF, Sox 2 and Notch, and this is presumably co-induced by Wnt. The borders of the prosensory area are determined by Hedgehog and Lmx1a. Now the prosensory area is genetically determined, and its borders are drawn. Cell increase is needed in the following step. Both Wnt and Notch are involved in the proliferation process by Ccnd 1 upregulation and p27 ^{kip1} downregulation, respectively (Zak et al. 2015). For differentiation, cell cycle withdrawal is a necessary prerequisite to keep a postmitotic state of the future hair cells. Increase in cyclin-dependent kinase inhibitors (e.g. p27 ^{kip1}) is responsible for this stage of cell cycle exit. This starts at the apex of the cochlear duct and progresses towards the base of the tube. The subsequent hair cell differentiation period is induced by the appearance of the transcription factor Math 1. This differentiation starts in the middle of the cochlear tube and continues into both the apex and the base direction. At the same time a mediolateral differentiation occurs in the sensory epithelium. Sox2 and Prox 1 inhibit the Math 1 effect, producing the supporting cells. "Inductive and inhibitory signalling create the correct cellular patterning of the organ of Corti. Much of this appears to be mediated by Notch signalling, which inhibits hair cell neighbours from adopting the same fate. Furthermore, the initial differentiation of the inner hair cells appears to direct the differentiation of other cell types, such as the pillar cells through FGF8 and Hey2 in a Notch-independent manner. Sprouty 2 (SPRY2) further restricts the differentiation of pillar cells" (Fig. 9.4/1; Kelly and Chen 2009; see also Kelly 2006 and Zak et al. 2015). Recent research indicates that Wnt via the Atoh-1 line produces hair cells, while these hair cells express Notch ligands, which block in the neighbouring cells Atoh-1 causing the differentiation in supporting cells (Zak et al. 2015).

The control of ion concentrations in endo and perilymph and the maintenance of the three-dimensional structure of the organ of Corti need cell junctions. Cell junctions in the cochlea are characterized by adherens junctions, gap junctions and tight junctions. Gap junctions are electrical or chemical coupling junctions. In the cochlea, an exchange of small molecules and signal transmission is indeed possible. The gap junctions contain connexins (connexin-26, 30, 31, etc.). Gap junctions are present in the epithelial structures of the cochlea (e.g. supporting cells and outer sulcus cells), where the gap junction is responsible for the circulation of potassium ions, and in the connective tissue of the cochlea (fibrocytes of the spiral ligament). Connexin-26 gene mutations are kept responsible for non-syndromic hereditary deafness. Tight junctions are present on the lateral walls of epithelial cells and prevent the motion of substances across the cellular wall or produces selective cellular permeability. These junctions keep the borders between endo and perilymph. Last but not least, the adherens junction forms the connection between the inner side of the cell with extracellular and often transmembrane proteins of another cell. The adherens junctions are involved in "cochlear development, growth of auditory neurons, immune mediation and planar cell alignment" (Wang et al. 2015).

The cochlea knows a blood–labyrinth barrier. Despite earlier studies, in which the blood–labyrinth was proposed to be an immune barrier, the cochlea contains resident macrophages, the predominant immune cells, T cell marker lymphocytes and T cell dominant immune cells in the endolymphatic sac. Cochlea T-marker lymphocytes can increase their population after immune challenge. Presumably, a larger differentiation of lymphocytes is present within the cochlea, but research is lacking. Nearly two-third of the macrophages resides in neural tissue; around the ganglia and along the nerve fibres. Ageing reduces the amount of macrophages (Hu et al. 2017).

9.4 Development of Neuronal Connections of the Cochlea

The connections of the cochlea are reciprocal (Fig. 9.4/1). The spiral ganglion cells connect to the inner (one row of cells) and outer hair cells (three rows of cells; see Fig. 9.4/1) and to the cochlear nuclei of the brainstem. The olivary-cochlear bundle, containing mainly brainstem (peri) olivary nuclei neuron axons' projects back to the inner and outer hair cells. The superior olive connections between inner and outer hair cells are different. The inner hair cells finally receive axosomatic and axodendritic connections while the outer hair cells are connected by the olivo-superior neurons in an axodendritic way. The spiral ganglion connections are early in development indifferent, but specialize in Type I for the inner hair cells and Type II connections for the outer hair cells (see Fig. 9.4/2, a). Note that Type I and Type II divisions are based on the connections to inner and outer hair cells, respectively. "Spiral ganglion neurons are charged with the difficult task of accurately capturing and encoding the features of each complex sound stimulus. How this is accomplished at the level of the circuit remains a mystery" (Appler and Goodrich 2011).

The bottom specializations of the hair cells, the so-called synaptic ribbons, coincide with the further adaptation of the olivo superior connections. The lateral olivary efferents connect to the inner hair cells, and the outer hair cells receive medial olivary efferents. These medial olivary efferents are cholinergic and bilateral. The contralateral lateral olivary efferents are GABA-erg or cholinergic with CGRP-colocalized homolateral and have effect on the primary afferent fibres, located directly beneath the inner hair cells (Warr and Guinan 1979). Functions of the medial descending auditory system are suppressing the cochlear gain. The lateral system effects are under discussion (for overview of the central auditory system see Nieuwenhuys et al. 2008). "The olivocochlear neurons thus form a system by which the brain can influence its own auditory input" (Nieuwenhuys et al. 2008, p. 744). "A vast body of evidence accumulated over more than 30 years has established that the auditory system continuously adjusts to changes in stimulus statistics and significance. These forms of plasticity occur at many different levels of the auditory pathway and over many timescales, and in most cases serve to optimize auditory performance" (Irvine 2017). Sound perception therefore can be individually different.

Type I spiral ganglion cells connect to only one inner hair cell, but the Type II spiral ganglion cells can connect several outer hair cells (Fig. 9.3/2, b and 3). Note that the inner hair cells can receive several different Type I connections, presumably important to detect soft sounds. The differentiation of the connection patterns is co-related to the Ca^{2+} signalling capacities of the neural cells involved (Mammano and Bortolozzi 2017). Spiral ganglion cells are also characterized by their spontaneous firing rates. Type I high firing rate spiral neurons are the most recognized in physiological studies (up to 60%). They are characterized by the lowest threshold of sound detection for all frequencies. "In contrast, spiral ganglion neurons with a low spontaneous firing rate can vary their response rates over a larger range of sound pressure levels, allowing them to better detect changes in sound volume"

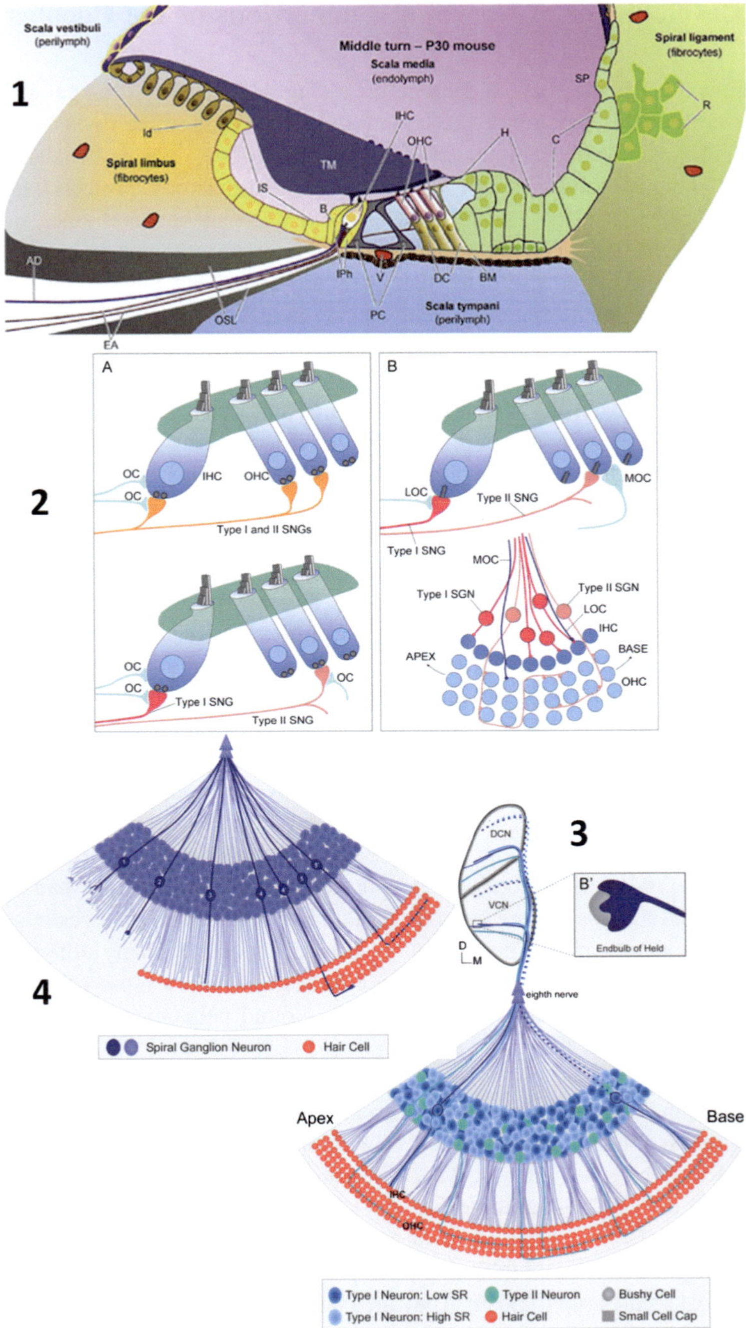
1
Scala vestibuli
(perilymph)
Middle turn – P30 mouse
Scala media
(endolymph)
Spiral ligament
(fibrocytes)
SP
R
IHC
Id
OHC
H
C
Spiral limbus
(fibrocytes)
TM
IS
B
AD
IPh
V
DC
BM
OSL
PC
Scala tympani
(perilymph)
EA
2
A
OC
IHC
OHC
OC
Type I and II SNGs
OC
OC
Type I SNG
OC
Type II SNG
B
LOC
Type II SNG
MOC
Type I SNG
MOC
Type I SGN
Type II SGN
LOC
IHC
APEX
BASE
OHC
3
DCN
B'
VCN
Endbulb of Held
D
M
eighth nerve
4
Spiral Ganglion Neuron
Hair Cell
Apex
Base
IHC
OHC
Type I Neuron: Low SR
Type I Neuron: High SR
Type II Neuron
Hair Cell
Bushy Cell
Small Cell Cap

◂**Fig. 9.4** In figure part 1 schematic diagram of a cross section of the mouse cochlear canal at postnatal day 30 is pictured: AD, afferent dendrites of Type I spiral ganglion neurons: B, inner border cell; BM, basilar membrane with Böttcher's cells; C, Claudius' cells; DC, (immature) outer phalangeal cells, also known as Deiters' cells; EA, efferent axons of olivocochlear neurons; H, Hensen's cells; HCR, hair cell region; Id, interdental cells; IHC, (immature) inner hair cell; IPh, inner phalangeal cells; IS, inner sulcus; OHC, (immature) outer hair cell(s); OSL, osseous spiral lamina; PC, (immature) inner (left) and outer (right) pillar cells; R, root cells; SP, spiral prominence; TM, tectorial membrane; V, vas spiralis (a blood vessel running in the tympanic layer of the basilar membrane just beneath the tunnel of Corti). 2: "Postnatal refinements of hair cell innervation patterns. **a** Immature mouse cochlea. Top: at birth, IHCs and OHCs can be innervated by the same spiral ganglion neurons (SGNs). Immature efferent fibres from olivocochlear neurons (OC) form both axosomatic and axodendritic contacts with IHCs, but not with OHCs. Bottom: by the end of the first postnatal week, Type I SGNs innervate exclusively IHCs, whereas Type II SGNs innervate OHCs, which receive also efferent axodendritic contacts. **b** Mature mouse cochlea. Top: synaptic ribbons acquire an elongated shape and each synaptic active zone contains a single ribbon both in IHCs and in OHCs. LOC: lateral olivocochlear efferents; MOC, medial olivocochlear efferents. Bottom: top view of the innervation pattern. Each Type I SGN contacts a single IHC; in contrast, each Type II SGN contacts multiple OHCs" (Mammano and Bortolozzi (2017) see also Bulankina and Moser (2012) and Rubel and Fritzsch (2002). 3; "Schematic view of a wedge from a flat mounted cochlea (bottom) and its connections with the cochlear nucleus complex (top). In the cochlea, peripheral projections are corralled in radial bundles that pass through the spiral lamina to the hair cells (red). Low spontaneous rate (dark blue) and high spontaneous rate (light blue) Type I neurons contact inner hair cells (IHC). Type II neurons (green) are positioned in the ganglia nearest to the hair cells and extend a projection past the inner hair cells and turn towards the base, with each projection contacting multiple outer hair cells (OHC) along its length. Information is conveyed to the cochlear nucleus by central axons, which bundle together to form the eighth nerve (double arrowhead). Upon entering the brainstem, individual axons bifurcate. The ascending projections terminate with bouton endings in the dorsal cochlear nucleus (DCN), while the descending projections target the ventral cochlear nucleus (VCN), where they form boutons with a variety of postsynaptic target neurons as well as unusual endbulb of Held synapses with bushy cells (B0). Within each division of the cochlear nucleus, auditory axons are tonotopically organized, such that high-frequency information from the base of the cochlea is processed dorsally (dotted lines) and low-frequency information from the apex is processed more ventrally (solid lines). In addition, the central projections from neurons with low spontaneous firing rates project more laterally than those with high spontaneous firing rates. Type II neurons project to the small cell cap that surrounds the cochlear nucleus complex (dark grey), as do some arbours from low spontaneous rate fibres" 4: "The development of peripheral processes in the cochlea. Diagram of the cellular events that occur as spiral ganglion neurons (blue) develop in the cochlea, with immature cells shown on the left (1) and progressively maturing towards the right (5). Soon after becoming postmitotic (1), the neurons become bipolar, with a central process extending into the eighth nerve and a peripheral process exploring the spiral lamina. Subsequently, neurons develop unbranched processes (2) that grow towards the edge of the organ of Corti, even before hair cells (red) have differentiated. Processes appear to wait while hair cell differentiation continues (3). Subsequently, Type I neurons (4) terminate beneath IHCs while Type II neurons (40) develop large growth cones that navigate past the IHCs and into the rows of OHCs, where they turn towards the base, resulting in a mature pattern of innervation by birth (5)" (Appler and Goodrich 2011). 1 and 2 are taken from Mammano and Bortolozzi (2017) see also Mammano (2013), 3 and 4 are from Appler and Goodrich (2011), all with permission

(Appler and Goodrich 2011). Their cellular synaptic locations are different. Inner hair cells are contacted by the low firing rate spiral Type I neurons halfway the hair cell, while high firing rate spiral Type I neurons reach the hair cell bottom.

Moreover, there is a topographically different projection of these spiral neurons within the cochlear nuclei (Fig. 9.4/3; see Nieuwenhuys et al. 2008).

The developing spiral ganglion cells are shaped by late expression of Ngn 1, of GATA-3 (in chicken and mouse) and of Lmx-1a, which also differentiate between vestibular and auditory ganglion neurons. Strong differences in expression of factors and genes are present in various experimental animals. The spiral ganglion cells send out their axons towards the developing organ of Corti, but are still mesenchyme and epithelium related early in development. They direct themselves towards the epithelial cells, later replaced by cuboid cells of the mature inner sulcus. This primitive layer of epithelial cells is named Kölliker's organ (Fig. 9.4/4). While hair cells begin their differentiation, the inner hair cells started earlier than the outer hair cells. This differentiation commences at the base and directs itself towards the apex. The outgrowth of the hair cells coincides with a radial organization of the spiral ganglion cells which will contact the inner row of hair cells or the hair cell of the outer rows (see Fig. 9.4/4, follow the numbers). While it looks that the hair cell differentiation induces the spiral ganglion organization, this is not the case. Without hair cell differentiation, the radial structure of spiral ganglion axons can be acquired (Appler and Goodrich 2011).

Within the cochlear nucleus, the bushy cells are present. Auditory axons project on these cells with the so-called endbulb of Held (inset Fig. 9.4/3). These endbulbs belong to the largest synapses (15–30 μm) present in the auditory system. The endbulbs are strong synapses by their effect. The bushy cells are capable by these synapses to fire more accurately to the temporal demands of acoustic stimuli.

9.5 Three Levels of Planar Polarity: Subcellular, Cellular and Tissue Polarity

The term planar polarity causes problems, since various definitions and approaches are present. "We suggest that when considering cellular organization, a planar polarization process is defined as one in which cell–cell communication causes two or more cells to adopt coordinated polarity. Moreover, such a process should only be referred to as evidence of 'planar cell polarity signalling' or requiring the 'planar polarity pathway' when shown to be dependent upon planar polarity proteins that mediate these polarized cell–cell interactions" (Goodrich and Strutt 2011). One should note that planar cell polarity has been originally studied in the fruit fly Drosophila and is later applied on other experimental animals, especially the cochlea of the mouse.

Polarity is considered: "The most fundamental characteristic of any pluricellular structure. The organism has at least two dissimilar poles, ends or surfaces" (Chandebois and Faber 1983). The study of polarity in a two-dimensional plane has been extremely fruitful due to the detection of the Frizzled/planar cell polarity and the anterior–posterior patterning system. Directional signals do have in several organisms the same underlying mechanism. The Frizzled transmembrane receptor is involved in Wnt-interactions but is believed to be independent in the planar cell polarity system and plays an important role in various cell types, including the otic/cochlear cells. Essential proteins are cell surface and cytoplasmic ones: Frizzled (Fz, 1, 2, 3, 6), Strabismus (Vang 1, 2), Flamingo (Celsr 1, 2, 3) and the cytoplasmic Dishevelled (Dvl 1, 2, 3), Prickle (prickle-like 1, 2) and Diego (Diversin) proteins (Fig. 9.5a–c, names in brackets are the mammal homologues). These proteins after their production in the rough endoplasmic reticulum are present in the cytoplasm. Like a submarine that comes to the surface, these proteins reach the luminal (cochlea) or apical (Drosophila) surface of the cell involved. Microtubules are involved in this process to bring Frizzled to the surface. Two types of groups are formed, a proximal and a distal one by activity of all six proteins (see Fig. 9.5c). Description of this process has been omitted here (see Zallen 2007). The induced local polarity by these proteins is not restricted to the cell, but can have effect on the tissue level or cell–cell interaction levels (Fig. 9.5, Deans 2013; Goodrich and Strutt 2011; Zallen 2007). Another polarity cascade is the Fat/Dachsous pathway important in the placing of actin-rich hairs in animal fur (see Goodrich and Strutt 2011). Hair cells of the cochlea contain cilia that are organized in a V or W form on top of the luminal surface (Fig. 9.5e). The height increases towards the point of the V and is called the staircase pattern. The V top is made by one large kinocilium. This structure "repeats" itself on all hair cells. All the V bundles are directed away from the cochlear neural side, that is abneural. Within the cochlea, the Prickle

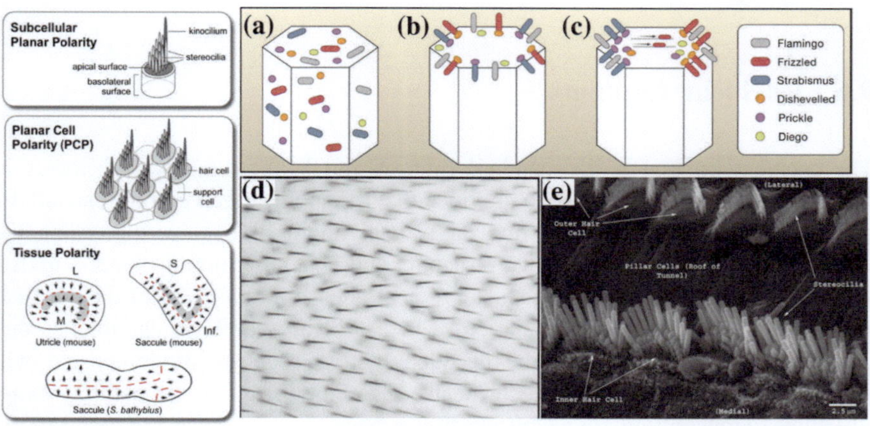

Fig. 9.5 Figures of planar polarity diagrams (see Deans 2013) showing the three levels of planar polarity, i.e. subcellular, cell and tissue [Lateral (L), medial (M), superior (S) and inferior (Inf.)]. Red is the Line of Polarity Reversal. **a**, **b**, **c** The planar cell polarity pathway as known from Drosophila wing epithelial cell. The proteins are present in the cytoplasm (**a**) proceed to the cell surface (**b**) and all six proteins (see column colour indication) reorganize on the surface in proximal and distal domains with the help of organized microtubules. **d** Is an example of planar polarity of the hairs of the fruit fly wing, and E shows planar polarity of the stereocilia of the cochlea. **a**, **b**, **c** and **d** are from Zallen (2007); **e** is taken from Raphael and Altschuler (2003). (All with permission of respective journals)

protein does not fulfil totally the expectations as explained in Fig. 9.5a–c. This is presumably related to the pluricellular situation in the mammalian cochlea compared to the unicellular presence of epithelial cells in the Drosophila wing. In rat and mouse cochlea, the Dishevelled 1, 2, 3 accumulates abneural, while Strabismus (Vang 1, 2) and Frizzled (Fz 3, 6) group at the neural side and presumably organize the polarity within the hair cell rows (Ezan and Montcouquiol 2013). The process is seemingly different from the one of the Drosophila wing. The relation of kinocilium polarity and the influence of the other ciliary polarity are still under discussion.

These interactions at the cellular level do have analogues at the tissue level. The vestibular maculae contain within the mature sensory epithelium opposite polarities. Within utricle and saccule maculae, the ciliary hair cell groups contain luminal opposite bundle orientations. The so-called Line of Polarity Reversal is the border between these opposite cilium distributions. While in the utricle, the polarity of the ciliary bundles is towards this line, in the saccule the ciliary bundles point away (Fig. 9.5 diagrams, red line). Finally, polarity is needed for the whole body composition, e.g. during growth to maturation along the anterior–posterior axis. Normally, these planar polarities for body development are combined, an antero-posterior body polarity with a dorsoventral body polarity is converted into bilateral symmetry of the mammalian body. Cell sociology already hypothesized the laws needed to understand this automation in animal development at cellular and tissue levels (see Chandebois and Faber 1983) but these embryological and anatomical studies are seemingly neglected by the molecular, cellular and genetic research.

9.6 Cochlear Implants

The cochlear implant is the standard solution for deafness and severe losses in hearing nowadays (Wilson 2017) provided that the cochlear nerve fibres can still be stimulated. Since the cochlear implant is a highlight of engineering, using the specific anatomy of the auditory structure and making running speech understandable attention is given to its history, electrodes and modelling of the cochlear morphology and function. Stimulation of the cochlea with electrodes implies that the electrode information can be decoded by the cochlear structures and will specifically reproduce what speech means, also emotional, by the auditory brain structures. Transmitting sound by electrodes became possible in the early 1950–1960s (see Sect. 9.1), but application of the correct human speech frequencies was a hard nut to crack. "The data indicated the limitation was due in particular to deterministic firing of neurons and failure to reproduce the normal fine temporo-spatial pattern of neural responses seen with sound. However, the data also showed the need for the place coding of frequency, and this meant multielectrodes inserted into the cochlea" (Clark 2015). Lots of these problems were solved by Clark from the University of Melbourne. In 1978, the formant-extraction speech code was developed. It allowed "severely profoundly deaf people to understand running speech" due to the fact that the place of stimulation induced timbre and was perceived as vowels (Clark 2015). It leads to the first multichannel implant approved by the US Food and Drug Administration in 1985 and in 1990 to the multichannel implant for children. By 2012, over 300,000 patients had received a cochlear implant worldwide. Figures for the USA are 58,000 cochlear implants in adults and 38,000 in children in 2013 (see NIH, National Institute on Deafness and other communication disorders, NIDCD). ***Severe*** auditory malformations are present in 1:10,000–1:20,000 newborns. These malformations are mainly unilateral in 70–90% of the children and right sided in 58–61% (Kösling et al. 2009 see also Sennaroglu et al. 2002). About 2–3 out of every 1000 children in the USA are born with a ***detectable level of hearing loss*** in one or both ears. This affects each year 8000–12,000 newborns in the USA. In the Netherlands for children between 6 and 12 years, auditory deafness and severe hearing loss are found in 0.74 per 1000 inhabitants, meaning that the Netherlands counts 12,500 deaf children only in this age category in 2017.

In the Netherlands, the centres that are involved in cochlear implants are working together (Cochlear Implant organization the Netherlands, CI-ON). It gives us the possibility to produce the required number on cochlear implants: nowadays, each year 550 cochlear implants are performed including 25–30% in children. In 2016, the 7000th cochlear implantation was done of which 2000 were in children (2000–2016, early years with lower numbers). Total bilateral implants concern 550 ones of which 80% were performed in children below 18 years.

Prelingual deaf children (deafness is present already before their language development) benefit from early implantation. If possible implantation should be done before one year of age but at least before the age of six. Language development and grammatical teaching are supported, and normal emotional development is strongly promoted by this early implantation (Boons et al. 2012a; Theunissen 2013). From bilateral implanted children, it is verified that their language is clearly improved (see Caldas et al. 2016; Boons et al. 2012b). Nevertheless, adequate stimulation is needed and parents are important in their practice of audition, in auditory-verbal education and in linguistic interactions on a daily basis (Ouellet and Cohen 1999).

Some aspects of cochlear implants are considered in this part. One is the congenital malformations of the inner ear and the other two modelling of the cochlea and properties of electrodes.

9.6.1 Congenital Cochlear Malformations

For the indication of the capacity and properties of the cochlea, one should reckon with: the cochlea in humans is capable to notice a frequency range of 20–20,000 Hz. (By the way, ageing reduces this frequency range.) The dynamic range is large, even over 100 decibel (dB). At threshold (20 μPa), the movement of stereocilia of 0.04 nm is still noticed. This distance is less than the diameter of the hydrogen atom. Frequencies that differ 10–20% can be resolved simultaneously. The minimal difference that is still noticed is 0.2–0.3%. Although miniaturization of electrodes and of equipment is a common technique; nowadays, still our technical capabilities are far beyond the natural possibilities of the cochlea.

The different malformations are congenital originate during the third to 7th gestational week (Fig. 9.6; see Sect. 9.2 for normal developmental topics). Complete labyrinth aplasia and complete cochlea aplasia are a consequence of mal development in the third gestational week. Nearly 4% of the patients are affected by aplasia. The common cavity congenital malformation is related to the 4th gestational week. The cochlea and vestibule make one cavity (Fig. 9.6). This is found in 25% of the patients. The incomplete Type I partition is responsible for 6% of the patients and contains absence of the modiolus and a cystic vestibule. The separation between cochlea and vestibule is present, and it is 5th gestational week related. The incomplete Type II partition, also known as the Mondini malformation, is responsible for 50–55% of the congenital malformations. The Mondini malformation is known by its cochlea with a normal basal turn and a total turn of 1.5, resulting in a cystic apex. A strong enlarged vestibular aqueduct and normal vestibular semicircular canals are present.

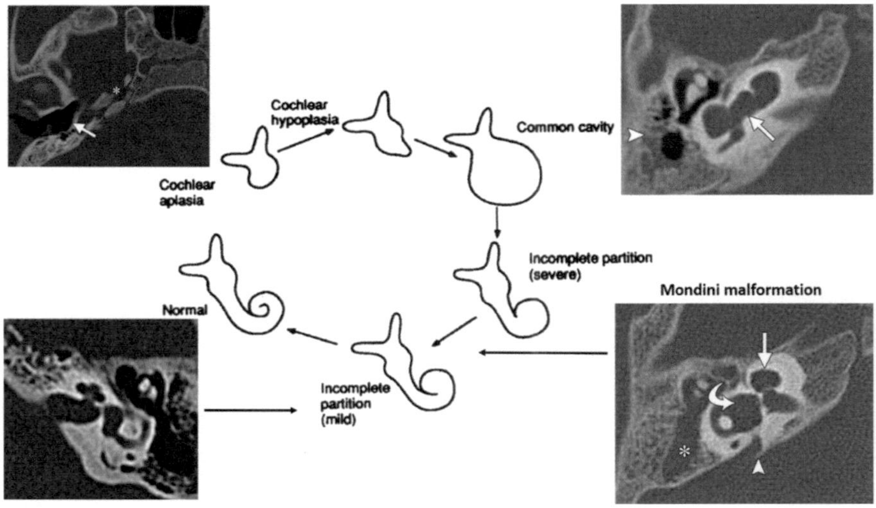

Fig. 9.6 Malformation cycle of the cochlea (Jackler et al. 1987, with permission) illustrated by four axial high-resolution CT images with partial texts taken from Joshi et al. (2012 with permission). "Complete labyrinthine aplasia. Inner ear structures are absent. The flat medial wall of the middle ear cavity is indicated by the white arrow and its hypoplastic petrous bone by an asterisk. A cystic cavity showing confluence of the cochlea and vestibule with no internal architecture is indicated by the straight arrow. An effusion (arrowhead) is seen in the mastoid air cells. Mondini malformation showing the absence of the modiolus from a cystic cochlear apex (straight arrow) formed by coalescent apical and middle turns. Dilatation of the vestibular aqueduct (arrowhead) and vestibule (curved arrow) and a coexistent middle ear effusion by the asterisk are also present. Incomplete partition is a mild malformation"

Cochlear hypoplasia is assigned to the 6th gestational week or somewhat earlier (Jackler et al. 1987 and Sennaroglu et al. 2002). It is found in 15% of the patient cases and is characterized by a small cochlea with less than one turn and often with deformed vestibule and semicircular canals (see Joshi et al. 2012; Yiin et al. 2011).

Here we used the Jackler et al. (1987) description that otherwise has been challenged (Papsin 2005). It is clear that cochlear stimulation by electrode implants is impossible in several congenital malformations. A solution is the so-called auditory brainstem implant (ABI, see below). Patient selection is important for a successful cochlear implant. The most suitable cases for cochlear implants can be found in the Type I to Type II patients or in Type II Mondini patients or with normal anatomy and connexin-26 deflection. Type II incomplete partition is often bilateral and goes with sensorineural hearing loss. Since the base of the cochlea is functional, high-frequency hearing is often conserved.

9.6.2 Electrodes and Model Studies

Early in the cochlear implant surgery, its clinical outcome was not very predictable. Even today a large group of patients achieve "poor" outcomes (Pisoni et al. 2017) that need a critical evaluation both psychologically and technically. One should note that cochlear implants cause significant vestibular disturbances in 1/3 of the implanted patients and are not age related, cause of deafness or the preoperative caloric result related (Huygen et al. 1995; Enticott et al. 2006; Cushing et al. 2008). The engineering approach by electrode and stimulation protocol studies as performed at the ENT department of my (EM) former academic hospital, the Leiden University Medical Center, are summarized in a restricted overview.

The electrode array that will be inserted in the cochlea should fulfil a series of conditions: it needs to reach a certain depth and should give an adequate coverage of the cochlear area. Moreover, it has to match the neurotonotopy of the basilar membrane. The electrode array to be brought in should not harm cochlear structures and fit into the anatomical characteristics of the (congenitally) damaged cochlea for optimal coverage of the neural fibres to be stimulated (Dhanasingh and Jolly 2017). A specific idea is that low tones are important for spatial hearing that means that the length of the electrode arrays should be appropriate for each individual cochlea (Mistrìk and Jolly 2016). A clear prerequisite is the presence of spiral ganglion cells. Areas missing these ganglion cells are known as "dead regions". Blood vessel damage is not excluded by the surgery and can be due to the implantation. In total, 21 different electrode arrays have been produced by firms in Sydney Australia, Innsbruck Austria, Valencia USA, Vallauris France and Hangzhou China that differ in total length, active stimulation length, angular insertion depth, buffer length and by their differences in length producing a different Greenwood's frequency distribution of the electrodes (Dhanasingh and Jolly 2017).

Model studies (Frijns et al. 1996, 2000; Briaire and Frijns 2005; Van Gendt et al. 2016) together with experimental animal and smart human cochlear implant research have been and still are a necessary prerequisite for improving the cochlear implant's effect (Wilson 2017). One should expect that electrode placing over an "as large as possible cochlear area (called the active insertion/stimulation length)" would be beneficial for speech perception. Insertion surpassing 1¾ turn in pitch matching experiments, deduced from model predictions (Kalkman et al. 2014), in unilaterally deaf cochlear implant recipients does not yield stimulation of more nerve fibres. It therefore is doubtful whether moving up the electrode array beyond the 1¾ turn should be sought.

Electrode arrays (12–22 electrodes) have gone through a development during the two decennia of cochlear implants (Fig. 9.7). One should wish the position of the electrodes as near as possible to the nerve fibres of the spiral cochlear ganglion. This placing of the contacts on the medial side of the array and positioning the implant against the medial wall of scala tympani (called the modiolar position) has been tested in model studies (Briaire and Frijns 2006; see Fig. 9.7/6). "The model predicts improved threshold, dynamic range and spatial selectivity for the peri-modiolar position at the basal end of the cochlea, with minimal influence of neural degeneration. At the apical end of the array (Sect. 9.2 cochlear turns), the dynamic range and the spatial selectivity are limited due to the occurrence of cross-turn stimulation, with the exception of the condition without neural degeneration and with the electrode array along the lateral wall of scala tympani" (Briaire and Frijns 2006).

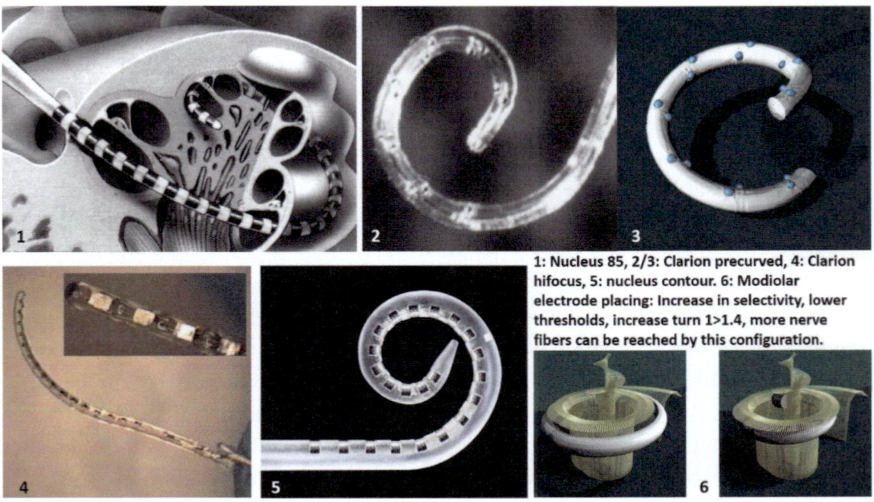

Fig. 9.7 A series of electrodes have been pictured along the cochlear implant timeline (1–5). From 1985 onwards, electrodes that should be pushed up in the tympanic duct, directly below the organ of Corti, have been adapted by containing more electrodes. An increase in turn and placement of electrodes at the peri-modiolar side (6) is favoured, since the nerve fibres spread towards the organ of Corti and centralize at the modiolar side. Pictures are by courtesy of Frijns (LUMC), and with permission of Cochlear Technology

Evoked compound action potentials (eCAPs) can be studied because all modern cochlear implants are supplied with a back telemetry system. The compound action potential amplitude is considered a measure of the number of excited nerve fibres. Moreover, the telemetry results show that 6 months after implantation eCAPs values decrease and remain afterwards stable (Christov et al. 2016). An adaptation period is clearly present.

A high current strength during stimulation with the cochlear implants occasionally produces lower eCAP amplitudes, which could be interpreted as less excited nerve fibres. The decrease of the amplitudes at high currents is attributed to saturation effects and stimulus artefacts. Model studies (Briaire and Frijns 2005) and experimental animal results show that increase in current strength normally does increase amplitude. However, at high stimulus levels this monotonous increase stops. "At high levels, the single unit contributions of fibres located centrally in the excitation area, close to the stimulating electrode, are atypical, without a clear negative peak in the single fibre action potential. The number of fibres with atypical responses also increases with stimulus level. Therefore, the overall evoked compound action potential amplitude decreases above a certain stimulus level" (Briaire et al. 2006; Westen et al. 2011). Thus, the evoked compound action potential can hardly be used as a measure of number of stimulated nerve fibres at high stimulation currents.

9.6.3 Anatomy of the Scala Tympani

The electrodes, embedded in a silastic carrier, are brought into the scala tympani. One of the concerns is that the pushing up of the electrode array will cause insertion trauma. Indeed, damage to the cochlear structures is of great concern (see Adunka and Kiefer 2006). The effects of damage are often observed as deterioration of hearing thresholds. Moreover, damage to cochlear structures could induce degeneration of spiral ganglion cells creating dead regions by rupture damage of the basilar membrane. It is self-evident that local narrowing of the scala tympani can induce insertion trauma (Zrunek et al. 1980). The silastic carrier contributes to safety of the implant because vertical stiffness and horizontal flexibility have been incorporated. It is claimed that upward excursion of the electrode array is opposed and basilar membrane damage reduced. Nevertheless, studies that report insertion trauma indicate predestined places: the base, at ± 180 and at 400 degrees. They are caused by surgical insertion techniques at the base and diameter differences between scala tympani and electrode array at the upper cochlear turns (see Verbist et al. 2009; Avci et al. 2014; van der Jagt et al. 2017). Lack of information on the spiral turning of the cochlea, its course, and steepness, and on irregularities present, and thus lack of knowledge of anatomical properties, should be kept to a minimum. Cochlear anatomy has been studied directed mainly by cross-sectional area of the cochlea and by the decrease of dimensions along the scala tympani (Hatsushika et al. 1990).

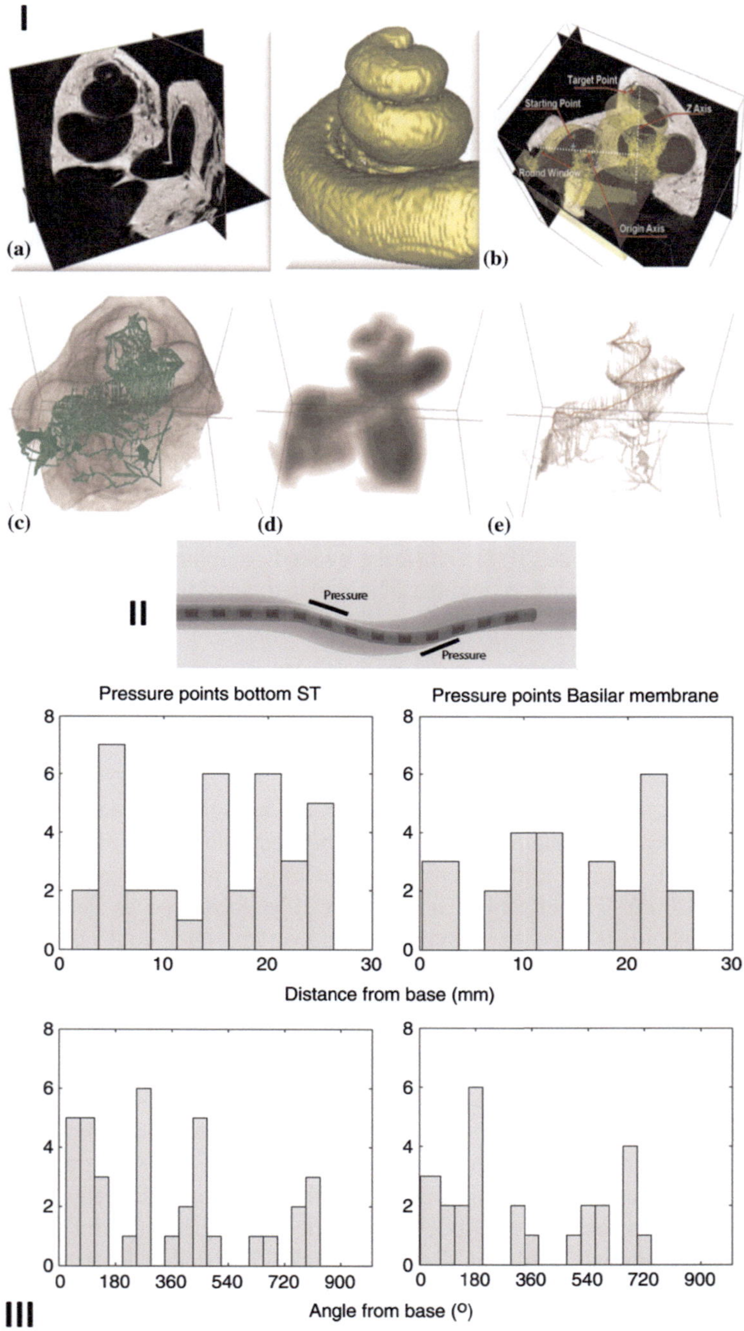
I
(a)
(b)
Target Point
Starting Point
Z Axis
Round Window
Origin Axis
(c)
(d)
(e)
II
Pressure
Pressure
Pressure points bottom ST
Pressure points Basilar membrane
Distance from base (mm)
Angle from base (°)
III

◂**Fig. 9.8** ***I***, **a**, Using a data set of eight microCT scan segments of the human cochlear canal to which is added a local coordinate system (**b**), side branches due to surface irregularities can be detected (**c**). By association of each point (voxel) with a distance to the closest surface point (**d**), a final central path of the cochlear tube will be produced (**e**) using wave-prop and modulation by the distance map. ***II*** demonstrates the pressures exercised by upward or downward movements of the electrode array by small tube deviations during implantation on the basilar membrane and scala tympani, respectively. ***III*** Histograms of the occurrence of pressure points along the cochlear tube's scala tympani (ST) and basilar membrane (figures are from Verbist et al. 2009 with permission)

Using eight human cochleae, the pressure points that can be encountered by insertion of the electrode array are found by the combination of cochlear morphology and mathematics (see Fig. 9.8). Confirmation of the change at the basal turn by a rise followed by a downward slope at start was affirmed. The basal turn will then narrow, and electrode array insertion will produce pressure on the basilar membrane in this area. Indeed, surgical adaptations to overcome such a pressure have been proposed. Pressure points are found at 180–225 degrees or 12–14 mm and at 405–450 degrees or at 18–22 mm distance from the round window (Fig. 9.8). This fine tuning of the slope changes of the scala tympani has consequences for the electrode array composition. Vertical stiffness of the electrode array is questionable and arrays with smaller electrodes are presumably passing the natural slope changes of the cochlear tube more easily. Although general local pressure points are found, individual differences are present supporting the need of an accurate estimation of these slope changes in each cochlear implant patient (Verbist et al. 2009).

9.6.4 Auditory Brainstem Implant (ABI)

In cases where deafness finds its origin in destruction of all or most of the spiral ganglion fibres or destruction of the cochlear/VIII cranial nerve by surgical treatment or due to neurofibromatosis NF2 (also called MISME syndrome) cochlear implantation is useless. Neurofibromatosis type 2 may be inherited or caused by a spontaneous mutation and is found with an incidence of 1:60,000. Neurofibromatosis causes tumours near the internal acoustic meatus, tumours that include the risk of damage of the cranial VIII nerve during removal. In such cases, an electrode mesh can be put on the cochlear nuclei of the brainstem, which are located on the brainstem surface. Adjustment of the electrodes is rather difficult due to the less distinctive tonotopy: the dorsal cochlear nucleus is mainly related to high tones while the ventral cochlear nucleus is associated with lower frequencies. The dorsal acoustic striae (becoming the degenerated incoming layer of spiral ganglion fibres in neurofibromatosis) cover partially the dorsal cochlear nucleus, while the ventral one is present partially below and to the side of the dorsal cochlear nucleus (Sennaroglu and Ziyad 2012). All these factors cause difficulties during surface stimulation of the cochlear nuclei (grouped as the selectivity hypothesis). The other hypothesis says that the electrodes hardly stimulate due to bypassing the specific auditory networks in the cochlear nuclei (bypass hypothesis). Children with a serious dysplasia of the cranial nerve VIII, with cochlear aplasia, hypoplasia, skull basal fracture, VIII nerve avulsion and cochlear ossification benefit from an auditory brainstem implant. However, additional handicaps (e.g. autism) do have direct effect on the function of the brainstem implants. Moreover, patients that underwent a tumour removal compared to non-tumour patients with the same auditory brainstem implants show a poorer modulation detection and speech understanding (Coletti and Shannon 2005). It is beyond dispute that auditory brainstem implants need improvement from an electric engineering point of view (Schmidt Goffi-Gomez et al. 2012). Neuroanatomical, neurophysiological and modelling studies of the cochlear brainstem nuclei are a prerequisite for the understanding of their functioning and to increase auditory brainstem implant effects. One should notice that damage of the cochlear nerve also induces defects of the olivocochlear reflex (suppression of cochlear gain by action on outer hair cells), the stapedius reflex and tensor tympani reflex (damping of ear ossicle vibration at high-intensity sounds) (see Nieuwenhuys et al. 2008).

9.6.5 Middle Ear Implants

Surprisingly, one of the arguments to use active middle ear implants is that conventional hearing aids are hardly used due to the short-comings of their design. The problems concern occlusion, discomfort and canal irritation by the tight-fitting ear mould. The problems with the sounds like distortion and feedback aggravate the "I leave my hearing aid in the drawer" (Carlson et al. 2014; Kochkin 2000). One should note that only 10.6% of the people using a hearing aid are satisfied, 27.1% are neutral, but 62.3% are dissatisfied (Kochkin 2000). The solution of the hearing aid problems is not simple. At one hand, the technical properties should be ameliorated, and on the other hand, hearing aids should be increasingly suited for the consumer's individual ear properties. Middle ear prostheses are used as an answer to these problems by bypassing the external acoustic meatus, and signal amplification is not needed.

Active middle ear implants are characterized by a transducer that acts as an input sensor with an output actuator. In fact, an electric signal is converted into mechanical vibration by a piezoelectric or an electromagnetic element. The contact electromagnetic system has an energizing coil and magnet brought together in one and the same construction that is connected either to the ossicle chain or to the round window membrane. The non-contact electromagnetic device has a magnet that is fixed to the tympanic membrane, ossicle chain or round window. An energizing coil is placed nearby. The magnetic field fluctuates that give a fibrillation in synchrony with the incoming electric signals produced by the sounds and speech registered by a microphone.

Middle ear prostheses actually benefit of ossicle measures (Fig. 9.9). Since such prostheses are implanted in patients with middle ear ossicle problems or hearing aid problems, a comparison between normal ($n = 10$), congenital aural stenosis ($n = 10$) and congenital aural atresia ($n = 10$) shows that the number, size and shape of the middle ear ossicles can differ and be affected diversely (see for normal development of ossicles Sect. 4.7.2 and Fig. 4.11).

The differences concentrate on the length of the malleus handle and the long process of the incus, which are larger in normal ossicles. The angles of the incudostapedial joint and between both the processing parts of the incus are smaller in normal ears (Fig. 9.9). Other differences are hardly present: "Anatomical parameters of the lower part, but not of the upper part, of the ossicular chain in these patients differ from those in normal ears. This may be explained by the hypothesis that the upper part of the ossicular chain originates from the first arch during embryonic development, whereas the lower part of the ossicular chain is derived from the second arch." (Li et al. 2017; for analogous measures see Kumar et al. 2018)

These distances are highly important because the ossicle length, shape and altered position determine the connection needed for the implant. Connections have been made with the incus, with the incus reconstructed with the stapes (note the incudomalleolar joint is transected), the stapes, both footplate and capitulum are used, and malleus together with the incus. A thin fibrillating plate has been placed between the disarticulated incudostapedial joint, but was not used clinically. Both the tympanic membrane and the round window membrane have been used for

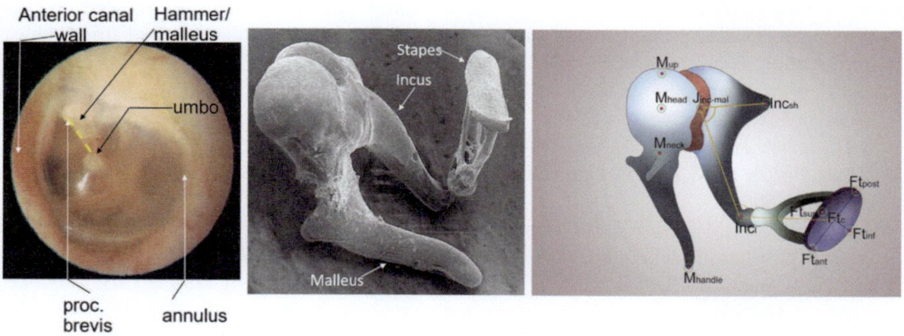

Fig. 9.9 View on the eardrum and scanning electron microscopic picture of the ear ossicles (adapted from Frijns: Ear surgery, with focus on cochlear implantation 2006). Right are the measures and angles of the human ossicles that are used by Li et al. (2017): the central points of the malleus head (Mhead), the malleus neck (Mneck), the incudomalleolar joint (Jinc-mal); the distal ends of the long process of the incus (Incl), the short process of the incus (Incsh), the malleus handle (Mhandle); the upper point of the malleus head (Mup); the anterior point of the stapes footplate (Ftant); the posterior point of the stapes footplate (Ftpost); the midpoint of the superior border of the stapes footplate (Ftsup); the midpoint of the inferior borders of the stapes footplate (Ftinf); the central point of the stapes footplate (Ftc); the angle between the short and long processes of the incus (Incsh Jinc-mal Incl); the angle of the incudostapedial joint (Jinc-mal InclFtc), with permission

application of vibrations (for overview see Carlson et al. 2014; Channer et al. 2011).

Measures of the inner ear cave with its structures are important for the surgical implantation of the transducers. The absence of differences in gender, size, age or pneumatisation of the mastoid in adult humans together with data which describe the middle ear quantitatively by CT secures the implantation surgery (Stieger et al. 2006).

Some of the devices caused the bankruptcy of the firm because the device was implanted but did not work in the long run, others performed well but development was not continue because research became too expensive and several small companies stopped because they were bought out by larger companies (Carlson et al. 2014). One should note that patients with a middle ear implant, but without support of the original firm, do have a serious problem.

References

Adunka O, Kiefer J (2006) Impact of electrode insertion depth on intracochlear trauma. Otolaryngol Head Neck Surg 135:374–382

Appler JM, Goodrich LV (2011) Connecting the ear to the brain: molecular mechanisms of auditory circuit assembly. Progr Neurobiol 93:488–508

Avci E, Nauwelaers T, Lenarz T, Hamacher V, Kral A (2014) Variations in microanatomy of the human cochlea. J Comp Neurol Res Syst Neurosci 522:3245–3261

Baker CVH, Bronner-Fraser M (2001) Vertebrate cranial placodes. Part I. Embryonic Induction. Dev Biol 232:1–61

Boons T, Brokx JPL, Frijns JHM, Peeraer L, Philips B, Vermeulen A, Wouters J, Van Wieringen A (2012a) Predictors of spoken language development following pediatric cochlear implantation. Ear Hear 33:617–639

Boons T, Van Wieringen A, Brokx JPL, Frijns JHM, Peeraer L, Philips B, Vermeulen A, Wouters J (2012b) Effect of pediatric bilateral cochlear implantation on language development. Arch Pediatr Adolesc Med 166:28–34

Briaire JJ, Frijns JHM (2005) Unrevaling the electrical compound action potential. Hear Res 205:143–156

Briaire JJ, Frijns JHM (2006) The consequences of neural degeneration regarding optimal cochlear implant position in scala tympani: a model approach. Hear Res 214:17–27

Briaire JJ, Westen AA, Frijns JHM (2006) Amplitude non-linearity of the electrically evoked compound action potential (eCAP) of the auditory nerve. Biomechanics 39:S435

Bruska M, Ulatowska-Blaszyk K, Weglowski M, Wozniak W, Piotrowski A (2009) Differentiation of the facio-vestibulocochlear ganglionic complex in human embryos of developmental stages 13–15. Folia Morphol Warsz 68:167–173

Bulankina AV, Moser T (2012) Neural circuit development in the mammalian cochlea Physiol 27:100–112

Caldas FF, Cardoso CC, de Souza Antunes, Chelminski Barreto M, Teixeira MS, da Silva Melo, Hilgenberg A, Serra LSM, Bahmad F Jr (2016) Analysis of electrically evoked compound action potential of the auditory nerve in children with bilateral cochlear implants. Braz J Otorhinolaryngol 82:123–130

Carlson ML, Pelosi S, Haynes DS (2014) Historical development of active middle ear implants. Otolaryngol Clin North Am 47:893–914

Chandebois R, Faber J (1983) Automation in animal development. In: Wolsky A, Karger S (eds) Monographs in developmental biology 16. Basel

Channer GA, Eshraghi AA, Liu XZ (2011) Middle ear implants: Historical and futuristic perspectives. J Otology 6:10–18

Christov F, Munder P, Berg L, Bagus H, Lang S, Arweiler-Harbeck D (2016) ECAP analysis in cochlear implant patients as a function of patient's age and electrode-design. Eur Ann Otorhinolaryngol Head Neck Dis 133S:S1–S3

Clark GM (2015) The multi-channel cochlear implant: Multi-disciplinary development of electrical stimulation of the cochlea and the resulting clinical benefit. Hear Res 322:4–13

Coletti V, Shannon RV (2005) Open set speech perception with Auditory Brainstem Implant? Laryngoscope 115:1974–1978

Cushing SL, Papsin BC, Rutka JA, James AL, Gordon KA (2008) Evidence of vestibular and balance dysfunction in children with profound sensorineural hearing loss using cochlear implants. Laryngoscope 118:1814–1823

Dhanasingh A, Jolly C (2017) An overview of cochlear implant electrode array designs. Hear Res 356:93–103

Deans MR (2013) A balance of form and function: planar polarity and the development of the vestibular maculae. Semin Cell Dev Biol 24:490–498

Dechesne CJ, Sans A (1985) Development of vestibular receptor surfaces in human fetuses. Am J Otolaryngol 6:378–387

Djourno A, Eyries C (1957) Prothese auditive par excitation electrique á distance du nerf sensoriel á l'aide d'un bodinage inclus á demeure. Presse Med 35:14–17

Ezan J, Montcouquiol M (2013) Revisiting planar cell polarity in the inner ear. Semin Cell Dev Biol 24:499–506

Enticott JC, Tari S, Koh SM, Dowell RC, O'Leary SJ (2006) Cochlear implant and vestibular function. Otol Neurotol 27:824–830

Fekete DM (1999) Development of the vertebrate ear: insights from knockouts and mutants. TINS 22:263–269

Frijns JHM, de Snoo SL, ten Kate JH (1996) Spatial selectivity in a rotationally symmetric model of the electrically stimulated cochlea. Hear Res 95:33–48

Frijns JHM, Briaire JJ, Schoonhoven R (2000) Integrated use of volume conduction and neural models to simulate the response to cochlear implants. Simul Pract Theory 8:75–97

Gisselsson L (1950) Experimental investigation into the problem of humoral transmission in the cochlea. Acta oto-laryng, Suppl, p 82

Goodrich LV, Strutt D (2011) Principles of planar polarity in animal development. Develop 138:1877–1892

Griffioen BKP (1988) Doktersgereedschap van toen. Reinier de Graaf stichting series nr 4:1–91

Hatsushika S, Shepherd RK, Tong YC, Clark GM, Funasaka S (1990) Dimensions of the scala tympani in the human and cat with reference to cochlear implants. Ann Otol Rhinol Laryngol 99:871–876

Helidonis ES (1993) The history of otolaryngology from ancient to modern times. Am J Otolaryn 14:382–392

Hillel AD (1983) History of stapedectomy. Am J Otolaryngol 4:131–140

Hu BH, Zhang C, Frye MD (2017) Immune cells and non-immune cells with immune function in mammalian cochlea. Hear Res. https://doi.org/10.1016/j.heares.2017.12.009

Huygen PLM, Hinderink JB, Van Den Broek P, Van Den Borne S, Brokx JPL, Mens LHM, Admiraal RJC (1995) The risk of vestibular function loss after intracochlear implantation. Acta Otolaryngol S 520:270–272

Irvine DRF (2017) Plasticity in the auditory system. Hear Res. https://doi.org/10.1016/j.heares.2017.10.011

van der Jagt AMA, Kalkman RK, Briaire JJ, Verbist BM, Frijns JHM (2017) Variations in cochlear duct shape revealed on clinical CT images with an automatic tracing method. Scie Rep 7:17566. https://doi.org/10.1038/s41598-017-16126-6

Jackler RK, Luxford WM, House WF (1987) Congenital malformations of the inner ear: a classification based on embryogenesis. Laryngoscope 97:2–14
Jeffery N, Spoor F (2004) Prenatal growth and development of the modern human labyrinth. J Anat 204:71–92
Joshi VM, Navlekar SK, Kishore GR, Reddy KJ, Kumar ECV (2012) CT and MR imaging of inner ear and brain in children with congenital sensorineural hearing loss. Radiogr 32:683–698
Kalkman RK, Briaire JJ, Dekker DMT, Frijns JMH (2014) Place pitch versus electrode location in a realistic computational model of the implanted human cochlea. Hearing Res 315:10–24
Kelly MC (2006) hair cell development: commitment through differentiation. Brain Res 1091:172–185
Kelly MC, Chen P (2009) development of form and function in the mammalian cochlea. Curr Opin Neurobiol 19:395–401
Kim JH, Rodriguez-Vazquez JF, Verdugo-Lopez S, Cho KH, Murakami G, Cho BH (2011) Early fetal development of the human cochlea. Anat Rec Hob 294:996–1002
Kochkin S (2000) Why *my hearing aids* are in the drawer: MarkeTrak V The Consumer's Perspective. Hear J 53:34–42
Kösling S, Ommenzetter M, Bartel-Friedrich S (2009) Congenital malformations of the external and middle ear. Eur J Radiol 69:269–279
Kumar IDV, Chaitanya DK, Singh V, Reddy DS (2018) A morphometric study of human middle ear ossicles in cadaveric temporal bones of Indian population and a comparative analysis. J Anat Soc India. https://doi.org/10.1016/j.jasi.2018.01.001
Li J, Chen K, Li C, Yin D, Zhang T, Dai P (2017) Anatomical measurement of the ossicles in patients with congenital aural atresia and stenosis. Int J Pediat Otorhinolaryngol 101:230–234
Lim R, Brichta AM (2016) Anatomical and physiological development of the inner ear. Hear Res 338:9–21
Mammano F (2013) ATP-dependent intercellular Ca(2+) signaling in the developing cochlea: facts, fantasies and perspectives. Semin Cell Dev Biol 24:31–39
Mammano F, Bortolozzi M (2017) Ca^{2+} signaling, apoptosis and autophagy in the developing cochlea: Milestones to hearing acquisition. Cell Calcium. https://doi.org/10.1016/j.ceca.2017.05.006
Mistrìk P, Jolly C (2016) Optimal electrode length to match specific cochlear anatomy. Eur Ann Otorhinolaryngol Head Neck Dis 133S:S68–S71
Nieuwenhuys R, Voogd J, van Huijzen C (2008) The human central nervous system. Springer, Berlin, Heidelberg
Ouellet C, Cohen H (1999) Speech and language development following cochlear implantation. J Neuroling 12:271–288
Papsin BC (2005) Cochlear implantation in children with anomalous cochlea-vestibular anatomy. Laryngoscope 115:1–26
Pisoni DB, Kronenberger WG, Harris MS, Moberly AC (2017) Three challenges for future research on cochlear implants. World J Otorhinolaryn-Head Neck Surg. https://doi.org/10.1016/j.wjorl.2017.12.010
Raphael Y, Altschuler RA (2003) Structure and innervation of the cochlea. Brain Res Bull 60:397–422
Rubel EW, Fritzsch B (2002) Auditory system development: primary auditory neurons and their targets. Ann Rev Neurosci 25:51–101
Schmidt Goffi-Gomez MV, Magalhães MT, Neto RB, Tsuji RK, de Queiroz Telles Gomez M, Bento RF (2012) Auditory brainstem implant outcomes and map parameters: report of experiences in adults and children. Int J Pediat Otorhinolaryngol 76:257–264
Sennaroglu L, Ziyad I (2012) Auditory brainstem implantation. Auris Nasus Larynx 39:439–450
Stieger C, Djeric D, Kompis M, Remonda L, Häusler L (2006) Anatomical study of the human middle ear for the design of implantable hearing aids. Auris Nasus Larynx 33:375–380
Streeter GL (1906) On the development of the membranous labyrinth and the acoustic and facialis nerves in the human embryo. Am J Anat 6:139–165

Streit A (2004) Early development of the cranial sensory nervous system: from a common field to individual placodes. Dev Biol 276:1–15

Theunissen (2013) Psychopathology in hearing-impaired children. Thesis, University Leiden

Timeline of hearing devices and early deaf education (2005–2009) Deafness in disguise, Washington University School of Medicine

Van Gendt MJ, Briaire JJ, Kalkman RK, Frijns JHM (2016) A fast stochastic and adaptive model of auditory nerve responses to cochlear implant stimulation. Hear Res 341:130–143

Verbist BM, Ferrarini L, Briaire JJ, Zarowski A, Admiraal-Behloul F, Olofsen H, Reiber JHC, Frijns JHM (2009) Anatomic considerations of cochlear morphology and its implications for insertion trauma in cochlear implant surgery. Otol Neurotol 30:471–477

Wang B, Hu B, Yang S (2015) Cell junction proteins within the cochlea: A review of recent research. J Otol 10:131–135

Warr WB, Guinan JJ Jr (1979) Efferent innervation of the organ of corti: two separate systems. Brain Res 173:152–155

Westen AA, Dekker DMT, Briaire JJ, Frijns JHM (2011) Stimulus level effects on neural excitation and eCAP amplitude. Hearing Res 280:166–176

Whitfield TT (2015) Development of the inner ear. Curr Opin Gen Devel 32:112–118

Wilson BS (2017) The cochlear implant and possibilities for narrowing the remaining gaps between prosthetic and normal hearing. World J Otorhinolaryngol-Head and Neck Surg. https://doi.org/10.1016/j.wjorl.2017.12.005

Yiin RSZ, Tang PH, Tan TY (2011) Review of congenital inner ear abnormalities on CT temporal bone. Br J Radiol 84:859–863

Zak M, Klis SFL, Grolman W (2015) The Wnt and Notch signalling pathways in the developing cochlea: Formation of hair cells and induction of regenerative potential. Int J Dev Neurosci 47:247–258

Zrunek M, Lischka M, Hochmairdesoyer I, Burian K (1980) Dimensions of the scala tympani in relation to the diameters of multichannel electrodes. Arch Otorhinolaryngol 229:159–165

Zwislocki JJ (1981) Middle ear cochlea and Tonndorf. Amer J Otolaryngology 2:240–250

Chapter 10
Posture and Motion of the Cervical Spine

Despite the still expanding research on the cervical spine, an explanation of the newborn cervical spine, of the mature head and neck motions, and of the biomechanical cause(s) of whiplash are incomplete. Presumably, this is due to the lack of notion of the interrelated anatomy of the constituting structures, to restricted studies on motion functions of the individual segmental and blocks of cervical vertebrae. The achondroplastic craniocervical junction is treated. The genetic research clarified the cartilage–bone development that presumably will ameliorate the development of achondroplastic children.

10.1 Craniovertebral Junction

10.1.1 Introduction: Rembrandt's Anatomical Lesson

Before the craniovertebral junction can be treated, three related problems concerning the skull base and its relation to the vertebral column have to be discussed. One is brought up by looking to Rembrandt's anatomical lesson of Dr. Deijman, the other is the foetalization theory of Bolk. The third is the rule of thirds. The first reminds to the ligamentous structure of the cervical column directly below the skull, the second indicates retention of foetal properties and the last one considers the cervical vertebral canal.

The anatomical lesson of Dr. Deijman has been overshadowed by the anatomical lesson of Tulip, both painted by Rembrandt (for an overview of the Amsterdam anatomical lessons; see Baljet 2000). Although less known, the position of the dissected body of Joris Fonteijn (a hanged criminal) has been in the centre of the discussion of dead body painting and not only in Deijman's anatomical lesson (Fig. 10.1; Six 1905; Middelkoop 1994a, b). Rembrandt had to paint Fonteijn's body extremely compact, due to the spectator's position he took.

E. Marani and C. Heida, *Head and Neck*,
https://doi.org/10.1007/978-3-319-92105-1_10

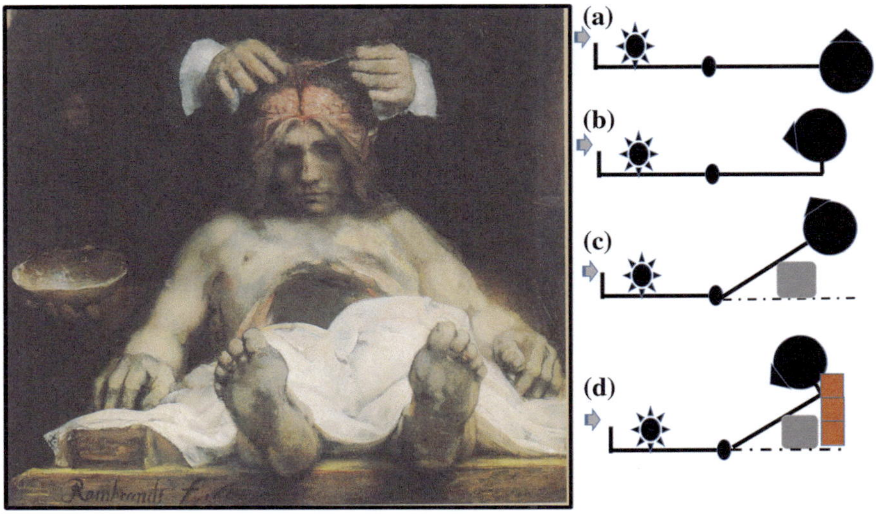

Fig. 10.1 Excision from "De anatomische les van Dr.Deijman" of Rembrandt, 1656, (Amsterdams Historisch Museum) and next to it the explanation of the position of the body of the hanged Joris Fonteijn (1633/34-1656; see text)

This position has been compared to the *Picta* of Matagna in Milan and has been related to Vesalius's portrait in *De humani corporis fabrica*. Hendrick Goltzius (1558–1617) has used this frontal view on the body for the dying Adonis and by Orazio Borgaianni (1574–1616) for the Lament of Christ. However, the position of Fonteijn's corpse is intriguing in its anatomical perspective. Rembrandt uses a low position of the spectator. One looks at the level of the feet towards the head of Joris Fonteijn and towards the praelector Dr. Deijman. The painting was heavily damaged, due to the fire of 8 November 1723, and only the portrait of Gijsbert Calkoen (1621–1664), the praelector's aid and master of the college of surgeons, of the bystanders survived it. Dr. Deijman's face was lost. His hands are still present. From the dissection room, anatomists know that from that spectator's position one cannot see the belly and the chest, particularly if a white cloth is draped over the legs (position A in Fig. 10.1). Vesalius recommends to decapitate the head if the brain should be dissected. van Mauden [1538–1597] (1647) explains it: "Because when it is decapitated you can easily and with fewer difficulties manipulate it. That is why the heads of decapitated humans are the best qualified, principally because they died recently and because they are haemorraghed (…) and can be obtained directly after the execution". And somewhat further on: "But if the head is still on the body put as much as possible square stones or a block below the neck and occiput till the head comes forward and is uplifted like it was placed on its base or a support" (see Middelkoop 1994a, b; see also van Hee and Lowis 2006)

Maer is het hooft noch aen het lichaam vast dan steeckt onder den necke ende onder het occiput soo veel viercante steenen ofte een blocxken tot dat het hooft so verre veurwaert helle ende opgeheven sy, gelijck het doen soude alst op sijnene basis of steunsel stonde.

It is still a matter of debate whether stones have been placed below the body to get the corpse in the painted position. The rigor mortis does not contribute to the raising of the head. In normal living, male humans one can bend the head (flexion), reaching the chin onto the chest, with an angle of 80–90° (position B in Fig. 10.1). By putting a block under the back of the corpse, the thorax will come up (position C in Fig. 10.1) and the spectator can look into the opened belly and into the chest space (the visceral organs were removed in the first part of the dissection). Since Joris Fonteijn was hanged he must have had a “Hangman’s fracture” (a rupture of C_2 isthmus; see Sect. “Compartments of Head and Neck” in Introduction). This supported the bending of the head (position D in Fig. 10.1) such that the frontal and parietal brain lobe of the diseased can be seen. To keep the head stable in that position, the square stones are needed, but also support of the back to overcome wobbling during dissection.

Thus, the male cervical vertebral column can (be) ben(d)(t) 90°, hyperextension and distraction (hanging) destructs the interarticular parts of C_2 and gives spondylolisthesis of the axis (disruption of the C_2–C_3 junction and vertebral sliding between C_2–C_3). Now note that C_0 (skull base joint with condyles occipitalis)–C_1 connection is involved and the relation between atlas and axis (C_1–C_2) remains. This is due to the strong connections between, atlas and upper part of the axis. These ligamentous reinforcements stop at the C_2–C_3 joint (Fig. 10.2).

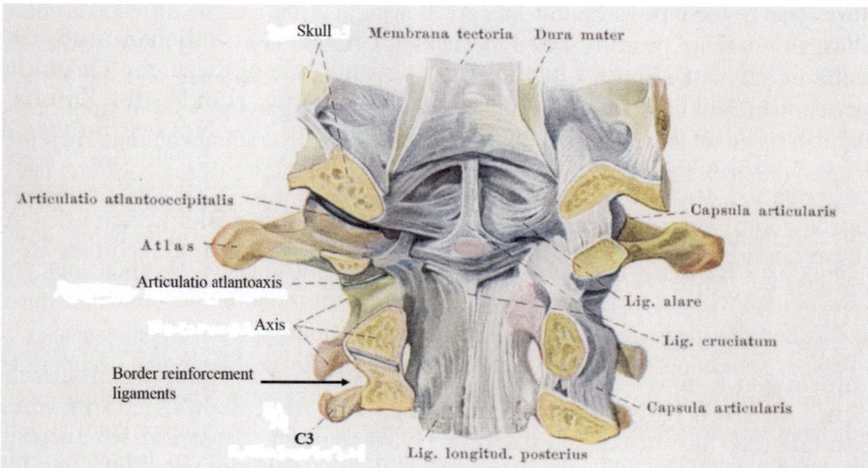

Fig. 10.2 Posterior reinforcement ligaments of the C_0–C_1–C_2 joints, viewed towards dorsal from inside the vertebral canal (adapted from Braus 1929)

10.1.2 Bolk's Foetalization Theory

Bolk (1866–1930) published in 1918 "Hersenen en Cultuur" (Brain and Cultivation) in which his "foetalisatieleer" (foetalization theory) was explained for the layman. Although his scientific publications were published in Dutch, German and English (see Origin of racial characteristics in man 1929), a restricted reaction appeared in scientific circles. The main idea of Bolk's theory is, without jumping to conclusions, that man is a sexually mature foetal ape. His argumentation was derived from comparisons between man and ape:

(1) The position of the foramen magnum: while in animals their foramen magnum is dorsally placed, it shifts more ventrally in the apes to arrive at the middle of the skull base in man. The foetal ape has its foramen magnum in the middle of the skull base, and it shifts caudodorsally during development and growth (see Fig. 7.17).
(2) The structure of the skull: Apes do have a brow ridge (supraorbital torus) that is absent in man. The foetal ape skull misses such a brow ridge, and its foetal anterior cranium compares to the mature human one. An analogous situation can be described for the orbit of both species.
(3) The hair covering of the body: while our hair is restricted to head, axilla and sexual organs, apes have a total hair covering of the body. However, the foetal ape has its hair restricted to the top of its head as in mature man.

There are more arguments that stem from pigmentation and from morphology of the mandible. Gavin Rylands de Beer (1899–1972) also puts the idea forward under the name paedomorphosis, the retention of juvenile features in the adult form (see Brigandt 2006).

Thus, the position of the foramen magnum varies with the species studied. Nevertheless, the biped man has this position right at the middle of the skull base, which is a unique property among vertebrates (Lieberman 2011). Moreover, the centre of gravity in humans is placed nearly above this foramen and the vertical cervical vertebral column, while in apes this centre of gravity is far in front of the foramen magnum and its horizontal or oblique cervical vertebral column. Therefore in apes, the neck ligament (ligamentum nuchae) and muscles of the back and neck, connected to the neck ligament, are reinforced to carry the head (Lieberman 2011). This ligament and its neck muscles play an important role in tracking down meningitis. Meningitis is caused by a viral (often not fatal) or bacterial infection (always fatal without treatment) of the membranes surrounding the brain and spinal cord called the meninges. Other microorganisms, fungi and protozoa, can also be involved. Symptoms are neck stiffness, headache and fever. Meningitis occurs in 16 million people with death in around 400,000 persons a year. The surviving group will have serious setbacks with long-term consequences, including cognitive deficits among deafness and epilepsy. Nuchal rigidity (inability to bring the chin towards the chest passively due to increased muscle tone and stiffness) concerns the flexion motion of the neck. Check is by tilting the head upward, while the patient is in the recumbent position. Tilting is unsuccessful (sometimes only in a certain amount) and as a consequence, the legs are pulled up (called the sign of Brudzinski). The neck stiffness is organized by the neck muscles that adhere onto the neck ligament that helps to keep the head upright (Bernsmeier et al. 1984; van der Meer and van 't Laar 2001).

10.1.3 The Rule of Thirds

The rule of thirds says: the cervical cord occupies one-third of the spinal canal, the dens of the axis occupies one-third and the remaining one-third is empty. What causes the one-third emptiness of the spinal canal. It appears to be in contradiction to the "coherence spatial" that was explained in Chap. 1. Spatial coherence stands for mutual influence and fitting of structures not for absence of structures by empty spaces. Let's try to explain this one-third emptiness.

By turning your head from extreme left to extreme right, you make an 180° turn. While your head rotates, the structures responsible for it have to contribute: the skull base with its joint to C_1, C_1 to C_2, and so on. This is done partially, but not equally (see Sect. 10.4), by the joints of each vertebra with the others above and below, together they reach this maximal turn. You should expect that each of the vertebrae circle exactly with its centre placed above the centres of the other vertebrae. However, the articulations between skull and the atlas contain a special form, called sandal joints or ellipsoid or egg joints. The form of the sandal joints forces the movement of the atlas into an ellipsoid and not into a circle. If you place an ellipsoid with the same middle diameter of the circle's diameter below it, the turning of the ellipsoid will reduce (or occupy parts of) the circular space of the spinal canal (see Fig. 10.3). Thus, the atlas with a sagittal diameter of 30.1 mm and a coronal diameter of 28.8 mm turns above the axis with a sagittal diameter of 19.0 mm and a coronal of 22.8 mm in an elliptic way. At 64° turn of the atlantoaxial complex, a narrowing of the central canal occurs even so that spinal cord compression happens (Mazzara and Fielding 1988). The empty space in the spinal canal is a necessary prerequisite to rotate the head and to overcome total oppression of the spinal cord between atlas and axis (see also Sect. 10.4 and Fig. 10.9)

10.2 Anatomy of the Cervical Spine

The anatomy of the craniocervical junction forces the description of the development of the rostral vertebral column, the development of the occipital skull base and the interaction between skull base and vertebral column. One should be aware of the gender differences. Passive flexion motion is larger in young (6–12 years) and adult females compared to young males (6–12 years) and adult males. Passive

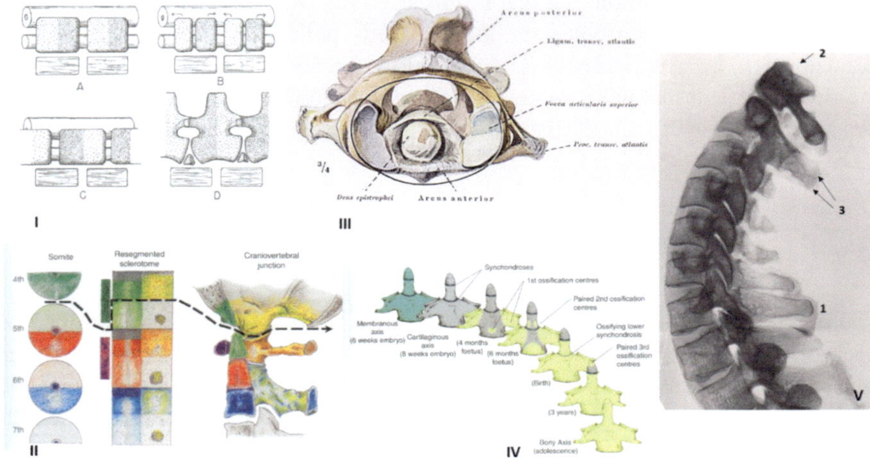

Fig. 10.3 ***I*** Ontogeny of the craniocervical junction. This diagram shows the mesenchyme contribution to the formation of the amniote vertebra. Tubes indicate spinal cord and notochord. Two sclerotomes related to their muscular portions are indicated (**a**). For the production of the vertebra, the sclerotome divides and upper and lower parts of the divided sclerotomes reunite (**b**). The vertebra shifts to its belonging musculature (**c**) and (**d** mature situation; from Romer 1962). ***II*** Shows somite contributions between the C0 and C1, and the separation between skull and spine. The border is dotted and involves the somites 4 and 5. Note that the border is placed differently between somite 5 and 6 by others (see Christ and Wilting 1992). ***III*** Demonstrates the position of the dens in relation of the elliptic turning of the skull (from Braus 1929). ***IV*** This figure of the dens development shows that 2½ somite contribute to the formation of the dens: "three developmental phases of the axis C2 and the three waves of ossification and note ages. ***V*** The mature cervical spine with vertebra prominens (1), processus articularis superior (2), split of the spinous process (3). (Figures ***II*** and ***IV*** are from Pang and Thompson 2014 with permission)

flexion diminishes with age (Seacrist et al. 2012). Impact loading of the cadaver female cervical spine compared to males shows that the segmental angles are significant larger at C2–C3, C4–C5, C5–C6 and C6–C7 levels in females. This fact is used to explain why higher incidence of whiplash-related complaints in female patients is noted secondary to rear-impact acceleration (Stempera et al. 2003; see also Sect. 10.4).

The seven cervical spine vertebrae are recognized by their lordosis, the small spinous processes that are split except the vertebra prominens (Fig. 10.3.V, 3), which is large (Fig. 10.3.V, 1). The cervical joint surfaces are nearly horizontal producing the typical mobility of head and neck. The transverse processes contain a foramen transversarium (with tubercle anterior and posterior) for the vertebral artery. Comparatively, the anterior part is a not "grown out" rib, also called costo-tranversarial process. The direct relation between artery and nerves causes by arthrosis obstruction of this artery. The transverse process of the atlas extends more to lateral. The superior processes of the atlas articulate with the condylar processes of the occipital bone (Fig. 10.3.V, 2). The atlas misses a corpus but the axis contains a dens (epistrophei) reaching into the hole of the missing corpus of the atlas (Fig. 10.3.III). Displacement of the dens is overcome by the transversal ligament. The dens is part of a synovial joint with facies articulares on the dens and on the lamina of the atlas (Fig. 10.3.III). The mutual cohesion between vertebrae and also the skull is organized by longitudinal ligaments: the anterior longitudinal ligament adheres to the ventral sides of the vertebral bodies and to the skull (skull part is called membrane atlanto-occipitalis anterior), dorsal side contains the ligamentum flavum that also reaches the skull (called membrane atlanto-occipitalis posterior). The vertebral canal contains the ligamentum longitudinalis posterior. Its contact with the skull is the tectorial membrane.

10.2.1 Development of the Cervical Vertebral Column

The embryology of the vertebrae is characterized by controversies and small and large adaptations of these controversial hypotheses can be discerned (for overview see Verbout 1976, 1985). Here we start with the "Neugliederung" hypothesis (Remak 1855; Kölliker 1859; His 1886) or called in English resegmentation of the somites. Left upper part of Fig. 10.3 gives a simplification of the resegmentation theory. Since the metamery of the muscles is in maturity shifted compared to the metamery of the sclerotomes, an explanation for such a shifting has been described from around 160 years ago on. Verbout's findings in the sheep showed that "from the beginning of development the vertebral bodies, their processes, and, even more important, the intervertebral discs, arise at their definite site. Phenomena such as shifting and migration of segment parts, or even a resegmentation of segment halves (in the sense of regrouping), do not take place" (Verbout 1985).

The early development of the vertebral column has been studied using gene expression (Pax 1 and 9; NKx3.2, N-cadherin, SHH, Sox 9 and IHH; overview in Scaal 2016), and with chick quail chimera's (Aoyama and Asamoto 2000 and references herein). The general conclusion of these studies, contradictory to Verbout's statement, is that a rostro-caudal gradient is present in the sclerotome and that resegmentation occurs. The caudal half of the sclerotome will reunite with the rostral half of its sclerotome directly below it. Due to the gene expression and chimera studies, the sclerotome not only knows a border in its middle, but also sclerotome concentrations: a ventral one, producing the vertebral body, a central one, responsible for the pedicle, a dorsal one, related to the neural arch and a lateral one, constituting the rib process (Christ et al. 2004). Thus, borders and compartments within the sclerotome are more intricate than known before. Two other conclusions are that resegmentation also involves the vertebral arches and the ribs and that the cervical vertebrae do arise from resegmentation, but the borders of the halves differ and thus do not coincide with the boundary between adjacent vertebrae. A lead in these boundary studies, using gene expression methods, is the production and the localization of the intervertebral disc (see Takahashi et al. 2013).

The occipital bone has been evolutionary developed by the fusion of vertebrae, and the several holes for the hypoglossal nerve remind to it (Fig. 4.9b). It means that the "Neugliederung" also holds for the transition skull to spine (Fig. 10.3.II). The discussions on the development of the dens seemingly agree on a 2½ somite that contributes to organize this structure in humans (Fig. 10.3.IV). "The primordia for the dens components are assembled during the membranous phase. Upper and lower dental synchondroses are shown as dense lines. First wave of ossification at the fourth foetal month consists of bilateral centres for the neural arches and a single centre for the centrum. Second wave at sixth foetal month consists of bilateral ossification centres for the basal dental segment. At birth, the basal dental centres should have integrated in the midline and begun to be fused to the centrum. Third wave of C2 ossification occurs from 3–5 years postnatal life at the apical dental segment, which does not become fused to the basal dens till the 6th–9th year and fully formed during adolescence" (Pang and Thompson 2014; Fig. 10.3.IV). The dens development thus continues after birth to adolescence.

10.2.2 Stiffness of the Cervical Spine

Low back pain and scoliosis directed research on the spinal column with the main tendency to study thoracic and lumbar areas. Cervical studies are clearly less present in literature, although manual therapy for cervical spinal manipulation and cervical mobilization are frequently described (Kranenburg et al. 2017). Severe and less adverse effects are clearly present using these therapeutic techniques by chiropractors, but are not a theme in this chapter. "Neck and head posture have been found to have a significant influence on the posture of the lower spine region during lifting and both an extended/upward gaze and a flexed/downward gaze have been hypothesized to lead to increased pain and/or overuse of the neck musculature. As a result, strength training recommendations have turned to the use of a retracted neck posture as being the safer posture to assume during lifting" (Hlavenka et al. 2017). This posterior position of the chin produces increased trunk and sternocleidomastoid activity and decreased spine flexion and decreased dorsal neck musculature activity during lifting. Head and neck position and movements therefore do have an extended effect over the whole spinal column and its adnexa like muscles and ligaments. The elasticity of the muscles and ligaments of head and neck plays an important role in accidents in motorsports or contact sports. Moreover, head and neck stabilization determines body balance, hearing and vision reliabilities, and the protection of the central nervous system. In other words, stiffness of the cervical spine is indissolubly connected to the "soft tissue" component of head and neck and together they influence the whole spinal column with its spinal muscles and spinal ligaments.

Seven main spinal column movements can be discerned: extension, flexion, right and left lateral bending, left and right rotation and axial compression. Stiffness of these seven movements of the vertebral column is needed for modelling and alleviates low back pain or scoliosis. These movements are co-determined by the intervertebral discs due to their compressibility, elasticity and thickness, by form and orientation of the zygapophysial joints, by the pressure of the joint ligaments and by the resistance and elasticity of the spinal column muscles and ligaments. Measurements can be performed on cadavers and in vivo, while load–displacements (correlating loading and torque) and/or range of motion (ROM, maximal possible displacement) are determined.

Stiffness can be calculated by the resonant frequency $\omega = \sqrt{S/m}$.

with ω the resonant frequency, S the stiffness and m the mass. The resonant frequency is constant in the elastic area, but will change to lower orders in the plastic area. Thus, stiffness is reduced in the plastic area. Stiffness is also methodologically measured by the vertebral column impedance (Ns/m, force *N* divided by velocity m/s) and angular frequency (rad/s).

Applying external mechanical loads, from which the moment can be calculated and redirected to the moment at the vertebrae, gives the total moments per vertebrae *segment* which makes it possible to calculate the stiffness at the vertebral segment if the distortions per vertebrae segment are known. Finally, the stiffness for axial rotation, axial compression, lateral bending and flexion and extension is found. The methodological approach is a motion capture system with its markers on the spinous processes of the vertebrae and the application of controlled loads. This is an intricate method, but as one of our students said: One needs a correct anatomical model for the determination of stiffness of vertebrae. A simple model will not suffice (Mongula 2007). This has been underlined by the study of Bell et al. (2016) on compressive loading methods: "The human cervical spine supports substantial compressive load in vivo arising from muscle forces and the weight of the head. However, the traditional in vitro testing methods rarely include compressive loads, especially in investigations of multisegment cervical spine constructs. Various methods of modelling physiologic loading have been reported in the literature including axial forces produced with inclined loading plates, eccentric axial force application, follower load, as well as attempts to individually apply and model muscle forces in vitro. The importance of proper compressive loading to recreate the segmental motion patterns exhibited in vivo has been highlighted in the previous studies. However, appropriate methods of representing the weight of head and muscle loading are currently unknown".

An effort to overcome the complex methodological approach is modelling the cervical spine as a single-joint system in the sagittal plane. The musculo-tendinous stiffness, expressed as "the series elastic component stiffness" that comprises tendon, connective tissue, and immanent muscle stiffness, has been studied. Energy is stored in this elastic component during isometric contractions and is freed earlier (first 15–30 ms) before structural change (e.g. muscle reflex activity) is started after sudden release of the system. The series elastic component therefore shows the elastic conduct of the studied system with minimum damping. This quick release method has been applied to the cervical spine. Subjects are strapped to a chair, and a headgear is placed on their head. Light-emitting diodes are placed on head and face with respect to the stabilized, strapped shoulder one. The headgear is connected to a wall mounted system for sudden release. The resulting musculo-tendinous stiffness values are higher and differ strongly from those earlier reported (28.38–216.28 Nm/rad and 22.57–41.25 Nm/rad, respectively). The musculo-tendinous stiffness versus the isometric cervical flexion torque showed a significance increase of the stiffness according to the torque, and the slope gives the stiffness index of the head–neck segment in flexion ($y = 3.35x + 43.97$; $R^2 = 0.74$). The centre of rotation placed by various authors or at C1 or C7 was found in this study below the C7 level. Some studies already placed the centre of rotation below C7, because many joints above and below T1 are involved in flexion (Portero et al. 2013). One should conclude that cervical spine stiffness is still under debate and that there is hardly agreement on musculo-tendinous stiffness nor on the centre of rotation. In fact, the cervical spine properties are a hard nut to crack.

Above was said that the intervertebral discs due to their compressibility, elasticity and thickness also determined stiffness. Few articles consider the intervertebral disc during loading of the head–neck complex. The approach in Yoganandan et al. (2017) is the fatigue response. Fatigue is the reaction of the intervertebral disc under axial and eccentric loads. After axial loading of parts of the cervical spine and a duration of cycles of this loading (in this article mimicking soldiers walking 19 km with their helmet and package with a total load of 150 N and around 10,000 cycles of footsteps), the intervertebral disc decreases in height due to this fatigue loading: C4–C5 nearly 5, 6% and C6–C7 6.2%. The viscoelasticity of the intervertebral discs also varies. However, these variances reach a plateau and remain constant during the rest of the cycles of footsteps. This is called "work hardening", which seemingly holds for all the spinal cord discs.

10.3 Craniocervical Junction in Achondroplasia

In Chap. 2, Sect. 2.3.2 on "sutures, fibroblast growth factor receptor and craniosynostosis" attention has been paid to the fibroblast growth factor receptor proteins (FGFRs). They are needed again to understand the changes present in the craniocervical junction in achondroplasia. The description of Armand Marie Leroi (2005) of achondroplasia's cause is by quality English: "Achondroplasia is caused by a mutation in a receptor for fibroblast growth factors. FGFs are the signalling molecules involved in the molecular clock regulating the near to the far axis of the foetal limb. After birth, however, FGFs, far from promoting the outgrowth of the limb, inhibit it. We know this because 99% of all cases of achondroplasia are caused by a mutation in which an amino acid (a glycine) at a particular location in the FGFR3 protein sequence (position 380) is replaced by another (an arginine). This mutation has the peculiar property of causing the FGFR3 molecule to become hyperactive. Nearly all of the mutations discussed cause a deficiency in the quantity of efficacy of some protein, often by causing it to be completely absent. If the protein is a signalling molecule like FGF, the disorder that we see is due to an absence of some critical piece of information that the cells require. The achondroplasia mutation is, however, different in that it occasionally causes the receptor to transmit a signal into the interior of the cell even if no FGF is bound to it. The effect is like a switch that spontaneously flips on when it should be off and that transmits a blast of unwanted information to the cells of the growing limb. If an excess of FGF signalling causes limbs to be unusually short, then the usual role of FGFs must be to act as a brake on the growth of the infant limb. They do this by limiting the rate at which the cells of the growth plate divide. The bones of achondroplastic children have growth plates that are only a fraction of the size they should be. They contain far fewer dividing chondrocytes than those of normal children, and fewer yet that swell and form cartilage" (Leroi 2005). The pelvis in achondroplasia, the small people, also named in vulgar tongue Lilliput, shows such an aspect. Achondroplasia is characterized by a retardation of the growth of cartilage and ossification during development, as explained above. Especially, the pelvis and the lumbar spine are important in determining this type of dwarfism. From the pelvic bones, the os ilium is square, while "the pubic and ischial bones are short and square and the overall pelvic configuration is short and broad" (Thomeer 1982). The inferior parts of the iliac bone are short, and "therefore, the sacrosciatic notch is short" (Thomeer 1982). Achondroplastic children that start to stand and walk produce a posterior tilting of the sacrum and reduced diameters of the pelvis are noted compared to normal pelvises. The tilting here does not produce an extra lordosis. The vertebral column becomes more oblique directed to the front. The sacral tilting overturns the pelvis further and brings the symphysis more inferior, so that the symphysis becomes the real lowest pelvic point (Marani and Koch 2014). Next to the retardation of limb growth in the small people, the skull base, foramen magnum and the craniocervical joint are also liable to this mutation. The spine in achondroplasia includes: "pseudoscalloping of the vertebrae, reduced pedicle length, and a reduced interpediculate distance that progressively narrows, going caudally". A narrowed subarachnoid space is often noticed (Wang et al. 1987). The skull features include a narrowed foramen magnum and shortening of the skull base and clivus (Thomeer 1982). "In the presence of a normal-sized nervous system, the deformities of the skull and spine may give rise to various neurological complications; for example, the narrowing of the foramen magnum may cause cervicomedullary compression with tetraparesis (weakness of all four limbs) and central apnoea (transitory cessation of breathing) as a result" (Van Dijk et al. 2007).

Until 1982, achondroplasia was surgically studied in small groups of up to 5–7 patients. The first extended overview on measurements of 31 achondroplastic patients (Thomeer 1982) studied basilar impression, petrosal angle, basal angle and clivus length. Basilar impression is the abnormally rostral presence of the upper cervical spine, such that it invades the foramen magnum. Confusion with platybasia (flattening of the skull base) is present in the older literature. The Chamberlain line is between hard palate and ophistion and is used (beside nowadays the McGregor and the McCrae lines) to estimate the smallest distance from the odontoid tip to this line. Positive diagnosis is made if the dens protrudes more than 2.5 mm above this line (Pang and Thompson 2014), while normal range indicated by Poppel et al.

(1953) is +3.5 to −3.5 mm above or below this line. Using the 2.5 mm border, 21 of the 31 achondroplasia patients present basilar impression. The other criterion gives 19 patients with basilar impression, which is 68% and 61%, respectively. General prevalence in symptom-free normal individuals is 1 in 3000, but in the achondroplasia series, it is 1 in 5 (Thomeer 1982).

The petrosal angle formed by the petrosal pyramid lines through the internal auditory canal showed this angle normally above or equal to 180°. The average angle in achondroplasia patients is 143° ± 18. Pyramids are directed: "upward and more medially and no correlation between the degree of basilar impression and petrosal angle was found" (Thomeer 1982). The basal angle is the angle between the line from nasion to the tuberculum sellae, and the second line is determined by the tuberculum sellae to the anterior rim of the foramen magnum or basion. While the basal angle normally lays between 130° and 135°, the achondroplastic studied group has a decreased basal angle with an average of 116°. The last part of the skull base that has been determined is the length of the clivus that has a normal range between 32 and 40 mm (Coin and Malkasian 1971). The average length is 25 mm in achondroplasts studied.

The conclusion is that in achondroplasia primary anterior basilar impression is present: "the anterior part of the cervical spine has migrated abnormally cephalad due to hypoplasia of the clivus" (Thomeer 1982). A second conclusion is that the whole skull base has been influenced. Moreover, the skull contains frontal bossing and the skull base is comparatively small. Suture lines are normal. A depressed nasofrontal junction also characterizes grown-ups (Thomeer 1982). In 1985 and 1989, Hecht et al. studied with tomography the foramen magnum in achondroplasia and followed the growth of the foramen magnum. Analogous results as described above originated.

Within the spinal cord of 40% of the adult achondroplasts studied at the University Medical Centres in Leiden and Groningen, a cervical high-intensity intramedullary (C1–C2, Fig. 10.4) lesion has been noted that did not correlate with compression of the spinal cord in flexion, retroflexion or neutral position of the craniocervical joint. No compression was found in the coronal plane either. The size of the cervical high-intensity lesion and the size of the cervical vertebral canal, nor the craniocervical junction at the site of the lesion could be related to compression of the C1–C2 spinal cord. The patient's complaints and the medical diagnosis for these lesions are described as mild. A clear explanation of the cervical high-intensity intramedullary lesions in the studied achondroplasts is unknown. It is speculated that such a lesion originates from compression of the cord at the craniocervical junction during their youth, thus in achondroplastic children (Fig. 10.5), which sometimes could only be demonstrated upon flexion and/or retroflexion of the head (Van Dijk et al. 2007; Brouwer et al. 2012).

Achondroplastic complications summarized for head and neck are (see Reina et al. 2014; Horton et al. 2007 and references herein):

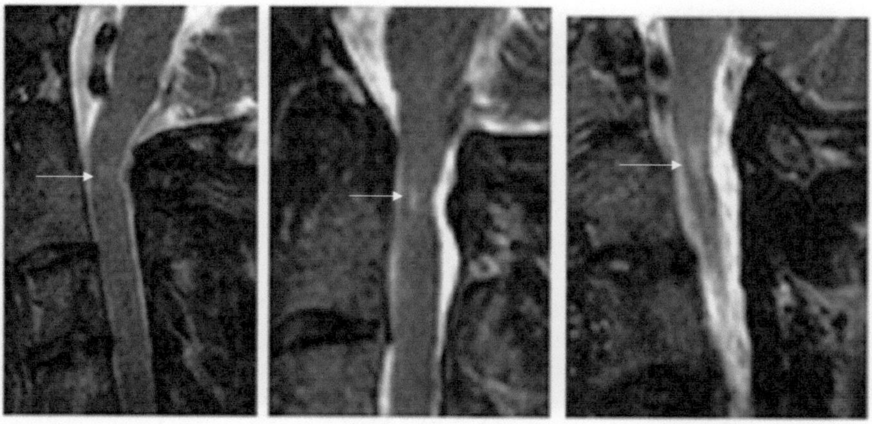

Fig. 10.4 Three examples of cervical high-intensity intramedullary lesions in achondroplasts. Arrows indicate the lesions at the C1–C2 level (courtesy Vleggeert-Lankamp with permission)

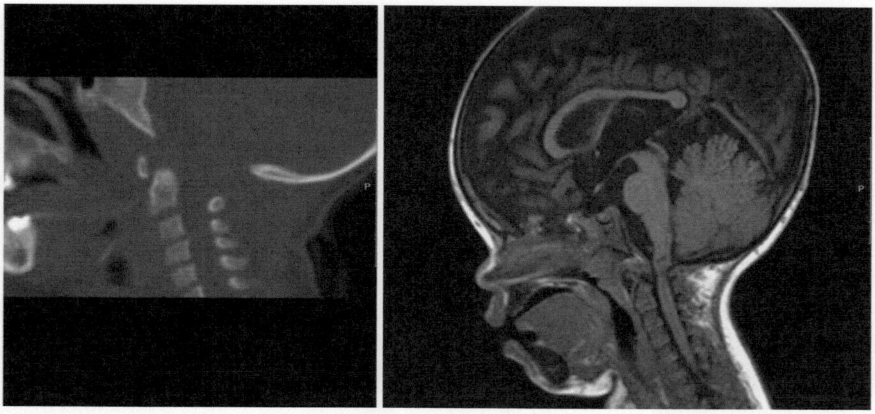

Fig. 10.5 CT and MRI of an achondroplastic girl of 2 years. Left the relation foramen magnum, clivus and cervical spine. Right compression of the spinal cord at the transition skull base, atlas and axis (C0–C1–C2) [courtesy Vleggeert-Lankamp with permission]

(1) The achondroplastic head supported by a small craniocervical joint can produce internal hydrocephalus, because the jugular foramen is also small and venous drainage is hampered increasing venous pressure.
(2) Megalencephaly (large brain) can be related to hydrocephalus or to dilated ventricles. Around 10% of the achondroplasia patients do have ventricular shunts in their youth to fight hydrocephalus.
(3) Cervical cord compression (Fig. 10.5) can induce lower limb hyper-reflexia and central apnoea. Attention was paid to the sudden death in achondroplastic

infants. Researchers have linked this to medullary cord compression (Hecht et al. 1987).

(4) The midface is underdeveloped with the consequences of short Eustachian tubes and a small pharynx. As a consequence, the ring of Waldeyer with tonsils and adenoids (see Sect. 7.5 and Fig. 7.23) occupies relatively much space. Otitis media is frequently present and is directly related to hearing loss (present in 40% of these patients). Speech problems have been related to midface hypoplasia, the small maxilla should be enlarged and the amount of teeth in the small mandible should be reduced.

(5) Sleep dysfunction, snoring and apnoea (already in 10% of the patients at the age of 4 years) are related to cord compression, small pharynx with large tonsils, jugular foramen compression by venous drainage on passing cranial nerves, inducing muscular mall function of the upper airways. The arousal response for apnoea is disturbed in achondroplastic children and supposed to be related to sudden unexpected death (Ednik et al. 2009).

Spatial coherence seems minimally disturbed by only one receptor's (FGFR-3) dysfunction, but the deviations from a normal skull and skull base extend their influence far further than has been recognized for this type of dwarfism at the start of its basic research: it concerns among others, respiration, cardiac function, speech, hearing, mastication and sleep (Horton et al. 2007; Afsharpaiman et al. 2013; Reina et al. 2014).

Some achondroplastic results are published that cannot be explained: A history of polyhydramnios (excess of amniotic fluid in the amniotic sac) was associated with a lower height, more severe shortening of the limbs, and higher frequency of left-handedness in achondroplasia (Latini et al. 2002). Moreover, the results of the Latini group indicate that "polyhydramnios is a frequent occurrence in the natural history of achondroplasia with a prevalence in their case series that is 22- to 64-fold higher than that reported for the general population" (Latini et al. 2002). The relation between polyhydramnios and achondroplasia stays an enigma.

Excavation of the nineteenth-century cemetery of the Dutch Middenbeemster village showed a skeleton that was suspected of achondroplasia. Thorough inspection of the skull base showed that no severe deviations from normal were present. It was decided that a case of hypochondroplasia was found, which could be confirmed by the offspring refound by studying the archives of the cemetery (Waters-Rits and Hoogland 2013). These results show that skull base deviations are also used as an important criterion to decide to achondroplasia.

10.3.1 The Chondrocyte in Achondroplasia

The chondrocyte is a bone marrow-derived cell in the developmental line "mesenchyme stem cell towards a condensed mesenchyme stem cell". The chondrocyte is the single cell type that constitutes cartilage. Chondrocyte differentiation into growth plate cartilage (the bone producer) or into articular cartilage ("joint" cartilage) is regulated by two transcription factors: RUNX2 for the growth plate and bone formation and SOX9 for articular cartilage. SOX9 inhibits so-called hypertrophic differentiation that is the initiation of the growth plate–bone development (Fig. 10.6). The cartilage homeostasis is organized by a series of signalling pathways: FGF, IHH, PTHrP, TGFβ, IGF, BMP and Wnt. The general overview of a cascade, started by these pathways, is explained by the IGF signalling pathway. The insulin growth factor (IGF) affects cells by their IGF-receptor and produces tyrosine kinase activity that brings intracellular IRS-1 activity (insulin receptor substrate-1). It produces a serine phosphorylation cascade with effects on glucose transport (via SHPTP2), protein synthesis (via PI-3 kinase and S6 nuclear protein phosphates), lipid synthesis (via SH2 and protein phosphatases), glycogen synthesis (via SH2 and MAP-kinase) and growth and gene expression (mainly via RAS complex: a family of 21 kDa GTP binding proteins, originally identified as oncogenes in certain types of cancer). In short, the other chondrocyte cascades will be indicated below. The fibroblast growth factor (FGF) signalling pathway is using heparin-binding proteins that interact with heparin sulphate proteoglycans present on the cell surface and is involved in proliferation and differentiation of cells. These factors are called pluripotent growth factors, due to their influence on a large variety of cell types.

Indian hedgehog homologue (IHH) is directly involved in chondrocyte differentiation and in enchondral ossification. [One other member of this family has already been encountered as the sonic hedgehog (SHH)]. Parathyroid hormone-related peptide (PTHrP) regulates endochondral bone development via the endochondral growth plate width. The transforming growth factor β (TGFβ) is a multifunctional cytokine and is involved in cell growth and cell differentiation due to specific receptors on the cell surface. Bone morphogenetic proteins (BMPs) are growth factors also known as cytokines. They induce production of cartilage and bone. "Wingless-related integration site" (Wnt), this strange name is also due to involvement of its detection by Drosophila genes, is a glycoprotein that needs lipid modifying to be effective in receptor binding and co-organizes cell proliferation and cell migration. It also plays an important role in the formation of the antero-posterior and dorsoventral axes of the embryo and as such it is important in the vertebral column construction.

Mesenchymal stem cells cultured in a hypoxia atmosphere influence positively the chondrocyte development, presumably due to induced lipid changes in culture, but also inhibits ossification (Leijten et al. 2012b). Articular cartilage can also persists cartilage because SOX9 remains active, and an escape towards bone replacement is blocked due to the expression of DKK1 (Dickkopf-related protein-1; inhibitor of the Wnt signalling pathway), FRZB (indirectly producing a Wnt inhibitor protein and antagonists) and GREM1 (Gremlin-1 is an inhibitor in the TGF β signalling pathway) (Fig. 10.6; Leijten et al. 2012a, 2013; Scholma et al. 2014; Schivo et al. 2016b).

The signalling pathway of FGFR-1 and -3 is presented in full in Fig. 10.7. The SHC protein is not only involved in the FGFR pathways, but is also used by the

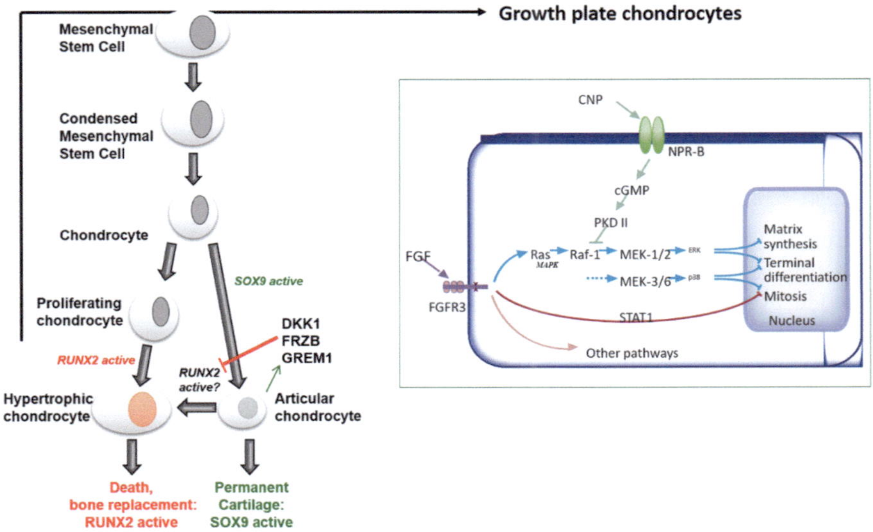

Fig. 10.6 Left: development of the articular cartilage (green) and growth plate cartilage (red). SOX9 starts the differentiation of the mesenchymal cells into chondrocytes. SOX9 is suppressed in the growth plate line, and RUNX2 is subsequently expressed. RUNX2 is responsible for the hypertrophic differentiation, start of enchondral ossification. In articular cartilage, SOX9 stays active. DKK1, FRZB and GREM1 are held responsible for SOX9 remaining activity (courtesy J. Post, see Schivo et al. 2016). Right: part of the signalling pathways of FGFR-3 in the growth plate chondrocytes (see also Fig. 10.7). FGFR-3 cascade inhibits chondrocyte proliferation by STAT-1, and MAPK legs of the signalling pathway. The CNP-NPR-B on its turn inhibits the MAPK leg of the FGFR-3 signalling pathway (Horton et al. 2007 with permission)

IGF pathway and this holds for several subdivisions of the above-enumerated signalling pathways. These interactions between signalling pathways have been modelled with Analysis of Networks with interactive MOdelling (ANIMO) "a powerful tool to formalize knowledge on molecular interactions. This formalization entails giving a precise mathematical description of molecular states and of interactions between molecules" (Schivo and Post 2016; Schivo et al. 2016b). Simulations can be carried out with ANIMO and one is demonstrated in Fig. 10.7 (right side) for the FGFR interaction with other signalling pathways. FGFR-3 does keep the chondrocyte in the enchondral ossification line (red, RUNX2 positive). The indicated square denotes the FGFR-3 effect with Wnt signalling (upper three red lines in the square): RUNX2 remains within this configuration. It is therefore understandable that in dysfunction of FGFR-3, the chondrocyte–endochondral line is weakened as is the case in achondroplasia.

On the other hand, since the signalling pathways are well known nowadays, use of this knowledge to enhance amelioration of achondroplasia is researched in experimental animals. Achondroplasia is also characterized by a decrease of parathyroid hormone-related peptide (PTHrP). Parathyroid hormone uses the same receptor as PTHrP. PTHrP inhibits the articulate chondrocyte line and supports

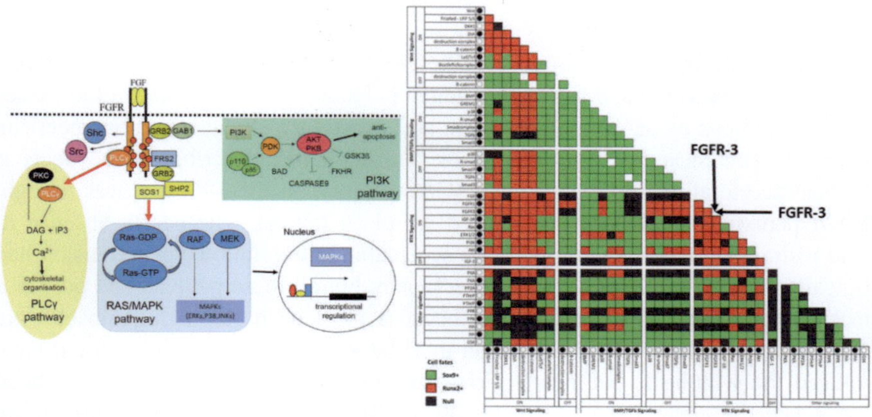

Fig. 10.7 Left side is the FGFR signalling pathway in full. "The downstream signal transduction pathways activated by phosphorylated FGFRs. A number of intracellular signal transduction proteins specifically bind these phosphorylated tyrosine kinase residues. The major direct substrates of activated FGFR kinase are FRS2 and PLC (the red arrows indicate their pathways). The cellular responses to the main downstream intracellular signalling pathways are indicated" (Moosa and Wollnik 2015 with permission). Right side shows network nodes for different signalling pathways with knockout and activation combinations (empty and full circles, respectively) and all other nodes in the RUNX2+ state. FGFR-1 and FGFR-3 keep the chondrocyte in the growth plate enchondral forming state (red blocks indicated by black arrows) combined with the RTK signalling. Combined with Wnt signalling FGFR-1 and FGFR-3 preserved the growth plate enchondral forming state (courtesy J. Post, see also Schivo et al. 2016b; Schivo and Post 2016 with permission)

enchondral ossification. Parathyroid hormone therefore is "a potential therapeutic agent for achondroplasia" (Ueda et al. 2007). The other possibility that is studied is activation of the C-type natriuretic peptide via the natriuretic peptide receptor B (NPR-B) (called the CNP-NPR-B signalling pathway; Figs. 10.6 and 10.7). Activation influences the MAPK pathway responsible for inhibition of proliferation, terminal differentiation and postmitotic matrix synthesis. FGFR-3 negatively influences both proliferation and terminal differentiation, i.e. inhibition of chondrocyte proliferation, which is needed for ossification. This means that dysfunction of FGFR-3 by mutation proliferation and differentiation is not inhibited. Influencing MAPK by the next step Raf-1, FGFR-3 function is mimicked. Trials with CNP have been started: "The strongest candidate therapy employs an analogue of C-type natriuretic peptide (CNP), which antagonizes the mitogen-activated protein (MAP) kinase pathway downstream of the FGFR3 receptor and may also act independently in the growth plate. Only the CNP analogue has reached clinical trials. Preliminary results of Phase 2 studies show a substantial increase in growth rate of achondroplastic children after six months of therapy with no serious adverse effects" (Klag 2015).

10.4 Whiplash Trauma, The Skull Base and Neck

In car accidents, mostly rear-impact crashes, external forces are exerted on head and neck in an unexpected and aggressive manner. These forces can damage the tissues in head and neck by severe compression, and the sudden changes in movement can extremely stretch and/or tear the tissues. The term that describes such effects is cervical acceleration and deceleration (CAD) injury. In the 1960s, it was thought that hyperextension of the neck caused whiplash trauma. The nearly absence of head restraints made such a hypothesis acceptable at the time. Hyperextension had logically also to be related to frontal neck pain, but all these patients had posterior neck pain. In fact, it is not illogical, because most head stabilizing muscles are placed dorsally. The federal obligation in 1969 to add head restraints should have reduced the whiplash occurrence. However, no significant reduction was noticed and it became clear that more than 30% of the head restraints were improperly adjusted. Moreover, only 10% of the cars tested had effective head restraints.

The anatomical and biomechanical research of head and neck whiplash can be subdivided into reproduction research for robotics and adapting research for ergonometric studies. Both strains of research are applied in whiplash-associated disorders (WAD). Generally, whiplash occurs 200–300 per 100,000 inhabitants in the normal population (Styrke et al. 2012) In car accidents, it concerns 50% neck trauma, 84% involves damage of soft tissue and in 10%, others indicate 20% (Fouquet and Doury-Planchout 2013), the trauma remains. During a collision, the car gets an acceleration and the driver will be pushed forward. The head remains behind due to its mass slowness, the head moves backwards and hyperextension occurs. This whiplash movement, as it is called, is also present in frontal and in lateral movements of the driver or passengers head in car accidents. The effects in lateral or frontal car accidents are comparable to rear-end accidents.

Another aspect in whiplash trauma is the aggression towards the muscles responsible to keep the head upright. The head stabilizing muscles are seriously affected during whiplash motion, expressed in the "cervical strain and sprain" saying. The group of stabilizing muscles of the head can be grouped in long and short muscles. Long ones are the trapezius and sternocleidomastoid, longissimus capitis and splenius capitis muscles. Short ones include semispinalis capitis and the rectus capitis muscles (see Fig 14.1). They all adhere to the occipital bone of the skull and find or on the cervical vertebrae or further away, scapula or clavicle, their grip to stabilize, rotate, extend or flex the head. Head movements are not exclusively extension or flexion but combinations of stabilizing, rotation, extension and flexion. It is in this balance of head and neck movements that something is going wrong and remains in the wrong after a car accident, which is noticed as a whiplash. One should note that vision disturbances, headache, dizziness and even unconsciousness are belonging to this phenomenon.

Theoretical and experimental studies have been carried out to understand the head and neck problems in order to safeguard the passengers of traffic vehicles that range from bike, motors, cars to airplanes. The first two finally brought up the helmet, the car got his airbags, but since the speed of an airplane is too high for any protection during impact the passengers are hardly saved from accident harm. Some attention has to be paid to motor accidents, especially those that cause nerve avulsions (see Chap. 11, Sect. 11.3.2). The protection measures seemingly indicate that we know how to overcome harm, but the cause of whiplash stays unknown. We take the review paper of 1996 by Ella Yeung for "Whiplash injuries" of the cervical spine as the start, because earlier papers on cervical spine are reviewed in it. Until 1996, the understanding of the whiplash injuries causes is unknown. Whether biomechanical, isolated tissue studies or human cadaveric research, the outcome of various studies can be diametrically opposed. On one side, forces after hyperextension are maximum and its greatest magnitude was found in the low cervical region (McKenzie and Williams 1971), while Yoganandan et al. (1989) found the whiplash injuries at the occiput–axis and the C5–C6 level. Cain et al. (1989) described fatal road accidents with the injuries at C2 and at C6. On the other hand, Ulbrich et al. (2014) could not find any trauma-related changes in the cervical spinal column using T2 weighted images. Earlier the same was noticed with X-ray studies (McRay 1960) or deviations from normal were related to the person's ageing and degenerative processes present (Rössler and Schmeisser 1957; Rothman and Simeone 1982). If no anatomical directed cause can be found the psychological one comes up: "Our review suggests that mobilization, manipulation and clinical massage are effective interventions for the management of neck pain. It also suggests that electro-acupuncture, strain-counter-strain, relaxation massage and some passive physical modalities (heat, cold, diathermy, hydrotherapy, ultrasound) are not effective and should not be used to manage neck pain" (Wong et al. 2015). The opposite conclusion is "Our study adds to the existing evidence that early intensive care is associated with slower recovery from whiplash-associated disorders (WAD), independent of expectation of recovery. The results have policy implications and suggest that the optimal management of WADs focuses on reassurance and education instead of intensive care" (Skillgate et al. 2016).

Siegmund et al. (2009) studied the literature and found that potential neck injury sites are facet joints, spinal ligaments, intervertebral discs, vertebral arteries, dorsal root ganglia (DRG), neck muscles and vertebral artery. The most striking is the DRG in this row. The neuroinflammation and glial cell activation in the DRG, and spinal cord has been considered a possible mechanism behind the transition from acute pain to chronic pain (Cairns et al. 2015). Moreover, central sensitization by pain is nowadays considered the base for whiplash dysfunctions. One should know that cross-sensitization also plays an important role. Cross-sensitization has been best studied in visceral organs (Fig. 10.8). "Visceral organ cross-sensitization, clinically known as referred pain, is the presence of pain in another organ and not in the pain inducing, sick organ. Heart failure correlates with pain in skin of the left arm and shoulder or lung, showing visceral–cutaneous interactions and visceral–visceral ones, respectively. Known relations are duodenum and oesophagus, lower and upper oesophagus, and colon and several other viscera. For the pelvic, it concerns colon and rectum, colon and urinary bladder/pelvic urethra, the uterus or prostate and upper urethra. Explanations are divers: A, one and the same DRG cell receives information from more than one organ; B, different DRG neurons project onto the same spinal neuron or interneuron; C, different DRG cells influence each other within the ganglion (Fig. 10.8). DRG neurons are isolated by satellite cells and mutual synapses are absent. Nevertheless, spike traffic over neurons can induce inhibition or excitation of neighbouring DRG neurons. The cell soma of the DRG that is needed for spike discharge is essential for this crosstalk" (Marani 2016). Thus, headache can be caused by pain in the neck. Neck pain also can induce psychological dysfunctions like emotional, cognitive and behavioural ones. Whiplash injuries can give long stay dizziness. All are explained by cross-sensitization.

Since a clear pathology cannot be found and discussions are still going on, other solutions are proposed of which the transient pressure changes within the vertebral canal during whiplash is intriguing. Figure 10.9 shows the assumed changes within the vertebral canal that is exercised on the spinal cord and its peripheral

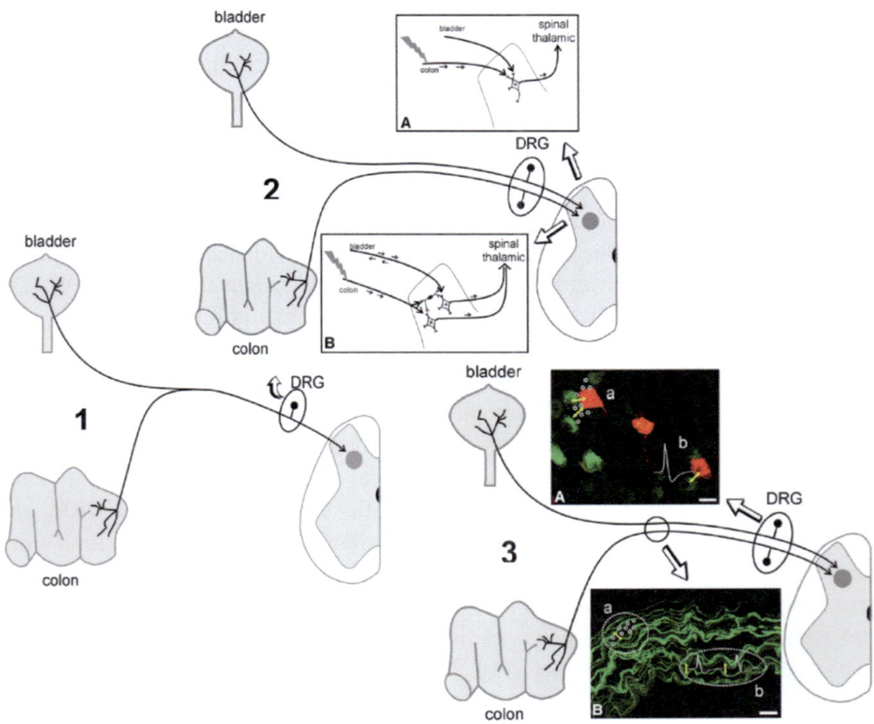

Fig. 10.8 Three explanations of visceral organ cross-sensitization (see text for explanation numbers). *1* One DRG cell receives information from more than one organ; *2* different DRG cells project on the same interneuron; *3* different DRG cells influence each other within the ganglion and *3* also shows influence of each other's axon or chemically (circles, a) or electrically (action potential, b). Colon DRG cells in green and bladder DRG cells in red. Figure is a compilation of those from Brumovsky and Gebhart (2010 with permission)

connections. During whiplash motion, volume changes within the spinal canal (Svensson et al. 1993) causing transient pressure gradients that damage the nervous tissue within the spinal canal. The vertebral canal is shortened during extension and elongated in flexion. Since the fluids within the spinal canal are incompressible, the volumes have to change. The volume adaptation can only be done by the venous plexus, since it has a connection with the intervertebral vein to outside the spinal canal and cerebrospinal fluid has a high flow resistance.

The arterial system can contribute, but is thought to have a minor influence due to its higher blood pressure. The venous blood can flow into both directions, inward (flexion, elongated spinal canal, increase in volume) or outward (extension, smaller spinal canal, volume decrease).

The hydrodynamic modelling by Yao et al. (2016) needs various suppositions. Nevertheless, the hydrodynamic model supports the damaging effect of the whiplash motion on the spinal cord. Perhaps such a damaging effect concerns also, by compression, the DRG's within the intervertebral foramina of the spinal canal (see Panjabi 2006), since vertebral neuroforamina C3–C7 decreased in extension and increased in flexion. For the measured area, height, and width only C5–C6 and C6–C7 did not change for area and height, respectively (Mao et al. 2015).

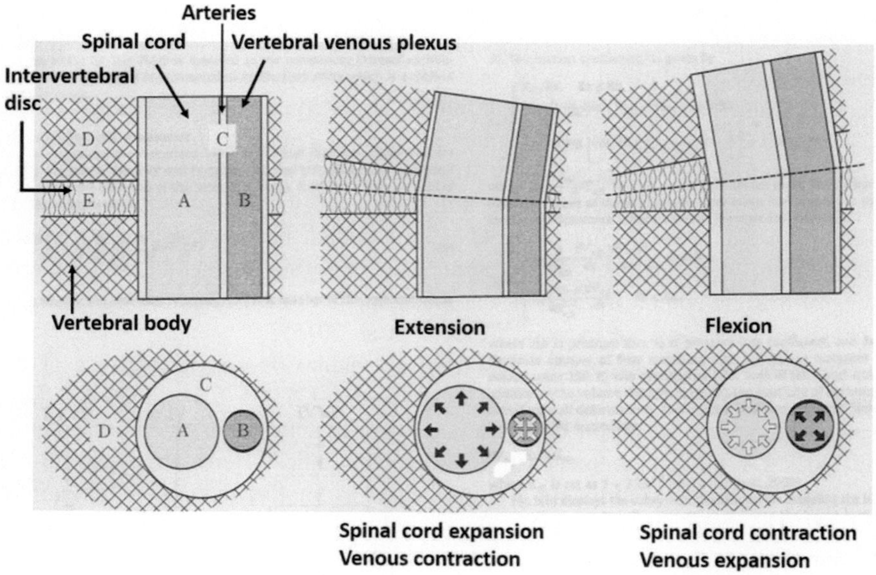

Fig. 10.9 Pressure changes during extension and flexion of the vertebral column within the spinal canal during whiplash. Adapted from Yao et al. 2016 with permission

10.5 Motion of the Cervical Spine

The whiplash studies showed that extravagant movements of the cervical spine are hardly explainable by researchers. This is presumably also due to the complicated research on the cervical spine motion. Early studies on the anatomy of the neck vertebrae in general say that each vertebra contributes to all movements, the more horizontal direction of the joints, compared to the thoracic ones, induces more indifferent motion types. During flexion, the higher vertebrae take a position in front of those below them producing an inverse stair topography, overhanging somewhat the intervertebral discs. During motion, this position inhibits partly the flexion and lateral flexion is reduced by the transverse processes (Braus 1929).

Biomechanically, the neck vertebrae are subdivided into segments. A segment is a unit consisting of two bony parts, the vertebrae, and their flexible connections, e.g. C5–C6. Others call a segment: Bewegungssegment, vertebral unit or basic vertebral unit (Van Mamaeren 1988, and references herein). Consequently, the neck spine is a chain of motion of seven segments. It is contested whether segment C7-Th1 belongs to the thoracic or cervical spine. Movement is possible between the adjacent segments, called segmental motion. However, it is also possible that motion is present between nonadjacent vertebrae, called motion within a block, e.g. C5–C7. Motion between skull and the seventh cervical spine is motion of the entire cervical spine. In general, cervical spine main relations to the separate segments or blocks are difficult to establish.

Studying the cervical spine a large intra- and intervariability is present. The inter variability in segmental range of motion is dependent of the position the person is in, i.e. lying or sitting position and the researcher's approach. Except for C2–C3 and C3–C4, the measured differences in motion between the results for anteflexion and those for retroflexion can be more than twice up to twelve more within a person. Thus, the range of segmental motion is not equal in all segments. C0–C1, C5–C6 and C6–C7 show a different range of motion in anteflexion and retroflexion, especially if performed immediately after each other (Van Mamaeren 1988). This has been confirmed recently: "Cervical motion segment contributions to head motion change over the full range of motion cannot be accurately characterized solely from endpoint data. The continuously changing segmental contributions suggest that the compressive and shear loads applied to each motion segment also change over the range of motion" (Anderst et al. 2015).

Nevertheless, strong ranges of motion for anteflexion and retroflexion of the neck can be noticed in C0–C1, C1–C2, and in C4–C5, C5–C6 and C6–C7. Less range of motion is found for C2–C3 and C3–C4. In anteflexion, the motion start is by block C4–C7 and segment C6–C7 starts this block. Hereafter block C0–C2 produces the largest contribution followed again by another motion contribution of C4–C7. The contribution C2–C4 is always in between the contributions of C5–C7. In retroflexion, the motion starts by contribution of C4–C7 followed by C0–C2 (Van Mamaeren 1988, analogous but somewhat different results, especially C1–C2, have been described by Anderst et al. 2015). Extracting a trend in these ranges of motion is rather dangerous. Nevertheless, our conclusion is that the motion ranges indicate strong contribution of the upper and lower cervical spine and less of the middle cervical spine, or upper and lower cervical spine contain a large variability during anteflexion and retroflexion, and a more stable action is produced by the middle cervical spine. Such a tripartition or regionalization of the cervical spine is also favoured in dogs and extended to mammalian species (Arnold et al. 2016). In other words, motion adaptation and execution are larger in upper and lower cervical spine than in its middle segment and that is why it is so difficult to catch it in numbers.

Buckling effect even increases the prediction difficulty of motion in the cervical spinal cord. First-order axial loading forces the extension of the vertebrae C1–C5 and flexion through T1. Second-order buckling produces flexion in C1–C3 and extension in C4 and C5 and again flexion in C6 through T1. Loading the cervical vertebrae places the motion segments easily in an extreme position. Something the athletic trainer should understand: "motion at the head is not a reliable indicator of the motion at various levels in the spine" (Swartz et al. 2005): injuries of the neck are waiting during training.

The following citation is from Strimpakos (2011): "It is obvious from the above that despite the widespread clinical use of cervical range of motion (ROM) and proprioception assessment the scientific literature reflects little agreement regarding the instruments as well as the methodologies utilized. The anatomical, biomechanical and physiological considerations in the assessment of physical parameters presented so far in this review have revealed the complicated nature of the cervical spine and the multiple issues a clinician or researcher have to take into account throughout its evaluation". One should note that emphasis is on the anteflexion and retroflexion of the cervical spine motion, leaving rotation and lateral flexion out (see, e.g., Lin et al. 2014). Nevertheless, in our interpretation cervical spine research hampered to give a sound basis for the motion function of the cervical spine and its segments or blocks in anteflexion and retroflexion, which is due to the complicated motion patterns present in the cervical spine.

10.6 Motion of Head and Neck in the Newborn Infant

The newborn baby shows in the first few hours of live typical motions of head and neck, but after several hours postpartum, newborn's behaviour is changed and is more predictable and less labile. These movements are developed during the foetal phase into a more constant pattern (Minkowski 1921). If the newborn infant is placed on its back, supine, it shows the typical flexion posture with the whole back on the underground and the head also slightly flexed. Limbs are semiflexed and legs are in slight abduction at the hips. If the head is to one side, arm and leg of that side are extended (so-called Fechterstellung or fencing reflex, official name is asymmetrical tonic neck reflex). Motility of the newborn is positively correlated with muscle tone (tonus). One should be aware of the fact that normal tone has only been found in around 80% of the newborns, around 5% shows hypertonicity and 15% hypotonicity. Normal motility is present in 65% of the newborns, 20% demonstrated hypermotility and around 15% had hypomotility (Willemse 1961). Not all newborns are diagnosed in the normal range for both tonus and motility and at the start of the research, all newborns studied belonged to a normal population of babies (Willemse 1961). Thus, variability within the normal population for tonus of the muscles and for motility is present: high tonus means strong motility and vice versa. Hypotonicity and/or hypomotility are sometimes the indication for serious syndromes.

Prechtl (1977) (co)founded the way of reflex testing in the newborn. From the 26 reflexes, tested 50% are related to head and neck. Resistance to movement tests involved the neck in 12.5% and in prone position tests, 33% were related to the head. Head and neck are important in testing neurological behaviour of the newborn infant.

By shaking the rump of the newborn in supine position one notices that the newborn's head moves into all directions (called ballottement), but not the extremities. In grown-ups shaking does induce extremity movements, but the head keeps steady (Willemse 1961). The craning of the neck muscles during ballottement gives nearly no reaction, but passive anteflexion of the head in the supine position produces a resistance, that suddenly disappears and the head goes in total in anteflexion. This is known as the clasp knife phenomenon. This neck stiffness is transitory, but is permanent in meningitis (see Sect. 10.1.2). Willemse (1929–2012) studied this anteflexion clasp knife phenomenon in newborns: 87% showed the phenomenon, and in 13%, it was absent. The 87% had normal to high tone of the muscles. With a low tone, nearly no resistance was noticed. Within a few days after birth, the neck muscles are strong enough to raise the head. It is the development of the labyrinthine righting reflex after the first few days that determine the extension of the head (Willemse 1961).

In the nineteen seventies and eighties, neonatal reflexes were studied with the intention to understand lateral asymmetries. Psychological literature opposes the medical results. Psychologists considered asymmetry of the reflex motility a consequence of the neurobehavioural organization of the newborn. Parental handedness was found to be related to lateral differences in neonatal head–neck turning and was caused by the dominant left hemisphere and is genetically based. Absence of asymmetries should indicate neurobehavioural dysfunction. For example, the asymmetry in the rooting reflex (touching the corner of the mouth, upper and lower lip, turning the head towards a stimulus, retroflexion of the head and trying to suck the stimulating finger) has shown that newborns with a high Apgar score (quality test for Appearance, Pulse, Grimace (stimulus reactions), Activity (muscle tone) and Respiration (one minute after birth) rooted more to the right than to the left. Premature infants demonstrate weak rightward head turning and no asymmetric rooting. It is supposed to be related to a higher incidence of neurobehavioural dysfunction (see Trehub et al. 1983 and references therein).

The clinician considers asymmetries in reflexes, motility or posture of the newborn an indication for neurological dysfunction. The opisthotonic posture (newborn with head retroflexed and both lower limbs extended) and head turning constantly to one side and an asymmetrical position are considered pathological (see Prechtl 1977). Moreover, Trehub et al. (1983) restudied the psychological results, and their conclusion was: "Our findings with respect to lateral asymmetries can be summarized as follows. We found marginal, largely nonsignificant differences favouring rightward responses in the overall incidence of only 2 of the 10 responses evaluated: spontaneous head turning and rooting". A polite way of saying asymmetries are not found.

Drug abuse of the mother does have influence on the newborn infant. Let us start with some information on pregnancies of teenagers in the USA and in Brazil. In the USA, 900,000 teenagers (15–19 years) become pregnant (3% of ±30 mil teenagers) and in Brazil 703662 (4% ±17 mil). By the way, birth per 1000 teenage females are 4 births in Japan, 4 in the Netherlands, 10 in Canada, 15 in the UK, 24 in the USA and 67 in Brazil for the years 2011–2015 (The World Bank, IBRD-IDA). Within the teenage group, drug abuse especially in Brazil is large (data unknown) but in the USA 2.1% of pregnant women use marijuana. From three large studies: Ottawa Prenatal Prospective Study, Maternal Health Practices and Child Development Study, and the Generation R study, it is known that foetal growth is retarded with 150 g in mothers who used the first two month cannabis. After constantly using cannabis, the foetal weight is reduced with 277 g, the head circumference is clearly less, and the aorta has a smaller inner diameter. Compared to tobacco, weight reduction is larger from the use of cannabis. The Ottawa study could not find the birth weight differences, but gestational age is lagging behind. Both the Ottawa and the Maternal health studies found: at 6 years more impulsivity and hyperactivity 9–12 years and impaired visuo-perceptual functioning in the Ottawa study and at 6 years, more impulsivity, hyperactivity, delinquency, at 10 years, more problems in abstract and visual reasoning in the Maternal one. Generation R started later and still has not reached this children's age (Huizink 2014). Thus, early effects during pregnancy and directly after birth are present, but the biggest impact is at older age.

References

Afsharpaiman S, Saburi A, Waters KA (2013) Respiratory difficulties and breathing disorders in achondroplasia. Pediatr Resp Rev 14:250–255

Anderst WJ, Donaldson WF, Lee JY, Kang JD (2015) Cervical motion segment contributions to head motion during flexion\extension, lateral bending, and axial rotation. Spine J 15:2538–2543

Aoyama H, Asamoto K (2000) The developmental fate of the rostral/caudal half of a somite for vertebra and rib formation: experimental conformation of the resegmentation theory using chick-quail chimeras. Mechan Dev 99:71–82

Arnold P, Forterre F, Lang J, Fischer MS (2016) Morphological disparity, conservatism, and integration in the canine lower cervical spine: Insights into mammalian neck function and regionalization. Mammal Biol 81:153–162

Baljet B (2000) The painted Amsterdam anatomy lessons: anatomy performances in dissecting rooms? Ann Anat 182:3–11
Bell KM, Yan Y, Debski RE, Sowa GA, Kang JD, Tashman S (2016) Influence of varying compressive loading methods on physiologic motion patterns in the cervical spine. J Biomech 49:167–172
Bernsmeier A, Schrader A, Struppler A (1984) Differentialdiagnose neurologischer Krankheitsbilder, 4th edn. Georg Thieme Verlag, Stuttgart, NY
Bolk L (1918) Hersenen en cultuur. Scheltema & Holkema's Uitgeverij, Amsterdam
Bolk L (1929) Origin of racial characteristics in man. Am J Phys Anthropol 13:1–28
Braus H (1929) Anatomie des Menschen, vol I. Bewegungsapparat. J, Springer, Berlin
Brigandt I (2006) Homology and heterochrony: the evolutionary embryologist Gavin Rylands de Beer (1899–1972). J Exp Zool 306B:317–328
Brouwer PA, Lubout CM, van Dijk JM, Vleggeert-Lankamp CL (2012) Cervical high-intensity intramedullary lesions in achondroplasia: Aetiology, prevalence and clinical relevance. Eur Radiol. https://doi.org/10.1007/s00330-012-2488-0
Brumovsky PR, Gebhart GF (2010) Visceral organ cross-sensitization- an integrated perspective. Auton Neurosci Basic Clin 153:106–115
Cain CMJ, Ryan GA, Fraser R, Potter G, McLean AJ, McCaul K, Simpson DA (1989) Cervical spine injuries in road traffic crashes in South Australia, 1981–1986. Austr NZ J Surg 59:15–19
Cairns BE, Arendt-Nielsen L, Sacerdote P (2015) Perspectives in pain research 2014: neuroinflammation and glial cell activation: the cause of transition from acute to chronic pain? Scand J Pain 6:3–6
Christ B, Huang R, Scaal M (2004) Formation and differentiation of the avian sclerotome. Anat Embryol 208:333–350
Christ B, Wilting J (1992) From somites to vertebral column. Ann Anat 174:23–32
Coin CG, Malkasian DR (1971) Clivus. Radiol of the skull brain I:348
van Dijk JMC, Lubout CMA, Brouwer PA (2007) Cervical high-intensity intramedullary lesions without spinal cord compression in achondroplasia. J Neurosurg Spine 6:304–308
Ednik M, Tinkle BT, Phromchairak J, Egelhoff J, Amin R, Simakajornboon N (2009) Sleep related respiratory abnormalities and arousal pattern in achondroplasia in early infancy. J Pediatr 155:510–515
Fouquet B, Doury-Planchout F (2013) Whiplash: actualités Revue rhumat monograph 80:67–71
van Hee R, Lowis S (2006) David Van Mauden (±1538–±1597), "Sworn medical Doctor and surgical Prelector of Antwerp", and his Book on Anatomy. Acta Chir Belg 106:130–135
Hecht JT, Nelson FW, Butler IJ, Horton WA, Scott CI jr, Wassman ER, Mehringer CM, Rimoin DL, Pauli RM (1985) Computerized tomography of the foramen magnum: achondroplastic values compared to normal standards. AM J Med Genet 20:355–360
Hecht JT, Francomano CA, Horton WA, Annegers JF (1987) Mortality in achondroplasia. Am J Hum Genet 41:454–464
Hecht JT, Horton WA, Reid CS, Pyeritz RE, Chakraborty R (1989) Growth of the foramen magnum in achondroplasia. Am J Med Genet 32:528–535
His W (1886) Untersuchungen über die erste Anlage des Wirbeltierleibes. Die erste Entwicklung des Hünnchens Ei, Vogel, Leipzig
Hlavenka TM, Christner VFK, Gregory DE (2017) Neck posture during lifting and its effect on trunk muscle activation and lumbar spine posture. Appl Ergonom 62:28–33
Horton WA, Hall JG, Hecht JT (2007) Achondroplasia. Lancet 370:162–172
Huizink AC (2014) Prenatal cannabis exposure and infant outcomes: Overview of studies. Progr Neuro-Psychopharmacol Biol Psych 52:45–52
Klag KA (2015) Horton WA (2015) Advances in treatment of achondroplasia and osteoarthritis. Hum Mol Genet. https://doi.org/10.1093/hmg/ddv419
Kranenburg HA, Schmitt MA, Puentedura EJ, Luijckx GJ (2017) Adverse events associated with the use of cervical spine manipulation or mobilization and patients characteristics: a systematic review. Musculoskelet Sci Pract 28:32–38

Kölliker A (1859) Ueber die Beziehungen der Chorda dorsalis zur Bildung der Wirbel der Selachier und einiger anderer Fische. Verh Phys Med Ges Würzburg X:193-242

Latini G, De Felice G, Parrini S, Verrotti M, Di Maggio G, Petraglia F (2002) Polyhydramnios: a predictor of severe growth impairment in achondroplasia. J Pediatr 141:274–276

Leijten JC, Bos SD, Landman EB, Georgi N, Jahr H, Meulenbelt I, Post JN, van Blitterswijk CA, Karperien M (2013) GREM1, FRZB and DKK1 mRNA levels correlate with osteoarthritis and are regulated by osteoarthritis-associated factors. Arthritis Res Ther 19. https://doi.org/10.1186/ar4306

Leijten JC, Emons J, Sticht C, van Gool S, Decker E, Uitterlinden A, Rappold G, Hofman A, Rivadeneira F, Scherjon S, Wit JM, van Meurs J, van Blitterswijk CA, Karperien M (2012a) Gremlin 1, frizzled-related protein, and Dkk-1 are key regulators of human articular cartilage homeostasis. Arthritis Rheum 64:3302–3312. https://doi.org/10.1002/art.34535

Leijten JC, Moreira Teixeira LS, Landman EB, van Blitterswijk CA, Karperien M (2012b) Hypoxia inhibits hypertrophic differentiation and endochondral ossification in explanted tibiae. PLoS ONE 7:e49896. https://doi.org/10.1371/journal.pone.0049896

Leroi AM (2005) Mutants. On genetic variety and the human body. Penguin Books, London

Lieberman DE (2011) The evolution of the human head. Belknap Press, Harvard University Press, London

Lin C-C, Lu T-W, Wang T-M, Hsu C-Y, Hsu S-J, Shih T-F (2014) In vivo three-dimensional intervertebral kinematics of the subaxial cervical spine during seated axial rotation and lateral bending via a fluoroscopy-to-CT registration approach. J Biomech 47:3310–3317

Marani E, Koch WFRM (2014) The Pelvis, structure, gender and society. Springer, Heidelberg

Marani E (2016) Dorsal root or spinal ganglion neurosci biobehav psychol: NRSM 04160

Mao H, Driscoll SJ, Li J-S, Li G, Wood KB, Cha TD (2015) Dimensional changes of the neuroforamina in subaxial cervical spine during in vivo dynamic flexion-extension. https://doi.org/doi.org/10.1016/j.spinee.2015.11.052

Mauden van D (1647) Bedienighe der anatomien. Amsterdam Danckertsz, Folio's 53–54 (first edition 1583 by Christophe Plantin)

Mazzara JT, Fielding JW (1988) Effect of C1–C2 rotation on canal size. Clin Orthop Relat Res 237:115–119

McKenzie JA, Williams JF (1971) The dynamic behavior of the head and cervical spine during whiplash. J Biomech 4:447–490

McRay DL (1960) The significance of abnormalities in the cervical spine. Am J Roentgenol 84:3–25

Middelkoop N (1994a) Rembrandt's anatomische les van Dr. Deijman. Ned Tijdschr Geneesk 138:2614–2618

Middelkoop N (1994b) De anatomische les van Dr.Deijman. Amsterdams Historisch Museum (ISBN 90-71361-06-3)

Minkowski M (1921) Sur les mouvements, les réflexes et les réactions musculaires du fœtus humain de 2 à 5 mois et leur relations avec le système nerveux fœtal. Revue Neurol 37(1105–1118):1235–1250

Mongula JE (2007) Bepaling van de stijfheden van de wervelkolom en romp. Master assignment: BW-229/BMT-021, University Twente

Moosa S, Wollnik B (2015) Altered FGRF signalling in congenital craniofacial and skeletal disorders. Semin Cell Dev Biol. https://doi.org/10.1016/j.semcdb.2015.12.005

Pang D, Thompson DNP (2014) Embryology, classification and surgical management of bony malformations of the cranio-vertebral junction. In: Di Rocco C, Aklan N (eds) Pediatric cranio-vertebral junction diseases. Advances in technical standards in neurosurgery, vol 40. Springer Cham, Heidelberg

Panjabi MM (2006) A hypothesis of chronic back pain: ligament sub-failure injuries lead to muscle control dysfunction. Eur Spine J 15:668–676

Poppel MH, Jacobson HG, Duff BK (1953) Basilar impression and platybasia in Paget's disease. Radiol 61:639–644

Portero R, Quaine F, Cahouet V, Thoumie P (2013) Musculo-tendinous stiffness of head-neck segment in the sagittal plane: an optimization approach for modelling the cervical spine as a single-joint system. J Biomech 46:925–930
Prechtl H (1977) The neurological examination of the full-term newborn infant. Clinics in developmental medicine, 2nd ed, no 63, pp 1–68
Reina V, Baujat G, Fauroux B, Couloigner V, Boulanger E, Sainte-Rose C, Maroteaux P, Le Merrer M, Cormier-Daire V, Legai-Mallet L, Zerah M, Di Rocco F (2014) Craniovertebral junction anomalies in achondroplastic children. In: Di Rocco C and Akalan N (eds) Pediatric craniovertebral junction diseases. Adv technical Standards Neurosurgery, vol 40. 295–312
Remak R (1855) Untersuchungen über die Entwicklung der Wirbeltiere. Reimer, Berlin
Romer AS (1962) The vertebrate body. W.B.Saunders Co, Philadelphia, London
Rössler G, Schmeisser A (1957) Die alternsabhängige Beweglichkeitsverminderung der Halswirbelsäule. Z Alternforsch 11:75–80
Rothman RH, Simeone FA (1982) The Spine. W.B.Saunders Co., Philadelphia
Scaal M (2016) Early development of the vertebral column. Siminars Cell Dev Biol 49:83–91
Schivo S, Post JN (2016) Computational modelling of complex protein activity networks. In: Prigent C (ed) Protein phosphorylation. Biochemistry, genetics and molecular biology. https://doi.org/10.5772/intechopen.69804
Schivo S, Scholma J, van der Vet PE, Karperien M, Post JN, van de Pol J, Langerak R (2016a) Modelling with ANIMO: between fuzzy logic and differential equations. BMC Systems Biology/BMC series. https://doi.org/10.1186/s12918-016-0286-z
Schivo S, Scholma J, Huang X, Zhong L, van de Pol J, Karperien M, Langerak R, Post JN (2016b) An echo in biology II: Insights in chondrocyte cell fate. Abstr Osteoarthr Cartil 24:S176
Scholma J, Schivo S, Karperien M, Langerak L, van de Pol J, Post JN (2014) An echo in biology: validating the executable chondrocyte. Abstr Osteoarthr Cartil 22:S157
Seacrist T, Saffioti J, Balasubramanian S, Kadlowec J, Sterner R, García-España JF, Arbogast KB, Maltese MR (2012) Passive cervical spine flexion: the effect of age and gender. Clin Biomech 27:326–333
Siegmund GP, Winkelstein BA, Ivancic PC, Svensson MY, Vasavada A (2009) The anatomy and biomechanics of acute and chronic whiplash injury. Traffic Inj. Prev. 10:101–112
Six J (1905) De ligging van het lijk in Rembrandt's anatomische les van Dr. Deijman. Oud Holland 23(1):37–40
Skillgate E, Côte´P, Cassidy JD, Boyle E, Carroll L, Holm LW (2016) Effect of early intensive care on recovery from whiplash-associated disorders: Results of a population-based cohort study. Arch Physic Med Rehabil 97:739–46
Stempera BD, Yoganandana N, Pintar FA (2003) Gender dependent cervical spine segmental kinematics during whiplash. J Biomech 36:1281–1289
Strimpakos N (2011) The assessment of the cervical spine. Part I: range of motion and proprioception. J Bodyw Mov Ther 15:114–124
Styrke J, Stalnacke BM, Bylund PO, Sojka P, Bjornstig U (2012) A 10-year incidence of acute whiplash injuries after road traffic crashes in a defined population in northern Sweden. PMR 4:739–747
Svensson MY, Aldman B, Hansson HA, Lövsund P, Seeman T, Suneson A,Örtengren T (1993) Pressure effects in the spinal canal during whiplash extension motion a possible cause of injury to the cervical spinal ganglia. In: International Research Council on the Biomechanics of Injury. Eindhoven, Netherlands, pp 189200
Swartz EE, Floyd RT, Cendoma M (2005) Cervical Spine functional anatomy and the biomechanics of injury due to compressive loading. J Athlet Train 40:155–161
Takahashi Y, Yasuhiko Y, Takahashi J, Takada S, Johnson RL, Saga Y, Kanno J (2013) Metameric pattern of intervertebral disc/vertebral body is generated independently of Mesp2/Ripply-mediated rostro-caudal patterning of somites in the mouse embryo. Dev Biol 380:172–184
Thomeer RTWM (1982) Achondroplasia. Management of neurological complications. Thesis, University of Leiden

Trehub SE, Corter CM, Shosenberg N (1983) Neonatal reflexes:A search for lateral asymmetries. In: Young G et al (eds) Manual specialization and the developing brain. Academic Press, NY

Ueda K, Yamanaka Y, Harada D, Yamagami E, Tanaka H, Seino Y (2007) PTH has the potential to rescue disturbed bone growth in achondroplasia. Bone 41:13–18

Ulbrich EJ, Anon J, Hodler J, Zimmermann H, Sturzenegger M, Anderson SE, Boesch C (2014) Does normalized signal intensity of cervical discs on T2 weighted MRI images change in whiplash patients? Injury, Int J Care Injured 45:784–791

Van der Meer J, van 't Laar (2001) Anamnese en lichamelijk onderzoek. Elsevier Gezondheidszorg, Maarssen

Van Mamaeren H (1988) Motion patterns in the cervical spine. Thesis University Maastricht

Verbout AJ (1985) The development of the vertebral column. Adv Anat Embryol Cell Biol 90:1–122

Verbout AJ (1976) A critical review of the "Neugliederung" concept in relation to the development of the vertebral column. Acta Biotheor [Leiden] 25:219–258

Wang H, Rosenbaum AE, Reid CS, Zinreich SJ, Pyeritz RE (1987) Pediatric patients with achondroplasia : CT evaluation of the craniocervical junction. Radiology 164:515–519

Waters-Rits AL, Hoogland MLP (2013) Osteological evidence of short-limbed dwarfism in a nineteenth century Dutch family: achondroplasia or hypochondroplasia. Int J Paleont 3:243–256

Willemse J (1961) De motoriek van de pasgeborene in de eerste levensuren. Uitg. Erven J Bijleveld, Utrecht

Wong JJ, Shearer HM, Mior S, Jacobs C, Côté P, Randhawa K, Yu H, Southerst D, Varatharajan S, Sutton D, van der Velde G, Carroll LJ, Ameis A, Ammendolia C, Brison R, Nordin M, Stupar M, Taylor-Vaisey A (2015) Are manual therapies, passive physical modalities, or acupuncture effective for the management of patients with whiplash-associated disorders or neck pain and associated disorders? An update of the Bone and Joint Decade Task Force on Neck Pain and its Associated Disorders by the OPTIMa Collaboration. SPINEE 56530. https://doi.org/10.1016/j.spinee.2015.08.024

Yeung E (1996) Whiplash injuries of the cervical spine. The relationship between mechanism of injury and neural tissue involvement. Physiotherapy 82:286–290

Yao H-D, Svensson MY, Nilsson H (2016) Transient pressure changes in the vertebral canal durin whiplash motion. A hydrodynamic modelling approach. J Biomech 49:416–422

Yoganandan N, Haffner M, Maiman DJ, Nichols H. Pinter FA, Jentzen J, Weinshei SS, Larson SJ, Sances A (1989) Epidemiology and injury biomechanics of motor vehicle related trauma to the human spine. In: Proceedings of Stapp Car Crash Conference, Society of Automotive Engineers, Warrendale USA pp 223–242

Yoganandan N, Umale S, Stemper B, Snyder B (2017) Fatigue responses of the human cervical spine intervertebral discs. J mech Behav Biomed Mat 69:30–38

Chapter 11
Neurologic and the Cervical Spinal Cord: Nerve, Plexus and Cord Regeneration

Currently, few residents are adequately trained in brachial plexus surgery, and fewer still become interested enough to pursue it. Thus, a focus on education is critical for the future
(Spinner and Shin 2004)

Using stab wounds and diving accidents, cervical spinal cord injuries are quantified and expenses of spinal cord injuries are given for the USA. The first main focus is on spinal cord structure, peripheral nerve fibres and muscle types. The history of spinal cord regeneration studies ends with modern spinal cord regeneration, which is hardly successful. Therefore, the realization of regeneration of nerves and of brachial plexuses is described together with the use of synthetic nerve grafts. This chapter ends with spinal cord stimulation for pain suppression.

11.1 Stab Wounds and Diving Accidents

In the introduction of this book, the story of the son of a craftsman of six or seven years of age was told. He was lifted up by the neighbour and grew refractory. The angry father threw a sharp Sadler's hammer into the pit of the neighbour's neck where it cut the marrow of the spine, whereof the neighbour expired. An analogous scissors stab wound was presented in an eight-year-old boy, due to the throw of the scissors as a dart by his little brother while playing (Zang and Yang 2016). It produced a Brown-Séquard-plus syndrome that ended happily in a complete recovery, due to the initial mild laceration of the spinal cord and ipsilateral spinal cord functional compensation. The Brown-Séquard syndrome consists of ipsilateral upper motor neuron paralysis (hemiplegia) and loss of proprioception with contralateral pain and temperature sensation deficits, mostly due to half-sided spinal cord damage. Several case reports are published in the literature on cervical spinal cord injury (e.g. Voelker et al. 2015) and separately on cervical Brown-Séquard syndrome (see Pouw et al. 2010), but combined reviews are restricted available.

E. Marani and C. Heida, *Head and Neck*,
https://doi.org/10.1007/978-3-319-92105-1_11

The most cited *stab wound* paper is an overview by Peacock et al. (1977) of 450 cases of spinal cord injury in South Africa. "The most commonly reported site for a stab wound is the thoracic spine, followed by the cervical and lumbar spines" (Zang and Yang 2016). A recent report on 51 spinal cord patients, also from South Africa, in which the knife or knife blade was still present in the patient's body showed that the blades were located cervical (18%), thoracic (74%), lumbar (4%) and sacral (4%) (Enicker et al. 2015). Both the Peacock overview and Enicker study concluded that the cervical spinal cord injury is the second in the row of stab wounds. It is a male problem in South Africa, since 88% concerns males and only 12% females, of which 4% is still caused by intimate partner violence. "Penetrating injuries commonly occur as a result of interpersonal violence which is a major contributor to the incidence of trauma in the Province of KwaZulu-Natal, South Africa, with the World Health Organization (WHO) identifying interpersonal violence as one of the leading causes of morbidity and mortality in young adults, and males in particular" (Enicker et al. 2015). Since the penetrating neck injuries are rare in Western countries, assessment and management can be problematic in trauma centres without a standardized protocol (Pakarinen et al. 2006).

Cervical spinal injuries due to diving bring about severe personal damage. Of all spinal cord injuries, 6–7% are caused due to diving. The following overview is a hotchpotch from the publications of Aito et al. (2005), Chan-Seng et al. (2013), Korres et al. (2006) and Borius et al. (2010) from France, Greece and Italy. These are the countries with a high amount of swimming pools, lakes, beaches and small rivers and many tourists. Diving in water that is less than twice your length can be dangerous, not to mention a depth of 1–1.5 m. In nearly 90%, others say 95%, of the cases, it concerns young healthy male amateur divers. The median age is estimated between 15 and 19 years in one article and in another 22 years. The use of alcohol or drugs is present in around 43% of the cases. 88% of the diving incidents occur in the summer months (from April to August). The cervical spinal cord injuries are found for 25% at the C1–C3 level, nearly for 50% at the C4–C5 and 1/3 for the C6–C7 level in diving accidents. As various publications say that mostly the lower cervical spine is the part involved in the injuries, the results for Italy confirm neurological motor deficits at C6 (40% of the studied cases) and vertebral injury at C5. Severity is unpredictable with estimated Asian A, B, C less than Asian D patients in France (see Sect. 11.1.1 for Asian explanation). These numbers are for Italy 54% patients with Asian A, 25% Asian B, 11% Asian C and 10% Asian D, thus a substantially higher percentage in the serious damage range. Diving accidents account for 20–40% of the neurological deficits of spinal injuries. In several patient cases with Asian A, recovery is nearly total, and others do not change in their severity, which holds up to Asian D, which can also ameliorate or stay unchanged. Death is not excluded in 2% of the cases.

To summarize: cervical spinal cord injuries in the cited literature are less frequently encountered than thoracic ones in stab wounds and diving. Nevertheless, a large group of (young) male patients acquire serious vertebral and spinal cord damage due to diving and stab wounds. Prediction of the result of treatment is difficult. One should note that spinal cord lesions by brute force of the back or the vertebral column can be present without noticing or even the presence of damage of ligaments, vertebrae or intervertebral discs (Mumenthaler 1973). Therefore, the incidence of spinal cord injury is difficult to determine (see part 11.1.3).

11.1.1 Explanation of Used Terminology

The spinal cord lesion terminology can be confusing:

Differences between neurological and rehabilitation definitions of spinal cord injury levels. "Doctors use two different definitions for spinal cord injury levels. Given the same neurological examination and findings, neurologists and physiatrists may not assign the same spinal cord injury level. In general, neurologists define the level of injury as the first spinal segmental level that shows abnormal neurological loss. Thus, for example, if a person has loss of biceps, the motor level of the injury is often said to be C4. In contrast, physiatrists or rehabilitation doctors tend to define level of injury as the lowest spinal segmental level that is normal. Thus, if a patient has normal C3 sensations and absent C4 sensation, a physiatrist would say the sensory level is C3, whereas a neurologist or neurosurgeon would call it a C4 injury level. Most orthopaedic surgeons tend to refer to the bony level of injury as the level of injury" (Young 2002).

American Spinal Injury Association (ASIA) impairment scale:

A = Complete: no motor or sensory function is preserved at the sacral segments S4–S5

B = Incomplete: sensory but not motor function is preserved below the neurological level and includes the segments S4–S5

C = Incomplete: motor function is preserved below the neurological level and more than half key muscles below the neurological level have a muscle grade less than 3

D = Incomplete: motor function is preserved below the neurological level, and at least half of the key muscles below the neurological level have a muscle grade of 3 or more

E = Normal: motor and sensory function are normal (Young 2002).

Key muscles:

C5, elbow flexors; C6, wrist extensors; C7, elbow extensors; C8, finger extensors (distal phalanx of middle finger); T1, finger abductors (little finger) (Young 2002).

Muscle strength:

0 = total paralysis; 1 = palpable or visual contraction; 2 = active movement, gravity eliminated; 3 = active movement against gravity; 4 = active movement against some resistance; 5 = active movement against full resistance; NT not testable (Young 2002).

11.1.2 Spinal Cord Structure and Organization

The classical structural organization of the spinal cord has been studied by its cytoarchitecture and by the ascending and descending fibre systems. The cytoarchitecture that has been frequently cited is Rexed's subdivision based on the Nissl colouration of spinal cord neurons on transversal sections (Rexed 1964, Fig. 11.1e). The division consists of seven layers (I–VII) from dorsal towards ventral and three neuron groups (also called layers) numbered VIII (commissural neurons and small motor neurons), IX (motor neurons) and X (neurons localized around the central canal and in the spinal cord commissures).

A structural organization that has been partly falling into oblivion is the motor pool approach by Romanes (1964). Van Gehuchten and Nelis (1899) and De Neef (1901) have brought the idea of motor pools forward. The lamina's VII–IX of Rexed are subdivided in motoneuron groups related to a single muscle or a functional muscle group that proceed over several spinal cord segments, the so-called motor columns. Large, medium, and small neurons within the columns are discerned. Medium and small neurons are scattered among the large motoneurons that

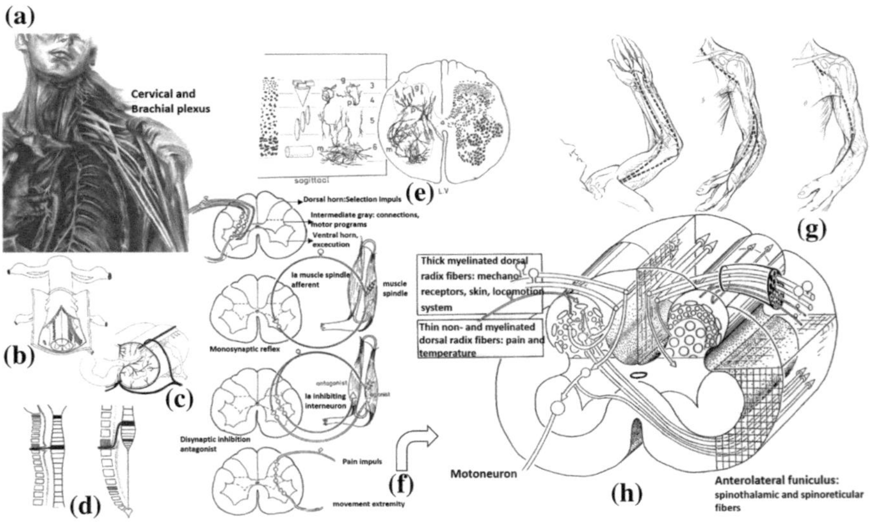

Fig. 11.1 Short review of the spinal cord structures and cord organization. **a** Ventral view of neck plexuses; **b** Relation dura mater, arachnoidea, pia mater and radiculi; **c** Blood vessel organization spinal cord; **d** Ascend of the spinal cord, nearly absent cervical, large at the conus medullaris (note filum terminale); **e** Spatial configuration of spinal cord neurons in longitudinal spatial and transverse sections; **f** Steering agonist and antagonist; **g** Trajectories of peripheral nerves of the brachial plexus. Left to right, n. medianus together with n. ulnaris, n. radialis and n.musculocutaneus. H, Interaction between **a** (δ, thick) and **c** (thin) fibres for the constitution of the spinothalamic (pain) tract. (A crayon drawing by miss Sofie van der Berg, dissection Willem Reychard, with permission Dept. Anatomy and Embryology, Leiden; **g** courtesy J.M.F. Landsmeer (†), all other figures with courtesy and permission of J. Voogd)

make the main criterion for the motocolumns. Due to the widespread of the dendrites of the motor columns, the cytological borders are always overstepped (Fig. 11.1e). Dorsal root, but also ventral root, fibre information has been subdivided in A, B and C types: A, large, thick myelinated fibres; B, medium to thin myelinated fibres, mainly belonging to the autonomous motor system; and C, exclusively thin myelinated or non-myelinated fibres. A-fibres are further generally divided in Aα (from muscle spindle/skin), Aβ (from skin), Aγ (joints) and Aδ (exteroception). The pain system has been connected with the interaction of Aδ and C fibres (Fig. 11.1f, h). In Table 11.1, the relation between the A, B, C fibres their diameter, action potential velocity and the relation with the physiological terminology (I–IV) is explained (Marani and Lakke 2014).

The large, long fibre systems are subdivided in ascending and descending systems to and from the cortex and brainstem. Descending systems are the cortical and upper brainstem ones: the pyramidal tract with its uncrossed and crossed parts (Schoen 1964; Kuypers 1964), the rubrospinal tract (Schoen 1964; Lakke and Marani 1991, *but see below*), involved in hand and foot steering. The interstitiospinal, tectospinal, vestibulo and reticulo spinal descending systems (grouped together as the fasciculus longitudinalis medialis and mostly located in the anterior and lateral funiculi, see Busch 1964) are important for posture and synergistic whole limb movements (Schoenen and Grant 2004). The long ascending systems concern the spinothalamic tract (Fig. 11.1h) and the dorsal funiculus, which continues, after the dorsal column nuclei, in the lemniscus medialis (Marani and Schoen 2005). The spinocerebellar fibres join from the ventral lateral funiculus into the mainstream at cervical levels. Here, the fibres are present in the dorsal part of the lateral funiculus (Marani and Schoen 2005). The smaller descending and ascending systems, e.g. spinoreticular and spino-olivary projections, can be found in Voogd (2004) and Marani and Schoen (2005), but are not discussed here. A special comment should be placed here: most publications concern experimental animal results, while these results are hardly supported by human results. The Schoen publications: Schoen (1964), Marani and Schoen (2005), Usunoff et al. (1997), and Voogd (2004) are all based on human results. By the way, man is not a rat. While the rat has a well-developed rubrospinal tract (Lakke and Marani 1991), humans “lack” a rubrospinal tract (Schoen 1964).

11.1.3 Epidemiology of Spinal Cord Injury

The USA has two main institutions that focus on spinal cord injury (SCI): “The Foundation for Spinal Cord Injury Prevention, Care and Cure (FSCIPCC) is a non-profit educational group dedicated to the prevention, care and cure of spinal cord injuries through public awareness, education and funding research. FSCIPCC is committed to improve the quality of care for persons with serious spinal cord injuries and to raising funds that support the search for a cure. FSCIPCC is comprised of victims of spinal cord injuries and their families, persons dedicated to the

Table 11.1 Nerve fibre and muscle types[α]

Aα-fibres; motoric and proprioceptive	Diam. (μm)	MCV (m/s)						
Somatic α-motor neuron	12–20	70–120						
DRG from annulospiral bag & chain muscle spindle (Ia sensory neurons)	12–20	70–120						
DRG from tendon organ (Ib sensory neurons)	12–20	70–120						
Aβ-fibres; proprioceptive and exteroceptive								
DRG from flower-spray chain muscle spindle (II sensory neurons)	5–12	30–70						
DRG from touch/pressure receptors in skin/synovia/ligaments (II sensory neurons)	5–12	30–70						
Myelinated:								
Aα; proprioception & somatic motor; 12–20 μm								
somatic motor								
proprioception; muscle spindle annulospiral		=Ia						
proprioception; Golgi tendon organ			=Ib					
Aβ; proprioception; 5–12 μm								
proprioception; muscle spindle flower spray		=II						
exteroception; touch & pressure			=II					
Aγ; motor to muscle spindle; 3–6 μm								
Aδ; exteroception; 2–5 μm								
exteroception; pain temperature (some) touch		=III						

(continued)

Table 11.1 (continued)

Aα-fibres; motoric and proprioceptive	Diam. (μm)	MCV (m/s)						
Unmyelinated								
B; pregangionic autonomic; <3 μm								
C dorsal root; exteroception; 0.4–1.2 μm								
C sympathetic; postganglionic sympathetic; 0.3–1.3 μm								
Summarized:								
Aα		myl	12–20 μm	Motor somatic muscle fibre				
Aα	Ia	myl	12–20 μm	Proprioception; muscle spindle annulospiral				
Aα	Ib	myl	12–20 μm	Proprioception; Golgi tendon organ				
Aβ	II	myl	5–12 μm	Proprioception; muscle spindle flower spray				
Aβ	II	myl	5–12 μm	Exteroception; touch & pressure				
Aγ		myl	3–6 μm	Motor somatic muscle spindle				
Aδ	III	myl	2–5 μm	Exteroception; pain temperature (some) touch				
B		unmyl	<3 μm	Motor autonomic preganglionic				

(continued)

Table 11.1 (continued)

Aα-fibres; motoric and proprioceptive	Diam. (μm)	MCV (m/s)						
C	IV	unmyl	0.4–1.2 μm	Exteroception; pain reflex responses				
C		unmyl	0.3–1.3 μm	Motor autonomic postganglionic sympathetic				
Muscle fibre types								
Type	Speed	Strength	Fatigability	ATPase	Colour	Motor unit	Motor neuron	Axon
I	Slow	Weak	Fatigue resistant	Low	Red	Slow	Small, very excitable	Small, fast
IIB	Fast	Strong	Fatigable	High	White	Fast, fatigable	Large, low excitable	Large, very fast
IIA	Fast	Intermediate	Fatigue resistant	High	Red	Fast, fatigue resistant	Large, low excitable	Large, very fast

[α]Note that these numbers are means from several publications [adapted from Marani and Lakke (2014) with permission] prevention, care and cure of SCI and professionals who provide free counsel to SCI victims" (http://www.fscip.org/ last visit on 13 February 2017). The second concerns NSCISC, The National Spinal Cord Injury Statistical Centre, which "supports and directs the collection, management and analysis of the world's largest and longest spinal cord injury research database" (see https://www.nscisc.uab.edu/ last visit 13 February 2017). NSCISC is part of the University of Alabama at Birmingham. Despite these institutions, Wyndaele and Wyndaele (2006) concluded that improved registration of SCI is a necessity and research with publication of the findings in many parts of the world is a must. Moreover, uniformity in methodology of spinal cord injury registration is missing, restricting comparison of data.

The 2016 data sheet of the NSCISC shows that 54 persons per million inhabitants of the USA do encounter spinal cord injury or nearly 17,000 new cases each year. This incidence relates to a prevalence of nearly 282,000 persons (range 243,000–347,000). The average age increased from 29 years in 1970s to 42 years currently. Males account for 80% of the new cases, mainly caused by vehicle accidents (38%), falls (30.5%) and acts of violence (13.5%, mainly gunshot wounds), together 82% of the spinal cord injuries. One-third concerns complete paraplegia and tetraplegia patients, and two-thirds are incomplete paraplegia and tetraplegia patients.

Knowing that spinal cord injury results are difficult to compare, Singh et al. (2014) have done a good effort, producing a meta-study "to identify relevant English literature published through June 2013". Within the USA, the highest incidence is found in Alaska (83 per million) and a good second is Mississippi (77 per million). The prevalence of spinal cord injury is the highest in the USA and the lowest in the Rhone-Alpes region in France together with Helsinki in Finland, and the USA prevalence is, of course, the highest in Alaska and Mississippi. It is remarkable that in several North America states spinal cord injury contribution is more related to sports than to acts of violence, in the USA, e.g. Utah 15.7% sports and 5.4% fire arms, and in Canada, British Columbia 17.9% sports and 1.5% violence.

Alaska has 43.9% ***cervical*** spinal cord injuries, Alabama 47.5%, British Columbia 49.7%, Manitoba 45–51.3% and Alberta at the top with 61.5% cervical spinal cord injuries. These cervical spinal cord injuries results are for Canada's London (Ontario) 75%, central Portugal region 51.2%, Taipei—Taiwan 46.8%, Russia's Novosibirsk 49%, Northwestern Kentucky and Southern Indiana 49.7%, Thessaloniki, Greece 48%, and Western Norway 52.4–57.8%. Few states and cities or regions studied by Singh et al. (2014) do have a lower percentage for cervical spinal cord injury, but the mainstream of this systematic search shows that cervical spinal cord injuries are more frequently reported locally than can be concluded from the more general studies. In a study on 122 children with spinal injuries, 60% were cervical ones, while "high" cervical injuries are more frequently encountered in younger children (Singh et al. 2011).

Spinal cord injuries have devastating effects: high cervical myelotomy gives death due to respiration insufficiency, and somewhat lower tetraparese or tetraplegia will occur. Thoracic damage of the spinal cord is characterized by complete or

Table 11.2 US expenses in 2015, in dollars, differentiated along the severity of the injury (with permission of **NSCISC**, see DeVivo 1997)

Severity of injury	Average yearly expenses (in 2015 dollars)		Estimated lifetime costs by age at injury (discounted at 2%)	
	First year	Each subsequent year	25 years old	50 years old
High tetraplegia (C1–C4) AIS ABC	$1,065,980	$185,111	$4,729,788	$2,599,411
Low tetraplegia (C5–C8) AIS ABC	$770,264	$113,557	$3,455,879	$2,125,674
Paraplegia AIS ABC	$519,520	$68,821	$2,312,846	$1,517,851
Motor functional at Any Level AIS D	$347,896	$42,256	$1,580,148	$1,115,312

Data Source: Economic Impact of SCI published in the journal *Topics in Spinal Cord Injury Rehabilitation*, Volume 16, Number 4, in 2011
ASIA Impairment Scale (AIS) is used to grade the severity of a person's neurological impairment following spinal cord injury

incomplete paraplegia. The direct costs are enormous for these US patients (see Table 11.2).

One should note that Table 11.2 gives only the direct costs. "The indirect costs such as losses in wages, fringe benefits, and productivity are averaged $72,047 per year, in 2015 in dollars" (NSCISC 2016 SCI data sheet). Total US ***direct*** costs per year are estimated 7.736 billion dollars (1997–2005), while lost productivity is calculated to be 5.5 billion. The amount of money that has to be spent to ***care and cure*** is a total of around 15 billion, thus indirect costs included (Berkowitz et al. 1998: note this figure is still without lost productivity). The spinal cord injury costs of patients justify fundamental and translational spinal cord research. This normally well-funded regeneration research can be carried out in specialized centres (e.g. The Miami project to cure paralysis) or at hospitals, medical faculties and universities. The USA now proposes a network of Spinal Cord Injury Research Centers of Excellence (Liverman et al. 2005) in which the existing centres of excellence (Miami project, Kentucky SCI Research Center, and Reeve-Irvin Research Center) will form the nodes.

11.2 History of Spinal Cord Regeneration

Let us start with the commentary of the Committee on Spinal Cord Injury of the Board on Neuroscience and Behavioral Health of the Institute of Medicine of the National Academies in their introduction to "Spinal cord injury, progress, promise and priorities": "Much remains to be learned about the basic biology of spinal cord injuries and the numerous potential therapeutic targets involved in the complex process of maintaining cell and tissue viability and promoting axonal growth and synaptic integrity that will result in improved and appropriate function in individuals with spinal cord injuries. Adding to the body of knowledge on neurological circuitry and mechanisms will be of benefit not only to improving function after spinal cord injury but also to developing therapies for other neurological diseases and conditions" (Liverman et al. 2005). Is this management swanking? Seemingly, history repeats frequently. Let us compare to the introduction of the Conference: "The enigma of central nervous regeneration" (Palm Beach, Florida, Feb. 4–6, 1970) and financially supported by the National Paraplegia Foundation from the Social Rehabilitation Service of the US Department of Health, Education and Welfare. The introduction by Alan Reich, President of the National Paraplegia Foundation says: "This greater awareness of our capabilities and increased optimism for meeting difficult human challenges have created renewed and more realistic hope that a cure for paraplegia will be found. The conclusion of the eminent scientists at the National Paraplegia Foundation conference that 'the problem of regeneration in the central nervous system, previously considered hopeless, is amenable to solution through basic research' has important meaning for those who, knowing there is hope, will be willing to work to advance the complex, multidisciplinary, basic neuroscientific effort required to achieve the breakthrough". We are now nearly 50 years and billions of dollars and euros later and there is no paraplegic patient, who raises out of his wheelchair due to regeneration research of spinal cord injuries. It is hard to say, especially for the paraplegic patients, but it is true.

Damage of the spinal cord (Fig. 11.2) contains three phases: the acute phase, in the first few days, causes direct mechanical damage to the spinal cord producing a situation of spinal shock, inducing the absence of spinal reflexes. The second injury phase, weeks to months, is characterized by a progressive loss of spinal cord parenchyma and increase in the lesion size. The chronic third phase contains the irreversible necrosis and apoptosis, giving degeneration of all structures in and around the lesion site. This brings forward a glial scar around the cystic lesion (Fig. 11.2). Although some regeneration can be noticed after the initial injury, this is in general abortive. The actions that contribute to the cystic cavity are posttraumatic ischaemia and oedema with loss of the blood–brain barrier. By these actions, uncontrolled excitatory amino acid release occurs together with calcium influx into the neurons, sodium influx in the mitochondria, collapse of oxidative metabolism, ATP production inhibition, cytochrome C release, free radical overproduction and lipid peroxidation. This results in invasion of blood cells, microglia activation, reactive astrogliosis, Wallerian degeneration, rupture of terminal clubs causing hydrolytic enzyme release, apoptosis of glial cells and invasion of fibroblasts (see Vroemen 2006). The cytoarchitectonic explanation of the events during spinal cord injury causes that one forgets a clearly present damage at the human anatomical level. This is damage or disruption of the anterior longitudinal ligament or of the ligament flavum, which both can be discerned by MRI. They are good predictors of bad or even lack of neurologic improvement (Martinez-Peres et al. 2017). How did research arrive at the neuroregenerative concepts and what kind of experiments emanates from this scheme of spinal cord injury effects?

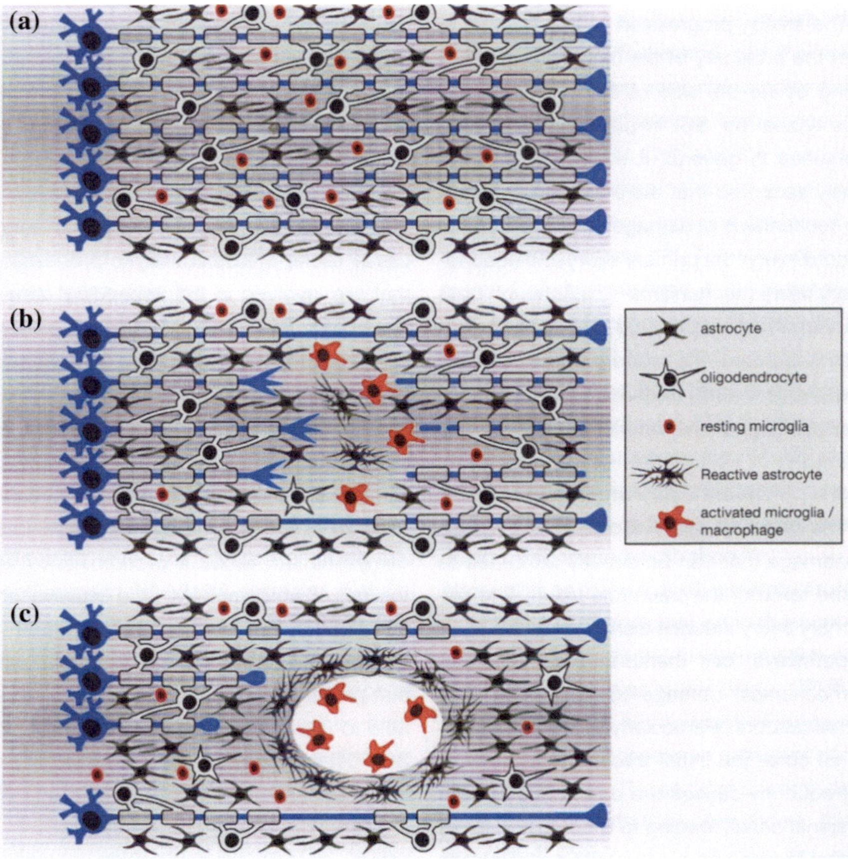

Fig. 11.2 Schematic diagram of spinal cord injury (courtesy M. Vroemen 2006 with permission). **a** Myelinated axons (in blue) are wrapped by oligodendrocytes. Astroglia produces structural and homeostatic aid. Microglia is inactive. **b** The traumatic insult disrupts axonal continuity resulting in activated astrocytes and microglia. Necrosis and gliosis occurs in the centre of the insult. Cellular destruction of the grey matter occurs too, but is not indicated. **c** this results in a cystic lesion cavity surrounded by reactive astrocytes and apoptosis of oligodendrocytes producing demyelinization and neuronal loss by lack of trophic support

The positive idea that restoration of the damaged spinal cord function was possible was supported already in 1849 by Brown-Séquard and later on by other scientists, e.g. Pavlov (1849–1936) in his publications in between 1920 and 1935. One should note that spinal cord and cranial nerve functions were just started to be understood (see, e.g. Arnold 1844, and his defence of Bell's law: dorsal roots sensory and ventral roots motoric). Accidental cases like the Philadelphia woman in which the damaged spinal cord was sewed together and survived for 24 years supported the positive idea on regeneration (but in the end the lady became a spastic paraplegic). The idea that spinal cord regeneration was impossible originated at the start of the previous age. Ramon y Cajal by his regeneration studies, which were published in *Degeneración y Regeneratión del Sistema Nervioso* (1914), favoured an abortive regeneration of the central nervous systems. In Russia, the positive attitude stayed present by Pavlov's ideas. First World War brought series of paraplegic soldiers and the earlier claims of restored function appeared to be false in lots of cases. Rehabilitation of nerve injuries and of brachial plexus avulsions, together with spinal cord injuries, was hardly increased in the period between both world wars (Lanska 2016; Eldar and Jelic 2003). Main target during and after the First World War was the restoration of limb functions by prostheses. Two main and important neurological findings in this period were: the Hoffmann–Tinel sign, twinkling sensations at the front of the outgrowing nerve, and the Guillain–Barré–Strohl syndrome, which is a muscle weakness with a rapid onset. It is caused by the immune system, being an autoimmune effect, and damages the nerves. It can be life-threatening by developing weakness of the breathing muscles requiring artificial respiration in about 15% of the patients (see Lanska 2009). "Little progress was made in the rehabilitation of central nervous system injuries during First World War—in part because of the horrific early mortality associated with such injuries at the time. For example, during First World War, early mortality within the first several weeks of SCI was approximately 80%, and the overall one-year survival was less than 10% and by some estimates only 3%" (Lanska 2009). This situation changed dramatically during Second World War, due to the Rusk (Howard A. Rusk 1901–1989) rehabilitation system named "The Convalescent Training Program" performed in so-called convalescent hospitals and increased antibacterial treatment of the patients also by the end of First World War introduced British technique of debridement (remove all dead and foreign material and to never primarily close wounds). "In Second World War, there were 2500 American service-connected combat paraplegics, and three-fourths of them were alive twenty years later. I might add parenthetically that of these survivors, 1400 were holding down jobs" (Rusk, see Lanska 2009). The mortality rate figures of spinal cord lesions for Russia were 62–80% in First World War and 52% in Second World War.

Whether one likes it or not, research on spinal cord injuries survived in Russia during the communistic era. To use William Windle's euphemistic expression: "In retrospect, the two decades since the 1954 Conference on Regeneration in the Central Nervous System have represented a period in which the interest in the problem flagged and the dogma of non-regeneration continued to be championed. Enthusiasm for support of research on central nervous regeneration died out in the USA. Fortunately, during this period Soviet scientists took up the search for a solution of the problem of regeneration in the mammalian spinal cord" (Windle 1976). Injustice has been done to the Russian researchers by American and Western investigators during the J. R. McCarthy pursuits (±1947–1960), scoffing that dogs with a thoracic myelotomy could only walk again after "*communistic* spinal cord injury treatment". Indeed several false claims were produced, but not only by Russian researchers. Western scientist also lived in an ivory tower.

It is useless to repeat here the history of all scientists that contributed in one or the other way to spinal cord injuries. It is far more interesting to see how new research topics (methods, themes and ideas) originated and how they were applied into spinal cord injury research (for somatosensory overview, see Snow and Wilson 1991). The start is *scar formation*. Scar formation is different in embryos and in foetuses compared to adult animals. Regenerating foetal axons can pass the scar, and mature ones not. The embryonic and foetal glial proliferation after injury does not produce an impenetrable scar. The inhibition of regenerative axons by the adult scar has been found in rats, cats, dogs, monkeys and humans (Windle 1955; Stepanian-Tarakanova 1959). It is reasonable to think that reduction of the scar could improve outgrowth of the regenerating axons. The technique used was biochemical and/or enzyme treatment of the mature scar. The scar in the mature spinal cord injuries belongs to the glial type and later on becomes a collagenous one, which irritates the spinal cord and is comparably thick. Thus, its constitution is time dependent. Therapy was done using Piromen® (a sterile, non-protein and non-anaphylactogenic bacterial polysaccharide complex) by Windle and Chambers (1950). Scar formation was claimed to be inhibited by this substance. The bacterial pyrogens were used both by American and Russian scientists, and both found a reduction of scar formation. Since Piromen® was also related to pituitary ACTH and was used in combination with ACTH in spinal cord regeneration studies (Mc Masters 1962), ACTH fractions were studied in spinal cord injuries and crush lesions by the Gispen group (Utrecht, The Netherlands) and others later on (see, e.g. Bijlsma et al. 1983; Houweling et al. 1998). Enzyme therapy by addition of hyaluronidase was thought to reverse the action of young fibroblasts in the scar area, which produce hyaluronic acid that together with water incorporates in the scar and increases the impenetrability. The enzyme was already known to have a positive permeability effect on keloid scars and on contractures. A claim of complete recovery of spinal cord transections after enzyme therapy in combination with bacterial pyrogenic substances (pyrogenal together with lidase and trypsin with proserine) was published (Matinian and Andreasian 1973). Beside the Russian researchers trypsin positive analogous results were reported by Freeman and colleagues (Magenis et al. 1952; Freeman 1952, 1954; see also Pettegrew 1976), but denied by Guth et al. (1978). Enzyme therapy with lydase, trypsin or elastase was definitely shown to dysfunction in regeneration after spinal cord transections by Kosel et al. (1979). The claim that reduction of the scar by biochemical or enzymatic therapy should enhance regeneration of spinal cord tracts became evidently not true. "Thus, even if we were able to eliminate the problem of getting regenerated fibres crossing a dense glial scar containing little extracellular space, we might still expect the growing fibres to experience orientational difficulties as a result of the complex glial environment of the spinal cord" (Guth and Windle 1970).

Clinical grafts have been used for treatment of Parkinson (Brundin et al. 2010) and other neurological deficits (Gash and Sladek 1988). "Dopaminergic neurons derived from the human embryonic brain were transplanted into striatum of patients with Parkinson's disease. Subsequent studies, in particular two that followed a double-blind, sham surgery, placebo-control design, showed variable and mostly negative results" (Brundin et al. 2010). Moreover, transplant-induced motor side effects (graft-induced dyskinesias) appeared in patients inhibiting graft transplantation. Far earlier grafting was applied in spinal cord injuries. The presence of spinal cord autotomy has been indicated as the cause of outgrowth arrest. Using the salami slice technique, non-respectfully said, but it is the removal of direct stump surfaces in spinal cord injuries after one week. Transplantation of autogenous sciatic nerve grafts in the gap should stimulate regeneration (see Fig. 11.6, left side; Kao 1974; Kao et al. 1977). The week was considered the time needed for autotomy to arrest. In half-sectioned spinal cords grafted with autogenous radial nerve parts supporting results were found (Wilson 1984). Kao claimed regeneration over the graft; thus, regenerated axons could reach the other side of the dog spinal cord (Kao et al. 1977). These results forced our neurosurgery group to check Kao's results. Repetition of Kao's results in cats showed that outgrowing axons could cross the first gap border only for 1–2 mm, but never the second one in transplants of 1 cm (Fig. 11.6). It meant that the claimed regeneration was still abortive and Kao's results were not repeatable. Moreover, the autotomy area could be shown to be localized after the preserved segment and was characterized by the lytic enzyme acid phosphatase. Behind this area, still within the metamorphic segment, succinate dehydrogenase positive axon endings (also called clubs) were present. Counting of the active terminal clubs (clubs with no dispersed reaction product, indicative for inhibition of enzyme activity) in the spinal cord stumps showed that the acid phosphatase clubs count was $\pm$ 200 per mm^2 at day 1 that decreased in day 3–7 after transection to 20 per mm^2. It stayed this amount at day 14. The results did not differ for both stumps. For succinate dehydrogenase, the counts were: day 1, 70–80; day 3, 50; day 7, 10; day 14, 10–15. The presence of succinate dehydrogenase (characteristic for mitochondria) behind the front of acid phosphatase, but later overgrown by the extending acid phosphatase area, indicates the termination of the capacity of the spinal tissue to regenerate (Voormolen et al. 1990).

Around the 1990s of the previous age, interest arose in the use of *embryonic implants* to heal spinal cord injuries. Chromatolysis of the rat rubrospinal neurons could be overcome if embryonic transplants or neurotrophic factors were added to hemisections of the thoracic spinal cord (Bergman et al. 1998). Addition of both reverses the axotomy-induced atrophy completely. Moreover, c-Jun expression was upregulated in these animals if both transplants and neurotrophins were used (Broude et al. 1999). Due to the hemisection doubt could exist whether the results were derived from the intact spinal cord part or not. This series ended with a publication in which foetal spinal cord transplants together with a continuous application of neurotrophins in a total transectioned cord were given. "Surprisingly, regeneration from supraspinal pathways and recovery of motor function were dramatically increased when transplants and neurotrophins were delayed until 2–4 weeks after transection rather than applied acutely. Axonal growth back into the spinal cord below the lesion and transplants was seen only in the presence of neurotrophic factors. Furthermore, the restoration of anatomical connections across the injury site was associated with recovery of function with animals exhibiting plantar foot placement and weight-supported stepping. These findings suggest that the opportunity for intervention after spinal cord injury may be greater than originally envisioned and that CNS neurons with long-standing injuries can reinitiate growth, leading to improvement in motor function" (Coumans et al. 2001). Stepping of the rat after total thoracic transection has nothing to do with regeneration, but with automatic leg movements if the rat pulls itself up over a rough underground or ladder. Your baby of 6 month can "walk" if you drag the baby hanging with his armpits in your hands and with the feet over the ground. The central pattern generators induce these leg actions. Recent renewed interest in addition of embryonic motoneurons in hemisectioned spinal cords was published (Nógrádi et al. 2011). Since the research set-up of hemisectioning is seriously criticized and the use of human aborted embryonic motoneurons is a large ethical problem, this research is presumably condemned to be restricted to animal research.

Within the central nervous system regeneration does occur. In fact, several unmyelinated connections do regenerate "well" with the catecholaminergic, serotoninergic and indolaminergic connections as the best known. Within the limbic loop of Papez (hippocampus, fornix, mammillary body, thalamus), the postcommissural fornix part does know axonal regeneration. It concerns one tract is unidirectional with a clear distinction between proximal and distal stumps after lesioning. Many studies on regeneration concern these systems (e.g. Guth and Windle 1970, on collateral sprouting as an enhancing or exponent of regeneration? See also Hagg 2006), but they hardly contribute to bridging anatomically and functionally the spinal cord injury. Nevertheless, these systems are used to sharpen our ideas on spinal cord regeneration (for an excellent overview, see Stichel-Gunkel 1997).

Neurotrophic interactions are mutual actions upon neural structures. "The nervous system uses also other rules than genetic ones to compose the organization of the nervous system for a specific species, via the so-called patterned migration. This also holds for the peripheral nervous system, where the origins of nerves are genetically coded, but the pathways towards innervated targets are also subjected to other rules. Once neurons are born in a germ layer, they migrate towards their place of destination in the central or peripheral nervous system. An axon develops to make contact with other neurons or targets. The axon length can reach one metre in the peripheral nervous system. For the direction that the axon will take and its velocity, the growth cone is the driver. Principles govern the growth cone: neurotro**p**ic factors in chemotaxis determine the direction of growth. A concentration gradient is needed, while such factors only work over a small distance. Neurotro**ph**ic factors in chemokinesis, on the contrary, determine the growth velocity of the axon. These factors are concentration dependent, but a gradient is unnecessary" (Marani and Lakke 2014). The early detection that nerve disruption causes: destruction of taste buds, degeneration of the lateral line organ, inhibition of salamander limb bud regeneration and muscle atrophy, all were brought back to neurotrophic chemical influences. Trophic factors cannot pass the blood–brain barrier, and in order to enhance regeneration, these neurotrophins have to be added into the spinal cord. This can be done by infusion or implantation of cells that produce in excess the desired factor. This is mostly done by genetic modification. The neurotrophins most used are nerve growth factor (NGF), brain-derived neurotrophic factor (BDNF) and glial cell-derived neurotrophic factor (GDNF, see below).

Enhancing regeneration is also done by the strategy of neutralization of *growth-inhibitory* molecules. The first detected growth-inhibitory molecules were NI 35 and NI250 by the Schwab group, and it was claimed that in combination with neurotrophin-3 (NT-3) outgrowth of the transected corticospinal tract was increased. This result was denied for sensory connections in the spinal cord (Oudega et al. 2000). The series of growth inhibitors has been increased. Myelin-associated glycoprotein (MAG), some extracellular matrix molecules, like tenscin-C, oligodendrocyte myelin glycoprotein (Omgp) and chondroitin-6-sulphate proteoglycan do show the growth inhibition property. MAG is known to promote outgrowth during the embryological and foetal period. It starts its inhibitory action far later, myelin has to be present, and this inhibition is neuron-type specific (see Filbin 2003). While MAG changes from proliferative to inhibitory around the first postnatal week, it is already long known that myelin lipid fractions stay changing till 6–8 months after birth (Edgar 1955).

Regeneration studies are induced by the growth inhibitors. "There are at least three important caveats to these regeneration studies. First, the contribution of regeneration to functional recovery is in most cases unknown. Second, the proportion of fibres that regenerate is always small. Third, the distance regrown is modest, certainly in the context of human spinal cord anatomy" (see Bradbury and McMahon 2006 for overview). However, several of these regeneration studies use the most tricky set-up, the dorsal column transection model and mostly carried out in the rat. Even if the most controllable and versatile retractable tungsten wire knife is used for lesioning, the extend of the lesions stays irrepressible. Lesions are mostly segmentally restricted to one or two segments but always concern both the corticospinal tract and the sensory primary axons in the dorsal column. It is impossible in the rat to lesion exclusively the corticospinal tract and its borders are frayed with the chance that parts of the tract are not lesioned. In regeneration studies, horizontal sections are used to show the expected traversing of the regenerating axons through the lesion. Nevertheless, a small deviation of the horizontal level already includes parts of the dorsal horn or intermediate grey in which tracts partially end or start. Moreover, if functional recovery is observed, this cannot be directly reduced to the elicited structural recovery. Incomplete injuries of the spinal cord open the possibility of compensation of spared connections. In relation to myelinated and unmyelinated axons and growth inhibitors, one should know that both the corticospinal tract and the dorsal column fibres do contain unmyelinated fibres. The incoming dorsal roots below and above the lesion contain thick and thin unmyelinated fibres that both ascend but also descend. The unmyelinated axons reach farther down than the myelinated ones. Incoming fibres go into the different Rexed laminas (see Sects. 11.1.2 and 11.1) and over several segments up and down. Already in 1970 (Guth and Windle), attention was given to collateral sprouting to indicate the intricated relation with regeneration. The discussion is back. "Myelin-associated inhibitors of axon growth, including Nogo, MAG and Omgp, have been the subject of intense research. A myriad of experimental approaches have been applied to investigate the potential of targeting these molecules to promote axonal repair after spinal cord injury. However, there are still conflicting results on their role in axon regeneration and therefore a lack of a cohesive mechanism on how these molecules can be targeted to promote axon repair. One major reason may be the lack of a clear definition of axon regeneration in the first place" (Lee and Zheng 2012, the citation is nearly their whole article abstract). Seemingly, everything is present to let you believe there is regeneration, but there is no regeneration. The two newly developed approaches, stem cell application and nano-techniques, have barely crystallized to be treated here in an historical overview.

11.3 Spinal Cord Regeneration

In general, there are two approaches to restore function of the damaged nervous system. Here, they are grouped under the term "Scientia restitutionis", science of restoration towards the original form and function:

(1) Regeneration of the nervous system. The nervous system can restore, with or without support, its connections and function.
(2) Artificial transfer of (electrical) information into the nervous system as applied in neurotechnology and functional electrical stimulation (FES). The overview of the amount of various implants and stimulator systems shows: cardiac- > 400,000, **spinal cord- > 130,000**, cochlear- > 80,000, deep brain- > 20,000, micturition- > 10,000, respiration stimulators > 1600 (Stieglitz and Meyer 2006).

In both cases, the interface with the human (central) nervous system is essential, which raises the question: "what do we know on interfacing with the ***human*** (central) nervous system? The answer is: too little to understand exactly what is going on and most positive effects are found empirically, but are hardly comprehensible except for cardiac FES. Moreover, the combination of morphometric and electrophysiological methods to measure regeneration outcome is few, meaning that ***functional*** regeneration is hardly studied in restoration experiments (see Vleggeert-Lankamp 2007). For a better understanding of the next parts, research is subdivided in studies related to:

(1) The peripheral nerve (Figs. 11.1f–h). Damage is known as neurapraxia (conduction block over one segment, axonal continuity and complete recovery; see ingenious studies of van Lier 1905), axonotmesis (interruption of axons while connective tissue (basal lamina) of the nerve remains intact; although slow, it may completely or nearly completely recover) and neurotmesis (nerve injury is a complete transection or disruption of nerve fibres that can only recover to a limited extent or not at all). This division of nerve damage is according to Seddon's classification (1943).

(2) The transitional region, entrance or exit area at the spinal cord for dorsal or ventral radiculi, respectively. Damage is known as avulsion (tearing nerve roots/rootlets out of the spinal cord) or as traction injuries, rupture of a plexus parts like ventral nerve, cord or trunk.
(3) The spinal cord, being an expression of the central nervous system. Damage is known as complete or incomplete tetraplegia or paraplegia.

11.3.1 Regeneration of Nerves

Start of this part is with the peripheral nervous system, because it has a rather good capacity for regeneration. If the regenerated, 12 weeks old, autografted rat tibial nerve is studied, the number of Aα-fibres in regenerating nerves did not increase in proportion to the number of Aβ-fibres (see Table 11.1). The increase in the number of fibres in the autografted nerve was caused exclusively by an increase in the number of Aα-fibres (see Fig. 11.3). Thus, at least the Aα-fibres branch extensively (amount control 700, regenerated nerve 3000) and the diameter is less. Some of the electrophysiological properties (charge displaced, firing threshold, refractory period) of the regenerating Aα- and Aβ-fibres were adjusted in proportion to the changes of their respective number and diameter, and some were not. Deviations of the observed from the predicted electrophysiological values were attributed to an upregulation of the $Na_v1.3$ channel in the Aα-fibres and a downregulation of the $Na_v1.6$ channel in the Aβ-fibres during regeneration. Deviation of the observed from the predicted value of the total charge displaced per monophasic action current of Aβ-fibres was attributed to repetitive firing of the regenerating Aβ-fibres (see Fig. 11.3). The inevitable conclusion is that although peripheral nerve regeneration occurs there are large differences in form and in function of the constituents of the regenerated nerve compared to the control situation (Vleggeert-Lankamp et al. 2004). "Functional motor recovery is obviously impaired if significant numbers of axons do not successfully reform functional connections with the muscle. Even if the numbers are adequate, erroneous cross-reinnervation may produce a suboptimal functional result: an originally 'fast' muscle may be reinnervated by axons previously innervating 'slow' muscle, and the result may be a mixed form with inefficient contraction" (Burnett and Zager 2004). Thus, restoration into the original form and function ("Scientia restitutionis") does not occur during peripheral nerve regeneration. Spinal nerve and spinal root stimulation is hampered since it is assumed that the distributions of axons in a peripheral nerve and spinal root are disorganized. In other words, it is unpredictable whether an (cuff) electrode will stimulate the desired motor axons or not. This issue has implications for the development of spatially selective recording (microneurography) and stimulation of the development of neural prostheses (see Deurloo et al. 1998, 2000, 2003).

In order to stimulate selectively a muscle on the level of the spinal roots, one should know the spatial pattern of motor fibres and their location in the root. In ventral spinal roots, less fibres are present than found in dorsal spinal roots and the

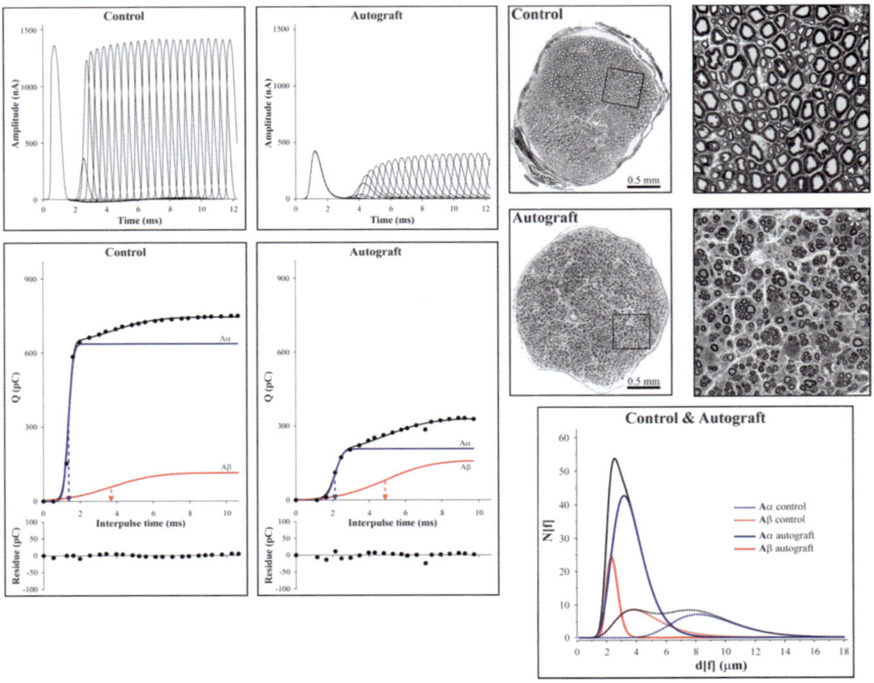

Fig. 11.3 Electrophysiology and morphometry evaluation of normal and regenerating sciatic nerve (see Vleggeert-Lankamp et al. 2004). Autograft Q/t curve shifted to the right, to longer interpulse times, the initial limb of the curve is less steep, Q_{max} is more than halved. Aα- and Aβ-fibre parameters: $Q_{max,\alpha}$ decreased, but stayed larger than $Q_{max,\beta}$; $Q_{max,\beta}$ remained the same. The mean refractory period $t_{50,\beta} \approx 3x\ t_{50,\alpha}$ as in control, but both are 1.5x the control value. Area of the regenerating nerve is comparable to control. Myelin sheath of the regenerating nerve fibres are thinner. Regenerating nerve fibres are frequently fasciculated within persisting Schwann cell basal laminar tubes. N is doubled and D halved. N_α 700 increased to 3000; N_β 1000 decreased to 800. Strong reduction of the diameters occurs: $D_{Fmax,\alpha}$ 9 → 3 μm; $D_{Fmax,\beta}$ 4 → 2 μm. Black lines in diagram N vs d are totals. (courtesy and with permission, C. Vleggeert-Lankamp)

density is slighter. Moreover for rat dorsal roots, a lumbar-level dependence is present, which is absent in ventral roots and no statistical differences were noted for different lumbar rat ventral roots (Prodanov and Feirabend 2007). Although left–right differences were found in the rat, no systematic prevalence of a side effect could be proven (Prodanov and Feirabend 2007). The internal neural fibre topography within ventral spinal roots was studied using cluster index functions, (local) clustering threshold functions, point processes and simulations in tracing experiments on gastrocnemic nerves in the albino rat (Prodanov et al. 2007). Some clustering of axons related to specific innervation of the gastrocnemic muscle could be demonstrated in the L6 ventral spinal roots. Together this indicates that internal neural fibre topography does exists in several ventral spinal roots but not in dorsal roots. Seemingly clustering in the ventral roots of axons could be established, but did not bring forth an electrophysiological stimulation solution.

In short: The study of the morphology and electrophysiology of the regenerated nerve denies a return to the original construction and function. The study of the normal distribution of the various types of axons in nerve and root shows some clustering of axons in the ventral root, but did not solve cuff or electrode placing problems.

11.3.2 Neurotization and Brachial Plexus Traction Injuries

Carlstedt et al. (2004) have extensively studied the transitional area between spinal cord and ventral and dorsal roots. The root–spinal cord junction is characterized by the splitting of the root in thinner microrootlets. The bulging of the spinal cord tissue into these rootlets is dome-shaped. The peripheral character of Schwann cells with collagen-rich extracellular space changes into a spinal cord oligodendrocyte rich setting with small extracellular spaces and serious reduction of collagen. The dorsal root starts a reorganization of the myelinated and unmyelinated axons (see Fig. 11.1h). Grouping for entrance into the dorsal funiculus by the thick myelinated fibres and of the small unmyelinated towards the substantia gelatinosa takes place.

The brachial plexus constructed from ventral roots for the upper limb itself is still under discussion. Table 11.3 demonstrates that the brachial plexus has been hardly considered a constant, which is important in avulsion research and clinics.

Avulsion is a traction lesion of dorsal and ventral roots, which tears roots or rootlets out of the spinal cord (see Fig. 11.4), mainly occurring in traffic accidents, resulting in a (partial) lamed arm. The types of mechanisms of cervical root avulsions have been described by Sunderland (1974). A partly functional recovery is possible after nerve reconstructive surgery using autologous grafts and nerve transfer, together called neurotization (Midha 2004; see also Holzer et al. 2002). This concerns mainly the musculocutaneous neurotization to produce elbow flexion with good results (accessory nerve transfer with MRC grade 3 or better outcome present in 65–83% of the patients, Midha 2004). However, overall results remain limited (Malessy et al. 1999a). In cats, it was demonstrated that after reimplantation of the ventral rootlets (Hoffmann et al. 1993, 1996; Han et al. 2007) or after reimplantation of an autologous transplant between spinal cord and ruptured nerve (Holzer et al. 2003), regeneration of the remaining motoneurons in the cervical spinal cord occurs. The motoneurons wait nearly three months to send their axon out again. After a month, several of the motoneurons start to die, and after three months, all are expired, which favours early surgical repair (Hoffmann et al. 1996; Holzer et al. 2000). Indeed, if an autologous transplant is offered rapidly, in which the motoaxons can regenerate, most remaining motoneurons survive. After 200 days, the normal amount of motoneurons sending out an axon is again reached in avulsed cats (Hoffmann et al. 1996). Thus, offering a mean for outgrowth is important. However, in general only nerve transfers or grafts for *ventral* roots are given, restoring motofunction partially, but not sensory function, if needed. The second problem is that the motoaxons have to reach the hand to bring hand

Table 11.3 A literature review of spinal nerve participation in the brachial plexus[α]

Literature	Brachial plexus
Vesalius (1543) 1964	R5–R9
Heitzmann (1886)	C5–C8, D1
von Lanz and Wachsmuth (1935)	C5–C8, T1
Piersol (1930)	C5–C8, D1
Paturet (1964)	C5–C8, D1
Anson (1971)	C5–C8, T1
Bonnel (1989)	C5–C8, T1
Herzberg (1989)	C5–C8, T1
Lang (1991)	C5–C8, T1
Cunningham (1877)	C5–C8, D1(*D2*)
Testut (1891)	C5–C8, T1(*T2*)
Bishop Harman (1900)	C5–C8, D1(*D2*)
Kerr (1918)	(*C4*)C5–C8, T1
Rouvière (1924)	(*C4*)C5–C8, T1
Toldt (1940)	(*C4*)C5–C8, T1
Bland (1987)	(*C4*)C5–C8, T1
von Lanz and Wachsmuth (1955)	(*C4*)C5–C8, T1(*T2*)
Zuckerman (1961)	(*C4*)C5–C8, T1(*T2*)
Gray (1973)	(*C4*)C5–C8, T1(*T2*)
Jachimonowicz (1925)	(*C3*)(*C4*)C5–C8, D1(*D2*)
Hirasawa (1931)	(*C3*)(*C4*)C5–C8, D1(*D2*)

[α][C = cervical, T = D = thoracic ramus dorsalis, R = "nerve group origin" (R5–R9 = C5–C8, T1) for references see Holzer et al. 2002]

movements back. Since the mature distance to bridge is around 50–60 centimetre (Fig. 11.1g), it will last one and a half year before the hand is reached by sufficient motoaxons to be effective. In the meantime, the non-functional hand develops contractures of muscles, tendons, skin and joints. The damaged nerve(s) disintegrate(s) over such a long time (Wallerian degeneration), because the nerve is not functional and therefore, the outgrowing motoaxon hardly can find back the muscle it belongs to. Wrong coaptation between axons and muscles is the consequence. The comparatively small distance to travel by regenerating axons can already be a problem in the rat. In lumbar root avulsion, compared to sacral and cervical avulsions, axons have to travel longer distances towards their target muscles. This results in prolonged denervation of the distal nerve and a diminished capacity to support regeneration (Eggers et al. 2010).

The main goal for brachial plexus reconstructive surgery is to get elbow flexion and shoulder function in grown-ups, since bringing up the hand with a functional stable shoulder makes a series of actions possible, needed for daily life. For babies, the goal for reconstructive surgery in cases of obstetric brachial plexus lesions (incidence 1–2 per 1000 births, 200–350 cases a year in the Netherlands) is first recovery of hand function, second is the elbow flexion, and third is the recovery of shoulder motions.

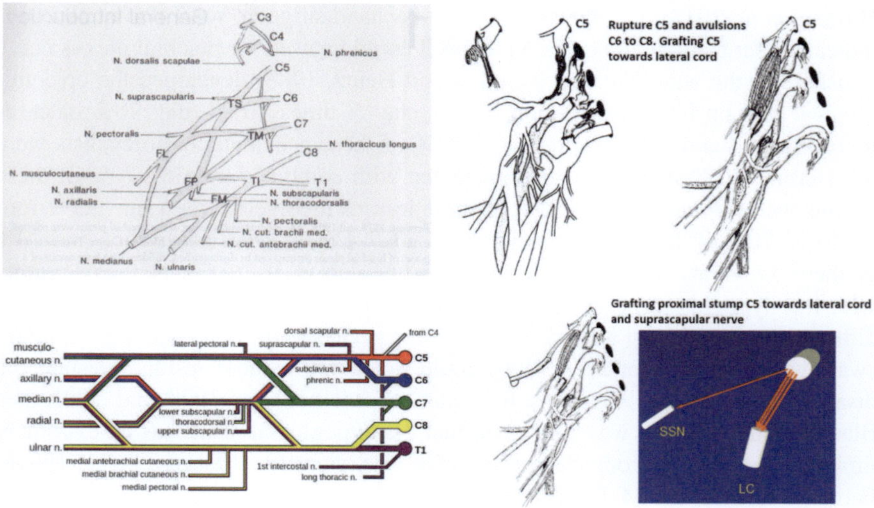

Fig. 11.4 Left shows the anatomy of the brachial plexus (courtesy RTWM Thomeer) with the cervical C5–C8 and thoracic T1 contributions to trunk and cord (TS, TM, TI: superior, medial and inferior trunk; FL, FP, FM: lateral, posterior and medial cord; Contributions of ventral roots to the upper limb nerves, from Selketderivative work: mcstrother (Marshall Strother see: https://commons.wikimedia.org/w/index.php?curid=8833181). The right side demonstrates C5 grafting in nearly total plexus brachialis lesion (from A. Narakas 1993 with permission) and the grafting towards both suprascapular nerve and lateral cord (M. Malessy with permission). Although the efficacy of nerve grafting to improve the outcome in severe brachial plexus traction lesions is no longer debated, the overall results are limited. Nerve transfer using the XI and XII cranial nerves in brachial plexus lesions is also less effective as originally was thought by hand surgeons and neurosurgeons

The study of Pondaag et al. (2005) belongs to the few critical analysis of the obstetric brachial plexus functional recovery after neurotization. Grafting the C5 to the suprascapular nerve is performed in order to reactivate the infraspinatus muscle to increase the external rotation of the arm. In cases where the C5 stump is overloaded with grafts for connections to other damaged brachial plexus subunits, nerve transfer of the suprascapular nerve occurs. In such cases, this nerve is connected to the distal branch(es) of the accessory cranial (XI) nerve. Both techniques have been compared by Pondaag et al. (2005): "The restoration of a fair range of true glenohumeral external rotation after neurotization of the suprascapular nerve in infants with obstetric brachial plexus lesions, whether by grafting of C5 or by nerve transfer of the accessory nerve is disappointingly low".

We avoid here the discussion whether ventral root reimplantation is better or inferior to nerve transplantation (see Su et al. 2013 and references herein).

Analogous is the story of nerve transfer of the hypoglossus nerve. Hypoglossal nerve transfer has been introduced already in 1905 in patients with facial paralysis and is presumably effective due to the small distance that is needed to be bridged. The technique is still in use and effective (Samii and Matthies 1997;

Hayashi et al. 2013). In 1988 Narakas (Swiss hand surgeon, see Taleb et al. 2013) indicated that the cranial nerves XI and XII could be used for brachial plexus nerve transfer. By the way, Narakas (Narakas and Hentz, 1988) developed the epidemiological rule on intraplexal lesion causes, the "7 times 70%" rule: 70% of cases involved road and street accidents, in 70% of these accidents, motorcyclists were the victims, 70% of these victims presented with combined injuries, 70% of these lesions were supraclavicular, 70% of these lesions involved at least one nerve root avulsion, 70% of these avulsions involved a C7–C8, T1 level nerve root, and 70% of these avulsions developed chronic pain" (Taleb et al. 2013). Narakas' results with the hypoglossal nerve transfer were considered employable and only slight difficulty with speaking and with eating were noticed, and none had any difficulty swallowing. Malessy et al. (1999b) could not confirm these results. Despite the disagreement between surgeons that was present between 1999 and 2003 (see Blaauw et al. 2006), it was concluded that the hypoglossal nerve has an extremely limited value for neurotization of brachial plexus parts (Malessy et al. 1999b; Ferraresi et al. 2002; Midha 2004).

The hypoglossal nerve transfer was also used in obstetric brachial plexus lesions. Here, the result was also considered negative: breast feeding was not possible anymore, and donor site morbidity was large: speech was retarded, anxiety of the parents increased, tactile sensations of the tongue were normal, although atrophy of the tongue occurred in single cases. "In all children, clear associated movements of the ipsilateral arm occurred while eating and drinking and during speech, but volitional movement of the arm was possible during speech, eating and drinking. Recovery of powerful volitional elbow flexion was achieved in four cases only (of the six). We do not believe that the sacrifice of such an important function as exerted by the hypoglossal nerve is balanced by the gain demonstrated in our series". (Blaauw et al. 2006). An escape route should be the use of half the hypoglossal nerve for nerve transfer in obstetric brachial plexus lesions (Al-Thunyan et al. 2015).

After more than an age of clinical research, the results of restoring function of the brachial plexus after lesions are distressingly meagre. Grafting and nerve transfer in most cases remain limited in its success to produce functional recovery. Both peripheral nerves and cord–nerve junctions show that regeneration is possible, but the understanding of morphology and physiology of these structures is seemingly inadequate for the targets primed in the clinical setting. "Both stimulation, sensing and surgery of peripheral nerves enhanced the interest in the fascicular construction of nerves. While Sunderland (1945; Sunderland and Ray 1948) favoured changing content and position of fascicles over their trajectory, others indicated a rather high constancy in content and position (Jabaley et al. 1980; Hallin and Wu 2001; Watchmaker et al. 1991). The clinical relevance of nerve repair also supports the study of fascicular organization (Steward 2003; Grant et al. 2004) for surgery and ultrasound or imaging techniques. Microneurography also uses fascicular organization for correct placement of the sensing needle (Hallin 1990). In general, mesoscopic studies over the length of nerves are difficult to perform and clearly distinguishing between sensory and motor areas within a fascicle is difficult due to the absence of sound markers" (Marani and Lakke 2014). The consequence is also that correct coaptation of fascicles is hardly possible. If fascicles are already hard to understand, what about smaller structures? Discord on several other peripheral nervous system topics can be found in the literature (see Marani and Lakke 2014), hardly supporting a unifying hypothesis on structure and function of the peripheral nervous system and its relation with the spinal cord. The way to an effective clinical approach passes inevitably by anatomy and physiology, but is seemingly obstructed by the false idea that we know already everything of the peripheral nervous system.

Earlier the Leiden Neurosurgery group resisted the transfer of animal avulsion research into a Leiden clinical set-up, based on the insufficient treatment strategy and insufficient research for avulsions. The arguments were given in a letter to the editor as a reaction on the paper of Carlstedt et al. (2000). "Even if root reimplantation were able to restore hand function, which apparently is not the case, this would ultimately result in a 'blind' hand. In the clinical presentation, dorsal rootlets are avulsed as well, and these are not amenable to repair. Furthermore, it must be kept in mind that, although probably less so in the ventral approach, any surgical manipulation of the spinal cord carries with it the risk of injuring long tracts. One of eight adult patients in the series presented by Carlstedt et al. (2000) reportedly suffered transient injury to the spinothalamic tract. These considerations, in combination with the reproducible satisfactory results of nerve transfers for biceps function, have kept us from giving into the temptation of performing root reimplantations in humans. The paper of Carlstedt et al. (2000) strengthens our opinion that, at the present time, clinical application of this technique is not warranted. Unfortunately, we do not share the overly optimistic future expectations presented in this paper". Thomeer et al. (2002).

The response of Carlstedt (see Thomeer et al. 2002) was: Professor Thomeer and his colleagues have reacted with a letter to the editor after our paper on intraspinal repair of brachial plexus lesions in humans was published in this journal. Although the reason for Professor Thomeer's letter is ambiguous, I am grateful I was given a chance to respond. His opinion is due in part to severe ignorance and in part to a difference in attitudes and philosophy regarding advancements in surgery. The intraspinal repair of severe brachial plexus lesions is based on meticulously conducted animal experiments. Based on our animal experiments in which intracellular techniques were used, we know that injured motor neurons regenerate and innervate the denervated muscles. We were cautious in our discussions regarding the possible origin of the regeneration. There is of course the plausibility of other intraspinal sources contributing to the recovery made possible by root reimplantation. More than a decade later Prof. Thomeer presented his experimental research on root avulsion injury. Unfortunately, nothing new was learned except his result supported our initial findings. This new surgical technique when applied to humans has led to an increase in surgical as well as basic science knowledge. It has been possible to show, for the first time in humans, that surgery can restore both motor and sensory function as well as alleviate pain after a spinal cord injury. That this surgery has been efficacious and has led to functional return in humans in beyond doubt. There are certainly reasons for optimism as future application of new molecular biology tools and treatments in concert with the intraspinal repair will further improve function. This will of course not be possible with the palliative approach advocated by Prof. Thomeer, which for instance will not give back function in the hand. More than 100 years ago, the first exploration of the intraspinal brachial plexus lesion was performed and comments and suggestions for its treatment were made. We know that the ambition of the courageous surgeons was later curtailed by other so-called nerve surgeons, who abandoned brachial plexus surgery. There was a setback for more than 50 years in knowledge of brachial plexus injury, repair, and recovery. I do hope that a return to the "dark ages" of nerve surgery will not repeat itself because of naysayers like Prof. Thomeer and his colleagues, Carlstedt (2002).

In 2011 Carlstedt published an article: "An overture to basic science aspects of nerve injury". From the text a series of citations are given:

(1) Neurotrophic factors and their receptors have been considered promising targets for therapeutic interventions; however, success has so far been limited.
(2) Scar formation is a major clinical problem in peripheral nerve repair. After a nerve injury, many axons fail to regenerate past the injury site, because they became trapped in scar tissue.
(3) Functional recovery may also be compromised by delayed repair as there is a deterioration of trophic and substrate support for regenerating axons in long-term denervated distal nerve stumps. Added to this is the time dependant atrophy of muscles and sensory receptors.
(4) Furthermore, axons are not accurately guided back to their correct target tissue: misalignment and profuse sprouting at the site of damage leads to regrowth to the wrong target.
(5) Attempts to improve outcome have been made with a variety of artificial implants such as conduits, scaffolds, electrodes, etc., but so far, nothing has been promising enough to be used routinely in peripheral nerve repair.

Seemingly, the arguments on which the Leiden Neurosurgery group refused performing root reimplantations in humans, in 2002, are still valid and are even extended and endorsed by Carlstedt in 2011: "De docta ignorantia" (Nicholas Cusanus ca. 1401–1464)

11.3.3 Synthetic Nerve Grafts

Synthetic nerve grafts are assumed to be useful in bridging nerve gaps since the limited quality and quantity of autologous donor nerves should be overcome by these nerve conduits. The slogan is "The nerve on the shelf". The production of nerve conduits belongs to the field of tissue engineering and several substances are applied. Nerve conduits can be subdivided in synthetic grafts (made from polyesters, polyols and polyurethanes) and biological ones (using polyesters, proteins and polysaccharides). Biodegradable and non-degradable conduits are made from:

silicone, collagen, gelatine, fibronectin, ethylene vinyl acetate, poly-L-lactide, polylactide-caprolactone, polyphosphoester, polysulphone, caprolactone and trimethyl carbonate to name a few. "An ideal nerve conduit needs to have properties like biocompatibility, biodegradability, flexibility, high porosity, compliance, neuroinductivity, neuroconductivity with appropriate surface, and mechanical properties. The nerve conduit may be designed in different ways; they could be cylindrical tubes with internal channels or matrix, porous walls, or cell incorporation, and the design may include bioactive agents" (Arslantunali et al. 2014, see also Vleggeert-Lankamp 2008, and some also holds for FES, see Prodanov et al. 2003). The biophysical and chemical properties of the often newly proposed nerve conduits are hardly studied. Several of the nerve conduit publications miss the mechanical properties like molecular weight, molecular weight distribution and intrinsic viscosities of the used polymers or biological substances. Thermal properties, tensile testing and wettability are seldom published together with the regeneration results (see Pĕgo et al. 2001, 2003 for correct additions).

For the biodegradable conduits the in vitro degradation properties should also be given. During the refereeing processes for several journals these omissions were clearly present and were also detected by the other referees. The next problem with nerve conduits is that *functional* regeneration is hardly studied. If morphometry shows a continuation between both stumps over the conduit, it is quickly concluded that the nerve conduit can be used as a replacement of the autograft. In Fig. 11.5 three tubes that are used as nerve conduits are shown from our own research:

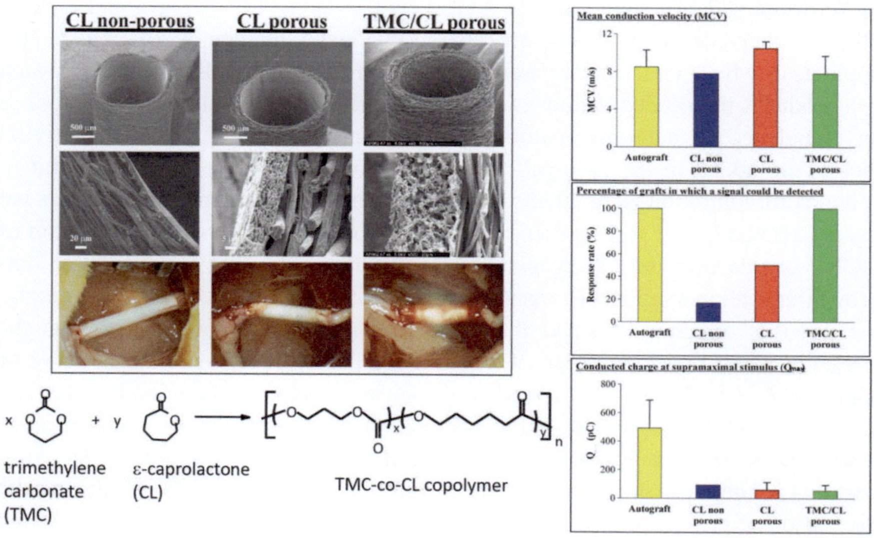

Fig. 11.5 Synthetic nerve grafts of caprolactone and in combination with trimethylene carbonate are shown in scanning electron microscopic photographs (overview and detail) together with the 6 mm implants for bridging the sciatic nerve. Comparison between synthetic tubes and autografts for mean conduction velocity, percentage of grafts with electrophysiological signals and conducted charge are given (courtesy C. Vleggeert-Lankamp and A. Pĕgo with permission)

porous and non-porous ε-caprolactone and a porous trimethylene carbonate combination with ε-caprolactone nerve conduit. While the mean conduction velocity is about equal compared to the autograft and the porous TMC/CL excels in constancy for signal transfer, the Qmax (the conducted charge) is less than 20% of the autograft, which is already around 50% of the normal nerve, indicating that signal transfer is present but presumably insufficient to return to a functional nerve. For the comparison between porous and non-porous TMC/LC grafts counts: "Current opinion holds that pores in synthetic nerve guides facilitate nerve regeneration. Solid factual support for this opinion, however, is absent; most of the relevant studies assessed only morphological parameters and results have been contradictory" (see Vleggeert-Lankamp et al. 2007). This study compared non-porous, macroporous and microporous conduits and based on bridging the gap, the amount of nerve fibres, the myelinization level, the electrophysiological response and the decrease in muscle cross-sectional area. The microporous nerve conduit was the most effective in regeneration of the rat sciatic nerve. Nevertheless, the conducted charge was still minimal. Despite the enormous efforts that are raised to increase the effectiveness of artificial nerve guides, the maximum gap that can be bridged is 2 cm (Arslantunali et al. 2014).

To keep faith in nerve conduit studies one should know that implanted medical devices progress in three stages: first there is speculation, than there is hope and finally there is promise (free after Dobelle 2000*), but only if the conditions of good research are fulfilled, which can be doubted for several nerve guide studies.*

The other research approach fills the nerve conduits or even the spinal cord with or Schwann cells, or BDNF and/or NGF gels (see review of Macaya and Spector 2012). Other neuromodulators are also used. Gene therapy is of course also applied, in which viral vectors are introduced in Schwann cells and the Schwann cell produces the needed therapeutic proteins. The lentiviral factors should produce an abundance of neuromodulators like NGF and GDNF or BDNF. Lentiviral factors are added in the autograft or in the Schwann cell filled nerve conduit or spinal cord. Lentiviral NGF effects are non-existent and lentiviral GDNF effects are mainly related to the pain system. However, "For motoneurons overexpression of GDNF causes trapping of regenerating axons impairing both long distance outgrowth and reinnervation of target muscles" (Tannemaat et al. 2008). Seemingly, the opposite effect is obtained than was targeted for. Another approach is the switching of or knocking down of negative regulators of regeneration in nerve or spinal cord, e.g. (Lingo-1) in spinal cord injuries present on oligodendrocytes and spinal cord neurons (Wu et al. 2013). DREZ (dorsal root entry zone) could be made permeable to regenerating sensory axons of the dorsal root ganglion by lentiviral vector (LV) delivery in the spinal cord of polysialyltransferase (PST) inducing the expression of polysialic acid (PSA) (a large carbohydrate attached to molecules on the cell surface; Zhang et al. 2007). This approach belongs to the fight against the "blind arm". However, the discussion of this article contains the following citation: "Although PSA can make the CNS environment less inhibitory to regenerating axons, overexpression of PSA at the injury site is not in itself sufficient to achieve successful axonal regeneration. As this study demonstrates, a conditioning lesion is necessary in order to stimulate the regeneration mechanism of DRG neurons". Is this "To repair the crack in your cup, I break off the ear"?

In 2015 a Dutch manifesto (a public declaration of policy by a body of individuals, Oxford Dictionary) appeared by the Dutch Brain Institute, its Department of Neuroregeneration, the departments of Neurosurgery and Neurology of the Leiden University Medical Centre and from Amsterdam the UniQure and Center for Neurogenomics and Cognition Research participated. The tendency in it is that the lentiviral approach is promising although the results are until now disappointing, but with pinpointing future areas of research and the start of a path for clinical studies, results will be potential in nerve regeneration, especially in plexus brachialis lesions. "Patients with a brachial plexus injury, a lesion which has a permanent negative impact, are a target group where gene therapeutic intervention could develop into a genuine adjuvant regenerative treatment strategy to further promote repair after neurosurgical intervention" (Hoyng et al. 2015). An ethical question appears: should research direct itself towards patients if we do not have a clear idea whether and how the lentiviral technique works; man as an experimental animal? This manifesto does nothing else than promoting a leap in the dark, which is sold as a pace forwards.

11.3.4 Damage and Stimulation of the Spinal Cord

To start with: the gene transfer by lentiviral and retroviral approach in the spinal cord does not work. The in vivo gene expression is not stable and in vitro differentiation of neurotrophins expression is in general lost (Vroemen et al. 2005). This result is not different from remarks above on lentiviral research. One can return to the study of the interaction between anatomy, physiology and clinics, which is the main topic of the rest of this chapter.

In 70% of the brachial plexus avulsions neurogenic pain is present, presumably due to damage of the spinal cord. Such pain occurs in the flail, deafferented limb soon after injury. The persistent pain (not in all patients) can be so awful that suicide is around the corner (Landers et al. 2015). Let's have a look into the fibre distribution exactly at the entrance area of the spinal cord. The incoming dorsal root and rootlets divides itself into large fibres for the dorsal columns and small myelinated and unmyelinated ones for the zone of Lissauer placed over the substantia gelatinosa (Fig. 11.1h). This zone of changing organization just on top of the spinal cord is called the dorsal root entry zone, always abbreviated as DREZ. The large and small dorsal root fibres dichotomize into ascending and descending primary branches. These fibres send out collaterals that terminate in the same or bordering cord segments (Fig. 11.1f, h). Collaterals of small, unmyelinated C fibres terminate on Rexed's laminae I (marginal cells) and II (substantia gelatinosa). Collaterals of small, thinly myelinated Aδ fibres terminate on Rexed's laminae II (substantia gelatinosa) and V/VI (intermediate grey substance). Collaterals of the largest (Aα) fibres end on Rexed's laminae V/VI (intermediate grey substance) and

IX (motor neurons). Large (Aβ and -γ) fibres end on Rexed's laminae II and III/IV (nucleus proprius). Contrary to all other afferent collaterals, which directly penetrate the dorsal horn, the collaterals of Aβ and -γ fibres make a loop around the superficial side of the dorsal horn before entering the latter medially (Marani 2017). DREZ lesioning to fight pain is also indicated as rhizotomy. Cutting the small fibres, which are localized more laterally in the DREZ, together with its Rexed layers (I–IV) reduces the corresponding dermatomal pain (Thomas and Sheehy 1983). "The only clear indication for this technique is for brachial plexus avulsion injuries, although it has been tried for pain resultant from many other central and peripheral nerve lesions. DREZ lesion is an open neurosurgical procedure requiring a laminectomy. A lesion is created affecting Rexed laminae I–IV. This can be difficult to identify when roots have been avulsed, so intraoperative neurophysiological techniques can be necessary to aid identification of the correct anatomy. Risks are: some ipsilateral loss of sensation and/or weakness may occur in as many as 60% of cases and be significant in up to 10%. For these reasons DREZ lesion is often attempted only as a last resort" (Eldridge 2010, see also Awad et al. 2013). Note that DREZ lesion not solely damages thin pain fibres, as claimed, but nearly all fibre types for that cord segment. Long-term follow-up of the pain relief showed a gradually declined over 5 years after surgery. The pain relief stayed present the first 3 years (Prestor 2006). In a study, in which the patients were under regular follow-up for 5–18 years, the pain relief rate was excellent or good in 80% of the cases in the early postoperative stage. A drop to 60% in the 5-year follow-up period was found and only 50% had still an excellent or good result in the 10-year follow-up (Chen and Tu 2006). A literature analysis showed that DREZ lesioning is an accepted technique for brachial plexus pain, but most articles don't contain sufficient injury descriptions and injury levels, control groups are absent and inclusion and exclusion arguments are often neglected (Robert et al. 2009). Therefore, a clear explanation of this brachial plexus avulsion pain is absent, although the outgoing motoaxons of the spinal motoneurons are closely related to the spinothalamic (pain) tract. One should note that in fact the segmental total dorsal horn (layers I–IV) is lesioned. This is clearly more than a simple DREZ lesion or rhizotomy.

In spinal cord stimulation the dorsal funiculi fibres are activated. The development of the dorsal funiculus needs now extra attention. During the development of the spinal cord a dorsal root bifurcation area can be distinguished above the DREZ. This bifurcation zone changes into the dorsal funiculus, due to its medial displacement. The shift of the bifurcation zone towards the position of the mature dorsally placed gracilis and cuneatus fasciculi supports and explains the construction of the human dorsal columns according to Litaert Peerbolte (1932; Fig. 11.7a; see Oudega et al. 1993; Miller and Basbaum 1975; Voogd, in Nieuwenhuys et al. 1998).

The more or less layered representation of the dermatomes in the dorsal columns is strongly favoured by Smith and Deacon (1984; see Fig. 11.5b; Glees et al. 1951) and is generally accepted. The discussion on the organization of the somatotopy in the dorsal columns is still important for the areas that can be reached by spinal cord stimulation. (By the way Smith and Deacon are rather selective in their staining indication for thoracic level, see their Fig. 11). This discussion on the structure of the dorsal funiculus is not without clinical significance, since small thoracic lesions next to the midline will suppress visceral pain and decreases analgesic treatment of patients with visceral cancers (Hirschberg et al. 1996; Willis et al. 1999). These midline lesions are made along the whole septum posterius (Fig. 11.7), which means in the Winkler organization that all segments are a little lesioned, while in the Litaert Peerbolte scheme the cervical segments stay unharmed. Indeed thoracic spinal cord lesioning produces degeneration within the rostral gracile nucleus (Fig. 11.7, see Marani and Schoen 2005). Moreover, these lesions are not without risk. Thoracic small lesions at the midline, do produce variable deficits in hind limb sensibility (responsible is the gracile nucleus) and movement. "In the summary of their review, Nathan et al. (1986) state that disturbances of mechanoreception attributed to lesions of the dorsal columns include 'lability of threshold, persistence of sensation, tactile and postural hallucinations and temporal and spatial disturbances. In man, lesions of the dorsal columns cause an increase in pain, tickle, warmth and cold' (see Berkley 1998). The better we know the ***human*** dorsal column's structure and the organization of its continuation into the lemniscus medialis (Marani and Schoen 2005), the better small lesions can be placed with a serious reduction of the risks, if such lesioning is really needed (see criticism Berkley 1998).

Spinal cord stimulation is used in cases of chronic intractable pain of neurogenic origin, e.g. brachial plexus avulsions. This spinal cord stimulation is now well established and applied in several clinics over the world. Electrical stimulation occurs by an electrode placed dorsomedially in the epidural space and thus above the dorsal funiculi (Fig. 11.6). It elicits a tingling sensation (paresthesia) in corresponding dermatomes. In spinal cord stimulation the large diameter cutaneous Aβ fibres in the dorsal columns are activated and do have an inhibiting effect on the transmission of pain signals by Aδ and C fibres from the corresponding dermatomes. The largest Aβ fibres can be activated up to a maximum depth of about 0.25 mm in the dorsal columns. No data were available on the distribution of these large fibres until the study of Feirabend et al. (2002) to examine the dorsal column's full spectrum of myelinated fibres with emphasis on the mediolateral distribution of these large fibres in the dorsal columns (Fig. 11.7). 'Holsheimer (1998) using the cable model (McNeal 1976; Struijk et al. 1992), together with a volume conductor model, representing anatomical structures and electrical characteristics of nerve membrane potentials, succeeded in producing correct predictions of stimulation situations for pain by spinal cord stimulation. Dorsal column stimulation in humans could be understood due to the study of the characteristics of the axonal distribution in human superficial layers of the dorsal columns (Feirabend et al. 2002), which are only reached by spinal cord stimulation (Holsheimer 1998, 2002). The most spectacular finding was that the amount of Aβ fibres in the superficial layer of the dorsal columns at T11 that can be reached by stimulation is minimal (4–5 fibres), still pain suppression did occur. The dermatome activated just below the discomfort threshold is likely to be stimulated by just a single fibre, suggesting that paraesthesia and pain relief may be effected in a dermatome by the stimulation of a single

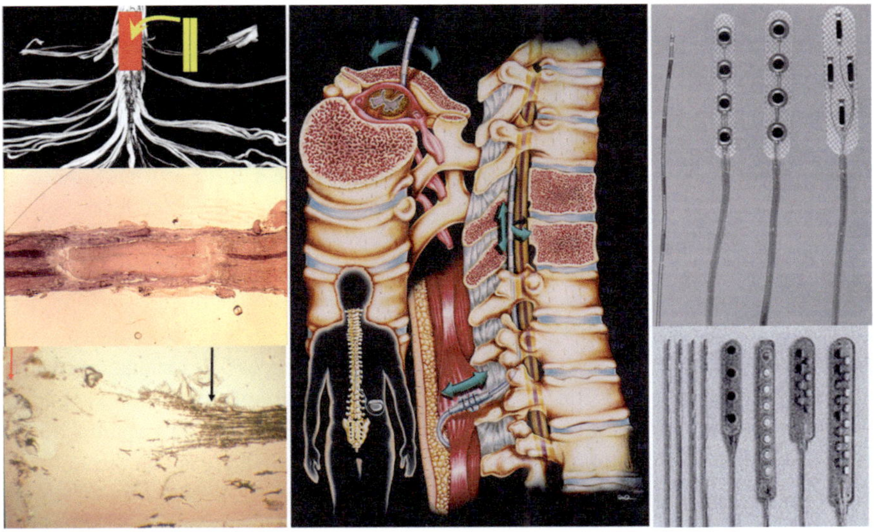

Fig. 11.6 Left panel shows spinal cord grafting by two autografts. Outgrowing axons only pass the first interface (left lower panel) but not the second interface. Black arrow first interface, red arrow is halfway autografts (© Marani). Middle panel gives an overview of spinal cord stimulation with epidural localization of the stimulation electrode (courtesy and with permission J. Holsheimer). Right panel shows the different types of electrodes used (Medtronic, The Netherlands), which are model-based optimized for percutaneous and surgical spinal cord stimulation (see Sankarasubramanian 2013)

large Aβ fibre' (Holsheimer 2002). Aβ fibres are held responsible for suppressing pain after stimulation due to synaptically suppressing the activity of spinal cord Rexed layers that transmit noxious information (Melzack and Wall 1965), caused by antidromic effects of the stimulation. Dorsal root fibres located near the electrode on the dura at a further away distance have the same possibility to be stimulated as dorsal column fibres, due to their lower threshold for excitation as compared to Aβ fibres in the dorsal columns. Placing of the electrode near the 'painful' spinal cord segment(s) is important to get orthodromic stimulation of Aβ fibres. The selectivity of a multielectrode cuff is dependent of the relative location of the electrode contacts, the nerve fascicles and the size and relative spacing of the neighbouring fascicles (Grill and Mortimer 1996), implicating that in nerve models the anatomical structure should be represented" (Marani and Lakke 2014) .

Recent developments concern motor cortex stimulation in order to suppress chronic pain (Fig. 11.8). At least it presumably gives the possibility to break free from lesioning. This technique needs optimization of the therapeutic efficacy, which is hampered since it is not known how electrically activated neural structures in the motor cortex can induce pain relief. It is unclear, which of the multiple neural elements parallel or perpendicular to the cortical layers (Fig. 11.8) are immediately excited by the electrical pulses in motor cortex stimulation.

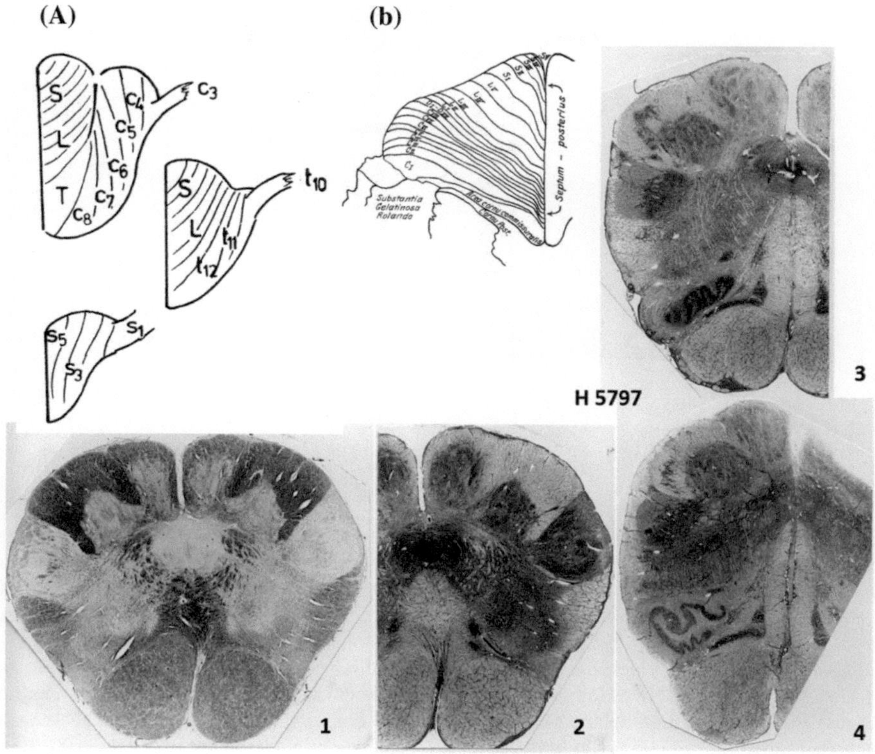

Fig. 11.7 Somatotopic organization in the dorsal columns. **a** Organization according to Litaert Peerbolte (1932). **B** Organization according to Winkler (1917) (See Marani and Lakke 2014). Four consecutive sections of the human nuclei gracilis and cuneatus are given. 1 is a Klüver section with the start of the lemniscus medialis by both fasciculus nuclei. Not earlier than in 4, the gracile nucleus reaches the sulcus area containing the projections for visceral pain (figs. are from Marani and Schoen 2005 with permission)

This depends on positioning of anodal and cathodal electrodes and the stimulation parameters, which is studied by computational modelling studies. The activation of neural elements in the precentral gyrus and anterior wall of the central sulcus has been studied by: (1) Calculating the stimulus-induced electrical field using a realistic 3D volume conductor model (Fig. 11.8, right upper part), and (2) simulation of the response of neural elements using compartmental neuron models including the axon, soma and dendritic trunk (Manola et al. 2005). The results show (Fig. 11.8) that electrode polarity and electrode position over the precentral gyrus and central sulcus have a large and distinct influence on the response of cortical neural elements to stimuli: cathodic and anodic stimulation at the same site results in the activation of different groups of local cortical nerve fibres and in bipolar stimulation both cathode and anode are active electrodes, selectively activating those axons which have the proper orientation in the cathodal and anodal field, respectively (Manola et al. 2007). This study does not solve the

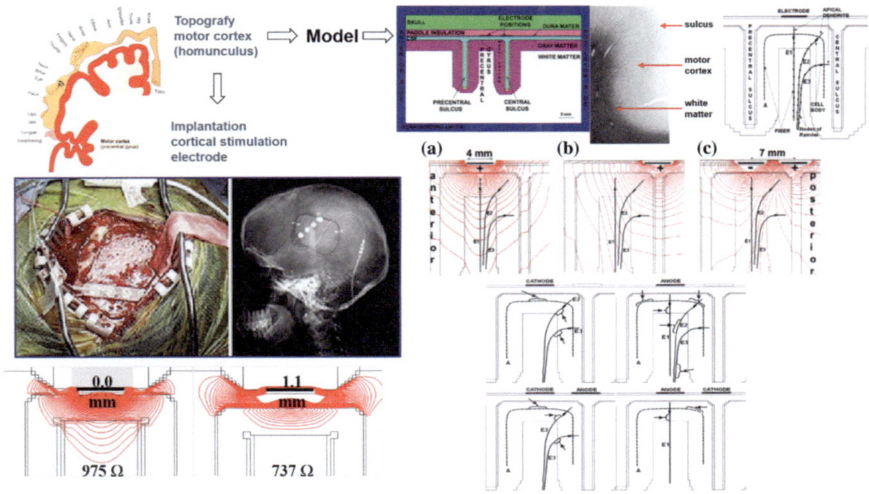

Fig. 11.8 Full model contains a nerve fibre parallel to the cortical laminae ('A'), a neuron in the crown of the precentral gyrus ('E1'), a neuron in the lip of the central sulcus ('E2') and a neuron in the wall of the central sulcus ('E3'). Iso-current density lines in the xy-plane through centre of cathode for different thicknesses of cerebrospinal fluid (CSF) layer between electrode and cortex are studied (left lower part). From monopolar cathodal stimulation (1 V), 55 equidistant lines (5–60 μA/mm^2) are drawn. Thickness of CSF layer and load impedance are indicated in each plot. A, B, C shows iso-potential lines for a stimulus of 1 V: **a** anodal stimulation centred on (Post Central Gyrus) PCG, **b** anodal stimulation centred on Central Sulcus (CS), **c** bipolar stimulation with cathode centred on PCG. Fifty iso-lines between 0.4 and 0.6 V in monopolar and 0.1 and 0.1 V in bipolar stimulation are shown (4.1 mV distance between lines). Lower right: sites of excitation in the "A" fibre and "E" neurons when stimulated by the electrodes having different positions and polarities. For each electrode configuration, only those neurons that are activated below 30 V are shown (courtesy L. Manola; see Manola et al. 2007 with permission)

question: "What are the cortical elements responsible for suppressing chronic pain"? However, it does give the scientific supported parameters for the settings of the cortical stimulation and is as such the "little" brother of the spinal cord stimulation.

The last in this series of research entities related to regeneration is the cell therapy approach, among them the mesenchymal stem cell approach. Although these stem cells can be transdifferentiated into Schwann cells, the real amount that change into the Schwann cell type is 60–70%. The undifferentiated stem cells (30–40%) are hardly contributing to nerve regeneration, presumably contribute to scar formation by in vivo differentiation, and reprogramming mutations are present in induced pluripotent human stem cells (Keilhoff and Fansa 2011). The use of mesenchymal stem cells for nerve regeneration still has to go a long way before it can be applied without serious risks.

11.3.5 A Sad Conclusion

Let us dwell more on efficacy of spinal cord injury therapies. In 2013, an article appeared with the title: "Demonstrating efficacy in preclinical studies of cellular therapies for spinal cord injuries-How much is enough?" The article was made by 27 authors of which 80% originated from institutions in USA and Canada and 20% from Europe (4), Australia (1) and Japan (1). Seemingly, the article was made for the US spinal cord injury market, since a series of European institutions that contribute to spinal cord injuries with cell therapies were excluded or at least could not participate. The introduction of the article starts as follows:

"Despite considerable progress over the past four decades in the medical, surgical, and rehabilitative care of individuals with spinal cord injury (SCI), treatment options to improve neurologic function are limited. As a result, SCI remains catastrophic for those affected. Numerous promising therapeutic approaches for SCI have been developed in the experimental setting. A growing number of these are emerging from the laboratory with the hope of being translated into human clinical trials (Tetzlaff et al. 2011; Kwon et al. 2011a, b). With this has come much optimism that efficacious treatments for human SCI will be established in the not-so-distant future" (Kwon et al. 2013). We have known this the whole time.

It is the next, but earlier manifesto for cellular therapy in research and clinical spinal cord injury one encounters in the literature. A questionnaire was distributed to participants of a meeting of these Spinal Cord Injury (SCI) scientists (45 invitations, 27 participants). "The purpose of this initiative was to seek perspectives on the level of evidence required in experimental studies of cellular therapies before proceeding with clinical trials of SCI. These included: the use of animal models, the use of injury models and mechanisms, the window for demonstrating efficacy, independent replication, defining 'relevant, meaningful efficacy' in preclinical studies, and the expectation of therapeutic benefits for cellular interventions". Questions like: "Should a cell therapy with evidence of efficacy in a rodent model of thoracic SCI be permitted to enrol cervical SCI patients in a trial?" are answered with no by 81% of the participants. Large animals and even primate studies should be carried out first, say most participants. However, the question "if regulatory/funding bodies required efficacy in primates, would this be too strict a bar for the SCI field?" is answered with yes by 78%. Hidden in the text of the introduction one notices that new therapies for spinal cord injury are hardly straightforward and some of the cellular therapies had to be terminated midstream. Still, the clinical trial governs the article by questions like "What are the criteria or parameters to start a clinical trial?". In the text of this chapter the counter arguments are given: cell therapy hardly works. Therefore, more attention should be given to clinical neuroprotection (Fawcett 2006) and rehabilitation interventions (Edgerton et al. 2006), since the spinal injury increases seriously after the initial damage.

The only possible conclusion is that ***our*** *research on regeneration failed. At present there is no universally accepted therapy for spinal cord injury. Moreover, the field is well known for scientific charlatans and fraud. The autograft technique* *for biceps reinnervation (and some other muscles) is what we are able to do clinically. The possibilities with just born babies look more promising. Regeneration science has not even reached the phase in which it can restore the minimal functions that belong to the cervical spinal cord yet, despite the enormous sums of money available and the huge research efforts. In this regeneration field we all need a new "neuro-logic".*

Such a needed breakthrough in spinal cord injury research is perhaps the identification of cyclic AMP that can modulate axonal regeneration due to its inhibitory role of growth inhibitors like MAG. Inhibition of the enzyme phosphodiesterase by Rolipram elevates the levels of cAMP (Nikullna et al. 2004). Moreover, increased levels of cAMP reduces the glial scarring that occurs after spinal cord lesioning. Dibutyryl cAMP (dbcAMP) addition is possible because it can pass the blood–brain barrier and has analogous effects like cAMP. The combination of Rolipram, dbcAMP together with Schwann cell addition gave "enhanced sparing of myelinated axons, increase in myelination of spinal cord axons by the added Schwann cells, and an overall increase in the number of axons within the grafts" (Hannila and Filbin 2008). The study of cAMP regulating genes to overcome myelin-mediated inhibition resulted in the pro-regenerative effects of arginase I (ArgI), interleukin (IL)-6, secretory leucocyte protease inhibitor (SLPI), and metallothionein (MT)-I/II (Siddiq and Hannila 2015). However, cAMP effects on spinal cord injury should be studied in the Kao paradigm, proving that both borders of the gap of 1 cm will be passed through a graft. Otherwise, the demonstrated regenerative capacity stays the same, bridging by a few axons for 1–3 mm, which has already been described several times and is still **abortive regeneration**.

References

Aito S, D'Andrea M, Werhagen L (2005) Spinal cord injuries due to diving accidents. Spinal Cord 43:109–116

Al-Thunyan A, Al-Qattan MM, Al-Meshal O, Al-Husainan H, Al-Assaf A (2015) Hemi-Hypoglossal nerve transfer for obstetric brachial plexus palsy: report of 3 cases. J Hand Surg 40. https://doi.org/10.1016/j.jhsa.2014.11.018

Arnold JW (1844) Ueber die verrichtung der Würzeln der Rückenmarksnerven. Physiologische und pathologische Untersuchungen zur Würdigung der Bell'schen Lehre. Verlag K. Groos, Heidelberg

Arslantunali D, Dursun T, Yucel D, Hasirci N, Hasirci V (2014) Peripheral nerve conduits: technology update. Med Devices: Evid Res 7:405–424

Awad AJ, Forbes JA, Jermakowicz W, Eli IM, Blumenkopf B, Konrad P (2013) Experience with 25 years of dorsal root entry zone lesioning at a single institution. Surg Neurol Int 13:4/64

Bergman B, Broude E, McAtee M, Kelley MS (1998) Transplants and neurotrophic factors prevent atrophy of mature CNS neurons after spinal cord injury. Exp Neurol 149:13–27

Berkley KJ (1998) Lesions of the dorsal column. Definitely disturbing. Pain Forum 7:129–134

Berkowitz M, O'Leary PK, Kruse DL, Harvey C (1998) Spinal cord injury: analysis of medical and social costs. Demos Medical Publ Inc, NY

Bijlsma WA, Schotman P, Jennekens FGI, Gispen WH (1983) Regeneration and protein synthesis in the spinal cord: The influence of ACTH-like peptides on cell-free protein synthesis. Neurosci Lett 38:297–302

Blaauw G, Sauter Y, Lacroix CLE, Slooff ACJ (2006) Hypoglossal nerve transfer in obstetric brachial plexus palsy. J Plastic Reconstr Aest Surg 59:474–478

Borius PY, Gouader I, Bousquet P, Draper L, Roux FE (2010) Cervical spine injuries resulting from diving accidents in swimming pools: outcome of 34 patients. Eur Spine J 19:552–557

Bradbury EJ, McMahon SB (2006) Spinal cord repair strategies: Why do they work? Nat Rev Neurosci 7:644–653

Broude E, McAltee M, Kelley MS, Bergman B (1999) Fetal spinal cord transplants and exogenous neurotrophic support enhance c-Jun expression in mature axotomized neurons after spinal cord injury. Exp Neurol 155:65–78

Brundin P, Barker RA, Parmar M (2010) Neural grafting in Parkinson's disease: problems and possibilities. Progr Brain Res 184:265–294

Burnett MG, Zager EL (2004) Pathophysiology of peripheral nerve injury: a brief review. Neurosurg Focus 16, 5:Article 1

Busch HFM (1964) Anatomical aspects of the anterior and lateral funiculi at the spinobulbar junction. In: Eccles JC, Schadé JP (eds) Organization of the spinal cord. Progr Brain Res II, pp 223–237. Elsevier, Amsterdam

Carlstedt T (2011) An overture to basic science aspects of nerve injuries. J Hand Surg Eur 36:726–729

Carlstedt T, Cullheim S, Risling M (2004) Spinal cord in relation to the peripheral nervous system. In: Paxinos G, Mai JK (eds) The human nervous system, second ed. Elsevier/Academic Press, Amsterdam, pp 250–263

Carlstedt T, Anand P, Halin R, Misra PV, Noren G, Severlis T (2000) Spinal nerve root repair and reimplantation of avulsed ventral roots into the spinal cord after brachial plexus injury. J Neurosurg 93(2 Suppl):237–247·

Chan-Seng E, Perrin FE, Segnarbieux F, Lonjon N (2013) Cervical spine injuries from diving accident: A 10-year retrospective descriptive study on 64 patients. Orthop Trauma Surg Res 99:607–613

Chen HJ, Tu YK (2006) Long term follow-up results of dorsal root entry zone lesions for intractable pain after brachial plexus avulsion injuries. Acta Neurochir Suppl 99:73–75

Coumans JV, Lin TT-S, Dai HN, MacArthur L, McAtee M, Nash C, Bregman BS (2001) Axonal regeneration and functional recovery after complete spinal cord transection in rats by delayed treatment with transplants and neurotrophins. J Neurosci 21:9334–9344

De Neef C (1901) Recherches expérimentales sur les localisations motrices médullaires chez le chien et le lapin Névraxe 2:71–111

Deurloo KEI, Holsheimer J, Bergveld P (2003) Fascicular selectivity in transverse stimulation with a nerve cuff electrode: a theoretical approach. Neuromodul 6:258–269

Deurloo KEI, Holsheimer J, Boom HBK (1998) Transverse tripolar stimulation of peripheral nerve: a modelling study of spatial selectivity. Med Biol Eng Comp 36:66–74

Deurloo KEI, Holsheimer J, Bergveld P (2000) Nerve stimulation with a multi-contact cuff electrode: Validation of model predictions. Arch Physiol Biochem 108:349–359

DeVivo MJ (1997) Causes and costs of spinal cord injury in the United States. Spinal Cord 35:809–813

Dobelle WH (2000) Artificial vision for the blind by connecting a television camera to the visual cortex. ASAIO J 46:3–9

Edgar GWF (1955) Myelination studied by quantitative determination of myelin lipids, with reference to the problem of demyelination. Thesis University Amsterdam

Edgerton VR, Kin SJ, Ichiyama RM, Gerasimenko YP, Roy RR (2006) Rehabilitative therapies after spinal cord injury. J Neurotrauma 23:560–570

Eggers R, Tannemaat MR, Ehlert EM, Verhaagen J (2010) A spatio-temporal analysis of motoneuron survival, axonal regeneration and neurotrophic factor expression after lumbar ventral root avulsion and implantation. Exp Neurol 223:207–220

Eldar R, Jelic M (2003) The association of rehabilitation and war. Disabil Rehabil 25:1019–1023
Eldridge P (2010) Neurosurgical techniques in the management of chronic pain. Anaest Intens Care Med 12:66–68
Enicker B, Gonya S, Hardcastle TC (2015) Spinal stab injury with retained knife blades: 51 consecutive patients managed at a regional referral unit. Injury Int J care injured 46:1726–1733
Fawcett JW (2006) Overcoming inhibition in the damaged spinal cord. J Neurotrauma 23:371–383
Feirabend HKP, Choufoer H, Ploeger S, Holsheimer J, van Gool JD (2002) Morphometry of human superficial dorsal and dorsolateral column fibers: significance to spinal cord stimulation. Brain 125:1137–1149
Ferraresi S, Garozzo D, Ravenni R, Dainese R, de Grandis D, Buffatti P (2002) Hemihypoglossal nerve transfer in brachial plexus repair: technique and results. Neurosurgery 50:332–335
Filbin MT (2003) Myelin associated inhibitors of axonal regeneration in the adult mammalian CNS. Nat Rev Neurosci 4:1–11
Freeman LW (1952) Return of function after complete transection of the spinal cord of the rat, cat and dog. Ann Surg 136:193–205
Freeman LW (1954) Return of spinal cord function in mammals after transecting lesions. Ann N Y Acad Med Sci 58:564–569
Gash DM, Sladek JR (1988) Transplantation into the mammalian CNS. Progr Brain Res 78, Elsevier, Amsterdam
Glees P, Livingstone RB, Soler J (1951) Der intraspinale Verlauf und die Endigungen der sensorischen Würzeln in den Nucleus Gracilis und Cuneatus. Arch Psych Zeitschr Neurol 187:190–204
Grant GA, Goodkin R, Maravilla KR, Kliot M (2004) MR Neurography: diagnostic utility in the surgical treatment of peripheral nerve disorders. Neuroimag Clin N Am 14:115–133
Grill WM, Mortimer JT (1996) Quantification of recruitment properties of multiple contact cuff electrodes. IEEE Trans Rehabil Eng 4:49–62
Guth L, Bright D, Donati EJ (1978) Functional deficits and anatomical alterations after high cervical spinal cord hemisection in the rat. Exp Neurol 58:511–520
Guth L, Windle WF (1970) The enigma of central nervous regeneration. Exp Neurol Suppl 5:1–43
Hagg T (2006) Collateral sprouting as a target for improved function after spinal cord injury. J Neurotrauma 23:281–294
Hallin RG (1990) Microneurography in relation to intraneural topography: somatotopic organization of median nerve fascicles in humans. J Neurol Neurosurg Psychiatry 53:736–744
Hallin RG, Wu G (2001) Fitting pieces in the peripheral nerve puzzle. Exp Neurol 172:482–492
Han L, Kan S, Yuan J (2007) Experimental study on reimplantation of ventral root into spinal cord after brachial plexus avulsion. Zhongguo Xiu Fu Chong Jian Wai Ke Za Zhi. 21:948–952
Hannila SS, Filbin MT (2008) The role of cAMP signalling in promoting axonal regeneration after spinal cord injury. Exp Neurol 209:321–332
Hayashi A, Nishida M, Seno H, Inoue M, Iwata H, Shirasawa T, Arai H, Kayamori R, Komuro Y, Yanai A (2013) Hemihypoglossal nerve transfer for acute facial paralysis. J Neurosurg 118:160–166
Hirschberg RM, Al-Chaer ED, Lawand NB, Westlund KN, Willis WD (1996) Is there a pathway in the posterior funiculus that signals visceral pain? Pain 67:291–305
Hoffmann CFE, Marani E, Oestreicher AB, Thomeer RTWM (1993) Ventral root avulsion versus transection at the cervical 7 level of the cat spinal cord. Rest Neurol Neurosci 5:291–302
Hoffmann CF, Marani E, van Dijk JG, vd Kamp W, Thomeer RT (1996) Reinnervation of avulsed and reimplanted ventral rootlets in the cervical spinal cord of the cat. J Neurosurg 84:234–243
Holsheimer J (1998) Computer modeling of spinal cord stimulation and its contribution to therapeutic efficacy. Spinal Cord 36:531–540
Holsheimer J (2002) Which neuronal elements are activated directly by spinal cord stimulation. Neuromodulation 5:25–31
Holzer CA, Feirabend HK, Marani E, Thomeer RT (2000) Ultrastructural and quantitative motoneuronal changes after ventral root avulsion favor early surgical repair. Arch Physiol Biochem 108:293–309

Holzer CAJ, Marani E, Lakke EAJF, Thomeer RTWM (2002) Repair of ventral root avulsions of the brachial plexus: a review. J Periph Nerv Syst 7:233–242

Holzer CA, Marani E, van Dijk GJ, Thomeer RT (2003) Repair of ventral root avulsion using autologous nerve grafts in cats. J Periph Nerv Syst 8:17–22

Houweling DA, Bär PR, Gispen WH, Joosten EAJ (1998) Spinal cord injury: Bridging the lesion and the role of neurotrophic factors in repair. Progr Brain Res 117:455–471

Hoyng SA, de Winter F, Tannemaat MR, Blits B, Malessy MJA, Verhaagen J (2015) Gene therapy and peripheral nerve repair: a perspective. Front Mol Neurosci 8:32. https://doi.org/10.3389/fnmol.2015.00032

Jabaley ME, Wallace WH, Heckler FR (1980) Internal topography of major nerves of the forearm and hand: a current view. J Hand Surg 5:1–18

Kao CC (1974) Comparison of healing process in transected spinal cords grafted with autogenous brain tissue, sciatic nerve and nodose ganglion. Exp Neurol 44:424–439

Kao CC, Chang LW, Bloodworth JMB (1977) Axonal regeneration across transected mammalian spinal cords: an electron microscopic study of delayed microsurgical nerve grafting. Exp Neurol 54:591–615

Keilhoff G, Fansa H (2011) Mesenchymal stem cells for peripheral nerve regeneration. a real hope or just an empty promise. Exp Neurol 232:110–113

Korres DS, Benetos IS, Themistocleous GS, Mavrogenis AF, Nikolakakos L, Liantis PT (2006) Diving injuries of the cervical spine in amateur divers. Spine J 6:44–49

Kosel KC, Wilkinson JM, Jew J, Itaya SK, Beckwith K, Williams TH (1979) Enzyme therapy and spinal cord regeneration. A fluorescence microscopic evaluation. Exp Neurol 64:365–374

Kuypers HGJM (1964) The descending pathways to the spinal cord, their anatomy and function. In: Eccles JC, Schadé JP (eds) Organization of the spinal cord. Progr Brain Res II, pp 178–202. Elsevier, Amsterdam

Kwon BK, Okon E, Hillyer J, Mann C, Baptiste D, Weaver LC, Fehlings MG, Tetzlaff W (2011a) A systematic review of non-invasive pharmacologic neuroprotective treatments for acute spinal cord injury. J Neurotrauma 28:1545–1588

Kwon BK, Okon EB, Plunet W, Baptiste D, Fouad K, Hillyer J, Weaver LC, Fehlings MG, Tetzlaff W (2011b) A systematic review of directly applied biologic therapies for acute spinal cord injury. J Neurotrauma 28:1589–1610

Kwon BK, Soril LJJ, Bacon M, Beattie MS, Blesch A, Bresnahan JC, Bunge MB, Dunlop SA, Fehlings MG, Ferguson AR, Hill CE, Karimi-Abdolrezaee S, Lu P, McDonald JW, Müller HW, Oudega M, Rosenzweig ES, Reier PJ, Silver J, Sykova E, Xu X-M, Guest JD, Tetzlaff W (2013) Demonstrating efficacy in preclinical studies of cellular therapies for spinal cord injury-How much is enough? Exp Neurol 248:30–44

Lakke EAJF, Marani E (1991) Prenatal descent of rubrospinal fibers through the spinal cord of the rat. Comp Neurol 314:67–78

Landers ZA, Lee SK, Mancuso CA, Wolfe SW (2015) The psychological impact of brachial plexus injury. http://peripheralnerve.org/meeting/abstracts/2015/27

Lanska DJ (2009) Historical perspective: neurological advances from studies of war injuries and illnesses. Ann Neurol 66:444–459

Lanska DJ (2016) The influence of the two world wars on the development of rehabilitation for spinal cord injuries in the USA and Great Britain. In: Tatu L, Bogousslavsky J (eds) Frontiers of neurology and neuroscience, war neurology, vol 38, pp 56–67

Lee JK, Zheng B (2012) Role of myelin inhibitors in axonal repair after spinal cord injury. Exp Neurol 235:33–42

Litaert Peerbolte M (1932) De loop van de achterwortelvezels, die in de achterstrengen van het ruggemerg opstijgen. Thesis Leiden University, The Netherlands

Liverman CT, Altevogt BM, Joy JE, Johnson RT (2005) Spinal cord injury: progress. The National Academies Press, Washington DC, Promise and Priorities

Macaya D, Spector M (2012) Injectable hydrogel materials for spinal cord regeneration: a review. Biomed. Mater 7. https://doi.org/10.1088/1748-6041/7/1/012001

Magenis TP, Freeman LW, Bowman DE (1952) Functional recovery following spinal cord hemisection and intrathecal use of hypochlorite treated trypsin. Fed Proc 11:99
Malessy MJ, van Duinen SG, Feirabend HK, Thomeer RTWM (1999a) Correlation between histopathology findings in C5 and C6 nerve stumps and motor recovery following nerve grafting for repair of brachial plexus injury. J Neurosurg 91:636–644
Malessy MJA, Hoffmann CFE, Thomeer RWTM (1999b) Initial report on the limited value of hypoglossal nerve transfer to treat brachial plexus root avulsion. J Neurosurg 91:601–604
Manola L, Roelofsen BH, HolsheimerJ Marani E, Geelen J (2005) Modelling motor cortex stimulation for chronic pain control: electrical potential field, activating functions and responses of simple nerve fibre models. Med Biol Eng Comput 43:335–343
Manola L, Holsheimer J, Veltink P, Buitenweg JR (2007) Anodal vs cathodal stimulation of motor cortex: a modelling study. Clin Neurophysiol 118:464–474
Marani E (2017) Dorsal Root or Spinal Ganglion. In: Reference Module in Neurosci Biobehav Psychol, ISBN 9780128093245
Marani E, Lakke EAJF (2012) Peripheral nervous system topics. In: Mai JK, Paxinos G (eds) The human nervous system, 3rd edn. Elsevier/Academic Press, Amsterdam
Marani E, Schoen JHR (2005) A reappraisal of the ascending systems in Man, with emphasis on the medial lemniscus. Adv Anat Embryol Cell Biol 179:1–76
Martinez-Peres R, Munarriz PM, Paredes I, Cotrina J, Lagares A (2017) Cervical spinal cord injury without computed tomography evidence of trauma in adults: Magnetic resonance imaging prognostic factors. World Neurosurg 99:192–199
Matinian LA, Andreasian AS (1973) Enzyme therapy in organic lesions of the spinal cord. BRI Publications (1976), University California
McMasters RE (1962) Regeneration of the spinal cord in the rat. Effects of Piromen and ACTH upon the regenerative capacity. J Comp Neurol 119:113–125
McNeal DR (1976) Analysis of a model for excitation of myelinated nerve. IEEE Trans Biomed Eng 23:329–337
Melzack R, Wall PD (1965) Pain mechanism: a new theory. Science 150:971–978
Midha R (2004) Nerve transfers for severe brachial plexus injuries: A review. Neurosurg Focus 16 (5):Srticle 5
Miller J, Basbaum AI (1975) Topography of the projection of the body surface of the cat to cuneate and gracile nuclei. Exp Neurol 49:281–290
Mumenthaler M (1973) Neurologie. G. Thieme Verlag, Stuttgart
Narakas AO (1993) Lesions found when operating traction injuries of the brachial plexus. Clin Neurol Neurosurg 95:S56–S64
Narakas AO, Hentz VR (1988) Neurotization in brachial plexus injuries. Clin Orthop Relat Res 237:43–56
Nathan PW, Smith MC, Cook AW (1986) Sensory effects in man of lesions of the posterior columns and of some other afferent pathways. Brain 109:1003–1041
Nikullna E, Tidwell JL, Dai HN, Bregman BS, Filbin MT (2004) The phosphodiesterase inhibitor rolipram delivered after a spinal cord lesion promotes axonal regeneration and functional recovery. PNAS 101:8786–8790
Nógrádi A, Pajer K, Marton G (2011) The role of motoneuron transplants to restore the lost motor function of the injured spinal cord. Ann Anat 193:362–370
Oudega M, Lakke EAJF, Marani E, Thomeer RTWM (1993) Development of the rat spinal cord: Immuno- and enzyme histochemical approaches. Adv Anat Embryol Cell Biol 129:1–166
Oudega M, Rosano C, Sadi D, Wood PM, Schwab ME, Hagg T (2000) Neutralizing antibodies against neurite growth inhibitor ni-35/250 do not promote regeneration of sensory axons in the adult rat spinal cord. Neurosci 100:873–883
Pakarinen T-K, Leppaniemi A, Sihvo E, Hiltunen K-M, Salo J (2006) Management of cervical stab wounds in low volume trauma centres: Systematic physical examination and low threshold for adjunctive studies, or surgical exploration. Injury. Int J Care Injur 37:440–447
Peacock WJ, Shrosbree RD, Key AG (1977) A review of 450 stab wounds of the spinal cord. SAfr Med J 51:961–964

Pěgo AP, Poot AA, Grijpma DW, Feijen J (2001) Copolymers of trimethylene carbonate and ε-caprolactone for porous nerve guides: synthesis and properties. J Biomat Sci-Polym 12:35–53
Pěgo AP, Poot AA, Grijpma DW, Feijen J (2003) Biodegradable elastomeric scaffolds for soft tissue engineering. J Controll Rel 87:69–79
Pettegrew RK (1976) Trypsin inhibition of scar formation in cordotomized rats. Anat Rec 184:501
Pondaag W, de Boer R, van Wijlen-Hempel RS, Hofstede-Buitenhuis SM, Malessy MJA (2005) Obstetric brachial plexus lesions: a framework for therapy. Neurosurg 57:530–537
Pouw MH, van de Meent H, van Middendorp JJ, Hirschfeld S, Thietje R, van Kampen A, EM-SCI study Group, Hosman AJF (2010) Relevance of the diagnosis traumatic cervical Brown-Séquard-plus syndrome: an analysis based on the neurological and functional recovery in a prospective cohort of 148 patients. Spinal Cord 48:614–661
Prestor B (2006) Microcoagulation of junctional dorsal root entry zone is effective treatment of brachial plexus avulsion pain: Long-term follow-up study. Croat Med J. 47:271–278
Prodanov D, Feirabend HKP (2007) Morphometric analysis of the fiber populations of the rat sciatic nerve, its spinal roots, and its major branches. J Comp Neurol 503:85–100
Prodanov D, Marani E, Holsheimer J (2003) Functional electric stimulation for sensory and motor functions: progress and problems. Biomed Rev (Varna) 14:23–50
Prodanov D, Nagelkerke N, Marani E (2007) Spatial clustering analysis in neuroanatomy: applications of different approaches to motor nerve fiber distribution. J Neurosci Methods 160:93–108
Ramon y Cajal S (1928) Degeneration and regeneration of the nervous system, vol 1, 2, Oxford University Press, London
Rexed B (1964) Some aspects of the cytoarchitectonics and synaptology of the spinal cord. In: Eccles JC, Schadé JP (eds) Organization of the spinal cord. Progr Brain Res II. Elsevier, Amsterdam, pp 58–92
Robert R, Perrouin-Verbe B, Albert T, Bussel B, Hamel O (2009) Chronic neuropathic pain in spinal cord injured patients: What ist the effectiveness of surgical treatments and excluding central neurostimulations? Ann Phys Rehab Med 52:194–202
Romanes GJ (1964) The motor pools of the spinal cord. In: Eccles JC, Schadé JP (eds) Organization of the spinal cord. Progr Brain Res II. Elsevier, Amsterdam, pp 93–119
Samii M, Matthies C (1997) Management of 1000 vestibular schwannomas (acoustic neuromas): the facial nerve – preservation and restitution of function. Neurosurg 40:684–695
Sankarasubramanian V (2013) Model-based stimulation optimization for chronic pain suppression using percutaneous and surgical leads in spinal cord stimulation (SCS). Thesis University Twente
Schoen JHR (1964) Comparative aspects of the descending fibre systems in the spinal cord. In: Eccles JC, Schadé JP (eds) Organization of the spinal cord. Progr Brain Res II. Elsevier, Amsterdam, pp 203–222
Schoenen J, Grant G (2004) Spinal cord: connections. In: Paxinos G, Mai JK (eds) The human nervous system. Elsevier/Academic Press, San Diego, pp 233–249
Seddon HJ (1943) Three types of nerve injury. Brain 66:237–288
Siddiq MM, Hannila SS (2015) Looking downstream: the role of cyclic AMP-regulated genes in axonal regeneration. Front Mol Neurosci. https://doi.org/10.3389/fnmol.2015.00026
Singh A, Goyal N, Gupta DK, Mahapatra AK (2011) An overview of spinal injuries in children: Series of 122 cases. Indian J Neurotrauma (IJNT) 8:25–32
Singh A, Tetreault L, Kalsy-Ryan S, Nouri A, Fehlings MG (2014) Global prevalence and incidence of traumatic spinal cord injury. Clin Epidemiol 6:309–331
Smith MC, Deacon P (1984) Topographical anatomy of the posterior columns of the spinal cord in man. Brain 107:671–698
Snow PJ, Wilson P (1991) Pasticity in the somatosensory system of developing and mature mammals- The effects of injury to the central and peripheral nervous system. In: Ottoson D (ed) Progress in sensory physiology 11. Springer, Berlin, Heidelberg
Spinner RJ, Shin AY, AT (2004) Brachial plexus surgery, the "orphan drug," whose time has come. Neurosurg Focus 16, 5, Introduction

Stepanian-Tarakanova AM (1959) Traumatic diseases of the spinal cord. Medgiz, Moscow, pp 154–174
Steward JD (2003) Peripheral nerve fascicles: anatomy and clinical relevance. Muscle Nerve 28:525–541
Stichel-Gunkel CC (1997) The role of microenvironment in axonal regeneration. Adv Anat Embryol Cell Biol 137:1–81
Stieglitz T, Meyer JU (2006) Neural implants in clinical practice. Biomems, Springer, Berlin
Strother M, Selketderivative work: mcstrother - Brachial_plexus.svg, CC BY-SA 3.0, https://commons.wikimedia.org/w/index.php?curid=8833181
Struijk JJ, Holsheimer J, van der Heide CG, Boom HBK (1992) Recruitment of dorsal column fibers in spinal cord stimulation:influence of collateral branching. IEEE Trans Biomed Eng 39:903–912
Su H, Yuang Q, Qin D, Yang X, Wong W-M, So K-F, Wu W (2013) Ventral root re-implantation is better than peripheral nerve transplantation for motoneuron survival and regeneration after spinal root avulsion injury. BMC Surg 13:21–25
Sunderland S (1945) The internal topography of the radial, median and ulnar nerves. Brain 68:243–298
Sunderland S, Ray LJ (1948) The intraneural topography of the sciatic nerve and its popliteal divisions in man. Brain 71:242–273
Sunderland S (1974) Mechanisms of cervical root avulsions. J Neurosurg 41:704–714
Taleb C, Nectoux E, Awada T, Liverneaux P (2013) The destiny of an ace: Algimantas Otanas Narakas (1927–1993). J Brachial Plex Peripher Nerve Inj 8:6. https://doi.org/10.1186/1749-7221-8-6
Tannemaat MR, Eggers R, Hendriks WT, de Ruiter GCW, van Heerikhuize JJ, Pool CW, Malessy MJA, Boer GJ, Verhaagen J (2008) Differential effects of lentiviral vector-mediated overexpression of nerve growth factor and glial-cell line derived neurotrophic factor on regenerating sensory and motor axons in the peripheral nerve. Eur J Neurosci 28:1467–1479
Tetzlaff W, Okon EB, Karimi-Abdolrezaee S, Hill CE, Sparling JS, Plemel JR, Plunet WT, Tsai EC, Baptiste D, Smithson LJ, Kawaja MD, Fehlings MG, Kwon BK (2011) A systematic review of cellular transplantation therapies for spinal cord injury. J Neurotrauma 28:1611–1682
Thomas DGT, Sheeny JPR (1983) Dorsal root entry zone lesions (Nashold's procedure) for pain relief following brachial plexus avulsion. J Neurol Neurosurg Psych 46:924–928
Thomeer RTWM, Malessy MJA, Marani E, Carlstedt T (2002) Nerve root repair (Letters). J Neurosurg 96(1):138–139
Usunoff K, Marani E, Schoen JHR (1997) The trigeminal system in man. Adv Anat Embryol Cell Biol 136:1–106
Van Gehuchten A, Nelis C (1899) La localisation mortice médullaire est une localistaion segmentair. J Neurol (Brux) 4:301–310
Van Lier EH (1905) Ueber Nervenkompression. Beitrage klin Chirurgie, XLVI 3:711–721
Vesalius A (1964) Nervorum in universam. In: De humani corporis fabrica libri Septum Facsimilé (no. 379) Culture et Civilisation, Bruxelles, 1543, pp 344–349
Vleggeert-Lankamp CLAM (2007) The role of evaluation in the assessment of peripheral nerve regeneration through synthetic conduits: a systematic review. J Neurosurg 107:1168–1189
Vleggeert-Lankamp CLAM (2008) The role of evaluation methods in the assessment of peripheral nerve regeneration through synthetic conduits: a systematic review. J Neurosurg 107:1168–1189
Vleggeert-Lankamp CLAM, van den Berg RJ, Feirabend HKP, Lakke EAJF, Malessy MJA, Thomeer RTWM (2004) Electrophysiology and morphometry of the Aα- and Aβ-fibre populations in the normal and regenerating rat sciatic nerve. Exp Neurol 187:337–349
Vleggeert-Lankamp CLAM, de Ruiter GCW, Wolfs JFC, Pego AP, van den Berg RJ, Feirabend HKP, Malessy MJA, Lakke EAJF (2007) Pores in synthetic nerve conduits are beneficial to regeneration. J Biomed Mat Res, A 80:965–982
Voelker A, von der Hoeh NH, Gulow J, Heyde C-E (2015) Cervical spondylodiscitis with epidural abscess after knife stab wounds to the neck: a case report. Injury Int J Care Injur 46:1684–1689

Voogd J (2004) Cerebellum and precerebellar nuclei. In: Paxinos G, Mai JK (eds) The human nervous system. Elsevier/Academic Press, Amsterdam, pp 321–392

Voogd J, Nieuwenhuys R, van Dongen PAM, ten Donkelaar HJ (1998) Mammals. In: Nieuwenhuys R et al (eds) The central nervous system of vertebrates, vol 3. Springer Berlin

Voormolen JHC, Thomeer RTWM, Marani E, De Beer F, Bots GTAM (1990) Enzyme histochemistry of the spinal cord after experimental transection (Th 9) in the cat. Eur J Morphol 28:404–417

Vroemen M (2006) Cellular therapy after spinal cord injury using neural progenitor cells. Thesis Leiden

Vroemen M, Weidner N, Blesch A (2005) Loss of gene expression in lentivirus- and retroviris-transduced neural progenitor cells is correlated to migration and differentiation in the adult spinal cord. Exp Neurol 195:127–139

Watchmaker GP, Gumucio CA, Crandall RE, Vannier MA, Weeks PM (1991) Fascicular topography of the median nerve: a computer based study to identify branching patterns. J Hand Surg 16:53–59

Willis WD, Al-Chaer ED, Quast MJ, Westlund KN (1999) A visceral pain pathway in the dorsal column of the spinal cord. PNAS 96:7675–7679

Wilson DH (1984) Peripheral nerve implants in the spinal cord in experimental animals. Paraplegia 22:230–237

Windle WF (1955) Comments on regeneration in the human central nervous system. In: Windle WF (ed) Regeneration in the central nervous system. Springfield, Ill, C.C. Thomas, pp 265–271

Windle WF (1976) Foreword. In: Enzyme therapy in organic lesions of the spinal cord. BRI Publications. Univ California

Windle WF, Chambers WW (1950) Regeneration in the spinal cord of the cat and dog. J Comp Neurol 93:241–257

Winkler C (1917) Handboek der Neurologie I. De Erven F, Bohn, Haarlem

Wu H-F, Cen J-S, Zhong Q, Chen L, Wang J, Den DYB, Wang Y (2013) The promotion of functional recovery and nerve regeneration after spinal cord injury by lentiviral vectors encoding for Lingo-1 shRNA delivered by Pluronic F-127. Biomat 24:1686–1700

Wyndaele M, Wyndaele J-J (2006) Incidence, prevalence and epidemiology of spinal cord injury: what learns a worldwide literature survey? Spinal Cord 44:523–529

Young W (2002) Spinal cord injury levels and classification. www.Sci-info-pages.com/levels.html

Zang X-Y, Yang Y-M (2016) Scissors stab wound to the cervical spinal cord at the craniocervical junction. Spine J 16:e403–e406

Zhang Y, Zhang X, Wu D, Verhaagen J, Richardson PM, Yeh J, Bo X (2007) Lentiviral-mediated expression of polysialic acid in spinal cord and conditioning lesion promote regeneration of sensory axons into spinal cord. Molecul Therap 15:1796–1804

Chapter 12
Cranial Nerves and Cervical Spinal Nerves

The historical drive for the study of the brain and its cranial nerves was the localization of the soul's interaction site with the brain: a dualistic approach. The primitive organization of the cranial nerves has been used to support the historical studies of Vesalius, Willis and Soemmerring for the understanding of our nowadays subdivision of the cranial nerves. Willis's use of the sheep's brain and its consequences for brain topography is discussed. The transition from cord to brainstem gives an intravertebral reorganization that is considered. The XIIIth (nervus terminalis) and XIVth (nervus intermedius) cranial nerves are treated. Gustatory and salivary pathways in relation to crocodile tears and the evolution of the vagus–accessory–hypoglossal complex are described. The trigeminal complex is considered by its evolutionary aspects, masticatory reflex, its ganglion, and its sensory and pain information transmission. The vagus nerve is considered by its topography, recurrent nerve and nerve stimulation.

12.1 Introduction

12.1.1 Dualism

In this chapter, the dualistic drive for studying the brain and its cranial nerves will be encountered regularly. For both Leonardo da Vinci (1452–1519) and Thomas Willis (1621–1675), their ambition was the detection of the localization of the soul's interaction site with the brain. "The soul was considered a non-physical substance in causal relation with the body. Mental reality (Res cogitans) is different from the physical world (Res existensa), since the physical world obeys laws in a causal closed system. Mind or soul is outside this system. It does not obey the causal closed system laws; e.g. mind or soul enjoys uncaused choice" (gathered from Smith-Churchland 1986). Descartes (1596–1650) placed the location of the

E. Marani and C. Heida, *Head and Neck*,
https://doi.org/10.1007/978-3-319-92105-1_12

interaction between body and soul within the pineal gland of the brain. The pineal gland should function as "a valve regulating the flow of animal spirits" (Berhouma 2013). Anatomy already showed that the pineal gland is situated at the ventricular wall side and therefore could not function as a ventricular valve for the animal spirits (Berhouma 2013). Despite the criticism on Descartes proposal (Berhouma 2013), most other scientists placed the contact site of the soul in the brain or considered it brain related. Here, we need some words in defence of Descartes: the third ventricle has an extension into the pineal gland, called the recessus pinealis (Villiger 1945), and as such the gland could have a valve function, although a minimal one. But the most important consequence is that Descartes indicated a brain structure that anatomists could examine. You could take the seat of the soul, as other scientists called it, in your hands after opening of the skull.

The world population in general advocates dualism since dualism is strongly interwoven with religion. Religion could and can be an obstacle in studying the brain. However, most scientists and philosophers deny dualism. (One should be aware, although denied, still implicitly dualistic ideas hold in a lot of reasoning in science.) Different types of dualism are discerned. Here, two are mentioned. *Substance dualism*: mind is not identical to the brain. Mental states are dependent on another substance, independent of the body or brain. *Property dualism* (*emergentism*): brain produces mental states, but these mental states cannot be reduced to brain states. Mental states are an extra property of matter or brain. To give an example: water has the property wet. However, one water molecule is not wet. The mutual contact between series of water molecules brings forward the property of being wet. Wet is an emerging property that cannot be reduced to a property of one water molecule. However, the dipole property of one water molecule organizes that several water molecules have to arrange in such a way that the property wet originates.

The substance dualist vision has "a special problem with interpretation, namely how to reconcile the results with their conception of the mind or soul as a non-physical substance in causal interaction with the nervous system" (Smith-Churchland 1986). Property dualism is roughly indicated by Smith-Churchland (1986) as a "subjective experience that has a character and quality uniquely and irreducibly mental…even if the mind is the brain". Although subjectivity cannot be coupled clearly to network functions of the brain until now, the scientific expectation is that with enough knowledge we can. Or in other words: the materialist expectation is that mind with its subjective experience of man will be reduced to brain network states within time. Nevertheless, there is a serious problem, even for the materialist: no general accepted definition of mind or soul exists. How then can we have knowledge (described as true justified belief) if we cannot even define mind!

The notion mind happily contains subnotions like memory, intelligence, awareness. Theoretical subnotion definitions of, for example, intelligence exist, thinking of IQ that we even can measure. Thus, one can make "operational" definitions (to design experiments) of subparts of the notion mind. Execute your observations or measurements and compare with theoretical definition or operational definition. Due to this approach, psychologists can test IQ and indeed assess memory capacity (important in diagnosis of a lot of neurodegenerative diseases). It is beyond dispute that memory is centralized in the hippocampus as demonstrated for amnesic patients (Press et al. 1989). Thus, we know the place in the brain where it can go wrong. All this information on subnotions of mind will bring us closer to understanding and defining mind (or soul as ancient scientists termed it).

The start of this chapter brought up intelligence, and the first subject one will think of is artificial intelligence. Its definition is "Thinking by, or by means of, machines". The consequence is that thinking is not a purely human property anymore, if somebody succeeds to let machines think. One should ask "What is thinking"? Let us follow the "operational" approach as described above, because there is also no good definition of thinking. Its operational definition used is "Manipulation of views (notions) or of mental words". Ask yourself: Do you think in pictures or words? Do you think in your own language? How do you think in another language? In addition, do not need your mental notions a representation in reality?

Alan Turing (1912–1954) tried to make a machine that could think called Pilot ACE (Automatic Computing Engine). Steps in its development made were time consuming, but research is progressing enormously nowadays. "Turing's discovery about the power of machines grew out of his interest, as a mathematician, in mathematics. But he then applied it to psychology (his paper of 1950 was addressed to the question "Can machines think?), drawing the conclusion that every human thought, provided it could be expressed in *language*, could be mimicked by his universal machine if it was suitable programmed" (Pratt 1987). How did you think, you said above?

The operational thinking approach not only tried to understand the inherent properties of subnotions of our mind, but some of its mental functions are attempted to be placed in machines: the thinking computer. "However, we have not got a clue about how to get a machine to feel, for example, sad" (Pratt 1987).

Thinking inevitably brings us to the separation between man and animal: Do animals think? Here, we turn to Thomas Willis. (Willis will turn up several times in this chapter.) The distinction between man and animal forced at least Willis to discern three types of soul or mind: sensitive, rational and immortal soul. The first two are from Aristotle (384–324 BC), the third is also Aristotle's, but strongly changed by Christian theology. The immortal soul was possessed by man and was lacking in animals: "he thought that the immortal soul had an implicit place in the human brain; the anatomical structure of humans indicated, he thought, a unique capacity for metaphysical contemplation that animals did not possess" McNabb (2014). One should note that mind (nous) is a component of the soul for Aristotle: "the part of the soul by which it knows and understands". Willis intermingles soul and mind in his text: it even reads as being identical items.

Willis dedicated his Cerebri anatome: "to unlock the secret places of Man's mind" to the Archbishop of Canterbury, because Gilbert Sheldon (1598–1677), before as Bishop of London, helped Willis to become professor in Oxford in 1660.

Moreover, the Archbishop together with others decided over the Licensing Act of 1662 permitting the publication of books on medicine and natural science. "There is no suggestion, however, that this provided any motivation for Willis's dedication" (Feindel 1964). Willis ideas became accepted by the church and fitted his studies into his own belief. (One should remember that during the Cromwellian period Anglican services were held at Willis's Oxford lodgings.) The dualistic drive of "finding" the soul or mind forced the study of the brain and its cranial nerves, but on the other hand could be dragged if the results were not in harmony with the church and/or belief.

12.1.2 Promorphology of Cranial Nerves

The conservatism in the nervous system makes it possible to discern groups of head structures and their specialized cranial nerves for most vertebrates:

(1) Somatic motoric nerves: eye and tongue muscle nerves (III, IV, VI, XII)
(2) Branchial nerves (terminalis, [III], V, VII, IX, X, XI)
(3) Special somatic sensory nerves (I, II, VIII)

In the primitive spinal cord, the dorsal and ventral nerves are separated, meaning that each has its own exit and is independent of the other (Fig. 12.1). Comparison between the primitive dorsal and ventral roots does not result in any substantial correspondence. Moreover while dorsal roots normally are afferent, efferent fibres leave the spinal cord also via the dorsal root in the primitive situation. Originally both Bell (1774–1824) and Magendie (1783–1855) demonstrated that dorsal radiculi are sensory and ventral ones are motoric, called the law of Bell and Magendie. Still ventral root afferents and dorsal root efferents are described, even in humans (Coggeshall 1980; Schalow 1992; Phillips et al. 2000).

A second aspect is the volume taken by the nervous system in the head:

> "By definition elements that are subordinate with respect to the property presence will have no or hardly any influence on the pattern when they appear. However, when do we say they have appeared? Apparently, the answer can only be given arbitrarily, depending on the definition of the element and the level of organization. Conversely, dominant elements have a strong influence on early development, and stage and time may differ.
>
> … Therefore, in the head the central nervous system seems to be dominant in all stages, followed by sense organs and then the pharyngo-oral cavity. First, the head is mainly determined by the developing brain, the facial area develops later on. In fact, the brain probably determines all the elements following in ontogeny, as is observable for the development of the eye and the ear. The changes are that this also holds for the nasal cavity and the mouth….. To all appearance the presence of sensory cells and muscles depends on the nervous tissue (Dullemeijer 1974).

In summary: the cranial nerves as an expression of the nervous system influence most organs in head and neck in their development. Cranial nerves hardly show

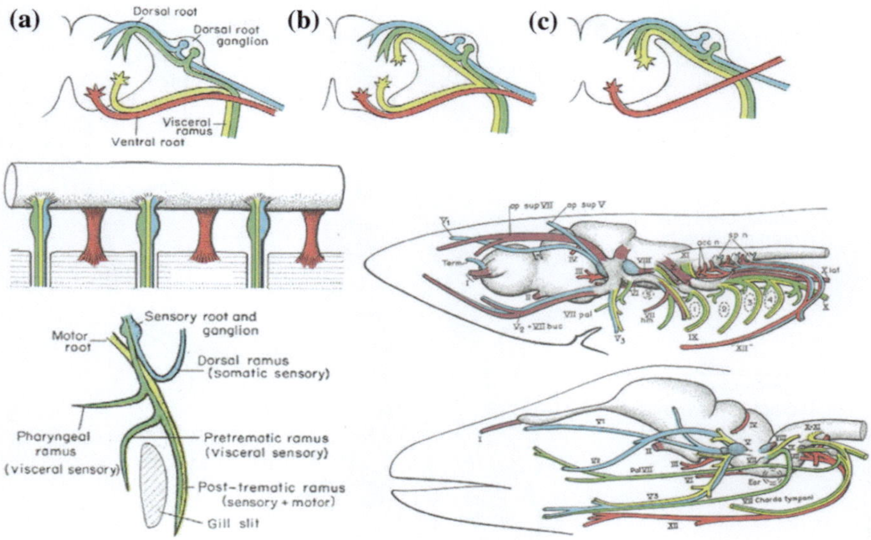

Fig. 12.1 These diagrams are all from Romer (1962, with permission) and show the nerve components in cranial and spinal nerves. Blue is somatic sensory; green, visceral sensory; yellow visceral motor and red somatomotor. Brown indicates the special somatic sensory systems In the upper part figure **a** is the mammalian situation: dorsal incoming sensible and ventral outgoing motoric, **b** is a more primitive situation present in lower vertebrates, in which visceral motor fibres leave via the dorsal root, and **c** is the primitive situation: ventral root exclusively maintained motoric fibres. The ventral and dorsal roots in the primitive situation have their proper sites for leaving the spinal cord (part figures below the **a**, **b** and **c** characterization, lamprey left side). The scheme of a typical branchial nerve in relation to the gill slits is shown. The cranial nerves are colour mapped for the shark's (Squalus, upper) and the lizard's (Anolis, lower) brain. Several details and abbreviations are omitted.

correspondence with combined spinal ventral and spinal dorsal root properties. Cranial nerves can contain solely afferent, solely efferent or a mixed appearance.

12.2 Survey of History

The cranial nerves count 12 nowadays, which was not always the case. Till the studies of Samuel Thomas Soemmering ([1775–1830]; Hildebrand 2011), we had nine or ten cranial nerves (Fig. 12.2). Even now the amount of cranial nerves is increasing by a "zero" cranial nerve and publications on a 13th and 14th cranial nerve (Bordoni and Zanier 2013). The amount is certainly not fixed, but for students kept on 12. Order has also been changed: the optic nerve was lost in the Jenenser Nomina Anatomica, since it was considered an extension of the diencephalon, being a brain tract and not a nerve. Thus, the olfactory nerve was I and the zero cranial nerve became cranial nerve II. After this "failure" of the Nomina Anatomica Committee, the situation was restored in 1955 by the Anatomical Congress in Paris

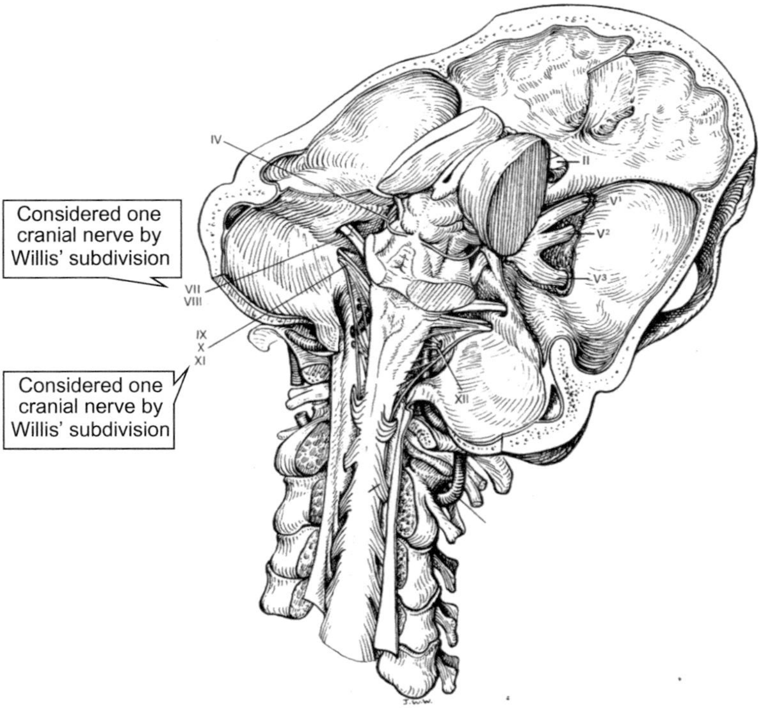

Fig. 12.2 Dorsal view on an opened skull with vertebral column, spinal cord and brainstem to show spinal and cranial nerves in the head and neck region. Cranial nerves VII and VIII, and IX, X and vagal XI were considered one cranial nerve in Willis's subdivision. Willis suboccipital nerve constituting the tenth pair of cranial nerves had already been placed among the cervical nerves by others (figure is adapted and with permission, J.M.F. Landsmeer†)

(Winkelmann 2014). Focus is on the accessory cranial nerve in this historical survey: you hardly can miss the olfactory or optic cranial nerves, but the characteristics of the accessory nerve ask for a special research quality as we will see.

The ancient amount of nine or ten cranial nerves comes from Thomas Willis (1621–1675), who described the circle of blood vessels at the brain base. He co-founded the Royal Society and was one of the first elected fellows. "It is difficult to decide how much of the important research in comparative anatomy associated with the name Willis we owe to him. It is probable that all the dissections were the work of others, and even the text is perhaps not wholly his own" (Cole 1975). Cole's statement seems rude and superfluous. However, this remark is clearly supported, true in a more positive way: "Following the common practice of the seventeenth century, Willis's research into anatomy, and the 'anatomy of the brain' (the translation of Cerebri anatome) in particular, was a collaborative effort. A number of individuals deserve credit for the result of the anatomy. Ralph Bathurst (1620–1704) may have helped revise the anatomy of the brain. Thomas Millington (1628–1704), who succeeded Willis's Professorship and later became

President of the Royal College of Physicians, conversed and helped analyse Willis's research. Christopher Wren (1632–1723) participated in the development of procedures to adequately dissect the brain, and he drew many of the illustrations for Willis's text. In like regard, Richard Lower (1631–1691, we will encounter him later on) also illustrated the anatomy and contributed to Willis's research. Other notables in Willis's circle include his apprentice Robert Hooke (1635–1703) who was an assistant to Willis and probably upon his recommendation later worked under Robert Boyle (1627–1691)" (McNabb 2014; see also Feindel 1964; and Donaldson 2010, for a more extensive description).

Thus, pinpointing to whom we owe the ten cranial nerve maps, so strongly defended by the Anglo-Saxons, after Soemmerring's proposal, is unclear, and Willis got the assets. Willis's Cerebri anatome: Cui accessit nervorum descriptio et usus (Anatomy of the brain with description and use of the nerves) contains not only the human brain but also the sheep's brain description. The comparative aspect will be treated later on.

First attention is paid to the history of the description of the cranial nerves to understand the development towards twelve cranial nerves and how functions were coupled to cranial nerves. Modern anatomy of the brain and its cranial nerves start with Leonardo da Vinci (1452–1519), since for surgeons, till into the Renaissance, the Hippocrates (c.460-c.370BC)/Galen (AD 130–200) rule was not to pass the dura mater, which was obeyed even in Arabic times (see Albucasis [936–1013], ±1000). At the turn of the fifteenth to sixteenth century, Leonardo da Vinci worked on the anatomy of the human body. Together with Marcantonio della Torre (1481–1511), he wanted to produce a "Treatise on anatomy" but never published it, due to the sudden death of della Torre from the plague. "The anatomical drawings of Leonardo da Vinci span the years between 1487 and 1513, but only in the latter half of this time did Leonardo have detailed knowledge gained through dissection. Da Vinci's source of information in the early years could only have been gleaned from textbooks, and, as a result, Leonardo spent much of his time producing drawings which tried to explain the conflicting viewpoints present in these books" (Morris 1986).

Leonardo was fascinated by the soul and its location within the body. Using frog experiments, Leonardo found that the soul must be located in the brain. The soul or "senso comune" was placed by Leonardo just above the chiasma of the optic nerves within the frontal part of the third ventricle after thorough skull, brain, cranial nerves and eye studies (see Del Maestro 1998). His search for the localization of the soul forced Leonardo to study the brain with its cranial nerves (Fig. 12.3). One can recognize the olfactory and optic cranial nerves, but also the three cranial nerves for the eye muscles (RL 19052r). By Leonardo's typesetting in his figures at least ten cranial nerves (KK 6287v; Fig. 12.3) can be discerned. It is clear that the facial nerve and the subdivisions van de trigeminal nerve are depicted. The cervical spinal nerves (RL19034v) are given, and the brachial plexus (RL19021v; RL19040 v, r) is depicted as a mathematical construct. Leonardo thus knew the general innervation that belongs to head and neck, not totally but enough to understand the basic organization of the peripheral nervous system of the upper part of the body.

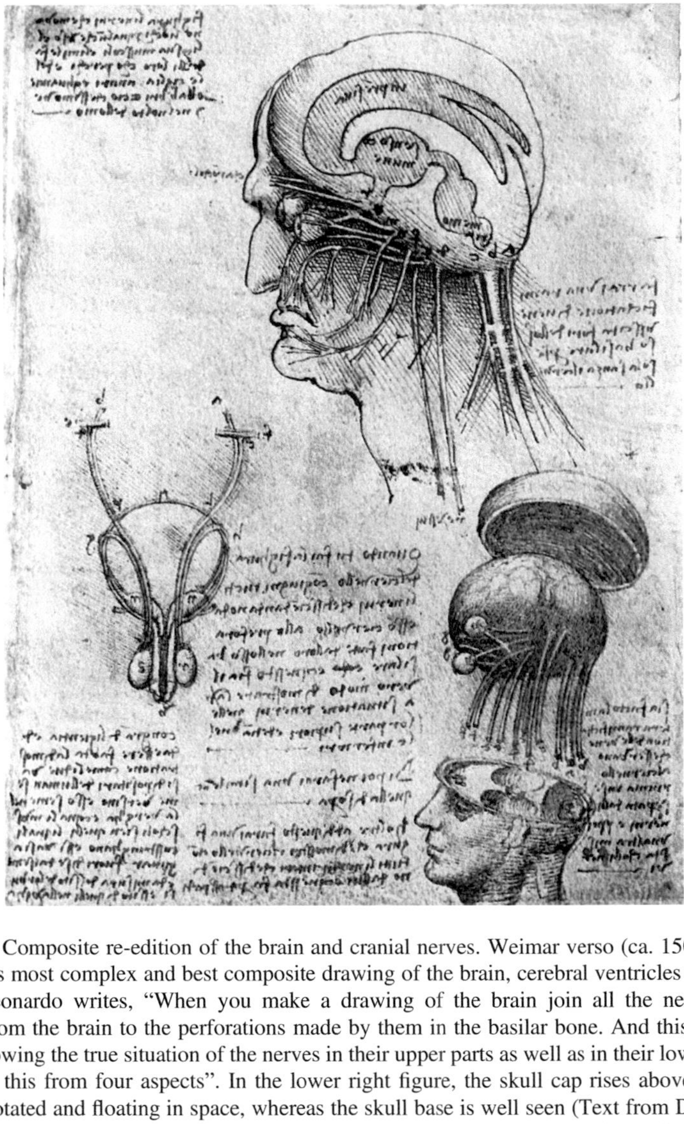

Fig. 12.3 Composite re-edition of the brain and cranial nerves. Weimar verso (ca. 1508). This is Leonardo's most complex and best composite drawing of the brain, cerebral ventricles and cranial nerves. Leonardo writes, "When you make a drawing of the brain join all the nerves which descend from the brain to the perforations made by them in the basilar bone. And this is the true way of showing the true situation of the nerves in their upper parts as well as in their lower parts … and do all this from four aspects". In the lower right figure, the skull cap rises above the brain, which is rotated and floating in space, whereas the skull base is well seen (Text from Del Maestro 1998; Reproduced with permission from Kunstsammlungen zu Weimar, Weimar, Germany)

Leonardo's anatomical results were not disseminated in the scientific world and became known far after his death.

We owe to Vesalius ([1514–1564]; 1543) a detailed macroscopic map of the peripheral nervous system including several of the cranial nerves. The map of the 'nervous man', the view on the base of the brain and extra details of the peripheral nervous system in several drawings (see Fig. 12.4), together with the spinal and brainstem origin of spinal and cranial nerves, laid the basis of the anatomy of the

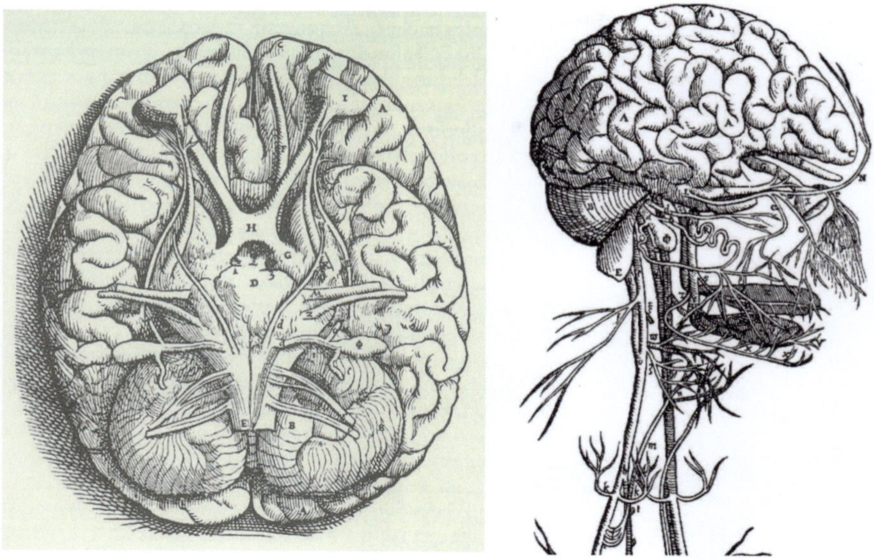

Fig. 12.4 Pictures of Vesalius (1543): Left: Base of the brain with its cranial nerves. Right: Lateral view on the brain and its main nerves towards head and neck. A, cerebrum; B, cerebellum; D and E, dorsal side of the medulla oblongata; F, smell organ(clearly the tractus olfactorius, Note that a bulbus olfactorius is not indicated); G, (optic tract) branch into the cerebrum; H, (chiasma) crossing of the eye nerves; I, tunica of the eye; K, second cranial nerve; L, third cranial nerve (note that it appears from the cerebrum); M, large part of the third cranial nerve; N and Q, small part of third cranial nerve (together with b and X they represent our three branches of the trigeminal nerve) Z, fourth cranial nerve. The brainstem nerves IX to XII are hardly indicated in the figures. Φ presumably the combination of the sup. and inf. ggl of IX and X or a wrongly placed ganglion N V (Gasseri) (Figures from author's facsimile specimen of Vesalius' Humani corporis fabrica)

peripheral nervous system. The quality of the pictured dissections and the discussions of Galen's conception of those days in book four 'Nervis proprius' make Vesalius studies the starting point of modern macroscopy of the peripheral nervous system. Data interpretation was restricted, and the descriptions of the cranial nerves were limited. The successors and contemporaries of Vesalius extended the knowledge of the peripheral nervous system: Bartolomeo Eustachi (1500/1514/1520–1574), innervation of muscles, Gabriele Fallopio (1523–1562) brain and peripheral nervous system. However, the variability present in the peripheral nervous system was hardly recognized.

Blasius's (1625–1692) research on the spinal cord of 119 animals showed the anterior and posterior roots present as a recurrent phenomenon of the spinal cord (Pearce 2008). Huber (1707–1778) produced the first correct description of the spinal nerves. Although nearly forgotten Huber's figures are still comparable to the recent representations found in handbooks (Naderi et al. 2004; Rengachary et al. 2008). The simple numbering from 1 till 31 of the spinal nerves stayed present next

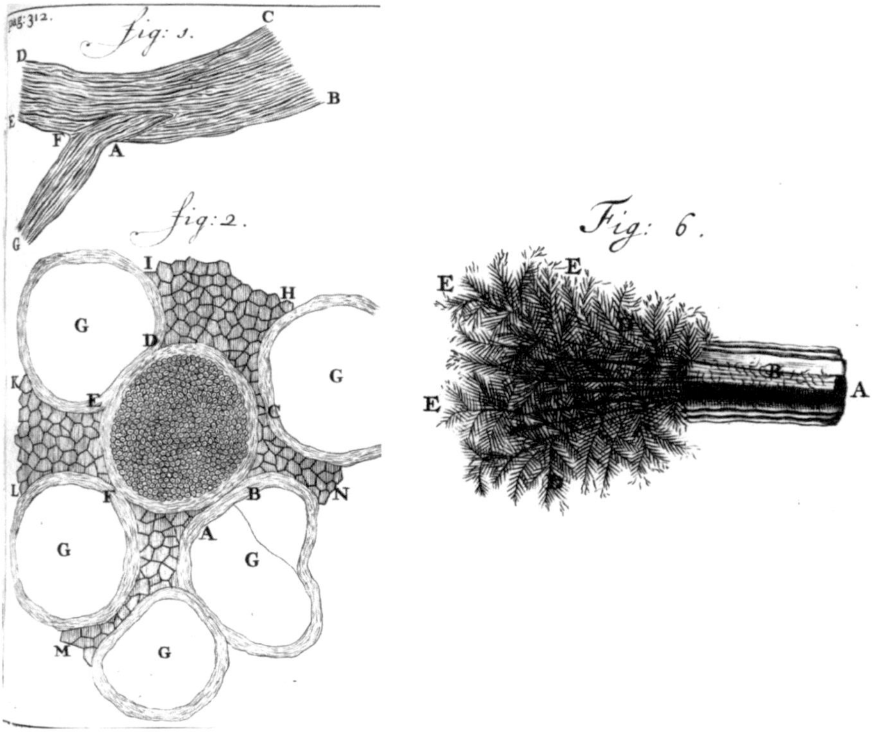

Fig. 12.5 Left: Microscopic picture of Anthony van Leeuwenhoek's longitudinal and transverse sections of nerve published in Letters of Antoni Van Leeuwenhoek, Royal Netherlands Academy of Arts and Sciences, 1939–2010 (Library Boerhaave museum, Leiden, with permission). Right: Bidloo's squash preparation of a nerve, studied with Van Leeuwenhoek's microscope (Leiden University Library, with permission)

to the regional subdivision in C, T, L, S, Co till around 1920 (e.g. Bolk 1910)" (Marani and Lakke 2012).

It lasted long before the content of nerves and their function became known. Microscopy of the vertebrate peripheral nervous system was initiated by Van Leeuwenhoek (1632–1723; see Schierbeek 1951) and Bidloo (1649–1713). Part of Van Leeuwenhooks' picture is still the front page of Neuroscience. However, it depicts only one fascicle in which myelinated fibres are pictured (Fig. 12.5), although Van Leeuwenhoek named them "hollow tubes". He had already shown several fascicles and the perineurium that surrounded a fascicle. A longitudinal section shows the splitting of a nerve branch from its main branch. Bidloo (1685) knew the work of Van Leeuwenhoek and published in his Anatomia a squash preparation of a nerve studied with Leeuwenhoek's microscope (Fig. 12.5). Despite the fame of Van Leeuwenhoek, the extensive diffusion of Bidloo's Anatomia, which was plagiarized by Cowper, this research was forgotten for over nearly a century (Hildebrand 2011; Marani and Koch 2014). Hooke (1635–1703), by his

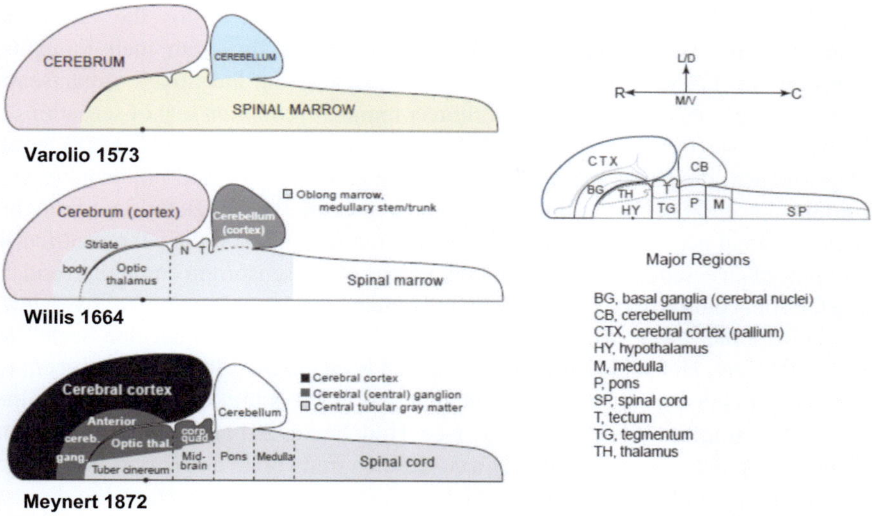

Fig. 12.6 Segmental subdivisions of the brain by Varolio, Willis and Meynert leading to the general nowadays accepted brain partition in ten parts (right part of the figure). This figure is a selection and adaption of Figs. 2 and 3 of Swanson (2000 with permission)

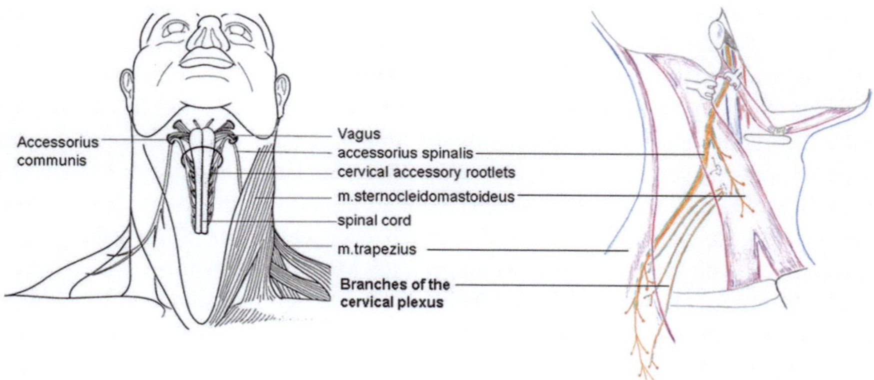

Fig. 12.7 Frontal and lateral view of the accessory nerve, innervating the mm. sternocleidomastoideus and trapezius (lateral view © E. Marani)

microscopic research of cork, introduced the term "cell" as used in modern times, although a change of definition occurred. However, the link between "cells" and "hollow tubes" was not recognized in those days.

Samuel Thomas Soemmerring's profound research on brain and nerves determined finally the acceptance of the twelve cranial nerves in the scientific circles of those days. Soemmerring's time was characterized by Goethe (1749–1832), they had an ambivalent relation with each other, but also by Samuel Hahnemann (1755–1843) founder of the homeopathy. Soemmerring appreciated to the so-called philosophische Arzte (philosophic physicians) and, in line with their thoughts, published in 1796 "Ueber das Organ der Seele" (on the soul's organ/seat). Soemmerring's attraction to the sensorium commune ("common seat of sensations") or localization of the soul originated from Ernst Platner's (1744–1818) book "Physiological questions", a mediator between exact science and philosophy, and from the detection of the origin of the trigeminal cranial nerve. Both ideas brought: 1) most cranial nerves were related to the ventricular wall, and thus, the ventricular fluid played an essential role (the medium uniens of the sensorium commune) and 2) here the soul could react on the body by influencing the cranial nerves (for overview, see: Hildebrand 2005, 2011). Soemmerring dedicated his "Organ der Seele" to Immanuel Kant. His publisher asked Immanuel Kant (1724–1804) for a comment to be placed as epilogue. Kant's argumentation was that it is impossible to localize the seat of the soul in a body or brain (see Hildebrand 2011). Note that also for Soemmerring the seat of the soul was a co-drive to study brain and nerves. He placed the soul in the fluids of the brain ventricles. Kant, in his epilogue, started to discuss the possibility that fluid could be organized. In fact, he taught fluid dynamics (Capozzi 2014).

One should look at Kant's end of the epilogue in Soemmerring's "Organ der Seele":

Über die eigentliche Aufgabe, wie sie nach Haller'n vorgestellt wird, ist hiermit nicht aufgelöst, sie ist nicht bloss physiologisch sodern sie soll auch zum Mittel dienen, die Einheit des Bewustseins seiner selbst (welche dem Verstande angehört) im Raumes verhältnisse der Seele zu den Organen des Gehirns (welches zum ausseren Sinne gehört) mithin den Sitz der Seele, als ihre locale Gegenwart vorstellig zu machen, welches eine Aufgabe für die Metaphysik, für diese aber nicht allein unauflösig, sodern auch an sich widersprechend ist. – Denn wenn ich den Ort meiner Seele, d.i. meines absoluten Selbt's, irgendwo im Raume anschaulig machen soll, so muss ich mich selbst durch eben denselben Sinn wahrnehmen wodurch ich auch die mich zunächst umgebende Materie wahrnehme; so wie dieses geschieht, wenn ich meinem Ort in der Welt als Mensch bestimmen will, nämlich dass ich meinem Körper ausser mir betrachten muss- Nun kann die Seele sich nur durch inneren Sinn, den Körper aber (es sei inwendig oder äusserlich) nur durch äussere Sinne wahrnehmen, mithin sich selbst slechterdings seinem Ort bestimmen weil sie sich zu diesem Behuf zum Gegenstand ihrer eigenen äusseren Anschauung machen und sich ausser sich selbst versessen müsste; welches sich widerspricht-

Die verlangte Auslösung also der Aufgabe vom Sitz der Seele, die der Metaphysik zugemuted wird führt auf eine unmögliche Grösse ($\sqrt{-2}$); und man kann dem, der sie unternimmt, mit dem Terens zurufen: nihilo plus agas, quam si des operam, ut cum ratione infanias; indess es den Physiologen, dem die blosse dynamische Gegenwart, wo möglich bis zur unmittelbaren verfolgt zu haben genügt auch nicht verargert werden kann, den Metaphysiker zum Erfass des noch Mangelnden aufgefordet zu haben.

Kant shows himself a dualist: the soul exists but is no substance or matter. His adept Ludwig Heinrich Jakob summarized already the point of view in a book, containing a chapter entitled "Von der Gemeinschaft der Seele und des Körpers (1791; on the community of the soul and the body): "This community, i.e. wanting to search for the seat of the soul, means to occupy oneself with a totally useless work. For a place is a relation in space needed and the latter can therefore pertain only a material substance. However, since the soul is not an object of the outer sense, the soul is no material substance; thus, it follows that no place can be attributed to it, and the question of the seat of the soul is completely senseless. The soul is everywhere she acts and produces changes. In this sense, therefore, the body, with which she is in immediate community, is her place, because she acts in and through it" (translation is by Capozzi 2014). Note that Jakob indicates still a substantial site for the soul: the whole body, in which he does not differ from Soemmerring. So, Soemmerring was warned: note the dates Jakob 1791 and Organ der Seele 1796.

Moreover, Kant's call-out in Latin is from Terentius, in his comedy Eunuchus, in which the slave Parmeno gives his master Phaedria a life's lesson, because he is in love with the girl Thais. She refuses him first, but later on indicates that he can visit her and Phaedria doubts whether he accepts her offer or not:

Proin tu, dum est tempus, etiam atque etiam hoc cogita, Ere, quae res in se neque consilium neque modum habet ullum, eam rem consilio regere non potes. In amore haec omnia insunt vitia: iniuria, suspiciones, inimicitiae, indutiae, bellum, pax rursum: incerta haec si tu postules ratione certa facere, nihilo plus agas, quam si des operam, ut cum ratione insanias.

The end is taken over by Kant and says: "you do presumably nothing else than that you exert logical arguments to be mentally disturbed", which hardly can be seen as polite denial of Soemmerring's research results. Just before it, Kant uses in the text: "it (the solution of the place of the soul) conducts to an impossible quantity $\sqrt{-2}$" that in itself is already an offence. So the epilogue is very destructive for Soemmerring's effort to localize the "Seele" in the brain; nowadays, one should say mind (for a more extended historical description, see Marino 1994 and Hildebrand 2005). The question stays whether Soemmerring was a "materialist" or not. If he was, his efforts are not different at all from our neuroscientists who claim that the network structure of the brain inherently produces mind. Soemmerring went to the ventricles (sure, nowadays considered a simple idea) and not to network structures, but the basic approach is the same: "mind or soul is brain".

There is one other aspect of Soemmerring's studies that need to be mentioned: "In later years, both as a physician in Frankfurt and then as a scientific member of the Bavarian Academy in Munich, Soemmerring emphasized 'electricity' as the major explanatory concept for nervous action. He was happy to find that the rapidity of electrical conduction explained the visible phenomena of nerve function better than models invoking 'wave transmission in humidity', which suggested a comparatively slow process. Though Soemmerring was more interested in the nature of electrical nerve transmission and tended to produce explanations by analogy, linking the physical phenomena to physiological observations, he did make use of contemporary work on 'animal electricity' by Luigi Galvani (1737–1798) in Italy". Haller with his theory of "irritability and sensitivity" induced the ideas on muscle force and nervous force; in fact, the motor–sensory system was described. Pfaff ([1773–1852], 1795) by his book "On animal electricity and activation" (Ueber Thierische Elektricität und Reizbarkeit) brought at least electricity in the public domain, coupled to the controversial and vivid discussions on mechanistic materialism opposing the vis vitalis. Although Soemmerring was already interested in Galvani and his electricity during his student's stay in London (1778), his attention was not only due to its function in the nervous system but also due to decapitation: "The stimulation experiments on bodies of executed people posed the question whether after decapitation the head of the executed still feels his existence, of which Soemmerring was convinced, or not" (translated from Hildebrand 2011). Soemmerring's guillotine paper was published in 1795. The "mind" question treated in his "Organ der Seele" strongly engaged Soemmerring into decapitation so fervently discussed in the scientific circles of those days (for overview, see Hildebrand 2011; Mann 1999).

12.2.1 Continuation of Survey of the History

The idea of the cell as used in modern terminology stems from Oken (1779–1851), although Van Leeuwenhoek already had depicted the cell nucleus in the blood corpuscles of fish. The idea that the plant cell was the essential unit of these organisms came from Schleiden (1804–1881). Moreover, he favoured the presence of a nucleus in each cell. Schwann (1810–1882) extended it to animals and gave a firm base to the cell doctrine: the animal or plant is entirely structured by cells and by substances secernated by these cells. Cells have their own life, but are subjected to the organism as a whole. That the axon indeed was an elongation of the cell body we owe to Koelliker (1817–1905), who also described for the first time neuronal networks in a modern sense. Haller (1708–1777) detected the irritability of muscle, leading to contraction, while another force directed from the nervous system by the peripheral nervous system could induce such irritability. Haller discerned the inherent muscular force from the nerve force. 'He was able to show that the tissues are not themselves capable of sensations, but that the nerves are the channels or instruments of this process' (Singer 1931). In fact, Haller discerned the motosensory system.

Luigi Galvani (1737–1798) stimulated the nerve and could prove the contraction of the muscle by it, and Volta (1745–1827) proved this was an electrical phenomenon. Bois-Reymond (1818–1896) showed that an action potential over nerve fibres always induced a change of the electrical properties along that nerve fibre. He also showed that the contraction of a muscle by nerve stimulation was caused by a change of the chemical composition within the muscle. Both Bell (1774–1824) and Magendie (1783–1855) demonstrated that dorsal radiculi are sensory, while ventral ones are motoric.

Fixation of nervous tissue greatly enhanced microscopy of the nervous system. Ruysch (1638–1731) had already introduced ethanol fixation for his studies of the human body. Neuronal tissue, however, asks for specialized fixation and colouration techniques, since myelin is hard to treat. 'Normally the presence of myelin was demonstrated with the osmic-acid technique. Before formalin came in use, nervous tissue was fixated in solutions of potassium dichromate and stained with Carmine red. The introduction of the Weigert method (1884) caused a tremendous revolution in the study of the white matter. The addition of the differentiation fluid of Pal (the so-called Lustgarten fluid) cleared up the tissue in between the myelin sheaths, thus further improving the contrast of the Weigert method' (Marani et al. 1987).

The trajectory of nerve fibres could be followed in the peripheral nervous system due to Waller (1816–1870) who described degeneration from the place of the lesion to the periphery, now known as Wallerian degeneration. The introduction of the Golgi technique by Golgi (1843–1926) in 1873 and the studies of Ramon y Cajal (1852–1934) raised the question whether the nervous system was a 'diffuse reticular network' or cellular entities that contact each other? Both Golgi and Cajal were co-recipients of the Nobel Prize for physiology or Medicine in 1906. The Nobel lecture of Cajal, who created and defended the neuron being a cellular entity, finally finished the debate, firmly establishing the so-called neuron doctrine for the nervous system too (Culberson 1985; review in Clarke and O'Malley 1996).

The neuron doctrine was scrutinized several times (e.g. Bullock 1959; Shepherd 1972; Gold and Stoljar 1999; Bullock et al. 2005). The discovery of specialized connections of the motoneuron and the mitral cell, the description of electrotonic coupling, dendro-dentric connections, changing ideas on neuronal network functions all contributed to doubts to a strict 'neuron doctrine'. However, till now the neuron doctrine has never been overthrown" (Marani and Lakke 2012).

12.3 Thomas Willis and Comparative Anatomy of the Brain

Thomas Willis "Anatomy of the brain and nerves" not only contains the human brain but also the sheep's brain. In fact, the human brain is regularly depicted next to the sheep's brain: Willis's Fig. 1 (Fig. 12.8) is the picture of the ventral human brain, and his Fig. 2 (Fig. 12.12) is the sheep's ventral brain side. This repeats in Fig. 5 (Fig. 12.8) with the inside of the human skull, and Willis's Fig. 6 is the inside of the sheep's skull. The Figs. 7 and 8 (Fig. 12.12) are also pictures of the sheep's brain. Sheep brain neuroanatomy has been frequently studied after Willis even for human brain instruction. One of the oldest printed instruction guides found by this author was by J.F. Burkholder from 1904, announced as: the anatomy of the brain: a manual for students and practitioners of Medicine. The brain of the sheep (Ovis aries) is being selected for description and illustration because of its availability and *its practical identity* with the human brain for laboratory use. Ranson and Clark (1959) in their anatomy of the nervous system also use sheep macroscopic neuroanatomy.

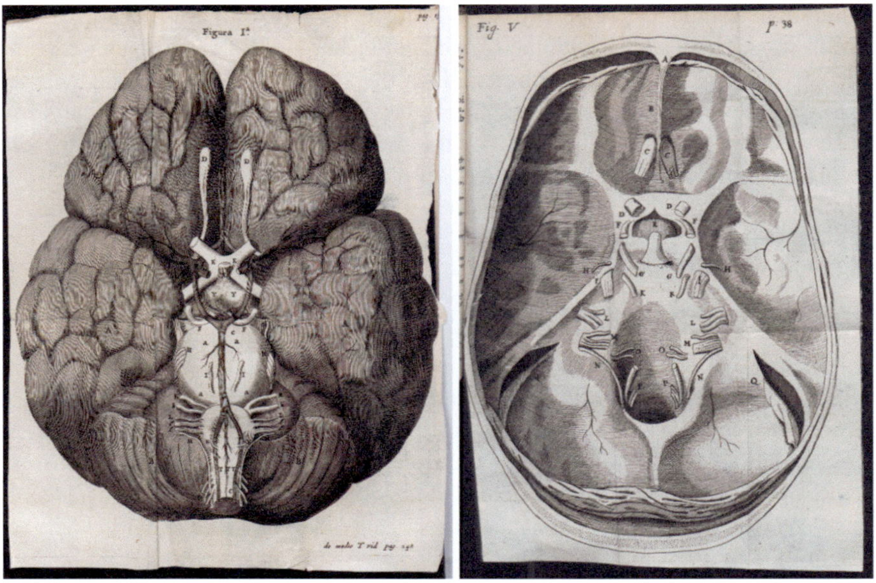

Fig. 12.8 Willis (1664) drawings of the brain base and the skull base. Brain base: AAAA, the anterior and posterior lobes of the brain quadripartite or divided in four parts; BB, the cerebel or little brain in the hinder of the head; CC, the long marrow or pith; DD, the smelling nerves or the first pair: EE, the optic or seeing nerves the second pair; FF, the moving nerves of the eyes the third pair; GG, the pathetic nerves of the eyes or the fourth pair; HH, the fifth pair of nerves; II, the sixth pair of nerves; KKKK, the auditory or hearing nerves and their two processes on either side of them the seventh pair; LLllll, the wandering pair, or the eight pair consisting of many fibres; MM, the spinal nerves coming from afar to the origin of the wandering pair; NN, the ninth pair consisting also of many fibres (tending downwards, grow together into one trunk) which arises a little above the process of the hinder part of the head; OO, the tenth pair tending downwards (rest of the letters are blood vessel indications). Skull base: A, the hollowness of the bone of the forehead; B, The close or mound of the cribiforme or sieve-like bone; CC, The mammillary processes, which are much thinner, and endued with less open cavity, than in the four-footed beasts, endued with more excellent sense of smelling; DD, the optic nerves, being far separated, go out of the skull otherwise than in most brute beasts; E, the pituitary glandula or kernel with top of the tunnel inserted into it; FF, the carotidick arteries shewing themselves nigh its sides; GG, the moving nerves of the eyes going out of the skull; HH, the pathetic nerves, bid under the dura mater, go out from the skull at the same hole with the former; II, the fifth pair of nerves hid under the dura mater; KK, the sixth pair stretched under the dura mater, and go out also at the same hole with the third and fourth pair; LL, the seventh pair entering with a double process the stony bone; MM, the eight or wandering pair seen to grow together with an accessory nerve of many fibres, NN, as it goes out of the skull; NN, the accessory nerve to the wandering pair; OO, the ninth pair; PP, the pair tending downwards, hid under the dura mater, where the vertebral artery ascends; QQ, lateral or side-bosom; R, the anterior branches of the carotids go away united, moving forward into the fissure or cleft of the brain; S, the posterior branches of the carotids united, and meeting with the vertebral trunk; TTT, the vertebral arteries and their three branches ascending; V, the branches of the vertebral growing together into one trunk; WW, the place designed where the vertebrals and the carotids are united, and either branch ascends to the choroidal infolding; X, the tunnel; YY, two glandula's or kernels placed behind the tunnel; aaaa, the annulary protuberance, which being sent from cerebel, embraces the stock of the long marrow (courtesy Leiden University and Walleus Library with permission)

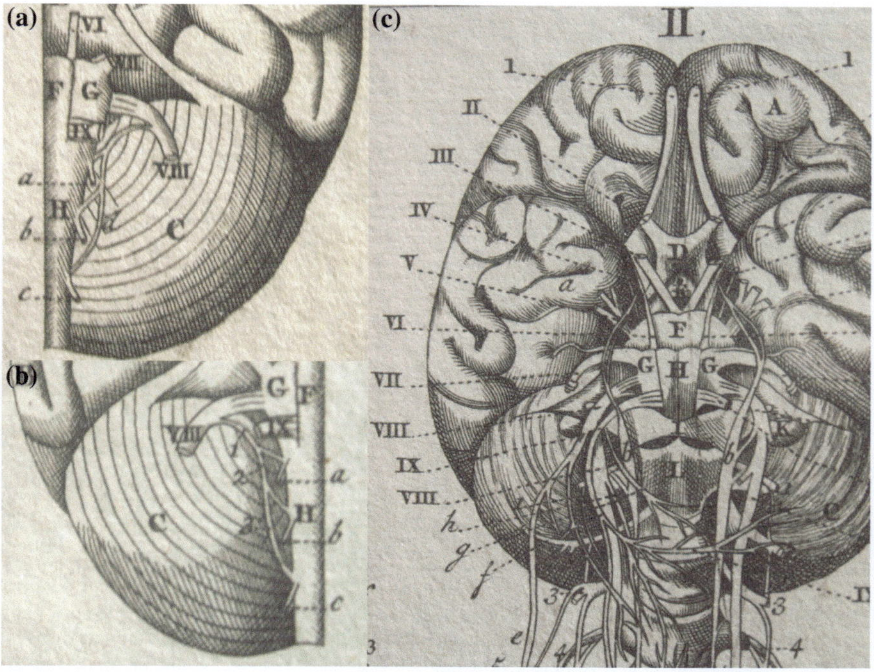

Fig. 12.9 Cranial nerve figures of Eustachius from "Ontleedkundige platen van B. Eustachius" (Bonn 1798): **a** a, b, c are the first, second and third cervical nerve and d the spinal recurrent nerve, connecting to the VIII cranial nerve. **b** a, b, c are the first, second and third cervical nerve with 1 the spinal recurrent nerve connected by 1 to the first cervical nerve, 2 connection with the second cervical spinal nerve and 3 the third cervical nerve connection. **c** All cranial nerves are labelled by Roman figures (Author's collection)

If the brain's description of Willis is placed in the row of brain models (Swanson 2000; Fig. 12.6), then the subdivision of the trunk or brainstem's end starts with Willis. This partition of the trunk's end is clearly based not only on his sheep's results, but also on the localization and direction of the optic nerve, hence optic thalamus. The same holds for the striated body, of which the picture of the sheep's dissection in his Fig. 8 (Fig. 12.13) corroborates this statement.

The cortical surface with its gyri and sulci, called by Willis (1664) gyrations, rollings or turnings and folds, is present in both the human and sheep hemispheres. Willis description is anatomical sound:

> The pia mater … is easily pulled away; and then the turnings and folds being opened and separated one from another, the substance of the brain is seen to be ploughed, or laid as it were with furrows; out of which arise banks or ridges of broken crevices, not in a direct series, but cross-wise; so that in the bottom of every furrow, a convolution arising from the right side, is drawn to the right, and so by turns the inequalities of the whole brain are variegated in this order. (Willis 1664)

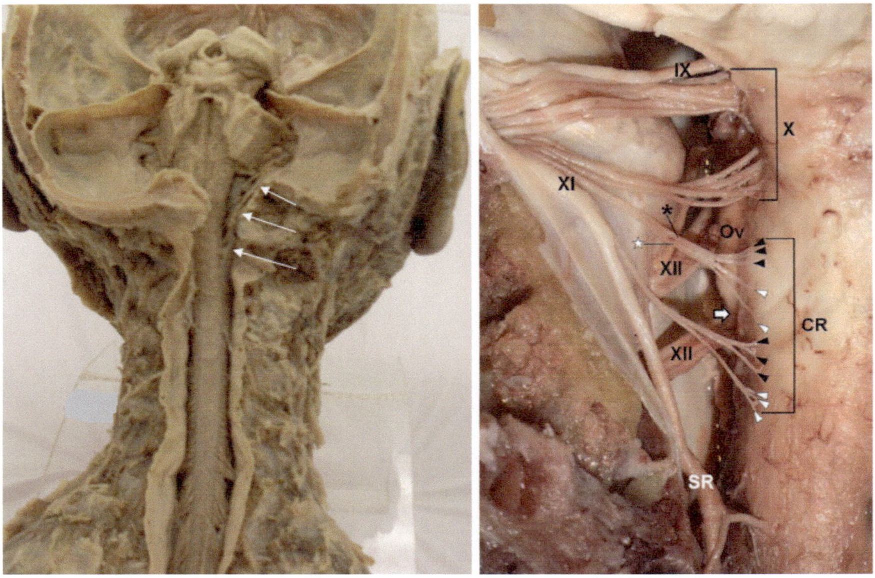

Fig. 12.10 Spinal(SR) and cranial (CR) accessory nerve. Left shows the spinal accessory nerve intravertebrally. Note that the trajectory is against the inside of the os occipitale (courtesy Dept. anatomy, Erasmus University, Rotterdam). Right demonstrates the origin of the cranial accessory nerve in relation to the hypoglossal (XII) and vagus (X) cranial nerves at the bulging of inferior olive (white arrow indicates its caudal margin). "Morphology of the filaments and rootlets of the CR. Black and white arrowheads indicate different rootlets. The uppermost (black*) and next (white*) rootlets of the CR are formed by the union of three and two filaments, respectively" (Liu et al. 2014 with permission)

In the next paragraph, it is indicated that the "brain is so framed, both for the more plentiful reception of the spirituous aliment, and also for the more commodius dispensation of the animal spirits for some uses". The convolutions of the brain hemispheres are clearly related to mind. Since man and sheep contain gyrencephalic (Fig. 12.14) and not lissencephalic brains, Willis could only state that animals had mind too but humans had in his ideas more and larger turnings:

> Hence these folds or rollings about are far more and greater in a man than in any other living creature, to wit, for the various and manifold actions of the superior faculties; but they are garnished with an uncertain, and as it were fortuitous series, that the exercises of the animal function might be free and changeable, and not determined to one. Those gyrations or turnings about in four footed beasts are fewer, and in some, as in cat, they are found to be in a certain figure and order: wherefore this brute thinks on, or remembers scarce any thing but what the instincts and needs of nature suggest. (Willis 1664)

Figure 12.14 shows how well the "turnings" are present in the sheep. White and grey substance (see Fig. 12.14) does not differ from man in construct, which was noted and described by the Willis group during their research. The only escape was:

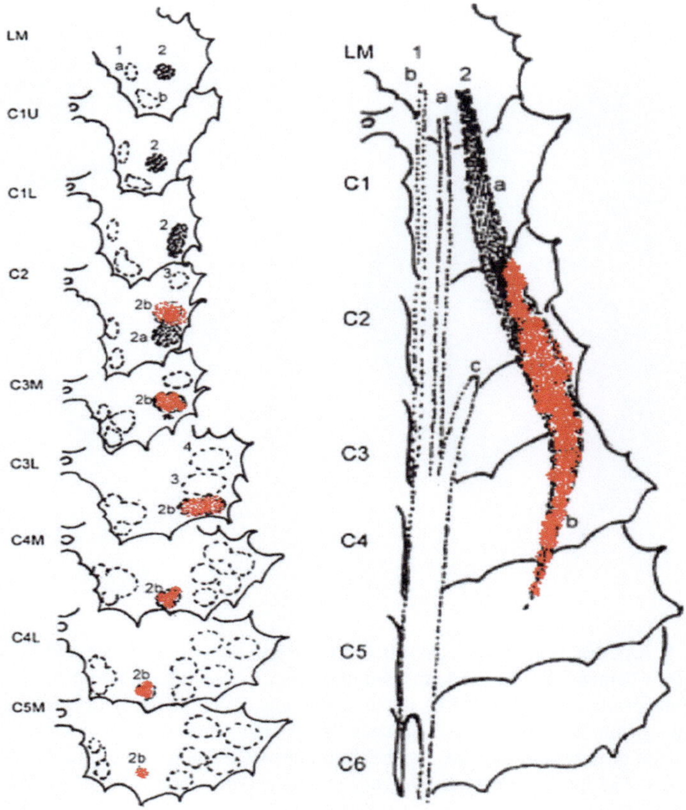

Fig. 12.11 Left contains cross sections through the cervical spinal cord grey matter (C, cervical; M, middle; L, lower and LM, lower medulla). Right: Segmental extent of motor neuron pool of the m. sternocleidomastoideus (black) and m. trapezius (red). (Figures adapted from Routal and Pal 2000)

the human gyri are larger and man has more. Willis supported the idea that cortical layers are the places of action in both cerebellar and hemispherical cortex:

> Whilst after this manner, for the performing the acts of motions and sense, we suppose the animal spirits to be expatiated within the oblong marrow; we affirm that they are not there begotten, but only exercised. For indeed, they being created only in the brain and cerebel, as they proceed from this or that, they perform offices, either of a merely involuntary function, or else of a spontaneous, as shall be shewed more largely hereafter. (Willis 1664)

While Willis did animal experiments for his anatomy of the brain (e.g. ligation of both vagus nerves in a dog), the study did not notice the different localization of the motor cortex in man and sheep. In man localized perpendicular and halfway the hemisphere (gyrus precentralis), at the contrary in the sheep it is placed in the length at the lateral side of the frontal part of the hemisphere (area gigantopyramidalis: 2 in Fig. 12.14). Three years later, Pathologiae Cerebri was published by Willis, in

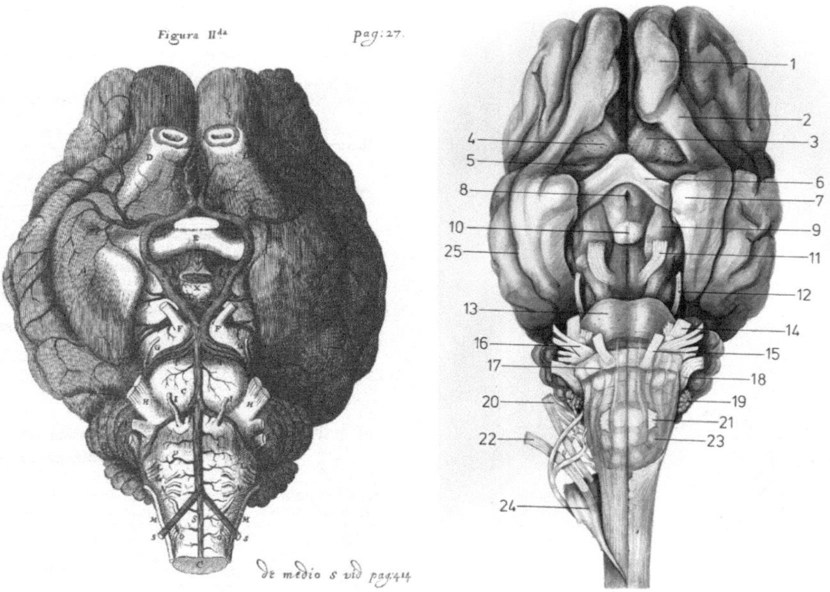

Fig. 12.12 Left part is from Thomas Willis (1664), Right part from Marani and Weyns (1984). Left; AA, Two hemispheres of the brain, without brain lobes, different from that in Man; BB, lateral portions of the cerebellum; CCC, medulla oblongata; DD, olfactory or smelling nerves cut off nigh to the mammillary processus, that their cavities may appear; E, the coalition of the optic nerves; FF, motory nerves of the eyes, or the third pair; GG, The pathetic (trochlearis) nerves or fourth pair; HH; The fifth pair: the trunk of which nerve is presently divided into two branches; II, the sixth pair; KK, kk The seventh pair, or the hearing nerves, on either sides of which are two processus; LL, The eight or wandering pair, the origin of which is made up of very many fibres seeming to grow together; MM, A nerve coming out of the spinal marrow to the beginning of the wandering pair; NN, the ninth pair, consisting also of divers fibres arising distinctly, which afterwards grow together into one trunk; OO, the tenth pair tending downwards: PP, the trunk of the carotidick artery cut off, where it is divided into the anterior and posterior branch; QQ, the inosculation of the anterior branches; R, The posterior branches of the carotids united, and meeting with the vertebral trunk; SSS, The vertebral artery ascending with a triple branch; TT, where the carotids meet the vertebral, and on either side many arteries ascend to the choroidean infolding; V, the coalition of the Vertebral branches into the same trunk; W, The tunnel; X, A white glandula or kernel placed behind it (courtesy Leiden University and Walleus Libraries with permission). Right:1, bulbus olfactorius; 2, tractus olfactorius lat I; 3, tractus olfactorius med; 4 area perforata; 5, diagonal band of Broca; 6, tractus opticus II; 7, uncus; 8, infundibulum; 9, pedunculus cerebri; 10, corpus mammillare; 11, nervus oculomotorius III; 12, nervus trochlearis IV; 13, pons; 14; nervus trigeminus V; 15, nervus abducens VI; 16, nervus facialis VII; 17, corpus trapezoides; 18, nervus statoacusticus VIII; 19, plexus choroideus of Bochdalek; 20, nervi glossopharyngeus and vagus (IX and X); 21, nervus hypoglossus XII; 22, nervus accessorius XI; 23, oliva inferior; 24, leptomeninx (© E. Marani)

which a series of brain diseases are described, including motor ones. One should ask whether Willis was aware of the importance of his ideas on cortical layers or was he forced in these ideas due to the distinction between animal and human soul?

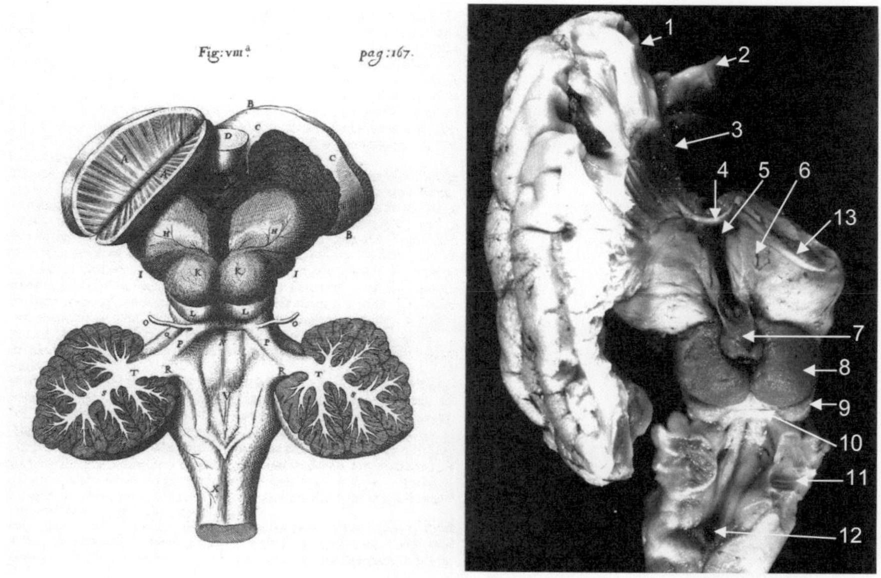

Fig. 12.13 Left is eight figure of Thomas Willis (1664), right side is figure from Marani and Weyns (1984). AA, the chamfered body cut in two in the midst, that its marrowy chamferings may appear. B The other chamfered body whole covered with choroidal infolding with the extreme portion of the callous body CC flicking to the same; CC, The brim or extremity of callous body cleaving to the chamfered body; D, the basis of the fornix; E, the right wing of the choroidal infolding; F, the passage of the veins being stretched out from the fourth bosom, which being presently forked, constitutes the veinous portion of either wing of the choroidal infolding; under the beginning of this passage, very much beset with fibres and sanguiserous vessels, the pineal glandula lyes hid; G, the hole or chink leading to the tunnel; HH, the chambers of the optic nerves; II, the medullary processes, or the ways of passage which lead from the medullar stock into the orbicular protuberances; KK, the buttock-form orbicular protuberances; LL, the lesser protuberances called Testes; M, the meeting together of the processes ascending obliquely from the Testes into the cerebel; N, the hole of the lower ventricle lying under the orbicular protuberances; OO, the pathetic nerves of the eyes; PP, the medullary processes, which being sent from the cerebel towards the oblong marrow, campass about its stock, and constitute the annular ringy protuberance; RR, the lowest or third processes of the cerebel, which being inserted to the medullar trunk, become additional cords or strings of it; SS, the medullar ramifications or branchings of the cerebel; TT, the middle marrows of either cerebel in which its three medullary processes, constituting either little foot of it, grow together; V, the ditch constituting the fourth ventricle in the medullary trunk; X, the extremity of the oblong marrow about to end in the spinal (courtesy Leiden University and Walleus Libraries with permission). 1, corona radiate; 2, bulbus olfactorius; 3, capsula interna & corpus striatum; 4, commissura anterior; 5, roof of third ventricle; 6, thalamus; 7, corpus pineale; 8, colliculus superior; 9, colliculus inferior; 10, velum medullare anterius; 11, cerebellar peduncles; 12, fourth ventricle arrow on its obex; 13, stria terminalis (© E. Marani)

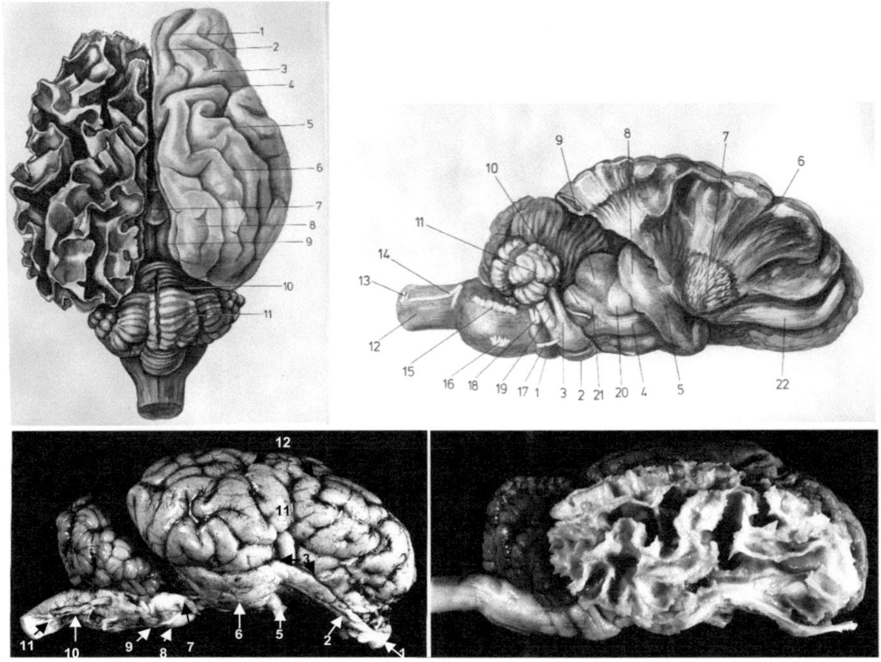

Fig. 12.14 All figures are from Marani and Weyns (1984). Left top: 1, sulcus coronalis; 2, area gigantopyramidalis (motor cortex); 3, premotor cortex; 4, sulcus ansatus; 5, parietal, sensory cortex; 6, sulcus supra sylvius; 7 corpus pineale; 8 sulcus endomarginalis; 9, colliculus superior; 10, cerebellar fissure prima; 11 cerebellar vermis. Right top: 1, corpus trapezoids; 2, pons; 3, brachium pontis; 4, pedunculus cerebri; 5, tractus opticus; 6, corona radiate; 7, capsula interna & corpus striatum; 9, colliculus superior; 10, colliculus inferior; 11, cerebellum (paraflocculus); 12, medulla oblongata; 13, spinal radix C1; 14, nervus accessorius; 15, nervus glossopharyngeus and vagus; 16, nervus hypoglossus; 17, nervus abducens; 18, nervus facialis and statoacusticus; 19, nervus trigeminus; 20, corpus geniculatum mediale; 21 brachium colliculus inferior; 22 tractus olfactorius; Left lower figure: 1, bulbus olfactorius; 2, tractus olfactorius lateralis; 3, sulcus rhinalis; 4, insula; 5, chiasma opticum; 6, uncus; 7, nervus trigeminus; 8, pons; 9, corpus trapezoids; 10, spinal part nervus accessorius; 11, nervus hypoglossus (© E. Marani)

12.3.1 The Accessory Nerve

The terminology as used in most of the classical anatomy books (Kopsch 1920; Hafferl 1969; see Figs. 12.2, 12.7 and 12.10) is: the accessorius spinalis originates from cervical accessory rootlets, placed in between the spinal cervical ventral and spinal cervical dorsal rootlets and dorsal to the denticulate ligament, while the accessorius vagi comprises the rootlets from the medulla oblongata. Together they constitute the accessorius communis, which is enclosed by the dura. Passing the foramen jugulare and in between the ganglion jugulare and nodosum, the accessorius vagi leaves the main trunk to participate in the nervus vagus. This branch is called the ramus internus. The ramus externus (=accessorius spinalis) innervates the sternocleidomastoid and trapezius muscles. Branches of the cervical plexus join the ramus externus. The spinal accessory nerve is a branchial arch nerve and thus leaves the spinal cord from the lateral line. This is in between the ventral and dorsal roots.

Thomas Willis gives the first correct description of the accessory nerve in 1664 (see Fig. 12.8): Chap. XXVIII. "Of the spinal nerve an accessory to the wandering pair (=nervus vagus); also the nerve of the diaphragma", in which Willis states:

> We have already shown, that the nerve of the wandering pair in the beginning is made up of numerous fibres: to which is joined another noted nerve arising from afar, and being ingrassed with them, goes forth together with them out of the skull. Concerning this nerve, because the beginning and distribution of it being very irregular, have not as yet been noted by other Anatomists, it may seem worth our labour to make a little more diligent inquiry. Therefore if we would search into the beginning of this nerve, that is found beginning with a sharp point in the inside of the spinal marrow nigh the sixth or seventh vertebrae of the neck. But being increased in its ascent, is nowhere inserted into the medullary trunk, but in its whole tract on both sides leans on its side, to which is knit by certain admitted fibres towards the superficies. This arising up from the neck after this manner, and being carried within the skull, is joined to the fibres of the wandering pair, and is ingrassed with them seeming to grow together in one trunk, and goes out with them at the same hole from the skull; which being passed through, the spinal nerve presently departing from the trunk of the wandering pair, is at length reflected outward. This stranger or travelling nerve after a short commerce having left its companion, is carried upon the muscles of the neck, to which it imparts some shoots, and is inoculated with a certain shoot of the tenth pair; but from thence being carried on further, it goes alone by a long passage, till it comes to the scapular muscle, on which it is almost wholly bestowed. This nerve is found constantly, not only in Man and four-footed beasts, but also in fowls and fishes; and in these seems to be delineated instead of arms, and for the moving of their wings and fins. (Willis 1664)

In the speculation of the function of this spinal accessorius nerve, Willis imputes an involuntary or autonomous function:

> That besides the spontaneous motions wherewith the neck and arms are wont to be imployed with the previous intention of doing this or that thing, those parts also, before any other member, are affected with pathetic and sudden motions, according to the force of the passions, the animal not being conscious of it. For almost all living creatures do not only turn about their necks at any noise to behold whatever might cause fear; but they being any ways affrighted in the twinkling of an eye fly away, their feet, wings, fins or other part answerable to them, being set into a rapid motion. The neck and arms are pliable or observant, not only to fear, but in like manner to other passions. For brute animals, as well as Man, being puffed up with pride or anger (as we have elsewhere intimated) have their necks swollen, and their crests lifted up: but as to Man, his hands and arms are so obsequious to the passions, and almost to all the conceptions of the brain, that they are continually agitated in the doing of any other thing with various gestures, whereof we are ignorant or not willing them: we scarcely think or speak any thing but at the same time the hands flung out here and there; and whil the tongue hesitates or sticks, or the words at it were stick between the jaws, the right hand is exercised, as if by its gesture it were endeavouring to draw out more swiftly the sense of the mind. Truly that these parts, to wit, the hands and arms, do so nearly conspire with the affections of the brain and heart in their motions in some measure in all living creatures, but more eminently in Man, the cause seems to be this nerve's coming from spinal marrow to the beginning of the wandering pair, and communicating with its nerves, and receiving from them as it were the note or private mark of the involuntary function. (Willis 1664)

Here, some attention has to be paid to Eustachius's (1500/1514/1520–1574) cranial nerve results (Fig. 12.9): "Any impartial investigation of priority must start from the fact that many of the discoveries which have been and still are attributed to Willis were anticipated by Bartholomeus Eustachius, a hundred years before him. Eustachius' Anatomical Tables were produced in 1552 but were only published in 1714. They contain accurate illustrations of ten cranial nerves, including the trochlear, abducens, spinal accessory and the sympathetic nerves with their visceral ramifications" (Hierons and Meyer 1962). Eustachius thus described most of the cranial nerves correctly including the spinal accesorius (see Fig. 12.9). It is said that Willis knew Eustachius results and therefore could produce the correct anatomy of the accessory nerve. Although it is claimed that there is enough historical antidote to deny such a "fraud" by Willis (for argumentation see Hierons and Meyer 1962), some remarks on Eustachius's results should be made. In Eustachius's "Opuscula Anatomica", published in 1564 and exactly hundred years earlier, one of the parts is "On the movement of the head" (De motu capitis). Eustachius clearly separated text and comments. Most of the comments concerned Galen's publications and are text page and rule of the Opuscula related. Comment 240.r2 and 240.r30 concerned the nervous system. Eustachius treated the cervical spinal nerves in relation to the neck vertebrae and remarks their strange appearance and localization and speaks of lateral foramina. The first and second vertebrae and the appearance of the spinal nerves are compared to Galen's result (Galenus "De usu, part lib I3.cap5).

"The nerves of the second connection of the neck do not leave by the foramina like the third and fourth and also not by the perforated second spondylus as is the case with the first, because it was impossible from the sides as already indicated for the first. He could not appear from any other part, since the first spondylus is closely surrounded. Thus on the only possible place, nature created space on both sides of the spinal cord, in between the first and second spondylus, by which the second nerve harmless emerges and will not be damaged by its own movement".

Emergunt nervi secundae colli coniugationes, non ex lateralibus foraminibus, ut in tertia ac quarta coniugatione, neque ex ipso perforato secondo spondylo quomodo est primus. Nam ex lateralibus non poterat propter causam, quam in primo attulimus. A nulla vero eius parte alia emergere poterat, quod ipsi primus spondylus esset circundatus. Ubi igitur solum licuit, ibi natura locum quendam comparauit, in utraq; scilicet spinarum parte, inter primum ac secundum spondylum, per quem secunda nervorum coniugatio egreditur, quae nihil ab ipsorum motu laeditur.

In 240r30 he discusses the distribution of the cervical spinal nerves over the muscles.

Willis certainly knew Eustachius's Opuscula Anatomica, and therefore, the argument of late publication of his plates and Anatomical Tables hardly counteracts the fraud hypothesis, since any anatomist should study the neck spinal nerves after these remarks. Eustachius' comment forces the discovery of the spinal accessory, so to say.

A good second is Bidloo (1685). A concise description of the components of the spinal accessory nerve inside the skull is also given by Bidloo (1685). Bidloo did not discern the facial nerve; as a consequence, his tenth nerve is equal to our accessory nerve.

The tenth pair of the nerves, called the "marrow pair", starts between the bones of the first vertebra, out of the sides of the spinal column, to which it was attached and ascends by very thin filaments. Its branches supply the neck muscles, although the lowest one goes to the scapula muscle (trapezius). Although this pair appears to be joined to the previous and next pairs, it was not joined with these, even in the exit to the aperture of the bone: the individual offshoots always remain separated from each other, which one can see more clearly in a head that has been boiled.

> Het tiende paar der zenuwen bij uitneementheid het Mergpaar genoemd, zijn begin tussen het eerste Rugwervelgebeente, uit de zijden van het Ruggemerg, aan het welk het in opklimmen met zeer dunne vezelkens gehecht werd, neemende, deeld zijn takken den spieren van de nek toe; toch de laatste ontfangd de schouderblads spier. Alhoewel dit met de vezelen van de voorgaande en volgende paren tot een schijnt te wellen, zo werd nochtans zelf in den uitgang door het gat van het gebeente, met dezelve niet verenigd, maar dezelve afscheuten blijven altijd van den ander gescheiden; het welk men met minder moeite in een afgekookt hoofd zien kan. (Bidloo 1685)

"The discussion in the literature nowadays concerns the number of segments that contribute in the spinal canal to the accessorius spinalis and the cervical nerves that join the ramus externus (see Figs. 12.7 and 12.10). Tumours present in the cranial nerve trajectory ask for a dissection of neck structures, and, if possible, sparing of the accessory complex overcomes shoulder dysfunction by the loss of innervation of trapezius and sternocleidomastoideus. The spinal segments that are thought to contribute to the accessorius spinalis range from five (C1–C5, Daube and Sandok 1978) to one (C1, Kopsch 1920). The cervical nerves, by their ventral roots, that participate in the ramus externus, are also at variance: C2–C4 (Flower 1861), C2 and C3 (Caliot et al. 1990).

The accessory spinalis anatomy has been brought up again recently. The junction of the bulbar and spinal parts has been denied again, based on microscopic inspection of the rootlets (Lachman et al. 2002; Ryan et al. 2007). Therefore, it should be considered that the accessory nerve exists only of a part of the spinal nerve trunk that originates from an intravertebral ascending accessory cervical plexus and forced by the ascensus to leave by the skull. This view was already given by Willis and Bidloo at the end of the seventeenth century, defended, e.g., by Van Gehuchten and Mohant (Gehuchten Van and Molhant 1911) and is well known see, e.g. Brodal (1969)" (Marani and Lakke 2012).

Thus, the contribution of the cervical roots to the spinal accessory nerve is unpredictable and affirmation of what is already known of this nerve is seemingly in vogue.

12.3.1.1 The Intravertebral Connections

In general, spinal segments C1–C4 are considered to contribute (Fig. 12.10), but accessory fibres can be added from C5 up to C7 (Hovelacque 1927; Lang 1976).

The exit line is oblique in such a way that spinal dorsal rootlets of C1 and sometimes C2 can join the accessory fibres (see Hovelacque 1927), which have been claimed to be present for C1 in 50% of the cases or more (see Tubbs et al. 2007). Moreover, indications for ganglionic cells in the accessory spinal nerve are present (see Hovelacque 1927). Interconnections between spinal cervical ventral and between spinal cervical dorsal roots are frequently present (Perneczky and Sunder-Plassmann 1980), which means that spinal cervical nerves can contain fibres of different spinal segments, when they join extracranially the accessory spinal nerve. Presumably, this is to be held responsible for the great variability in extracranial anastomoses between cervical spinal nerves and the spinal accessory nerve.

The spinal cord and brainstem nuclei involved in the accessory nerve have been studied by Pearson (1938) in human embryos, by Hopkins et al. in rat, dog, pig and monkey (see, e.g. Bieger and Hopkins 1987; Hopkins and Armour 1998; Hopkins et al. 1996), rabbit, Kitamura et al.(1989) and by Yoshida et al. (1984, 1985) in monkeys, all indicating that the origin of the bulbar accessory part is related to the nucleus ambiguous (thus vagus nerve), while the spinal accessory originates from dorsolateral situated motoneurons at the spinal level and in the medullary area, not reaching the ambiguous nucleus proper.

These motor neuron pools have been studied in rabbit, sheep and monkey. The human embryological and mature results contain differences, due to the developmental aspect, but two motor neuron groups can be differentiated with some overlap at C2/C3 in all studied animals. Upper part of the neuron pool steers the m. sternocleidomastoideus, and the lower group of neurons relays information to the m. trapezius (see Fig. 12.11; Routal and Pal 2000).

Motor cell groups or columns in the ventral horn of the spinal cord were described by Romanes (see overview 1964). These columns can be related to muscle groups and sometimes even to large single muscles. These motor columns are conservative in their appearance and are most clearly discerned in the cervical and lumbo-sacral enlargements: “Thus there is considerable similarity in the arrangement of motor cell groups in man, sheep, cat, dog, *Galeopithecus* (flying-lemur, remark ours) etc. …, a fact that would seem to suggest that the morphology of the limbs was of more importance in determining the arrangement of the cell groups rather than functional adaptations” (Romanes 1964). Moreover, in the acardiacus these motor columns develop normally in the lumbo-sacral enlargement supporting this autonomous organization of motor neurons in relation to muscles and less to the large ascending and descending spinal cord tracts (Romanes 1947).

12.3.1.2 The Extracranial Connections

The plexus accessorio-cervicalis, also more widely defined and then called spinal accessory nerve plexus, was extensively restudied around the 1930s (Foerster 1933; Nishizuka 1934; Freeman 1935; Pearson 1938; Straus and Howel 1936). The interest was directed to the neck dermatome organization, variants or interconnecting branches of cervical nerves, intra- or extradurally, or to the constitution of the accessorio-cervical plexus itself. Renewed interest in between 1970 and now derives from surgical neck dissections (Brown 2002), especially in the supraclavicular triangle. "The term spinal accessory nerve plexus may be defined as the spinal accessory nerve with all its intra- and extracranial connections to other nerves, principally cranial, cervical and sympathetic" (Brown 2002).

Recent anatomical literature shows few deviations from these earlier published cervical results, although contributions of C3–C4 are also described for the sternocleidomastoid muscle (Caliot et al. 1989) and percentages can vary.

Radical neck surgery, as said, gave a renewed impetus to the study of the relationships of the spinal accessory plexus (Brown et al. 1988). Interrelations were found for the great auricular nerve, the trigeminal nerve, facial nerve, the phrenic nerve, brachial plexus and hypoglossal nerve that are considered directly or indirectly connected with the spinal accessory plexus. Moreover, evidence was gathered that the accessory nerve is not purely motoric (Brown et al. 1988). "The spinal accessory nerve in man is not a purely motor nerve. It appears to have within it non-proprioceptive sensory and nociceptive fibres. Our study suggests that these fibres are inherent to the nerve" (Bremner-Smit et al. 1999).

The variability of the trajectories of branches of the spinal accessory plexus in relation to trapezius and sternocleidomastoid muscles have been extensively described by Lanz and Wachsmuth (1955).

12.3.2 *Willis' Colliculi, Corpus Striatum and Optic Thalamus*

From the dissection room, one knows that the spinal contribution of the accessory nerve is difficult to detect in the human brain due to the detaching of the brain from the spinal cord, mostly at C1, leaving most of the spinal contribution within the spinal canal. The rest of the spinal accessory is destructed due to the removal of the brain from the skull. Also from experience, the spinal accessory nerve in sheep is easier to detect, since the bending at the transition spinal cord–medulla oblongata is notably less and more of the spinal contribution stays on the brainstem: nevertheless still, it is often torn off. The guess is worthwhile that the Willis group's attention was also attracted by the sheep's localization of the spinal accessory nerve, Eustachius description (see Sect. 12.3.1) and that it was worked out for man afterwards.

The same holds for the striated body. Figure 12.13 shows Willis's sheep's figure and our dissection preparation of the sheep. Willis's Chap. 10: "A description of the brain, properly so called, and the explication of the use of its parts" shows regularly the comparison with the animal brain: "In some animals, the substance of the brain is divided in two parts." and "Those gyrations and turnings about in four footed beasts are fewer" (in which Willis found the prove for his immortal soul to be present in man and not in animals). Therefore, reading Willis's brain anatomy should be done from a comparative, neuroanatomical perspective.

A small intermezzo is needed: the ancient anatomists studied and named the brain structures from their own daily experience: best known is the naming "mammillary bodies" mirroring the view on the female breasts in a Bavarian "dirndel" dress. The cerebellar tonsil remembers the insight of the throat with an inflamed tonsil and the colliculus superior and inferior were compared to the buttocks or nates with its hanging testes below it (Fig. 12.13). Niels Stensen already remarked that using these "household" terms for neuroanatomical structures indicates that the user presumably does not know the real meaning of these expressions (Kardel and Jacobeaus 1994).

In the literature especially published before and after the tercentenary edition of "Anatomy of the brain" of 1965, Willis has been imputed with detection of nearly all the brain's structure. A lead to Willis human neuroanatomical detections can be found with Burdach's historical overview according to Meyer (1966): the first description of the corpus striatum, the anterior commissure, the pyramids, the olives and their relation to the lamina quadrigemina, the cerebellar peduncles and their relation to the spinal cord and cortex all are Willis's inventions. The first description of the corpus striatum is, to start with, untrue. It was found and described by Vesalius (1543), also indicated by Meyer (1966). The anterior commissure was known already, and Willis confused the stria terminalis with the anterior commissure (Meyer 1966). In his Fig. 7, the stria terminalis is indicated (G) according to Meyer (1966), but the text is: the transverse medullary process knitting the two streaked bodies one to another. Thus, Willis considered it a connection between both corpora striata and not a bundle of fibres from the amygdala towards septal and hypothalamic areas. Olivary connections are to the cerebellum, being the climbing fibres, not to the lamina quadrigemina. The cerebellar peduncles do not have a direct connection with the cortex, the superior cerebellar peduncle projects to the red nucleus and to the thalamus not to the cortex, the other two to the cerebellum, and so on and so on. Accentuating the neuroanatomical greatness of Willis by mistakes harms Willis reputation.

Function of nates and testes (also called by Willis the four-mole hills, Latin use now is lamina quadrigemina) stays unknown as Willis clearly indicated. Nevertheless, Willis *speculated* that cerebellar information (acts and progress of passions and natural instincts) had to pass the colliculi or protuberances as he called them (Willis was uneasy with the old terminology of nates and testes):

> When in a young one newly born the stomach crys out for hunger, the instinct of this is carried by the passage of the nerves to the cerebel, and from thence by the medullary processes to these protuberances; and the spirits there inhabiting form the idea of the impression and carry it to the brain, wherein presently, without any previous knowledge or experience, such kind of conceptions of the soul are stirred up, that every living creature seeks out the mothers breasts and sucks. (Willis 1664)

This indicates that Willis thought reflex-like or autonomic actions to pass by the colliculi (his protuberances) and by-passing the main route by which:

> forces of sensible things are carried, their acts might be greatly confounded by the mutual meeting or gathering together of the animal Spirits. (Willis 1664) (each of the three souls its own tract?)

Moreover, the collicular pathway, in fact the surface of the colliculi, is in the brains of calf, sheep and some other animals larger than in man, Willis noticed it, seemingly indicating that this reflex pathway is animal related and less man associated.

It was well known that the optic nerve, chiasma and optic tract were related to vision and that the follow-up of the optic tract brings one into the brainstem end. Thus, trailing the optic nerve brings everyone via chiasma and optic tract towards the thalamus, hence optic thalamus: hardly a surprising result (see top right Fig. nr. 5 of 12.14).

> For the optic and smelling nerves creep through the superficies of the brain by a long passage and windings about, that they might be inserted below this part of the oblong marrow. (Willis 1664)

However, Willis let the olfactory tract end in or near the thalamus.

Enticements are seemingly everywhere: "It is difficult not to read into Willis's description and conclusions the inklings of modern neurophysiology. When he states "the active Memory consists in the regurgitation or flowing back of the Spirits from exterior compass of the Brain towards its middle, on might be tempted to compare this with the concept put forward by Penfield in which the role of the temporal lobe in memory is established by means of its deep subcortical connections to the brainstem. But there is probably no real basis for such historical extrapolations, no matter how attractive they might appear" (Feindel 1964).

The humours had to be explained for their passage from the brain to the body and vice versa. The most logical approach should have been to use the area perforata (nr 4 in Fig. 12.12) directly located below the basal ganglia with its corpus striatum and ventricle and interspersed with small blood vessels. However, another approach is chosen:

> … that the translations of humors in living creatures are easily performed through places that seem impervious or unpassable; for while the pores and passages in all parts of a living creature are dilated by spirit an heat, they transmit the rain of the serum as through the fine texture of a woolen cloth. This plainly appears from arthritick distempers, in which the ferrous latex creeps by degrees through the nervous bodies, and passing through very small spaces, makes a falling down of humors sometimes upon these parts, and sometimes upon others; so that it is obvious, that the membranes and nervous processes drink in the ferrous humors like sponges, and then by a light compression render them by heaps, as is manifest in tooth-ach; for as often as a bodkin or instrument is put into a hollow tooth, clear water will come out plentifully. Indeed, in the body of a living creature the passage of humors are not only made through open passages and channels; but thinner and more watry latex creeps through the solid and smooth bodies of nerves, as also the fibres and the membranes, as through the holes of a filter, and so is transferred through imperceptible straits from place to place. (Willis 1664)

A second way of spreading humours, according to Willis, is from the ventricles towards the pituitary and mammillary bodies, thus the dorsal brain side. From here humours should be "carried away" by nerves or membranes (pia and dura mater are meant). Note that there is only one cranial nerve (IV) that leaves the brain dorsally. The cranial nerves, the neurons and fibres were seen as a transport system that functioned and could be related to diseases whatever the ideas on humour. Moreover, the back ground idea shows that neurons, fibres and nerves are grouped together functionally, which is hardly recognized in the literature. Remember: the link between "cells" and fibres or "hollow tubes" was not recognized, also not by Willis.

Studying Willis results, one should be surprised that (to give one example) the rhinocortex and its connections are hardly researched and thus not described in the parts on the brain's ventral side (Fig. 12.12). The sheep's rhinal sulcus delineating hemispherical cortex from the rhinocortex is so clear and impossible to miss (Fig. 12.14) together with the olfactory tracts around the area perforata that one asks: how biased was the Willis group or Willis by the "spirits" and souls. Was it interest in the brain and did they used the spirit conception to approach nervous tissue as Leonardo did, or was the drive a religious defend and consequence action against atheistic approaches born and carried out during and after the Cromwellian period? Due to the glorification of Willis's "Anatomy of the brain", called ancestor worship or hagiography, a series of questions are not treated that are a necessary prerequisite to judge the value of Willis's Cerebri anatome. This assertion is partially confirmed: "The early seventeenth century saw an unprecedented increase in learning and scientific advancement. Certain Puritan sects held a radical apocalyptic view that the end of the world was imminent and would be followed by a one thousand year reign by Christ. This doctrine spread to all religious groups and became the dominant concern among English academics. The study of God in the natural world was encouraged, to gain deeper knowledge of the Creator and thus prepare the reformers for the coming of Christ. This accelerated the growth of 'natural philosophy' as a discipline, so that by the Restoration of the monarchy in 1660 scientific publishing had reached new heights. The mechanistic philosophies typified by Cartesian dualism had begun to permeate continental philosophy and were rapidly absorbed in England. They carried atheistic elements, and theologians such as Henry More responded with antiatheist writings. More and his fellow Platonist Ralph Cudworth were strongly critical of medicophilosophical work with any atheistic implications" (O'Conner 2003).

Thus, historically the brain and cranial nerve studies were suffused with a spiritual atheistic antidote in Willis's research and the "inclination to regard pons, pyramids and other structures as reservoirs for animal spirits" (Meyer 1966) was hardly contradicted in these earliest studies. Scientists still regularly try to make Willis ideas to be conceptually sound, e.g. "Willis gave a structured account of diseases of the nervous system as it was known in his day. His account was largely derived from personal observations and not from traditional authorities and was based around his concept of the animal spirits, a fictitious entity in many ways analogous to the present day idea of the nerve impulse. This concept allowed him to develop a pathology of the animal spirits which embraced the whole content of the clinical neurology and psychiatry of his times" (Eadie 2002). That the animal spirit is not comparable with the nerve impulse is in the sentence above: brainstem parts functioned as reservoirs for animal spirits; thus, the animal spirit is not electrically based (you cannot store electricity in the mammalian brain), which is inherently to Willis Christian philosophy, by which one does not say that Willis's neurological approach is not worthwhile studying.

History remains the interpretation of passed facts and of ideas passed on. Foster's (1836–1907) interpretation of Willis's results (1901) is rather destructive: "Lower as we have already seen was a real man of science, with a clear penetrating mind, with a genuine love of truth for truth's sake, a worthy mate of Boyle, of Hooke and of Mayow. Willis was of a different type: love of truth was in him less potent than love of fame. Mixing with and indeed in daily intercourse with the band of exact inquirers, who at Oxford and London were striving to establish the new philosophy and advance by experiment natural knowledge, Willis caught up their phrases and thinking himself one of them, attempted to expound in their fashion the physiology of the nervous system. However, this method, when he was left to himself, and deprived of the aid and guidance of Lower, was in reality wholly different from theirs. They made exact observations and careful experiments and, guided by the dry light of reason, drew conclusions with caution, and expounded them with brevity, using words only as expressing the meaning of things. Willis's mind was of rhetorical art, he loved words as words, looked upon an illustration as an argument, and when discovered an analogy thought he had found a proof. Hence when we come to examine the view he put forward, we find that while they are expounded with a certain philosophic air, which perhaps goes far to explain the influence they had in their time, they do not themselves form any real solid contribution to knowledge. They are indeed to a large extend the views of Descartes, modified by more exact anatomical knowledge, occasionally by sound physiological deductions, in which we may probably trace the influence of Lower and other of Willis's contemporaries, but chiefly and especially by certain conceptions and certain modes of expression, which appear to be entirely Willis's own" (Foster 1901). By the way, Foster was professor of physiology at the Trinity College in Cambridge. His textbook on physiology was famous as was his history of physiology, from which the citation above originated. Foster's most famous student at Cambridge was Charles Scott Sherrington who went on to win the Nobel Prize in 1932.

A second critical approach of Willis work came from his contemporary Niels Stensen (1638–1686) in his famous "Discours sur l' anatomie du cerveau (held in 1665 and published in 1669): Monsieur Willis nous donne un systeme tout à fait particulier" [Willis gives us a very special system (of the nervous system)]. Stensen's sarcasm in his introduction also concerned Willis according to Foster (1901): "There abounds indeed a rich plenty of men to whom everything is clear. Such dogmatizing with the utmost confidence, make up and publish the story of the brain and the use of its several parts with the same assuredness as if they had mastered with their actual eyes the structure so admirable a machine and penetrated into the secrets of the great artificer" (Stensen 1669). This is substantiated by the French critical text on Willis anatomical considerations.

The text on Willis by Mettler and Mettler (1947) is even more harmful: "Thomas Willis (1622–1675) had one of the most fashionable and lucrative practices in London. He was something of a dandy and, while neither original nor erudite, he had the good fortune to secure the help of two men, each about a decade his junior, Richard Lower (1631–1691) and Christopher Wren (1632–1723), in the preparation of the work (Cerebri anatome 1664) upon which his reputation as an anatomist is based. Lower was a capable anatomist and Wren a good draughtsman, but the 13 small plates in this work are stiff and unnatural and the text lacks sequential development. The physiological conceptions which Willis held were, moreover, purely imaginary, and he offered no objective proof for the majority of his conclusions. Willis instituted a much-needed reclassification of the cranial nerves. Charles Bell whose unfamiliarity with the anatomic literature was exceeded only by his ambition, made the error of attributing the 'discovery' of the accessory nerve to Willis. This structure was familiar to Eustachius, whose teachings, though not his plates were widely known. Volcher Coiter also spoke of the accessory nerve in unmistakable terms. The so-called circle of Willis was also previously well known. Willis's account of the autonomic ganglia was something of a compromise between the notions of Galen and Descartes. In fact all his conceptions of the function of the neural system were Cartesian modifications of Galenic opinions". … and so on.

Such a damaging approach of Willis also inhibits the study of series of questions that are a necessary prerequisite to judge the value of Willis's Cerebri anatome.

12.4 The XIIIth and XIVth Cranial Nerves

The scientific content of spinal nerves and cranial nerves was known around 1900. Studying form gave basic results, but physiology produced misty dreams until Bois-Reymond's studies on action potentials. Nevertheless, the distribution of the spinal and cranial nerves in the end was in general understood, with cranial nerves that have their neurons within the central nervous system. Moreover, the so-called vago-accessory complex has been studied in the cat, dog, macaque, chimpanzee and man for its composition of myelinated and unmyelinated fibres. The conclusion is that the roots, stem and branches of this complex do not show basic differences. Thus, the cranial nerves X and cranial and spinal XI and their intravertebral and extravertebral branches are basically, hardly at variance in the mammalian species (Botár et al. 1937). Thus, the most difficult area IX, X and XI was well described and the VII to VIII problem, still waiting, will be treated here.

Still the question arises, do we have 12 cranial nerves?

First the synonyms for the studied extra cranial nerve structures are given:

(1) The *nervus terminalis* has also been called: zero nerve, nulla nerve and cranial nerve XIII (Bordoni and Zanier 2013).

(2) The *nervus intermedius* has collected the names: nerve of Wrisberg, the intermediate nerve, sensory and parasympathetic part of the facial nerve, portio intermedia, Sapolini's nerve and cranial nerve XIV (Bordoni and Zanier 2013; Alfieri et al. 2010; for overview see Tubbs et al. 2013).

Thus, one is dealing with two nerve bundles one related to the olfactory cranial nerve (nervus terminalis), the other has relations to the facial and vestibular–cochlear cranial nerve complex (nervus intermedius). Their naming is a hopeless muddle, because the nervus terminalis is not considered a cranial nerve and the nervus intermedius is considered a branch of the facial nerve, according to the Nomina Anatomica.

Making cranial VII simple: the nervus intermedius is the sensible branch of the facialis nerve and has a ganglion, the genicular ganglion. The facialis nerve proper is the motor branch. They join in or at the edge of the brainstem inducing the likeliness of two cranial nerves.

12.4.1 The Nervus Terminalis

Here, we partially reproduce the abstract of the review of Vilensky (2014, see Fig. 12.15): "The nervus terminalis (NT; terminal nerve) was clearly identified as an additional cranial nerve in humans more than a century ago, yet remains mostly undescribed in modern anatomy textbooks. The nerve is referred to as the nervus terminalis because in species initially examined its fibres were seen entering the brain in the region of the lamina terminalis. It has also been referred to as cranial nerve 0, but because there is no Roman symbol for zero, an N for the Latin word nulla is a better numerical designation.

This nerve is very distinct in human foetuses and infants but also has been repeatedly identified in adult human brains. The NT fibres are unmyelinated and emanate from ganglia. The fibres pass through the cribriform plate medial to those of the olfactory nerve fila. The fibres end in the nasal mucosa and probably arise from autonomic/neuromodulatory as well as sensory neurons. The NT has been demonstrated to release luteinizing-releasing luteinizing hormone and is therefore thought to play a role in reproductive behaviour".

Since the "nulla" nerve is considered a mixed nerve, the contributing ganglia are found adhered to the dura of the cribrose area, but the localization of the other nerve cells within the brain is speculative till now.

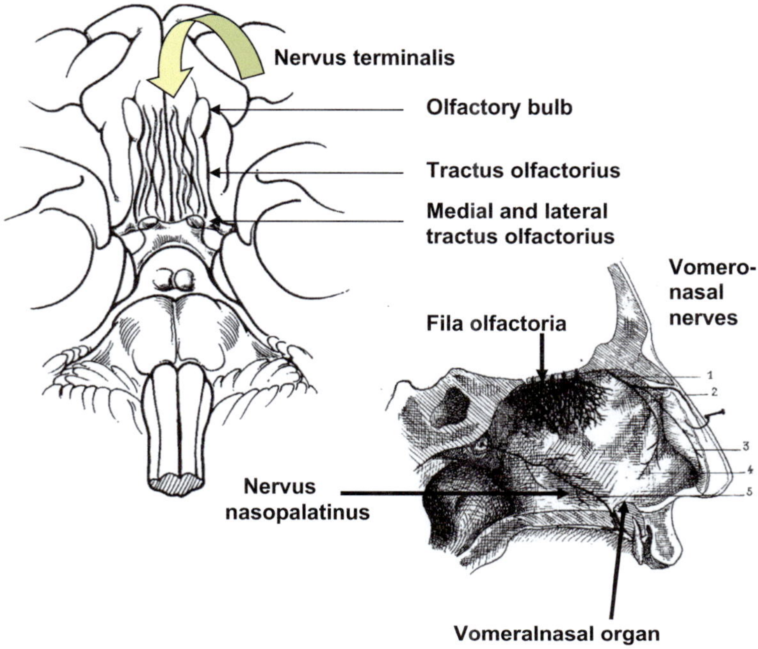

Fig. 12.15 "Zero" cranial nerve or nervus terminalis. Left: Nervus terminalis contacts the organ of Jacobson or vomeronasal organ by the vomeronasal nerves. Topographically placed in between both olfactory tracts and bulbs, it enters the brainstem. Right: Midsagittal section of the nasal cavity: 1, Le nerf nasal interne; 2, le nerf naso-lobaire; 3, La branche externe du filet ethmoidal; 4, la branche interne du filet ethmoidal; 5, Le nerf naso-palatin (English translation in the figure). Left upper figure, changed after Vilensky (2014), lower right figure adapted from Hovelacque (1927)

12.4.2 The Nervus Intermedius

"Of note, in tribute to his teacher, Soemmerring attached the eponym "nerve of Wrisberg" to the nervus intermedius, which was first described by Wrisberg in 1777" (Davis et al. 2013; see also Alfieri et al. 2010).

The intermedius nerve shows origins at variance (Fig. 12.16), inducing all kinds of speculations: branch of VIII (still in 20% of the cases) and not of VII, independent cranial nerve (that can have four or more roots) till a composite nerve. On 43 brainstems were found: a single bundle in 23.8%, two bundles in 30.95%, three in 27.38%, four in 10.71% and five in 7.14%, demonstrating its variability (Alfieri et al. 2012).

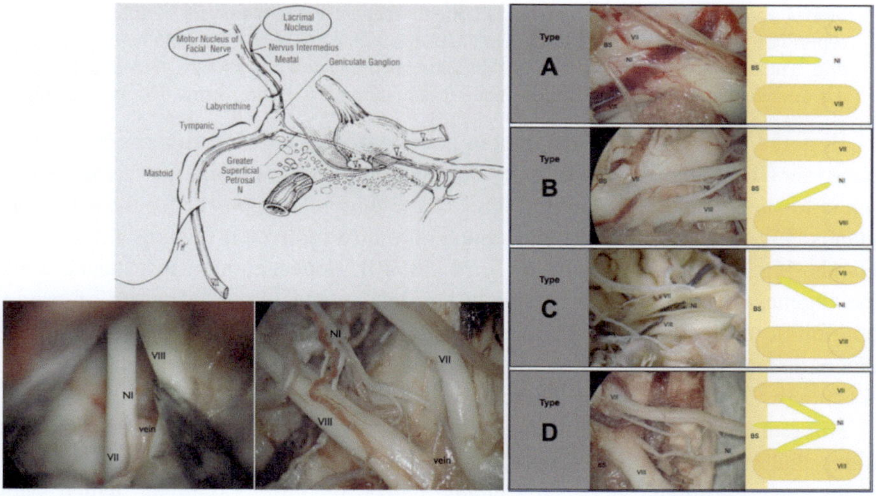

Fig. 12.16 Nervus intermedius: upper figure shows the localization of the facial sensory and motoric branches near the trigeminal ganglion (Jordan 1993), the lower two figures demonstrate both branches next to each other (left) and the nervus intermedius existing of several fascicles. Scheme at the right shows the origin of the nervus intermedius from the brainstem (A), from the statoacusticus (B), from the facialis (C) and from all three: brainstem, VII and VIII (D; Alfieri et al. 2012 with permission)

12.4.2.1 Crocodile Tears

"Crocodile tears syndrome is a lacrimal hypersecretion disorder characterized by excessive tearing with gustatory stimulation while eating, drinking or smelling food. Damage of the nervus intermedius is held responsible" (Nakamizo et al. 2012). In vestibular schwannoma operations, the surgeon needs to reach the tumour that is between brainstem and meatus acusticus interna around the cranial nerves VII and VIII, including the nervus intermedius (see Fig. 12.16). Damage in these operations of the facialis complex is not excluded, including the nervus intermedius.

Let us have a more extensive approach into the relation between tears and salivation: the first description of the use of the naming "crocodile tears" with a nearly appropriate explanation is from F.A. Bogorad (1928), and his name is coupled to the syndrome.

Tales on crocodiles that whine are numerous. In general, a crocodile can wet his nictitating membrane, also called the second eyelid, and teardrops can be noticed because the lacrimal glands are producing a proteinaceous fluid. This led to spectacular stories from ancient Roman Claudio Eliano via Byzantine times into Shakespeare (Murube 2005):

> If that the earth could teem with woman's tears
>
> Each drop she falls would prove a crocodile.
>
> Out of my sight!

> Stories and expressions related to hypocritical weeping express two main different concepts. The first is that weeping is done in order to sweeten and attract a person before killing him. The second concept is somewhat different, as the victim has already been killed, and the weeping starts when the victim is being eaten, perhaps to feign sorrow. (Murube 2005)

Bogorad's syndrome is an uncommon consequence of recovery from facial paralysis (Bell's palsy) where faulty regeneration of this cranial nerve causes sufferers to shed tears while eating. Fibres in the *intermediate nerve* for the salivatory glands (Fig. 12.17 passing by 5, 4 and 3) are also redirected towards the lacrimal glands (Fig. 12.17 now passing by 5, 34, 38, 40). Hence, eating causes tears. Since the regenerated fibres, innervating the lacrimal gland, are cholinergic of origin, botulinum toxin administration within the gland, indeed reduce lacrimation of the crocodile tear syndrome. Botulinum injections are effective for the duration of some six months. Then, the hyper-lacrimation reappeared.

Afferent taste sensation from the anterior two-thirds of the tongue is conveyed by fibres that run in the lingual nerve, a branch of the mandibular nerve. The taste fibres leave the lingual nerve in a bundle named the chorda tympani, which leads these fibres to the intermediate nerve: damage to the intermediate nerve can change taste perception.

Their cell bodies are placed within the geniculate ganglion at the first bend of the facial nerve and enter the intermediate nerve (in Fig. 12.17 the gustatory pathway 3, 4, 5, 8).

"The *Homo sapiens* is the only animal that in its etiological behaviour has at present been proved to have developed lacrimal psycho-emotional hypersecretion in some specific emotional states of mind. Psychic weeping has been proved to occur as a normal function only in humans" (Murube 2009). Note that these cortical or psychological effects are mediated via the parasympathetic system (Fig. 12.18). Moreover, weeping is a combined action between the cranial nerves V (ophthalmicus), sensory, afferent effect, and VII (intermedius), efferent, motor effect and can be a combined action of parasympathetic and sympathetic systems (Fig. 12.18). One could state that there are two loops involved: one parasympathetic, the main one, and one sympathetic, additional one, both activated by local stimuli from cornea and conjunctiva. Addition of sympathetic to the parasympathetic autonomic systems increases the lacrimal secretion. Relaying of sensory and motor innervation occurs within the brainstem, but sympathetic effects, via the ramus descendens of the trigeminal nerve, pass over the thoracic spinal cord to reach the sympathetic chain. One should note that the sympathetic effects are mediated by the sympathetic plexus caroticus after the superior cervical ganglion, by which this innervation, via the canalis pterygoideus, by the head blood vessels connects to those of and around the eye, a rather diffuse efferent system (not indicated in Fig. 12.18). The sympathetic system not only activates blood vessels of the lacrimal glands but presumably also increases the amount of additional substances in the tears.

Psychologically, we cry with tears coming out of both eyes. Reflex moistening of the eye is both sided. Single eye weeping is only known if the cornea or conjunctiva of one eye is stimulated (dust in one eye); then, tears appear in one eye as is the case with crocodile tears (Sect. 12.4.2.1). So where is the connection

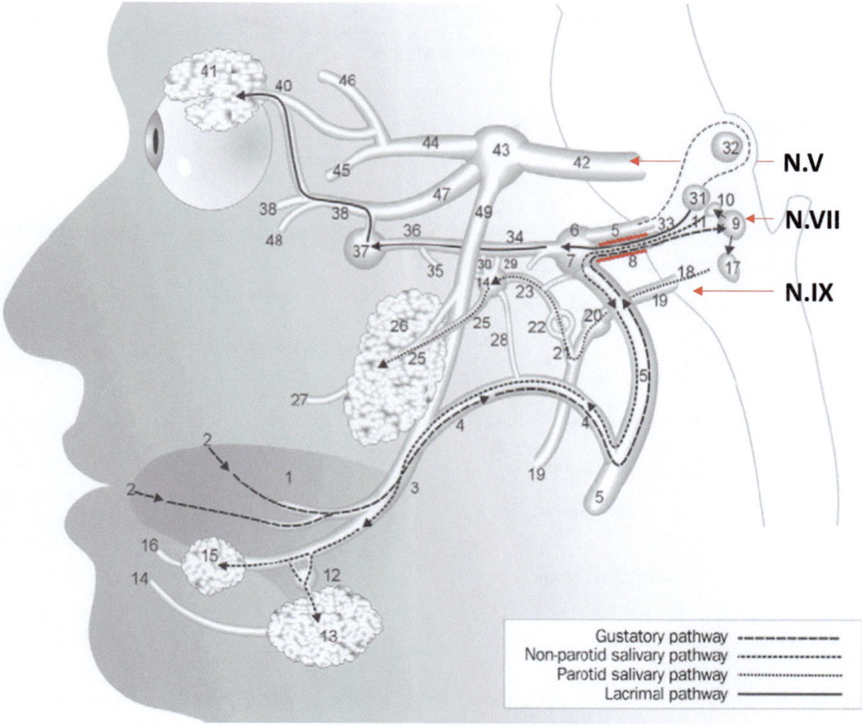

Fig. 12.17 Gustatory and salivary pathways related to the intermediate nerve (adapted from Murube 2005). In red are indicated: nervus intermedius and cranial nerves V, VII and IX. "Diagram containing the anatomy of the afferent gustatory, and efferent secretory fibres of the lacrimal and salivary glands; 1, Lingua; 2, fibrae gustationis; 3, nervus lingualis; 4, chorda tympani; 5, nervus facialis (VII); 6, ganglion geniculi; 7, genu nervi facialis; 8, nervus intermedius Wrisbergi; 9, nucleus solitarius; 10, nucleus salivarius superior (et lacrimalis); 11, fibrae salivariae non parotideae; 12, ganglion submandibulare; 13, glandula salivalis submandibularis; 14, ductus submandibularis Whartoni; 15, glandula sublingualis; 16, ductuli sublinguales et ductus Rivini; 17, nucleus salivarius inferior; 18, fibrae salivariae parotideae; 19, nervus glossopharyngeus (IX); 20, ganglion petrosum inferiore; 21, nervus tympanicus; 22, plexus tympanicus; 23, nervus petrosus superficialis minor; 24, ganglion oticum; 25, nervus auriculotemporalis; 26, glandula parotidea; 27, ductus parotideus Stenoni; 28, ramus communicans cum chorda tympani; 29, nervus sphenoidalis internus; 30, ramus communicans Bidderi; 31, nucleus nervi facialis; 32, nucleus nervi abducentis; 33, fibrae lacrimatoriae; 34, nervus petrosus superficialis major; 35, nervus petrosus profundus major; 36, nervus canalis pterygoidei (sive vidianus); 37, ganglion pterygopalatinum Meckeli; 38, nervus zygomaticus; 39, ramus communicans cum nervo lacrimale; 40, nervus lacrimalis; 41, glandula lacrimalis; 42, nervus trigeminus (V); 43, ganglion trigeminale sive semilunare Gasseri; 44, nervus ophthalmicus; 45, nervus nasociliaris; 46, nervus frontalis; 47, nervus maxillaris; 48, nervus infraorbitalis; 49, nervus mandibularis" (Murube 2005 with permission)

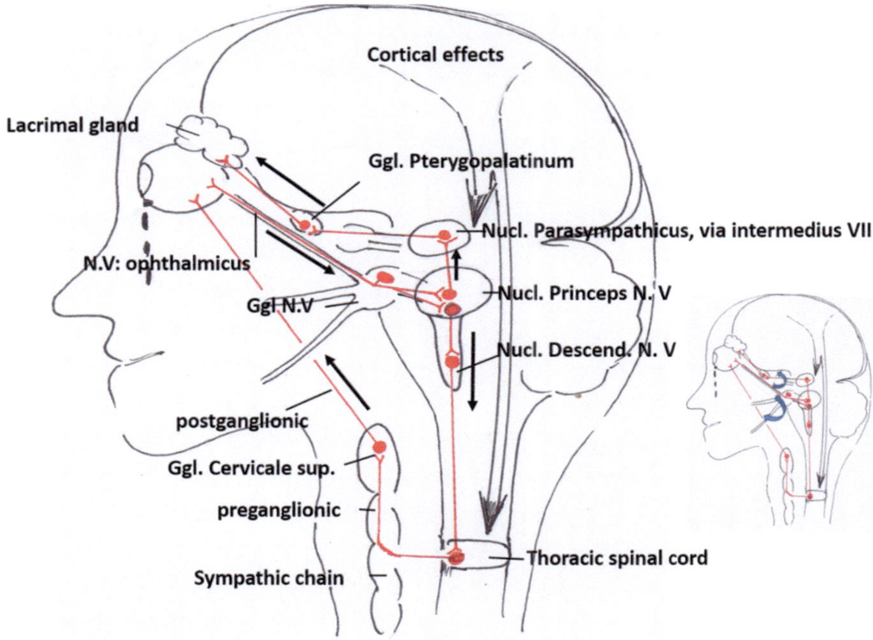

Fig. 12.18 Parasympathetic and sympathetic loops related to tear production. The parasympathetic loop uses the nervus intermedius and the sympathetic loop passes over the ramus and nucleus descendens of the trigeminal nerve into the thoracic spinal cord, by which the sympathetic chain is reached. The superior cervical ganglion has its effects diffusely via the sympathetic innervation of the blood vessels

between left and right lacrimal system? Of course in the brainstem. Monkey research showed that the ciliary ganglion can only play a role one sided in the lacrimal tear production but a left–right connection is organized by the ganglion pterygopalatinum (Van der Werf et al. 1996). How this connection is established is unknown.

A second surprising result is that both the parasympathetic, cholinergic effective, and the sympathetic system, adrenergic effective, can inhibit tear production by their neurotransmitters that normally increase lacrimal function (Dartt 2009).

12.5 Evolution of the Vago-Accessory Complex(X–XI) and the Hypoglossal Nerve (XII)

Fossil fishes are the lead in this part. Animal systematics discerns three superclasses of fishes, and restrictions remain to these superclasses to keep things simple and avoid discussions on evolutionary trends (see Romer 1962): I Cyclostomi (fishes without jaws), II Elasmobranchiomorphi (fishes with jaws and without lungs or airbladder) and III Teleostomi (fishes with jaws and with lungs or airbladder). I is also known as agnathi, and II and III are also called the gnathostomes. Examples of living cyclostome are lampreys and hagfishes. Elasmobranchiomorphi contains among others rays, sharks and dogfish. Teleostomi are best known by ray-finned fishes like the ocean sunfish and oarfish to steal the largest (2300 kg) and the longest (11 m), respectively, from Wikipedia. Sturgeon and salmon also belong to this superclass.

Some Cyclostomi lived already 500–600 million years ago, Elasmobranchiomorphi started to appear 400 million years back, and Teleostei since 350 million years. We need not to look twice for 50 million years. Bone appeared earlier than most scientists thought and lime-impregnated skeletons, together with living fossils are useful in the study of the brain and its cranial and spinal nerves. Not the nervous tissue survived, of course, but the bone, brain space and the channels by which the nerves had to leave the brain skull. After these years, spaces and channels are filled with sand and/or with rock that can be visualized (Fig. 12.19).

Since the basic morphology of the brain is rather conservative, which is also confirmed by Hox-genes (Schilling and Knight 2001), the main brain parts tel-, di-, mes-, met- and myelocephalon can be recognized in the fossil fish brain too. If these parts are known, then the cranial nerves can be distinguished, often in relation to their target organs: vestibulum, eye, jaw muscles and other structures. The outcome is at all times that fossil fishes contain ten cranial nerves (Stensiö 1963). The most caudal one is the vagus nerve (X) that has its branches to all segments of the body (Fig. 12.20). Although this clear vagus segmental nature is lost in humans, our vagus travels also through a large part of our body. Missing the hypoglossal cranial nerve (XII) is understandable since the tongue is a late development. By opening the fish mouth together with gill activity, enough low pressure, resulting into an inward water stream, can be organized to get food particles into the oesophagus.

The fossil fishes do have a spino-occipital bone, which can be longer or shorter in proportion to the length of the myelencephalon. The myelencephalon contains at its end the so-called spino-occipital nerves that differ from our spinal nerves by a separate dorsal and ventral root leaving by their own channels. This spinal nerve construction is reminiscent of the lamprey brain. Our spinal nerve joins the ventral and dorsal root to form one spinal nerve that passes through one foramen intervertebrale.

Reconstructions of the neurocranium with its channels are made by grinding perpendicular to the length axis of the fossil brain, and photographs are made each time after thin parts were removed. Afterwards, three-dimensional reconstructions can be made. Problem is that blood vessels have their own channels or are co-localized with the cranial nerve or their branches. In general, the vagus nerve stays the most caudal cranial nerve in the representatives of the three fish superclasses. This situation stays present in amphibians like frogs and toads grouped together as Anurans (missing a tail during maturity). The accessory cranial nerve appears in the Stegocephalians, a group of fossil animals belonging to the amphibians but possessing a tail. Others indicate the subclass Crossopterygian

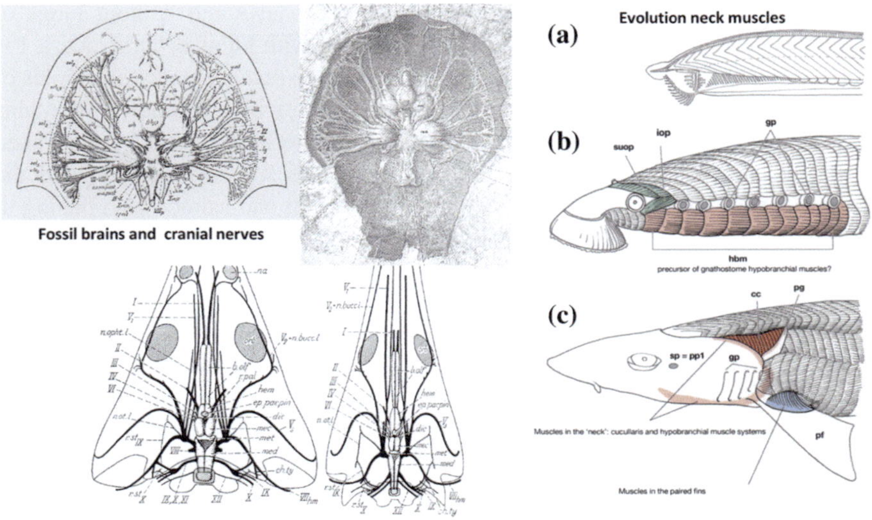

Fig. 12.19 Left side fossil cranial nerves: left upper part is the cephalic shield of an agnathe (Kiaeraspis auchenaspidoides), with its drawing (nervous system is visible due to sand and rock that filled the spaces): a.xx are all blood vessels (e.g. a.fac, canal for facial artery); bu, canals for nerve branches supplying infraorbital sensory lines; fs.hyp, hypophysial fossa; icf, infraorbital sensory line; l.lg, otico-orbital canal, for fibres of the n. ophtalmicus lateralis and n. buccalis lateralis and widening is space for ganglia of said nerves; nd, canals for the separate dorsal and ventral roots of occipital nerves; III-IX+X cranial nerve channels (text adapted from Stensiö 1963). Lower left: "Attempted restorations of the brains of two Triassic Stegocephalians, dorsal views of left Lyrocephalus euri and right Aphaneramma sp": ch.ty, chorda tympani; dic, diencephalon,; ep.par.pin, dorsal appendage of diencephalon (pineal organ, parietal organ or both); ey, eyeball; hem, hemisphere; mec, mesencephalon, med, myelencephalon; met, metencephalon; na, nasal opening; n.ophth.l n ophthalmicus lateralis; n.o.t.l, n. oticus lateralis; orb, orbital opening; r.st supratemporal branch of IX or X; I-X cranial nerves; V1 profundus nerve; V2+n.bucc.l, n.maxillaris V and n.buccalis lateralis; V3, n.mandibularis; VIIhm, n. hyoideomandibularis VII (Stensiö 1963). Right side fish neck muscles. "**a** Amphioxus. Segmented myotomes are present along the entire axis. **b** Lamprey. Somitic muscles are restricted primarily in the postotic levels. In the lamprey, however, some rostral myotomes move rostrally to cover the head as supraoptic (suop) and infraoptic muscles (iop). Light red colouring indicates the hypobranchial muscle of the lamprey (hbm), which morphologically resembles the gnathostome hypobranchial muscles lying ventral to the gill pores (gp). These coloured muscles have been specified specifically in cyclostomes. **c** In the shark anatomy, which represents the basic gnathostome body plan, rostral myotomes form both hypobranchial muscles and cucullaris muscles (cc). The latter arise as an element to bind the neurocranium, pharynx, and pectoral girdle (pg). Similar to the hypobranchial system, the cucullaris also possesses connective tissues derived from the neural crest, which represents a major component of the vertebrate 'neck' according to the definition of Matsuoka et al. (2005 with permission). pf, paired fins; sp = pp1, spiracle or the first pharyngeal pouch" (Kuratani 2008 with permission)

(lobe-finned fish, e.g. the still living bichir or Polypterus) being the first (Campos et al. 2011); nevertheless, Stegocephalians are evolutionary a subgroup born from the Crossopterygians and are the forerunners of the tetrapods or quadrupeds. This

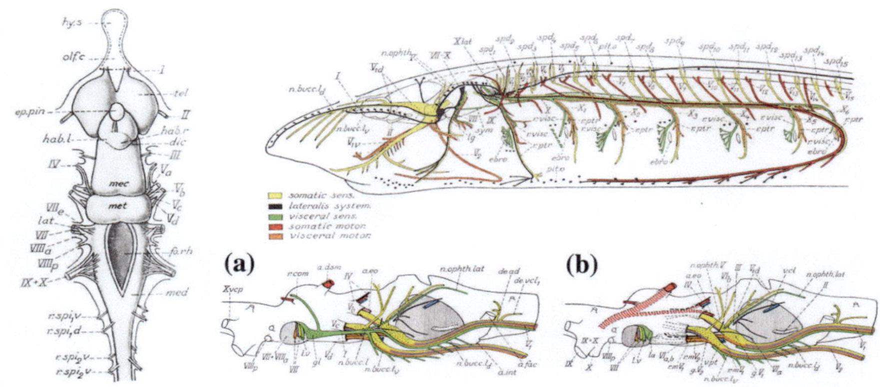

Fig. 12.20 Evolutionary aspects of the trigeminal system. Left upper part, reconstructed brain and cranial nerves of an Osteostraci-form (belonging to the cyclostomes) in a dorsal view: dic, diencephalon; hy.s, hypophysal duct; ep.pin, pineal organ; fo.rh, fossa rhomboidalis; hab, habenular ganglia left and right; lat, preotic lateralis root or roots; mec, med, met, mesencephalon, myelencephalon, metencephalon; olf.c, position of nasal cavity; r.spi.d and r.spi.v dorsal and ventral roots of two spino-occipital nerves; tel, telencephalon and hemispheres. I till IX+X cranial nerves (from Stensiö 1963). Right upper figure: Petromyzon (sea-lamprey) cranial nerves and their composition: ebro, external branchial openings; lg, lateralis ganglion associated with the ganglionic complex of trigeminus I and trigeminus II; n.bucc.l dorsal (d) and ventral (v) of n buccalis lateralis; n.ophthv, n.ophthalmicus superficialis V (dorsal ramus of trigeminus II) represented by two ascending twigs; pit.o., pit organs; r.ptr, post-trematic branches of VII and IX and of X 1–6; r.visc. visceral branches of VII, IX and of X 1–6; sp.d 1–15, dorsal roots of spinal nerves 1–15; I till X cranial nerves; V1d, dorsal branch of trigeminus I (=profundus nerve); V1v, branchial trunk of trigeminus I (=r.maxillaris V); V2, branchial trunk of trigeminus II (=n.mandibularis V); Xlat, ganglion lateralis of vagus complex; X1-6, branchial trunks of vagus nerve. Lower right: both **a** and **b** contain the distribution of the trigeminus complex of Mimetaspis hoeli (belonging to the superclass Cyclostomi, class Osteostraci), lateral view. Note colours are different from upper figure: eye muscle nerves black, ganglia coarsely stippled; somatic sensory components yellow, viscerosensory components violet; visceromotor components bright red; lateralis components bright green, arteries dark red. a.dsm, dorsal artery for skull roof; a.eo artery of orbital cavity; a.fac, canal for facial artery; a.int, canal for internal carotid artery; de.ad, de.vcl1 dorsal preorbital branches of branchial trunk of trigeminus I; gl, ganglion of lateralis fibres for the trigeminus nerves (this ganglion may also very well have been situated in the orbital cavity) gV1, ganglionic complex of trigeminus I; gV2, ganglionic complex of trigeminus II; l, unbranched prefacialis lateralis nerve; la, prefacial lateralis root; l.v. root of the lateralis nerve associated with the n. facialis; n.bucc.l, n. buccalis lateralis- ld and lv dorsal and ventral branches of n.bucc.l; n. ophth.lat, n.ophthalmicus lateralis (for vestigial supraorbital sensory line); n.ophth.V dorsal branch of trigeminus I(=n.ophthalmicus superficialis V); r.com, anterior part of n.oticus lateralis; r.mV1, r. mV2 probable visceromotor roots of trigeminus I and trigeminus II, respectively; v.cl, vena capitis lateralis (foremost part; v.pt, possible middle cerebral vein; I-IX+X cranial nerves; V1 branchial trunk of trigeminus I; Vb, Vc, Vd, canals for the roots of trigeminus I and II; V1d, dorsal branch of trigeminus (= profundus nerve) V2, branchial trunk of trigeminus II; VI a, b, ab, abducens nerve; VII, somatosensory, visceral sensory, and visceromotor roots of facialis nerve; VII+VIII, acustico-fasialis foramen; VIIIp, canal for a posterior branch of the acoustic nerve

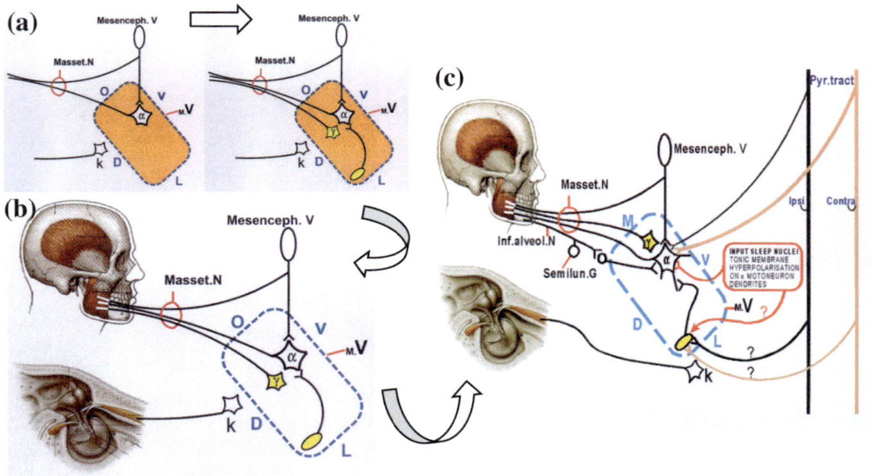

Fig. 12.21 Motor circuitry of jaw-muscles. **a** Comparison of previous and new situation for neurons present in the trigeminus motoneuron nucleus: extra are interneurons (yellow) and γ neurons (see Usunoff et al. 1997). **b** Shows the loops involved including the group K for eardrum muscles. **c** Gives the pyramidal effects on the jaw-jerk reflex, problematic connections included. mV: motor trigeminal nucleus, Masset N: nerve towards m. masseter, portio minor present as branch of the mandibular nerve, MesencephV: ramus mesencephalicus. Directions are v, ventral, d, dorsal etc. (see Marani and Usunoff 1998)

extinct group contains all pre-Jurassic salamander like amphibians and is strongly related to the reptiles. Reptiles, birds and mammals are amniotic "egg" animals and are tetrapods. The accessory nerve of the Stegocephalians is part of one cranial nerve in which the branches IX, X and XI are grouped near the myelencephalon. The hypoglossus as a cranial nerve arrives also with the Stegocephalians (for overview see Stensiö 1963, and see also Campos et al. 2011).

The other approach is: do not look into the evolution of the cranial nerves as done above, but use the evolution of the muscles innervated by the accessory cranial nerve (Fig. 12.19). Note that in this description the arguments derived from genetic modification of muscles have been omitted.

The XI cranial nerve innervates the sternocleidomastoideus and trapezius muscle that in evolution is in fact one muscle: the cucullaris muscle. The cucullaris muscle has been forcefully developed in the jaw-possessing fishes, the gnathostomes, for a better neck movement, a necessary prerequisite for terrestial live. Comparison between the cyclostomes and gnathostomes gives: "Thus, apart from the absence of the jaw and paired fin-related muscles, loss of the cucullaris muscles is one of the most conspicuous anatomical features that differentiate cyclostomes from gnathostomes. Instead, cyclostomes possess somite-derived muscles in the head, which is reminiscent of amphioxus. However, the resemblance is superficial, and these muscles represent, at least in the lamprey, typical trunk muscles that originate in the postotic region, as in gnathostomes, and which moved secondarily into the preotic domain to cover the head. The latter behaviour of myotomes is not encountered in the gnathostome developmental process" (see Fig. 12.19; Kuratani 2008).

The end of this story is that based on the study of fossil fish cranial nerves and on the evolution of neck muscles, the accessory nerve evolved in the gnathostomes, presumably within the Crossopterygian–Stegocephalian-line halfway the Devonian or a 375 million years ago. Earlier fishes had 10 cranial nerves, later 12 cranial nerves, since the hypoglossus nerve evolved around the same time.

12.6 Cranial Nerve V: The Trigeminus Complex

The trigeminal central nervous complex has been described together with Kamen Usunoff (Usunoff et al. 1997). The motor trigeminus co-organizes the jaw-jerk loop (see Figs. 12.21 and 12.23), and the involvement of γ neurons and interneurons in this loop has long be debated. Arguments are produced that the jaw-jerk reflex can be influenced by the γ loop (see below Sect. 12.6.1; Marani and Usunoff 1998). The structure of the constituting nuclei of the human central trigeminal complex contains differences compared to other mammals: a supratrigeminal nucleus is absent in man that is present in cats. Be aware that the cat scheme is automatically used for man (see Lazarov 2007). Moreover, the human mesencephalic trigeminal nucleus reaches far more rostral than assumed before (Usunoff et al. 1997; see also Lazarov 2000).

For a jaw-jerk or masseter reflex, one needs a jaw together with its muscles and the above described cyclostomes are agnathi and thus could not innervate jaw muscles. Presumably, their trigeminal system indicates a primitive situation (see Fig. 12.20). The main results from the fossil fish studies for the trigeminal complex are surprising: "In the Osteostraci (being cyclostomes) … there existed two complete cranial nerves anterior to the facialis nerve, two nerves, which according to their position from in front backwards, have been termed trigeminus I and trigeminus II, respectively. Each of these two nerves was represented by a dorsal ramus made up of solely somatic sensory fibres and a mixed branchial trunk consisting of somatic sensory, visceral sensory and visceromotor fibres" (Stensiö 1963).

Somatic motor fibres for jaw muscles are seemingly absent. The results as described by Stensiö (1963) show: "As a result of the changes which then took place in the foremost part of the branchial apparatus by the evolution of a rasping tongue, a sectorial or biting mouth or true jaws, however, the two original trigeminus nerves joined more or less completely to form the nerve currently known as the trigeminus. As has long been assumed: *the trigeminus hence comprises two*

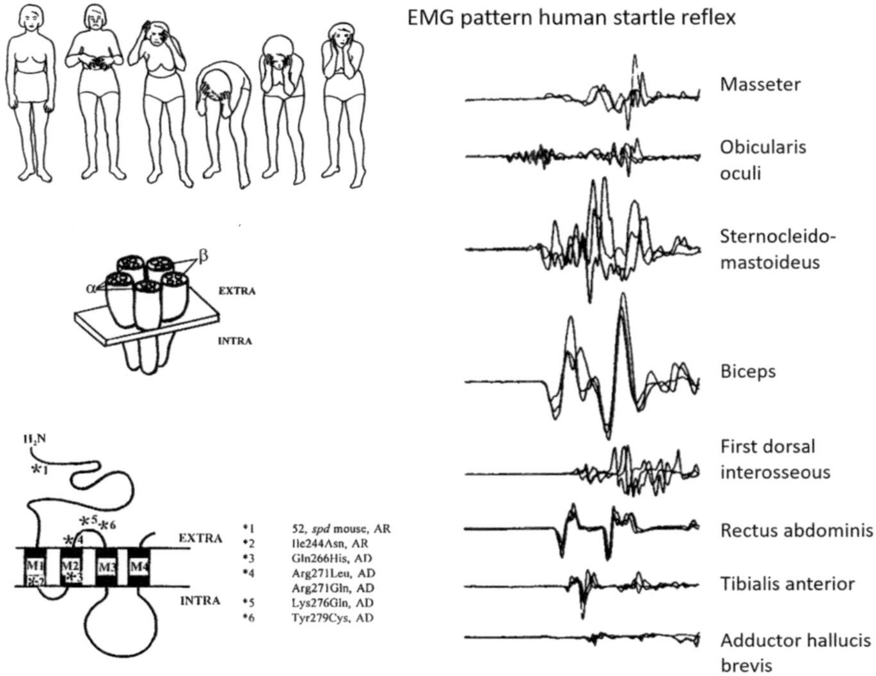

Fig. 12.22 Startle disease or hyperekplexia, a consequence of change of the glycinergic receptor. Upper figure shows the separate movements of the startle reaction (from Suhren et al. 1966). The glycine receptor is built from three α and two β subunits. The α1 subunit mutations are presented. Most mutations are found extracellular and 1 is located at the N-terminal at amino acid 52. Right side figure shows the human EMG pattern for the startle reflex. The startle reflex is located in the caudal brainstem and the latencies of the muscles increase both upward and downwards from the startle "generator" (Brown et al. 1991 with permission). Figures are adapted from De Koning-Tijssen (1977)

segmental nerves: trigeminus I and trigeminus II. Under these circumstances the homologies of the principal trigeminus branches in lower vertebrates in general would be roughly as follows:

n. ophthalmicus profundus = dorsal ramus of trigeminus I
n.maxillaris V = branchial trunk of trigeminus I
n.ophthalmicus superficialis V = dorsal ramus of trigeminus II
n.mandibularis V = branchial trunk of trigeminus II" (Stensiö 1963)

Note that profundus and superficialis are in most mammals simply the ophthalmicus, in which the profundus branch = n. nasociliaris and the n.ophthalmicus superficialis = n. frontalis after branching (see Sect. 12.6.2 *and Fig.* 12.20*).*

Thus, passing evolutionary along the Elasmobranchii and the Teleostei towards the tetrapods, two dorsal and two ventral roots (Fig. 12.20) fuse to produce the trigeminal cranial nerve. The question thus is whether the subdivision of the

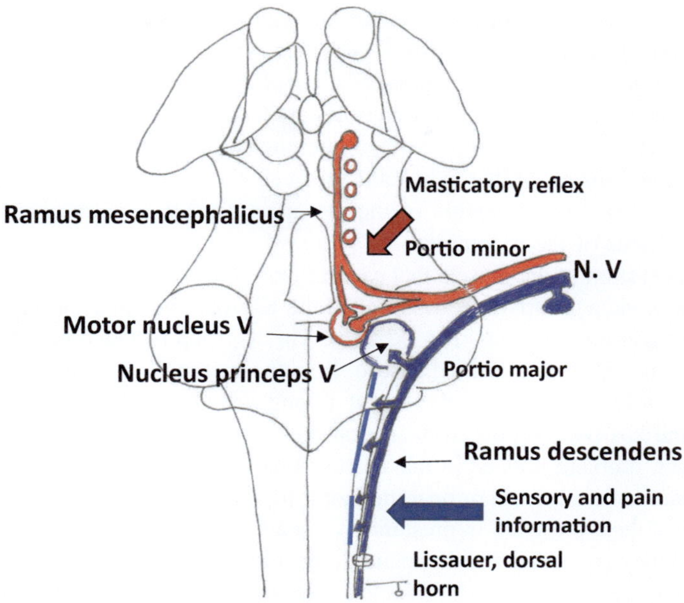

Fig. 12.23 Schematic representation of the trigeminal portio minor (red) and portio major (blue) entering the brainstem with its branches and the principal brainstem nuclei involved in orofacial movements and orofacial sensibility and pain. The trigeminal primary afferents that innervate intra- and peri-oral sensory receptors have their cell body in the trigeminal ganglion (blue line, portio major), while those innervating jaw muscle spindles (red line, portio minor) are located within the brainstem in the mesencephalic trigeminal nucleus. Mesencephalic information also contains part of the periodontal afferents innervating the teeth. The trigeminal motoneurons are depolarized and fire during closure (mesencephalic and motor activity along the portio minor constitute the masticatory reflex). Trigeminal motoneurons are hyperpolarized during mouth opening. Lateral to the motor nucleus is localized the sensory complex (blue nuclei) comprised of the main trigeminal sensory princeps nucleus and the spinal nucleus or ramus descendens of the N. V which both receive massive inputs from primary afferents that are related to pain and somatosensory information. The descending part of the portio major is continuous into the area of Lissauer in the spinal cord dorsal horn (changed after J. Voogd)

trigeminal nerve in a portio major and a portio minor (jaw-jerk reflex part) mirrors the original quadripartite subdivision.

12.6.1 *The Trigeminal Reflex*

The masticatory reflex is, just like the knee jerk reflex, considered a monosynaptic unilateral reflex and passes over the portio minor (Fig. 12.23). The difference is topographical, mainly in its localization of the sensory ganglion neurons. These neurons are not placed outside the brainstem but inside and distributed along a long caudo-rostral line, called the ramus mesencephalicus. So stretch of, for example, the

m. masseter is noted by muscle spindles and changes in the joint capsules by the Golgi tendon sensors/organs. Muscle spindles will instantly activate the motor neurons, via the intrabulbar ganglion cells, which let the masseter muscle contract. The motor neurons of all masticator muscles are located in the motor nucleus of the trigeminal nerve, called the motor trigeminal nucleus (Fig. 12.21a and 12.23). The Golgi tendon sensors of the jaw joint, as well the sensory information of the periodontium and skin receptors of the jaw area, send their information into the princeps nucleus of the trigeminus.

The innervation of the muscle that can influence the eardrum (the tensor tympani muscle) goes via a group of neurons, called the K-group, which is closely related to the motor trigeminal nucleus (Fig. 12.21b). The K-group is added here because an additional innervation to the digastric (mouth opening) and masseter (mouth closing) muscles is found in the rabbit (Donga et al. 1992) for these K-group neurons and has been supposed in several other mammals, including man.

The main problem noticed in masticatory function was the presence of γ motor neurons and interneurons in or near the motor trigeminal nucleus (see Usunoff et al. 1997). Now their presence is presumably established (Lund 1991; Masuda et al. 1997), and they are added and compared to the old organization in Fig. 12.21a. The interneurons are located in a perinuclear rim around the trigeminal motonucleus also called border zone or Regio h.

Chewing movements can be made voluntary. Therefore, the voluntary brain *activating* messages have to reach the motor trigeminal nucleus. They do, via the ipsilateral and contralateral pyramidal tract (Fig. 12.21c: see also Usunoff et al. 1997). The contralateral connection is seemingly dominant. To be honest, while our own research (Usunoff et al. 1997; see also Cruccu et al. 1989) found pyramidal degeneration, after large cortical cerebral vascular accidents lesions, on the human trigeminal motoneurons, Hatanaka et al. (2005) doubts such a direct steering by the cortex of the monkey trigeminal motoneurons. They argue for pyramidal influence on the interneuronal rim or border zone around the trigeminal motonucleus.

During sleep, the masticator muscles relax, hence snoring. Best known is alcohol consumption that induces relaxation (Baranyi and Chase 1984). This *relaxation* is mediated by inhibition of the trigeminal motor neurons. This tonic membrane hyperpolarization (Chase et al. 1980) is caused by both the reticular formation (Chase et al. 1984) in man as by the supratrigeminal nucleus in animals (your cat can snore!). The main discussion concerns whether this effect is directly on the dendrites of the trigeminal motor neurons or whether interneurons are also involved or not.

The inferior alveolar nerve (Fig. 12.21), part of the portio major, also brings sensory information to trigeminal sensory neurons. During sleep, the activity of these neurons is suppressed, which should also bring lower activity towards the trigeminal motor neurons. So, how is this relaxation organized?

"The suppression of motoneuron activity by postsynaptic inhibition during active sleep (AS or rapid eye movement (REM) sleep) is produced by glycinergic postsynaptic processes" (Morales et al. 2006). Thus, the neurotransmitter responsible for relaxing the trigeminal motoneurons during sleep is known. Two glycinergic groups of neurons in the brainstem reticular formation are discerned: a magnocellular group that influences several brainstem motoneurons, including the trigeminal motoneurons, and a group around the nucleus ambiguous responsible for relaxation of the larynx and pharynx muscles (Morales et al. 2006). The receptor going with glycine can be mutated at several places. Dysfunction of the glycinergic receptor causes a startle reflex disease or hyperekplexia (Fig. 12.22).

The syndrome contained an abnormally severe startle reaction to unexpected stimuli, in particular auditory ones. Two forms are discerned a major form and a minor form. The major form not only showed the severe startle reaction but also has a temporary generalized stiffness, inducing the patients to fall (see De Koning-Tijssen 1997). "The distribution and latency of the startle response in different muscles made it possible to localize the startle response in humans in the caudal brainstem, probably in the medial bulbopontine reticular formation [the magnocellular group]. The efferent pathway of the reflex could be detected by the differences in latency of the EMG activity of the different muscles (Fig. 12.22). The first muscle activity (not counting the orbicularis oculi muscle [presumably normal part of the blink reflex, remarks ours]) was measured in sternocleidomastoid muscles (31–85 ms. after auditory stimulation), followed by the mentalis, masseter, trunk and limb muscles. Latencies to the hand and foot muscles were disproportionally long" (De Koning-Tijssen 1997). From these latencies, it was concluded that the efferent pathway coincided with the efferent part of the spinobulbospinal reflex.

The masticatory reflex is unilateral, and to be effective, both the left and right reflex must have the possibility to adapt to requirements and to each other. Therefore, some words have to be spent on paired systems before we can go on: "The branchial system, although it appears as a paired system—the paired branchial arches of early stages—it is nevertheless related to a single unpaired cavity, the pharynx, and in later stages to the nose, the mouth and the larynx. In fact, we face the most intriguing problem, the innervation or central representation of a system, which is *related* to an unpaired system, the pharynx, within an essentially bilateral central nervous system. Structures involved are: the muscles of mastication, the facial musculature, including some muscles relating to the floor of the mouth, and one of the small muscles of the middle ear, the pharyngeal musculature and the laryngeal musculature. It is clear that these systems, although they show a high degree of bilateral structural independence, nevertheless contribute to an unpaired functional system. Thus, the tempero-mandibular joints are structurally independent of each other to a high degree. Functionally, however, mastication is an unpaired – though not symmetrical – mechanism, which often shows unilateral predominance" (Landsmeer, unpublished). Unilateral predominance because normally chewing is alternated, left side followed by right side, and so on.

This brings us back to the question: Where is the connection between left and right? Now this question is not only valid for lacrimal glands or masticatory muscles. It concerns nearly all branchial muscles within head and neck.

Let us resume the study of the trigeminal reflex and its masticatory muscles to get an answer. Two systems have been treated until now: contraction that works by acetylcholine and relaxation using glycine. The occlusive force of the masticatory muscles can reach 150–200 lb (=700–900 N). What kind of movements is possible with the mandible? To close the mouth, bringing both jaws together, the mandible moves upward due to action on both sides of the m. temporalis, m. masseter and m. pterygoideus medialis. The mandible can be brought forward. This is an action of the superficial fibres of the m. masseter and m. pterygoideus medialis, which assist the m. pterygoideus lateralis muscle. The reversed movement is done by the deep m. masseter fibres and the fibres at the back of the m. temporalis. Trituration of food is mainly done by the m. pterygoideus lateralis muscles. Thus for movements of the jaw, the masticatory muscles should be subdivided in superficial and deep (mm. masseter and pterygoideus medialis) and anterior and posterior (m. temporalis). Note that in the temporalis muscle already five functional compartments are discerned (Chang et al. 2013). Subdivisions of the masseter muscle are deep, intermediate and superficial. Subsets are described, based on endplate distribution, resulting in a finer partition of the masseter muscle (Widmer et al. 2006). Moreover, what we see as two muscles: masseter at the outside of the jaw and pterygoideus medialis at its inside; in fact, construct a noose around the mandible edge. These muscles have to work on both sides of the left and right parts of the mandible simultaneously. Therefore, at each jaw side there exists still a lateral and medial half, which asks for its own coordination that should be synchronized for the whole jaw's left and right movements.

Masticatory movements are directed by the brainstem's motor trigeminal neurons (see Lazarov 2000, 2007), and since chewing is a rhythmic occupation, trigeminal motor activity should contain such a rhythmicity, for chewing is a sequence of (partially) opening and closing the mouth. Your mother taught you to keep your mouth closed during eating; in fact, she asked to keep your lips together not the jaws. "As in other cyclic vital functions (e.g. respiration and locomotion), the basic rhythmic activity of these muscles is determined by a neuronal network referred to as central pattern generator. By definition, central pattern generators are capable of producing such rhythmic network activity even in the absence of rhythmic inputs from descending or sensory afferents" (Morquette et al. 2012). Several types of chewing can be discerned that are modified by the sensory input. The masticatory central pattern generator is capable to start and keep its rhythmicity without sensory information, and sensory information is considered the modulator of the rhythmicity in such a way that different types of chewing and the final swallowing are made possible. The second large input that adapts the chewing is of course the brain cortex using the pyramidal or corticofugal system (see Fig. 12.21c), but several other areas are involved too: hypothalamus, amygdala, several brainstem areas, cerebellum and the basal ganglia (for Parkinson-related masticatory effects see Usunoff et al. 2002).

Sensory information of mastication goes to two sensory nuclei (Fig. 12.23): the ramus mesencephalicus that is involved in the masticatory reflex and to the nucleus princeps of the trigeminus and to its ramus descendens, an extension of the princeps nucleus that behaves more as a dorsal horn of the spinal cord and related to head and neck pain included toothache.

The information that goes into the masticatory reflex loop (portio minor) is in itself already rhythmic and is derived mainly from masticatory muscle spindles. Moreover, presynaptic inhibition and antidromic firing change the information that reaches the trigeminal motor neurons. All these modulations of the sensory information, among others, organize that chewing is adapted to the food hardness. These mesencephalic trigeminal neurons are capable of producing oscillatory rhythms and do impose such rhythm to other mesencephalic neuronal neighbours by gap-junctions, which is a prerequisite for rapid biting and extra force for chewing hard food.

The neurons present in the nucleus princeps of the trigeminus are responsible for the production of rhythmic bursting that produce series of action potentials to initiate and keep muscle contraction. Normally bursts are calcium ion dependent, and calcium channel blockers, nickel or calcium chelators all reduce and some even suppress the bursts (see Fig. 12.24). However, trigeminal sensory bursts are less calcium ion dependent because these bursts are also present in several experimentally calcium free or sodium-ion-manipulated situations. A sodium persistent current (I_{NaP}) plays an important role in these bursts. In general, I_{NaP} is a current that is slowly inactivated and therefore ends the burst. Moreover, the repetition of bursts occurs in these nucleus princeps neurons presumably by a hyperpolarization activated current (I_h).

Brown (1911, 1914) proposed the half centre model (Left, Fig. 12.24; Morquette et al. 2012) for extensor and flexor muscles. In masticatory movements, they are opener and closer muscles. Fatique and interneurons produced its rhythmicity. This model is extended, and for the masticatory movements, premotor neurons cross-lined the influence towards the trigeminal motor neurons, producing rhythmicity. Such model approach of rhythmicity also answers the question: Where is the connection between left and right? Within the brainstem: the two independent left and right masticatory systems connect with interneurons and premotor neurons each other at brainstem level (and of course also on cortical level). Thus, the innervation or central representation of a system, that is *related* to an unpaired system, the mouth and pharynx, within an essentially bilateral central nervous system has chosen to keep the mutual influence of the bilateral innervation within the central nervous system at various levels (see for brainstem, Nakamura and Katakura 1995, for cortex, Nordstrom 2007 and for trigeminal lemniscus projections, Marani and Schoen 2005).

The recorded trigeminal activity of currents and receptors leads to an organizational model of the masticatory functions that can simulate trigeminal actions (Right, Fig. 12.24). Jaw-closing information goes to the princeps nucleus. This jaw information is negatively calcium dependent, in which glia plays an important role. By a decrease in extracellular calcium, an increase in the sodium current is noted via an I_{can} (calcium-activated non-specific cation current). A decrease in extracellular calcium induces depolarization, and this depolarization can induce a plateau potential needed for bursting via I_{NaP} (see Heida et al. 2008). Moreover, this decrease also slowly affects the potassium currents reducing the burst activity. Incoming stimulation has to be high frequent to change the extracellular calcium

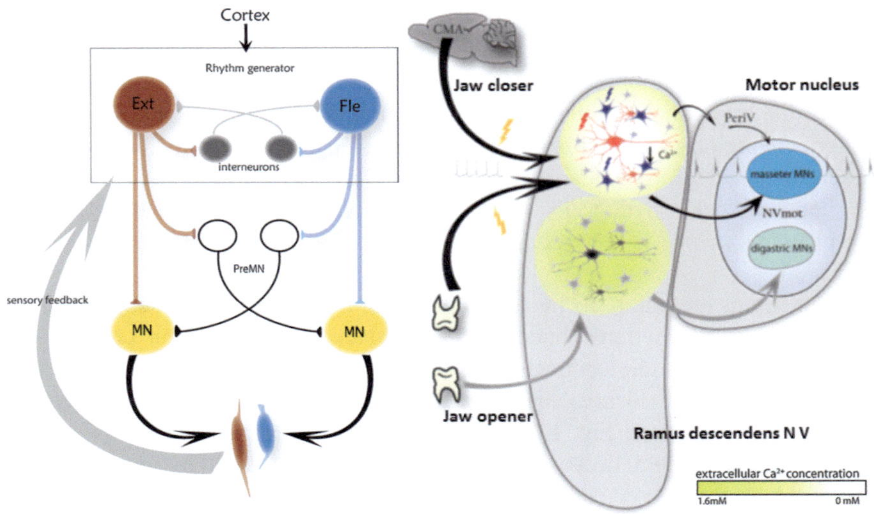

Fig. 12.24 Left: "Schematic illustration of half-centre models that produce alternating rhythmic activity between flexors (Fle in blue) and extensors (Ext in red). Flexor and extensor neurons have projection to motoneurons innervating muscle (also in blue and red) implicated in the rhythmic movement. In Graham Brown's model (box), mutual inhibition through inhibitory interneurons (in grey) insures alternating activity between two half centres (flexor in blue and extensor in red). Developing fatigue in the inhibitory neurons would be responsible for the switch between the two half-centres. In later models, the CPG circuitry is divided into two levels: a first one (within the box) that generates the basic alternating rhythm and a second one composed of premotoneurons that determine the envelope of the motoneuronal burst pattern. Cortical and sensory inputs are useful to trigger or correct the pattern"(Morquette et al. 2012). Right: "Model of the masticatory CPG. Populations of neurons in the dorsal half of the nucleus princeps would be functionally segregated in domains (yellow circles) defined by the boundaries of astrocytic syncitia. Adjacent neurons (cells in red and black) and astrocytes (cells in blue and grey) within these domains receive roughly the same topographically organized sensory and cortical inputs. In domains receiving weak afferent input (lower circle), I_{NaP} is poorly activated because of insufficient neuronal depolarization, on the one hand, and of insufficient astrocytic activation, on the other. In domains receiving greater inputs or maybe coincident inputs from the cortex and the sensory fibres, a more efficient astrocytic activation would decrease the extracellular Ca^{2+} and favour activation of I_{NaP}. This condition combined with the common incoming input would synchronize the population of neurons encompassed within the astrocytic syncytium and momentarily turn this population into the rhythm generator driving premotoneurons or motoneurons directly. The population driving the rhythm could constantly change depending on cortical inputs and sensory feedback. The bursting pattern and frequency of this population could be adjusted by sensory feedback, since sensory fibres provide depolarizing inputs that would determine the membrane potential of the bursting neurons, until depolarization exceeds I_{NaP} activity range reverting firing to a tonic pattern" (Morquette et al. 2012 with permission)

concentration. The princeps neurons transmit their information in a bursting manner towards the trigeminal motor neurons responsible for jaw closing. Jaw-opening information passes via a calcium-dependent information stream towards the trigeminal motor neurons responsible for jaw opening. One should note that the model is unilateral and contra- and homolateral connections are absent.

12.6.2 The Trigeminal Branches

Although each of the three main branches of the trigeminal nerve is important, here the maxillaris and mandibularis nerves are treated due to their importance in toothache.

Both nerves have to organize a network, called plexus, of small nerves that have to reach each of our teeth (Fig. 12.25). The endings have to enter the tooth roots, and other branches reach the periodontal receptors. The nerves thus have to enter the upper and lower jaws to reach the teeth. Upper jaw contains the branches of the nerves maxillaris: rami alveolaris superior that constitute the plexus dentalis superior containing the rami dentales et gingivales and the same construction is present by the branches of the mandibular nerve: rami alveolaris inferior and so forth.

Now look to the branches of the three main nerves: each branch is subdivided in three main sub-branches. If one looks to what is innervated by these three main sub-branches, then the following is discerned.

There is in all branches present (Fig. 12.25):

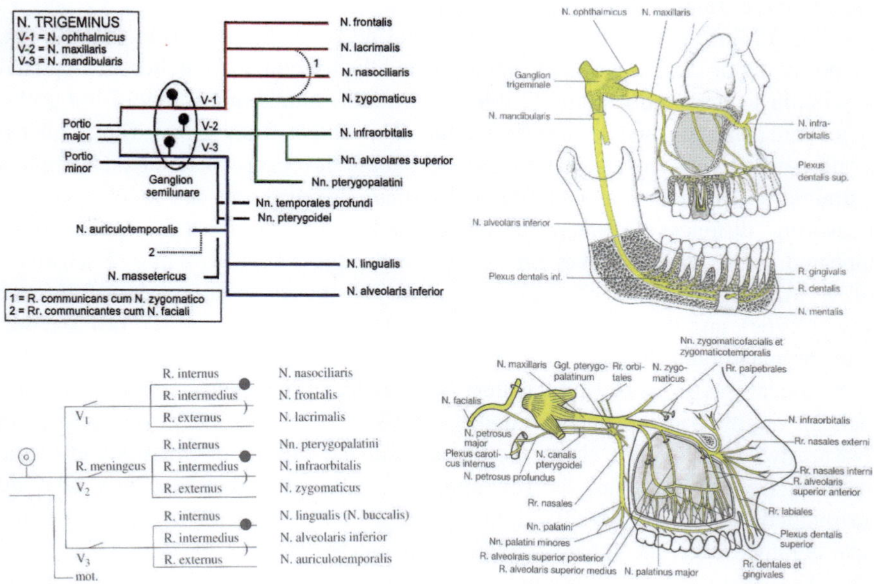

Fig. 12.25 Topographic overview of the main branches of the trigeminal portio major (coloured) and portio minor (black, adapted from Feirabend 2002). Right upper figure shows the innervation of the teeth by branches of the maxillaris and mandibularis. Upper second molar is longitudinally cut. Lower left shows the trigeminal subbranches in their comparison. Lower right figure shows the interrelation between the maxillary nerve and the ganglion pterygopalatinum in man (these three figures are from Samandari and Mai 1995 with permission)

(1) a ramus internus that is related to mucous layer of the cavities of the head and is always associated with a ganglion,
(2) a ramus intermedius that innervates the frontal skin of the head and always passes by a foramen or canalis to reach its target,
(3) a ramus externus responsible for the innervation of the lateral skin of the head.

Nevertheless, it is the ramus intermedius by its branches that innervate the teeth. Toothache therefore is mediated both over the maxillary and mandibular nerves into the brain (portio major) to reach the nucleus princeps and the ramus descendens of the trigeminal nerve (Fig. 12.23). It is a simple neuroanatomical statement, supported by an enormous amount of literature, both neuroanatomical and physiological, but very unpleasant and very painful for the carious patient.

12.6.3 The Trigeminal Ganglion and Its Cave

The trigeminal ganglion is stored in a separate cave, called Meckel's cave, on the skull base. It is an enlargement of the dura over the petrous bone that creates this cave (Fig. 12.26).

The ganglion Gasseri is large (broad 14–18 mm and long 6 mm, Fig. 12.26a), and the three trigeminal branches if compared to the cranial nerves III, IV and VI (see Fig. 12.26b) belong to the largest cranial nerve(s) (branches). In Fig. 12.26a, the portio major (5 till 9 mm), the ganglion and the trigeminal branches are depicted in relation to the portio minor (2 mm). The portio minor hides behind the portio major and ganglion and is therefore difficult to detect from above. Since microscopic surgery is also applied to the skull base for causes of, e.g., trigeminus neuralgia and tumours near this area, landmark localizations are ameliorated by measuring distances between these landmarks. Several of these publications appeared in the literature over time (see Hovelacque 1927; Lanz und Wachsmuth 1979; Arslan et al. 2012). The ganglion is placed within Meckel's cave (Fig. 12.26c), and the portio minor enters the foramen ovale (Fig. 12.26d) together with the mandibular branch.

If one places a sheet (dura mater) over a line (plica petroclinoidea ant. and incisura tentorii) and pushes with the fist (N.V) into the sheet, you will push forward two layers of the sheet: the one at your fist and the one that is part of the overhanging portion. Thus, the trigeminal subdivisions within Meckel's cave are surrounded by a double layer of dura (Fig. 12.27 left). Since the extension into the cave is rather large, the subarachnoidal space enters Meckel's cave too (Fig. 12.27 left). The topographical consequence is: the trigeminal ganglion is placed on the anterior aspect of the petrous bone, in a reduplication of the dura mater. The three branches run in dural sheets each directed to its skull base channel: ophthalmic to the superior orbital fissure, maxillary to the foramen rotundum and mandibularis to the foramen ovale. Directions are ophthalmic straight forward, maxillary horizontal and mandibular vertically. Essential trigeminal neuralgia is also called the suicide

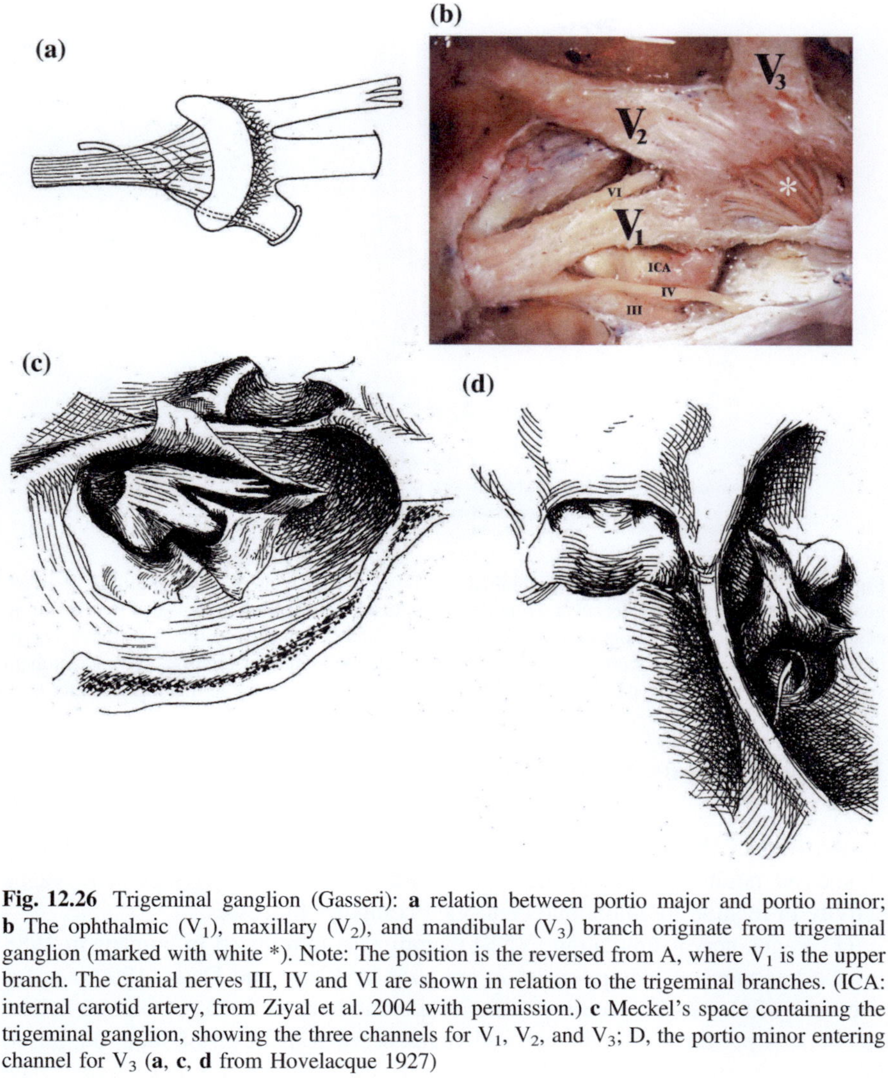

Fig. 12.26 Trigeminal ganglion (Gasseri): **a** relation between portio major and portio minor; **b** The ophthalmic (V_1), maxillary (V_2), and mandibular (V_3) branch originate from trigeminal ganglion (marked with white *). Note: The position is the reversed from A, where V_1 is the upper branch. The cranial nerves III, IV and VI are shown in relation to the trigeminal branches. (ICA: internal carotid artery, from Ziyal et al. 2004 with permission.) **c** Meckel's space containing the trigeminal ganglion, showing the three channels for V_1, V_2, and V_3; D, the portio minor entering channel for V_3 (**a**, **c**, **d** from Hovelacque 1927)

disease due to the unbearable pain that can come with it. Causes are: neurovascular compression, its principal cause, Chiari's malformation, multiple sclerosis or a lesion of the cerebellopontine angle stretching the root. The neurovascular compression is mainly caused by the arteria cerebelli superior (Fig. 12.27 left), and this compression destructs the nerve's myelin, easily triggering unwanted action potentials over the involved branch leading to pain. Microvascular decompression is applied if medication works badly. Both the a. cerebelli anterior inferior and tumours around the carotis interna can induce this trigeminal compression.

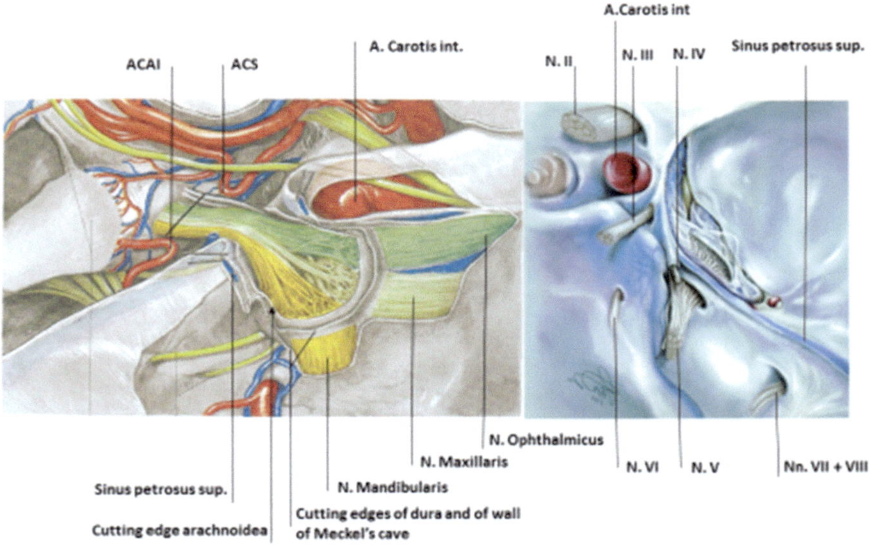

Fig. 12.27 Meckel's cave (cavum trigeminale, left 1979) with content: postganglionic radiculi, ganglion and cranial nerves. Its subarachnoidal space, cisterna trigemini, is indicated by the cutting edge of the arachnoidea (Lanz und Wachsmuth 1979) and its relation to the eye cranial nerves (right; Sabanci et al. 2011 with permission). ACAI, a. cerebelli anterior inferior; ACS, a. cerebelli superior. Nervi petrosi are passing below Meckel's cave, but above the a. meningea media

12.7 The Vagus Nerve (X)

At the start of this part on the vagus nerve, two expansions in its research area should be pointed to: vagus nerve stimulation and the parasympathetic vagus activity in relation to autonomic controlled computer processes.

(1) Vagus nerve stimulation (VNS) therapy was introduced in humans in 1988 for the treatment of therapy-resistant epilepsy by Penry and Dean (1990; see for overview of functional electric stimulation Prodanov et al. 2003). A bipolar helical electrode is placed around the cervical vagus nerve at the level of about the 5th to 6th cervical vertebra (Fig. 12.28), which is stimulated in a regular cycle. The preliminary results indicated a reduction in seizure frequency and a decrease in the duration and severity of seizures, without serious complications or mechanical failures (Wilder 1990). From then on clinical research extended and VNS therapy was finally approved as a treatment for medically refractory epilepsy in Europe in 1994 and in the USA and Canada in 1997. In the meantime, its application has been extended to several other diseases, namely depression, migraine, Alzheimer's disease, eating disorders and multiple sclerosis. The mechanisms behind VNS effectiveness remain poorly understood. Similarly, electrical engineering is very relevant to VNS, not only for the development of electrodes and stimulation, but also for

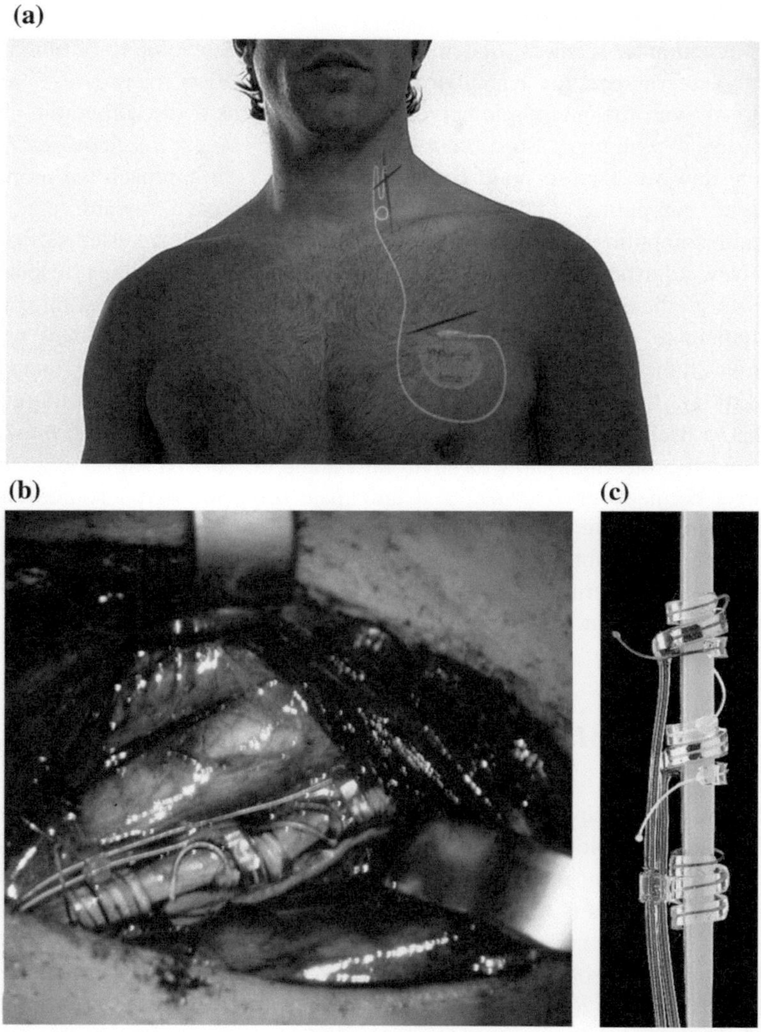

Fig. 12.28 VNS system. **a** Position of the VNS system projected on a human body. The location of the required incisions is marked. **b** Intra operative picture of electrode placed around the vagus nerve. Location of the head is on the right side of the picture. **c** The three helical parts of the electrode in detail. From bottom to top: anchor, anode and cathode (courtesy Jeroen Beekwilder, Tjemme Beems† with permission and Cyberonics)

sensing electrical signals along the vagus nerve with respect to cardiovascular function and disease, again opening new fields of clinical applications.

(2) The vagus autonomic activities attracted IBM's attentions: the claim is made by "*IBM's Perspective of the State of Information Technology*" (Horn 2001) that autonomic controlled processes in the human body are a good example to learn

from and to apply such processes in computing to solve a series of problems now arising in computer sciences. Statements such as: "In the evolution of humans and human society automation has always been the foundation for progress" and the comparison with the autonomic nervous system "It is as if the autonomic nervous system says to you, *don't think about it – no need to. I've got it all covered.* That is precisely how we need to build computing systems—an approach we propose as autonomic computing" should stimulate a new approach toward, e.g., e- or tele-health computing. "It is time to design and build computing systems of *running themselves*, adjusting to varying circumstances and preparing their resources to handle most efficiently the workloads we put upon them. These autonomic systems must anticipate needs and allow users to concentrate on what they want to accomplish rather than figuring how to rig the computing systems to get them there" (Horn 2001). Computing systems need to "know itself", must configure and reconfigure itself, have to optimize its workings, perform self-healing, must be an expert in self-protection, know its environment and the context surrounding, cannot exist in a hermetic environment and anticipate the optimized resources needed while keeping its complexity hidden, which is all indicated in the report of Horn (2001). One can doubt that all these properties are executed by the vagus nerve. Nevertheless, it describes autonomic functions of the nervous system from the engineering viewpoint.

12.7.1 Unified Modeling Language

Due to the clinical application of vagus stimulation, renewed interest in vagus anatomy and physiology occurred. The first problem concerns the language to exchange information: neural engineering needs profound knowledge of the nervous system, and this is used to analyse, design and test new ways of neurostimulation, e.g. vagus nerve stimulation. The need for a framework to unify experimental and theoretical results in neuroscience has been advocated by, for instance, Eliasmith (2003) in which a computational framework for modelling neural behaviour and dynamics was presented. Two research efforts in this direction are neuroScholar (Burns 2001), and the Neural Open Mark-up Language (NeuroML; Goddard et al. 2001). The neuroScholar system is a "computational knowledge management system with the objective of providing non-computational neuroscientist with a method to manipulate the published literature". The scope of the system is "to fully delineate the neural circuitry involved in a particular behaviour" Burns (2001). The NeuroML is an open standard for "describing models of neural systems and processes in a form that is suitable for use by simulation software tools". This language has been developed with the aim of supporting neuroscientists to collaborate in the modelling of neuronal networks. For such collaboration, it is necessary to have technologies in place that support development, exchange and discussion of computational models and their components, including databases (Goddard et al. 2001). Furthermore, NeuroML integrates existing neuron models and simulators, e.g. NEURON (Carnevale and Hines 2006), GENESIS (Bower and Beeman 1998), NeuroSim (Revest 1995) and JSim (JSim 2009). "Hence, NeuroML is not a neuroscientific application or application component itself. Also, because NeuroML is based on XML, it is intended for computer systems and not for communication between neuroscientists" (Van Beijnum et al. 2010).

A recent proposed new framework for the modelling of the peripheral nervous system is using the Unified Modeling Language (UML) and was applied to the connections of the vagus nerve (Van Beijnum et al. 2010). UML is not a programming language, but resides at a higher conceptual level. Its building units are simple and understandable for the neuroscientist, and it allows the construction of integrated neuroanatomical–physiological models, combining macroscopic with microscopic and ultrastructural views. As an example, a model of the vagus upper bodily connections is given below (Fig. 12.29).

12.7.2 The Vagus Topography

The vagus nerve leaves the skull by the foramen jugulare, but does this with several other cranial nerves. The combination of IX, X and XI near the vena jugularis interna, the carotis, cranial nerves XII and VII and this all around the processus styloideus stays an intriguing situation for the surgeon. The hypoglossus nerve leaves the skull by the hypoglossal canal in the base of the occipital condyle. The facialis uses the foramen stylomastoideum (Fig. 12.30).

The cranial nerves IX, X, XI, XII reach in this manner a neck space that is called spatium parapharyngeum. This space is bordered towards the outside of the neck by the so-called styloid fan. This screen is formed by three styloid muscles, which are enclosed by the same fascia. This fascia extends medially to the pharyngeal wall. It incorporates the posterior belly of the m digastricus and adheres to the sheath of the sternocleidomastoid muscle. The screen has to be perforated by the external carotid artery; the internal carotid stays within the spatium parapharyngeum. "The external carotid artery, arising from the common carotid, passes through the screen between the stylohyoid and digastric laterally and styloglossus medially. Just before passing the artery sends a branch, the superior thyroid artery, to the larynx and thyroid. The space in front of the screen and behind the mandible is known as the parapharyngeal space and is occupied by the parotid salivary gland, which therefore relates to the structures which lie between the wall of the pharynx and the mandibular ramus" (Landsmeer, unpublished).

The vagus nerve possesses two ganglia: the ganglion jugulare or superius and the ganglion nodosum or inferius. At the level of the ganglion jugulare, a ramus

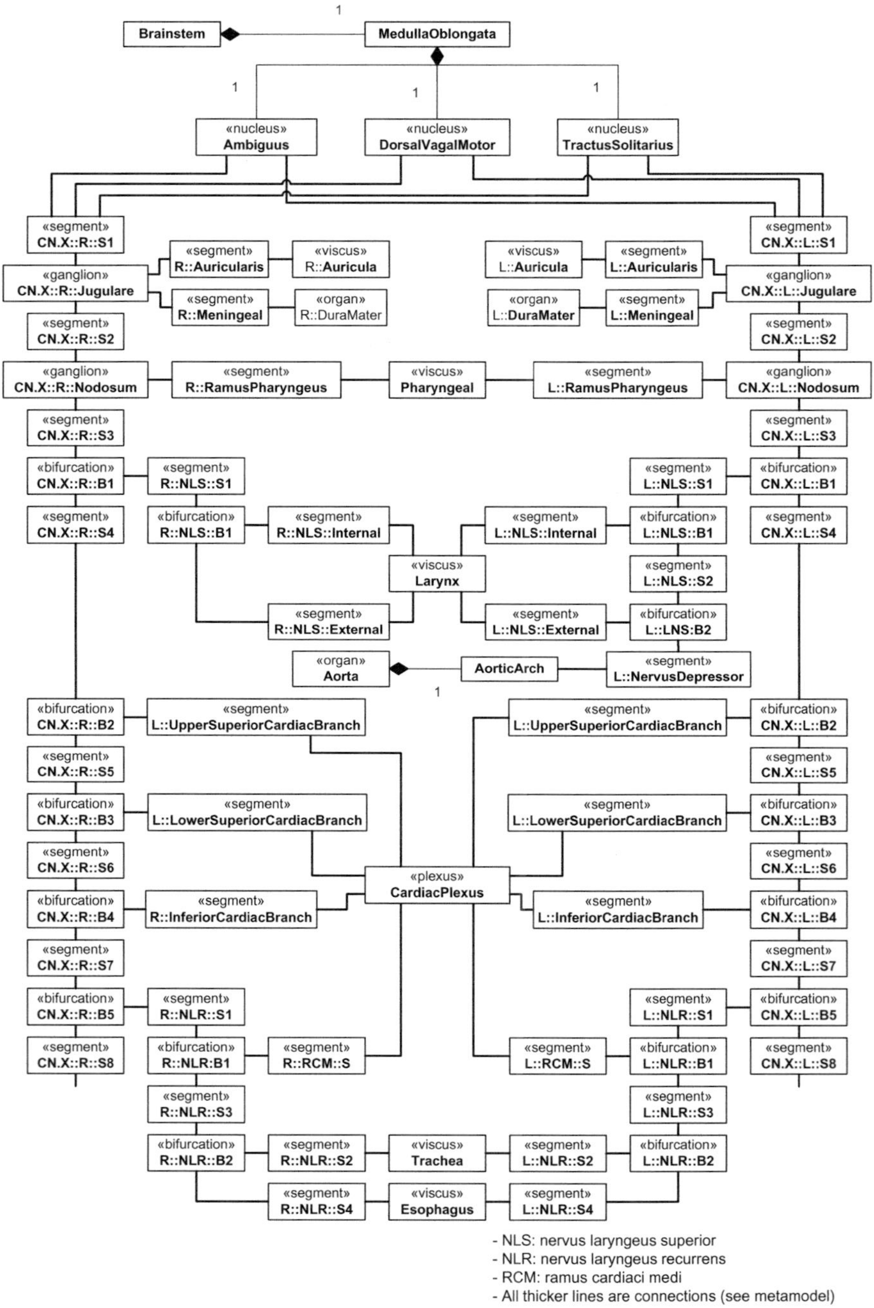

Fig. 12.29 Vagus branches of head and neck and cardiac plexus in the Unified Modeling Language (Van Beijnum et al. 2010)

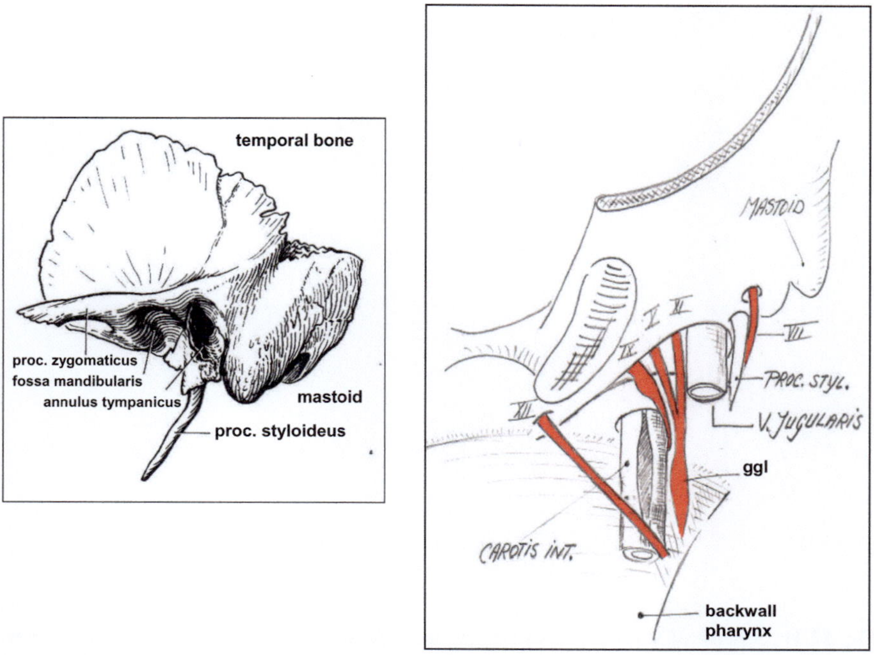

Fig. 12.30 Left shows temporal bone from lateral with the clear processus styloideus. Cranial nerves located in and near the foramen jugulare are shown from below under in the right part. (© E. Marani)

auricularis branch off that is responsible for the somatosensoric innervation of the central part of the shell of the ear and of the external auditory canal (not indicated in Fig. 12.31). The rami pharyngei (Fig. 12.31) are responsible for somatosensory innervation of mucous membranes of trachea, oesophagus and epiglottis but also for viscerosensible information: the taste from the epiglottis. Branchiomotor and somatosensory branches of the vagus nerve that innervate larynx and pharynx are the n. laryngeus superior and n.laryngeus recurrens (Fig. 12.31). The recurrens leaves the vagus nerve at the level of the a. subclavia (right side) or the arcus aortae (left side).

As long as the vagus descends in the neck, by which it stays a long time within the parapharyngeal space, it adheres to the internal carotid and the vena jugularis interna within the carotid sheath.

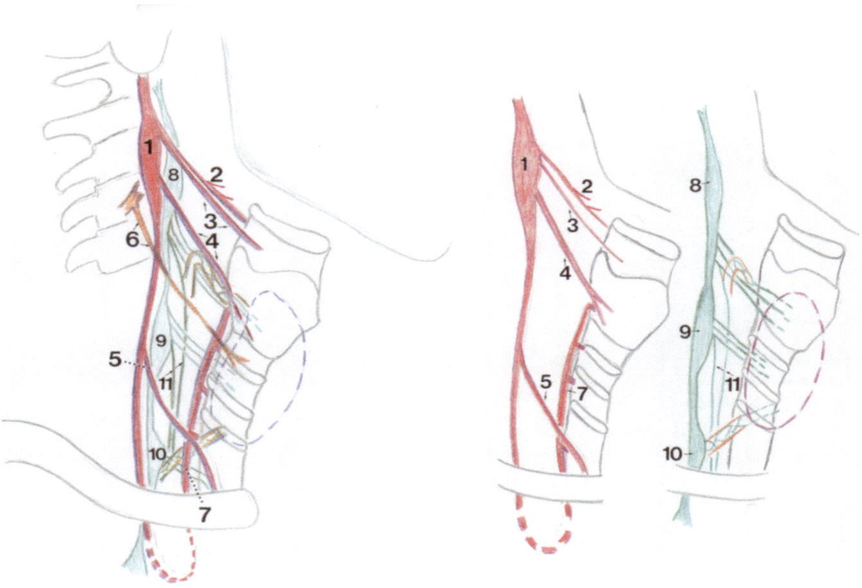

Fig. 12.31 Scheme of vagus nerve overlaying the sympathetic chain of collateral ganglia (left). Both the vagal nerve connections and sympathetic chain are separately drawn (right), both are lateral views (© E. Marani). 1, ganglion nodosum (inferior vagus ganglion); 2, pharyngeal nerve; 2, ramus internus and 3, ramus externus, are separately branched of from the ganglion (sometimes a separate r. thyreoideus can be discerned, variability is rather large); 4, nervus laryngeus superior; 5, ramus cardiacus, medius, branches of at level C7; One cm above this branch runs the r. descendens cervicalis (6), which anastomoses with the plexus thyroideus. The recurrens nerve (7, its end is named nervus laryngeus inferior) parallels the trachea till the thyroid. The sympathic chain contains three ganglia in this area: ggl cervicale superior (8), ggl cervicale medius (9) and ggl cervicale inferior (10). Each ganglion leaves a branch that innervates the thyroid gland. The branches of the ggll cervicale superior and inferior use the aa. thyroidea superior and inferior, respectively, to reach the gland. Branches known as rami cardiaci are derived from the ggl cervicale superior and medius and sometimes join the vagus nerve

References

Albucasis (±1000) On surgery and instruments. A definitive edition of the Arabic text with English translation and commentary by MS Spink and GL Lewis (1973), Wellcome Institute of the History of Medicine, University Press, Oxford

Alfieri A, Fleischhammer J, Peschke E, Strauss C (2012) The nervus intermedius as a variable landmark and critical structure in cerebellopontine angle surgery: an anatomical study and classification. Acta Neurochir 154:1263–1268

Alfieri A, Strauss C, Prell J, Peschke E (2010) History of the nervus intermedius of Wrisberg. Ann Anat 192:139–144

Arslan M, Deda H, Avci E, Elhan A, Tekdemir I, Tubbs RS, Silav G, Yilmaz E, Baskaya MK (2012) Anatomy of Meckel's cave and the trigeminal ganglion: Anatomical landmarks for a safer approach to them. Turkish Neurosurg 22:317–323

Baranyi A, Chase HM (1984) Ethanol-induced modulation of the membrane potential and synaptic activity of trigeminal motoneurons during sleep and wakefulness. Brain Res 307:233–245

Berhouma M (2013) Beyond the pineal gland assumption: A neuroanatomical appraisal of dualism in Descartes' philosophy. Clin Neurol Neurosurg 115:1661–1670

Bidloo G (1685) In: Someren et al (eds) Anatomia humani corporis. Amsterdam

Bieger D, Hopkins DA (1987) Viscerotopic representation of the upper alimentary tract in the medulla oblongata in the rat: the nucleus ambiguus. J Comp Neurol 262:546–562

Bogorad FA (1928) The symptom of crocodile tears. Vrachebnoe delo 11:1328–1330 (see for translation: Bogorad FA, Sekersen A (1979) J Hist Med Allied Sci (1979) XXXIV(1):74–79 https://doi.org/10.1093/jhmas/xxxiv.1.74

Bolk L (1910) De segmentale innervatie van romp en ledematen bij den mensch. Erven F, Bohn, Haarlem

Bonn A (1798) Ontleedkundige plaaten van B. Eustachius. J.B. Elwe, Amsterdam

Bordoni B, Zanier E (2013) Cranial nerves XIII and XIV: nerves in the shadow. J Multidiscip Health 6:87–91

Botár J, Popjak G, Bense A (1937) Die Fasern des Nervus Vagoaccessorius beim Menschen und bei Säugetieren. Acta Litterarum ac scientiarum Reg Uni Hung Francisco-Iosephinae. Tome III, Fasc 2:102–229

Bower JM, Beeman D (1998) The book of genesis: exploring realistic neural models with the general neural simulation system, 2nd edition. Springer

Bremner-Smit AT, Unwin AJ, Williams WW (1999) Sensory pathways in the spinal accessory nerve. J Bone Joint Surg 81:226–228

Brodal A (1969) Neurological anatomy, 2ed edn. Oxford University Press, NY, London

Brown H (2002) Anatomy of the spinal accessory nerve plexus: relevance to head and neck cancer and atherosclerosis. Exp Biol Med 227:570–578

Brown H, Burns S, Kaiser W (1988) The spinal accessory nerve plexus, the trapezius muscle and shoulder stabilization after radical neck cancer surgery. Ann Surg 208:654–661

Brown P, Rothwell JC, Thompson PD, Britton TC, Day BL, Marsden CD (1991) The hyperekplexias and their relationship to the normal startle reflex. Brain 114:1903–1928

Brown T (1911) The intrinsic factors in the act of progression in the mammal. Proc R Soc Lond B Biol Sci 84:308–319

Brown TG (1914) On the nature of the fundamental activity of the nervous centres: together with an analysis of the conditioning of rhythmic activity in progression, and a theory of the evolution of function in the nervous system. J Physiol 18–46

Bullock TH (1959) Neuron doctrine and electrophysiology: a quiet revolution has been taken place in our concepts of how nerve cells act alone and in contact. Science 129:997–1002

Bullock TH, Bennett MVL, Johnston D, Josephson R, Marder E, Fields RD (2005) The neuron doctrine. Science 310:791–793

Burkholder JF (1904) The Anatomy of the brain: A manual for students and practioners of Medicine. G.P. Engelhard & Co, Chicago

Burns GAPC (2001) Knowledge management of the neuroscientific literature: the data model and underlying strategy of the neuroscholar system. Philos Trans: Biol Sci 356(1412):1187–1208

Caliot Ph, Bousquet V, Midy D, Cabanie P (1989) A contribution to the study of the accessory nerve: surgical implications. Surg Radiol Anat 11:11–15

Caliot P, Bousquet V, Midy D (1990) Surgical anatomical study of the spinal nerve. Rev Laryngol Otol Rhinol (Bord) 111:275–278

Caliot P, Dumont D (1986) A note on the anastomoses between the hypoglossal nerve and the cervical plexus. Surg Radiol Anat XX:75-79

Campos D, Rieger A, Mohr H, Ellwanger JH, Borba AM Jr (2011) Anatomy and evolution of accessory nerve: cranial or spinal origins? – A review. J Morphol Sci 28:222–227

Chang Y, Cantelmi D, Wisco JJ, Fattah A, Hannam AG, Agur AM (2013) Evidence for the functional compartmentalization of the temporalis muscle: a 3-Dimensional study of innervation. J Oral Maxillofac Surg 71:1170–1177

Capozzi M (2014) Kant, Soemmerring and the importance of the sense of hearing. Lexicon Philosophicum LPh 2, ISSN 2283-7833, http://lexicon.cnr.it/

Carnevale NT, Hines ML (2006) The NEURON Book. Cambridge University Press

Chase HM, Chandler SH, Nakamura Y (1980) Intra cellular determination of memebrane potential of trigeminal motoneurons during sleep and wakefulness. J Neurophysiol 44:349–358

Chase HM, Enomoto S, Hiraba K, Katoh M, Nakamura Y, Sahara Y, Taira M (1984) Role of medullary reticular neurons in the inhibition of trigeminal mototneurons during active sleep. Exp Neurol 84:364–373

Clarke E, O'Malley CD (1996) The human brain and spinal cord. Norman Pub, San Francisco, CA

Coggeshall RE (1980) Law of separation of functions of the spinal roots. Physiol Rev 60:716–755

Cole FJ (1975) A history of comparative anatomy. Dover Publ Inc, NY, From Aristotle to the eighteenth century

Culberson JL (1985) Golgi, Cajal and the 1906 Nobel prize. Occasional paper of the Cajal Club

Cruccu G, Berardelli A, Inghilleri M, Manfredi M (1989) Functional organization of the trigeminal motor system in man. A neurophysiological study. Brain 112:1333–1350

Dartt DA (2009) Neural regulation of lacrimal gland secretory processes: relevance in dry eye diseases. Prog Retinal Eye Res 28:155–177

Daube JR, Sandok BA (1978) An approach to anatomy, pathology and physiology by systems and levels. Little Brown, Medical Neurosciences, Boston

Davis MC, Griessenauer JC, Bosmia, Tubbs RS, Shoja MM (2013) The naming of the cranial nerves: a historical review. wileyonlinelibrary.com: https://doi.org/10.1002/ca.22345

De Koning-Tijssen MAJ (1997) Hyperekplexia: Startle disease. Thesis University Leiden

Del Maestro RF (1998) Leonardo da Vinci: the search for the soul. J Neurosurg 89:874–887

Donaldson IML (2010) Cerbri anatome: Thomas Willis and his circle. J R Coll Physicians Edinb 40:277–279

Donga R, DuBuc R, Kolta TA, Lund JP (1992) Evidence that the musculatory muscles receive a direct innervation from cell group K in the rabbit. Neuroscience 49:951–961

Dullemeijer P (1974) Concepts and approaches in animal morphology. Van Gorcum & Co, Assen, The Netherlands

Eadie MJ (2002) A pathology of the animal spirits – the clinical neurology of Thomas Willis (1621–1675) Part I – Background, and disorders of intrinsically normal animal spirits. J Clin Neurosci 10:14–29

Eliasmith C (2003) Neural engineering: unraveling the complexity of neural systems. IEEE Can Rev 13–15

Eustachius B (1564) Sancto severinatis medici ac philosophi Opuscula anatomica. Quorum numerum argumenta aversa pagina indicabit. Venetiis MDLXIIII, Vincentius Luchinus excudebat

Feindel, W (1964) The origin and significance of Cerebri anatome. In: Thomas Willis The Anatomy of the brain and nerves, Tercentenary edn 1664–1964, vol I. Mc Gill University Press, Montreal, pp 5–60

Feirabend HKP (2002) Syllabus Anatomie van het hoofd-halsgebied en de zintuigen. Dept Neurochirurgie and Fysiologisch Laboratorium, Leiden

Flower WH (1861) Diagram of the nerves of the human body. John Churchill, London

Foerster O (1933) The dermatomes in man. Brain 56:1–39

Foster M (1901) Lectures on the history of physiology during the sixteenth, seventeenth and eighteenth centuries. Cambridge

Freeman W (1935) Motor and sensory overlap in the spinal roots. Ann Surg 101:133–137

Gehuchten Van A, Molhant M (1911) Contribution a l'etude du nerf pneumogastrique chez l'homme. Bull Acad Royal Med Belgique XXV:859–900

Goddard NH, Hucka M, Howell F, Cornelis H, Shankar K, Beeman D (2001) NeuroML: Model description methods for collaborative modelling in Neuroscience. Philos Trans: Biol Sci 356 (1412):1209–1228

Gold I, Stoljar D (1999) A neuron doctrine in the philosophy of neuroscience. Behavior Brain Sci 22:809–869

Hafferl A (1969) Lehrbuch der topographisen Anatomie. Springer, Berlin

Hatanaka N, Tokuno H, Nambu A, Inoue T, Takada M (2005) Input–output organization of jaw movement-related areas in monkey frontal cortex. J Comp Neurol 492:401–425

Heida T, Marani E, Usunoff KG (2008) The subthalamic nucleus. Part II: Modelling and simulation of activity. Adv Anat Embryol Cell Biol 199:1–88

Hierons R, Meyer A (1962) Some priority questions arising from thomas willis' work on the brain. Proc Roy Soc Med 55:287–292

Hildebrand R (2005) Soemmerring's work on the nervous system: a view on brain structure and function from the late eighteenth century. Anat Embryol 210:337–342

Hildebrand R (2011) Samuel thomas soemmerring werke: vol 4 Hirn und Nervenlehre. Akademie der Wissenschaften und der Literatur, Mainz; Schwabe Verlag, Basel

Hopkins DA, Armour JA (1998) Brainstem cells of origin of physiologically identified cardiopulmonary nerves in the rhesus monkey (Macaca mulatta). J Auton Nerv Syst 68:21–32

Hopkins DA, Bieger D, deVente J, Steinbusch WM (1996) Vagal efferent projections: viscerotopy, neurochemistry and effects of vagotomy. Prog Brain Res 107:79–96

Horn P (2001) IBM's perspective on the state of information technology. Citeulike 1512356

Hovelacque A (1927) Les nerfs craniens. Presses Universitaires de France, Paris, Anatomie Macroscopique

Jakob LH (1791) Grundriβ der Erfahrungs-Seelenlehre, Halle, Hemmerde-Schwetschke

Jordan DR (1993) The nervus intermedius. Arch Ophthalmol 111:1691–1692

JSim (2009), The Physiome Project. http://physiome.org/jsim/index.html (Last visited on February 5, 2009)

Kardel T, Jacobeaus H (1994) Steno: Life-science-philosophy. Acta hist Sci nat med 42

Kitamura S, Nishiguchi T, Ogatu K, Sakai A (1989) Neurons of origin of the internal ramus of rabbit accessory nerve: Localization in the dorsal nucleus of the vagus

Fr Kopsch-Rauber (1920) Lehrbuch der Anatomie. Thieme, Leipzig

Kuratani S (2008) Evolutionary developmental studies of cyclostomes and the origin of the vertebrate neck. Dev Growth Differ 50:S189–S194

Lachman N, Acland R, Rosse C (2002) Anatomical evidence for the absence of a morphologically distinct cranial root in the accessory nerve of man. Clin Anat 15:4–10

Lang J (1976) Kopf: Gehirn und augenschaedel Teil B, in Lanz von T. Wachsmuth W Praktische Anatomie, Springer, Berlin

Lanz and Wachsmuth (1938) 1955, 1972, 1979, 2004) Praktische Anatomie: Kopf-uebergeordnete Systemen, Kopf-Gehirn und Augenschädel, Hals, Arm, Bein und Statik, Bauch, Rücken. Springer, Berlin, Heidelberg, NY

Lazarov NE (2000) The mesencephalic trigeminal nucleus in the cat. Adv Anat Embryol Cell Biol 153:1–97

Lazarov NE (2007) Neurobiology of orofacial proprioception. Brain Res Rev 56:362–383

Liu H-F, Won H-S, Chung I-H, Kim I-B, Han S-H (2014) Morphological characteristics of the cranial root of the accessory nerve. Clin Anat 27:1167–1173

Lund JP (1991) Mastication and its control by the brain stem. Crit Rev Oral Biol Med 2:33–64

Mann G (1999) Soemmerring, Lichtenberg und die Ärzte im Streit um die Guillotine. In Samuel Thomas Soemmerring Werke: Organ der Seele. Band 9 pp 105–130. Akademie der Wissenschaften und der Literatur, Mainz Schwabe & Co Verlag, Basel

Marani E, WFRM Koch (2014) The pelvis. Structure, gender and society. Springer Heidelberg, NY, pp 1–558

Marani E, Lakke AEJF (2012) Peripheral nervous system topic. In: System Human Nervous (ed) Mai JK. Elsevier, Academic Press, Amsterdam Boston, Paxinos G, pp 82–140

Marani E, Roeling TAP, Velders GA, Lakke EAJF (1987) The Jelgersma collection. In: Marani E (ed) Progress in neuroscience and neuroregulation, neuroregulation group. ISBN: 90-71803-01-5

Marani E, Schoen JHR (2005) A reappraisal of the ascending systems in man, with emphasis on the medial lemniscus. Adv Anat Embryol Cell Biol 179:1–76

Marani E, Usunoff KG (1998) The trigeminal motonucleus in man Arch. Physiol Biochem 106:346–354
Marani E, Weyns A (1984) Hersenatlas van het schaap met een ontleedkundige handleiding. Depts. Anatomy and Embryology, Leiden, Netherlands and Veterinary Anatomy and Embryology, Anvers, Belgium
Marino L (1994) Soemmerring, Kant and the organ of the soul. In: Poggi S, Bossi M (eds) Romanticism in science: science in Europe, 1790–1840 Boston studies in the philosophy of science, vol 152, pp 127–142
Masuda Y, Morimoto T, Hidaka O, Kato T, Matsuo R, Inoue T, Kobayashi M, Taylor A (1997) Modulation of jaw muscle spindle discharge during mastication in the rabbit. J Neurophysiol 77:2227–2231
Matsuoka T, Ahlberg PE, Kessaris N, Kessaris N, Iannarelli P, Dennehy U, Richardson WD, McMahon AP, Koentges G (2005) Neural crest origins of the neck and shoulder. Nature 436:347–355. http://www.nature.com/nature/journal/v436/n7049/full/nature03837.html-a1
McNabb J (2014) Thomas Willis: The faculties and his two cognitive frameworks. Brain Cogn 91:131–137
Mettler CC, Mettler FA (1947) history of medicine, a correlative text, arranged according to subjects. The Blakistone Co Philadelphia, Toronto
Meyer A (1966) Karl Friedrich Burdach on Thomas Willis. J Neurol Sci 3:109–116
Morales FR, Sampogna S, Rampon C, Luppi PH, Chase MH (2006) Brainstem glycinergic neurons and their activation during active (Rapid Eye Movement) sleep in the cat. Neuroscience 142:37–47
Morris AG (1986) On the sexual intercourse drawings of Leonardo da Vinci. S Afr Med J 69:510–513
Morquette P, Lavoie R, Fhima MD, Lamoureux X, Verdier D, Kolta A (2012) Generation of the masticatory central pattern and its modulation by sensory feedback. Prog Neurobiol 96:340–355
Murube J (2005) Crocodile tears. Ocul Surf 3:69–72
Murube J (2009) The tear apparatus of animals: do they weep? Ocul Surf 7:121–127
Naderi s, Türe U, Pait TG (2004) History of the spinal cord localization Neurosurg Focus 16:E15
Nakamizo A, Yoshimoto K, Aman T, Mizoguchi M, Sasaki T (2012) Crocodile tears syndrome after vestibular schwannoma surgery. J Neurosurg 116:1121–1125
Nakamura Y, Katakura N (1995) Generation of masticatory rhythm in the brainstem. Neurosci Res 23:1–19
Nishizuka T (1934/35) Ueber den Plexus accessorocervicalis des Menschen. Arb Anat Inst Kais Uni Kyoto Serie A 4:6–13
Nordstrom MA (2007) Insights into the bilateral cortical control of human masticatory muscles revealed by transcranial magnetic stimulation Arch Oral Biol 52:338–342
O'Conner JPB (2003) Thomas Willis and the background to Cerebri Anatome. J R Soc Med 96:139–143
Pearce JMS (2008) The development of spinal cord anatomy. Eur Neurol 59:286–291
Pearson AA (1938) The spinal accessory nerve in human embryos. J Comp Neurol 68:243–266
Penry JK, Dean JC (1990) Prevention of intractable partial seizures by intermittent vagal stimulation in humans: preliminary results. Epilepsia 31(Suppl 2):S40–s43
Perneczky A, Sunder-Plassmann M (1980) Intradural variant of cervical nerve root fibres. Acta Neurochir 52:79–83
Pfaff ChH (1795) Ueber thierische Elektricität und Reizbarkeit. Ein Beytrag zu den neuesten Entdeckungen ueber diese Gegenstände. Crusius, Leipzig
Phillips LH, Park TS, Shaffrey ME, Shaffrey CL (2000) Electrophysiological evidence for afferent nerve fibres in human ventral roots. Muscle Nerve 23:410–415
Platner E (1794) Questiones physiologicae Leipzig
Pratt V (1987) Thinking machines. The evolution of artificial intelligence. Basil Blackwell, Oxford
Press GA, Amaral DG, Squire LR (1989) Hippocampal abnormalities in amnesic patients revealed by high-resolution magnetic resonance imaging. Nature 341:54–57

Prodanov D, Marani E, Holsheimer J (2003) Functional electric stimulation for sensory and motor functions: Progress and problems. Biomed Rev 14:23–50

Ranson SW, Clark SL (1959) The anatomy of the nervous system. Its development and function, 10th ed. W.B. Saunders Co Philadelphia, London

Rengachary SS, Pelle D, Guthikonda M (2008) Contributions of Johann Jacob Huber to the surface anatomy of the spinal cord and meninges. Neurosurgery 62:1370–1374

Revest P (1995) NeuroSim for Windows. Trends Neurosci 18:556–559

Romanes GJ (1947) Notes on the spinal cord in acephalic fetuses. J Anat (Lond) 81:391–392

Romanes GJ (1964) The motor pools of the spinal cord. In: Eccles JC, Schadé JP (eds) Organization of the spinal cord. Progr Brain Res 11:93–116

Romer AS (1962) The vertebrate body. Third ed, W.B. Saunders Co, Philadelphia, London

Routal RV, Pal GP (2000) Location of the spinal nucleus of the accessory nerve in the human spinal cord. J Anat 196:263–268

Ryan S, Blyth P, Duggan N, Wild M, Al-Ali S (2007) Is the cranial accessory nerve really a portion of the accessory nerve? Anatomy of the cranial nerves in the jugular foramen. Anat Sci Int 82:1–7

Sabanci PA, Batay F, Civelek E, Al Mefty O, Husain M, Abdulrauf SI, Karasu A (2011) Meckel's Cave. World Neurosurg 76:335–341

Samandari F, Mai JK (1995) Funktionelle Anatomie fuer Zahnmediziner, Band I and II, Quintessenz Verlags-GmbH, Berlin

Schalow G (1992) Ventral root afferents and dorsal root efferent fibres in dog and human lower sacral nerve roots. Gen Physiol Biophys 11:123–131

Schierbeek A (1951) Antoni van Leeuwenhoek. Zijn leven en werken, Vol 1 + 2. De tijdstroom, Lochem

Schilling TF, Knight RD (2001) Origins of the antero-posterior patterning and Hox gene regulation during chordate evolution. Phil Trans R Soc Lond B 356:1599–1613

Shepherd GM (1972) The neuron doctrine: A revision of functional concepts. Yale J Biol Med 45:584–599

Singer C (1931) A short history of biology. Oxford Clarendon Press

Smith Churchland P (1986) Neurophilosophy. Toward a unified science of the mind-brain. Bradford book, MIT Press Cambridge Massachusetts

Soemmerring STh (1796) Ueber das Organ der Seele. Nicolovius, Koenigsberg

Soemmerring, S Th (1795) Sur le supplice de la guillotine. Magasin encyclopédique tom III: 468–477, ou Journal des sciences, des lettres et des arts 3 (1795): 463–77. See also: Samuel Thomas Soemmerring Werke: Organ der Seele. Band 9 pp 155–252. Akademie der Wissenschaften und der Literatur, Mainz, Schwabe & Co Verlag Basel

Stensen N (1669) Discours sur l'anatomie du cerveau. Opnieuw vertaald. Originele tekst en Nederlandse weergave. In: Niels Stensen Onderzoeker van het zenuwstelsel 1638–1686. J.G. Vugs (1968) Universitaire Pers Leiden

Stensiö E (1963) The brain and the cranial nerves in fossil, lower craniate vertebrates. I. Mat.-Naturv. Klasse Ny Serie No. 13. Universitetsforlaget, Oslo, pp 1–120

Straus WL, Howel AB (1936) The spinal accessory nerve and its musculature. Quat Rev Biol 11:387–405

Suhren O, Bruyn GW, Tuynman A (1966) Hyperekplexia, a hereditary startle syndrome. J Neurol Sci 3:577–605

Swanson LW (2000) What is the brain? TINS 23:519–527

Tubbs RS, Loukas M, Slappey JB, Shoja MM, Oakes WJ, Salter EG (2007) Clinical anatomy of the C1 dorsal root, ganglion and ramus: a review and anatomical study. Clin Anat 20:624–627

Tubbs RS, Steck DT, Mortazavi MM, Cohen-Gadol AA (2013) The nervus intermedius: a review of its anatomy, function, Pathology, and role in neurosurgery. World Neurosurg 79, 5/6:763–767. http://dx.doi.org/10.1016/j.wneu.2012.03.023

Usunoff KG, Marani E, Schoen JHR (1997) The trigeminal system in man. Adv Anat Embryol Cell Biol 136:1–126

Usunoff KG, Itzev DE, Ovtscharoff WA, Marani E (2002) Neuromelanin in the human brain: a review and atlas of pigmented cells in the substantia nigra. Arch Physiol Biochem 110:257–369

van Beijnum BJ, Widya IA, Marani E (2010) Modeling the vagus nerve system with the Unified Modeling Language. J Neurosci Methods 193:307–320

Vesalius A (1543) Humani corporis fabrica, libri septem. Basileae

Vilensky JA (2014) The neglected cranial nerve: nervus terminalis (Cranial Nerve N). Clin Anat 27:46–53

Villiger L (1945) Gehirn und Rueckenmark. B. Schwabe Co, Basel

van der Werf F, Baljet B, Prins M, Otto JA (1996) Innervation of the lacrimal gland in the cynomolgous monkey: a retrograde tracing study. J Anat 188:591–601

Widmer CG, English AW, Morris-Wiman J (2006) Developmental and functional considerations of masseter muscle partitioning. Arch Oral Boil 52:305–308

Wilder BJ (1990) Summary and conclusions. Epilepsia 31:S60

Willis T (1664) Cerebri Anatome, cui accessit nervorum description et uses. London Fleher, Martyn & Allestry also consulted: Thomas Willis (1681) Anatomy of the brain and the description and use of the nerves. Translated by S. Pordage, printed for Dring, Harper, Leigh, Martyn, Fleet-street

Winkelmann A (2014) Letter to the editor: response to "The neglected cranial nerve: nervus terminalis (Cranial nerve N). Clin Anat 27:806–807

Yoshida Y, Mitsumasu T, Miyazaki T, Hirano M, Kanaseki T (1984) Distribution of the motoneurons in the brain stem of monkeys, innervating the larynx. Brain Res Bull 13:413–419

Yoshida Y, Mitsumasu T, Hirano M, Kanaseki T (1985) Somatotopic representation of the laryngeal motoneurons in the medulla of monkeys. Acta Otolaryngol 100:299–303

Ziyal IM, Sekhar LN, Özgen T, Söylemezoğlu F, Alper M, Bes M (2004) The trigeminal nerve and ganglion. An anatomical, histological, and radiological study addressing the transtrigeminal approach. Surg Neurol 61:564–574

Chapter 13
Sleep and Head and Neck

> *Cetacean sleep: An unusual form of mammalian sleep. We hypothesize that three factors – the need to come to the surface to breathe, more efficient monitoring of the environment and thermogenesis – may have been important in the evolution of the observable cetacean sleep phenomenology: sleeping while in motion, uni-hemispheric slow wave sleep, absence of REM sleep, and sleeping with one eye open* (Lyamin et al. 2008)

The overview of sleep and sleep disorders in The Netherlands is used to demonstrate the prevalence of "social jet lag", insufficient sleep and sleep disorders. A restricted overview on neuroanatomical connections related to the sleep–wake cycle is given and synapse remodelling during sleep is indicated. Head and neck do move during sleep, which is studied and neck myoclonus is described. The position of the head seems directly related to the removal of "brain waste".

13.1 Overview of Sleep and Sleep Disorders in the Netherlands

Sleep, sleep disorders and its associated health have been described extensively for The Netherlands (Kerkhof 2017). Sleep quality is directly related to the circadian system that under normal conditions regulates our day–night rhythm. Humans will not always stick to their rhythmicity due to shift work or weekend/daily pleasure. Irregularity, thus, is forced on the normal day and night rhythm. The consequence is that the restorative and adaptive functions of our sleep system are disturbed even so that sleep disorders arise. Sleep timing showed that the Dutch population had an average weekend time shift (i.e. later bedtimes in the weekend), sometimes called the "social jet lag", of 28 min. The youngest age group (18–24 years) with 40.3 min and the oldest age group (55–70) showed 16.4 min for weekend time shifts. There is a strong gender difference for the sleep onset times. Males go clearly later to bed and this difference disappears only in the age group of 55–70 years.

E. Marani and C. Heida, *Head and Neck*,
https://doi.org/10.1007/978-3-319-92105-1_13

However, the weekend shift size had no gender difference. The overall prevalence of insufficient sleep (i.e. actual sleep at least one hour less than sleep need) was 43% of this studied representative population of 2089 Dutch individuals aged 18–70 years. The occurrence of insufficient sleep was significantly higher for females (45.9%) than for males (40.2%), and reached maximal values during midlife (35–44). As compared to "sufficient sleepers", insufficient sleepers reported worse sleep quality and had higher levels of fatigue and worse daytime functioning.

As estimated on the basis of a global index of disturbed sleep, the prevalence of some form of sleep disorder was higher for females (34.8%) than for males (29.2%). In particular, among young females (18–24 year), disturbed sleep was very prominent (prevalence of 52.7% vs. 31% for males). Prevalence of insomnia is 8.2% and is higher for females compared to males. Circadian rhythm sleep disorder showed a prevalence of 5.3%, parasomnia of 6.1% and restless leg syndrome of 12.5%. Sleep breathing disorders (including obstructive sleep apnoea) demonstrated a prevalence of 7.1% with 8.1% in males and 5.6% in females.

Generalization of these results shows that females score high in insufficient sleep with higher levels of fatigue and worse daytime functioning, and the number showing sleep disturbances is worryingly high. The prevalence of sleep disorders is high in the Dutch population. By the way, Gallup polls calculated that western population sleeps 6.7 h per night, while 50 years ago it was 7.6 h, a reduction of 12%. Nowadays, 16% enjoys only 5 h sleep, while earlier this was 3% of the population. A publication of the same tenor for these differences, now compared to 20 years ago, has been published by the American Academy of Sleep Medicine and the Sleep Research Society (Luyster et al. 2012).

13.1.1 Opiates and Sleep

"Analgesic-induced overdose fatalities in New York City increased almost 700% during the interval from 1990 to 2006. In September 2014, the National Institutes of Health (NIH) Office of Disease Prevention sponsored a workshop entitled "The Role of Opioids in the Treatment of Chronic Pain". The NIH announcement also noted that chronic pain is a major public health problem affecting about 20–30% of the world's population. The problem of prescription opiate abuse is exacerbated by a 300% increase over the past 20 years in clinical prescriptions written for opiates. The number of deaths due to prescription opiates has exceeded the total number of deaths caused by cocaine and heroin. Since 2005 in the state of Massachusetts, annual deaths attributed to opiate overdose have outnumbered those caused by motor vehicle accidents" (Lydic et al. 2017). The above citation shows that pain management fails and induces severe problems also for sleep. Opiates taken or given to reduce pain undeniable affect sleep. Opiates disrupt the organization both for sleep and for wakefulness. At least it decreases the REM sleep periods and presumably also the deeper non-REM sleep duration, while light non-REM sleep periods are longer. Total sleep time is found to be reduced. Benzodiazepines induce

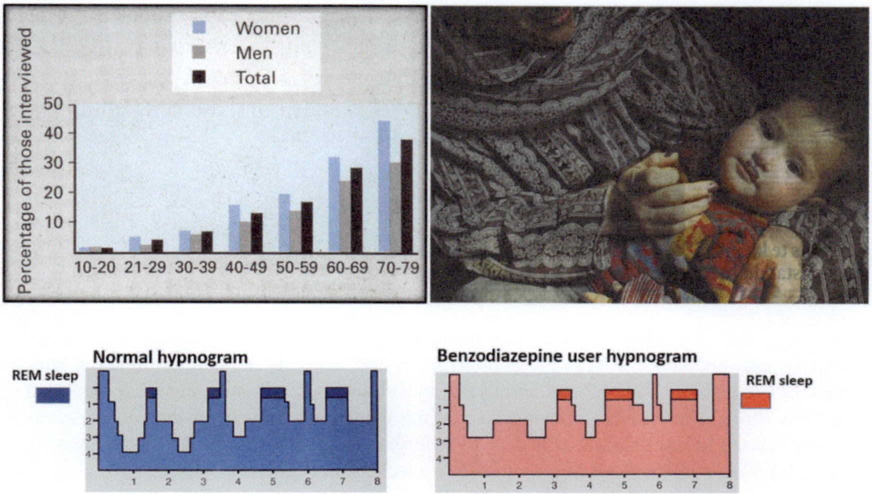

Fig. 13.1 Effect of opiates. The use of soporifics until 1979 for woman and man shows a modest increase compared to nowadays (see text), next the Chinese mother that gives her baby some opium on the upper lip to get it asleep. Benzodiazepines in drug abuse are combined with certain opioids to enhance the euphoric effects. Benzodiazepines do have analogous effects as opiates on sleep, non-REM and REM reduction in the early hours (diagrams from Shapiro 1993; courtesy G. A. Kerkhof with permission)

the same effects and are often combined with opiates to enhance the euphoria (see Fig. 13.1). Opiates induce so-called torpor-like states of the user, which means that lethargy and reduced mental and physical activity are present. Torpor states are easily confused with sleep. Wakefulness is reduced in these cases, and thus, the relation with sleep induction is changed. Although tolerance for wakefulness develops, the REM and non-REM effects do not change. The psychosocial consequences of torpor states and of the concomitant changed sleep patterns are unknown (Lydic et al. 2017). From China, during the opium wars (1839–1842 and 1856–1860), when Britain enforced the use of opium, it is well known that Chinese economy and social structures were endangered by opium smoking.

13.1.2 Cerebellar Sleep Research

A new research approach in sleep studies is the participation of the sleep–wake cycle in cerebellar function. Since motor functions also reside in the cerebellum, cerebellar sleep physiology, the sleep interactions between cortex and cerebellum and cerebellar consolidation of memory during sleep are studied. Moreover, there exists an interaction between cerebellar malfunction and the sleep–wake cycle and sleep disorders. During sleep, changed spike frequencies of the Purkinje cells are noted. In non-REM sleep, these spike frequencies decrease, while during REM sleep they increase. Within both the cerebellar fastigial nucleus (brainstem related) and dentate nucleus (thalamus and cortex related) altered sharp slow wave-like activities are found. Indications are found that sleep has a positive effect on cerebellar episodic memory formation. Correctly, it is stated that the cerebellum is still an "Uncharted Land" in sleep research (Canto et al. 2017).

13.2 Neuroanatomy of Sleep

> As suggested by Hess (1954), the aminergic system controls ergotrophic (energy expending) functions, whereas the cholinergic system controls trophotropic (energy conserving) functions. Because the aminergic and cholinergic effectors modulate neuronal activity through processes occurring both at the membrane and within the cell, it seems likely that the robust differences in sensory information processing that distinguish waking from sleeping may be explained by the same mechanisms. Hobson (1990)

Daylight is noticed by the eye and is transferred by the optic nerve to a small nucleus localized directly above the chiasma of the optic nerve, the suprachiasmatic nucleus (SCN). This photic retinal stimulation modulates the SCN via the retino-hypothalamic tract (Marani et al. 1990). The SCN, together with other brain parts, works as a pacemaker for the sleep–wake cycle. Within the retina, non-photic receptors (melanopsin receptors) are present that co-produce light information. By a crossroad via the latero-geniculate complex, light information will still reach the SCN even by damage of the optic nerve. Using hypothalamic nuclei, basal forebrain and preoptic/ventrolateral area, cortex and striatum are influenced. By the preoptic/ventrolateral area, the midreticular formation is co-governed in which the dorsal serotoninergic raphe nuclei induce the cortical sleep information by reduction of serotoninergic activity and influences the sleep inducer nuclei: parabrachial and lateral tegmental area. The locus coeruleus, responsible for the noradrenergic Non-Rapid Eye Movement (non-REM) sleep (Fig. 13.2). This starts the changes in the thalamic nuclei. During REM sleep the locus coeruleus activity is inhibited by GABA activity presumably from the periaquaductal gray. The serotoninergic raphe nuclei, by their decrease of serotonin, activate the cholinergic brainstem system, pedunculopontine and parabrachial nuclei, co-maintaining the REM sleep. These brainstem nuclei do influence thalamus, cortex and the striatum, which also connects to the cortex (see, Moruzzi 1972; Mesulam 1995; Sakai 1991; Steinbusch 1981; Rosenwasser 2009; Brown et al. 2012). A mutual influence induces a series of alternate periods of REM and non-REM sleep. Next to the inducers of sleep, wakefulness is maintained by the so-called ascending reticular activating system. This dorso–meso-brainstem–hypothalamus–basal forebrain system keeps and maintains the "thalamo-cerebral cortex unit" awake, using partly the same neuroanatomical structures and connections (Reinosos-Suàrez et al. 2011). These results are also based on animal research which Hobson (Hobson 1980) among

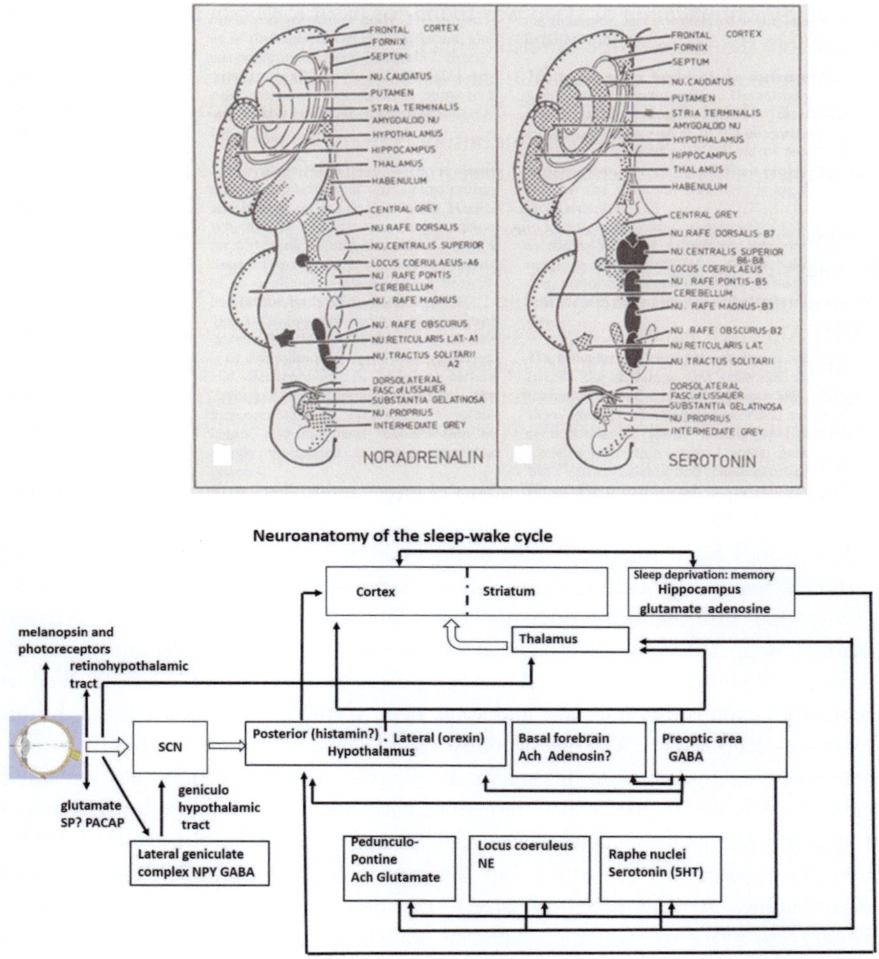

Fig. 13.2 Upper part related to sleep–wake gives a schematic overview of the neuroanatomical connections cycle (see Rosenwasser 2009; Brown et al. 2012; Nieuwenhuys et al. 2008). Excluded are the afferent connections of the SCN. Lower picture shows overview of the chemical neuroanatomy of noradrenalin and serotonin. Black are positive nuclei and stippling indicates projection areas (courtesy J. Voogd 1981 with permission). 5HT, serotonin; Ach, acetylcholine; GABA, gamma-aminobutyric acid; NE, noradrenalin; NPY, neuropeptide-Y; SCN. Suprachiasmatic nucleus; PACAP, pituitary adenylate cyclase activating peptide; SP, substance P

others, summarized for sleep-related lesions and stimulation experiments. Hypersomnia can be induced by the cerveau isolé, posterior hypothalamus lesion and midbrain reticular formation stimulation. Insomnia was noticed after lesions or stimulation of basal forebrain area, anterior hypothalamus, and mediopontine pre-trigeminal area and raphe nuclei. Vestibular lesions decreased REM in desynchronized sleep. Mesencephalic reticular formation showed desynchronization and arousal. Synchronization of sleep was found for basal forebrain area, thalamus and presumably the some of the brainstem nuclei.

Neurochemistry of noradrenalin and serotonin distributions in the brain contributed to the understanding of the effect of these chemical interconnections (Fig. 13.2). There exists a good conformity between lesion and stimulation results and the neurochemical distribution of neurotransmitters involved in the sleep–wake cycle. This sleep–wake balance does not belong to the lemniscal sensory conduction system (Marani and Schoen 2005) nor to the long descending motor system (Kuypers 1987). It belongs to the vaguely defined core of the brainstem that rests after subtraction of all sharply defined nuclei and fibre systems (Nieuwenhuys et al. 2007). Although anatomically vaguely defined in a physiological study on sleep position and sleep stage on collapsibility of the pharynx in patients with sleep apnoea, no significant effect was found by head-body position on sleep stages, but collapsibility of the upper airways is reduced by changing position from supine into a lateral sleep position. Seemingly, the REM and non-REM stages and its alternations are conservative and difficult to change, even during oxygen desaturation (Penzel et al. 2001).

Sleep does have important effects on brain morphology and on function of course. Synapse remodelling during sleep is far out the most exciting. "General activity and information processing while an animal is awake drive synapse strengthening. This is counterbalanced by weakening of synapses during sleep" (Science 2 February 2017). Serial scanning electron microscopy has been used to reconstruct axon–spine interface and spine head volume in the mouse brain during the wake–sleep cycle. A large decrease in interface size after sleep was found. Extensive changes occurred among weak synapses, whereas strong ones remained stable (De Vivo et al. 2017). The immediate early gene *Homer1a* produced changes in synaptic glutamate receptors during the sleep–wake cycle. "In awake animals, *Homer1a* accumulates in neurons but is excluded from synapses by high levels of noradrenaline. At the onset of sleep, noradrenaline levels decline, allowing *Homer1a* to move to excitatory synapses and drive synapse weakening" (Diering et al. 2017). Moreover, sleep deprivation reduces spatial memory function of the hippocampus by reduced glutamate receptor expression and increases in adenosine effects, which impairs hippocampal plasticity (Kreutzmann et al. 2015).

Striatal influence on sleep–wake cycle can be shown in Huntington's disease (HD). This genetic-based neurodegenerative disease is characterized by caudate nucleus (part of the striatum) degeneration, resulting in motor dysfunction, cognitive decline and psychiatric disturbances (Walker 2007). Huntington's sleep–wake cycle displays increased abnormal motor activity during wake and non-REM stages and a nearly absence of this disturbed motor activity in the REM period (Piano et al. 2015, 2016) and "sleep disturbances are among the earliest non-motor symptoms in HD" (Lazar et al. 2015).

13.3 Head Movements During Sleep

Figure 13.3 shows husband, wife and cat sleeping together as published in "The ethology of sleep" by Hobson (1979). During the first half of the night, a deep sleep reduces head and body movement (Hobson et al. 1978). Each 90 min a change of position can be noted. The second half of the night/morning sleep is lighter and more head and body movements are observed. Sleeping together needs

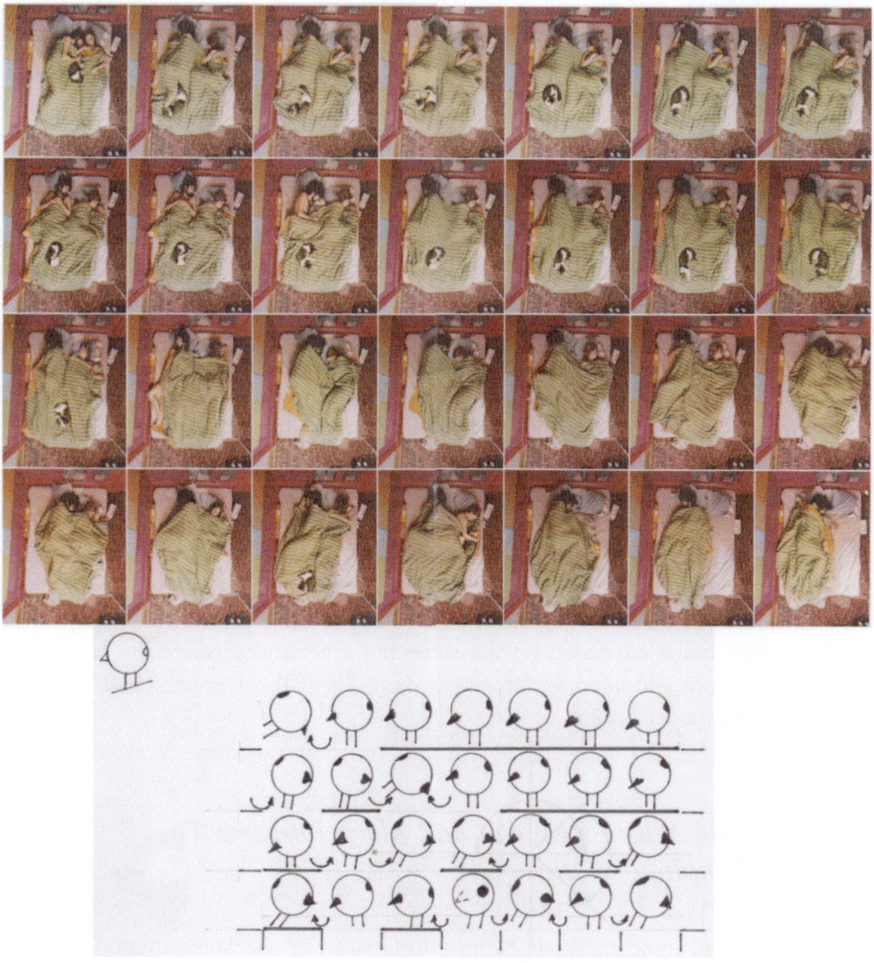

Fig. 13.3 Sleep together, pictures of a man and woman with their cat in bed. Each 15 min a time-lapse picture was made, starting left upper corner (pictures represent 7 h registration). Studying the head movements of the man (lower picture, nose and back head are indicated) shows head turnings (arrow indicates direction of head turning) after long and short periods of bodily rest (black lines). Time-lapse bed pictures are from "De ethologie van de slaap" by courtesy of J.A. Hobson and with permission

synchronisation of body positions and is seemingly induced here by lovemaking. The cat changes three to four times its position on one position change of his bosses. The sleep cycle of the cat is 30 min. The cat leaves when the position changes of husband and wife increases towards the morning.

Studies carried out before, during and after the Second World War in Zürich, Swiss, researched the cat's behaviour during sleep. The free-moving cat implanted with electrodes in the diencephalon could be brought into sleep by stimulation of the ventral part of the massa intermedia. This stimulation organizes the slowly laying down of the cat by its typical well-known behaviour looking for the most comfortable position. Caressing or touching the cat's skin is followed by short pupil dilatation (see Sect. 15.7) and raising and turning of the head. After it, the cat normally goes back to sleep again. Turning of the head can be restricted to head and upper part of the body alone, while awake. However, strong head turning will be followed regularly by turning of the whole body (Hess 1949). (The back position is directly followed by the standing position.) Thus, head turning occurs by stimuli during sleep and can induce turning of the whole body.

People that sleep well have longer periods of immobility of head and body than bad sleepers. Older people should be categorized as bad sleepers since sleep diminishes, including REM sleep, and immobility can disappear almost totally. Sleep posture influences the frequencies of sleep apnoea periods in obstructive sleep apnoea patients (van Kesteren et al. 2011). Generally, it is well known that head position during sleep is vital in several disease conditions. For example, sudden infant death happens frequently in the prone position and it is advocated that babies sleep in the supine position (Dwyer and Ponsonby 2009; Yiallourou et al. 2008).

Head turning in infants, left-side or right-side turns, during the earliest days after birth shows a preference for right-side head turns (Turkewitz and Creighton 1974). This preference has been related to lateral preference at later age. Theories borrowed from this research are that head turning effects hand preference and that head and hand preference are based on the same brain networks. Since the correlation between head turn and hand preference is weak, these theories presumably should go to the waste-paper basket.

13.3.1 Brain Waste Removal and Sleep Posture: The Glymphatic Pathway

"The rodent brain removes waste better during sleep or anaesthesia compared with the awake state. Animals exhibit different body posture during the awake and sleep states, which might affect the brain's waste removal efficiency. We investigated the influence of body posture on brain wide transport of inert tracers of anesthetized rodents. The major finding of our study was that waste, including soluble amyloid β (Aβ), removal was most efficient in the lateral position (compared with the prone position), which mimics the natural resting/sleeping position of rodents. Although our finding awaits testing in humans, we speculate that the lateral position during sleep has advantage with regard to the removal of waste products including Aβ, because clinical studies have shown that sleep drives Aβ clearance from the brain" (Lee et al. 2015). In Chap. 7, bulk flow transport of therapeutics (Sect. 7.4.3) has been treated, in which the interstitial fluid around cortex cells can displace substances if fluid pressure gradients are present. During sleep, the paravascular spaces increase their volume that lowers the resistance to paravascular inflow and enhances the exchange between interstitial fluid and the cerebro-spinal fluid. Consequently, bulk flow increases and substances can be "swiftly" displaced including waste. Body posture changes the hemodynamic and intracranial pressure that can also induce extra waste removal from the brain. Sleep related to fluctuations of Aβ in the extracellular space (hallmark for Alzheimer) has been described by Kang et al. (2009) and Lee et al. (2015). From the three positions studied in rats by the Benveniste group, prone, supine and right lateral decubitus, the glymphatic transport is the most efficient when sleeping in the lateral position (Lee et al. 2015). The above parts of this chapter demonstrated that head position from baby till grown-ups is important for sleep function. Publications on body position during sleep have been used to explain relations between woman and man or one's personality (Dunkell 1977; Domino and Bohn 1980). "Indeed, most of the literature on sleep positions (and personality) is based on anecdotal observations, personal impressions and clinical speculations" (De Koninck et al. 1983). Both articles are grouped under speculations. This chapter does not deal with the speculations on sleep positions and personality.

In Sect. 4.6, cervical reflexes of the foetus were described, indicating that movement of the head causes movement of the limb to which the head is turned. Turning the head to the right gives abduction of the right arm and an adduction of the left arm. Labyrinth reflexes can be activated in the foetus. Spatial head movements also distinguish symmetric bilateral movements of the extremities. Thus, while skull base and skull cap are still in the middle of their development, the head can change its position and bring about reflexes to extremities and trunk. In a quick, presumably inadequate study of head movements during sleep, it became clear that head turning always precedes bodily movements during the non-REM sleep. The amount of turnings during sleep is rather variable from 18 in REM sleep behaviour disorder and insomnia till 8 in parasomnia. The last compares to sleep turnings during normal sleep.

13.3.2 Neck Myoclonus at Onset and During Sleep

The term myoclonus was introduced by Friedreich in 1881, while earlier Dubini in 1846 mentioned already these movement disturbances. Due to the EEG application in the 1930s, it became possible to associate myoclonus with nervous system electrical activity (Tassinari et al. 1998). The definition of myoclonus is taken from the NIH-Myoclonus fact sheet (2012):

> Myoclonus describes a symptom and not a diagnosis of a disease. It refers to sudden, involuntary jerking of a muscle or group of muscles. Myoclonic twitches or jerks usually are caused by sudden muscle contractions, called positive myoclonus, or by muscle relaxation, called negative myoclonus. Myoclonic jerks may occur alone or in sequence, in a pattern or without pattern. They may occur infrequently or many times each minute. Myoclonus sometimes occurs in response to an external event or when a person attempts to make a movement. The twitching cannot be controlled by the person experiencing it.

Neck myoclonus is encountered during REM sleep, although REM sleep is mainly characterized by atonia. Head jerks during sleep have a characteristic appearance: a sudden twitch-like dorsal or ventral flexion or version of the head to one side. In a study on sleep disorder patients, 50% had neck myoclonus during the REM sleep. In fact, a mean of one head jerk per hour was found. The younger the higher the head jerk index is (the mean index is calculated for every recording and given per hour of REM sleep per individual patient). Younger than 45 years gives a myoclonus index of 2, while for the group of patients older than 45 years ± 0.5 myoclonus was registered per hour. In non-REM sleep, myoclonus was nearly absent (index of 0.04). The mean duration of a neck jerk was 0.6 ± 0.4 s. Although it concerned an exclusive patient study, the authors think that neck myoclonus is a physiological phenomenon (Frauscher et al. 2010).

Myoclonus can also be noted at the wake–sleep transition. Involuntary jerks can arise when falling asleep. Four types of myoclonus are discerned: cortical, subcortical (brainstem or reticular) propriospinal and peripheral (Shibasaki 1996; Cassim and Houdayer 2006; Dijk and Tijssen 2010). This sleep-onset type of myoclonus belongs to an axial myoclonus of propriospinal origin and did not have a cortical or reticular origin (Vetrugno et al. 2001). It arises in axial muscles and presumably proceeds by a multimeric propriospinal propagation (it advances up and down the spinal cord; Montagna et al. 2006; see also Antelmi and Provini 2015). From the above-provided literature, only a few involved anatomical muscular structures are mentioned, besides the general term axial muscles: splenius capitis muscle and the trapezius muscle, mainly due to electrode placing.

Electrophysiology indicates that "muscle twitches are caused by phasic glutamatergic input (Brown et al. 2012). These twitches, particularly in developing animals, have been suggested to be important in the refinement of spinal cord connectivity (Petersson et al. 2003). In intracellular recordings from lumbar motoneurons, twitches are accompanied by brief depolarizing events that can be blocked by the non-NMDA glutamate receptor antagonist kynurenate, but not the selective NMDA receptor antagonist APV. Similarly, muscle twitches in masseter muscles were blocked by an AMPA/kainate receptor antagonist (CNQX) into the trigeminal motoneuron pool. Thus, twitches during REM sleep are due to bursts of excitatory, glutamatergic AMPA/kainate receptor-mediated synaptic potentials" (Brown et al. 2012). A good candidate for this glutamatergic input is the pedunculopontine nucleus. This nucleus contains N-type neurons that are glutamatergic (Pahapill and Lozano 2000) and mainly active during REM sleep (Garcia-Rill et al.

2015). This peculiar nucleus is also involved as a possible target in deep brain stimulation (Lourens et al. 2011).

The generation of myoclonus, which can lead to a diagnosis, is difficult to establish. Myoclonus may arise from anywhere in the nervous system, central and/or peripheral. The types of myoclonus are based on general features for the cortex in the human brain and for its subcortical causes holds, "a topography for which there are no positive diagnostic arguments" (Cassim and Houdayer 2006).

To summarize this chapter: sleep studies show a decrease of the sleep time and an increase in sleep disorders in the Western world. Head and body movements and positions are important during normal sleep and in sleep disorders. Head position during REM sleep is related to brain waste removal and head position is influenced by myoclonus at the sleep onset and during sleep. The neurodegenerative Huntington's disease can change movements during non-REM sleep but not during REM sleep. Morphological structures or site of generation related to sleep myoclonus are hardly provided in the literature. Cerebellar relations with the sleep–wake cycle are in the highlights of sleep research.

References

Antelmi E, Provini F (2015) Propriospinal myoclonus: The spectrum of clinical and neurophysiological phenotypes. Sleep Med Rev 22:54–63

Brown RE, Basheer R, McKenna JT, Strecker RE, McCarley RW (2012) Control of sleep and wakefulness. Physiol Rev 92:1087–1187

Canto CB, Onuki Y, Bruinsma B, van der Werf YD, De Zeeuw CI (2017) The sleeping cerebellum. Trends Neurosci 40:309–323

Cassim F, Houdayer E (2006) Neurophysiology of myoclonus. Neurophysiol Clin 36:281–291

De Vivo L, Bellesi M, Marshall W, Bushong EA, Ellisman MH, Tononi G, Cirelli C (2017) Ultrastructural evidence for synaptic scaling across the wake/sleep cycle. Science 355 (6324):507–510

Diering GH, Nirujogi RS, Roth RH, Worley PF, Pandey A, Huganir RL (2017) Homer1a drives homeostatic scaling-down of excitatory synapses during sleep. Science 355(6324):511–515

Dijk JM, Tijssen MA (2010) Management of patients with myoclonus: available therapies and the need for an evidence-based approach. Lancet Neurol 9:1028–1036

Domino G, Bohn SA (1980) Hypnagogic exploration: Sleep positions and personality. J Clin Psychol 36:760–762

Dunkell S (1977) Sleep positions. The night language of the body, William Morrow, NY

Dwyer T, Ponsonby AL (2009) Sudden infant death syndrome and prone sleeping position. Ann Epidemiol 19:245–249

Frauscher B, Brandauer E, Gschliesser V, Falkenstetter T, Furtner MT, Ulmer H, Poewe W, Högl B (2010) A descriptive analysis of neck myoclonus during routine polysomnography. Sleep 33:1091–1096

Garcia-Rill E, Luster B, Mahaffey S, Bisagno V, Urbano FJ (2015) Pedunculopontine arousal system physiology – Implications for insomnia. Sleep Sci. https://doi.org/10.1016/j.slsci.2015.06.002

Hess WR (1954) Diencephalon: autonomic and extrapyramidal functions. Grune & Stratton, New York

Hess WR (1949) Das Zwischenhirn. Benno Schwabe & Co, Basel

Hobson J A (1980) Toward a cellular neurophysiology of the reticular formation: Conceptual and methodological milestones. In: Hobson JA, Brazier MAB (eds) The reticular formation revisited. IBRO monographs 6, Raven Press NY
Hobson JA (1979) De ethologie van de slaap. Hoffmann-La Roche, Mijdrecht
Hobson JA (1990) Sleep and dreaming. J Neurosci 10:371–382
Hobson JA, Spagna T, Malenka R (1978) Ethology of sleep with time-lapse photography: Postural immobility and sleep-cycle phase in man. Science 201:1251–1253
Kang JE, Lim MM, Bateman RJ, Lee JJ, Smyth LP, Cirrito JR, Fujiki N, Nishino S, Holtzman DM (2009) Amyloid-β dynamics are regulated by Orexin and the sleep-wake cycle. Science 326:1005–1007. https://doi.org/10.1126/science.1180962
Kerkhof GA (2017) Epidemiology of sleep and sleep disorders in The Netherlands. Sleep Med 30:229–239
De Koninck J, Gangnon P, Lallier S (1983) Sleep positions in the young adult and their relationship with the subjective quality of sleep. Sleep 6:52–59
Kreutzmann JC, Havekes R, Abel T, Meerlo P (2015) Sleep deprivation and hippocampal vulnerability: changes in neural plasticity, neurogenesis, and cognitive function. Neurosci 309:173–190
Kuypers HGJM (1987) Some aspects of the organization of the output of the motor cortex. In: Motor areas of the cerebral cortex. Ciba Found Symp 132 Wiley, Chichester, pp 63–82
Lazar AS, Panin F, Goodman AO, Panin F, Goodman AOG, Lazic SO, Lazar ZI, Mason SL, Rogers L, Murgatroyd PR, Watson LPE, Singh P, Borowsky B, Shneerson JM, Barker RA et al (2015) Sleep deficits but no metabolic deficits in premanifest Huntington's disease. Ann Neurol 78:630–648
Lee H, Xie L, Yu M, Kang H, Feng T, Deane R, Logan J, Nedergaard M, Benveniste H (2015) The effect of body posture on brain glymphatic transport. J Neurosci 35:11034–11044
Lourens MAJ, Meijer HE, Heida T, Marani E, van Gils SA (2011) The pedunculopontine nucleus as an additional target for deep brain stimulation. Neural Netw 24:617–630
Luyster FS, Strollo PJ, Zee PC, Walsh JK (2012) sleep: a health imperative. Sleep 35:727–734
Lyamin OI, Manger PR, Ridgway SH, Mukhametov LM, Siegel JM (2008) Cetacean sleep: an unusual form of mammalian sleep. Neurosci Biobehav Rev 32:1451–1484
Lydic R, Keifer JC, Baghdoyan HA, Craft R, Angel C (2017) Opiate action on sleep and breathing. In: Kryger MH, Roth T, Dement WC (eds) Principles and practice of sleep medicine, 6th Edition. Elsevier, New York, pp 250–259
Marani E, Choufoer H, Van der Veeken J (1990) The neurofilament architecture of the rat suprachiasmatic nucleus. Eur J Morphol 28:279–288
Marani E, Schoen JHR (2005) A reappraisal of the ascending systems in Man, with emphasis on the medial lemniscus. Adv Anat Embryol Cell Biol 179:1–76
Mesulam MM (1995) Cholinergic pathways and the ascending reticular activating system of the human brain. Ann NY Acad Sci 757:169–179
Montagna P, Provini F, Vetrugno F (2006) Propriospinal myoclonus at sleep onset. Neurophysiol Clin 36:351–356
Moruzzi G (1972) The sleep-waking cycle. Ergeb Physiol 64:1–165
Nieuwenhuys R, Voogd J, van Huizen C (2007) The human central nervous system. Springer, Heidelberg
Nieuwenhuys R, Voogd J, Van Huijzen C (2008) The human central nervous system. A synopsis and atlas, Springer, Heidelberg
NIH-Myoclonus fact sheet (2012) "Myoclonus Fact Sheet", NINDS, NIH Publication No. 12–4793; https://www.ninds.nih.gov/Disorders/Patient-Caregiver-Education/Fact-Sheets/Myoclonus-Fact-Sheet. Accessed 27 Jan 2017
Pahapill PA, Lozano AM (2000) The pedunculopontine nucleus and Parkinson's disease. Brain 123:1767–1783
Piano C, Losurdo A, della Marca G, Solito M, Calandra-Buonaura G, Provini F, Bentivoglio AR, Cortelli P (2015) Polysomnographic findings and clinical correlates in Huntington disease: a cross-sectional cohort study. Sleep 38:1489–1495

Piano C, Mazzucchi E, Bentivoglio AR, Losurdo A, Buonaura GC, Imperatori C, Cortelli P, Della Marca G (2016) Wake and sleep EEG in patiënts with Huntington disease: an eLORETA study and review of the literature. Clin EEG Neurosci. https://doi.org/10.1177/1550059416632413

Penzel T, Möller M, Becker HF, Knaack L, Peter J-H (2001) Effect of sleep position and sleep stage on the collapsibility of the upper airways in patients with sleep apnea. Sleep 24:90–95

Petersson P, Waldenstrom A, Fahraeus C, Schouenborg J (2003) Spontaneous muscle twitches during sleep guide spinal self-organization. Nature 424:72–75

Reinosos-Suàrez F, de Andrés I, Garzón M (2011) Functional anatomy of the sleep-wakefulness cycle: wakefulness. Adv Anat Embryol Cell Biol 208:1–130

Rosenwasser AM (2009) Functional neuroanatomy of sleep and circadian rhythms. Brain Res Rev 61:281–306

Sakai K (1991) Physiological properties and afferent connections of the locus coeruleus and adjacent tegmental neurons involved in the generation of paradoxical sleep in the cat. Prog Brain Res 88:31–45

Shapiro CM (1993) ABC of sleep disorders. BMJ Publishing group, London

Shibasaki H (1996) Overview and classification of myoclonus. Clin Neurosci 3:189–192

Steinbusch H (1981) Distribution of serotonin immunoreactivity activity in the central nervous system of the rat – cell bodies and terminals. Neuroscience 6:557–618

Tassinari CA, Rubboli G, Shibasaki H (1998) Neurophysiology of positive and negative myoclonus. Electroencephal Clin Neurophysiol 107:181–195

Turkewitz G, Creighton S (1974) Changes in lateral differentiation of head posture in the human neonate. Devel Psychol 8:85–89

van Kesteren ER, van Maanen JP, Hilgevoord AA, Laman DM, de Vries N (2011) Quantitative effects of trunk and head position on the apnea hypopnea index in obstructive sleep apnea. Sleep 34:1075–1081

Vetrugno R, Provini F, Meletti S, Plazzi G, Liguori R, Cortelli P, Lugaresi E, Montagna P (2001) Propriospinal myoclonus at the sleep-wake transition: A new type of parasomnia. Sleep 24:835–843

Voogd J (1981) Neuroanatomy of sleep-waking regulation. Boerhaave cursus: Slaap als verschijnsel en als stoornis. Ed Kamphuis HAC et al. pp 17–24

Walker FO (2007) Huntington's disease. Semin Neurol 27:143–150

Yiallourou SR, Walker AM, Horne RSC (2008) Prone sleeping impairs circulatory control during sleep in healthy term infants: implications for SIDS. Sleep 31:1139–1146

Chapter 14
Peripheral Nerve and Blood Supply, and Lymphatic Drainage of Head and Neck

The sensitivity of ultrasound to identify metastatically involved necks is very high when compared to palpation. The specificity however, is low since images do not allow differentiation between benign and malignant nodes (Baatenburg de Jong and Rongen 1990)

This chapter relates the laminae of the cervical deep fascia to nerves and blood vessels. Atherosclerosis is linked to the bends of the arteries and its herbal solutions are disputed. Topography of the large arteries is associated with Blalock-Taussig operations and situs inversus. The Sloan-Kettering lymphatic maps are correlated with metastases, especially squamous cell carcinomas and lymphatic node distributions. Attention is given to lymphatic development and lymph vessel pump actions. Human development of the venous system explains its independence. The venous pterygoid plexus is described together with its damping effect on masticatory pressure. The blood vessels need to pass from weak and transformational areas (front of the neck) into bony-related areas. This forces blood vessels into trunks and fans.

14.1 Dorsal Cervical Roots and Rami

The presence of a first dermatome (C1) in Man is still debated. The dermatome C1 is placed differently by various authors, either at the back of the head or half way. A number of authors have even argued that there is no C1. Herpes zoster area descriptions also deny a C1 dermatome. The discussion in experimental animals is likewise unclear. Its presence was demonstrated in monkey and cat. Nevertheless, the dermatome fields were differently localized, but cutaneous branches of cat C1 were later denied. Great variability in the presence of C1 dorsal roots, the uptake in the accessory nerve or in the hypoglossal nerve is responsible for the confusing data. Moreover, dorsal cervical roots anastomose within the cervical spinal canal. The discussion on C1 dorsal roots is still going on in recent literature. In 20–27% of the

E. Marani and C. Heida, *Head and Neck*,
https://doi.org/10.1007/978-3-319-92105-1_14

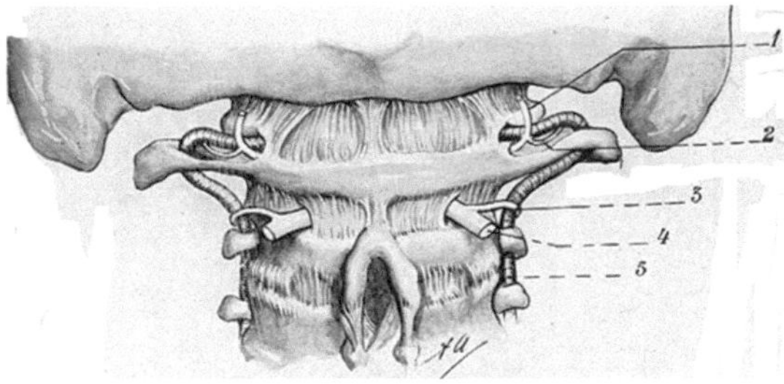
1
2
3
4
5

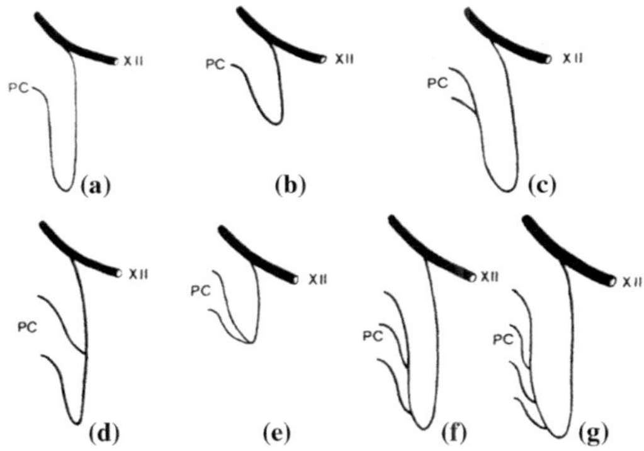
XII
PC
(a)
(b)
(c)
(d)
(e)
(f)
(g)

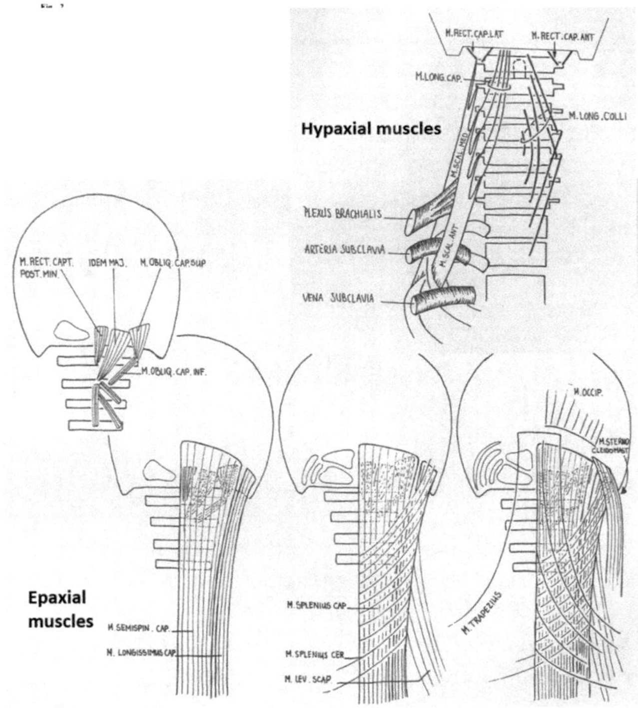
Hypaxial muscles
M.RECT.CAP.LAT
M.RECT.CAP.ANT
M.LONG.CAP.
M.LONG.COLLI
M.SCAL.MED
PLEXUS BRACHIALIS
ARTERIA SUBCLAVIA
M.SCAL.ANT
VENA SUBCLAVIA
M.RECT.CAPT. POST.MIN.
IDEM MAJ.
M.OBLIQ.CAP.SUP
M.OBLIQ.CAP.INF.
M.OCCIP.
Epaxial muscles
M.SEMISPIN.CAP.
M.LONGISSIMUS CAP.
M.SPLENIUS CAP.
M.SPLENIUS CER.
M.LEV.SCAP.
M.TRAPEZIUS

◂**Fig. 14.1** Upper figure shows the relation of the C1 and C2 cervical nerves with the vertebral column: 1, C1 dorsal ramus; 2, C1 ventral ramus; 3, C2 ramus ventralis; 4, C2 dorsal ramus; 5, vertebral artery (from Hovelacque 1927). The ansa cervicalis variability is shown in the middle figure: **a** 40%, **b** 27%, **c** 11%, **d** 9%, **e** 8%, **f** 1%, **g** 1% (see Caliot and Dumont 1986 with permission). The epaxial muscles are subdivided into superficial ones (m. levator scapulae, mm. splenius capitis and splenius cervicis) and erector spinae (medial, intermedial and lateral groups). The medial group contains mm. semispinalis cervicis and capitis, the intermedial one encloses mm. longissimus cervicis and capitis, while the lateral group encompasses the mm. iliocostales and the mm. scaleni. The relation mm. scaleni and the hypaxial muscles (m. longus colli and capitis together with the mm. recti capiti) and both the anterior and posterior hiatus of the scaleni muscles are associated respectively to the v. subclavia and the brachial plexus together with the a. subclavia (courtesy N. M. Gerrits, Dept. anatomy, Erasmus University, Rotterdam)

cases, C1 dorsal roots are absent. No connections of existing dorsal rootlets with the spinal accessory nerve were found in 6–27% of the studied cadavers, but connections with the spinal accessory nerve were found in 46–74% in 250 human cases. Therefore, in nearly half to three-quarters of the cases C1 dorsal roots are present in humans (see Fig. 14.1). Moreover, racial differences appear to be present, since in Japanese subjects C1 dorsal roots are present and connected to the spinal accessory nerve in nearly 93% of the studied cases. Since dorsal roots of C1 and C2 together with trigeminal sensory input are presumably involved in tension headaches, whiplash, head neuralgia and temporomandibular diseases, nociceptive studies on dorsal horn neurons at the upper cervical spinal cord and spinal trigeminus nucleus caudalis were carried out in experimental animals. Essential, in this respect, is the bordering of the trigeminal nucleus caudalis, both at its upper and lower level (Usunoff et al. 1997; Waite and Ashwell 2012). The craniofacial input studies to the upper cervical dorsal horn in rats, demonstrated that by constant borders, as measured from the outside, the C1 dorsal horn area is located adjacent and mediocaudally to the mandibular trigeminal area. Therefore, it is expected that in humans, as in animals, sensory C1 information is also relayed by the caudal trigeminal sensory system. Electric and pain stimulation indeed show receptive fields of nociceptive dorsal horn neurons indicating C1 dermatome areas (Moskowitz 2008). If a C1 ganglion is present, its number of ganglion cells is low (20% of C2 ganglion cells in the cat) as are its ganglionic fibres. Sharing of a ganglion with the spinal accessory nerve and small aberrant ganglia can occur. In 28.5% of the cases, a dorsal root ganglion associated with the C1 root was found. Dorsal rhizotomies at C1 in humans do give degeneration in the posterior columns and occipital neuralgia is treated with C1 rhizotomies (see Marani and Schoen 2005). Dorsal rami of the cervical roots also innervate neck and back muscles (Fig. 14.1). The epaxial muscles are innervated by the dorsal rami. These epaxial muscles are divided into two longitudinal running groups: a medial and a lateral one. The dorsal rami follow this division: they have medial and lateral branches. The medial longitudinal muscle group is more developed in the head and neck region. The so-called transversospinalis group is subdivided into semispinalis and multifidi muscles and both are innervated by the medial branch. The longissimus and iliocostal muscles are innervated by the lateral branch of the dorsal cervical rami (see Fig. 14.1). It is in general the medial branch of the dorsal rami that is responsible for

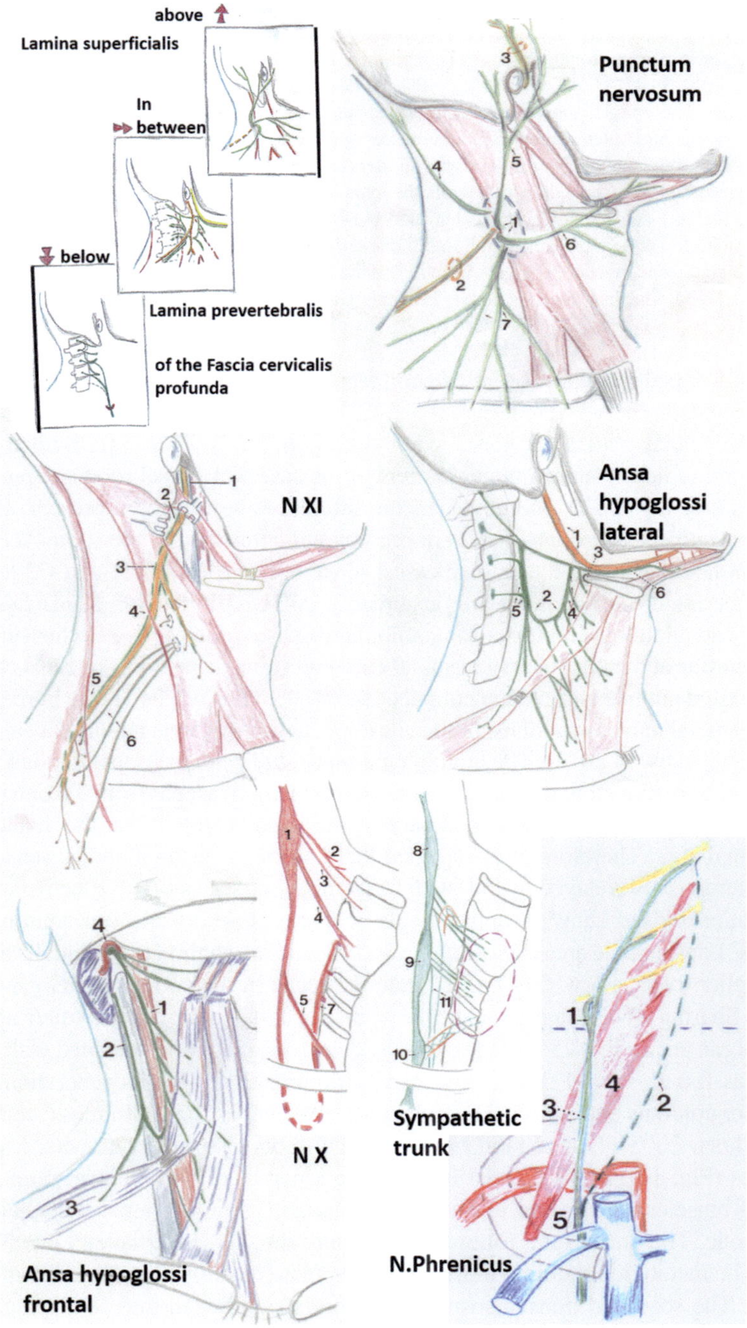
above
Lamina superficialis
In
between
below
Lamina prevertebralis
of the Fascia cervicalis
profunda
Punctum
nervosum
1
2
3
4
5
6
7
N XI
Ansa
hypoglossi
lateral
1
2
3
4
5
6
7
8
9
10
11
N X
Sympathetic
trunk
Ansa hypoglossi
frontal
N.Phrenicus

◂**Fig. 14.2** Rami ventrales cervicis responsible for skin innervation are above the lamina superficialis of the fascia cervicalis profunda and pass through this fascia at the **punctum nervosum** (1) half way the posterior edge of the sternocleidomastoid muscle. The cervical, spinal and skin branches are from rostral to caudal: N. occipitalis minor (C2 or C3, or from sling C2 + C3), retroauricular and adjacent to occipitalis major area (4); N. auricularis magnus (from C2 + C3; often C3; sometimes from sling C3 + C4) (5), retromandibular; N. transversus colli = n cutaneous colli (C2 + C3 or C3 + C4) anterior submental (6); N. supraclavicularis (C3 + C4, regular only C4), between clavicular and mm trapezius and deltoideus (7). The origin of the n. auriculotemporalis is indicated (3). The trajectory of the **N XI** starts at the jugular foramen leaving the jugular vein (1) to go on between digastricus and sternocleidomastoid muscles (2) and passes below the sternocleidomastoid muscle to innervate this muscle from the back. Continuation is found below the punctum nervosum (4) in the supraclavicular triangle to reach and innervate the trapezius (5). Cervical plexus branches (6) reach the sternocleidomastoid (C2) and trapezius (C3 + C4) muscles. The **ansa hypoglossi** is depicted lateral and frontal. The N XII (1) in its trajectory to the tongue combines with the C1–C2 branch (6) towards geniohyoid and thyrohyoid muscles and passes above the lateral end of the hyoid (3). The superior root or ramus descendens hypoglossi (4), inferior root or ramus descendens cervicalis (5) and the ansa cervicalis (2) innervate the infrahyoidal muscles. The frontal view of the ansa shows the relation with the a. carotis sup. (1), the jugular vein (2) and the omohyoid muscle (3). The relation between the a. sternocleidomastoideus that rides over the C1–C2 branch is indicated (4). The **phrenic nerve** arises from C3 to C4 or C4 to C5, has to pass through the prevertebral lamina (1 stippled line) to reach the anterior scalenus muscle (4), where it is placed within a double sheet of the phrenic fascia (3). After leaving the muscle it descends between a. and v. subclavia (5). In 5% of the cases a n. phrenicus accessories is present (2), which joins the main branch above the large vessels. The **N X** placed in front of the prevertebral lamina contains the ganglion nodosum (1) that contains the rami pharyngici (2). The internal ramus (3) and the external ramus (4) of the n. laryngeus superior branch separately from the ganglion nodosum. The cardiac ramus branches of at C5 height (5). One centimetre above this nerve the ramus descendens cervicalis (6) is present that anastomoses with the plexus thyroideus. The n. recurrens, its end is called the n. laryngeus inferior parallels the trachea (7). The **sympathetic trunk** knows three ganglia just in front of the prevertebral lamina: ggl. cervicale superior (8), ggl. cervicale mediale (9) and the ggl. cervicale inferior (10). Each contributes to the innervation of the thyroid. Some use and parallel the a. thyroidea superior and inferior to reach the thyroid. The ggl. cervicale superior gives cardiac rami (11) that can join the ramus cardiacus of the vagus. (All figures © E. Marani)

the cutaneous innervation. The dorsal rami of the cervical roots do innervate their dermatomes (see also part on C1 dermatome). According to some authors, only the rami dorsalis of C7 and C8 lack sensory fibres (see Marani and Lakke 2012 for an overview and references).

14.1.1 The Ventral Cervical Rami: The Cervical Plexus

The description of the cervical plexus will follow the fascia layers of the neck: the area dorsal of the deep cervical, in between the deep and superficial and above the superficial cervical lamina (see Fig. 14.2). The cervical plexus is considered to be constituted from the ventral roots of spinal nerves C1 to C4, strictly speaking, while the brachial plexus takes up C5–C8. Since there is overlap between cervical and brachial plexus, branches of C3 and C4 can participate in the brachial plexus, while brachial plexus nerves can join in their trajectory the nerves originating from cervical plexus (Lanz und Wachsmuth, Hals 1955).

After leaving the vertebral column, cervical ventral rami innervate the levator scapulae and rhomboid muscles by the dorsoscapular nerve. Literature indicates that for the levator C2–C5 and for rhomboid C3–C6 can be involved with a variance from one to several cervical ventral rami. Traumatic brachial plexus lesions together with anatomical dissections established that the levator scapulae is innervated by segmental branches of C3 and C4, while the rhomboid muscle is also innervated by two segmental branches from C4 to C5 (Malessy et al. 1993 and references herein).

Recent anatomical dissections indicate that the origin of the dorsoscapular nerve from one segmental ramus is present in 18% of the cases, from a common trunk with the long thoracic nerve in 30%, from C4 in 28% and from C4 to C5 in 23% of the cases, while one segmental ramus is taken as being usual for this nerve. The origin from simply one cervical ventral ramus (C5) is only retained by two authors; all other publications (12 references) favour a multiple origin (see Marani and Lakke 2012).

The phrenic nerve (Fig. 14.2) originates according to Bolk (1910) either from C3 to C4 or from C4 to C5. Lanz and Wachsmuth (1955) give C4 the main contribution with additions of C3 and C5. The variability of the morphology of the phrenic nerve is large. In 97% of the cases a phrenic nerve and in 3% only an accessory one is present. Nevertheless, in 28.6% of cases an accessory phrenic nerve is found together with the phrenic nerve. Higher percentages found in the literature are around 44%. The segments and percentages for the phrenic nerve reported are: C3 + C4 in 2.5%, C3 + C4 + C5 in 1.5%, only C4 in 43.5%, C4 + C5 in 52% and C5 alone in 0.5% of 200 cases. If the accessory phrenic nerve was included in this enumeration even C6 was involved. Lanz and Wachsmuth (1955) already showed that accessory fibres can in 6% of the cases be added by the caudal brachial plexus (C8 and Th1 in 5%). Moreover, addition can be organized via the ramus cervicalis of the ansa hypoglossi (2%), and along cardiac nerves (3%). However, the phrenic mnemonic: "C3, C4, C5 keep you alive" is now under discussion presumably by the high incidence of accessory fibre distributions (see Marani and Lakke 2012 and references therein). The trajectory of the phrenic nerve and the accessory phrenic nerve (see Figs. 14.1 and 14.2) are extensively described by Lanz and Wachsmuth (1955).

The article: "Ansa Hypoglossi or Ansa Cervicalis? That's the question…" by Olry and Haines (2002), ends: "As far back as in 1835, Christophore Ernest Bach claimed that the ansa cervicalis contained only fibres arising from the cervical plexus. It took one hundred and twenty years to acknowledge that he was right…" The ansa (Fig. 14.2) indeed contains cervical fibres. Renewed interest in the ansa came from ansa hypoglossi-to-recurrent laryngeal nerve anastomosis in cases of cricothyroid reinnervation and in total laryngeal transplantation. The ansa cervicalis can be a source for nerve transfer. The reviews on the ansa cervicalis show seven different morphological forms of the ansa cervicalis, in which the double classic form is found in only 40% of the cadavers (Fig. 14.1). Absence of the ansa is also reported. An even more refined classification is described in 7 main groups with 21 subgroups based on the ansa's inferior root. Seemingly Lanz and Wachsmuth (1955) belong to lumpers, indicating only four main variations from their general pattern, while others should be considered splitters, trying to catch all variations for a "good" clinical reference (for overview see Marani and Lakke 2012). *Variability is the only suitable word for the topographical results of the cervical roots.*

All cutaneous sensory cervical spinal nerves for the ventral innervation of head and neck originate in the "punctum nervosum" (Fig. 14.2) placed halfway the posterior border of the sternocleidomastoid muscle. The cutaneous nerves pass here through the lamina superficialis to reach their subcutaneous fields. Although the superficial cervical plexus block has been introduced in the 1980s (see Pandit et al. 2003), no anatomical studies on the cutaneous cervical spinal nerves originated. Therefore, the description of Lanz and Wachsmuth (Hals 1955) still holds. A recent poster (Schulze et al. 2016) studying ultrasound-guided intermediate cervical plexus block in regional anaesthesia shows that the lamina prevertebralis is impenetrable for anaesthetics, but the lamina superficialis is not anaesthetically tight. Although the cranial accessory nerve has been described in Chap. 12, its neural constitution is of clinical importance (Fig. 14.2). Leaving the jugular foramen (Chap 4), this cranial nerve has intradurally taken up fibres from C1. Passing over the jugular vein its direction is towards lateral. Uptake of cervical fibres C2–C4 in the cranial nerve or these fibres paralleling the accessory for sternocleidomastoid and trapezius innervation can be noted (Fig. 14.2). In the last case, a plexus accessorio-cervicalis is discerned (Brown 2002; see also Nishizuka 1934). The sympathetic trunk contains the axons of the second sympathetic neurons and its ganglia in the head and neck region lodge the third sympathetic neurons (the first neurons are in the brainstem). These neurons reach via the internal and external carotid artery skin, eyes and other head and neck organs and are vasomotoric, sudomotoric and oculosympathetic in origin (see Gibbins 2012 for overview).

14.2 Blood Vessels, Nerves and Their Trajectory Through the Head

The head is different in its anatomical construction from the rest of the body. The segmental subdivision of the body is absent in the head. The consequence is that blood vessels and nerves don't show conformity in their trajectories. Even the head veins in general don't follow the passageways of the head arteries. Plexus formation is absent in the head and the restricted topographic possibilities forces anastomoses. Main subdivisions that can be recognized (see Fig. 14.3a, b) are an area for the eye with the cranial nerves II, III, IV and VI. A trigeminal area that reaches from the eye area to the mouth and a sublingual area containing the nerves from the jugular foramen (IX, X, XI) and XII. This sublingual area finds *topographic* contact with the neck structures (e.g. plexus cervicalis, see above). In a layer above (ventral of) the trigeminal area, the facialis innervated mimic muscles are located and they extend over the eye and sublingual area. While the nerves mainly show a (oblique) dorsoventral direction, the large blood vessels of the head all do have a caudo-rostral or, of course, a rostrocaudal direction. This deep (over or under)

(a)

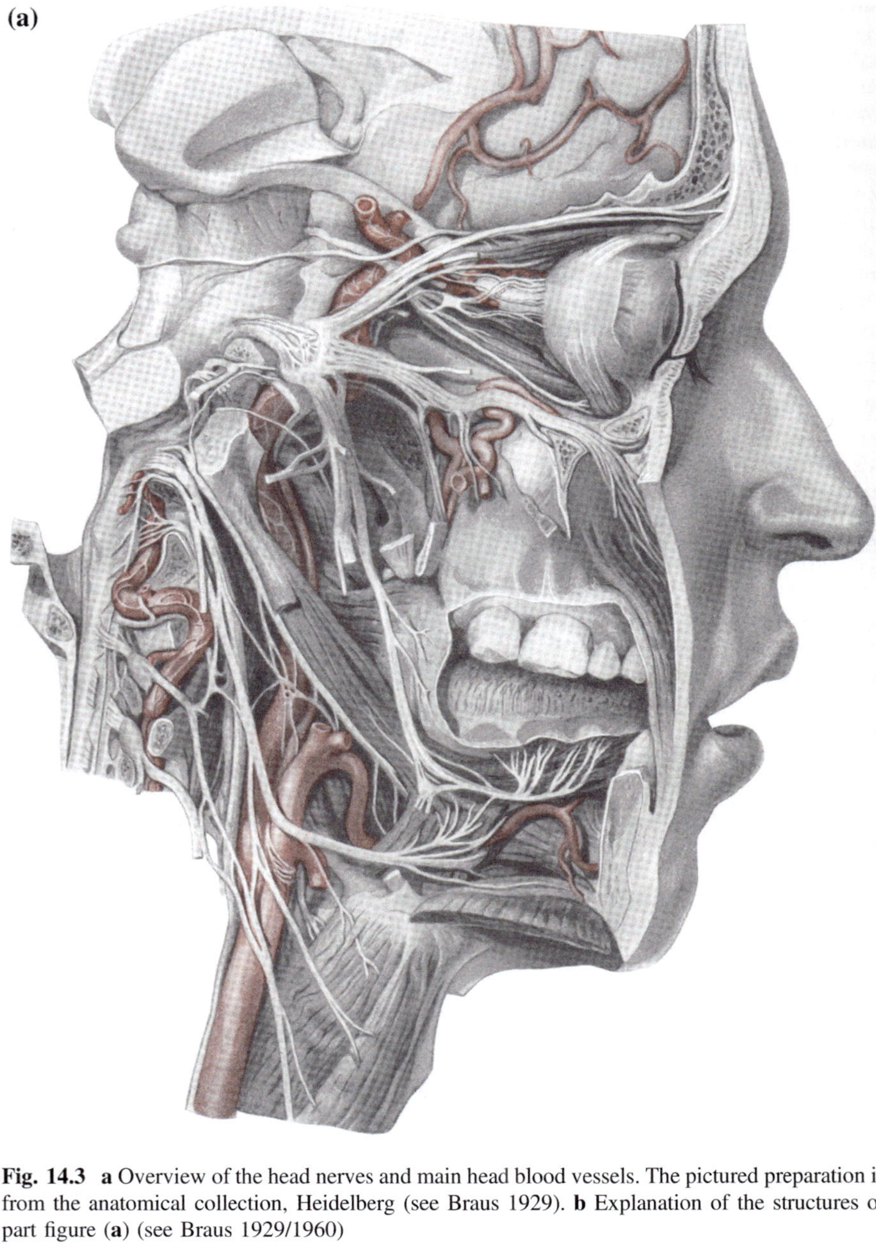

Fig. 14.3 **a** Overview of the head nerves and main head blood vessels. The pictured preparation is from the anatomical collection, Heidelberg (see Braus 1929). **b** Explanation of the structures of part figure (**a**) (see Braus 1929/1960)

(b)

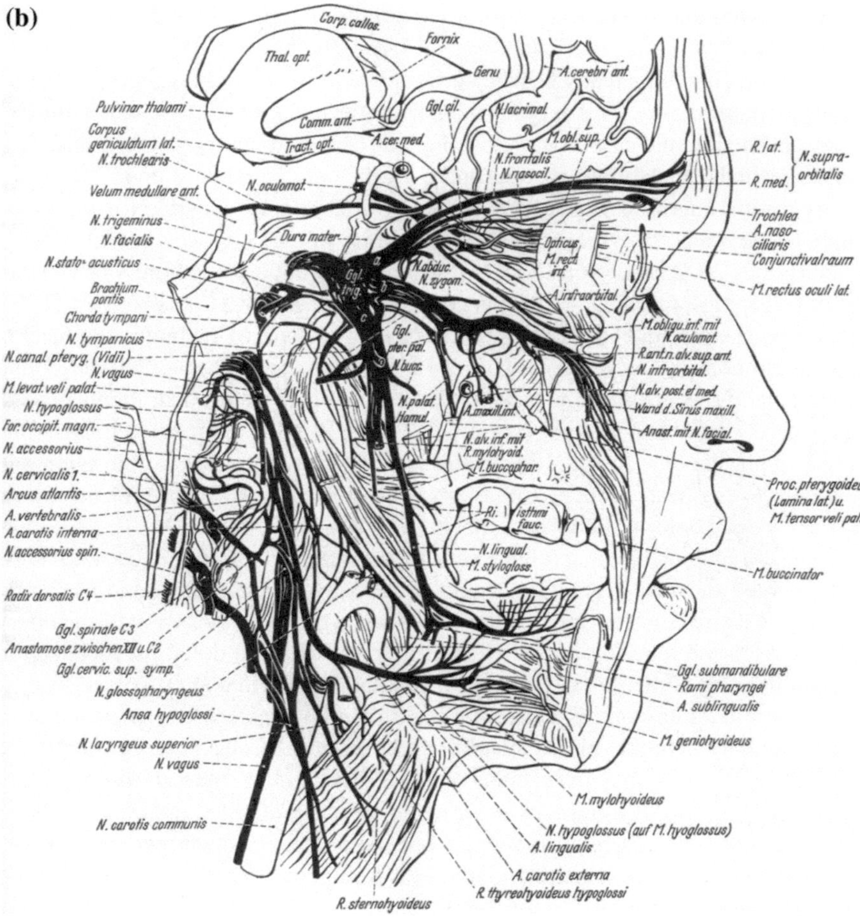

Fig. 14.3 (continued)

crossing of blood vessels with nerves is a typical macroscopic phenomenon of the head (see Fig. 14.3a, b) and is easily lost in the overwhelming anatomical data.

The two-layer construction of the nerves as exemplified by trigeminus and facialis cranial nerves repeats itself in the blood vessels. In general, the spreading branches of the a. carotis externa (aa. facialis, angularis and occipitalis) produce most of the superficial layers of blood vessels, although contributions from other vessels can be present due to the enormous variability of these head vessels. Branches of these blood vessels contribute to the deeper layers and to the muscles of the face. The external carotid branches are generally described as anterior, posterior and medial. Anterior localized is aa. thyroidea sup, lingualis and facialis, while posterior branches are aa. sternocleidomastoidea, occipitalis, auricularis posterior and the medial ones aa. pharyngea ascendens and maxillaris (interna). The external carotid end branches are the a. temporalis superficialis (see Figs. 14.3a, b and 14.5) and the a. maxillaris (see Sect. 14.4).

The carotis interna already has an anastomosis with the arteria vertebralis before reaching the head area. This type of mutual connections remains present even into the brain. The internal carotid passes along, but deeper, to the external carotid till it reaches medially the styloid process and uses the carotid canal to enter the skull.

The carotis interna is together with the a. vertebralis responsible for the blood supply of the brain. "The brain constitutes only 2% of the body weight (1.3–1.5 kg) but consumes 20–25% of your body's resting energy budget, regardless of whether you are sleeping, watching TV or puzzling over this sentence" (Lieberman 2013). The head makes around 10% of your body (± 7 kg) including the brain. The head in total should consume even more than 20–25%, presumably over 30% energy and sleeping or TV watching now does make a difference for the extrabrain tissues. Don't forget, the visual system, including its brain parts, is a big oxygen user. One should emphasize that nearly 15% of the total blood supply goes to the brain and thus more to the head in total. *Exact* figures on energy consumption of the whole head and neck could to the best of our knowledge not be found.

A rete mirabile or "the marvellous net", first described by Galen (Rocca 2014), is the branching out of a single artery into a number of smaller vessels that finally reconstitute a single or few larger vessels and it is a direct continuation of the artery that generated the rete. Retia mirabilia are best known from dolphins, localized in between the ribs of the thorax and around the spinal cord that are involved in flow and pressure-damping effects during diving. It can be compared to an arteriolar or arterial capillary network, but existing of larger vessels. It is also used to supply the brain of artiodactyls that have either an involuted or no internal carotid artery (Fukuta et al. 2007). Although its presence is seriously doubted in humans, a kind of false carotid "rete mirabile" does exist, but should be named carotid anastomoses and is defined as an arterial network at the cavernous portion of the internal carotid artery. The meningeal branches of the external carotid produce these anastomoses just around the entrance of and in the carotid canal. Connections between the cavernous internal carotid and middle meningeal arteries at the skull base are more common (58.3%) than not in humans. In animals, the main artery that produces a real rete mirabile is the maxillary artery. "These connections in humans most likely function in times of internal carotid artery ischemia that is chronic in nature and, in the past, may have been misidentified as rete mirabile" (Tubbs et al. 2015). Moreover, it is claimed that these anastomoses are not involved in internal carotid artery stenosis (Tubbs et al. 2015).

14.2.1 Atherosclerosis

Those working in pathology or anatomy do know that the large blood vessels of the human head and neck are characterized by atherosclerotic plaques. Students are completely astonished if they feel the large carotid arteries of older cadavers or the circle of Willis, with their totally calcified walls, mirroring plastic tubes. Squeezing in these large arteries causes the cracking of the calcified walls. The risk factors for the carotid atherosclerotic disease, responsible for 25% of all stroke cases, are well studied. Nevertheless, a series of risk factors for atherosclerotic disease are badly explained: left–right variation in carotid atherosclerotic disease in the same individual, interracial differences in location of the plaques along the large blood vessels and sex differences (Phan et al. 2012). The idea that carotid anatomy and geometry play also a role in the development of atherosclerotic plaques is not improbable. Measures of the internal and external carotid bifurcation showed that the following factors were independently associated with internal carotid artery (ICA) stenosis: ICA radius at the bifurcation (radius of circle that can be placed at ICA, external carotid and common carotid branching area), ICA angle (angle between ICA and common carotid), age, hypertension, diabetes, male sex and ever smoker. Traditional factors like age, diabetes, hypertension and even smoking are not an astonishing outcome. However, absence of stenosis is found in patients with a mean ICA angle of 23.3° and those with an angle of more than 31.3° had ICA stenoses. Abnormal hemodynamic flows are seemingly induced by the carotid anatomy and geometry and its asymmetry is presumably responsible for the left–right individual variability (Phan et al. 2012). Thus, the relation between common carotid, internal and external carotids, as expressed in radius and angle, co-rules the stenosis in the internal carotid and it is to expect also in the other two vessels. To repeat: It's the anatomy stupid!

The SAPPHIRE trial in which carotid endarterectomy is compared to carotid stenting with an emboli protection device for stroke anticipation was needed, because stenting in the internal carotid has been regularly attacked by vascular clinicians. "In observing these early carotid stenting efforts, the majority of vascular surgery's national leaders, along with their societies, did not serve their constituency well. Rather than approaching the emerging carotid stent method as a potential adjunct to their care of the patient with carotid disease—therefore in their domain—and integrating it into the investigational process going forward, a pitched, all-or-nothing strategy to limit its use and demonize its practitioners was chosen instead" (Gray 2004). After one year, the outcome of the trial was still not believed. The results of the 3 years follow-up show: "In our trial of patients with severe carotid artery stenosis and increased surgical risk, no significant difference could be shown in long-term outcomes between patients who underwent carotid artery stenting with an emboli protection device and those who underwent endarterectomy" (Gurm et al. 2008). This result was already known after the one-year follow-up and one should condemn the clinical attitude and to describe it as: "They are invading my territory, so less money will be earned". This is at the patient's cost, since stenting is less invasive for the patient, and less hospital admission time is needed. The risk of both clinical methods for the treatment of carotid atherosclerotic disease is equal. The discussion has not ended and payment and "turf" issues are still involved (see Safian 2017). [Turf issues arise from competition between medical specialties, and they involve competition for patients and for the dollars that accrue from the care of these patients (Veith 1998)].

At the basis of stenosis is often atherosclerosis and thus the fatty acid/lipid metabolism. The normal fatty acid metabolism (Fig. 14.4) involves hydrolysis and re-esterification. This can occur four times or more before fatty acid residues are finally oxidized in the muscle. The four "generations" of triglyceride (TG in Fig. 14.4) have different compositions and finally the liver produces TG from (among others) free fatty acids. A necessary prerequisite is a sufficient insulin signal. TG is organized within a droplet surrounded by a monomolecular layer of phospholipids. Stabilization of the phospholipid layer is caused by some free cholesterol and the apo-molecules (B, E and C). The hepatic end product is called very low-density lipoprotein (VLDL, Fig. 14.4). Apo C enhances the uptake by lipoprotein lipase in *capillary* endothelium, in fat and in muscle tissue. Restoration of the triglycerides into fatty acids and glycerol by hydrolysis is the main action of lipoprotein lipase. Apo E organizes the next action that is uptake by the hepatocyte B-E-receptor of the triglycerides and consequently they are metabolized by the hepatocyte. The reduced droplets now contain mainly low-density lipoprotein (LDL) and are further reduced due to the Apo B receptors by nearly all types of tissue. Unspecific uptake is also possible of these last droplet rests by endocytosis. Thus "peeled off" chylomicrons (fat droplets) are circulating within the bloodstream (Gevers Leuven 1987). At places within the arteries where the laminar flow is disturbed by branching off or bends of the blood vessels, the cell lining of endothelial cells reacts differently. It is normal that the endothelial cells of large arteries are selective in their uptake (endocytosis), transport through the cell and the

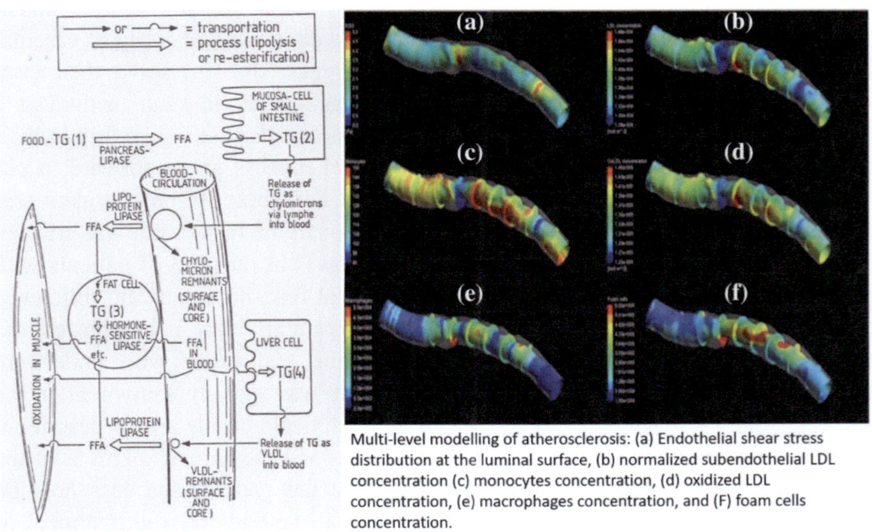

Fig. 14.4 Left: Diagram of the transportation of triglyceride and of free fatty acids (courtesy and permission Gevers Leuven [1987], Gaubius Institute and LUMC). FFA, free fatty acids; TG, triglyceride; VLDL, very low-density lipoprotein. Right multilevel model of atherosclerotic characteristics (with permission Sakellarios et al. 2017)

delivery at the other side of the cell (exocytosis). These "peeled off" chylomicrons are usually not taken up, but the endothelial cells exposed to low time-average shear stress in the arteries do start to take up the LDL containing peeled off chylomicrons, now called apolipoprotein B (apo B) containing lipoproteins (LP's). Presumably the coating (glycocalyx) of these shear stressed endothelial cells is changed and thickened and contribute to the reduction of extravasation of LP's into the subendothelial space. Thus, LP's are directed into the intima of the larger blood vessels. The increase in modified LP's attracts inflammatory monocytes and they start to macrophage the LP's and become lipid-loaded foam cells within the intima. These macrophages are co-responsible for plaque initiation or plaque progression that involves (extra) inflammatory endothelial and smooth muscle activation. The plaque disturbs the blood flow above and below the inflammatory area and as a consequence LP's enter the intima in these upper and lower regions and so on. Finally, the artery is atherosclerotic over a large distance. The ongoing development of the atherosclerotic plaques destroys the cap of the plaque that still made a border between blood and the plaque. The destruction of the endothelial-lined cap induces thrombosis that can impede blood flow (see Tabas et al. 2015; Libby et al. 2011).

A proof-of-concept numerical model study of the mechanisms of atherosclerotic plaque growth extends and confirms recent results (Sakellarios et al. 2017). The Italian researchers used a human reconstructed left circumflex coronary artery for this multilevel modelling approach. Its first level consists of the modelling of blood flow and endothelial shear stress computation. The second level contains the modelling of LDL and HDL. To this second stage belonged monocyte transport through the endothelial membrane to vessel wall. The third level encompassed the modelling of LDL oxidation, macrophages differentiation and foam cells formation. All modelling levels integrate animal (pigs) and human experimental physiological results. Measuring up the baseline computational results with the changes in arterial wall thickness, lumen diameter and plaque components using follow-up data gave the following results: "endothelial shear stress and LDL concentration have a good correlation with the changes in plaque area whereas the introduction of the variables of oxidized LDL macrophages, and foam cells as independent predictors improve the accuracy in predicting regions potential for atherosclerotic plaque development. Advanced computational models can be used to increase the accuracy to predict regions which are prone to plaque development" (Sakellarios et al. 2017). However, the confirmation that regions with a low endothelial shear stress or disturbed flow are more prone to exhibit disease progression is remarkable. In fact, areas with high endothelial shear stress and areas with low endothelial shear stress both induce atherosclerosis. The low shear stress induces more a fibrous type of plaques (see also Goubergrits et al. 2002).

14.2.2 Herbal Medicine

The use of raw plant extracts to cure atherosclerotic problems has frequently been promoted. The placebo effect or the psychodynamic component in the overall therapeutic effect is stressed in herbal medicine. Phytotherapy is the medical use of pure (raw) plant extracts that are applied in prevention as a medicine and should be discerned from phytopharmacology and chemical taxonomy: chemical knowledge of plants for medical practice. Chemical taxonomy (Hegnauer 1981) serves regular medicine still day by day to invent new medicaments. Plant extracts are studied for their chemical substances and for their use in medicine. Purification and their effective concentrations are studied in phytopharmacology, which is contradicted in phytotherapy. In this phytotherapy part a choice is made. There are so many different approaches, together with a plethora of recipes that it asks for regulation. In the series "Le livre de poche pratique", a book written by Dr. Jean Valnet (1920–1995, military surgeon, carrier of the Medaille de Bronze pour travaux scientifique [1954], officier de la Legion d'Honneur et commandeur des Palmes académique) entitled: "Phytothérapie: Traitement des maladies par les plantes" (editions 1972, 1975, 1979, 1983) has been used extensively. Since 1985 his recipes have been produced by "Laboratoire Cosbionat", which is still one of the main distributers of phyto-extracts in France.

Table 14.1 shows the seven herbs that are good medical remedies for atherosclerosis, according to Valnet: common lady's mantle, hawthorn, birch, mistletoe, walnut, sage and meadowsweet. Since herbs in phytotherapy do have several medical actions the other main effects as described by Valnet are also given in this table. The "one plant extract for one typical disease" rule is not found in phytotherapy. So there are main effects and side effects (in the table indicated with other effects). What someone indicates as side effect will be considered by others as amain effect.

PR of phyto-extract firms and even Wikipedia will show different main effects of the same herb. Double blind placebo studies have shown that the claimed phytotherapy effects of several plants are not present. Besides these seven herbs, there are several others that do have the capacity to reduce atherosclerosis: artichoke, asparagus, thistle, sundew, curcuma, greater celandine, sweet briar, lily of the valley, nettle, passionflower, dandelion, rosemary, bramble, lime-tree and a few more. To it belongs a life-regime of nitrogen- and chloride poor meals, little meat and a lot of fruit, green vegetables, garlic, onions, watercress, artichoke, cherry, strawberry and prunes. For body treatment, baths of seaweed are recommended. "*Alternative medicine cannot guarantee healing, methods are not transparent and the chance is rather large that "healing" has to be attributed to placebo effects. Evident, it is the old Galenic-Hippocratic framework that governs alternative and herbal medicine*" (Marani and Koch 2014).

Table 14.1 Herbal medicine and atherosclerosis (see Phytothérapie by Valnet 1983)

Name: fr +eng	Latin name	Effect of herb	Chemistry	Other effects
Alchemille	Alchemilla vulgaris	Detergent, astringent, stomachic, prevention fibromas	Tanins, salicylic acid, phytosterols	Wound healing, regul, menstruation
Lady's mantle	Rosaceae			
Aubepine	Crataegus oxyacantha	Vasodilatation, antispasm, cardiovascular function (+)	Tannins, flavonoids, phenolic acids, proanthocyanidins, managanese	Hypnotic, antipyretic, astringent
Hawthorn	Rosaceae			
Bouleau	Betula alba	Diuretic, elemination of chloride	Resins, betulin, phytochemicals	Depurative, stimulant, sudoriferous
Birch	Fagales			
Gui	Viscum album	Vasodilatation, hypotension, anti-cancer	Alkaloids, glucoside, resin, folate, anthocyanins, tannin, vit B, magnesium	Antispasm, diuretic, decongestion
Mistletoe	Loranthaceae			
Noyer	Juglans nigra	Astringent, antidiabetic, antiscrofulous, vermifuge	Tanin, juglandin, juglon, naphtoguinine, vit C	Excitation liver, antiseptic, stomachic, depurative
Walnut	Juglandeae			
Sauge	Salvia officinalis	Antiseptic, antibacterial, relaxant, stomachic, heatburn	Tannin, oleic acid, ursonic-, ursolic acid, carnosol, caffeic acid, nicotinamide	Depression, memory loss Alzheimer, diarrhea
Sage	Lamiaceae			
Ulmaire	Filipendula ulmaria	Diuretic, astringent, diaphoretic, anti-inflammatory	Salicylic acid, iron, calcium, flavone glycosides, tannins	Fevers, astringent, sleep, sweating, choleretic
Meadowsweet	Rosaceae			

14.2.3 Vascular Anastomosis

Successful transplantation of organs and successful carotid endarterectomy is co-determined by blood vessel reconnection called vascular anastomosis. Its history starts with Vincenz Czerny (1842–1916). His attention and methods to increase asepsis during his operations on stomach and digestive tract were clinically generalized. He started a cancer clinic (1906–1908) where he applied his asepsis methods: Instituts für Experimentelle Krebsforschung (later it became Deutschen Krebsforschungszentrums, DKFZ) in Heidelberg, known by the people as the Samariterhaus (Samaritan house). Asepsis is a prerequisite for successful vascular anastomosis. Moreover, it is Czerny who introduced the double suture using silk threads, the precursor method for blood vessel anastomosis. Then came a sordid affair: two scientists working together on blood vessel anastomoses and one of the two gets the Nobel Prize, the other not and of course, since the other is an American, the Prize (or the primary credit for this work) should have gone to the American. The Nobel Prize winner was a Frenchmen, you see. Alexis Carrel (1873–1944) was a French surgeon, eugenicist and biologist. He has been named: the "father of Transplantation", "father of Vascular Anastamoses" and "father of Experimental Vascular Surgery". His experiments brought advances sorely needed in vascular surgery. His first main target was to perfect the anastomosis of blood vessels. In which he succeeded. "Charles Claude Guthrie (1880–1963) was an American physiologist. A major part of the work that was the basis for Carrel being awarded the Nobel Prize was performed jointly by Carrel and Guthrie from spring 1905 to summer 1906 at the Hull Laboratory at the University of Chicago, resulting in the publication of 21 jointly written papers" (Sade 2005).

The drive for Alexis Carrel's clinging to vascular anastomosis has been described by Levin (2013): "In 1894, Francois Carnot, the popular president of France, was assassinated in Lyons. He was stabbed in the abdomen, and when surgeons were unable to stop the massive haemorrhage, he died on the operating table. The wound had divided his portal vein, a problem at that time beyond anyone's ability to solve. Carnot was delivering a political speech. In the audience was Alexis Carrel, a 20-year-old medical student. The horrifying event was an epiphany that changed his life. With a leap of faith, he believed he could solve such a problem and was determined to do so". Living in Lyons, centre of lace industry and his mother owning a lace factory, in which he worked during his medical studies doing embroidery and stitching with fine needles and silk threads. After his studies in 1900, he joined the Lyons' anatomy department and started performing successfully vascular anastomoses in experimental animals. His publications attracted other scientists and he accepted in 1904 to work at the University of Chicago together with Claude Guthrie. Kidney, thyroid, spleen, ovary and adrenal gland transplantations became possible, also due to the asepsis approach, and even a whole dog's head was transplanted on another living dog (Lamba et al. 2016). Coronary artery bypass grafting, mitral-valvulotomy and transplantation of a heart–lung block were performed (Sade 2005). By the way, Carrel was also known for Carrel-Dakin's solution in the treatment of infected war wounds, embryo cultures, and he made a start with the heart–lung machine by improving pump functions, but these are beyond the scope of this chapter (for more information see Levin 2013; Lamba et al. 2016; Sade 2005). Carrel's method existed of three retaining sutures to form a triangular shape. Along each of the sides of the triangle the upper and lower vessel walls were stitched together by fine threads and using fine needles. The method protected against postoperative haemorrhages, embolism and prevented strictures along the sutures.

The interspecies variations in branching of the main blood vessels from the mammalian aortic arch are large (see Fig. 14.5). Differential growth is held responsible for these variations. Comparison of the embryonic situation (Fig. 14.5a), in which the main vessels of pairs of arches 3 and 4 each contain the right and left carotids and the left and right subclavian arteries, respectively, to the different mature mammalian situations shows: one main trunk for carotids and subclavian arteries (Fig. 14.5b), a main trunk with left subclavian separated

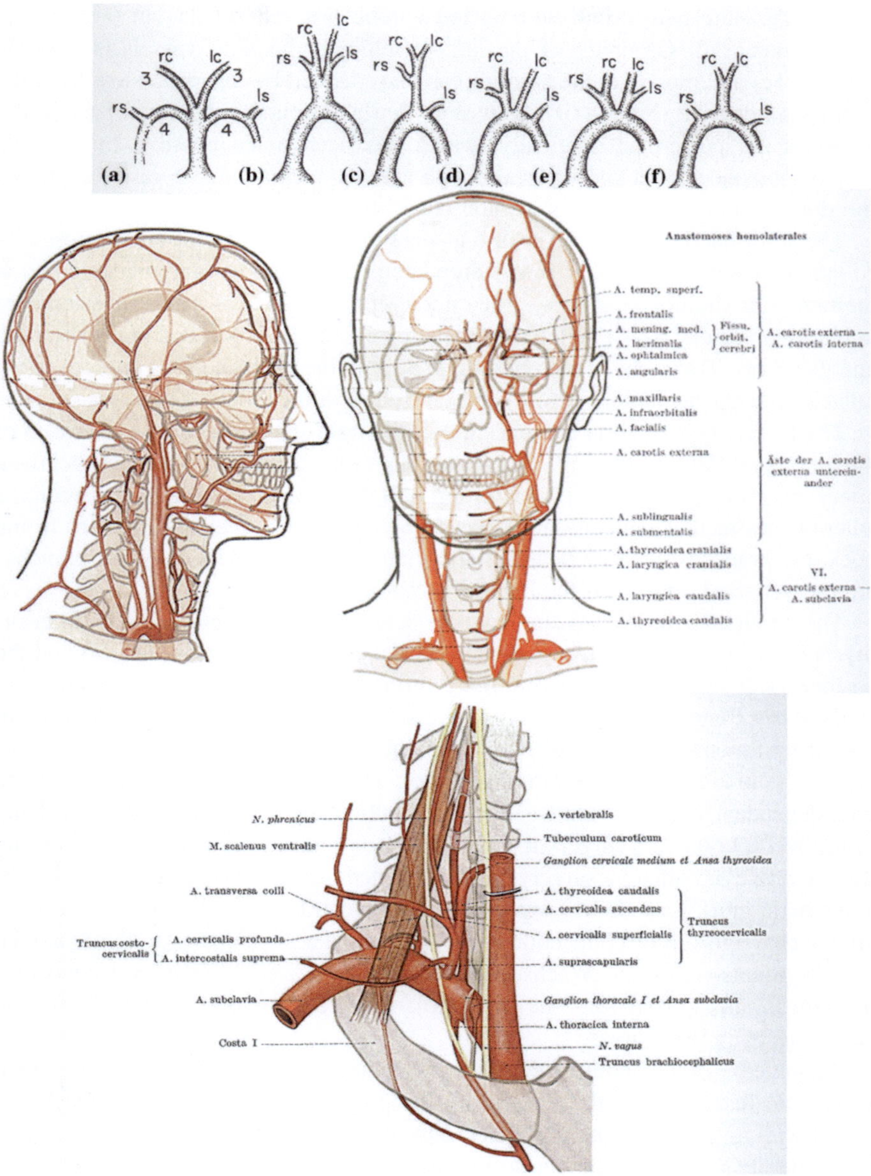

Fig. 14.5 Upper figure shows the diagrams of the variations in the branching of the main blood vessels from the mammalian aortic arch. A, embryonic situation, third arch pair producing the carotids from the ventral aortic trunk. The right fourth arch is later in development reduced beyond the point of departure of the subclavian artery (see text; figure is from Hafferl 1933). Middle figure on head and neck anastomoses, indicated blackish, is adapted from Lanz und Wachsmuth (1955, Figs. 22 and 23) and lower figure on the branches of the subclavian artery is also from Lanz and Wachsmuth (1955, Fig. 294). All figures with permission

(Fig. 14.5c), the main trunk without left carotid and left subclavian (Fig. 14.5d, human situation). Grouping of the left vessels and the right vessels is possible (Fig. 14.5e) and moreover the carotids can be separated by one main stem from the subclavian arteries (Fig. 14.5f). Now note that anastomoses of most of the aortic branches are performed surgically, which has been made possible by Carrel's method and one should know that although vascular techniques are vastly improved (no doubt about that), Carrel's method is still basic.

The left and right parts of head and neck do contain natural anastomoses by collaterals reaching the other side (contralateral anastomoses) and anastomoses at the same side (homolateral ones). The middle of Fig. 14.5 shows these anastomoses in blackish vessels. Note that the contralateral ones are mainly found in neck, mouth and eye areas. The homolateral ones are present in the neck and are vertebral artery related (already mentioned before) and are found in the frontal skull area.

The human type of aorta branches shows the right subclavian artery to branch of the brachiocephalic trunk (Fig. 14.5 lower picture). The branches of the subclavian artery leave this artery before it passes through the scalenus opening in between the anterior and medial scalenus muscles and before it goes over the first rib. In the neck, the branches of the subclavian artery are single or grouped. Single branches are the vertebral artery and the internal thoracic artery.

The grouped arteries are the thyrocervical trunk branches: (caudal (inferior) thyroid, ascending cervical, superficial cervical and suprascapular arteries) and the costocervicalis trunk branches (profound cervical and supreme intercostal arteries). In the *early* Blalock-Taussig operation the subclavian artery is cut to make a shunt with the pulmonary artery to increase pulmonary blood. Sometimes, the cut is even before the branching of the vertebral artery. The operation is to increase flow in duct-dependent cyanotic heart defects like pulmonary atresia (called "blue baby syndrome"). In modern surgery it is only temporarily used while the child is waiting for corrective or palliative surgery with a modified method. Nowadays surgeons use a synthetic graft. One should expect severe circulation problems in arm and neck after such a rigorous disconnection of the subclavian artery. However, this is not the case in infants (Lanz and Wachsmuth, Hals 1955, p. 23). A good explanation for the arm is still looked for. The best explanation for the neck and head are the anastomoses in these areas.

An increase of metabolism produces a widening of the blood vessels. The blood vessels do have a tonus that is kept by the proper smooth muscles of the blood vessel wall and by the autonomic nervous system. If organs are metabolic active extra blood is directed into these organs. The resting organs with a far lower metabolic activity are subject to a decrease of blood intake or flow. This distribution of blood over the various organs is sympathicus dependent. The threshold for sympathicus constriction is increased in the metabolic active organs. Thus, the total peripheral resistance of blood vessels is caused by different subresistances of tissues and organs that are floating from organ contribution to organ contribution by their respective different metabolism over time. This effect of collateral sympathetic vasoconstriction, causing adequate blood distribution over the human body, is mainly organized by the smaller arterioles and not by the large elastic vessels. Thus, the large elastic vessels function as rapid-transit passageways for blood. By their elasticity they act as a pressure reservoir, which is needed in the relaxing phase of the heart for keeping a more constant blood transport. However, the elasticity still contributes to *transport*.

Kartagener syndrome is a type of primary ciliary dyskinesia. It is also characterized by situs inversus totalis (mirror-image reversal of internal organs) in 50% of the patients. In most cases, it is caused by mutations in different genes determined as an autosomal recessive inheritance for the production of a large cilia protein complex called dyenin. Concern of this part is the situs inversus. In non-Kartagener syndrome situs inversus totalis has a prevalence of 1 in 10,000 and is characterized by the total inversion or mirroring of the thoracic and abdominal organs. It concerns a genetic autosomal recessive alteration in chromosome 14, influencing the development of the left–right distribution during development. Note that in the Kartagener syndrome it concerns a mutation that changes the cilia in their movement effectivity. Heart abnormalities occur in situs inversus in 3–5% of the cases, which is far higher than in normal cases (1%). Nevertheless the aorta arch and its branches in situs inversus are reported to be normal, although mirrored. Seemingly the large elastic vessels contain and remain in their mutual spatial coherence despite left–right changes. This is supported by the mal-position of the great arteries, which is "Abnormal spatial relations of the aorta and pulmonary trunk to each other. Each of the abnormally related great arteries arises above the anatomically correct ventricle" (Perloff 2011). Although the aorta and pulmonary trunk are differently related, the main branches of the aorta are hardly deviated from the normal situation (see Ali et al. 2016; Schallert et al. 2013; Van Praagh 2016). *Although the aortic arch interspecies difference in mammals is large, the human intraspecies aorta construct is invariable and is hardly changed by aberrations of the heart. This conservatism assures the blood transport function.*

The situs inversus is infrequently detected in humans. It is mainly noticed if (digital) visual systems are used for the patient's examination like X-ray, CT or NMRI. The Museum of Pathological Anatomy in Padua has a teratological collection in which a situs inversus (Fig. 14.6) is present. The pathological specimen was fixated using tannin tissue preservation and colours have been injected into arteries and veins (Zanatta et al. 2014). The aorta arch and the heart apex show the situs inversus and dextrocardia, respectively (see arrows in Fig. 14.6). By removal of the sternum both the claviculae hang free and point to characteristics of the situs inversus. For comparison reasons an analogous specimen from the Museo La Specula, Florence, is shown in Fig. 14.6 with the normal topography. The enigma is that only 50% of the Kartagener's syndrome and also the half of the conjoined twins (see Chap. 3) do possess a situs inversus. Science knows that the left–right organization is dependent of an organizer of mesoderm cells. These mesoderm cells do contain cilia that sweep from right to left, presumably inducing the left–right organization of our body by putting on genes that enhance the left–right organization. These messengers produced by the genes can be tricked and situs inversus can be purposely made in chicken embryos. But why 50% of the cases? We know that Kartagener's syndrome can be caused by various mutations of the dynein cilia

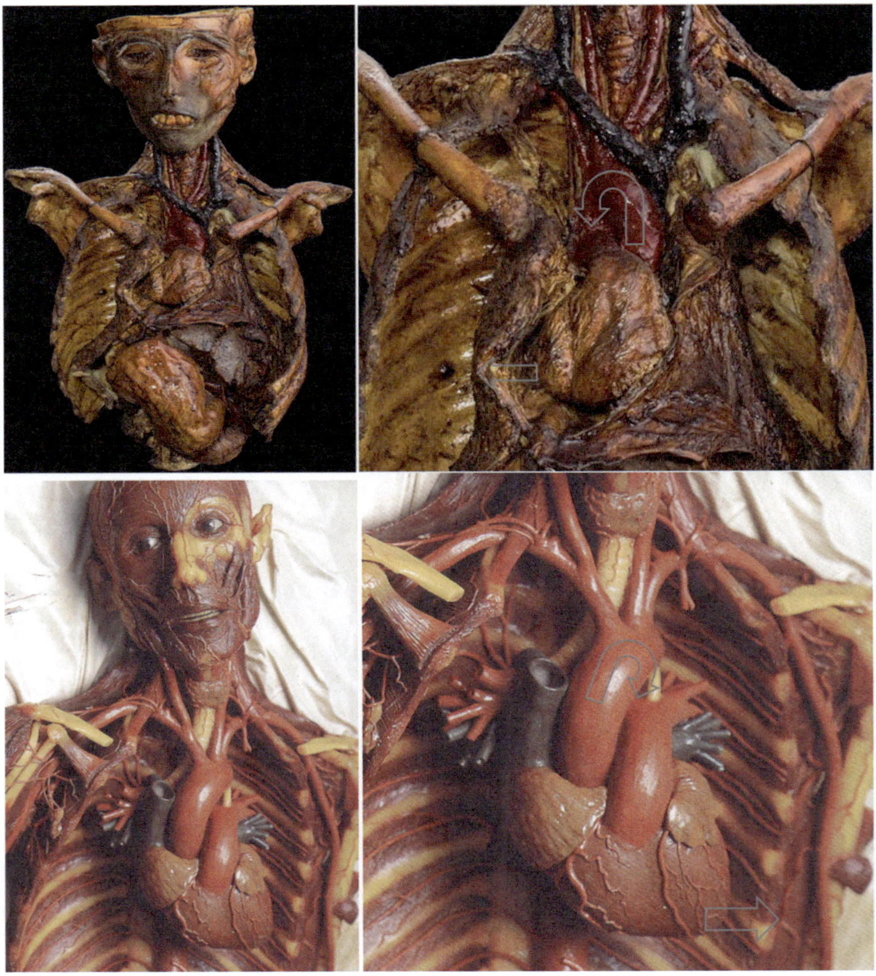

Fig. 14.6 Upper two figures concern situs inversus with dextrocardia in a mummy of the Museum of Pathological anatomy of the Padua Medical School. Arrows indicate aorta and heart apex directed towards the right (Zanatta et al. 2014, with permission). Lower two figures are from the Encyclopedia Anatomica (1999) of the Museo La Specola (XXV.446) in Florence (with oral permission). Arrows indicate aorta and heart apex directed towards the left. Note the position of the right a. subclavia from the a. brachiocephalica and branches towards the neck

protein complex having different effects on the cilia activity. Leroi (2003) relates it to the probability that when symmetry always fell to the left, it now has an even probability of falling to either side due to the destruction or changing of the messengers determining left.

14.3 Lymphatic System

14.3.1 Development of the Lymphatic System

Blood vessels induce the development of the lymphatic system. Only after the invasion of blood vessels into developing tissues and organs, will lymphatic vessels emerge. This also holds for wounded areas. This idea was set forth by Florence Rena Sabine (1871–1953; 1902). The idea was not generally accepted, but after her defence in "The origin and development of the lymphatic system", it was incorporated in most handbooks on this topic and became a guiding principle for basic and clinical studies (Oliver 2004). The lymphatic sacs are found to be the sprouting areas of the lymphatic vessels. They are considered primitive hearts that are thought to initiate the flow of the lymphatic chyle (but see Sect. 14.3.4 active pump organization). Lymph nodes develop and keep their position at main entrances of the rump. Extremities are mainly provided with lymphatic vessels, and nodes are few. Morphological studies showed that the lymphatic primordia originate at the angulus venosus (see Fig. 14.10), the area in between the superior and inferior cardinal veins in the 42 days old embryo (11 mm CRL). The two first primordia that are bilateral present are: a jugular lymphatic sac and an axillar one just below each other. They keep an open connection to the veins. Later on these two primordia fuse but the open connection stays present. Several other places produce primordia that form lymphatic parts: thoracic lymphatic internal plexus, paratracheal, lumbal and iliac ones. A series of primordia produce the thoracic duct that is localized along the aorta. Not all lymphatic parts are bilaterally shaped. Exceptions are the mesenteric lymphatic and the subtracheal lymphatic plexuses (see van Limborgh and van der Putten 1981; van der Putten 1975). Genetic studies confirmed that embryonic veins change into lymphatic vessels by an unknown external factor. Lymphatic vessel endothelial hyaluronan receptor 1 (Lyve1) and expression of the transcription factor prospero-related homeobox 1 (Prox1), and presumably the Sox 18 transcription factor are responsible for the appearance of both and determine the early development of the lymphatic vessels. Prox1 negative vessels remain blood vessels, while Prox1 positive vessels develop endothelial cell types for lymphatic vessels (see Fig. 14.9, I). Lymphatic endothelial cells that are Prox1 positive are also suspected of forming the lymphatic sacs by leaving the vessels and regrouping into a lymphatic sac, a process called budding (Oliver 2004). The budding process is dependent on the production of mesoderm cells characterized by lymphangiogenic growth factor Vegf-c (=vascular endothelial growth factor-c). This Vegf-c attracts Prox1/Lyve-1-positive cells because these cells have receptors for it (Schulte-Merker et al. 2011). A cascade of factors is involved in the final staging of the lymphatic vessels (Fig. 14.9, I), but why sac formation occurs only at certain places is unknown. Moreover, the exclusive venous origin of the lymphatic system is debated. Alternative origins for the lymphatic endothelium next to the main venous one are indicated in the paper of Semo et al. (2016).

14.3.2 Lymphatic Maps and Metastases

Indications for neck operations are medial and lateral neck cysts and fistulas, infections, primary tumours (sarcoma or lymphoma) and cervical lymph node metastases (inter alia metastatic squamous cell carcinoma). Head and neck cancer incidence worldwide surpasses 550,000 cases per year. Head and neck metastases are in their extension nearly always guided along the lymphatic system. However, "cutaneous melanomas of the head and neck metastasized to clinically predict nodal groups in 73.3% of patients in this series. This leaves a significant 26.7% of nodal metastases outside of the arbitrarily defined regions" (Suton et al. 2012). Thus, melanomas of head and neck use in a quarter of the patient cases extralymphatic pathways for metastasizing. Squamous cell carcinomas use also the lymphatic system, but invade the soft tissues and bone or cartilage at the same time. One should note that head and neck accounts for only 20% of the body's volume. Nevertheless, it contains 40% of the body's lymph nodes. Moreover, head and neck cancers accounts for around 300,000 deaths worldwide per year. As a consequence of the need for effective tumour surgery, an adequate topography of the tumour spread and of the areas involved or to which the tumour could be directed was needed. These maps are based on lymph node distribution (Fig. 14.7b, c). Several different levelling systems have been published. Each do have their advantages and only one, the well-known Memorial Sloan-Kettering Cancer Center's levelling system is indicated in Fig. 14.7c. Picturing the exterior of the neck for the projections of different lymphatic/metastatic regions on the surface helps to locate the surgical extension needed for the specific extension of the metastasis. Besides the complicated and variable anatomy of head and neck that influences surgery, staging of these cancers gives difficulties. "The staging of head and neck cancers has recently been updated to address both the poor prognosis in the presence of extracapsular extension of nodal disease and the increased prevalence of HPV-mediated oropharyngeal cancers" (Kulzer and Branstetter 2017).

Squamous cell carcinomas can develop into large swellings (Fig. 14.7d) and the extended removal of the tumour that is needed ask for an extremely good knowledge of the local anatomy in order to spare (or not to spare) structures such as the accessory nerve.

There are several approaches to describe the lymphatic system. The topographic description is given here, while above the surgical approach of the lymphatic system was given. The lymphatic system can be considered an extra vessel system that takes up tissue fluid and also fatty compounds from the digestive tract. The content of the total lymphatic fluid is 4–5 litres that passes with a velocity of 1 cm/min through lymphatic capillaries, nodes and vessels.

The lymphatic system contains cells: lymphocytes and monocytes and is involved in the immunological response. There are regional differences in uptake. The lymphatic vessels of the digestive tract can take up fat. This fatty fluid is called chyle or chylus. The lymphatic fluid is the result of the filtration of blood fluid (leakage to provide cells with the necessary substances to survive) out of the blood

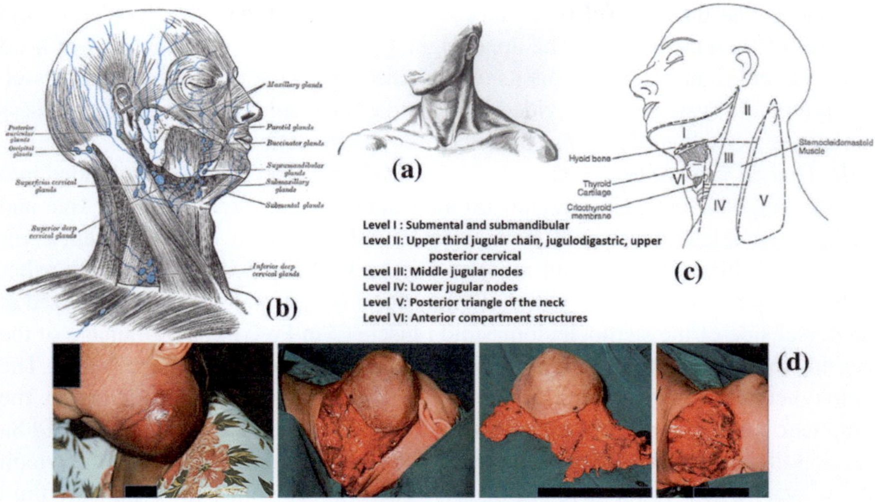

Fig. 14.7 **a** Shows the main palpable structures form mentis to clavicula, m.sternocleidomastoideus, m.omohyoideus, m.trapezius, thyroid, hyoid, cricoid and trachea (courtesy J.M.F. Landsmeer †). The borders of the supraclavicular triangle are explicitly visible. **b** Gives the overview of the main lymphatic nodes in head and neck (Gray's anatomy, 25th edition). **c** Clinical subdivisions of the neck for surgery from the Memorial Sloan-Kettering Cancer Center's levelling system. This system knows a refinement by subdivision of the areas I, II and V in subareas A and B. **d** Demonstrates the extent of a squamous cell carcinoma before, during and after the removal operation. Note that several areas of the levelling system are involved (courtesy C. Meeuwis, Erasmus MC, with permission)

vessels and the reabsorption by the further on located venous capillaries. The difference of leakage minus reabsorption is tissue fluid that will be taken up by the lymphatic capillaries. If this system was not present a person would increase in volume by 5 litres a day and would "explode" after a few days. This extra tissue fluid is reabsorbed by pinocytosis (drinking of the cell and by which IgG is taken up with the neonatal Fc receptor) and pressure-driven fluid uptake. Physiological pressure laws do explain this fluid reuptake. The structure of lymphatic capillaries and of vessels is comparable to the venous system. Capillaries join to larger vessels that will be the afferent vessels of a lymphatic node. The node is surrounded by a capsule and contains a ring vessel (subcapsular sinus) in the periphery in most cases related to the outer capsule. This ring vessel gives branches that are guided by trabecula (perinodular vessels) around the cortex of the lymphatic node. Capillaries from the vessels penetrate the cortex of the node and reassemble vessels in the trabecula to leave the node as the efferent lymphatic vessel. The node itself gets arterial and venous vessels to support the tissue of the node. The lymphatic vessels contain valves, since the internal pressure is often too low to bring the fluid from lower places to the collecting higher located sacs (the main sac is the cisterna chili) and ducts. The main vessel is the thoracic duct, which brings the lymphatic fluid into the left subclavian vein. The upper right quarter of the body (right half of head and neck and right arm) contains its own connection to the right subclavian vessel. The lymphatic system is subdivided into a superficial and deep part, which possess connecting vessels to direct the fluid towards the thoracic duct.

In a lateral view of head and neck (Fig. 14.8a), the superficial part is divided into several lymphatic areas, in principle on the position of the internal jugular vein and the m. sternocleidomastoidus, since most lymphatic nodes organize themselves along and behind this vein and muscle. The back of the head and neck, together with the supraclavicular area will drain into those lymph nodes that are localized at the dorsal side of the sternocleidomastoid muscle (2 in Fig. 14.8a). Drainage of the area in front of the sternocleidomastoid muscle is related to two facial areas. The region between masseter and sternocleidomastoid muscles will connect to the lymphatic cranial nodes, ventral of the sternocleidomastoid muscle (in Fig. 14.8a n's. 5, 3 till 1). The real facial area, orbit, nose and mouth will connect to the lymph nodes located next to the myeloid muscle. These nodes gather the lymphatic fluid and they connect to the main lymphatic nodes in front of the sternocleidomastoid muscle (in Fig. 14.8a, n. 1). From Fig. 14.8a in comparison to 14.8b it must be clear that two superficial passages are present either via the deep nodules towards the ductus thoracicus (left side) or the main lymphatic vessel at the right side. If metastatic spread has not passed the sternocleidomastoid border the surgical approach can be restricted to the supraclavicular triangle with tumours at the back of the neck. The deep drainage nodules are present below the sternocleidomastoid muscle and below the internal jugular vein. The bottom of the mouth and submandibular gland area will connect to the superficial lymph nodes and to those nodes below the sternocleidomastoid muscle (Fig. 14.8c).

The tongue, pharynx and larynx and the rear of the mouth will send their lymphatic connection to the retro pharyngeal lymph nodes (located in fossa or spatium parapharyngeus; Fig. 14.8d, n. 1), but also to retrotracheal nodes in the thorax (Fig. 14.8d, n. 2).

14.3.3 Funaoka's Lymphatology and Mechanobiology of Lymphatic Vessels

The anatomy department of the Kyoto Imperial University published research volumes in series A, peripheral nervous system to D lymphatic research from 1930 to 1944. The forerunner of the Kyoto Imperial University was the Seimi-kyoku or chemistry school established in 1869 and replaced by the Third Higher School. At the foundation of the Imperial University System, the Third Higher School was partly integrated into the Kyoto Imperial University (1897) and its college of

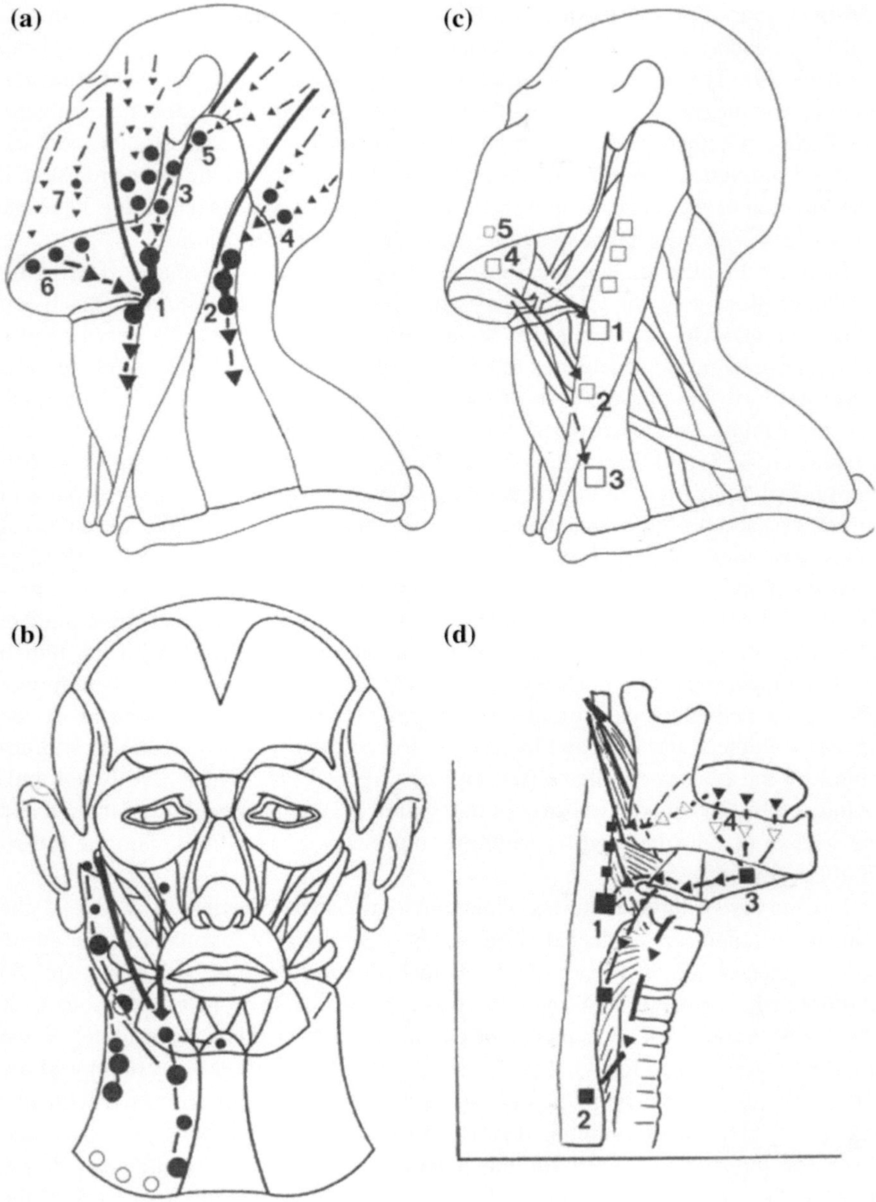

Fig. 14.8 Superficial and deep lymphatic drainage of head and neck. **a** 1 n.l. cervicalis sup, 2 n.l. cervicalis prof median group, 3 n.l. cervicalis prof superior group, 4 n.l. occipitales, 5 n.l. auriculares posterior, 6 n.l. submentales. **c** 1 n.l. infrahyodei, 2 n.l. prelaryngei, 3 n.l. pretracheales, 4 n.l. submandibulares, 5 n.l. submentales. **d** 1 n.l. cervicales prof superior group, 2 n.l. cervicales lower group, 3 n.l. submandibulares, and 4 n.l. submentales (n.l. lymph nodes)

medicine started in 1899. An enormous effort was undertaken by Seigo Funaoka and his collaborators of the Kyoto University to understand the lymphatic system (1930–1944). The pressure within the rabbit lymphatic system, their experimental animal, was measured, the uptake of several substances by the lymphatic peripheral capillaries, together with the arrival in the thoracic duct, was studied and the pressure difference between the vasa afferens and efferens of the nodes registered. The function of the valves within the lymphatic system was clear to these Japanese researchers. The metastases spreading lymphatic roads were studied.

In order to measure the pressure, new methods were developed. The rabbit lymphatic fluid contained 1925/mm^3 white blood cells and 358/mm^3 red blood cells. Lymphocytes constituted 84%, mononuclear leucocytes 2%, erythrocytes 15% and polynuclear leucocytes 0.2% of the cell population. The red blood cells were of course the consequence of the needle method to collect the lymph fluid. Specific gravity was determined for the fluid in the vasa afferens (1.0114) and efferens (1.0128) and their water content studied. Most physical measurements that were possible, just before the Second World War, were applied. The resorption of substances administered subcutaneously was studied together with the chemical substances present in normal lymph fluids to understand filtration difference between blood plasma, interstitial and lymph fluid. Antibody and antitoxin production of the body as it was mirrored in the lymph fluid was another topic studied. The enzymes lipase, amylase, peptidase and catalase in relation with the lymph fluid and their cells were determined. The innervation of the lymphatic system was discovered and electrical stimulation applied. However, the mechanism of the uptake of intercellular fluid and the driving force of the lymph fluid stream was not found by these research efforts (see Funaoka 1930–1944). The active lymph pulsating activity by smooth muscles in the wall of lymphatic vessels was missed and the passive activity induced by extremity muscles was used to explain the lymph fluid stream.

In a survey of three centuries of tumour lymphatics (Sleeman et al. 2009), the Funaoka studies are neglected. The series of Funaoka's "lymphatologie" were known since Notes on Books of The British Medical Journal, 1937 (Jan., p. 21) described the content as follows: "Lymphatologie (Heft 5) gives an account of about fifty researches in lymph carried out at the Anatomical Institute of the Imperial University at Kyoto. The present section deals with the physiology of the lymph glands and biochemical and immunological problems connected with this system. The various researches therefore cover a wide variety of problems, and Professor Funaoka, the head of the department, is to be congratulated on the manifold activity he has inspired". The explicit citation of Funaoka's work at the start of Hellman's chapter (1943) on lymphatic vessels in Bargmann's overview makes missing it nearly impossible.

14.3.4 Active Pump Organization of the Lymphatic System

In general, filtration of fluid and substances out of the blood capillaries is larger than the reuptake or resorption back into these capillaries. The excess of filtered fluid makes up the lymph. The question is: how is this abundant fluid taken up from the tissues and transported within the lymphatic system? The volume of lymph fluid is dependent on the capillary pressure, of the colloid-osmotic pressure of the blood, of the passage quality of the capillary for proteins, of the pressure difference between osmotic pressures, of the tissue and the type of capillary, of the permeability of the capillary wall and of the tissue pressure, together with physical and chemical changes within the interstices. In short: flushing the extravascular space depends on the pressure distributions in, and properties of the blood vessels that are present in tissue or organ. The Starling equation describes this fluid filtration. The calculations show that the net filtration pressure is 10 mmHg and the net reabsorption pressure is −9 mmHg. Not all fluid is reabsorbed since net filtration pressure is higher than the net reabsorption pressure.

The extrafiltered fluid is resorbed by the finest, dead-end lymph capillaries called initial lymphatics. Increase of the interstitial fluid volume stretches the interstitial anchoring filaments that are connected to tissue elements. This increases the volume of the lumen of the initial lymphatics. The slight pressure difference that originates pushes interstitial fluid into the initial lymphatics. The endothelial cells of the initial lymphatics allow only a one-way passage of fluid, always into the capillary lumen. From the initial lymphatics, the lymph goes into the so-called collecting lymphatics. The collecting lymphatics contain valves, inducing a one-way stream and their walls contain smooth muscle cells, responsible for contractions of compartments (called lymphangions) between two valves (Fig. 14.9, I). Several collecting lymphatics unite and end in the lymph node.

To produce a lymph fluid stream the lymphangions smooth muscles have to contract, the upstream valve should be closed and the down-stream valve be opened. Moreover, if two lymphangions next to each other both contract their lymph stream is seriously hindered. Thus, synchronization of the contractions of the one-way lymphangions is a prerequisite for efficient fluid displacement (see Fig. 14.9, I, III, IV; von der Weid and Zawieja 2004: Munn 2015). Several contraction schemes are proposed (Fig. 14.9, IV). In general, Ca^{2+} is held responsible for the synchronization. Figure 14.9 contains a compilation of article figures that relate exclusively to the smooth muscle actions in collecting lymphatics. The original figure texts are added and some are extended. Valve development has been neglected, not because it is less important, but because attention is focused mainly on functionality of the mature lymphatic system.

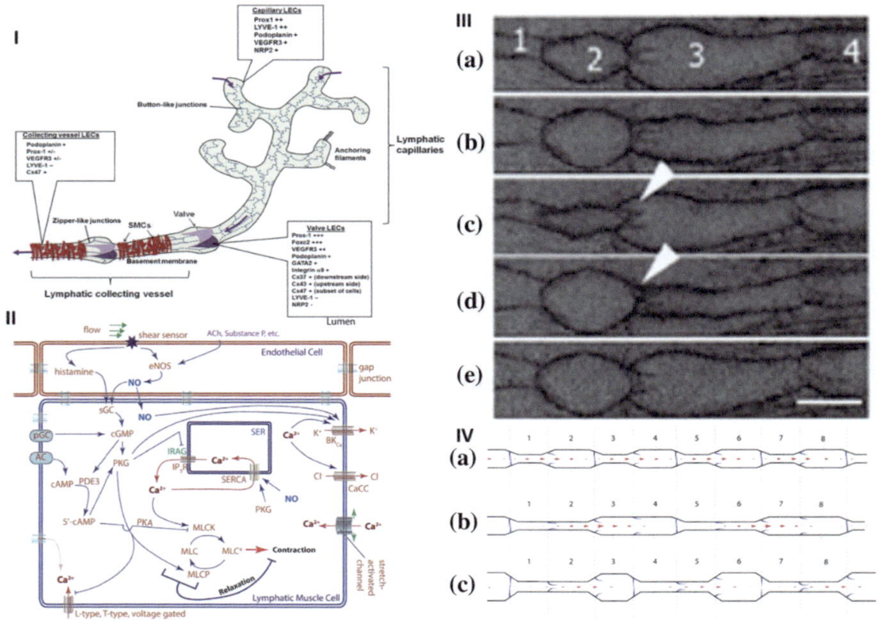

14.4 Veins of the Head and Neck

The diagrams of the development of the human venous system show that at the 16 mm stage the veins of the head and neck form an independent entity (Fig. 14.10c). The venous drainage of head and neck is along the right-sided superior vena cava. This isolation of the upper venous system causes its own medical problems, e.g. persistent left superior vena cava ending into the coronary sinus (Fig. 14.10c/16 mm) or trigeminal neuralgia [Voormolen 1998] by arterial but also by venous compression mainly within the neurocranium. Venous compression was present in 13% of the trigeminal neuralgia (Hong et al. 2011). The most effective therapy for trigeminal neuralgia microvascular decompression of the trigeminal nerve root or percutaneous radiofrequency rhizotomy is still under discussion. “At present both sets of protagonists seem to be resisting a randomized trial. Regrettably, a strong belief in their preferred technique deeming it unethically to withhold it from the patient precludes a move to “evidence-based medicine” in the treatment of trigeminal neuralgia” (Voormolen 1998; see also Obermann 2010)

Infections towards the dura can be caused by the plexus pterygoideus (Fig. 14.10b). Smaller veins of the plexus pterygoideus are related to and pass through the foramen ovale, foramen lacerum and foramen Vesalii. Moreover, the medial meningeal vein connects to this plexus. The plexus pterygoideus has connections by the fissure orbitalis inferior with the orbita (Fig. 14.10b, v ophthal. inf.) and with the lacrimal gland. It collects the muscle veins of the pterygoid, masseter and alveolar areas and those from the palate.

◄**Fig. 14.9** Pump function of the lymphangions. ***I*** "Schematic representation of the organization and of the molecular identity of the lymphatic vasculature. The blind-ended lymphatic capillaries capture fluids, proteins and cells from the interstitial space. They are constituted of oak leaf-shaped cells that are connected to the extracellular matrix by anchoring filaments and that display button-like intercellular junctions. Lymphatic endothelial cells (LECs) of the collecting lymphatic vessels that transport the lymph have a basement membrane and exhibit zipper-like intercellular junctions. Collecting vessels are covered by smooth muscle cells (SMCs) that possess intrinsic contractile activity ensuring lymph propulsion. They contain intraluminal valves to prevent backflow. The LECs of capillaries, collecting vessels and valve-forming cells exhibit differences in their molecular identity. Their respective expression profiles are indicated in boxes" (Vittet 2014 with permission): Cx, connexins; Foxc2, forkhead box C2; LEC, lymphatic endothelial cell; LYVE-1, lymphatic vessel endothelial hyaluronan receptor; podoplanin, muscin type of o-glycosylated glycoprotein is a specific lymphatic lymphangions marker and upregulated a marker for squamous cell carcinoma's; Prox-1, prospero-related homeobox domain 1; SMC, smooth muscle cell; VEGFR3, vascular endothelial growth factor receptor 3. ***II*** "Control of calcium kinetics. Calcium enters the cytosol through ion channels (L-type, T-type, voltage-gated, stretch-activated) in the plasma membrane or smooth endoplasmic reticulum (SER; SERCA, IP_3R) and acts through myosin light-chain kinase (MLCK) to phosphorylate MLC, allowing formation of the myosin–actin cross-bridges and cell contraction. Stretch-activated channels can also allow calcium to enter the cell. Calcium-activated chloride channels (CaCCs) can enhance depolarization during STD generation. Endothelial cells produce endothelial-derived relaxing factors (EDRFs) such as histamine and NO in response to fluid shear. The EDRFs act on adjacent muscle cells through soluble guanylyl cyclase (sGC), cyclic GMP, protein kinase G (PKG) and myosin light-chain phosphatase (MLCP) to reduce intracellular Ca^{2+} and can dephosphorylate MLC, causing relaxation" (Munn 2015 with permission). ***III*** "Mechanism and modulation of lymphatic smooth muscle contractile activity. **a** Successive images of a lymphatic vessel of the guinea-pig mesentery under intraluminal perfusion. Four chambers are displayed which underwent sequential contractions. In **c** for example, chamber 2 that has been filled by chamber 1 is now contracting and filling chamber 3. The unidirectional valve (arrowhead) visible between chambers 2 and 3 is open. In **d** chamber 2 is filled by chamber 1, while chamber 3 is contracting, moving its lymph content forward. The valve is now closed to prevent backfilling. Scale bar 100 m" (von der Weid and Zawieja 2004 with permission). ***IV*** "Examples of lymphangion coordination. **a** Contractions might alternate in adjacent lymphangions. **b** It is also possible that two or more adjacent lymphangions contract and relax together. c If there is asymmetric contraction/relaxation in terms of the number of in-phase adjacent lymphangions, then the stroke volumes must not be the same (e.g. lymphangions 1 and 2 will have lower stroke volume than lymphangion 3)" (Munn 2015 with permission)

The plexus pterygoideus can be considered a venous rete mirabile (see above, Sect. 14.2) that is incorporated in the normal paths of drainage of the head (Buckmund et al. 1971). Placed in between the caput and the processus coronoideus of the mandible, below the zygomatic arch, interspersed with arterial branches of the transversa facie and maxillary artery and surrounding the mandibular nerve it lies between the temporalis and a small part of the masseter muscle and the lateral and medial pterygoid muscle. During mastication, the plexus is pressed by the bellies of the masticatory muscles and above it was explained that rete mirabile is effective in flow and pressure-damping effects. Thus, undisturbed venous drainage of large areas of the head, including the meninges of the brain, is organized by this plexus construction that causes the damping of the pressures exerted by the masticatory muscles.

The front of the neck area is characterized by the possibility of transformation. Caused by the absence of large bony structures, except the vertebral column of course at the back, the vessels must have the ability to cope with such transformations. These transformations are often induced by the muscles in these areas and passive movements of head and neck (think of niddle-noddle during sleeping in a chair). This movable area continues behind the mandible. Mouth bottom and skull muscles miss this transformation capacity, due to their relations to the bony skull. Before vessels enter such bony skull related structures, a fan of vessels leave the main large blood vessels. This trigonum caroticum region contains at the level of the carotid division both an arterial and venous trunk, each which fans out. The arterial one contains ventrally from rostral towards caudal a. facialis, a.lingualis, a.thyroidea cranialis and dorsally a.occipitalis, a. retroauricolaris, a.sternocleidomastoidea. The variability of the venous fan is large. Here, the truncus thyreolinguofacialis is considered with the v.retromandibularis, v.facialis, v. lingualis and v.thyreoidea cranialis entering this truncus (note vv.pharyngica and laryngica are branches). Both the arterial and venous fans are localized in the area bordered by the digastric and omohyoid muscles (Fig. 14.11 carotid sheath and trigonum submandibulare).

In between viscero and neurocranium a small cleft, the fossa pterygopalatine, is present. The a. maxillaris goes into this cleft. The a. maxillaris is one of the two end arteries of the external carotid (the other one is the a. temporalis superficialis) and gives the a.tympanica, a. meningea media, a. alvolaris inferior, a. masseterica, a. temporalis profunda and a. buccalis before entering the fossa. It makes a circle

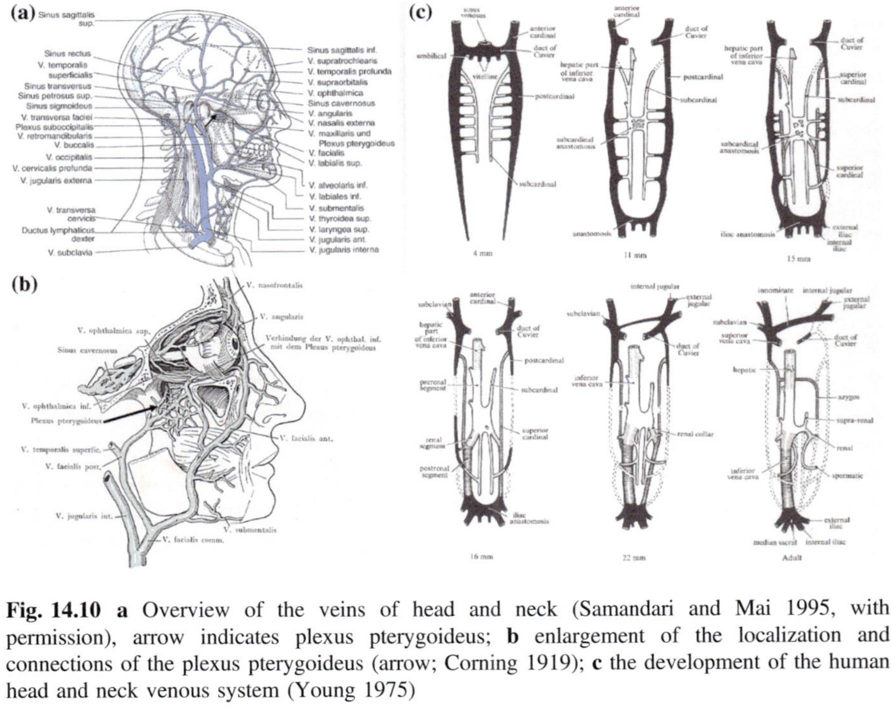

Fig. 14.10 **a** Overview of the veins of head and neck (Samandari and Mai 1995, with permission), arrow indicates plexus pterygoideus; **b** enlargement of the localization and connections of the plexus pterygoideus (arrow; Corning 1919); **c** the development of the human head and neck venous system (Young 1975)

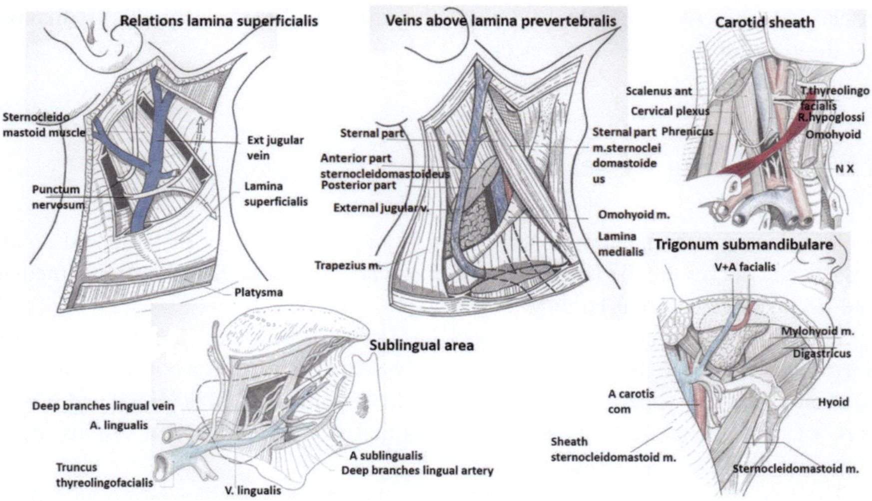

Fig. 14.11 Topography of neck veins above and in between the lamina superficialis and the prevertebral lamina, and in the sublingual area and the submandibular trigone. In the carotid sheath figure, the omohyoid muscle is indicated in dark red to show its landmark function. Arteries are pictured in red and veins in blue (figures are adapted in colour and in text from Perlemuter and Waligora 1985)

within this fossa and gives blood vessels that descend (a. palatina descendens, a. alveolaris superior), one blood vessel that goes to the orbit (a. infraorbitalis) and it ends in three more or less ascending vessels (a. sphenopalatine, a. nasalis posterior and a. canalis pterigoidea). Thus, deep skull areas are reached by branches of the a. maxillaris. They have left the artery before the fossa pterygopalatine or in the fossa where the a. maxillaris ends due to the splitting in a blood vessel fan reaching orbid, palatum and nasal areas.

Like most venous systems in the extremities a superficial and a deep venous net is present in the neck. The official terminology (v. jugularis superficialis and v. jugularis superficialis ventralis paramediana and sternocleidomastoidea) is left. The superficial veins are indicated with the terms dorsalis exterior and ventralis exterior. The dorsal exterior branch passes over the sternocleidomastoid muscle and splits into veins responsible for the drainage around the ear, mouth bottom and connects to the intern jugular vein near or behind the clavicle below the sternocleidomastoid muscle. It has a direct relation to the punctum nervosum (Fig. 14.11 left upper two). The submental veins do collect blood from the skin areas. They connect to the ventral exterior vein that goes down until it reaches the incisura of the sternum and connects to the internal jugular vein (not depicted in Fig. 14.11). Thus the arterial system stays in the carotid layer to reach the muscles and organs with the various vessels. The venous system divides into a deeper part that organizes the ventral drainage of mouth bottom and skull areas. The superficial venous system can be divided into a ventral and a dorsal part. The dorsal part of the superficial system (the dorsal branches of the external jugular vein: v. cervicalis- v. occipitalis) is mainly for drainage of the back of the neck.

14.5 Tapering Vessel Tapes

A good lead for the anatomy and presumably also for surgery of large and small vessels and nerves are the laminae of the deep cervical fascia: the superficial, medial and deep or prevertebral laminae.

The venous drainage in head and neck has been differentiated in a superficial and deep system. In the extremities, both venous systems supply the whole extremity. In head and neck, the superficial system is mainly responsible for the lateral and dorsal parts of the neck as exemplified by the occipital vein. Seemingly, a venous topographical division is present.

The finest branches of the venous system abandon the parallelization with their arteries and there is no equivalent topography between arteries and veins with the cutaneous or superficial distribution of the nerves. Vascular-nerve bundles are less frequently encountered in head and neck, although the carotid sheath belongs to the largest vascular-nerve bundles.

The plexus pterygoideus has been considered a venous rete mirabile and has its drainage into the deep venous system (v. facialis ant and post, Fig. 14.10b). These smaller veins that together constitute a venous network are responsible for the damping effect of the masticatory pressure on the venous co-drainage of orbit and masticatory muscles.

The arterial system knows an intraspecies constancy of its large transport vessels, although a great variety can be noticed for the interspecies constructions. These large vessels induce atherosclerosis by bends that have an angle larger than ± 25–30°.

The lymph nodes are so variable in number and topography that it must drive the surgeon crazy. Nodes are already difficult to detect with digital systems and the smaller ones are easily missed. Moreover, the lymph node density is high(est) in head and neck. Whether the lymphatic nodes are filters or culture chambers for cancer cells, the surgeon has to restrict its cancer area that he wants to clean in a safe manner. Therefore, maps of head and neck are developed that presumably gives a 60–70% certainty that the surgeon removes the cancer. Chemotherapy should do the rest.

References

Ali AL, Sicari R, Fest P (2016) Anatomically corrected malposition of the great arteries: review of the literature. Progr Pediat Cardiol 43:113–117

Baatenburg de Jong RJ, Rongen RJ (1990) Ultrasound examination of the head and neck. Thesis Erasmus University, Rotterdam

Bolk L (1910) De segmentale innervatie van romp en ledematen bij den mensch. Erven F. Bohn, Haarlem
Brown H (2002) Anatomy of the spinal accessory nerve plexus: relevance to head and neck cancer and atherosclerosis. Exp Biol Med 227:570–578
Braus H (1929/1960) Anatomie des Menschen. Part III, 1st and 2nd ed. Springer, Berlin
Buckmund H, Grusche A, Schmidt-Vanderheyden C (1971) Venous pattern in normal lateral serial angiograms of the carotid artery. Neuroradiol 3:20–26
Caliot P, Dumont D (1986) A note on the anastomoses between the hypoglossal nerve and the cervical plexus. Surg Radiol Anat 8:75–79
Corning HK (1919) Lehrbuch der topographischen Anatomie. Verlag J.F. Bergmann, Wiesbaden
Encyclopedia Anatomica, Museo La Specola Florence (1999) Benedikt Taschen Verlag, Köln. ISBN 3-8228-7613-5
Funaoka S (1930–1944) Arbeiten aus der dritten Abteilung des Anatomischen Institutes der Kaiserlichen Universität Kyoto, Serie D Lymphatologie, Heft 1 to 9
Fukuta, Kudo H, Sasaki M, Kimura J, bin Ismail D, Endo H (2007) Absence of carotid rete mirabile in small tropical ruminants: implications for the evolution of the arterial system in artiodactyls. J Anat 210:112–116
Gevers Leuven JA (1987) Metabolic and endocrine aspects of cardiovascular risk factors. University Leiden, Thesis
Gibbins I (2012) Peripheral autonomic pathways. In: Mai JK, Paxinos G (eds) The human nervous system. Elsevier/Academic Press, Amsterdam, pp 141–185
Gray WA (2004) A cardiologist in the carotids. J Am Coll Cardiol 43:1602–1605
Goubergrits L, Affeld K, Fernandez-Britto J, Falcon L (2002) Atherosclerosis and flow in carotid arteries with authentic geometries. Biorheology 39:519–524
Gurm HS, Yadav JS, Fayad P, Katzen BT, Mishkel GJ, Bajwa TK, Ansel G, Strickman NE, Wang H, Cohen SA, Massaro JM, Cutlip DE, forthe SAPPHIRE Investigators (2008) Long-Term results of carotid stenting versus endarterectomy in high-risk patients. N Engl J Med 358:1572–1579
Hafferl A (1933) Das Arteriensystem. In: Ellenberger W et al. (eds) Handbuch der vergleichenden Anatomie der Wirbeltiere. Berl, pp 563–684
Hegnauer R (1956–1981) Chemotaxonomie der Pflanzen, part 1–13. Birkhäuser, Basel
Hellman T (1943) Lymphgefässe, lymphknötchen und lymphknoten. In: Blutgefäss und lymphgefäss apparat Inner sekretorische Drüsen. Springer, Berlin
Hong W, Zheng X, Wu Z, Li X, Wang X, Li Y, Zhang W, Zhong J, Hua X, Li S (2011) Clinical features and surgical treatment of trigeminal neuralgia caused solely by venous compression. Acta Neurochir 153:1037–1042
Hovelacque A (1927) Anatomie des nerfs craniens et rachidiens et du systeme grand sympathique. Libr, Octave Doin, Paris
Kulzer MH, Branstetter BF (2017) Chapter 1. Neck Anatomy, Imaging Based Level Nodal Classification and Impact of Primary Tumor Site on Patterns of Nodal Metastasis, Seminars in Ultrasound, CT, and MRI. http://dx.doi.org/10.1053/j.sult.2017.05.002
Lamba N, Holsgrove D, Broekman ML (2016) The history of head transplantation: A review. Acta Neurochir 158:2239–2247
Lanz and Wachsmuth (1938, 1955, 1972, 2004) Praktische Anatomie: Kopf-uebergeordnete Systemen, Kopf-Gehirn und Augenschädel, Hals, Arm, Bein und Statik, Bauch, Rücken. Springer, Berlin, Heidelberg, NY
Leroi AM (2003) Mutants. On genetic variety and the human body. Penguin Books, London
Levin SM (2013) Alexis Carrel's historic leap of faith. J Vasc Surg 61:832–833
Libby P, Ridker PM, Hansson GK (2011) Progress and challenges in translating the biology of atherosclerosis. Nature 473:317–325
Lieberman D (2013) The story of the human body. Penguin books, London
Limborgh van, van der Putten (1981) De embryonale ontwikkeling van de grote lymphe vaten bij de mens. Ned T Geneesk 125:769–775

Malessy MJA, Thomeer RTWM, Marani E (1993) The dorsoscapular nerve in traumatic brachial plexus lesions. Clin Neurol Neurosurg 95:S17–S23

Marani E, Lakke EAJF (2012) Peripheral nervous system topics. In: Mai JK, Paxinos G (eds) The human nervous system 3th ed. AP/Elsevier, Amsterdam, pp 82–140

Marani E, Koch WFRM (2014) The Pelvis, structure, gender and society. Springer, Berlin, Heidelberg

Marani E, Schoen JHR (2005) A reappraisal of the ascending systems in Man, with emphasis on the medial lemniscus. Adv Anat Embryol Cell Biol 179:1–76

Moskowitz MA (2008) defining apathway to discovery from bench to bedside: the trigemino-vascular system and sensitisation. Headache 48:688–690

Munn LL (2015) Mechanobiology of lymphatic contractions. Semin Cell Dev Biol 38:67–74

Nishizuka T (1934) Ueber den Plexus accessoriocervicalis des menschen. In: Untersuschungen ueber das periphere Nervensystem. Heft 4, Serie A. Arbeiten Anatomischen Institutes der Kaiserlichen Universität Kyoto, pp 6–13

Obermann M (2010) Treatment options in trigeminal neuralgia. Ther Adv Neurol Disord 3:107–115

Oliver G (2004) Lymphatic vascular development. Nat Rev Immunol 4:35–45

Olry R, Haines E (2002) Ansa hypoglossi or ansa cervicalis? That is the question… J History Neurosci 11:302–304

Pandit JJ, Dutta D, Morris JF (2003) Spread of injectate with superficial cervical plexus block in humans: an anatomical study. Brit J Anaesth 91:733–735

Perlemuter L, Waligora J (1985) Tête et Cou. Cahiers d'Anatomie 7/8 3th ed. Masson Paris

Perloff JK (2011) The cardiac malpositions. Am J Cardiol 108:1352–1361

Phan TG, Beare RJ, Jolley D, Das G, Ren M, Wong K, Chong W, Sinnott MD, Hilton JE, Srikanth V (2012) Carotid artery anatomy and geometry as risk factors for carotid atherosclerotic disease. Stroke 43:1596–1601

Van Praagh R (2016) What determines whether the great arteries are normally or abnormally related. Am J Cardiol 118:1390–1398

Putten van der SCJ (1975) The development of the lymphatic system. Adv Anat Embryol Cell Biol 51:1–65

Rocca J (2014) Galen. Encycl Neurol Sci 392–393

Sabin FR (1902) On the origin of the lymphatic system from the veins, and the development of the lymph hearts and thoracic duct in the pig. Am J Anat 1:367–389

Sade RM (2005) Transplantation at 100 years: Alexis Carrel, pioneer surgeon. Ann Thorac Surg 80:2415–2418

Safian RD (2017) Carotid artery revascularization. The known knowns and the known unknowns. J Amer Coll Cardiol 69:2276–2278

Sakellarios AI, Räber L, Bourantas CV, Exarchos TP, Athanasiou LS, Pelosi G, Koskinas KC, Parodi O, Naka KK, Michalis LK, Serruys PW, Garcia-Garcia HM, Windecker S, Fotiadis DI (2017) Prediction of atherosclerotic plaque development in an *in vivo* coronary arterial segment based on a multilevel modeling approach. IEEE Trans Biomed Eng 64:1721–1730

Samandari F, Mai JK (1995) Funktionelle Anatomie für Zahnmediziner. Band I. Quintessenz Verlag Berlin

Schallert EK, Danton GH, Kardon R, Young DA (2013) Describing congenital heart disease by using Three-Part Segmental notation. www.rsna.org/education/search/RG

Schulte-Merker S, Sabine A, Petrova TV (2011) Lymphatic vascular morphogenesis in development, physiology, and disease. J Cell Biol 193:607–618

Schulze M, Wree A, Seidel R (2016) Anatomical study of ultrasound-guided intermediate cervical plexus block in regional anaesthesia. 111th annual meeting Anatomische Gesellschaft, Göttingen

Semo J, Nicenboim J, Yaniv K (2016) Development of the lymphatic system: new questions and paradigms. Development 143:924–935

Sleeman J, Schmid A, Thiele W (2009) Tumor lymphatics. Semin Cancer Biol 19:285–297

Suton P, Lukšić I, Müller D, Virag M (2012) Lymphatic drainage patterns of head and neck cutaneous melanoma: does primary melanoma site correlate with anatomic distribution of pathologically involved lymph nodes? Int J Oral Maxillofac Surg 41:413–420

Tabas I, Garcia-Cadeña G, Owens GK (2015) Recent insights into the cellular biology of atherosclerosis. J Cell Biol 209:13–22

Tubbs RS, Walker AM, Demerdash A, Matusz P, Loukas M, Cohen-Gadol AA (2015) Skull base connections between the middle meningeal and internal carotid arteries. Childs Nerv Syst. https://doi.org/10.1007/s00381-015-2763-x

Usunoff KG, Marani E, Schoen JHR (1997) The trigeminal system in Man. Adv Anat Embryol Cell Biol 136:1–126

Valnet J (1983) Phytothérapie: Traitement des maladies par les plantes. Le livre de poche pratique. Maloine S.A, Paris

Veith FJ (1998) Turf issues: How do we resolve them and optimize patient selection for intervention and ultimately patient care? J Vas Surg 28:370–372

Vittet D (2014) Lymphatic collecting vessel maturation and valve morphogenesis. Microvas Res 96:31–37

Voormolen JHC (1998) Treatment for trigeminal neuralgia. Arch Physiol Biochem 106:335–337

Waite PME, Ashwell KWS (2012) Trigeminal sensory system. In: Mai JK, Paxinos G (eds) The human nervous system. Elsevier/Academic Press, Amsterdam, pp 1110–1143

von der Weid P-Y, Zawieja DC (2004) Lymphatic smooth muscle: the motor unit of lymph drainage. Int J Biochem Cell Biol 36:1147–1153

Young JZ (1975) The life of mammals, their anatomy and physiology. Clarendon Press, Oxford

Zanatta A, Zampieri F, Rippa Bonati M, Frescura C, Scattolin G (2014) Situs inversus with dextrocardia in a mummy case. Cardiovasc Pathol 23:62–64

Chapter 15
Head and Neck Reflexes

The reflex is considered a simple automatic reaction mechanism of the nervous system. Research that started in the previous century showed its premotor circuitry to be far more extended than expected. Brainstem reticular formation plays an important role in the transfer from the afferent reflex leg towards the efferent reflex leg in head and neck reflexes. Moreover, the divergent areas provided for by head and neck reflexes extend the proper head and neck area. German research on the diencephalic influence on these reflexes has been restricted incorporated in the head and neck reflexes in the literature.

15.1 Introduction

The concept of the reflex dates from Descartes (1596–1650), who described in his "Treatise of Man" (1664) the reflex as an involuntary motor response, reactive to a peculiar and specific stimulus, which involves a "simple" spinal pathway. The spiritus animalis move from the central nervous system by the nerve tubes to the muscles for contraction. Descartes speaks of esprits réfléchis not of reflexes. We owe to Haller (1708–1777) the notion of two forces, one in muscles for contraction and one from outside the muscle transferred by the nerves and organized in the central part of the nervous system. Although serious disagreement between Haller and Whytt (1714–1766) existed, the concept of the stimulus arose and was accepted. Unzer (1727–1799) distinguishes afferent and efferent connections. Galvani (1737–1798) discovers that electricity can be guided by the nerves and postulates that in animals electricity is needed for the nerve to contract a muscle. The presence of ganglia in the afferent leg was well known (e.g. Willis, Haller and Unzer). Magendie (1783–1855) finally unites the literature results with his own experiments. He couples afferent as being *sensible*, and the central part or premotor neurons with the efferent *motoric* leg. Although Bell (1774–1842) claims it too, his fraud by reprinting an old text with unmentioned changes, so it looked like he

E. Marani and C. Heida, *Head and Neck*,
https://doi.org/10.1007/978-3-319-92105-1_15

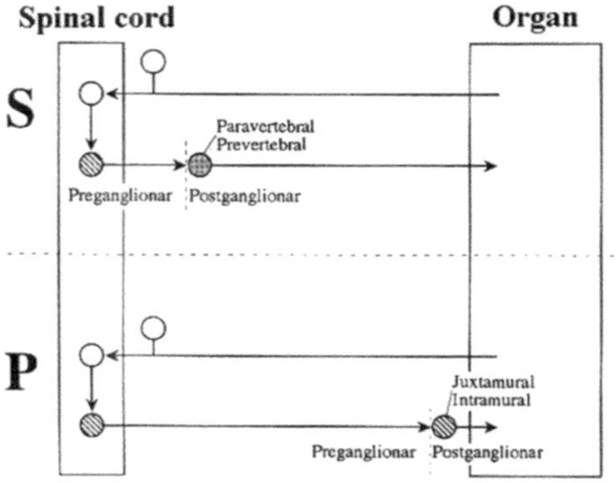

Fig. 15.1 Autonomic system. General scheme of the sympathetic (S) and parasympathetic (P) system together with neurotransmitters and anatomical nomenclature © E. Marani

described the reflex earlier, is well known. Hall (1790–1857) explicitly describes the afferent leg, spinal cord and efferent leg and their interconnections (Brunia 2015). Thus, around 1850, nearly 200 years after Descartes, the description of the reflex has found its completion.

Note that we distinguish somatic and autonomic reflexes, whose bases can be different. Somatic reflexes concern striated muscles. Autonomic reflexes (Fig. 15.1) involve smooth muscles and glands. A reflex can be an open (cornea reflex) or closed (knee tendon reflex) loop. It starts at the afferent side by muscle, gland or skin receptors. Tickling is noticed at the periphery of the autonomic and somatic system and transferred by the sensory neuron to the central nervous system, called the afferent leg of the reflex (by the way, this part of the loop is hardly shown as being present in the autonomic part of the nervous system). Here, in the central nervous system (spinal cord or brainstem), action potentials will be relayed directly or indirectly to a motor neuron, which causes responses in, e.g. muscle or gland, called the efferent leg of the reflex. So, the efferent leg of the reflex loop activates one (or more) effector(s). The somatic system uses a direct connection from the motoneuron to the effector. The autonomic system uses a two-step system. The autonomous motoneuron in the central nervous system transfers its message to a second neuron. These neurons are situated outside the central nervous system and can still be influenced before the message reaches the effector. A reflex always minimally concerns a sensory or sensible neuron in the afferent leg and one somatomotor neuron or two in both the autonomic parasympathetic and sympathetic system in its efferent leg (Fig. 15.1).

15.2 The Somatomotoneuron Recruitment

Special attention is paid to the somatomotoneuron and its recruitment, which is by its size principle. Motor units, the motoneuron and its connected muscle fibres, are recruited and the more units are recruited the more force is executed by the muscle. This motoneuron in both the spinal cord and the brainstem organizes the connection with the striated muscles and is responsible for the accurate activation of the striated muscles. Moreover, the somatomotoneuron can be steered by higher centres (cortex or subcortical centres) and is part of the reflex loop. Somatomotoneurons are localized in the anterior horn. In the myotatic reflex, the motoneurons receive input from the Golgi tendon organs and from the nuclear bag fibre and nuclear chain fibre. Here we restrict ourselves to the nuclear bag fibres: Ia fibres = A alpha fibres. The myotatic reflex is exemplified for the knee muscles and the same holds for other muscles in the extremities.

Sherrington laid the base for the study of reflexes. He could measure the reflex by manipulating the afferent leg and measuring the efferent leg outcome. He found that in the motoneuron pool facilitation or occlusion is present (Fig. 15.2a). There exists a discharge zone and a subliminal fringe zone. Each afferent fibre ended on a localized group of cells in the motoneuron pool. Each incoming neurotransmission gives an excitatory or inhibitory postsynaptic potential (EPSP or IPSP, respectively) Depending on the background excitation reaching the pool, each new input could result in a separate discharge zone with a surrounding subliminal fringe of cells that were excited but not discharged. Overlapping of subliminal fringes leading to a summation of PSPs results in firing in a zone of convergence. Overlapping of two discharge zones could also result in occlusion. "Unfortunately, the organization of input that Sherrington inferred is not consistent with a number of recent observations on recruitment of motor neurons and is contradicted by direct evidence of how afferent terminals of group Ia fibres are distributed" (Henneman 1974).

The place of the incoming PSP determines the amplitude of the PSP. This is schematically given in the model of Fig. 15.2b. Each incoming Ia = Aα gives an EPSP, but note that (1) each incoming Ia fibre from a muscle will project on all motoneurons belonging to that muscle and (2) the Ia fibre will descend in planes perpendicular to the dendritic trees (Fig. 15.2c–e).

Each dendritic tree gets all Ia fibres of that muscle. Each Ia fibre synapses only once on one dendrite of each alpha motoneuron involved, via a collateral. Whether small or large the density of synaptic terminals is suggested to be in proportion to their surface area, in such a way that each motoneuron has an equal density of Ia endings. Thus, the total amount of EPSP distributed over dendrites is equal per surface area (Fig. 15.2e). So recruitment of motoneurons cannot depend on the input, because preferential distribution of Ia fibres is ruled out. What then makes recruitment possible? The solution is in the motor neuron membrane showing the following properties: (1) the cell membrane contains resistance (the channels), and they are organized in parallel, (2) a large cell contains more parallel resistances than a small cell; thus, the total resistance R is smaller in large cells. (3) According to Ohm's law, the electric potential V over a resistance is equal to the current I through the resistance times the resistance: V = I.R. To generate an action potential, a constant difference towards a certain threshold must be overcome: V = c for all motoneurons independent of its size. This means that a smaller neuron needs less current to reach the threshold. Therefore, by a low level of activity of Ia fibres,

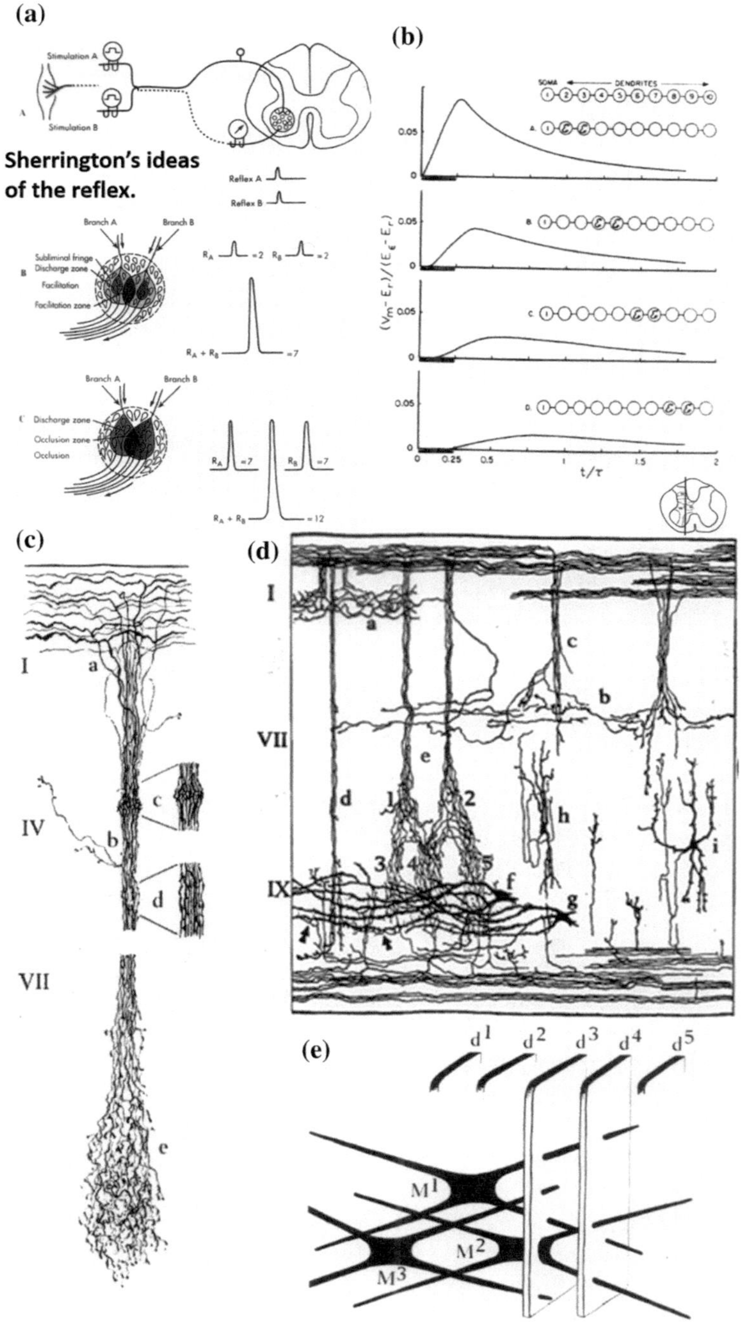
(a)
Stimulation A
Stimulation B
Sherrington's ideas
of the reflex.
Reflex A
Reflex B
Branch A
Branch B
Subliminal fringe
Discharge zone
Facilitation
Facilitation zone
Discharge zone
Occlusion zone
Occlusion
(b)
SOMA
DENDRITES
t/τ
(c)
I
IV
VII
(d)
I
VII
IX
(e)
M1
M2
M3

◄**Fig. 15.2** **a** Sherrington's sublimal fringe and occlusion zone. **b** Model to show EPSP effects as a function of time and depending on the time constant τ of the fibre and dendritic position, **c** incoming primary afferent fibre collaterals into a single microbundle, a coalesces of collaterals of various sizes, b "plumbers joint" formation, c intrafascicular specializations, d terminal neuropil field, e its compression into the rostrocaudal dimension. **d** Sagittal section through the spinal cord showing that various motoneurons receive several incoming Ia fibres: a longitudinal neuropil of dorsal horn, b of the intermediate nucleus, e microbundles of long (short c) primary afferent collaterals, f and g motoneuron somata with overlapping sagittal-running dendrites. **e** Model to show that several bundles produce a sagittal terminal domain (d1–d5). The longitudinally running dendrites of motoneurons (M1–M3) pierce these fields in varying amounts, dependent on the position of each apparent cell body. (Figures are from Henneman 1974 [**a**, **b**]; Scheibel and Scheibel 1969 [**c**, **d**, **e**] with permission; Rall 1964 [**b**])

small neurons are recruited, at higher levels of activity large motoneurons are activated, and at inhibition, first large cells stop firing and at higher inhibition the small neurons (see Henneman 1974 for a short overview).

15.3 Monosynaptic Reflexes of Head and Neck

The best-known monosynaptic reflex in the head and neck area is the jaw jerk reflex. A tap on the lower jaw with a slightly opened mouth closes the mouth and opens it again involving masseter and temporalis muscle. The afferent leg passes via the mandibular nerve of the trigeminal complex and the efferent leg uses the same nerve. The incoming fibres reach the princeps nucleus, but by its collaterals to the trigeminal motor nucleus the monosynaptic reflex loop is organized (see Sect. 12.6). At the moment that you eat and bite in your currant bun, you cannot stop the action, even if a small stone is present. Often you damage the grinder or other teeth.

The hiccoughing (hiccup) reflex, a brief strong inspiratory movement accompanied by glottis adduction, has been considered a monosynaptic reflex that passes over the phrenic nerve with both its afferent and efferent legs. Its central part is still under discussion. The phrenic nerve originates according to Bolk (1910) or from C3 and C4, or from C4 and C5. Lanz and Wachsmuth (1955) give C4 the main contribution with additions of C3 and C5. "The C4 segment is almost indispensable because it was present in all except one case (99.5%)" (Banneheka 2008; see also Marani and Lakke 2012). The reflex has been extended towards the medulla of the brainstem due to research in cats (Kondoa et al. 2003; Oshima et al. 1998; Arita et al. 1994). "Hiccup could be due to a referred gastrointestinal reflex elicited by abdominal inputs. The afferent pathway consists of the phrenic nerve, vagal nerve, cervical plexus (C2–C4) and sympathetic fibres of the lower thoracic segments (Th6–Thl2), and the efferent pathway consists of the motor part of the phrenic nerve (diaphragm), vagus nerve (musculature of the larynx and oesophagus) and sympathetic fibres of the cervical (C5–C7) and thoracic segments (Thl–Thll) innervating the auxiliary respiratory musculature" (Fetter and Kennard 2003).

Numerous neural elements organize a well-coordinated contraction at supraspinal level with a polysynaptic control. The hiccup-evoking site is localized somewhat above the ambiguus nucleus", in the medullary reticular formation lateral to the nucleus ambiguus on the right side" (Kondoa et al. 2003; Oshima et al. 1998; Arita et al. 1994; for the reticular formation subdivision see Fig. 8.11). Clinical experience supports its location in the medulla oblongata (Fetter and Kennard 2003). Native medicine has plenty of tips to control hiccups: drinking water, stop respiration, blow a balloon. A small, half-page article by Kumar uses the "Gag-reflex" (pharyngeal reflex): just reach out digitally to the base of the tongue and depress it slightly, and even intractable hiccups disappears" (Kumar 2005). Fight the reflex with a reflex!

15.4 Polysynaptic Vagus Reflexes

In the previous section, we saw that the reflexes present in head and neck show a strong divergence for the area to be effected, from head to diaphragm. Several polysynaptic reflexes are treated now, starting with baroreceptor reflexes present in the carotid body at the carotid sinus, which also show a rather large topographical connection distribution, as a consequence of the development of the heart. Baroreceptors are stress-sensitive, also called pressor receptors, and are present in the high- and low-pressure parts of the circulation. The arterial circulation has its stress receptors in the carotid sinus body and the aortic arch (Armour and Ardell 2004). Two side branches of the glossopharyngeal and of the vagus nerve, the sinus nerve and the aortic depressor nerve, respectively, carry information into the brainstem towards cardio-regulatory centres of the medulla oblongata. The signals are relayed by myelinated and unmyelinated fibres. Within the medulla, the information reaches the solitary tract. Connections from the solitary tract to the ventrolateral superficial reticular formation influence the so-called C1 adrenergic cell group.

The C1 adrenergic cell group reaches the thoracic intermediolateral cell column. These preganglionic fibres induce vasoconstriction. The solitary tract neurons do influence the dorsal vagus nucleus, which is responsible for the slowing down of the heart rate. The A5 noradrenergic cell group is also reached by the neurons of the ventrolateral superficial reticular formation. The A5 neurons do have an inhibitory effect on the thoracic intermediolateral cell column, inducing vasoconstriction (see Nieuwenhuys et al. 2008). Thus, in the central nervous system, aminergic transmission plays an important role in the control of the circulation. The sympathetic relations are by descending spinal connections that reaches the intermediolateral nucleus, which is connected by transverse strands with its intermediomedial nucleus around T3 level to reach the heart via the sympathetic trunk (Fig. 15.3).

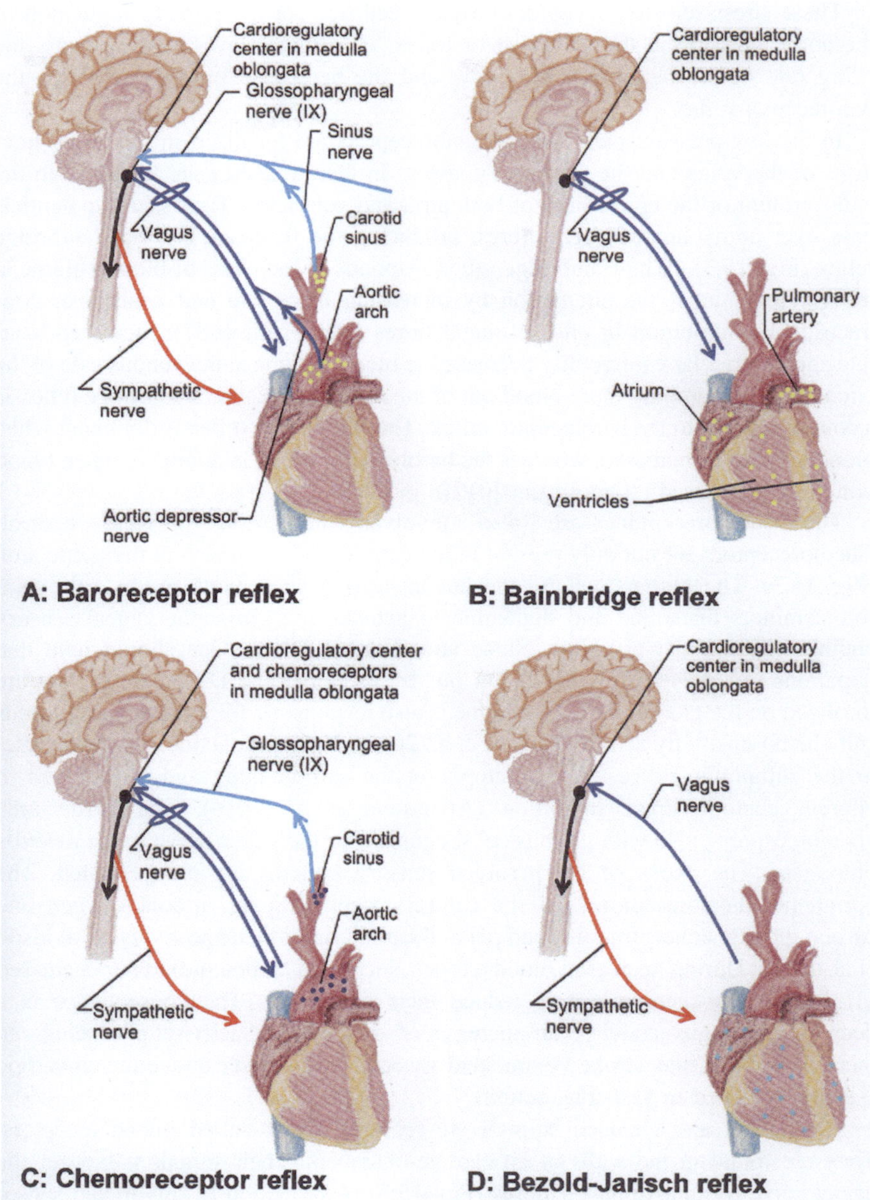

Fig. 15.3 **a** Connection diagrams of the reflexive control in the baroreceptor reflex with its feedback signals from cardiovascular receptors, **b** the Bainbridge reflex also called atrial reflex, **c** the chemoreceptor reflex and (**d**), the Bezold–Jarisch reflex (courtesy and with permission Ordelman 2012)

These stress-sensitive receptors provide feedback for the primary regulation of the aortic pressure in the baroreceptor reflex. At a rather low blood pressure, the firing rate of the baroreceptors is low and the heart rate increases through the baroreceptor reflex.

In the low-pressure part, the mechanoreceptors are localized in the atrial junctions of the veins and the pulmonary artery. In the atria, they are localized in the endocardium or the epicardium of both atria and ventricles. Their action potentials pass over unmyelinated vagus fibres, producing the feedback for the Bainbridge reflex (Fig. 15.3). "The Bainbridge reflex responds to increases of blood volume at the venous side of the circulation by increasing heart rate and ventricular contractility via inhibition of efferent vagal fibres (Hakumaki 1987). Increased heart rate and ventricular contractility decrease the blood volume at the venous side of the circulation by pumping more blood out of the right atrium. The Bainbridge reflex is a counterbalance to the baroreceptor reflex. The Bainbridge reflex is dominant when blood volume is increased, whereas the baroreceptor reflex is dominant when blood volume is decreased" (Ordelmans 2012).

The chemoreceptors are also involved in blood pressure control. Chemoreceptors are not only present in the carotid body, but also in the aortic arch (Fig. 15.3). The glomus cells in the human carotid sinus body produce the neurotransmitters histamine and dopamine to activate the glossopharyngeal sensory endings (Lazarov et al. 2009; Nurse and Piskuric 2013). One should note that dopamine not only causes excitation but by its dopamine D2 receptors that are localized on the glomus cells, dopamine is also responsible for inhibition of glomus cell chemosensitivity activity (Wakai et al. 2015). NO (nitric oxide) is a transmitter in the autonomic nerve endings supplying the carotid body and is involved in efferent chemoreceptor inhibition (Atanasova et al. 2016). The aortic arch chemoreceptors, also with glomus cells, connect via the vagus nerve to the medulla oblongata. The axons of the involved sensory systems are unmyelinated. The peripheral chemoreceptors and the central receptors sense arterial oxygen and carbon dioxide concentrations, and pH and carbon dioxide, respectively, (Davidson et al. 1976). During decreased blood oxygen, increased carbon dioxide or decreased pH levels, the chemoreceptors reduce their firing rate. The consequence is a decrease of vagus activity, an increase of sympathetic activity producing, an increase in heart rate, stroke volume and vasoconstriction (see above the aminergic systems involved in vasoconstriction).

Mechanical and chemical stimuli are received by so-called mixed receptors. They are found in the walls of all cardiac chambers. Their signals pass over the vagus nerve by unmyelinated fibres (Fig. 15.3). A reduction of this mixed sensor information results in an increase of the sympathetic system by the Bezold–Jarisch reflex and blood pressure and heart rate are increased. The Bezold–Jarisch reflex is complementary to the baroreceptor reflex (see Campagna and Carter 2003).

Resistant hypertension can be relieved by unilateral carotid body resection. According to Narkiewicz et al. (2016), it lowers the blood pressure with 26 mmHg in 57% of the patients operated upon, the other 43% were insensitive to the treatment.

15.5 Polysynaptic Eye Reflexes

The accommodation reflex organizes the focussing of the object in the visual field. The retina projects to the lateral geniculate body that projects on the visual cortex by the optic radiation. The cortex projects to the pretectal olivary nucleus that informs the Edinger–Westphal nucleus to effect the ciliary ganglion to steer the ciliary muscle for adequate focussing (Fig. 15.4).

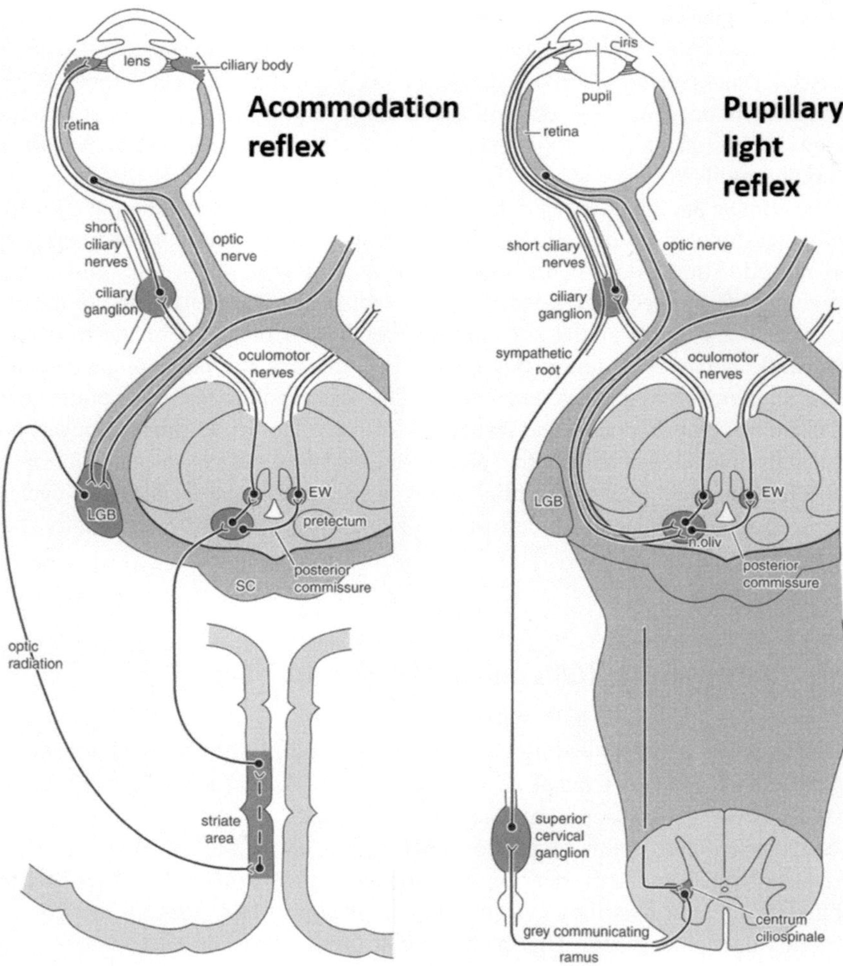

Fig. 15.4 Accommodation reflex and pupillary light reflex, note that only the temporal connections are shown. EW, nucleus of Edinger–Westphal; LGB, lateral geniculate body; n. oliv, pretectal olivary nucleus (courtesy and permission J. Voogd from Nieuwenhuys et al. 2008)

If a small bundle of light is shone in one eye, in both eyes the pupil will contract. It is called the pupillary light reflex. The retina of one eye send its light information into occipital cortices, but by passing the mesencephalon collaterals end into both the pretectal olivary nuclei, which are connected via the posterior commissure to the Edinger−Westphal nuclei (note medial, nasal field and lateral, temporal fields of the retina of the same eye project to left and right hemispheres and brainstem areas, respectively). Both the pretectal olivary nuclei projects to the Edinger–Westphal nucleus of both sides and via the oculomotor nerve the ciliary ganglion gives activation of the sphincter pupillae muscle, causing contraction of both pupils (Fig. 15.4; for an extended overview see McDougal and Gamlin 2015).

The pupillary light reflex is mostly used in the clinical setting, e.g. during general anaesthesia. It stays present if somebody is unconscious. Pupil dilatation and or constriction (mydriasis and myosis, respectively) indicates the depth of unconsciousness. If the pupillary reflex cannot be elicited, serious damage is present in the brainstem. This reflex is also used by the police, because drug abuse inhibits dilatation of the pupils and they stay in the constricted state, known as "pinpoint pupils".

The corneal and blink reflex induces closure of the eye by touching the cornea or analogous stimulations like strong bright light, while heavy emotions seemingly are also related. Stimulation of the supraorbital area also produces this blink reflex. Perceiving of an object approaching the eye swiftly also induces closing of the eye. The nasociliary nerve or the supraorbital nerve, both branches of the trigeminal ophthalmic nerve (V1), relays this information towards the princeps nucleus and descending branch of the trigeminal nerve. The trigeminal interneurons project onto the facial nucleus producing the ipsilateral R1 response, while ruber impulses and cortical impulses reach the pontine and medullary blink centres responsible for the bilateral component (R2) of the blink reflex. The axons of the facial motor nucleus contracts bilaterally the orbicularis oculi muscles of both eyes briefly (see Nieuwenhuys et al. 2008).

15.6 Polysynaptic Reflexes of Mouth and Nose

Sneezing reflex and swallowing reflex are treated here. The lacrimation reflex is described in Chap. 12: Cranial nerves (see Sect. 12.4.2.1 and in Figs. 12.17 and 12.18).

Stimulation of receptors by dust or other irritants in the mucous covering of the nose (see Chap. 7) induces sneezing and produces somatosensory impulses over the ophthalmic (V1) or maxillary (V2) trigeminal divisions to the princeps nucleus and its ramus descendens part. The interneurons connect to ventral respiratory parts: ventral inspiration (rostral) and expiration (caudal) respiratory groups, which are located the near the nucleus ambiguous. Inspiratory neurons project to the cervical C3, **4**, 5 motoneurons of the phrenic nerve, while expiratory neurons that innervate abdominal muscles project to intermediate grey and motoneurons of thoracic levels

(T1–T12; see Nieuwenhuys et al. 2008). A deep inspiration is followed by building up pressure against a closed glottis that results into a heavy exhalation. The closure of the oropharyngeal isthmus (see Sect. 6.6.2 and 7.4.2) brings the stream through the nose. The coughing reflex is essentially similar to the sneezing reflex due to receptors in larynx and trachea without closing the oropharyngeal isthmus. One should note that the pacemaker for respiration (so-called Pre-Bötzinger cells and Bötzinger complex) has to adapt its effect during coughing and sneezing.

The swallowing reflex consists of three motor phases: the oral or first phase, during which mastication occurs and a second pharyngeal phase in which the food bolus is displaced due to pharyngeal muscle constriction; in the third or oesophageal phase the larynx is elevated and sealed. Swallowing is initiated by the food bolus present in the mouth. Receptors in this area activate the sensory receptors of the wall that activate the branches of the glossopharyngeal nerve. From the tractus solitaries, the incoming information is distributed towards the hypoglossal, facial and trigeminal motor neurons for the first oral phase. The premotor neurons are localized in the dorsomedial reticular formation. Projections from this reticular area reach the ambiguus nucleus together with solitary information innervating the pharyngeal muscles (tongue, mouth floor muscles, strap muscles, pharynx and larynx muscles) responsible for the second phase of swallowing. The oesophageal motoneurons are activated by the neurons of the solitary tract. Higher centres are involved in steering the swallowing process: the caudal pontine tegmentum (cat), the bilaterally in ventral M1, the ventral premotor area, inferior parietal lobule and the right anterior insula (human by PET studies). Cerebellum and striatum are also involved (Nieuwenhuys et al. 2008). The control of orofacial movements are widespread through the central nervous system. In case non-food-related stimuli are present in the oropharynx, the "gag" reflex induces a violent retching that can end up in vomiting.

15.7 W.R. Hess and Vegetative Functions

Nobel prize winner and professor of physiology at the University of Zürich, Walter Rudolf Hess (1881–1973) focussed his work on autonomic networks for circulation and heart rate, respiration, eye functions, especially pupil functions, and sleep (Hess 1949). Using implanted electrodes in the free-moving cat with stimulation of diencephalic structures he showed several autonomic functions in this brain area. In Sect. 4.7.1 and Fig. 4.10, attention was paid to the diencephalic relation between thalamus, subthalamus and hypothalamus during development. Note that the topographic relations stay nearly unchanged during maturity in the diencephalon, only the thalamus increases enormously. Hess's results on sleep were slowly accepted in the scientific world. The area of adynamy in the lateral hypothalamus is now considered related to the sleep–wake cycle (see Reinoso-Suárez et al. 2011). Moreover, Hess considered the "bodily reactions during sleep-manifestations of an active state and in contrast to anaesthesia not a passive state with clearly diminished ergotropic (sympathetic) functions" (Akert 1999), a rather modern vision of sleep. "Hess's work appeared almost exclusively in the German language. Hess's name therefore remained virtually unknown in the Anglo-Saxon literature for a long time. Even after the wars, one rarely finds Hess's original papers and monographs quoted" (Akert 1999).

Cat vegetative syndromes studied by Hess showed among others:

(1) Pupil constriction is thalamus related, and pupil dilatation is linked to hypothalamic activity. A rather sharp border is noticed between both areas.
(2) Blood pressure drop is rostro-lateral hypothalamic related, while blood pressure rise is related to activity in the posterior hypothalamus and around the diencephalic entrance of the aquaduct.
(3) Forced increase in respiration (paroxysmal tachypnoea) relates to the superior zone around the hypothalamic fornix and the lateral hypothalamus. A normal increase of respiration is related to the subthalamic region towards the posterior commissure. A decrease of respiration is anterior hypothalamic related.
(4) Panting is elicted in the hypothalamus, like salivation. Salivation was also activated within the nucleus ventralis thalami.
(5) Sniffing was activated within the lateral hypothalamus. Licking of the cat was present after stimulation of the thalamus.
(6) Defecation and micturition were activated after stimulation of the septum pellicidum and the hypothalamus. Thalamus stimulation never gave defaecation or micturition in the cat.

For several of the reflexes described above like pupil, circulation and sneezing reflexes and for those not described (vomiting, defensive responses, increased motor activity passing over to flight), diencephalic areas are present that are involved in these reflexes. They are hardly incorporated in their reflex functions in the literature as is done for Hess's results regarding sleep. Hypothalamic stimulation (pupil dilatation) indeed abolishes Edinger–Westphal activity for pupil constriction (Sillito and Zbrozyna 1970a, b). Thalamic haemorrhage induces hypertension in humans (Tokgoz et al. 2013). Accordingly, diencephalic damages are described for several other reflex functions of head and neck. Clinical representation of diencephalic deflections are well known, but restrictedly described in neuroanatomy in their relation to the head and neck reflexes.

References

Akert K (1999) Walter Rudolf Hess (1881–1973) and his contribution to neuroscience. J History Neurosci 8:248–263

Arita H, Oshima T, Kita I, Sakamoto M (1994) Generation of hiccup by electrical stimulation in medulla of cats. Neurosci Lett 175:67–70

Armour JA, Ardell JL (2004) Basic and clinical neurocardiology. Oxford University Press, USA

Atanasova DY, Dimitrov ND, Lazarov NE (2016) Expression of nitric oxide-containing structures in the rat carotid body. Acta Histochem 118:770–775

Banneheka S (2008) Morphological study of the ansa cervicalis and the phrenic nerve. Anat Sci Int 83:31–44

Bolk L (1910) De segmentale innervatie van romp en ledematen bij den mensch. Erven F, Bohn, Haarlem

Brunia K (2015) Het brein van farao tot fMRI. Uitgeverij Eburon, Delft, The Netherlands

Campagna JA, Carter C (2003) Clinical relevance of the Bezold-Jarisch reflex. Anesthesiol 98:1250–1260

Davidson NS, Goldner S, Mccloskey DI (1976) Respiratory modulation of baroreceptor and chemoreceptor reflexes affecting heartrate and cardiac vagal efferent nerve activity. J Physiol (Lond) 259:523–530

Fetter M, Kennard C (2003) Chapter 18: Hiccup, cranial nerves and brain stem neurological disorders: course and treatment, 2nd ed. Elsevier Science (USA)

Hakumaki MO (1987) Seventy years of the Bainbridge reflex. Acta Physiol Scand 130:177–185

Henneman E (1974) Principles governing distribution of sensory input to motor neurons. In: Schmitt FO, Worden FG (eds) The neurosciences third study program. MIT Press, Cambridge, MA

Hess WR (1949) Das Zwischenhirn. Syndrome, lokalisationen, Funktionen. B Schwabe & Co Verlag Basel

Kondoa T, Toyookac H, Aritaa H (2003) Hiccup reflex is mediated by pharyngeal branch of glossopharyngeal nerve in cats. Neurosci Res 47:317–321

Kumar A (2005) Gag reflex for arrest of hiccups. https://doi.org/10.1016/j.mehy.2005.08.003

Lanz and Wachsmuth (1938, 1955, 1972, 2004) Praktische Anatomie: Kopf-uebergeordnete Systemen, Kopf-Gehirn und Augenschädel, Hals, Arm, Bein und Statik, Bauch, Rücken. Springer Verlag, Berlin, Heidelberg, NY

Lazarov NE, Reindl S, Fischer S, Gratzl M (2009) Histaminergic and dopaminergic traits in the human carotid body. Resp Physiol Neuro boil 165:131–136

Marani E, Lakke EAJF (2012) Peripheral nervous system topics. In: Mai JK and Paxinos G (eds) The human nervous system. Elsevier, pp 82–140. ISBN 9780123742360

McDougal DH, Gamlin PD (2015) Autonomic control of the eye. Compr Physiol 5:439–473. https://doi.org/10.1002/cphy.c140014

Narkiewicz K, Ratcliffe LEK, Hart EC, Briant LJB, Chrostowska M, Wolf J, Szyndler A, Hering D, Abdala AP, Manghat N, Burchell AE, Durant C, Lobo MD, Sobotka PA, Patel NK, Leiter JC, Engelman ZJ, Nightingale AK, Paton JFR (2016) Unilateral carotis body resection in resistant hypertension. Saf Feasibility Trial. https://doi.org/10.1016/j.jacbts.2016.06.004

Nieuwenhuys R, Voogd J, van Huijzen C (2008) The human central nervous system. Springer, Berlin, Heidelberg

Nurse CA, Piskuric NA (2013) Signal processing at mammalian carotid body chemoreceptors. Semin Cell Dev Biol 24:22–30

Ordelman S (2012) Sensing and stimulation of the vagus nerve for artificial cardiac control. Thesis University Twente

Oshima T, Sakamoto M, Tatsuta H, Arita H (1998) GABAergic inhibition of hiccup-like reflex induced by electrical stimulation in medulla of cats. Neurosci Res 30:287–293

Rall W (1964) Theoretical significance of dendritic trees for neuronal input-output relations. In: Reisa RF (ed) Neural theory and modeling. Stanford University Press, pp 73–79

Reinoso-Suárez F, de Andrés I, Garzón M (2011) Functional anatomy of the sleep-wake-fulness cycle: Wakefulness. Adv Anat Embryol Cell Biol 208:1–130

Scheibel ME, Scheibel AB (1969) terminal patterns in cat spinal cord. III. Primary afferent collaterals. Brain Res 13:417–443

Sillito AM, Zbrozyna AW (1970) The activity characteristics of the preganglionic pupilloconstrictor neurones. J Physiol 211:767–779

Sillito AM, Zbrożyna AW (1970) The localization of pupilloconstrictor function within the mid-brain of the cat. J Physiol 211:461–477

Tokgoz S, Demirkaya S, Bek S, Kasikci T, Odabasi Z, Genc G, Yucel M (2013) Clinical properties of regional thalamic hemorrhages. J Stroke Cerebrovas Dis 22:1006–1012

Wakai J, Takayama A, Yokoyama T, Nakamuta N, Kusakabe T, Yamamoto Y (2015) Immunohistochemical localization of dopamine D2 receptor in the rat carotid body. Acta Histochim 117:784–789

Chapter 16
Head and Neck During Puberty and Ageing

The bones of the skull can be considered as produced by an extended lamina over the outer surface of the cranial table, bone being resorbed at the inner surface of the table. This is a method of increase that would hardly be thought of as practicable for a human artefact- to increase the size of a box by thickening the wall and removal of material from the inside!
(Young 1974)

The bony and soft tissue during facial development is prepubertaly determined. At 6 years of age, 90% of the total growth has passed. Attention is paid to the development of the groove between neuro and viscerocranium, the pterygopalatine fossa and fissure. Bodily fat and facial fat behave differently during puberty and ageing. The relation between superficial muculo-aponeurotic system (SMAS) and facial subcutaneous fat is considered. Thermogenetic brown fat is localized in the neck, and its leptin and uncoupling protein 1 relations are discussed. Bodily bone mineral content can change during puberty and ageing. Skull mineral content behaves differently, and osteoporosis is low in the cranium. Ageing of the face is primarily expressed in subcutaneous fat reduction in head and neck. The reduction of the "nervus accessory fat" (NAF) in the supraclavicular triangle is considered.

16.1 Main Facial Development Is Pre-Pubertal

The human face is characterized by diminution during evolution. Modern *Homo sapiens* have a shorter anterior–posterior and shorter super-inferiorly face compared to archaic *Homo*. The reduction can reach 15–25 per cent. Causes are unknown but speculations contain smaller embryologic condensation of mesenchyme in skull structures and growth reduction by diminished reaction on hormones and growth factors. Higher resorption of bone is another suggested possibility (see Lieberman 2011). In short, increase of the neurocranium coincides with decrease of the viscerocranium with consequences for the midface construct (see Sect. 7.4.1). The

E. Marani and C. Heida, *Head and Neck*,
https://doi.org/10.1007/978-3-319-92105-1_16

face is not a fixed structure during evolution nor individually either. It will adapt to different situations (see Sect. 7.1 facial attractiveness and physiognomy).

Most people think that our face is chiefly shaped during puberty. "Puberty can be defined as a maturational process of the hypothalamic–pituitary–gonadal axis, resulting in growth and development of the genital organs and concomitantly physical and psychological changes towards adulthood leading to the capacity to reproduce. The development of the sex characteristics encompasses the development of vulva, uterus, vagina epithelium (secondary), breasts, sexual hair and fat composition (tertiary) in girls and enlargement of penis, scrotum, prostate, seminal vesicles (secondary), growth of facial and sexual hair and vocal cords (tertiary) in boys" (Delemarre van de Waal 1984). Puberty is from 9 to 12/13 years and adolescence from 12/13 to 16/17 years (note that racial, individual and gender ranges are large). The growth and increase of weight of the reproductive organs in man start no earlier than 10 years of age (autopsy data Tanner 1962; Delemarre van de Waal 1984). Figure 16.1 shows four colour pictures of the face development of a girl. The constitution of the face at the age of 9 years already shows the contours of the mature face.

Head circumference (i.e. neurocranium circumference) at one year of age is 47 cm, at 9 years 53 cm and at 20 years 57 cm for Dutch males. The figures for Dutch girls are 46, 52 and 55.5 cm, respectively (TNO groeidiagrammen, see also Mellinger 1940). If size is attained in per cent of total growth, the head and brain will have reached, at the age of 9 years, 95% of the total growth. Its main growth is from birth until 6 years and attained 90% of its total increase at this age (Fig. 16.1; Tanner 1962). Lanz and Wachsmuth (1985) also inform us that the total skull length of the 6-year-old child has been increased by 75% compared to the newborn. *Thus, before puberty the constitution of the head has been organized and is seemingly less puberty dependent.*

During the post-natal development, the mandible increases markedly (Fig. 16.1) and the sulcus nasolabialis deepens. Facial width-to-height ratio (FWHR), defined as the width of the face divided by the upper facial height, has been used to prove that sexual dimorphism is also present in the skull. It is not (Kramer 2017). This sexual skull dimorphism has been thought to relate the FWHR to certain types of behaviour, which is weakly based and looks like it revives physiognomy.

One important aspect concerns the pterygomaxillar fissure and its continuation into the pterygopalatine fossa. Since the neurocranium (sphenoid) increases its volume and the viscerocranium (maxilla) lengthens the groove between both, which forms this fissure and fossa, it also lengthens and increases in height. Maxilla, palatinum and from the sphenoid, the medial and lateral lamina of the processus pterygoideus grow together but leave a canal known as the canalis pterygopalatinus, which opens within the fossa pterygopalatinus (see Sect. 6.1 and Fig. 6.2). One should note that the maxillary artery is located within the fossa and provides the nasal cavity, hard and soft palate, lacrimal gland, some eye muscles and teeth with their gingiva with blood. Also located in this fossa is the pterygopalatine ganglion that has a parasympathetic secreto-motor function and is related to the greater petrosal nerve of the facial cranial nerve. In summary, the maxillary nerve that enters

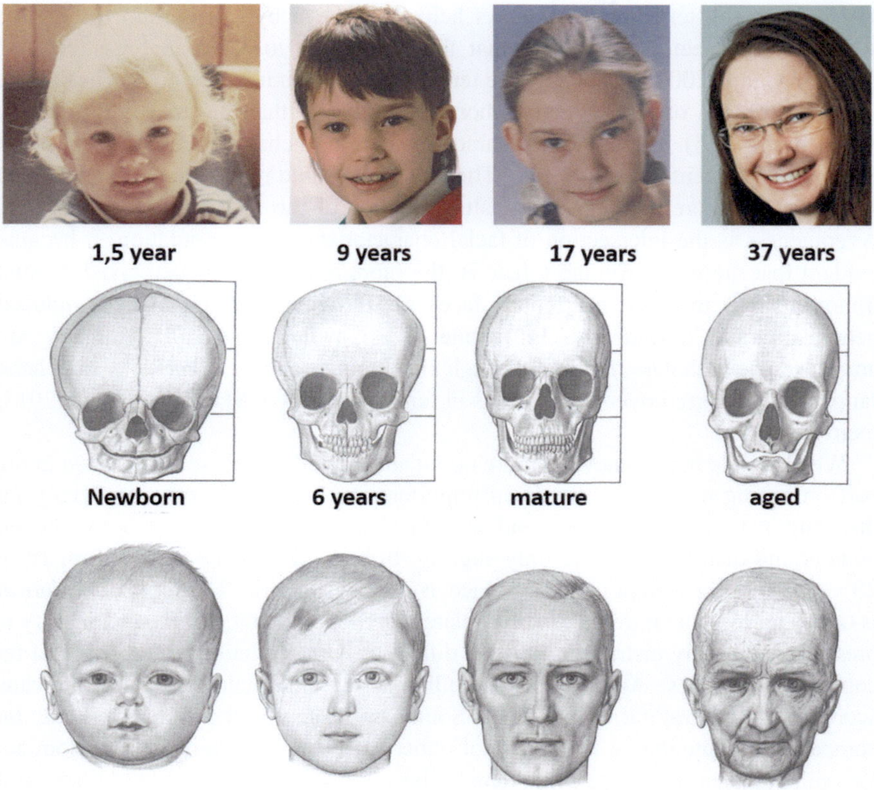

Fig. 16.1 Upper series: Individual female face development from 1.5 years until maturity. Note that the increase of the visceral skull at the age of 9 years has already been attained (with permission). Skull development from newborn to old age. Frontal view of skull and face (adapted from Lanz and Wachsmuth 1985)

the fossa via the foramen rotundum leaves it by the infraorbital fissure, giving off seven branches in the meantime and contacts the pterygopalatine ganglion via the pterygopalatine nerves. Increase of the length and height of the pterygopalatine fossa influences postnatally all these structures.

16.2 Bodily Fat and Facial Fat

While the newborn is characterized by its strongly curved cheeks due to buccal fat pads, these fat pads disappear early during face development (Fig. 16.1). Some of the explanations for the fat pads are factual and others amusing, respectively: "the distribution of atmospheric pressure and the prevention of the drawing in of the cheek and buccinator muscle between the gums during the act of suckling" (Crelin

1973) or "a full-term infant's cheeks help keep the baby's tongue in the middle when sucking seems to indicate that the purpose is to assist in breastfeeding" (Watson Genna 2008). Especially the female face gets an increase of the cheek bone around the age of 10/11, even more pronounced than in males (Lanz and Wachsmuth 1985). The feminine characteristic of high cheekbones is attractive and is race-independent (see Sect. 7.1). There are noticeable marks that influence mate selection. They are, among others, attractiveness of the face and fat distribution. Averageness is the intersection of facial characteristics of a population. It became evident that the more average a face is, the more attractive it is perceived. Sexual dimorphism is the fact that female faces are different from male ones, induced genetically and hormonally. In female faces, averageness and symmetry are attractive. Nevertheless, femininity, high cheek bones and full cheeks by the cheek fat pads, is preferred over averageness in female faces (see Marani and Koch 2014, their Chap. 11).

We elaborate here somewhat more on fat and fat pads, because it is needed in the part on ageing of the face and is an important issue in plastic surgery. Bodily fat distribution is different in men and women after puberty. Before puberty, the fat content and distribution differ only slightly. In boys, during the period from 10 to 20 years, ***free-fat*** weight increases twice as much as in girls. The ***fat accumulation*** is opposite. In general, females accumulate twice as much fat, but race variability is present. The bodily distribution is also different. Females build up less visceral fat compared to males. Again race variability is present: male African Americans accrue less visceral fat, while male Southeast Asians gather more visceral fat compared to white males. At the end of puberty, the male contains more abdominal fat, called the android type, while females have more adipose tissue at buttocks and femoral regions, called the gynoid type (Tschernof 2009). These contour differences play an important role in gender recognition. Moreover, the waist-to-hip ratio, determined by subcutaneous fat at buttocks and femoral regions, is also important for mate selection (Marani and Koch 2014).

The subcutaneous fat of face and neck needs a special notion of facial layers for its understanding. The face and neck can be subdivided into a skin layer, subcutaneous fat, the superficial musculo-aponeurotic layer or system (SMAS), deep spaces with deep fat, and finally the deep fasciae together with the periost layer. While in apes and fur animals, fibro-muscular tissue, called panniculus carnosus, separates superficial fat (panniculus adiposus) from the other deep layers, in man and pig this layer is clearly interrupted. In man, remnants are the platysma, the galea aponeurotica (=epicranial aponeurosis) on the vault (see also 7.1.5 and Fig. 7.10) and some ligamentous structures. The mimic muscles co-organize the SMAS. The facial superficial fat is organized in longitudinal or semicircular pouches (Fig. 16.2 left) with in between fine septa that can form the retaining ligaments. The fat content can vary, mainly near septa and skin, in quantity and quality. The cheek is also organized by two fat storages with, in between them, the masseter ligament. During ageing, this masseter septum stays present and is recognized as a groove.

Deep fat pouches are mainly present around and behind the eye lids and the cheeks. Suborbicularis oculi fat (SOOF), deep fat below and lateral of the

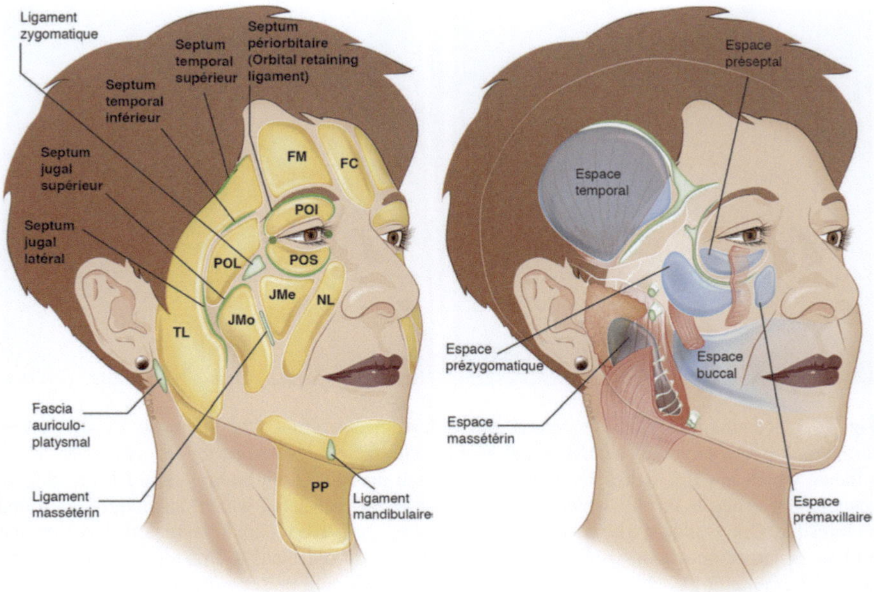

Fig. 16.2 We took over the remarkable French figures on facial fat and facial spaces without changing figure texts (figures are from Sinna et al. 2017, with permission). Left: superficial pouches that contain facial fat (yellow), together with their retaining ligaments or re-enforcements (green). FC, central frontal pouch; FM, middle frontal pouch; JMe, medial jugal pouch; JMo, middle jugal pouch; NL, nasolabial pouch; PL, lateral periorbital pouch; POI, inferior periorbital pouch; POS, superior periorbital pouch; PP, preplatysmal fat; TL, tempero-lateral pouch. Right: spaces in the face that can be approached surgically

orbicularis muscle, and retro-orbicularis oculi fat (ROOF), just extending above the eyelid over the eye brow, are discerned. Deep medial cheek fat is placed at the transition of the jugal to labial area, below the levator labii superioris and levator anguli oris. It is located next to the infraorbital foramen. The buccal fat pad (Fig. 16.3) stretches from over and under the temporalis muscle, below the masseter muscle and can peep out in between the masseter and buccinators muscles. In it are the parotid duct and the facial branch for the buccinator muscle (see Sinna et al. 2017; Niamtu 2010; Mertens et al. 2016; Gierloff et al. 2012).

The buccal fat pad is frequently operated upon (Matarasso 2006). Its position is such that a small incision into the cheek inside, a so-called mucosal incision, is often sufficient. Since the end of the parotid duct is easily recognized opposite the first molar, placing of the injection with local anaesthesia is easy. For other places for removing or grafting fat spaces to enter facial areas are known (see Fig. 16.2): the Merckel space located below the epicranial aponeurosis, named the temporal space with facial nerve branches in it, below the lower eye lid the preseptal space is recognized, with caudally the premaxillar (medial) and prezygoma (lateral) spaces. The buccal space has been treated in Fig. 16.3 describing the buccal fat pad.

Fig. 16.3 Rembrandt: Old man seated and old woman reading. "In his early Leyden years, Rembrandt made numerous studies of old people, with their faces marked by the years. Old age is a theme which has fascinated Rembrandt all through his life. Often he dressed up his models in costumes of brocade and fur which were not worn at that time, but which, apparently, he found picturesque. One wonders whether this painting, dating from 1651, should be regarded as a regular portrait or rather, in a more general way, as a study of an old man. In this old woman painting (1655), Rembrandt has chiefly expressed the reader's concentration: an old woman engrossed in her book. The source of the light illuminating her face is invisible, but is reflected by the pages of the book". Rembrandt painted the same light theme regularly in 1631, presumably his mother, 1647, portrait of Jan Six reading, 1656, Titus reading, and in 1658 twice, man with manuscript and portrait of Lieven van Coppenol. White arrow buccal fat pad, yellow arrow cheek fat pad, light blue arrows capaciously orbit and periorbital fat pad, blue arrow nasolabial fold. (Text and paintings adapted from catalogue Rembrandt 1669/1969, Rijksmuseum Amsterdam)

16.3 Neck Brown Fat

There are two types of fat, white fat and brown fat. The above treated facial fat is all white fat. Fat in the neck area is not only white fat, but the neck also contains brown fat. This brown fat, for example in bats, is directly related to thermal regulation of the bats body, which is needed in winter time during flight at cold temperatures. The bat's brown fat makes 20–25% of the total fat content and is located interscapular, jugular, squamo-occipito-cervical and in the carotid bodies (O'Farell and Schreiweis 1978). So, brown fat makes a minor portion of total fat, and we know its existence already from the 1960s of the previous age on. Humans do contain brown fat and its research reoccurred more recently, because brown fat (officially named brown adipose tissue) is thought to give clues in fighting obesity (Scheele and Nielsen 2017). Radiographic studies combined with anatomical ones show its human presence: interscapular, around the large neck blood vessels and neck muscles and also both trachea and oesophagus do contain brown fat (Okuyama et al. 2018; Bhatt et al. 2017; Avram et al. 2005). At least this brown fat tissue is reported at the nape of the neck in children and its content diminishes towards maturity (Okuyama et al. 2018; Bhatt et al. 2017). The brown colour originates from the high content of capillaries, a multiple of the capillaries present in the white adipose tissue. Cold activates brown fat. Brown fat is characterized by the uncoupling protein 1 (UCP1) for its non-shivering thermogenesis. This protein uncouples the proton gradient of the production of ATP with the consequence that energy normally stored as ATP now becomes available for heat production.

We start now with considering the puberty-related leptin hormone for white fat and afterwards return to brown fat, because they are interrelated. Leptin is a hormone secreted by adipose tissue that is involved in the signalling of fat metabolism towards the brain. In the beginning considered as an anti-obesity hormone, it later became clear that it is involved in "inflammation, angiogenesis, haematopoiesis, immune function, and most importantly, reproduction" (Mosgos et al. 2002). Leptin is not only produced in fat tissue, but also in the hypothalamus among other tissues. Leptin can be noticed by its receptors in the anterior pituitary, ovary and testis. Leptin production or its inhibition occurs by hormones: insulin increases leptin production as oestrogens do, while androgens suppress leptin production. Leptin increases gonadotropin-releasing hormone secretion and other hypothalamic prohormones and neurotransmitters, but also anterior pituitary hormones like LH and FSH. It is obvious that leptin is related to gonadal steering, due to its influence on gonadotropic hormones. "What is the data linking leptin concentrations and pubertal development? Several large studies in children have shown a strong correlation of serum leptin levels with adiposity as determined by body mass index. In general, girls have higher leptin levels than boys, although some investigators note that the concentrations are indistinguishable at the same relative fat mass. As pubertal development begins in boys, or just before its onset, there is a marked relative rise in leptin levels, whether reported cross-sectionally or longitudinally" (Rogol 1998). Leptin levels in boys have their highest peak just before puberty or at the start of puberty. Testosterone appearance coincides with a decrease in leptin levels towards baseline levels. In girls, a consequent rise of leptin levels occurs during puberty, and a direct relation between the leptin increase and menarche was established. This gender difference is called the sexual leptin dimorphism. "Leptin appears to be a necessary but not sufficient factor for the initiation of puberty in humans" (Mosgos et al. 2002). The interaction between leptin and the gonads is still unclear and is certainly multi-factual (part leptin text has been taken from Marani and Koch 2014, pp. 502–503).

Sexual dimorphism is also important in energy metabolism, and sex hormones are involved in brown fat thermoregulation in both males and females (Quarta et al. 2012). The addition of leptin results in effects on brown and white fat, activation of the Jack/STAT pathway (transcription pathway involved, among others, in adipogenese and sexual dimorphic growth), increase in the activity of target genes and the increase in glucose utilization and lipolysis (Siegrist-Kaiser et al. 1997). On the gene level, UCP1 and leptin are reciprocally related. High temperatures give high mRNA leptin levels and low UCP1 mRNA levels. Low temperatures show the reverse. Brown fat histochemistry demonstrated at high temperatures high levels of leptin and at low temperatures high levels of UCP1 were detected (Cancello et al. 1998). The cooperation between leptin and UCP1 goes further: leptin can reduce white fat by a mechanism that involves UCP1 of brown fat (Commins et al. 2001).

Thus, brown and white fat do have a reciprocal balance involved in puberty, and brown fat is reduced during maturity. *For head and neck, the brown fat localization is exclusively in the neck, some restrict it to the nape of the neck. Brown neck fat contributes to puberty and low temperature is a driver for brown fat thermogenesis that is related to puberty in an unknown way.*

16.4 Bone Mineral Content, Osteoporosis, and Sports

"It has been suggested recently that body fat is an important determinant of bone mass in premenopausal and postmenopausal women, although other investigators consider weight more important" (Rico et al. 1994). Bone is made out of 15% organic material (mainly collagen) and 85% of hydroxyapatite (calcium phosphate and calcium carbonate). These minerals determine the bone mineral content, which is the main determinant of the body mass. Dual-energy X-ray absorptiometry is used to determine the bone mineral content and can directly be calculated to body mass. The weight of the postpubertal females (14–18 years old) correlated directly with body mass and not with body fat. Bone mineral content is also important in osteoporosis (WHO 1994), which is the loss of the density of the bone due to extra loss of calcium and phosphate. It is known mainly in elderly women, due to the reduction in oestrogens after menopause. Using the same absorptiometric technique, several studies were directed to the effect of sports on the bone density and bone mass in boys and girls. The reason: "Bone health later in life may rely on the bone mass accumulation during growth" (Vicente-Rodriguez et al. 2004). Female handball players were studied: "Compared to the controls, handball players attained better results in the physical fitness tests and had a 6 and 11% higher total body and right upper extremity lean mass. The handball players showed enhanced bone content and bone density in the lumbar spine, pelvic region, and lower extremity" (Vicente-Rodriguez et al. 2004). The same results were found for pubertal soccer players for legs and pelvis. Moreover, this increase in bone density and mass could be linked to their increased hormonal metabolism (Nebigh et al. 2009). Thus, sports are beneficial for the pelvic bone mass and pelvic bone density in both boys and girls. If mature soccer players are compared to controls, pelvic bone mass, in fact "lower body" (pelvis and legs) mass, is significantly higher. *This does not hold for head and arms* (Karlsson et al. 2001). Nevertheless, exercise knows its limit even for league soccer players: "we conclude that, in national league soccer, the bone mineral density needed to attain a bone strength commensurate with that of duration of activity is achieved by 6 h of exercise per week. Beyond this, additional exercise confers no higher bone mineral density" (Karlsson et al. 2001). The conclusion is that exercise before and during puberty enhances the bone mineral content. Continuing exercise even during ageing keeps the bone mineral content high and reduces osteoporosis. The lower body part, pelvis and legs, benefits most of these exercises. *Thus, upper body parts (head and arms) behave differently compared to lower body parts (pelvis and legs) for bone mineral content during puberty and adolescence.*

As to osteoporosis of the skull the following citation is from Science Daily, Science News (dec. 2009): "Scientists at Queen Mary, University of London, have uncovered fundamental differences between the bone which makes up the skull and the bones in our limbs, which they believe could hold the key to tackling bone weakness and fractures". In fact, the skull excludes osteoporosis; the genes involved in limb and skull bone seemingly behave differently.

16.5 Ageing of the Face

There is no good definition of an aged human. Since the authors don't like it to be entangled in an ethical discussion on age limits (65 or 75?), here elderly are defined as those humans that are older than 70. Above this age, subcutaneous fat diminishes and skin starts to be harder. In 2014, the ageing issue has already been discussed related to the pelvis (Marani and Koch 2014). The problem remains intriguing. A short summary of the start of the chapter is repeated here (pp. 517–518): "Construction and restoration of robots are carried out from the outside. Construction and restoration in the living being occurs inside the body and are immanent properties. Such regenerative capacity, whether on DNA, cellular, or tissue level is inherent to the biological world. The turnover of structures, like cells or tissues, is rather high in the living being. There is a constant renewal. The turnover of the total protein content in the human body lasts 100 days. So, at least three times in a year, our body is totally renewed. For certain tissues, it is even faster". In 2003, IBM published "the vision on autonomic computing" that concerned the information and communication technology (ICT) interest in the (autonomic) nervous system of vertebrates in general and man specifically. It originates from an emerging field known as autonomic computing. This term was first introduced by IBM in 2001 (Kephart and Chess 2003), and it has inspired many ICT researchers to propose and evaluate new closed-loop architectures for managing (i.e. monitoring and controlling) information and communication components and networks (Parashar and Harir 2007). Autonomic computing aims at computing systems that can manage themselves. Still updates have to be found elsewhere, and only a limited amount of bits are used for program restoration and checks within the PC or server. The inherent programs of restoration in living systems, like those in computers, become worn out over the years of use. Restoration of form and function in the elderly is still ongoing, but no longer at optimal possibilities. This is called ageing. So, in the elderly inherent restoration is still ongoing, but its quality is reduced. "The rate-of-living theory of ageing postulates that longevity is inversely proportional to metabolic rate. Although the theory is flawed in its strict sense its circumstantial features reflect some important implications of the oxidative damage and mitochondrial theories of ageing. There is no doubt that a variety of metabolism-related features have an impact on ageing-dependent processes" (Heininger 2002). Ageing has not only been related to the damage by oxidative stress and dysfunction of mitochondria, but also to programmed cell death and malfunction of DNA repair (Heininger 2002; Harries et al. 2011).

Discussions on terminology are present in the literature: Ageing is defined as: "the inherent decline over time, from the optimal fertility and viability of early maturity that may precede death and may be preceded by other indications, such as sterility". Senescence in their formulation (in Gene Ontology; de Luce 1996) is a subcategory and not a synonym of ageing and is defined as: "the process that occurs in an organ near the end of its active life that is associated with the dismantling of cell components and membranes, and an overall decline in metabolism" (de Luce 1996). "Whatever the philosophical and scientific discussions and research, here we stick to the simple definition that ageing is the wearing out of the living being" (Marani and Koch 2014).

From a total other viewpoint, the contrary has been supported: one should notice that ageing is a lifelong process. It concerns a series of biological, socio-economic and psychologic processes. Ageing is individual and therefore nearly always different within the elderly. Ageing is not phasing out life. Ageing contains stability and in certain cases also growth (Mittelstrass et al. 1994).

So, two definitions oppose: ageing is not phasing out life and ageing is the wearing out of the living being. It does not matter how one considers ageing, it always leads to death. Nevertheless, the heterogeneity of the ageing process is badly understood. Some are frail early, others are fit even to old age over 90s.

Let us return to the Rembrandt paintings (Fig. 16.3). Both man and woman do have buccal fat pads that extend at the lower parts of the mandible called jowling. These buccal fat pads are clearly separated from the cheek fat pads in both paintings. In Sect. 2.6, the arguments against Leonardo da Vinci "oldest" miniature-scale sculpture of the skull showed that the orbital form and space change during ageing, which has been underpinned by Ilankovan (2014): "Skeletal ageing therefore results in curvilinear distortion of the orbital aperture in that the supramedial upper orbit and inferolateral lower orbit curve progressively with age. The loss of bony volume and projection of the lateral aspects of the orbit can cause a loss of support, and a variation in volume of the overlying soft tissue, which will contribute to its descent and bunching". In both Rembrandt's portraits, the eyes are capaciously placed within the orbit, indicating roomy orbital spaces. Fat pouches below the eyes are clearly extending and the nasolabial sulcus deepens to name a few of the ageing characteristics of these elderly faces. Preplatysmal fat is changed and alveolar height reduced with the consequence that the chin gets its typical "witch's chin" (see the female in Fig. 16.3). The differences are accentuated by resorption of bone in the mandible and decrease of maxillary angles. The mimetic muscles decline in their function. Mid- and lower face changes are related to SMAS alterations. These changes altogether do show a different face, and their modifications are typical for ageing and are directly recognized by our surroundings as belonging to elderly people.

16.6 Ageing of the Neck

"The mechanism of ageing of the neck is not well understood. Ptosis of the skin as a result of the effect of gravity is one explanation, and antero-posterior changes in volume can lead to the false appearance of descent of skin, but the change in the cervicomental angle is a key to the anatomical changes" (Ilankovan 2014). The oncologic surgeon regularly has to operate in the lateral parts of the neck. They know that the supraclavicular triangle on both sides contains fat, in which the accessory nerve and the cervical nerve branches (see Sect. 12.3.1) hide in their trajectory to trapezius and sternocleidomastoid muscles. Tumours in this area should be removed, if possible with sparing of the accessory nerve to keep a stable shoulder. Here we call this supraclavicular white fat area: nervus accessory fat (NAF). Note that this white fat area changes into brown fat at predisposed places.

Before more attention is paid to the NAF situation in the elderly, one should note that the body fluid contributes in the newborn nearly 70–80% of the total weight. This is reduced to 45–50% in the elderly. Water or fluid loss contributes seriously to the altered appearance of the aged person and not only fat reduction. Figure 16.4 clearly shows the extreme NAF reduction in the comparison between a young and an aged person. This supraclavicular fat reduction has no continuation into the skin fat jowls, which overhang the submandibular trigone. It therefore should be considered a typical NAF reduction. Left and right platysma do have a split area in which platysma fibres are absent. Due to their separate adherences on the mandible and the thin connective area in between both platysma muscles, they express themselves as columns diverging slightly within the elderly's front of the neck. The lateral sides of the platysma columns are accentuated by the strong NAF reduction (see Fig. 16.4). Skin elastic fibres produce the skin tone that is lost "due to intrinsic ageing secondary to genetic and natural corporal changes and also extrinsic ageing as a result of exposure to external factors. The end result is the development of fine rhytids and deep furrows, coarse texture, changes in pigmentation, and growth of telangiectasias" (dilatation of skin/surface capillaries; Ko et al. 2017). The web is full of techniques to reduce the sagging, crepey, baggy and scrawny neck, double chin, which is increased fat rise, and the turkey neck: not forgetting the botox injections. In fact, it is slackening of the skin collagen and of the skin elastic fibres together with platysma dysfunction. Fat increase is also possible during ageing as shown by the double chin by the submental, preplatysmal fat increase.

In an aged neck, the digastric and omohyoid muscles sometimes can be clearly noticed due to the neck fat reduction. One should also note that the sternocleidomastoid muscle lies buried directly under the skin and is remarkable recognized in head movements as a cable pushing up the skin like a boat's sail. As a consequence, the carotid fossa is also placed more directly under the skin and carotid palpation is easier due to the reduced subcutaneous fat.

Neck muscles to keep the head steady are losing force, and the typical thin back of the neck appears. Neck muscle atrophy with fatty infiltrates has been

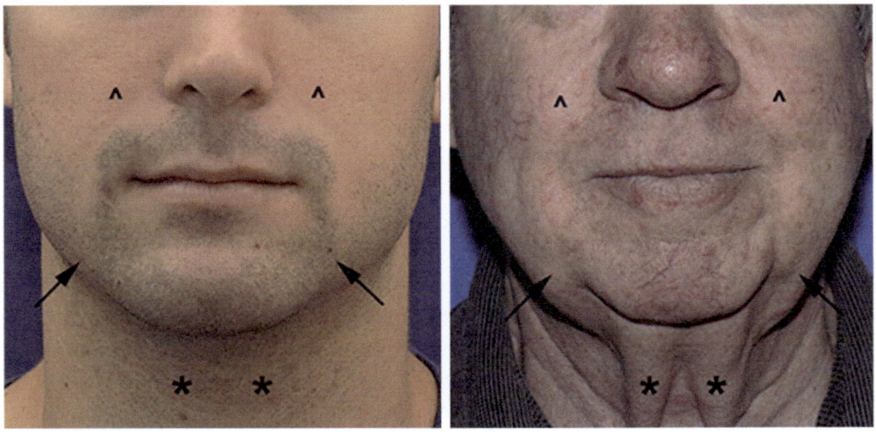

Fig. 16.4 "Comparison of nasolabial folds (caret), jowling (arrow) and platysma bands (asterisk) in a youthful **a** versus aged **b** face. Note the worsened appearance of nasolabial folds, and absence of jowling and platysmal bands, in the young face" (Ko et al. 2017, with permission)

demonstrated often in relation to headache. Cervicogenic headache is connected to ageing and to neck muscle dysfunction (Uthaikhup et al. 2017). The relationship of cervicogenic headache and neck muscles or cervical spine needs a critical note: "Despite a growing body of literature on cervicogenic headache and an increasing acceptance that headaches can originate from the cervical spine, there remains considerable controversy and confusion concerning all aspects of this topic" and later on in the same conclusions: "The significance of radiological findings and advanced diagnostic testing is unclear. Evidence to support treatment with surgery and injections consists mainly of case series without controls or standardized follow-up" (Haldemann and Dagenais 2001).

Foreward-head posture (scholar's or reading neck) produced by tightness in the pectoralis and sternocleidomastoid muscles is regularly registered in elderly people. It produces hyperactivity of the trapezius muscle, which soon leads to head sinking during meals.

The main point is the lack of a definition of an old person, although we start knowing lots of processes that contribute to ageing. We all recognize an old face and neck and determine it as being old. An analogous situation occurs in the skull of the youngest among us. We still think that puberty makes the face, however it is already formed.

One of the other, less important, remarks is that the amount of folk and scientific nick names around the head and neck during puberty and ageing is high: scholar's or reading neck, witch's chin, baggy or turkey neck, youngsters sitting in front of the PC with a chicken head, and the readers can easily produce several more. Together with several scientific names like cervical migraine and the 3rd occipital headache for the cervicogenic headache, all these names indicate that our knowledge is still insufficient and seemingly we hide this lack of knowledge behind nick names.

References

Avram AS, Avram MM, James WD (2005) Subcutaneous fat in normal and disease states. 2. Anatomy and physiology of white and brown adipose tissue. Am Acad Dermatol. https://doi.org/10.1016/j.jaad.2005.05.015

Bhatt PS, Dhillo WS, Salem V (2017) Human brown adipose tissue- function and therapeutic potential in metabolic disease. Curr Opin Pharmacol 37:1–9

Cancello R, Zingaretti MC, Sarzani R, Riquier D, Cinti S (1998) Leptin and UCP1 are reciprocally regulated in brown adipose tissue. Endocrinol 139:4747–4750

Commins SP, Watson PM, Frampton IC, Getty TW (2001) Leptin selectively reduces white adipose tissue in mice via a UCP1-dependent mechanism in brown adipose tissue. Am J Physiol Endocrinol Metab 280:E372–E377

Crelin ES (1973) Functional anatomy of the newborn. Yale University Press, London

Delemarre-van de Waal HA (1984) Central regulation of human puberty. Thesis University Amsterdam

Gierloff M, Stöhring C, Buder T, Wiltfang J (2012) The subcutaneous fat compartments in relation to aesthetically important facial folds and rhytides. J Plast Recon Aest Surg 65:1292–1297

Haldemann S, Dagenais S (2001) Cervicogenic headaches: a critical review. Spine J 1:31–46

Harries LW, Hernandez D, Henley W, Wood AR, Holly AC, Bradley-Smith RM, Yaghootkar H, Dutta A, Murray A, Frayling TM, Guralnik JM, Bandinelli S, Singleton A, Ferrucci L, Melzer D (2011) Human aging is characterized by focused changes in gene expression and deregulation of alternative splicing. Aging Cell 10:868–878

Heininger K (2002) Aging is a deprivation syndrome driven by a germ–soma conflict. Ageing Res Rev 1:481–536

Ilankovan V (2014) Anatomy of ageing face. Br J Oral Maxil Surg 52:195–20212

Karlsson MK, Magnusson H, Karlsson C, Seeman E (2001) The duration of exercise as a regulator of bone mass. Bone 28:128–132

Kephart JO, Chess DM (2003) The vision of autonomic computing. IEEE Comput 36(1):41–50

Ko AC, Korn BS, Kikkawa DO (2017) The ageing face. Surv ophthalmol 62:190–202

Kramer RSS (2017) Sexual dimorphism of facial width-to-height ratio in human skulls and faces: a meta-analytical approach. Evol Hum Behav 38:414–420

Lanz von T, Wachsmuth W (1985) In: Lang J, Kopf Teil A (eds) Uebergeordnete systeme. Springer, Berlin

Lieberman DE (2011) The evolution of the human head. Belknap Press, Harvard University Press, Cambridge, MA

de Luce J (1996) Mythology. In: Birren JE (ed) Encyclopedia of gerontology, vol II. Academic Press, NY, pp 187–195

Marani E, Koch WFRM (2014) The pelvis, structure, gender and society. Springer, Berlin, Heidelberg

Matarasso A (2006) Managing the buccal fat pad. Aesthetic Surg J 26:330–336

Mellinger WJ (1940) The canine fossa. Arch Otolaryng 31:930–937

Mertens A, Foyatier JL, Mojallal A (2016) Quantitative analysis of midface fat compartment mass with aging and body mass index, anatomical study. Ann Chir Plast Esth 61:798–805

Mittelstrass J, Baltes PB, Gerok W, Häfner H, Helmchen H, Kruse A, Mayer UK, Staudinger UM, Steinhagen-Thiessen E, Wagner G (1994) Wissenschaft und Altern. In: Baltes B (ed) Alter und altern: ein interdisziplinärer Studientext zur Gerontologie. de Gruyter, Berlin, pp 695–721

Mosgos S, Chan JL, Mantzoros CS (2002) Leptin and reproduction: a review. Fertil Steril 77:433–442

Nebigh A, Rebai H, Elloumei M et al (2009) Bone mineral density of young boy soccer players at different pubertal stages: relationships with hormonal concentration. Joint Bone Spine 76:63–69

Niamtu J (2010) Management of cervicofacial fat. In: Cosmetic facial surgery, 2nd ed. Elsevier Health Sciences. Amsterdam

O'Farell MJ, Schreiweis DO (1978) Annual brown fat dynamics in Pipistrellus hesperus and Myotis californicus with special reference to winter flight activity. Comp Biochem Physiol 61:423–426

Okuyama C, Ushijima Y, Kubota T, Yoshida T, Nakai T, Kobayashi K, Nishimura T (2018) ^{123}I-Metaiodobenzylguanidine uptake in the nape of the neck of children: likely visualization of brown adipose tissue. J Nuclear Med 44:1421–1425

Parashar M, Harir S (2007) Autonomic computing: concepts, infrastructure and applications. CRC Press, Boca Raton

Quarta C, Mazza R, Pasquali R, Pagotto U (2012) Role of sex hormones in modulation of brown adipos tissue activity. J Mol Endocrinol 49:R1–R7

Rico H, Revilla M, Alvarez del Buergo M, Ruiz-Conterras D (1994) Determinants of total body and regional bone mineral content and density in postpubertal normal women. Metabolism 43:263–266

Rogol AD (1998) Editorial: Leptin and puberty. J Clin Endocrinol Metabol 83:1089–1090

Scheele C, Nielsen S (2017) Metabolic regulation and the anti-obesity perspectives of human brown fat. Redox Biol 12:770–775

Siegrist-Kaiser CA, Pauli V, Juge-Aubry CE, Boss O, Pernin A, Chin WW, Cusin I, Rohner-Jeanrenaud F, Burger AG, Zapf J, Meier CA (1997) Direct effects of leptin on brown and white adipose tissue. J Clin Invest 100:2858–2864

Sinna R, Herlin C, Garson S, Dast S, Delay E (2017) Bases anatomiques du tégument facial appliquées à la chirurgie du rajeunissement facial. Ann Chir plast esth 62:355–364

Tanner JM (1962) Growth at adolescence. Blackwell, Oxford

Tschernof A (2009) Sex differences in energy balance, body composition, and body fat distribution. In: Tsatsoulis A et al (eds) Diabetes in women: pathophysiology and therapy. Humana Press, Springer, Berlin

Uthaikhup S, Assapun J, Kothan S, Watcharasaksilp K, Elliott JM (2017) Structural changes of the cervical muscles in elder women with cervicogenic headache. Musculoskelet Sci Pract 29:1–6

Vicente-Rodriguez G, Dorado C, Perez-Gomez J, Gonzalez-Henriguez JJ, Calbet JAL (2004) Enhanced body mass and physical fitness in young female handball players. Bone 35:1208–1215

Watson Genna C (2008) Supporting suckling skills in breastfeeding infants. Jones & Bartlett Publishers, Burlington, Massachusetts

World Health Organization (1994) Assessment of fracture risk and its application to screening for postmenopausal osteoporosis: report of a WHO study group. WHO Technical Report Series No.: 843. Geneva: WHO. http://whqlibdoc.who.int/trs/WHO_TRS_843.pdf

Young JZ (1974) An introduction to the study of Man. Oxford Paperbacks/Oxford University Press, London

Chapter 17
Head and Neck Jerks and Tremor

In a number of motor disorders, involuntary short (myoclonic) or rhythmic (tremor) motor actions in head and/or neck occur. In this chapter, a summary is included on Parkinson's disease, cervical dystonia, neck myoclonus and one of the progressive myoclonic epilepsies, lipofuscinosis. Unconscious motor activities during REM sleep behavioural disorder and the consequences of the absence of sensory corticofugal modulation are treated. A reduced ability to stabilize the head has consequences for gaze stabilization and sensorimotor function. In contrast, myoclonus can be sensitive to sensory activity, while sensory information can also be used to improve motor function. Therefore, (visual) cueing to overcome motor symptoms of Parkinson's disease is shortly described. The conclusion of this chapter concerns modulation of sensory information.

17.1 Parkinson's Disease

Parkinson's disease is the second most common neurodegenerative disease after Alzheimer's disease. While of most neurodegenerative diseases, the cause is still unknown. Progress is made in Parkinson's disease due to Braak's hypothesis, which states "that sporadic Parkinson's disease is caused by a pathogen that enters the body via the nasal cavity, and subsequently is swallowed and reaches the gut, initiating Lewy pathology in the nose and the digestive tract" (Rietdijk et al. 2017).

Parkinson's disease is characterized by loss of predominantly dopaminergic neurons and the presence of intracellular α-synuclein aggregations. The loss of dopaminergic- and catecholaminergic-related neurons is caused by a complicated interaction between toxic α-synuclein aggregates, mitochondrial dysfunction, oxidative stress, impaired intracellular calcium homoeostasis, altered iron metabolism, free radical formation, glutamate excitotoxicity and neuroinflammation. High concentrations of neuromelanin ("degradation" product of catecholamines) are related to atrophy and loss of pigmented nerve cells in the locus coeruleus

E. Marani and C. Heida, *Head and Neck*,
https://doi.org/10.1007/978-3-319-92105-1_17

(noradrenerg) and substantia nigra (dopaminerg). The disease is more common in men than in women. Its incidence varies and depends on ethnicity, environmental, lifestyle and genetic factors (Usunoff et al. 2002).

The cardinal symptoms of Parkinson's disease are: tremor of resting muscles (in 70–100% of patients), stiffness or rigidity (in 90–99%), slowness of movements or bradykinesia (in 77–89%) and absence or loss of voluntary movements (akinesia). A resting or parkinsonian tremor is absent in animals, and no experimental or hereditary animal form can replicate this resting tremor, that is in humans at a frequency of 3–8 Hz (see Lowrie and Garosi 2016).

There is no cure for the disease, and drug treatment is the main method to control disease symptoms by restoring dopamine levels in the brain. Over time, the benefits of drugs frequently diminish and patients start to experience on–off fluctuations: "on" episodes during which a new dose of medication starts to take effect become shorter and the wearing "off" periods start sooner. In addition, drug-related side effects like dyskinesia, hallucinations or delusions, and impulsive and compulsive behaviour may occur. Deep brain stimulation, the continuous delivery of electrical pulses in a specific area of the brain, can effectively reduce symptoms in Parkinson's disease. Its mechanisms, however, are still badly understood, and therefore, the pathophysiological mechanisms of the disease and the effects of electrical stimulation within the neuronal circuits involved are extensively studied using single neuron, neuronal network and neural mass models (Heida et al. 2008, 2011; Heida and Modolo 2017).

Parkinson's disease is determined by the invariable presence of Lewy inclusion bodies in the surviving neurons of the substantia nigra after autopsy. Lewy bodies contain proteins, free fatty acids, sphingomyelin, polysaccharides and aromatic α-amino acids and α-synuclein (Usunoff et al. 2002; Goedert 1997). Different divisions are made: a familial, mainly by genetic aberrations, and a sporadic, in which the environmental factors predominate. Another division is into idiopathic Parkinson's disease (accounts for 75–80% of autopsy cases) and Parkinson's multiple system degenerations (accounts for 20–25% of the autopsy cases). The multiple system degenerations contain Parkinson-plus syndromes (e.g. progressive supranuclear palsy, corticobasal degeneration, multiple system atrophy, Pick's disease and Alzheimer's disease) and secondary parkinsonian syndromes (e.g. postencephalitic parkinsonism, vascular parkinsonism, toxic and drug-induced parkinsonism; for an extended overview, see Usunoff et al. 2002). Alzheimer's disease is characterized by senile plaques, neurofibrillary tangles and granulovacuolar degeneration of the human cortex. Alzheimer's disease is included within the Parkinson-plus syndromes. This is due to the prevalence of rigid-akinetic extrapyramidal signs that can range from 33 to 90% within Alzheimer populations. Moreover, parkinsonian symptoms in Alzheimer patients are two to four times more frequent than in age-matched controls (see Jellinger 1989; 1996). At least one can state that Alzheimer's disease contains typical parkinsonian traits.

Although Parkinson's disease is considered a movement disorder, a number of non-motor symptoms may develop including mood disorders (e.g. anxiety, depression), cognitive changes, orthostatic hypotension, loss of smell, constipation, pain and fatigue. In addition, around 60–95% of the Parkinson population suffers from sleep disturbances with REM sleep behaviour disorder (RBD), one of the common ones.

17.2 Unconscious Motor Actions

On 17-12-2002, the court of Utrecht (the Netherlands, see Judgement, 2002) decided to acquit a young lady accused of inflicting grievous bodily harm to her girlfriend. She attacked her girlfriend in the night during her sleep with a knife. The violence occurred during the attacker's sleep too. The arguments for the acquittal are found in the rapport of the expert Prof. Dr. G. Kerkhof of which the main reason is: "*The widely held concept that the brain stem and other more 'primitive' neural brain structures primarily participate in elemental/vegetative rather than behavioral activities is inaccurate. There are overwhelming amounts of data documenting that extremely complex emotional and motor behaviours can originate from these more primitive structures – without involvement of 'higher' neural structures such as the cortex*" (Mahowald et al. 1992). The court's decision was based on a lower consciousness of the young lady during sleep and the complex motor behaviour that still can be performed during sleep. Aggression is such a known motor behaviour during sleep.

In the REM sleep behaviour disorder (RBD), the REM atonia is absent (see Chap. 13) and humans are active and sometimes even aggressive during this period (see the example above). This REM sleep behaviour disorder is a good risk indicator for obtaining Parkinson's disease often with a time lag of over 10 years (Iranzo et al. 2013). The disease is more common in men, but whether this is biased or not is unknown. Reliable prevalence estimates are still discussed. Estimates till now are: 0.38–0.5% in the normal population, 2% in those older than 60 years and 6% older than 70 years (Peever et al. 2014). The disorder is clearly related to lower brainstem damage (pons and medulla; see Postuma et al. 2012 and Ferini-Strambi 2011).

Blumberg and Plumeau (2016) "challenge the assumption that only motor cortex produces muscle twitches during REM sleep, thus calling into question the related assumption that motor cortex is primarily responsible for all of the pathological movements of REM sleep behavioural disorder. Moreover, motor cortex is not even necessary to produce complex behaviour; for example, stimulation of some brainstem structures can produce defensive and aggressive behaviours in rats and monkeys that are strikingly similar to those reported in human patients with REM sleep behavioural disorder" (Fig. 17.1a, c). The statement of Blumberg and Plumeau (2016) that the red nucleus, as one of the brainstem nuclei that is involved in sleep muscle twitching in one-week-old **rats** and young **cats** and **monkeys** by steering limb motoneurons, is supported by neuroanatomical studies for the rat (Lakke and Marani 1991). These studies show the presence of the rat rubrospinal tract over the whole spinal cord on postnatal day four. However in Chap. 6, Sect. 6.6.3 on palate myoclonus it was established that the rubrospinal tract in

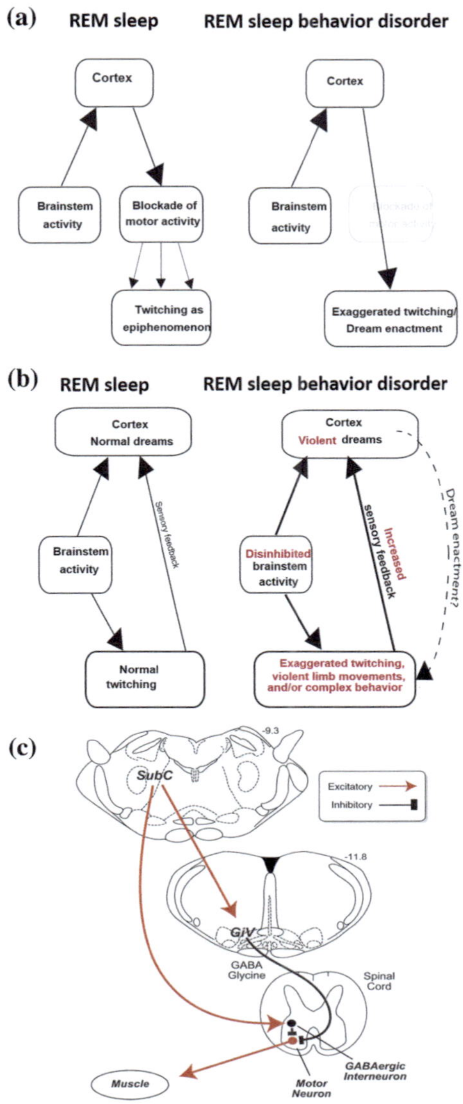

humans is absent and that red nucleus connections are exclusively related to the inferior olive and thus to the cerebellar circuits. Therefore, human red nucleus effects on limb motoneurons are excluded and the twitching disorder seems mainly to originate in other brainstem areas and presumably also in the cortex.

Sensory systems send their information towards the human cortex (Marani and Schoen 2005), but the cortex exercises control over the new incoming information. Some are allowed to ascend, others diminished or stopped. The best-known situation is the thalamo-cortical circuitry, but it holds for most brainstem sensory

◂**Fig. 17.1** **a** Initially it was suggested that the increase of muscle twitching that is seen during sleep in REM sleep behaviour disorder results from the absence or insufficient motor blockade in the brainstem. **b** Alternative models of the relationship between brainstem and cortical activity. "In normal REM sleep, brainstem activity provides parallel ascending activation of the cortex and descending activation of the skeletal muscles, the latter producing twitching. Sensory feedback (i.e. reafference) from twitching limbs provides another source of stimulation to the cortex. In REM sleep behavioural disorder, degeneration of sleep-related neural circuits may lead to disinhibition of brainstem structures that control motor behaviour (e.g. red nucleus, superior colliculus), resulting in exaggerated twitching, violent limb movements and/or complex behaviour, with concomitant effects on dream mentation. Motor cortex activity may occasionally be engaged, especially during "dream enactment" (dashed line)" (Blumberg and Plumeau 2016). **c** Schematic overview of the lower brainstem descending circuitry producing muscle atonia during REM sleep. The descending subcoeruleus (SubC) glutamatergic projections excite glycinergic neurons of the brainstem reticular formation. These glutamatergic axons project also to the medullary ventral reticular gigantocellular nucleus (GiV). The inhibitory output from the GiV inhibits spinal motoneurons, producing muscle atonia. Moreover, a direct SubC glutamatergic projection to the spinal cord is also present. SubC axons synaps directly on inhibitory interneurons of the ventral horn (**a** and **c** are adapted from Blumberg and Plumeau 2016; **b** is from Brown et al. 2012 with permission)

nuclei. Thus not only are ascending sensory systems present but also descending sensory control systems (for an overview of corticofugal modulation of somatosensory, visual and auditory systems, see Nuñez and Malmierca 2007; Alitto and Usrey 2003). Patterns of movement are encoded in so-called sensory engrams of motor movements. It is not the motor cortex alone that controls the rundown of motor patterns but also the sensory parts of brain and brainstem. This sensory modulation contributes to REM sleep atonia. By the way for extremely rapid motor patterns, the sensory control is too slow. For example, motor engrams are responsible for rapid typing.

After the initiation of sleep by both the locus coeruleus and surrounding area and the serotoninergic areas, these and other areas have to keep sleep going on by inhibiting motoneurons in the spinal cord (Fig. 17.1b). The subcoeruleus nucleus border (in Fig. 17.1c) is ill-defined. This nucleus contains noradrenergic neurons and enkephalinergic neurons. Their axons descend bilaterally with a preponderance to the ipsilateral side. The ventral gigantocellular reticular nucleus sends its GABAergic (40%) and glycinergic (15%) connections to all ipsilateral levels of the spinal cord. All levels receive substance P containing fibres from this nucleus. Other nuclei that are involved are the colliculus superior (visual system related), the laterodorsal tegmental nucleus (by its Barrington part miction related) and the pedunculopontine nucleus (Brown et al. 2012). For example, the pedunculopontine area is directly involved in leg movements and is considered a locomotion induction site (Garcia-Rill 1991; Benarroch 2013) and is well studied (Lourens 2013; Usunoff et al. 2003). Stimulation of the cholinergic part of this nucleus produces stepping as well as a series of other rhythmic events in the cat, and therefore, this area is considered rhythmogenic.

During sleep leg movements should be absent, but the reports on somnambulists are well known. Seemingly, the failure of sensory control and activation of sensory engrams open the possibility for motor systems to be activated and rundown; hence sleep-walking may occur, which produces sensory increased feedback. The disinhibited brainstem areas give cortical motor systems the possibility to be active and influence motor brain stem areas, and the occupations of the "sleeping" cortex (dreams) are expressed in the motor activity (Fig. 17.1c). This activity can be violent behaviour or simply heavy limb movements. In REM sleep behaviour disorder lesions of brainstem areas are at the base of this violent behaviour. Thus, awakening of local brainstem areas is presumably possible and are thought responsible for motor activities during sleep. Other scientists support a central control over this local awakening.

17.3 Myoclonus: Involuntary Short Motor Actions

A "belly dance" syndrome occurred in a pregnant woman. "A 32-year-old pregnant woman (gestational age 37 weeks) with unremarkable past medical history presented with a ten-day history of involuntary myoclonal abdominal contractions". Note that upper and lower parts of the rectus abdominal muscles can tilt the pelvis and the oblique muscles can contribute to it, inducing rolling one's hips, called belly dancing (Moreside et al. 2008). "A full gynaecological and neurological examination excluded complications. The presence of involuntary, rhythmic and bilateral contractions of the abdominal muscles was found. Recordings showed bursts of muscular activity restricted to thoracic myomeres, without extension to lumbar or cervical segments. The movements did not cease during sleep. A whole-spine MRI was performed, which was normal. The patient underwent caesarean delivery under general anaesthesia, to prevent foetal distress. Surprisingly, myoclonus disappeared immediately and definitively after the delivery. At follow-up evaluation, six months after the onset, the patient was still asymptomatic. Spinal segmental myoclonus is a special type of myoclonus involving muscles innervated by one or two contiguous spinal segments; in case thoracic segments are affected. This myoclonus type may generate abdominal movements resembling a "belly dance"" (text was adapted from Taga et al. 2017).

Myoclonus is seemingly not only possible in the oral cavity by the soft palate (see Sect. 6.6.3). "Myoclonus is a hyperkinetic movement disorder caused by an abrupt muscle contraction (positive myoclonus) or interruption of muscle activity (negative myoclonus)" (Zutt et al. 2017). In fact in all subdivisions of the nervous system myoclonus can originate with the duration time of the muscle contractions varying from 20 to 250 ms. It may cause jerks; however, not all people with jerks do have myoclonus, since patients with a functional movement disorder also produce jerks (spasms, motor tics and startle syndromes; Kojovic et al. 2011). Shivering can produce myoclonic activity. The different types of myoclonus can be subdivided along their site of origin (Fig. 17.2). The distribution on a diagnosed group of patients shows that cortical myoclonus is found in 34%, subcortical myoclonus in 11%, spinal/segmental myoclonus in 6%, peripheral myoclonus in

2% and functional jerks in 47%, indicating that cortical myoclonus is the most occurring myoclonus (Zutt et al. 2017). Cortical myoclonus is stimulus sensitive, typically to touch, but also to visual stimuli (Shibasaki and Neshige 1987). Cortical myoclonus is often related to damage or dysfunction of both cortices, for example produced by lack of oxygen (Fig. 17.2). However, single hemispheric cortex myoclonus does appear without obvious cortex damage (Yoshida et al. 2015). Transient myoclonus has been described in elderly (Hiraga et al. 2014). The difference between functional jerks, myoclonus and tics are difficult to establish without polymyography and accelerometry as shown by the following citation: "tics are repetitive stereotyped movements involving predominantly the eyes, face, neck and shoulders. Some tics are very simple, causing pure jerking movements" (Apartis and Vercueil 2016). Epilepsy produces myoclonus, and from the progressive myoclonic epilepsy only one is treated here, the neuronal ceroid lipofuscinoses (see Marani and Lazarov 2017). Lipofuscin is the most common pigment that is present not only in neurons, but also in other tissues. It accumulates with age and is therefore regarded as the wear and tear pigment and the granula are of lysosomal origin. Lipofuscinosis is characterized by high concentrations of lipofuscin granula. The disease is subdivided along genetic codes, numbered *CLN1–CLN14,* and is characterized by deficiency of tripeptidyl peptidase I (Dimitrova et al. 2017). All of the lipofuscinose diseases show more or less myoclonic jerks and mostly with dementia, ataxia, visual impairment and cognitive deficits (see Patel and Frucht 2015; Marani and Lazarov 2017).

Neck myoclonic activity is present during sleep disorders. In 205 mixed sleep disorder patients nearly 55% had neck myoclonus during REM sleep. Young people < 45 years do have more neck myoclonus (2 per h) compared to older persons > 60 years (1 per 2 h). REM sleep behaviour disorder patients had a higher neck myoclonus, nearly 3 per h. These neck myoclonus were directly related to ventral or dorsal head movements or combinations of these movements (see also

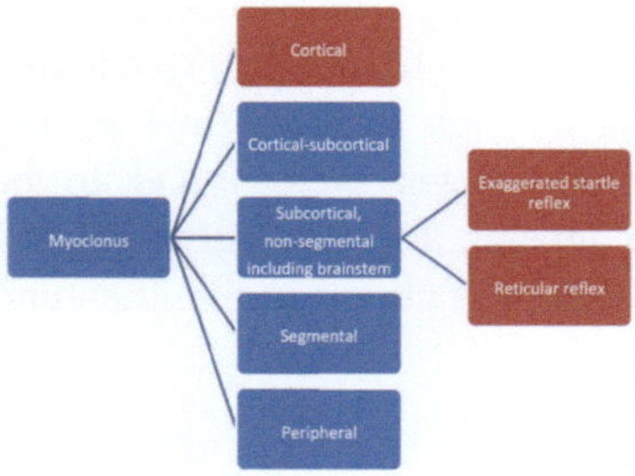

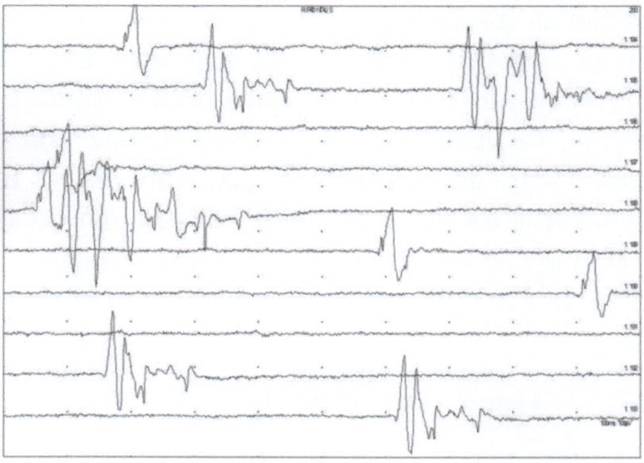

Fig. 17.2 Subdivision of myoclonus by their site of origin. Predominant types in postanoxic myoclonus are demonstrated in red (Ong et al. 2017). Right side shows hemifacial spasm characterized by very short and regular myoclonic EMG activity

Chap. 13). However, simultaneous movements of other body parts did almost not occur. Surface EMG could detect neck muscle activity related to the myoclonus in 94% of the cases (Frauscher et al. 2010).

Myoclonus is present in most types of parkinsonism like idiopathic Parkinson, Lewy body degeneration disease (all types of Parkinson's disease characterized by Lewy bodies), corticobasal degeneration, multiple system atrophy and several others. The myoclonus in idiopathic Parkinson's disease can also be induced by medication. Although not as frequent as the tremor, myoclonus is clearly present in Parkinson's disease (Truong and Bihayadisiri 2007).

17.4 Tremor: Involuntary Rhythmic Motor Actions

"In subcortical myoclonus, the onset of the EMG correlate tends to be less abrupt, and its duration is much more prolonged. Among others, rhythmic oscillation of the soft palate with or without associated movements of the extremities, which has long been called "palatal myoclonus", is especially associated with a long EMG burst. Since it is constantly rhythmic, the term "palatal ***tremor***" was introduced by Deuschl et al. (1994)" (Tassinaria et al. 1998). The above citation shows that the difference between repetitive myoclonus and tremor is gradual. Duration and rhythmicity are seemingly criteria that are involved in the distinction between myoclonus and tremor. The definition of essential tremor is even more difficult at present (Louis 2018). Subjects of discussion are: "the age of onset" of essential tremor, the relation of dystonia and essential tremor, the cerebellar signs like ataxia (mild is acceptable, but a more sever lack of muscle control or coordination may point to another disease) and whether or not non-motor features, like cognitive difficulties, psychiatric and sensory features, and sleep dysregulation, are linked to the primary disease (Louis 2018). Moreover, it is proposed that when Parkinson's patients show tremor of the head, this should result from trunk or limb tremor. This tremor is transmitted to the head and is not a genuine head tremor (Roze et al. 2006). Another hypothesis is that a central generator of tremor is present that influences the cranial and spinal motor neurons in an even manner, since resting tremor frequencies in lips, jaw, tongue and index finger were found to be remarkably uniform (Hunker and Abbs 1990). In general, a unified definition and hypothesis for the explanation of head and neck tremor is absent in the literature, which opens the possibility for reconsideration.

17.5 Head (in)Stability

What keeps the head steady? Sensory information of the head position is brought into the nervous system mainly by the neck muscles and neck ligaments (see Figs. 10.2, 14.1 and 14.2) and, of course, by the vestibular apparatus. Moreover, myoclonic and tremor movements of the head and neck involve the activation of the neck muscles. Not all neck muscles are involved in the same quality and quantity; e.g. obliquus capitis inferior is mainly engaged in cervical dystonia (a painful condition in which the neck muscles contract involuntarily causing the head to twist or turn to one side or causing the head to uncontrollably tilt forward or backward, Schramm et al. 2017). A short survey of the long and short muscles of the neck is recapitulated (see Chap. 14): the long erector spinae muscle is affixed to the occiput as the semispinalis capitis muscle and to the mastoid as the longissimus capitis, which constitutes part of the sacrospinalis muscles. To this erector spinae complex belongs semispinalis cervicis, longissimus cervicis and the transversospinal and intertransverse muscles. The latter two are short muscles. The splenius capitis and splenius cervicis enwrap the erector spinae. Thus, very long muscles that originate at the lumbar level that reach the occiput and very short muscles at the cervical level all contribute to the stability of the head. At the posterior side the trapezius muscle and at the anterior side the sternocleidomastoideus muscle both also contribute either directly or by their antagonistic actions (see Figs. 14.1 and 14.2). The neuroanatomical substrate concerns the interstitial nucleus of Cajal and the nucleus of Darkschewitsch (Fig. 6.11). The relationship between the neck muscle responses during interstitial nucleus of Cajal manipulation is complex and not extensively treated here (see below and Farshadmanesh et al. 2008). Together with the reticular formation, these structures constitute a neural integrator that organizes the head steadiness.

The mechanisms to stabilize the head and gaze make use of neural integrator circuits. Our visual system allows us to assimilate information from our surroundings. It enables us to focus both eyes on a stationary target, to smoothly follow objects in motion, and to jump from position to position to rapidly scan the environment, which involves different types of eye movements, i.e. vergence, pursuit and saccadic movements, respectively. It is hypothesized that neural integrator circuits are responsible for holding the eyes still in eccentric gaze positions. Velocity commands to overcome orbital viscous forces to move the eyes to a new position are integrated in the rostral medulla and the midbrain interstitial nucleus of Cajal controlling horizontal and vertical saccades, respectively. However, controlling gaze also requires controlling head movements. In analogy to controlling gaze, it is hypothesized that keeping the head aimed at a stationary target involves similar neural integrator circuits. All of these integrators, however, are inherently leaky, causing the eyes to drift to a null position near the midline, and causing the head to deviate from a steady eccentric position on the trunk. Feedback mechanisms are required to compensate for this drift. A cerebellar feedback loop improves gaze holding, while impairment of this system, e.g. by cerebellar lesions, causes gaze-evoked nystagmus, a drift of the eye that is only present for certain directions of gaze. Head movements analogous and highly correlated to the gaze-evoked nystagmus of the eyes may result from inactivation of the interstitial nucleus of Cajal suggesting a partial overlap in neural integration circuits. But head movements are far more complex than eye movements, since they are controlled by numerous neck muscles that allow complex movements in several planes and around several joints. Visual feedback can improve the function of the head integrator. In contrast, eye movements can compensate for head movements, and it was found that proprioception from neck muscles also has an important influence on the head integrator, as well as input from the cerebellum. The nature of the proprioceptive feedback, being it either negative velocity feedback or positive position feedback, is, however, still unknown. Head velocities observed during balance corrections depend to a large extent on the mass-viscoelastic properties of the head and neck without significant involvement of vestibulocollic reflexes (Allum et al. 1997). This mechanical dampening in the neck may have a similar effect as that of a central neural integrator.

Cervical dystonia has previously been associated with defects in basal ganglia, or sometimes the cerebellum or proprioceptive defects. However, it may very likely be related to an abnormal feedback to the midbrain head neural integrator. This would explain the fact that defects within different anatomical structures cause similar symptoms. It would also explain why some treatments have not been successful or have shown unpredictable outcomes like deep brain stimulation of the internal part of the globus pallidus or the subthalamic nucleus. The concept of a neural integrator suggests that stimulation of the interstitial nucleus of Cajal or its cerebellar inputs may result in a larger clinical benefit (for overviews, see Shaikh et al. 2013, 2016).

Head instability can even present as a disease. In nowadays southern Sudan and Uganda, and earlier in 1962 in Tanzania a so-called nodding disease was noticed. It concerns a progressive disease characterized by head nodding, mental retardation and stunted growth. Although the cause of the disease is still discussed, nearly 93% of the young patients (with a mean of 14 years, and start of the disease between 6 and 10 years) were infected by the filarial worm Onchocerca volvulus that is spread by Simulium, called the black fly. This fly reproduces near fast streaming rivers. "The characteristic clinical feature of nodding disease is a paroxysmal spell where the head nods forward repeatedly in a seemingly unresponsive affected child. These nodding episodes represent a form of epilepsy during which generalized electrodecrement is seen on electroencephalography and paraspinal dropout on electromyography" (Colebunders et al. 2015). The disease is subdivided into nodding and nodding plus, the plus mostly by nodding with epileptic-like seizures. The inductor of the nodding was food, and familial clustering is high (84%). MRI study showed that in several patients (40%) central nervous system lesions were present, described as hippocampal sclerosis or encephalo-pathology (Winkler et al. 2008). The antifilarial antibodies are suspected to attack also neuroproteins, explaining the mental retardation that is inherent to this illness (Idro et al. 2016). Nevertheless, a viral infection is also encouraged (Colebunders et al. 2015) and even the consequences of prior measles infection followed by poor nutrition (Spencer et al. 2016). The disease has been related to myoclonic epilepsy. Tone loss of neck muscles causes the head nodding, and tone losses of the upper extremities are reported in 60% of the studied cases. It is a pity that the literature on nodding disease hardly indicates the causal relation between spinal innervation of the long and short neck muscle groups. Hippocampal sclerosis can directly influence memory or limbic functions but hardly spinal motor functions related to neck muscles.

17.6 The Impact of Head Instability in Elderly and Parkinson's Patients

Decline of sensorimotor function in elderly has a direct effect on their mobility; for example, it causes a reduction in their step length, which is strongly related to a fear of falling. The degraded sensorimotor function and the adjusted gait pattern both have an indirect effect on head stability. The fear of falling in elderly is strongly related to decrease in step length but has nearly no direct influence on head stability. A reduced step length has a large influence on pelvic stability; it increases pelvic instability and has a less but still significant effect on upper body attenuation. It is mainly the pelvic (in)stability that is directly related to the (in)stability of the head. Absence of head stability makes you feel unsteady and induces falls (Menz et al. 2007; Veneman et al. 2008). One should note the following: "falls in older people are a major public health problem, with one in three community-dwelling people aged over 65 years falling each year. Approximately two-thirds of falls in older people occur while they are walking, and it is therefore not surprising that a number of studies have reported significant differences in gait patterns between older fallers and non-fallers. These differences include reduced velocity and step length, increased variability in step length and cadence, reduced peak hip extension and less rhythmic accelerations of the upper body. These findings suggest that, despite adopting a more conservative basic gait pattern, older people who fall have difficulty maintaining the stability of the body" (Menz et al. 2007).

Step length and step cadence in Parkinson patients are also seriously altered. Falls present in the studied Parkinson population reached 50% and recurrent falls 25% (Bloem et al. 2001). Freezing of gait, the total stop of stepping or of its start, is well known in Parkinson's disease. The exact cause of freezing of gait is not known, but it has been suggested that it may be related to a temporary exacerbation of defective mechanisms of central motor control that interact with each other (Nieuwboer 2008). Moreover, freezing of gait and falls is related. Since freezing of gait is a major cause of falls, it also has a devastating effect on quality of life (Bloem et al. 2004). One should know that while control of gaze is coordinated by eye movements (see Chap. 7 and Sect. 8.7) and head stability contributes to it, still eye and head movements can be distinguished as being generated by separate control systems (Shaikh et al. 2013). A comparison of fixed and free gaze shows that if a Parkinson patient has to fix to a point during walking his/her step width increases, which is not the case if he/she can look freely around (Reed-Jones and Powell 2017). This is the reverse approach from what we have seen above. Thus, head stability is not only related to muscle control and vestibular information, but is also related to pelvic/hip stability and gaze control. Moreover, head stability is phonation-related. Essential head tremor increases in tremor amplitude during or immediately after phonation. Moreover, patients that were classified with absence of head tremor produced head tremor with phonation in 5% of the studied essential tremor cases (Wright et al. 2014).

17.7 Cueing: Tricking the Parkinsonian Motor System

The definition of cueing in psychology is: "a stimulus, either consciously or unconsciously perceived, that elicits or signals a type of behaviour". In learning theory, it is "a pattern of stimuli to which an individual has learned or is learning to respond". In contrast, Parkinson's patients have difficulty to perform learned movements automatically, and start to rely more on external stimuli to perform these movements. It is well known that patients with Parkinson's disease are more dependent of optic information for crossing obstacles and keeping their postural stability than are healthy controls (e.g. Vitório et al. 2013, 2014). Various examples exist of Parkinson's patients being able to perform movements with more ease under specific conditions. The "university storyteller" uses the example of the Parkinson patient in bed, while downstairs somebody cries "Fire, fire". The Parkinson patient can leave the bed, come down and leave the house without any rigid-akinetic problems, and when all is safe all Parkinsonian traits are back. Similarly, the painted staircase example, which has been published by Janssen et al. (2016), describes a Parkinson patient that is able to climb stairs without any signs of gait disturbances and freezing of gait. Relatives painted the staircase on the floor of the patient's room to provide a similar condition as actually climbing stairs in order to overcome freezing of gait. However, when walking just next to the painted staircase freezing of gait returned.

The use of external cues such as a continuous rhythm of a metronome or music, stationary lines or bars on the floor or laser lines projected on the floor are strategies for Parkinson's disease patients to improve speed, cadence and stride length variability, and may prevent or reduce FOG (Young et al. 2016; Palacios-Navarro et al. 2016 and Ginis et al. 2017; Nieuwboer 2008 and on the Web: Parkinson's patient visual cueing gait training with partial weight relief, Ludwig Maximilians University, Munich). Analysis of articles on cueing training for the reduction of freezing of gait, indeed, shows an improvement of gait and of upper limb movements. However, although cueing has a positive effect on Parkinson symptoms "findings were not univocal, and long-term consolidation and transfer of the effects appear to be hampered" (Ginis et al. 2017).

Several defects in motor control have been identified, such as an ineffective scaling of movement amplitude, a distorted internal rhythm generation and impaired bilateral coordination, which can be partly overcome by using cues. Visual cues have been suggested to provide spatial information that could aid in scaling of movements, while auditory cues may provide an external rhythm that overrules the defective internal rhythm. In addition, Parkinson patients use more brain activity to perform automatic tasks and shift to a more attention-demanding way to perform them. Thus, not just the stimulus itself, but also the attention that is paid to the actual performance of the movement may have a significant impact on the effectiveness of the cue. For example, based on several studies it was found that three-dimensional (3D) visual cues, that give the impression that you have to step over an object, may be more effective than flat (2D) cues. This may be due to the activation of alternative motor programs which might be better preserved, or the cues may trigger the patient to make a more conscious and larger weight shift onto the stance leg to be able to lift the other leg high enough to step over the object. Moreover, 3D cues may induce a stronger activation of the visual areas of the cortex that may result in a stronger activation of neuronal pathways that compensate for the dysfunctional circuits of the basal ganglia.

Visual and auditory cues may easily be applied within a patient's home; however, they are only sporadically present in the outside world. Currently, novel technologies such as smart glasses in combination with augmented reality and methods for real-time object recognition and motion tracking are being tested to provide an alternative way to benefit from cueing anywhere and whenever required. The fact that older people in general are increasingly using computers, the Internet and mobile technologies suggests that smart glasses are easily adopted, and according to a survey in which 62 Parkinson patients participated, patients are especially enthusiastic about its potential to restore confidence and independence. However, since freezing of gait is highly variable in Parkinson patients and is influenced by a variety of factors including the environment, task demands, medication and attention, increased insight into the pathophysiological mechanisms underlying parkinsonian motor symptoms and especially freezing of gait, as well as the mechanisms through which cues become effective, is required for these tools to have the most optimal and long-lasting beneficial effect for each individual patient (Zhao et al. 2015, 2016; Janssen et al. 2017).

Seemingly, Parkinson patients and elderly need selective attention for effective motion in which head position plays an important role. "Selective attention is defined as the cognitive function that allows the focusing of processing resources onto the relevant sensory stimuli among the environmental information available at a given moment, while other irrelevant sensory information is largely ignored" (Nuñez and Malmierca 2007). The paradox is that selective attention, for which a stable head position is needed, can improve walking and stepping, but an instable head position can cause stress or raise emotion and then disturbs motion or increases tremor severity. For selective attention, not only sensory suppressing of distractive stimuli, by, e.g. fixed gaze, is needed, but also increase of sensory neural processing that can be organized by, e.g. cueing. The sensory component is undeniable in REM sleep behaviour disorder, myoclonus, tremor and cueing. Main studies on modulation of sensory information are found in auditory research and not in somatosensory system studies, which should be beneficial for Parkinson's patients and elderly.

References

Alitto HJ, Usrey WM (2003) Corticothalamic feedback and sensory processing. Curr Opin Neurobiol 13:440–445

Allum JH, Gresty M, Keshner E, Shupert C (1997) The control of head movements during human balance corrections. J Vest Res 7:189–218

Apartis E, Vercueil L (2016) To jerk or not to jerk: a clinical pathophysiology of myoclonus. Revue Neurol 172:465–476

Benarroch EE (2013) Pedunculopontine nucleus: functional organization and clinical implications. Neurology 80:1148–1155

Bloem BR, Hansdorff JM, Visser JE, Giladi N (2004) Falls and freezing of gait in Parkinson's disease: A review of two interconnected, episodic phenomena. Movem Disord 19:871–884

Bloem BR, Grimbergen YAM, Cramer M, Willemsen M, Zwinderman AH (2001) Prospective assessment of falls in Parkinson's disease. J Neurol 248:950–958

Blumberg MS, Plumeau AM (2016) A new view of dream enactment in REM sleep behaviour disorder. Sleep Med Rev 30:34–42

Brown RE, Basheer R, McKenna JT, Strecker RE, McCarley RW (2012) Control of sleep and wakefulness. Physiol Rev 92:1087–1187

Colebunders R, Post R, O'Neill S, Haesaert G, Opar B, Lakwo T, Laudisoit A, Hendy A (2015) Nodding syndrome since 2012: recent progress, challenges and recommendations for future research. Trop Med Int Health 20:194–200

Deuschl G, Toro C, Valls-Sole J, Zeffiro T, Zee DS, Hallett M (1994) Symptomatic and essential palatal tremor. I. Clinical, physiological and MRI analysis. Brain 117:775–788

Dimitrova MB, Atanasova DY, Lazarov NE (2017) Tripeptidyl peptidase I and Its role in neurodegenerative and Tumor Diseases. In: Chakraborti S, Dhalla NS (eds) Pathophysiological aspects of proteases. Springer Nature Singapore Pte Ltd. https://doi.org/10.1007/978-981-10-6141-7_6

Farshadmanesh F, Chang P, Wang H, Yan X, Corneil BD, Crawford JD (2008) Neck muscle synergies during stimulation and inactivation of the Interstitial Nucleus of Cajal (INC). J Neurophysiol 100:1677–1685

Ferini-Strambi L (2011) Does idiopathic REM sleep disorder really exist? What are the potential markers of neurodegeneration in idiopathic REM sleep disorder? Sleep Med 12:S43–S49

Frauscher B, Brandauer E, Gschliesser V, Falkenstetter T, Furtner MT, Ulmer H, Poewe W, Högl B (2010) A descriptive analysis of neck myoclonus during routine polysomnography. Sleep 33:1091–1096

Garcia-Rill E (1991) The pedunculopontine nucleus. Progr Neurobiol 36:363–389

Ginis P, Nackaerts E, Nieuwboer A, Heremans E (2017) Cueing for people with Parkinson's disease with freezing of gait. A narrative review of the state-of-the-art and novel perspectives. Ann Phys Rehabil Med. http://dx.doi.org/10.1016/j.rehab.2017.08.002

Goedert M (1997) The awakening of α-synuclein. Nature 388:232–233

Heida T, Marani E, Usunoff KG (2008) The subthalamic nucleus: Part II. Modeling and simulation of activity. Adv Anat Embryol Cell Biol 199:1–80

Heida T, Modolo J (2017) Models of deep brain stimulation. Scholarpedia. https://doi.org/10.4249/scholarpedia.33311

Heida T, Moroney R, Marani E (2011) Modeling and simulation of deepbrain stimulation in Parkinson's disease. In: Pereira M, Freire M (eds) Biomedical diagnostics and clinical technologies: applying high-performance cluster and grid computing. Med Inf Sci Reference, Hershey, NY

Hiraga A, Kamitsukasa I, Kuwabara S (2014) Isolated transient myoclonus in the elderly: An under-recognized condition. Clin Neurol Neurosurg 117:51–54

Hunker CJ, Abbs JH (1990) Uniform frequency of Parkinsonian resting tremor in the lips, jaws, tongue and index finger. Mov Disorder 5:71–77

Idro R, Opar B, Wamala J, Abbo C, Onzivua S, Mwaka DA, Kakooza-Mwesige A, Mbonye A, Aceng JR (2016) Is nodding syndrome an Onchocerca volvulus-induced neuroinflammatory disorder? Uganda's story of research in understanding the disease. Int J Inf Dis 45:112–117

Iranzo A, Tolosa E, Gelpi E, Molinuevo JL, Valldeoriola F, Serradell M, Sanchez-Valle R, Vilaseca I, Lomeña F, Vilas D, LLadó A, Gaig C, Santamaria J (2013) Neurodegenerative disease status and post-mortem pathology in idiopathic rapid-eye-movement sleep behaviour disorder: an observational cohort study. Lancet Neurol 12:443–453

Janssen S, Bolte B, Nonnekes J, Bittner M, Bloem BR, Heida T, Zhao Y, van Wezel RJA (2017) Usability of three-dimensional augmented visual cues delivered by smart glasses on (freezing of) gait in Parkinson's disease. Front Neurol 8:279

Janssen S, Soneji M, Nonnekes J, Bloem BR (2016) A painted staircase illusion to alleviate freezing of gait in Parkinson's disease. J Neurol 263:1661–1662

Jellinger K (1989) Pathology of Parkinson's syndrome. In: Calne DB (ed) Drugs for the treatment of Parkinson's disease. Handbook of experimental pharmacology, vol 88. Springer, Berlin, pp 47–112

Jellinger K (1996) The neuropathologic diagnosis of secondary parkinsonian syndromes. In: Battistin et al (eds) Parkinson's disease. Adv Neurol 69:293–303

Judgement Court Utrecht (2002) ECLI:NL:RBUTR:2002:AF2058

Lakke EAJF, Marani E (1991) Prenatal descent of rubrospinal fibers through the spinal cord of the rat. J Comp Neurol 314:67–78

Kojovic M, Cordivari C, Bhatia K (2011) Myoclonic disorders: a practical approach for diagnosis and treatment. Ther Adv Neurol Disord 4:47–62

Louis ED (2018) The evolving definition of essential tremor: What are we dealing with? Park Rel Dis 46:S87–S91

Lourens M (2013) Neural network dynamics in Parkinson's disease. Thesis Twente University

Lowrie M, Garosi L (2016) Classification of involuntary movements in dogs: Tremors and twitches. Vetinary J 214:109–116

Mahowald MW, Schenck CH, Rosen GM, Hurwitz TD (1992) The role of a sleep disorder center in evaluating sleep violence. Arch Neurol 49:604–607

Marani E, Schoen JHR (2005) A reappraisal of the ascending systems in Man, with emphasis on the medial lemniscus. Adv Anat Embryol Cell Biol 179:1–76

Marani E, Lazarov N (2017) Lipofuscine and lipofuscinosis. Encycl Neurosci Behavior Psychol. https://doi.org/10.1016/B978-0-12-809324-5.02594-3

Menz HB, Lord SR, Fitzpatrick RC (2007) A structural equation model relating impaired sensorimotor function, fear of falling and gait patterns in older people. Gait Posture 25:243–249

Moreside JM, Vera-Garcia FJ, McGill SM (2008) Neuromuscular independence of abdominal wall muscles as demonstrated by middle-eastern style dancers. J Electromyogr Kinesol 18:527–537

Nieuwboer A (2008) Cueing for freezing of gait in patients with Parkinson's disease: a rehabilitation perspective. Mov Disord 23:S475–S481

Nuñez A, Malmierca E (2007) Corticofugal modulation of sensory information. Adv Anat Embryol Cell Biol 187:1–75

Ong MT, Sarrigiannis PG, Baxter PS (2017) Postanoxic reflex myoclonus in a child and proposed classification of postanoxic myoclonus. Pediat Neurol 68:68–72

Palacios-Navarro G, Albiol-Perez G, Garcia IGM (2016) Effects of sensory cueing in virtual motor rehabilitation. A review. J Biomed Inf 60:49–57

Patel A, Frucht S (2015) Myoclonus: classification, clinical features, and genetics. In: LeDoux Ms (ed) Movement disorders, genetics and models, Second Edn. Academic Press, Elsevier, Amsterdam, pp 655–670. http://dx.doi.org/10.1016/B978-0-12-405195-9.00040-8

Peever J, Luppi P-H, Montplaisir J (2014) Breakdown in REM sleep circuitry underlies REM sleep behaviour disorder. Trends Neurosci 37(5):279–288

Postuma RB, Gagnon JF, Montplaisir JY (2012) REM sleep behavior disorder: from dreams to neurodegeneration. Neurobiol Disease 46:553–558

Reed-Jones RJ, Powell DW (2017) The effect of gaze stability on gait parameters in individuals with Parkison's disease. Neurosci Lett 655:156–159

Rietdijk CD, Perez-Pardo P, Garssen J, van Wezel RJA, Kraneveld AD (2017) Exploring Braak's hypothesis of Parkinson's disease. Front Neurol. https://doi.org/10.3389/fneur.2017.00037

Roze E, Coêlho-Braga MC, Gayraud D, Legrand AP, Trocello J-M, Fénelon G, Cochen V, Patte N, Viallet F, Vidailhet M, Pollak P, Apartis E (2006) Head tremor in Parkinson's disease. Mov Disord 21:1245–1248

Schramm A, Huber D, Möbius C, Münchau A, Kohl Z, Bäumer T (2017) Involvement of obliquus capitis inferior in dystonic head tremor. Park Rel Disorders. https://doi.org/10.1016/j.parkreldis.2017.07.034

Shaikh AG, Wong AL, Zee DS, Jinnah HA (2013) Keeping your head on target. J Neurosci 33:11281–11295

Shaikh AG, Zee DS, Crawford JD, Jinnah HA (2016) Cervical dystonia: a neural integrator disorder. Brain 139:2590–2599

Shibasaki H, Neshige R (1987) Photic cortical reflex myoclonus. Ann Neurol 22:252–257

Spencer PS, Mazumder R, Palmer VS, Lasarev MR, Stadnik RC, King P, Kabahenda M, Kitara DL, Stadler D, McArdle B, Tumwine JK, Members of the Oregon-Uganda Nodding Syndrome Research Team (2016) Environmental, dietary and case-control study of nodding syndrome in Uganda: a post-measles brain disorder triggered by malnutrition. J Neurol Sci 369:191–203

Taga A, Florindo I, Kiener A, Frusca T, Pavesi G (2017) Spinal segmental myoclonus resembling "Belly dancing" in a pregnant women. Clin Neurophysiol 128:2248–2249

Tassinaria CA, Rubbolia G, Shibasak H (1998) Neurophysiology of positive and negative myoclonus. Electroencephalo Clin Neurophysiol 107:181–195

Truong DD, Bihayadisiri R (2007) Myoclonus and Parkinson. In: Koller WC, Melamed E (eds) Handbook of clinical neurology 84, Parkinson's disease and related disorders, part II. Elsevier, Amsterdam, pp 549–560

Usunoff KG, Itzev DE, Ovtscharoff WA, Marani E (2002) Neuromelanin in the human brain: A review and atlas of pigmented cells in the substantia nigra. Arch Physiol Biochem 110:257–369

Usunoff KG, Itzev DE, Lolov SR, Wree A (2003) Pedunculopontine tegmental nucleus. Part I: cytoarchitecture, transmitters, development and connections. Biomed Rev 14:95–120

Veneman JF, Menger J, Van Asseldonk EHF, Van der Helm FCT, Van der Kooij H (2008) Fixating the pelvis in the horizontal plane affects gait characteristics. Gait Posture 28:157–163

Vitório R, Lirani-Silva E, Barbieri FA, Raile V, Stella F, Bucken Gobbi LT (2013) Influence of visual feedback sampling on obstacle crossing behaviour in people with Parkinson's disease. Gait Posture 38:330–334

Vitório R, Lirani-silva E, Pieruccini-faria F, Moraes R, Gobbi LTB, Almeida QJ (2014) Visual cues and gait improvement in Parkinson's disease: which piece of information is really important? Neurosci 277:273–280

Winkler AS, Friedrich K, Konig R, Meindl M, Helbok R, Unterberger I, Gotwald T, Dharsee J, Velicheti S, Kidunda A, Jilek-Aall L, Matuja W, Schmutzhard E (2008) The head nodding syndrome—Clinical classification and possible causes. Epilepsia 49:2008–2015

Wright BA, Michalec M, Louis ED (2014) Triggering essential head tremor with sustained phonation: a clinical phenomenon with potential diagnostic value. Parkin Rel Dis 20:230–232

Young WR, Shreve L, Quinn EJ, Craig C (2016) Auditory cueing in Parkinson patients with freezing of gait. What matters most: action-relevance or cue-continuity? Neuropsychol 87:54–62

Yoshida T, Kato T, Awaya T, Shibata M, Saito K, Ide M, Matsuhashi M, Heike T (2015) A boy with unilateral neck myoclonus of cortical origin independently on both sides. Seizure 24:131–133

Zhao Y, Heida T, van Wegen EEH, Bloem BR, van Wezel RJA (2015) E-health support in people with Parkinson's disease with smart glasses: a survey of user requirements and expectations in the Netherlands. J Park Dis 5:369–378

Zhao Y, Nonnekes J, Storcken EJM, Janssen S, van Wegen EEH, Bloem BR, Dorresteijn LDA, van Vugt JPP, Heida T, van Wezel RJA (2016) Feasibility of external rhythmic cueing with the Google Glass for improving gait in people with Parkinson's disease. J Neurol 263:1156–1165
Zutt R, Elting JW, van der Hoeven HJ, Lange F, Tijsen MJA (2017) Myoclonus subtypes in tertiary referral center. Cortical myoclonus and functional jerks are common. Clin Neurophys 128:253–259

Index

E. Marani and C. Heida, *Head and Neck*,
https://doi.org/10.1007/978-3-319-92105-1

MIX
Papier aus verantwortungsvollen Quellen
Paper from responsible sources
FSC® C105338

If you have any concerns about our products,
you can contact us on
ProductSafety@springernature.com

In case Publisher is established outside the EU,
the EU authorized representative is:
Springer Nature Customer Service Center GmbH
Europaplatz 3, 69115 Heidelberg, Germany

Printed by Libri Plureos GmbH
in Hamburg, Germany